Advances in Science, Technology & Innovation

IEREK Interdisciplinary Series for Sustainable Development

Advances in Science, Technology & Innovation (ASTI) is a series of peer-reviewed books based on the best studies on emerging research that redefines existing disciplinary boundaries in science, technology and innovation (STI) in order to develop integrated concepts for sustainable development. The series is mainly based on the best research papers from various IEREK and other international conferences, and is intended to promote the creation and development of viable solutions for a sustainable future and a positive societal transformation with the help of integrated and innovative science-based approaches. Offering interdisciplinary coverage, the series presents innovative approaches and highlights how they can best support both the economic and sustainable development for the welfare of all societies. In particular, the series includes conceptual and empirical contributions from different interrelated fields of science, technology and innovation that focus on providing practical solutions to ensure food, water and energy security. It also presents new case studies offering concrete examples of how to resolve sustainable urbanization and environmental issues. The series is addressed to professionals in research and teaching, consultancies and industry, and government and international organizations. Published in collaboration with IEREK, the ASTI series will acquaint readers with essential new studies in STI for sustainable development.

More information about this series at http://www.springer.com/series/15883

Amjad Kallel • Zeynal Abiddin Erguler
Zhen-Dong Cui • Ali Karrech
Murat Karakus • Pinnaduwa Kulatilake
Sanjay Kumar Shukla
Editors

Recent Advances in Geo-Environmental Engineering, Geomechanics and Geotechnics, and Geohazards

Proceedings of the 1st Springer Conference of the Arabian Journal of Geosciences (CAJG-1), Tunisia 2018

Springer

Editors
Amjad Kallel
Sfax National School of Engineering
University of Sfax
Sfax, Tunisia

Zeynal Abiddin Erguler
Kütahya Dumlupınar University
Kütahya, Turkey

Zhen-Dong Cui
China University of Mining
and Technology
Xuzhou, Jiangsu, China

Ali Karrech
The University of Western Australia
Perth, WA, Australia

Murat Karakus
University of Adelaide
Adelaide, WA, Australia

Pinnaduwa Kulatilake
Department of Materials Science
and Engineering
The University of Arizona
Tucson, AZ, USA

Sanjay Kumar Shukla
Edith Cowan University
Perth, WA, Australia

ISSN 2522-8714 ISSN 2522-8722 (electronic)
Advances in Science, Technology & Innovation
IEREK Interdisciplinary Series for Sustainable Development
ISBN 978-3-030-01664-7 ISBN 978-3-030-01665-4 (eBook)
https://doi.org/10.1007/978-3-030-01665-4

Library of Congress Control Number: 2018958941

This Springer imprint is published by the registered company Springer Nature Switzerland AG
The registered company address is: Gewerbestrasse 11, 6330 Cham, Switzerland

Preface

The physical and mechanical properties of geomaterials are inherently uncertain due to their anisotropic, heterogeneous, and extremely complex characteristics. This limitation was commonly observed in geotechnical projects particularly performed before 1960, with a vague understanding of its influence on natural and man-made structures. Therefore, a number of catastrophic failures (e.g., slope instabilities during construction of Panama channel and Vaiont dam, Malpasset disaster, earthquakes induced soil liquefaction, construction and landslide problems, etc.) have been recorded during and/or after implementation of many geotechnical projects. In spite of the negative consequences such as death tolls and destroyed properties (including whole villages, towns, and cities), the major disasters that occurred before 1960s are considered as important input parameters in the advancement of analytical and numerical approaches and models in geotechnics. In parallel with population increase, new engineering projects, particularly the construction of new residential areas, tunnels, dams, roads, underground openings, and mining activities, etc., were carried out in more complex and challenging geological units through advances in excavation and support system technology. The knowledge and experience gained from these engineering projects, the lessons learned from these natural disasters, as well as the wealth of data presented at international conferences or disseminated otherwise are instrumental to better understand the physical and mechanical behavior of geomaterials. In this spirit, the 1st Springer Conference of the Arabian Journal of Geosciences was organized during November 12–15, 2018 in Tunisia with further contributions on the issues of geo-environmental engineering, geomechanics and geotechnics, and geohazards.

This edited proceedings volume contains the best papers accepted for presentation at the 1st Springer Conference of the Arabian Journal of Geosciences (CAJG-1), Tunisia 2018 in the field of geo-engineering. In addition, it includes three keynotes entitled "*a new three-dimensional rock mass strength criterion*", "*new tools and techniques of remote sensing for geologic hazard assessment*", and "*land subsidence induced by the engineering-environmental effects in Shanghai China*".

The book has nine chapters covering the following main areas: (a) Applications in geo-environmental engineering including soil remediation, (b) Characterization of geomaterials using geological, geotechnical, and geophysical techniques, (c) Soil improvement applications, (d) Soil behavior under dynamic loading, (e) Recent research on expansive soils, (f) Analytical and numerical modeling of various geo-structures, (g) Slope stability, (h) Landslides, (i) Subsidence studies, and (j) Characterization and impacts of various geohazards.

Several books exist in the literature on each of the topics:

1. Geo-environmental engineering
2. Geotechnics
3. Geomechanics
4. Geohazards

However, it is almost impossible to find a single book that investigates all these areas. The suggested book provides a comprehensive coverage on the abovementioned topics. The papers have a good balance between theory and practice. Therefore, the content is useful for educators, researchers, and practitioners who would like to get a broad, comprehensive coverage in geo-engineering. Some papers report on pioneering developments and others provide state-of-the-art-techniques to address a wide variety of geo-engineering problems.

Sfax, Tunisia	Amjad Kallel
Kütahya, Turkey	Zeynal Abiddin Erguler
Xuzhou, China	Zhen-Dong Cui
Perth, Australia	Ali Karrech
Adelaide, Australia	Murat Karakus
Tucson, USA	Pinnaduwa Kulatilake
Perth, Australia	Sanjay Kumar Shukla
July 2018	

Acknowledgements

Our appreciation is extended to the authors of the papers for their hard and diligent work and producing high-quality contributions. We would like to thank the reviewers of the papers for their in-depth reviews and great efforts in improving the quality of the papers. Also, thanks are extended to Amjad Kallel who supervised and handled the evaluation process, to Sahbi Moalla who handled the submission and evaluation system for the ten conference proceedings volumes, and the publishing staff of Springer headed by Nabil Khélifi, Senior Editor for their efforts and contributions in completing this conference proceedings volume. All the above-mentioned efforts were very important in making this book a success.

About the 1st Springer Conference of the Arabian Journal of Geosciences (CAJG-1), Tunisia 2018

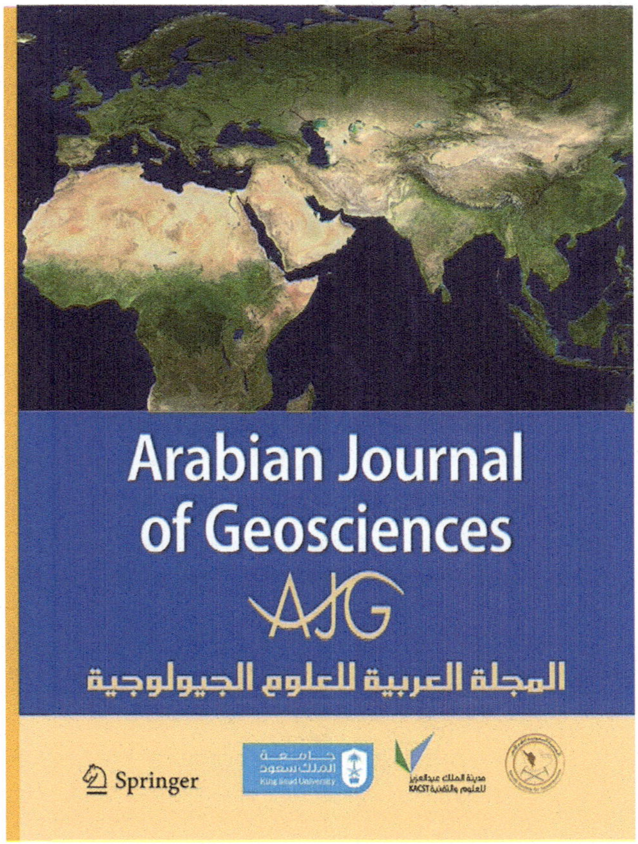

The *Arabian Journal of Geosciences* (AJG) is a Springer journal publishing original articles on the entire range of Earth sciences in partnership with the Saudi Society for Geosciences. The journal focuses on, but not limited to, research themes which have regional significance to the Middle East, the Euro-Mediterranean, Africa, and Asia. The journal receives on average 2000 submissions a year and accepts around 500 papers for publication in its 24 annual issues (acceptance rate 25%). It enjoys the participation of an editorial team of 100 international associate editors who generously help in evaluating and selecting the best papers.

In 2008, Prof. Abdullah Al-Amri, in close partnership with Springer, founded the Arabian Journal of Geosciences (AJGS). In this year, the journal celebrates its tenth anniversary. On this occasion and to mark this event, the Founder and Editor-in-Chief of the AJGS Prof. Al-Amri organized in close collaboration with Springer the 1st Conference of the Arabian Journal of Geosciences (1st CAJG) in Hammamet, Tunisia from November 12 to 15, 2018 (www.cajg.org).

The conference was an occasion to endorse the journal's long-held reputation for bringing together leading authors from the Middle East, the Euro-Mediterranean, Africa, and Asia who work at the wide-ranging fields of Earth sciences. The conference covered all crosscutting themes of Geosciences and focused principally on the following ten tracks:

- Track 1. Climate, paleoclimate, and paleoenvironmental changes
- Track 2. Geoinformatics, remote sensing, geodesy
- Track 3. Geoenvironmental engineering, geomechanics and geotechnics, geohazards
- Track 4. Geography, geoecology, geoarcheology, geotourism
- Track 5. Geophysics, seismology
- Track 6. Hydrology, hydrogeology, hydrochemistry
- Track 7. Mineralogy, geochemistry, petrology, and volcanology
- Track 8. Petroleum engineering and petroleum geochemistry
- Track 9. Sedimentology, stratigraphy, paleontology, geomorphology, pedology
- Track 10. Structural/petroleum/mining geology, geodynamics, marine geology

The dynamic four-day conference provided more than 450 attendees with opportunities to share their latest unpublished findings and learn the newest geoscience studies. The event also allowed attendees to meet and discuss with the journal's editors and reviewers.

More than 950 short contributing papers to the conference were submitted by authors from more than 70 countries. After a pre-conference peer review process by more than 500 reviewers, 700 papers were accepted. These papers were published as chapters in the conference proceedings by Springer.

The conference proceedings consist of ten edited volumes, each edited by the following group of Arabian Journal of Geosciences (AJGS) editors and other guest editors:

Volume 1. Patterns and Mechanisms of Climate, Paleoclimate, and Paleoenvironmental Changes from Low-Latitude Regions

Zhihua Zhang (AJGS Editor): Beijing Normal University, Beijing, China
Nabil Khélifi (AJGS Editor): Earth Sciences Editorial Department, Springer, Heidelberg, Germany
Abdelkader Mezghani (Guest Editor): Norwegian Meteorological Institute, Norway
Essam Heggy (Guest Editor): University of Southern California and Jet Propulsion Laboratory, Caltech, USA

Volume 2. Advances in Remote Sensing and Geo Informatics Applications

Hesham M. El-Askary (Guest Editor): Schmid College of Science and Technology at Chapman University, USA
Saro Lee (AJGS Editor): Korea Institute of Geoscience and Mineral Resources, Daejeon, South Korea
Essam Heggy (Guest Editor): University of Southern California and Jet Propulsion Laboratory, Caltech, USA
Biswajeet Pradhan (AJGS Editor): University of Technology Sydney, Sydney, Australia

Volume 3. Recent Advances in Geo-Environmental Engineering, Geomechanics and Geotechnics, and Geohazards

Amjad Kallel (AJGS Editor): ENIS, University of Sfax, Tunisia
Zeynal Abiddin Erguler (AJGS Editor): Dumlupinar University, Kütahya, Turkey
Zhen-Dong Cui (AJGS Editor): China University of Mining and Technology, Xuzhou, Jiangsu, China
Ali Karrech (AJGS Editor): The University of Western Australia, Australia
Murat Karakus (AJGS Editor): University of Adelaide, Australia

Pinnaduwa Kulatilake (AJGS Editor): Department of Materials Science and Engineering, The University of Arizona, USA
Sanjay Kumar Shukla (AJGS Editor): School of Engineering, Edith Cowan University, Perth, Australia

Volume 4. Exploring the Nexus of Geoecology, Geography, Geoarcheology and Geotourism: Advances and Applications for Sustainable Development in Environmental Sciences and Agroforestry Research

Haroun Chenchouni (AJGS Editor): University of Tebessa, Algeria
Ezzoura Errami (Guest Editor): Chouaïb Doukkali University, El Jadida, Morocco
Fernando Rocha (Guest Editor): University of Aveiro, Portugal
Luisa Sabato (AJGS Editor): Università degli Studi di Bari "Aldo Moro", Bari, Italy

Volume 5. On Significant Applications of Geophysical Methods

Narasimman Sundararajan (AJGS Editor): Sultan Qaboos University, Muscat, Oman
Mehdi Eshagh (AJGS Editor): University West, Trollhättan, Sweden
Hakim Saibi (AJGS Editor): United Arab Emirates University, Al-Ain, Abu Dhabi, UAE
Mustapha Meghraoui (AJGS Editor): Université de Strasbourg, Strasbourg, France
Mansour Al-Garni (AJGS Editor): King Abdulaziz University, Jeddah, Saudi Arabia
Bernard Giroux (AJGS Editor): Centre Eau Terre Environnement, Québec, Canada

Volume 6. Advances in Sustainable and Environmental Hydrology, Hydrogeology, Hydrochemistry and Water Resources

Helder I. Chaminé (AJGS Editor): School of Engineering—ISEP, Polytechnic of Porto, Portugal
Maurizio Barbieri (AJGS Editor): University of Rome La Sapienza, Italy
Ozgur Kisi (AJGS Editor): Ilila State University, Tbilisi, Georgia
Mingjie Chen (AJGS Editor): Sultan Qaboos University, Muscat, Oman
Broder J. Merkel (AJGS Editor): TU Bergakademie Freiberg, Freiberg, Germany

Volume 7. Petrogenesis and Exploration of the Earth's Interior

Domenico Doronzo (AJGS Editor): Consejo Superior de Investigaciones Cientificas, Spain
Emanuela Schingaro (AJGS Editor): Università degli Studi di Bari Aldo Moro—UniBa, Italy
John S. Armstrong-Altrin (AJGS Editor): The National Autonomous University of Mexico, Mexico
Basem Zoheir (Guest Editor): Benha University, Egypt and University of Kiel, Germany

Volume 8. Advances in Petroleum Engineering and Petroleum Geochemistry

Santanu Banerjee (AJGS Editor): Indian Institute of Technology Bombay, Mumbai, India
Reza Barati (AJGS Editor): The University of Kansas, Lawrence, KS, USA
Shirish Patil (Guest Editor): Saudi Aramco and King Fahd University of Petroleum and Minerals, Dhahran, Saudi Arabia

Volume 9. Paleobiodiversity and Tectono-sedimentary Records in the Mediterranean Tethys and Related Eastern Areas

Mabrouk Boughdiri (AJGS Editor): University of Carthage, Amilcar, Tunisia
Beatriz Bádenas (AJGS Editor): University of Zaragoza, Zaragoza, Spain
Paul Selden (AJGS Editor): University of Kansas, Lawrence, Kansas, USA
Etienne Jaillard (Guest Editor): Université Grenoble Alpes, France
Peter Bengtson (AJGS Editor): Universität Heidelberg, Heidelberg, Germany
Bruno R. C. Granier (AJGS Editor): Université de Bretagne Occidentale, Brest, France

Volume 10. The Structural Geology Contribution to the Africa-Eurasia Geology: Basement and Reservoir Structure, Ore Mineralisation and Tectonic Modelling

Federico Rossetti (Guest Editor): Università Roma Tre, Roma, Italy
Ana Crespo Blanc (Guest Editor): University of Granada, Spain
Federica Riguzzi (Guest Editor): National Institute of Geophysics and Volcanology, Roma, Italy
Estelle Leroux (Guest Editor): IFREMER, Unité Géosciences Marines, Plouzané, France
Kosmas Pavlopoulos (Guest Editor): Paris Sorbonne University Abu Dhabi, Abu Dhabi, UAE
Olivier Bellier (Guest Editor): CEREGE, Aix-en-Provence, France
Vasilios Kapsimalis (Guest Editor): Institute of Oceanography, Hellenic Centre for Marine Research, Anavyssos, Greece

About the Conference Steering Committee

General Chair

Abdullah Al-Amri: Founder and Editor-in-Chief of AJGS, King Saud University, Saudi Arabia

Conference Supervisor

Nabil Khélifi: Senior Publishing Editor, Springer Middle East and North African Program Springer, a part of Springer Nature, Heidelberg, Germany

Scientific Committee Chair

François Roure: Guest of Editorial Board of AJGS, IFP—Energies Nouvelles, France

Walter D. Mooney: Guest of Editorial Board of AJGS, US Geological Survey Western Region, USA

Local Organization Chair

Mabrouk Boughdiri: Associate Editor of AJGS, University of Carthage, Bizerte, Tunisia

Evaluation Chair

Amjad Kallel: Assistant Editor of AJGS, ENIS, University of Sfax, Tunisia

Publication Chair

Biswajeet Pradhan: Associate Editor of AJGS, University of Technology Sydney, Sydney, Australia

Essam Heggy: Guest of Editorial Board of AJGS, University of Southern California and Jet Propulsion Laboratory, Caltech, USA

Program Chair

Hakim Saibi: Associate Editor/Assistant Editor of AJGS, United Arab Emirates University, Al-Ain, Abu Dhabi, UAE

Domenico Doronzo: Associate Editor/Assistant Editor of AJGS, Consejo Superior de Investigaciones Cientificas, Spain

Communication Chair

Mohamed Ksibi: Guest of Editorial Board of AJGS, ISBS, University of Sfax, Tunisia

English Language Advisory Committee

Abdelmajid Dammak: ENIS, University of Sfax, Tunisia
Chokri Khalaf: FMS, University of Sfax, Tunisia
Dhouha Mabrouk: FLSHS, University of Sfax, Tunisia
Mohamed Elbahi: ENIS, University of Sfax, Tunisia
Sami Shami: ENIS, University of Sfax, Tunisia
Yasmine Basha: FLSHS, University of Sfax, Tunisia

Conference Manager

Mohamed Sahbi Moalla: Coordinator of AJGS, ISET, University of Sfax, Tunisia

Contents

About the Editors

Dr. Amjad Kallel is an Associate Professor of Environmental Geology. He holds a B.Eng. in Georesources and Environment (1998) from the University of Sfax (Tunisia) and an M.Sc. and a Ph.D. in Georesources and Environment (2004) from Hokkaido University (Japan). He joined Venture Business Laboratory (VBL) at Akita University, Japan (2005–2006) as a researcher focusing on refining and recycling technologies for the recovery of rare elements from natural and secondary sources. Back in Tunisia, he worked at the University of Gabes from 2006 to 2011, where he contributed to the elaboration of teaching programs at the Higher Institute of Water Sciences and Technologies of Gabes. Since 2011, he has joined the Sfax National School of Engineering (University of Sfax, Tunisia). There, he has also been involved in various research projects related to Environmental Geology and Environmental Geotechnics. He has organized many prestigious workshops, seminars, and international conferences. In 2016, he joined the *Arabian Journal of Geosciences* (Springer) as an Assistant Editor supporting the Editor-in-Chief. Currently, he is Executive Editor of the *Euro-Mediterranean Journal for Environmental Integration* (Springer).

Dr. Zeynal Abiddin Erguler is a Full Professor at Geological Engineering Department at Kütahya Dumlupınar University (Turkey). He holds a B.Sc. (1998), an M.Sc. (2001), and a Ph.D. (2007) in Geological Engineering from Hacettepe University (Turkey). His research interests mainly focus on rock mechanics, engineering geology, environmental geology, and soil mechanics. His current investigation is to understand and model thermo-hydro-mechanical behavior of shale rocks in the area of shale gas production. In addition to doing many researches and performing industry funded projects, he has also been teaching and supervising undergraduate and graduate students. In 2017, he joined the *Arabian Journal of Geosciences (AJGS)* as an Associate Editor responsible for evaluating submissions in the fields of rock mechanics, engineering geology, environmental geology, and soil mechanics.

Prof. Dr. Zhen-Dong Cui is the Dean of Institute of Geotechnical Engineering, China University of Mining and Technology. In July 2008, he obtained the Ph.D. from School of Civil Engineering, Tongji University, Shanghai, China. Since then, he had been a postdoctoral research fellow in the Hong Kong University of Science and Technology for 1 year. He joined Shanghai Institute of Geological Survey in 2009. In July 2010, he joined CUMT as an associate professor and was promoted to full professor in 2013. Supported by China Scholarship Council, from August 2015 to August 2016, as a visiting scholar, he researched and studied in the department of Civil, Environmental and Architectural Engineering, University of Colorado Boulder.

He won Nomination of 100 Excellent Doctoral Dissertations in China in 2011. He won Shanghai Excellent Doctoral Dissertations in 2010 and Excellent doctorate thesis of Tongji University in 2009. In 2015, he was awarded the third prize of Shanghai Natural Science. In 2003, he was awarded the second prize of Natural Science of the Ministry of Education. In 2008, he was awarded the second prize of Progress of Science and Technology in Shanghai. In 2012, he was selected as Qinglan Project for Outstanding Young Teachers of Jiangsu Province and in 2016, he was selected as 333 Talent Project in Jiangsu Province. In 2014 and 2017, he was twice selected as Young Academic Leader of China University of Mining and Technology. In 2015, he was awarded as Excellent Innovation Team Leader of China University of Mining and Technology.

His research interests focus mainly on the urban subway tunnel, the deformation of soft foundation, the soil dynamics, the centrifuge model test, and the artificial frozen soil. In the past 5 years, he took charge of National Natural Science Foundation of China (NSFC), Jiangsu Natural Science Foundation of China, Outstanding Innovation Team Project in China University of Mining and Technology, and Special Fund for China Postdoctoral Science Foundation. He published more than 60 papers, in which 37 English papers (indexed by SCI) have been published in *Engineering Geology*, *Natural Hazards*, *Computers and Geotechnics*, *Cold Regions Science and Technology*, *Environmental Earth Sciences*, *International Journal of Rock Mechanics and Mining Sciences*, etc. He published three books. He has applied for six national invented patents, among which four patents have been awarded. He is the specialized committee member of the soft soil engineering of the geotechnical engineering branch of the Civil Engineering Society of China, the member of international society for soil mechanics and geotechnical engineering (ISSMGE), and the member of international association for engineering geology and environment (IAEG).

Assoc./Prof. Ali Karrech has a Multidisciplinary B.Sc. in Engineering from Tunisia Polytechnic School (Tunisia, 2001), a M.Sc. in Mechanics of Solids and Structures (2004), and a Ph.D. in Materials and Structures (2008) from the Ecole Nationale des Ponts et Chaussées (Paris France). In addition, he received an Accreditation to Supervise Research in Engineering Sciences from the Ecole Normale Supérieure Paris-Saclay (Paris, France, 2013). Associate Professor Karrech is currently an Associate Professor and Graduate Research Coordinator at the School of Civil, Environmental, and Mining Engineering of the University of Western Australia. Prior to his appointment at the UWA, Assoc./Prof. Karrech was a Senior Research Scientist at the Commonwealth Scientific and Industrial Research Organisation (CSIRO) until July 2012. He had also held the position of Assistant professor at the Petroleum Institute of Abu Dhabi (2007–2009). He received several research awards especially from the Australian Research Council and The Research Council of Oman. Associate Professor Karrech's research interests are in the computational geomechanics for resource engineering applications. In particular, he focuses on poromechanics, plasticity, finite transformations, continuum damage mechanics, nonequilibrium thermodynamics, and multi-physics coupling. He is currently a member of Engineers Australia (MIEAust) and a member of the Australian Institute of Mining and Metallurgy (MAusIMM). In 2017, he joined the *Arabian Journal of Geosciences* (Springer) as an Associate Editor responsible for evaluating submissions in the fields of Civil, Environmental and Mining Engineering with particular focus on computational geomechanics for resource engineering applications.

Dr. Murat Karakus holds a B.Sc. in Mining Engineering (1991) from Hacettepe University (Turkey), and a Ph.D. in Geotechnical Engineering (2000) from University of Leeds in the UK. He is currently an Associate Professor in the School of Civil, Environmental and Mining Engineering and a Mining/Geotechnical Engineering Research Group Leader at the University of Adelaide, South Australia. His research interests focus on numerical analysis (continuum and discontinuum) in tunneling and rock mechanics. He has also been researching on strain burst analysis for deep tunnels and mining excavations. He has directed and involved many research projects including Australian Research Council grants. He has published several book chapters and over 80 papers in international Journals and Conferences. He is a member of Editorial board of three journals; *International Journal of Rock Mechanics and Mining Sciences*, *Tunnelling and Underground Space Technology*, *International Journal of Mining, Reclamation and Environment*. In 2017, he joined the AJGS as an Associate Editor responsible for evaluating submissions in the fields of Geo-environmental/Geotechnical Engineering.

Dr. Pinnaduwa Kulatilake is a Professor of Geotechnical Engineering and Director of Rock Mass Modeling and Computational Rock Mechanics Laboratories at the University of Arizona. He is a registered Professional Civil Engineer in California. He received his B.Sc. (in 1976) in Civil Engineering from the University of Sri Lanka, Peradeniya, MS (in 1978) in Soil Engineering from the Asian Institute of Technology, Bangkok, Thailand, and Ph.D. (in 1981) in Civil Engineering (with geotechnics emphasis) from the Ohio State University, USA. He has over 38 years of experience in rock mechanics & rock engineering associated with mining, civil and petroleum engineering, geotechnical engineering, and applications of probabilistic and numerical methods to geo-engineering. He has written over 240 papers and is a member of several technical committees. He has delivered over 30 keynote lectures and over 50 other invited lectures throughout the world on topics related to rock fracture network modeling, probabilistic geotechnics, mechanical and hydraulic properties of joints, rock slope stability, and mechanical and hydraulic behavior of rock masses. He has been a research paper reviewer for over 25 technical Journals and an editorial board member for *International Journal of Rock Mechanics and Mining Sciences*, *International Journal of Geotechnical and Geological Engineering*, *International Journal of Advances in Geological and Geophysical Engineering*, *Coal Science and Technology* and *Journal of Mining & Science*-Turkey. Currently, he is serving as an Associate Editor for *Arabian Journal of Geosciences*. He has taught short courses on stochastic fracture network modeling, rock slope stability analysis, Block theory, and rock joint roughness and aperture in Sweden, Mexico, Austria, USA, Canada, Hong Kong, Poland, Finland, Australia, South Korea, Sri Lanka, Egypt, Iran, Chile, China, Italy, Peru, and Tunisia. He has served over 20 years either as the primary or the sole examiner for the geological engineering professional exam conducted by the Arizona State Board of Technical Registration. He was a Visiting Professor at the Royal Institute of Technology and Lulea University of Technology in Sweden as part of his sabbatical leave. Also, he was a Visiting Research Fellow at the Norwegian Geotechnical Institute, for another part of his sabbatical leave. Due to the contributions he made on teaching, research, consulting and service activities, he was elected to the Fellow Rank of the American Society of Civil Engineers at the relatively young age of 45. In 2002, he received Distinguished Alumnus Award from the College of Engineering, Ohio State University and Outstanding Asian American Faculty Award from the University of Arizona in recognition of his achievements and contributions made to the advancement of his profession. In December 2005, Eurasian National University, Kazakhstan conferred him "Honorary Professorship". In August 2007, he organized and ran a successful International Conference on Soil & Rock Engineering in Sri Lanka. In January 2009, he organized and ran a high-quality International Conference on Rock Joints and Jointed Rock Masses in Tucson, Arizona. He was the guest editor for two special issues published in the Jour. of Geotechnical and Geological Engineering. He received "Kwang-Hua Visiting Professorship" for 2009–2010 from the College of Engineering, Tongji University, China. He was a Recipient of "Guest Professorship" from Wuhan University, China for 2010–2013. A few years back, he also received an award from the Chinese Academy of Sciences to spend a sabbatical in China as a Senior Visiting Professor. In 2013 and 2016, he received Peter Cundall awards.

Dr. Sanjay Kumar Shukla is an internationally recognized expert in the field of Civil (Geotechnical) Engineering. He is the Founding Editor-in-Chief of *International Journal of Geosynthetics and Ground Engineering*. He is also the Program Leader of Civil and Environmental Engineering at Edith Cowan University, Perth, Australia. He holds the distinguished professorship in civil engineering at five international universities in India and Fiji. His research interests include geosynthetics, ground improvement, and several other topics. He is an author of 10 books, and more than 200 research papers, including over 120 refereed journal papers. He is a fellow of Engineers Australia and several other professional bodies. He serves on the editorial boards of 12 international journals.

Part I
Keynote

A New Three-Dimensional Rock Mass Strength Criterion

Pinnaduwa H. S. W. Kulatilake, Mohammad Hadi Mehranpour, Ma Xingen, and Manchao He

Abstract

A new three-dimensional rock mass strength criterion was developed in this paper by extending an existing rock mass strength criterion. This criterion incorporates the effects of the intermediate principal stress, minimum principal stress and the anisotropy resulting from these stresses acting on the fracture system. In addition, the criterion has the capability of capturing the anisotropic and scale dependent behavior of the jointed rock mass strength by incorporating the effect of fracture geometry through the fracture tensor components. The new criterion was proposed after analyzing 284 numerical modeling results of the poly-axial, triaxial and biaxial compression tests conducted on the jointed rock blocks having one or two joint sets by the PFC^{3D} software. Some of these simulation results were compared with experimental ones to validate the developed PFC^{3D} model that was used for numerical modeling of jointed blocks. In this research to have several samples with the same properties, a synthetic rock material that is made of a mixture of gypsum, sand and water was used. To express the new rock mass strength criterion, it was also necessary to determine the intact rock strengths under the same confining stress combinations mentioned earlier. Therefore, the intact rock was also numerically simulated for all three compression tests and the intact rock strengths were found for 33 different minimum and intermediate principal stress combinations.

P. H. S. W. Kulatilake (✉) · M. H. Mehranpour
Rock Mass Modeling and Computational Rock Mechanics Laboratories, University of Arizona, Tucson, AZ 85721, USA
e-mail: kulatila@u.arizona.edu

M. Xingen · M. He
State Key Laboratory for Geomechanics and Deep Underground Engineering, China University of Mining and Technology, Beijing, China

Keywords

Discrete element method (DEM) • Particle flow code (PFC) • Rock mass strength • Polyaxial compression test • Intermediate principal stress • Fracture tensor

1 Introduction

Jointed rock masses are known as the combination of intact rock blocks and discontinuities. Due to the presence of complicated discontinuity geometry patterns, the inherent statistical nature of discontinuity geometrical parameters, and the variabilities and uncertainties involved in the estimation of discontinuity mechanical and geometrical properties, estimation of the jointed rock mass strength is difficult and challenging [1]. On the other hand, understanding the rock mass strength is crucial to design safe and economical structures in or on jointed rock masses.

Analytical, Empirical, and numerical are three available approaches to estimate the jointed rock mass strength [1, 2]. Limitations of the analytical and empirical approaches, and advantages of the numerical approaches in estimating rock mass strength are discussed in Kulatilake [2]. Because several parameters affect the strength of rock masses, numerous experimental tests are required to find the effect of these parameters on the strength of rock masses. That task is time consuming, very costly and impractical to perform in the field and laboratory. To solve this problem some researchers modeled rock masses with numerical modeling to propose new rock mass failure criteria. In this method, at first, a numerical model is calibrated with a limited number of experimental tests and physical modeling of the rock masses and then the calibrated model is expanded to more complicated situations with more diverse conditions [3–6]. Kulatilake et al. [1] and Wu and Kulatilake [6] used this procedure incorporating the 3DEC software to find the effect of the joint geometry parameters on the deformability properties of rock masses. To quantify the joint geometry

A. Kallel et al. (eds.), *Recent Advances in Geo-Environmental Engineering, Geomechanics and Geotechnics, and Geohazards*, Advances in Science, Technology & Innovation, https://doi.org/10.1007/978-3-030-01665-4_1

parameters, they used an extended form of the fracture tensor concept. Kulatilake et al. [4, 5] and He et al. [3] also extended the fracture tensor concept to fracture tensor components and developed new rock mass strength criteria.

Mehranpour et al. [7] used the same procedure based on experimental tests and PFC^{3D} [8] modeling on intact rock, jointed rock with one joint set and jointed rock with two non-orthogonal joint sets to develop new rock mass strength criteria in three dimensions. The new criteria considered the effect of all principal stresses in three dimensions and they are applicable for any type of rock mass, especially for rock masses which generally have non-orthogonal fracture systems. These criteria also show the anisotropic strength behavior of rock masses due to the orientation of joint sets. It should be mentioned that compared to other numerical methods, in the PFC^{3D}, macro parameter values are not directly used in the numerical model, and micro parameter values applicable between the particles should be calibrated using the macro property values, and then these calibrated micro parameter values are used in PFC^{3D} modeling. This paper presented one of the strength criteria developed in Mehranpour et al. [7] in a summarized form.

2 Procedures

In the conducted research to have several samples with the same properties a synthetic rock material that is made of a mixture of gypsum, sand and water was used. To develop a new rock mass strength criterion, first conventional experimental tests on the intact rock and the joint as well as the polyaxial compression tests on the intact rock and jointed samples were performed on the synthetic rock samples. It should be mentioned that polyaxial and triaxial compression tests were performed in the laboratory with a limited number of boundary stress conditions and joint set systems, because the experimental tests are expensive. Then, these experimental tests were simulated using PFC^{3D} and the numerical results were compared with the experimental results of synthetic intact rock and synthetic jointed rock blocks. If these two groups of results did not match, micro parameter values of PFC^{3D} were modified until very close results were obtained with an acceptable error. Accordingly, the micro properties of the PFC^{3D} model were calibrated based on the experimental test results; it turned out to be one of the challenging parts of this project. Afterwards, polyaxial, triaxial and biaxial compression tests for the intact rock and jointed rock blocks were simulated in the PFC^{3D} with different combinations of minimum (σ_3) and intermediate principal stresses (σ_2). After gathering the results from 284 numerical simulations, the development of a new rock mass failure criterion was initiated using the fracture tensor concept which was introduced by Oda [9] and developed into the fracture tensor components by Kulatilake et al. [1, 5]. Fracture tensor combines the joint orientation distribution, joint size distribution, joint density for each joint set and the number of joint sets by a second order tensor. Thus, the fracture tensor can show the anisotropy and scale effects of rock masses which are exhibited by the presence of joints. To express the new rock mass strength criterion, it was also necessary to determine the intact rock strengths under the same confining stress combinations mentioned earlier. Therefore, the intact rock was also numerically simulated for all three compression tests and the intact rock strengths were found for 33 different minimum and intermediate principal stress combinations. For details of the experimental and numerical procedures, the reader is referred to Mehranpour and Kulatilake [10] and Mehranpour et al. [7].

3 Results

Figures 12–15 of Mehranpour et al. [7] show the rock block strength values obtained for the synthetic jointed and intact rock models under different minimum and intermediate principal stress combinations. These figures indicate that for each combination of the minimum and intermediate principal stresses, the jointed rock blocks with 2 joint sets and 1 joint set have resulted in a lower strength compared to that of the synthetic intact rock and the jointed rock blocks with 2 joint sets have resulted in a lower strength value compared to that of the jointed rock blocks having 1 joint set with the same properties as the first joint set of the rock sample with 2 joint sets. This means that adding of joint sets to a sample under the same minimum and intermediate principal stress combination reduces the strength of the sample. Figure 12 of Mehranpour et al. [7] also shows that the intermediate principal stress has a significant effect on the synthetic intact rock strength and it can increase the intact rock strength up to about 25%. Increase of the intermediate principal stress while keeping the minimum principal stress constant, increases the strength of intact rock to a peak value and then the strength decreases. However, in Figs. 12–15 for each σ_3 level in the jointed rock models, the reduction of the strength after reaching the peak strength due to increase of σ_2 seems to be lower compared to that of the intact rock model. In some plots, even the strength reduction does not seem to exist especially for low σ_3 values and high joint set dip angles.

4 Development of a New Rock Mass Strength Criterion

By subjecting 150 synthetic jointed rock blocks to biaxial loading, Kulatilake et al. [5] developed a biaxial strength criterion for jointed rock masses. Based on extensive laboratory and numerical polyaxial test results on jointed coal blocks, He et al. [3] extended the Kulatilake et al. [5] criterion to the polyaxial compressive stress condition. The obtained results in the current study led to the following observations:

(a) Increase of joint set dip angles, in general, reduces the jointed rock block strength and increases F_{22} (the fracture tensor component in σ_2 direction) and F_{33} (the fracture tensor component in σ_3 direction). Thus, the increase of F_{22} and F_{33} reduces the jointed rock block strength.
(b) Increase of the minimum and intermediate principal stresses reduces the effect of joint shearing on the jointed rock block strength. Therefore, the increase of the minimum and intermediate principal stresses reduces the effects of F_{22} and F_{33}. However, this reduction for low minimum and intermediate principal stresses is relatively higher compared to high minimum and intermediate principal stresses.
(c) The effect of the minimum principal stress on the joints increases with decreasing angle between the dip direction angle of the joint set and the minimum principal stress direction. Thus, the increase of F_{33} increases the effect of σ_3 on the joints.
(d) The effect of the intermediate principal stress on the joints increases with the decreasing angle between the dip direction angle of the joint set and the intermediate principal stress direction. Thus, the increase of F_{22} increases the effect of σ_2 on the joints.

Based on the above-mentioned observations the following equation was proposed as a new rock mass strength criterion:

$$S_r = \frac{\sigma_J}{\sigma_I} = exp - \left[\frac{\lambda_3}{p_3\left(\frac{\sigma_3}{\sigma_c}\right)^{q_3}+1}F_{33} + \frac{\lambda_2}{p_2\left(\frac{\sigma_2}{\sigma_c}\right)^{q_2}+1}F_{22}\right] \tag{1}$$

where S_r is the strength ratio between the jointed rock mass strength, σ_J, under the minimum and intermediate principal stresses σ_3 and σ_2 and the intact rock strength, σ_I, under the same minimum and intermediate principal stresses, σ_c is the uniaxial compressive strength of the intact rock, λ_2, λ_3, p_2, p_3, q_2 and q_3 are empirical coefficients. It should be mentioned that if σ_I for the intended σ_3 and σ_2 combination is not available, based on the Mehranpour and Kulatilake [11] paper one of the three intact rock failure criteria out of the Modified Lade, Modified Wiebols and Cook and Mogi is recommended to represent the intact rock strength value. However, because in this research the intact rock strength for all minimum and intermediate principal stress combinations is available, it is not necessary to use intact rock failure criteria to estimate the intact rock strength.

It should be mentioned that if the joints have the same mechanical properties with isotropic behavior on the joint plane, the effect of σ_2 variation on F_{22} should be the same as the effect of σ_3 variation on F_{33}. Therefore, under this condition Eq. 1 can be simplified to Eq. 2 as follows:

$$S_r = \frac{\sigma_J}{\sigma_I} = exp - \lambda\left[\frac{F_{33}}{p\left(\frac{\sigma_3}{\sigma_c}\right)^{q}+1} + \frac{F_{22}}{p\left(\frac{\sigma_2}{\sigma_c}\right)^{q}+1}\right] \tag{2}$$

In Eq. 2, λ, p and q are empirical coefficients. In the conducted research because all the joints were saw cut, they have the same isotropic mechanical behavior on the joint plane. Thus, to fit the new rock mass strength criterion for the numerical modeling results and to find the accuracy of the new rock mass strength criterion Eq. 2 can be used. The best combination of the empirical coefficients was found by maximizing the coefficient of determination, R^2. The maximum R^2 was found to be 0.94, indicating a very strong fit. It results in the best values of 0.675, 3.16 and 0.6, for λ, p and q, respectively. Figure 1 shows the predicted strength values versus the strength values from the PFC^{3D} modeling for all 284 data points. It indicates that the suggested strength criterion (Eq. 2) is highly suitable to represent the PFC^{3D} data.

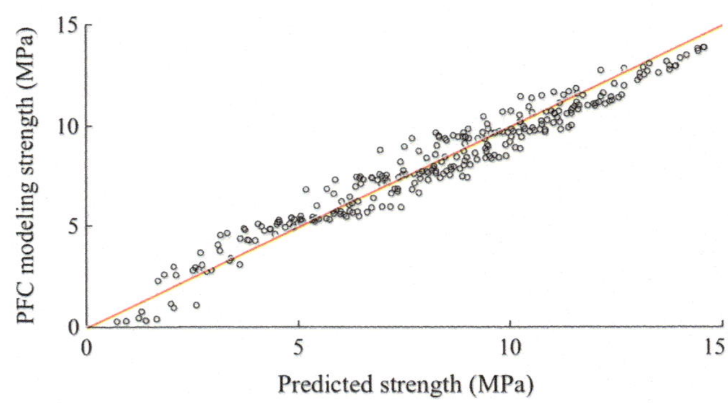

Fig. 1 Predicted strength values based on the new rock mass strength criterion based on Eq. 2 versus the strength values from PFC^{3D} modeling for all 284 data points from 12 different joint systems having different boundary conditions (R^2 = 0.94)

5 Discussion

The equations given in Sect. 4 to estimate the jointed block strength for synthetic rock are normalized with respect to the synthetic intact rock strength. Therefore, the equations are applicable for any rock mass. The equations allow estimating the normalized jointed block strength in any direction in three dimensions. By estimating the strength in different directions, the strength anisotropy and the minimum normalized jointed block strength can be estimated in three dimensions. The intact block strength can be estimated using one of the available intact rock strength criteria. To estimate the parameters of the intact rock strength criterion, it will be necessary to perform a few laboratory tests as usual. To apply the equations given for normalized jointed block strength for any rock mass, first, the fracture geometry data (number of fracture sets and orientation distribution, size distribution and intensity of each set) should be collected for the intended rock mass. These data will allow the calculation of the two fracture tensor components perpendicular to the direction jointed block strength is desired. These two fracture tensor components go into the normalized jointed block strength equation. The confining stresses should be applied based on the in situ stress system. For the time being the estimated coefficient values of the equations can be used to estimate the jointed block strength. It is important to note that these coefficient values depend on the ratios of joint mechanical property values to intact rock property values. This dependence should be investigated in a future research.

6 Conclusion

Based on the observations from synthetic rock testing (on intact, joints and jointed rock) in the laboratory, and the intact and jointed rock modeling results using PFC^{3D}, and the fracture tensor component concept, an existing rock mass strength criterion was extended to include the stress anisotropy and develop a new three-dimensional rock mass strength criterion (Eq. 1). The new criterion clearly showed the effect of the intermediate principal stress as well as the minimum principal stress and joint orientation on the rock mass strength. Because the developed jointed block strength criterion is expressed in a normalized form by dividing by the intact block strength, the normalized jointed block strength criterion is applicable for any rock mass. Guidelines are given to show how the developed strength criterion can be applied to field rock masses.

References

1. Kulatilake, P.H.S.W., Wang, S., Stephansson, O.: Effect of finite size joints on deformability of jointed rock at the three-dimensional level. Int. J. Rock Mech. Min. Sci. **30**(5), 479–501 (1993)
2. Kulatilake, P.H.S.W.: Physical, empirical and numerical modeling of jointed rock mass strength In: Feng, XT., Hudson, J. (eds.) Invited Book Chapter Rock Mech. Eng., CRC Press, Balkema, Taylor & Francis Group (2016)
3. He, P.F., Kulatilake, P.H.S.W., Liu, D.Q., He, M.C.: Development of a new three-dimensional coal mass strength criterion. Int. J. Geomech. **17**(3) (2016)
4. Kulatilake, P.H.S.W., Malama, B., Wang, J.: Physical and particle flow modeling of jointed rock block behavior under uniaxial loading. Int. J. Rock Mech. Min. **38**(5), 641–657 (2001)
5. Kulatilake, P.H.S.W., Park, J., Malama, B.: A new rock mass failure criterion for biaxial loading conditions. Geotech. Geol. Eng. **24**(4), 871–888 (2006)
6. Wu, Q., Kulatilake, P.H.S.W.: REV and its properties on fracture system and mechanical properties, and an orthotropic constitutive model for a jointed rock mass in a dam site in China. Comput. Geotech. **43**, 124–142 (2012)
7. Mehranpour, M.H., Kulatilake, P.H.S.W., Xingen, M., He, M.C.: Development of new rock mass strength criteria. Rock Mechanics and Rock Engineering (2018). https://doi.org/10.1007/s00603-018-1538-6.
8. Itasca Consulting Group Inc.: PFC manual, version 5.0, Minneapolis, (2016)
9. Oda, M.: Fabric tensor for discontinuous geological materials. Soils Found. **22**(4), 96–108 (1982)
10. Mehranpour, M.H., Kulatilake, P.H.S.W.: Improvements for the smooth joint contact model of the particle flow code and its applications. Comput. Geotech. **87**, 163–177 (2017)
11. Mehranpour, M.H., Kulatilake, P.H.S.W.: Comparison of six major intact rock failure criteria using a particle flow approach under true-triaxial stress condition. Geomech. Geophy. Geo-energ. Geo-resour. **2**(4), 203–229 (2016)

New Tools and Techniques of Remote Sensing for Geologic Hazard Assessment

Janusz Wasowski

Abstract

The use of earth surface remote sensing in geology is increasing thanks to the continuous development of sophisticated sensors and the improvements in digital image processing techniques. Here we focused on new remote sensing tools and techniques capable of delivering high-resolution data for geologic hazard investigations. These include airborne imaging systems such as UAVs (Uninhabited Aerial Vehicles) and LiDAR (Light Detection and Ranging), as well as new radar sensors onboard of Earth-orbiting satellites. We emphasize the applications of advanced synthetic aperture radar interferometry (InSAR) techniques referred to as multi-temporal interferometry (MTI). With the free imagery availability from the current (since 2014) European Space Agency (ESA) Sentinel-1 mission, MTI can now be more affordably exploited for long-term (years), regular (weekly-monthly), precise (mm) measurements of ground displacements over large areas (thousands of km^2). This, in turn, means improved detection and monitoring capability of landslide/slope instability, seismic and volcanic hazards.

Keywords

Remote sensing • Geologic hazard • UAV • LiDAR • Satellite InSAR

1 Introduction

Remote sensing is often defined as the process of obtaining information about an object, area or process via the analysis of data (typically images) acquired by terrestrial or air- or space-borne sensors. For background information on remote sensing principles and digital image processing and interpretation, we refer the readers to textbooks and manuals (e.g., [1–3]).

Systematic use of remote sensing in geology began in the 20th century, thanks to an increased availability of airphotos. Initially, black and white photography was used in reconnaissance geologic mapping [4]. Another important step was the launch of Earth Resource Technology Satellite-1 (ERTS-1) in 1972, which heralded the exploitation of space-borne sensors for mapping of the Earth's resources.

Here we discussed selected innovative remote sensing techniques and their applications in investigations of landslides/slope instabilities, as well as seismic and volcanic hazards. These hazards can affect wide areas and require synoptic (and possibly low-cost) information for their assessment. We also provide representative references on the uses of new remote sensing techniques in studies of specific geologic hazards.

2 Innovative Remote Sensing Techniques and Applications

2.1 Uninhabited Aerial Vehicles

UAVs, also referred to as Unmanned Aerial Systems (UAS), Remotely Piloted Aircraft Systems (RPAS) or simply drones, typically require a human operator on the ground [5]. Drones can have onboard various types of simple or sophisticated imaging sensors. However, they usually include light digital cameras used to gather very high-resolution (cm-dcm) images. Given the flexibility in survey scheduling, the UAVs are especially suitable for rapid assessment of geologic hazards during the emergency response phase (e.g., [6]). Furthermore, with the extended flight endurance (several hours), a day-long surveillance capability is secured for monitoring active geologic hazards (e.g., volcanos, landslides).

J. Wasowski (✉)
National Research Council, CNR-IRPI, Amendola 122i,
70126 Bari, Italy
e-mail: j.wasowski@ba.irpi.cnr.it

A. Kallel et al. (eds.), *Recent Advances in Geo-Environmental Engineering, Geomechanics and Geotechnics, and Geohazards*, Advances in Science, Technology & Innovation, https://doi.org/10.1007/978-3-030-01665-4_2

Importantly, being usually low-flying platforms, UAVs can acquire images also under low altitude cloud conditions. However, their application can be restricted (or even unfeasible) in the presence of strong winds. The aviation regulations can also significantly limit the use of UAVs.

While airborne LiDAR and satellites provide, respectively, (sub)regional and regional to global scale coverage, UAVs are best suited for acquisition of very high-resolution images over smaller areas and local-scale applications (e.g., 3D mapping, [7]; engineering geological surveys, [8]).

2.2 LiDAR

A good overview of LiDAR technology is given by Tratt [9]. LiDAR relies on a laser beam scanning which generates spatially "continuous" very high-resolution imagery (clouds of points) of the ground surface and associated natural and artificial features. One can distinguish between Terrestrial Laser Scanner (TLS) and Airborne Laser Scanner (ALS), or Airborne Laser Swath Mapping (ALSM), with the former being more suitable for local-scale applications, and the latter for regional studies. TLS and ALS can achieve, respectively, dcm and cm spatial resolutions and sub-cm and dcm measurement precisions.

ALS is exploited to produce high-resolution topographic maps and digital elevation models (DEM) for local to wide-area geo-hazard studies. Digital cameras are typically employed during airborne LiDAR surveys to acquire high-resolution optical images. Furthermore, surface change detection is provided through repeated TLS or ALS surveys. This approach has been used for landslide motion and soil erosion volume estimates [10] and volcanic activity assessment and monitoring (e.g., [11] and references therein). Importantly, good results can also be obtained in the presence of dense vegetation. For example, Cunningham et al. [12] reported an interesting case of mapping of seismogenic faults in forested mountainous terrain.

However, multi-temporal LiDAR has significant costs especially in case of airborne surveys. Regular repeated measurements are more feasible with TLS, which, nevertheless, requires the presence of a human operator during the scans. One important drawback is that the TLS and ALS instruments are expensive.

2.3 Satellite Multi-temporal Interferometry (MTI)

Conventional differential interferometry (DInSAR) and advanced MTI techniques e.g., Persistent Scatterers Interferometry—PSInSAR™/PSI and related approaches, as well as Small Baseline Subset—SBAS and similar methods [13], rely on radar imagery acquired by satellites periodically re-visiting the same area of interest. Using these techniques, we can obtain measurements of distance changes between the space-borne radar sensor and the ground surface features (e.g., human-made structures such as buildings, roads, but also rock outcrops and bare ground) that backscatter electromagnetic radiation emitted by the radar.

Where vegetation cover is scarce, MTI can provide precise (mm-cm resolution), high-density measurements (from tens/hundreds to thousands points/km^2) on slow (mm-dcm/year) deformations of the ground surface or human-made structures. Radar satellites offer regional to global coverage and nearly all-weather ("see" through the clouds) measurement capability. In the last 10 years, the MTI users have benefited from the improved spatial (from 3 to 1 m) and temporal (from 11 to 4 days) resolutions of the new generation radar sensors (COSMO-SkyMed constellation and TerraSAR-X). Since 2014, the applications of MTI in geologic hazard investigations can take advantage of the Sentinel-1 radar satellite mission of the European Space Agency. Sentinel-1 provides regular global-scale coverage, high temporal resolution (from 12 to 6 days) and, importantly, free imagery [14].

Radar interferometry has been exploited in studies of ground deformations related to earthquake and volcanic hazards since the 1990s, (e.g., [15]). An overview of the applications of MTI and other space-borne techniques in specific geohazard investigations, as well as relevant literature, can be found in a comprehensive ESA report [16]. The use of conventional InSAR and MTI in research oriented engineering geology investigations of landslides and unstable slopes has been thoroughly discussed in scientific literature (e.g., [17–19]).

3 Conclusion

New generation aerial and space-borne sensors and innovative remote sensing techniques are capable of delivering high-resolution imagery needed to produce detailed topographic maps and digital elevations models. The background topographic information can be frequently updated and represents essential input for geologic hazard mapping and assessment.

Furthermore, high-precision measurements of ground surface displacements can be delivered by repeated LiDAR surveys and MTI. Satellite radars are well suited for multi-scale (from regional to local scale) ground deformation monitoring because of wide/global-area coverage and regular schedule with increasing re-visit frequency (e.g., Sentinel-1 mission). Thanks to this, the detection of geologic hazards (e.g., active volcanos or unstable slopes) and their monitoring can be now more effective, especially when combined with suitable ground truth.

References

1. Drury, S.A.: Image interpretation in Geology, 3rd edn. Nelson Thornes, Cheltenham, UK (2001)
2. Khorram, S., van der Wiele, C.F., Koch, F.H., Nelson, S.A.C., Potts, M.D.: Principles of Applied Remote Sensing. Springer International Publishing, Cham, Heidelberg, New York, Dordrecht, London (2016)
3. Lillesand, T.M., Kiefer, R.W., Chipman, J.W.: Remote Sensing and Image Interpretation, 7th edn. Wiley, Hoboken, USA (2015)
4. Finkl, C.W. (ed.): Encyclopedia of Earth Sciences Series, pp. 1388–4360. Springer ISSN, Berlin (1984)
5. Barnhart, R.K., Hottman, S.B., Marshall, D.M., Shappee, E.: Introduction to Unmanned Aircraft Systems. CRC Press, Boca Raton, Fl, USA (2012)
6. Kerle, N.: Remote sensing of natural hazards and disasters. In: Bobrowsky, P. (ed.) Encyclopedia of Natural Hazards, pp. 837–847. Springer, Dordrecht, Heidelberg, New York (2013)
7. Nex, F., Remondino, F.: UAV for 3D mapping applications: a review. Appl. Geomatics **6**, 1–15 (2014)
8. Tziavou, O., Pytharouli, S., Souter, J.: Unmanned Aerial Vehicle (UAV) based mapping in engineering geological surveys: considerations for optimum results. Eng. Geol. **232**, 12–21 (2018)
9. Tratt, D.M.: Emerging technologies, LiDAR. In: Njoku, E.G. (ed.) Encyclopedia of Remote Sensing, pp. 177–185. Springer International Publishing, Cham, Heidelberg, New York, Dordrecht, London (2014)
10. DeLong, S.B., Prentice, C.S., Hilley, G.E., Ebert, Y.: Multitemporal ALSM change detection, sediment delivery, and process mapping at an active earthflow. Earth Surf. Proc. Land. **37**, 262–272 (2012)
11. Abellan, A., Marc-Henri Derron, M.H., Jaboyedoff, M.: Use of 3D point clouds in geohazards special issue: current challenges and future trends. Remote Sens. **8**(2), 130 (2016)
12. Cunningham, D., Grebby, S., Tansey, K., Gosar, A., Kastelic, V.: Application of airborne LiDAR to mapping seismogenic faults in forested mountainous terrain, southeastern Alps, Slovenia. Geophys. Res. Lett. **33**, L20308 (2006)
13. Hooper, A., Bekaert, D., Spaans, K., Arıkan, M.T.: Recent advances in SAR interferometry time series analysis for measuring crustal deformation. Tectonophysics **514–517**, 1–13 (2012)
14. ESA Homepage, https://www.esa.int/Our_Activities/Observing_the_Earth/Copernicus/Overview4, Last accessed 2018/06/01
15. Massonnet, D., Feigl, K.L.: Radar interferometry and its application to changes in the Earth's surface. Rev. Geophys. **36**, 441–500 (1998)
16. Bally, P. (ed): Satellite Earth Observation for Geohazard Risk Management. In: The Santorini Conference, Santorini, Greece, 21–23 May 2012. ESA Publication, STM-282 (2013). https://doi.org/10.5270/esa-geo-hzrd-2012
17. Colesanti, C., Wasowski, J.: Investigating landslides with space-borne synthetic aperture radar (SAR) interferometry. Eng. Geol. **88**(3–4), 173–199 (2006)
18. Wasowski, J., Bovenga, F.: Investigating landslides and unstable slopes with satellite multi temporal interferometry: current issues and future perspectives. Eng. Geol. **174**, 103–138 (2014)
19. Wasowski, J., Bovenga, F.: Remote sensing of landslide motion with emphasis on satellite multitemporal interferometry applications: an overview. In Davies, T. (ed.) Landslide Hazards, Risks and Disasters, pp 345–403. Elsevier Inc, Amsterdam (2014). http://dx.doi.org/10.1016/B978-0-12-396452-6.00011-2

Land Subsidence Induced by the Engineering-Environmental Effect in Shanghai, China

Zhen-Dong Cui

Abstract

Urban land subsidence is influenced by many factors, including the building loads, the changing of groundwater level, the construction of underground structures, etc. The study results indicate that the building itself experiences the maximum subsidence and there exists remarkable superimposition effect among the high-rise buildings. However, the land subsidence decreases dramatically with the distance increasing. The range of land subsidence caused by building loads is within 300 m through the in-site monitoring data.

Keywords

Land subsidence • In-site monitoring • High-rise buildings • Numerical simulation

1 Introduction

Land subsidence in the soft soil area has characteristics of the slow generation, the long duration, the wide ranges of affection, the complex mechanisms of generation and the difficult prevention [1]. Shanghai is one of the first cities which suffers serious land subsidence in China [2]. It was mainly due to the irrational withdrawal of groundwater, but from the 1960s, the withdrawal of groundwater was restricted. From the end of the 1970s, the pumping of groundwater was strictly controlled in the urban area of Shanghai. During this period, the 1-year average land subsidence was 3–5 mm and the land subsidence was controlled better. During the 1990s, the pumping of groundwater was strictly controlled in the urban area. However, the rate of land subsidence accelerated again in Shanghai [3].

Land subsidence caused by building loads was studied by some researchers. The additional stress caused by building loads in the foundation was the main reason for changing the stress state of shallow soil [4]. The land subsidence caused by building loads was mainly concentrated in the single building as well as the dense high-rise building group, which was studied by the theoretical calculation, the in-site monitoring, the laboratory tests and the numerical simulation.

In this paper, the land subsidence caused by four high-rise buildings including World Financial Center (SWFC), Bank of China Tower (BOC Tower), Shanghai Merchants Tower and Jinmao Tower, which is located in Lujiazui area of Shanghai, was studied by the in-site monitoring data. The numerical simulation was conducted to study the land subsidence caused by two high-rise buildings with relatively close distance, Jinmao Tower and World Financial Center.

2 In-Site Monitoring and Numerical Simulation

Shanghai is situated at the fore of the Yangtze River delta. The Quaternary deposit is thick and generally reaches 250–300 m in the urban area. The upper layer (about 0–150 m) is mainly of grey soil and is composed of soft-plastic clay and sandy soil. In order to better monitor the land subsidence in Shanghai, a large number of monitoring systems has been built up in the early days. Up to now, a comprehensive land subsidence monitoring systems in Shanghai, consisting of bedrock marks, layered marks and GPS stations have been set up.

Figure 1 illustrates the monitoring points around four buildings. Along Huayuanshiqiao Road, at the positions of 7, 12, 20, 30, 60, 100, 150, 200, 250 and 300 m away from World Financial Center, there are 10 monitoring points: A01, A02, A03, A04, A05, A06, A07, A08, A09 and A10, respectively. Along Dongtai Road, at the positions of 7, 12,

Z.-D. Cui (✉)
State Key Laboratory for Geomechanics and Deep Underground Engineering, School of Mechanics and Civil Engineering, China University of Mining and Technology, Xuzhou, 221116, Jiangsu, People's Republic of China
e-mail: czdjiaozuo@163.com

A. Kallel et al. (eds.), *Recent Advances in Geo-Environmental Engineering, Geomechanics and Geotechnics, and Geohazards*, Advances in Science, Technology & Innovation, https://doi.org/10.1007/978-3-030-01665-4_3

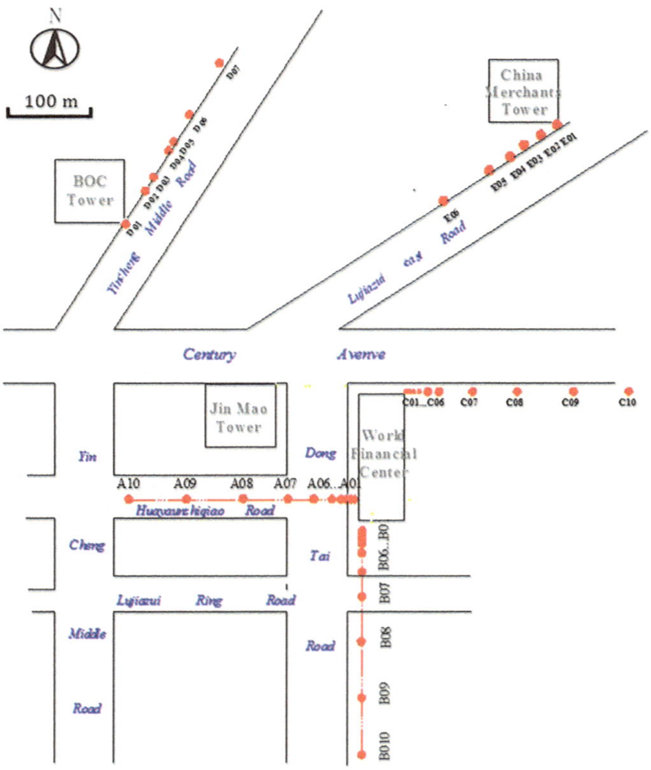

Fig. 1 The relative location and arrangement of monitoring points of four buildings

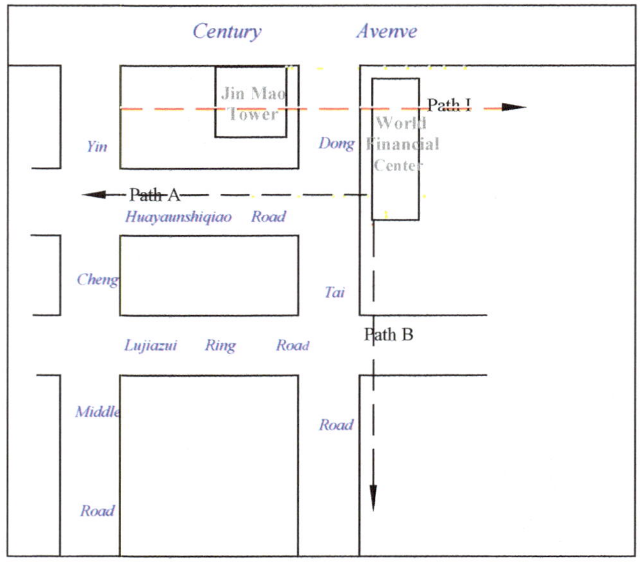

Fig. 2 Layout of buildings in numerical simulation model

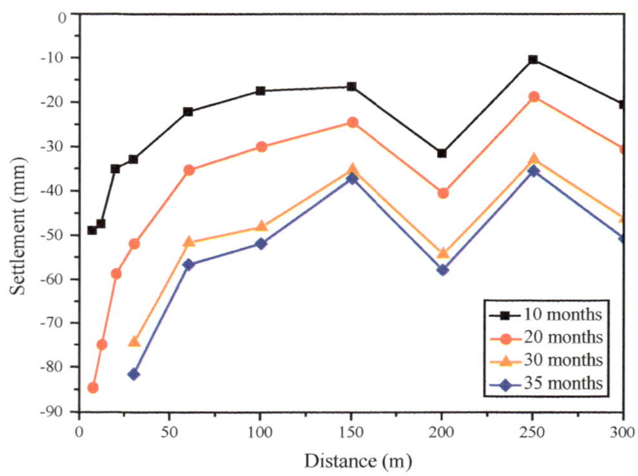

Fig. 3 Subsidence of A01–A10

20, 30, 60, 100, 150, 200, 250 and 300 m away from World Financial Center, there are 10 monitoring points: B01, B02, B03, B04, B05, B06, B07, B08, B09 and B10, respectively.

Due to the complexity of the actual project, it is difficult to accurately simulate the land subsidence in the study area. So only two super high-rise buildings were selected, which were Jinmao Tower and SWFC and the numerical simulations were conducted to study the land subsidence caused by two closely spaced high-rise buildings. Figure 2 illustrates the layout of buildings in the numerical simulation model.

3 Results

Figures 3 and 4 illustrate the variations of land subsidence with the distance away from World Financial Center at 10 months (October 2006), 20 months (August 2007), 30 months (June 2008) and 35 months (November 2008), respectively.

Figure 3 illustrates the variations of land subsidence with the distance away from the World Financial Center along Huayuanshiqiao Road. The point A01, which was the nearest point to the building, experienced the maximum land subsidence. With the distance increase, the land subsidence decreased. But at the point A08 (200 m), the land subsidence increased again, because the monitoring point A08 had the minimum distance from Jinmao Tower, which showed the superimposition effect of two high-rise buildings. In addition, the monitoring point A10 had greater land subsidence than the point A09, which may be affected by the traffic load of Yincheng Middle Road.

Figure 4 illustrates the variations of land subsidence with the distance away from World Financial Center along Dongtai Road. The subsidence of monitoring points from B01–B10 had the same varying regulation at 10 months, 20 months and 30 months. However, the subsidence of monitoring points from B01–B06 at 35 months was abnormal.

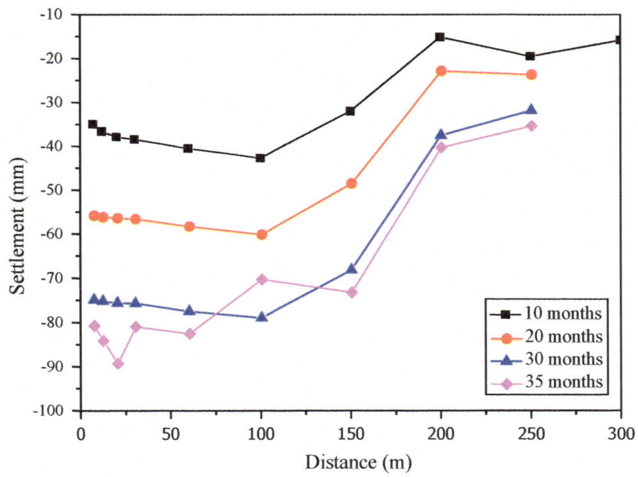

Fig. 4 Subsidence of B01–B10

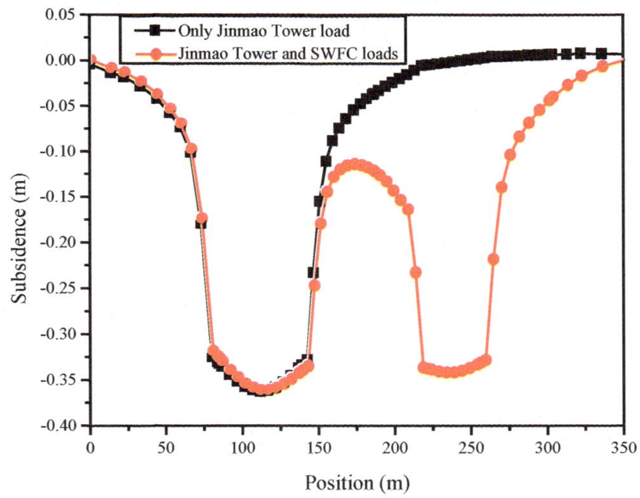

Fig. 6 The subsidence of path I

Figure 5 illustrates the variations of the additional stress with the different positions. When the load of Jinmao Tower was applied, the maximum additional stress was generated at the center of the base and gradually decayed to zero along the radius of the 150 m. After the Jinmao Tower and SWFC loads were both applied, the additional stresses in the bottom of the two building foundation centers had two extreme points. Meanwhile, the superposition effect of stresses existed in the central area of two high-rise buildings. Figure 6 illustrates the variations of land subsidence with different positions along Path I. The building itself experienced the maximum subsidence. The differential subsidence occurred in the building foundation and its surrounding ground. A comparative analysis of Figs. 5 and 6 showed that the distribution of soil settlement was determined by the distribution of the additional stress of the foundation under the flexible pile cap.

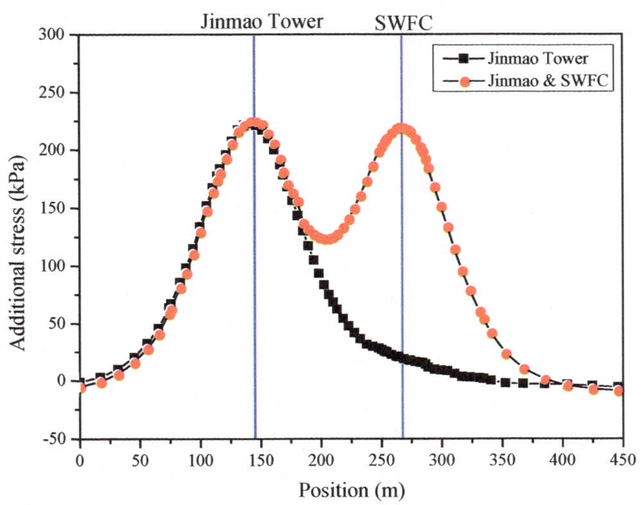

Fig. 5 Variations of additional stress with positions (path I, 100 m soil depth)

4 Discussion

Points A08 and B08 have the same settlement rules, which is mainly related to the construction process of the World Financial Center. Nevertheless, the settlement value of monitoring point A08 is larger than that of Point B08 due to the effect of the Jinmao Tower load. The settlement value of A08 is larger than that of B08 by 113.2% at the maximum and the minimum value is higher than 43.8%. This shows that the land subsidence caused by high-rise buildings has a remarkable superposition effect, which will cause the value of land subsidence to exceed the expected value. It is important to control the value of the loads in order to prevent excessive land subsidence.

5 Conclusion

Based on the high-rise building group, the rules of land subsidence caused by the engineering-environmental effect were studied.

(1) Building itself may experience a maximum settlement because of the remarkable subsidence superimposition effect between the high-rise buildings which exceeds the allowable values.
(2) The land subsidence decreases dramatically with the distance increase and the range of the land subsidence caused by the building load is about 300 m from the in-site monitoring.

(3) In the design stage, it is reasonable to control the distance between the dense high-rise buildings in order to avoid the settlement of regional ground over the allowable value and affect the normal use of other structures.

Acknowledgements This work was funded by National Key R&D Program of China (2016YFC0600903).

References

1. Cui, Z.D.: Study of the land subsidence caused by the dense high-rise building group in the soft soil area. Ph.D. thesis, Tongji University, Shanghai (2008)
2. Monjoie, A., Paepe, R., Su, H.Y.: Land subsidence in Shanghai (P.R. of China). Bull. Eng. Geol. Environ. **46**(1), 5–7 (1992)
3. Cui, Z.D., Tang, Y.Q., Yan, X.X., et al.: Evaluation of the geology-environmental capacity of buildings based on the ANFIS model of the floor area ratio. Bull. Eng. Geol. Environ. **69**, 111–118 (2010)
4. Tang, Y.Q., Cui, Z.D., Wang, X.H., et al.: Preliminary research on the land subsidence induced by engineering environmental effect of dense high-rise building group. Northwest. Seismol. J. **29**(2), 105–108 (2007)

Part II

Geo-Environmental Engineering: Identification and Assessment of Different Geo-Environmental Problems

Synchrotron X-Ray Absorption Spectroscopy Applications to Speciation of Metals in Soil

Tatiana Minkina, Dina Nevidomskaya, Tatiana Bauer, Saglara Mandzhieva, Abdulmalik Batukaev, Svetlana Sushkova, Victoria Shuvaeva, Inna Zamulina, and Victor Chaplygin

Abstract

Synchrotron radiation methods provide direct information on the chemical bonds formed by metals. The investigation of local structure of Haplic Chernozem saturated with Cu^{2+} and Pb^{2+} ions using X-ray absorption spectroscopy (XANES and EXAFS) was performed. As a result of the study, the reliable qualitative data on the nature of bonds formed by metal atoms have been obtained and the mechanism of their interaction with soil phases and all the chemical bindings occurring in the course of this process were determined.

Keywords

Synchrotron radiation • XANES • Cu^{2+} • Pb^{2+}

1 Introduction

Synchrotron-based X-ray absorption spectroscopy techniques are nondestructive methods for providing information for metals speciation in soils. X-ray consists of two methods —absorption fine structure (EXAFS) spectroscopy and absorption near-edge structure (XANES) spectroscopy. These methods can be used for understanding the transformations of metal compounds entered in soil and the possible environmental consequences [1]. In this study the coordination of Cu^{2+} and Pb^{2+} ions with layered silicates and humic acid in the Haplic Chernozem was investigated using synchrotron-based X-ray absorption spectroscopy.

T. Minkina (✉) · D. Nevidomskaya · T. Bauer · S. Mandzhieva · S. Sushkova · V. Shuvaeva · I. Zamulina · V. Chaplygin
Southern Federal University, B. Sadovaya str. 105, 344006 Rostov-on-Don, Russia
e-mail: tminkina@mail.ru

A. Batukaev
Chechen State University, Dudaeva blvd. 17, 366007 Grozny, Russia

2 Materials and Methods

The soil samples were selected from the depth 20 cm fallow lands covered by zonal soils (Haplic Chernozem [2]) of the Lower Don region (Russia). Characterization of some physical and chemical properties of the investigated soil are given in Table 1. Different dozes of nitrates and oxides Cu and Pb (2000 and 10,000 mg/kg) were added into the soil. After one year the soil samples of model experiments were analyzed.

The X-ray absorption spectra (EXAFS and XANES) were measured at the Cu K-edge (8985–8990 eV) and Pb L_{III}-edge (13,040 eV) at the Structural Materials Science beamline of the Kurchatov Center for Synchrotron Radiation (NRC "Kurchatov Institute", Moscow [3]).

XANES and EXAFS spectra were measured both in the transition and in the fluorescence yield modes in a step of 0.5 keV using Si (111) channel-cut monochromator, which provided an energy resolution of E/E $\sim 2 \times 10^{-4}$. Fluorescence photons were counted using an Si avalanche photodiode and for monitoring the incident beam intensity an ionization chamber was used. The exposition time in each point of the spectrum was 60 s. Energy calibration was achieved by means of spectra of copper and lead foils, which were used as reference samples were taken. For detailed analysis of the differences in the spectra the first derivatives of XANES spectra were calculated.

3 Results

It can be seen at the Fig. 1, that the experimental spectra of soil samples, contaminated by CuO, are very similar to the spectra of CuO. On the other hand, strong differences between the spectra of soils treated by $Cu(NO_3)_2$ and those of the initial copper-bearing compound indicate significant changes in environment of the Cu^{2+} ion, occurring in the soil. Copper nitrate is well-soluble in water; because of this,

A. Kallel et al. (eds.), *Recent Advances in Geo-Environmental Engineering, Geomechanics and Geotechnics, and Geohazards*, Advances in Science, Technology & Innovation, https://doi.org/10.1007/978-3-030-01665-4_4

Table 1 Physical and chemical properties of the investigated soil samples (Haplic Chernozem) in the 0–20 cm

C_{org} (%)	$CaCO_3$ (%)	*pHwater*	Exchangeable bases, cmol(+)/kg			Contents (%) of particle-size fractions		
			Ca^{2+}	Mg^{2+}	Na^{+}	Sand fractions (1.0–0.05 mm)	Silt fractions (0.01–0.001 mm)	Clay fraction (<0.001 mm)
3.7	0.4	7.6	31.0	6.0	0.06	9.8	62.1	28.1

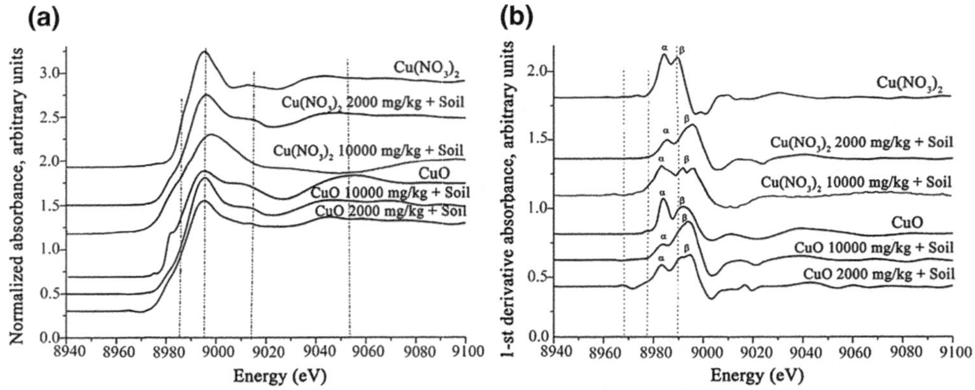

Fig. 1 Experimental spectra (**a**) and spectra of the first derivate (**b**) of X-ray absorption XANES for initial copper-bearing compounds [CuO and $Cu(NO_3)_2$] and contaminated soil samples

Cu^{2+} ions during the one year of incubation were sorbed by the soil and formed complexes with organic components [4].

The intensity of the α peak is controlled by the degree of bond covalence and characterizes the coordination environment and chemical bonds of the absorbed metal ion with its closest surroundings. With a decrease in the α peak energy, Cu complexes with soil components have predominantly the covalent character of the bond. The intensities of the α and β peaks in the experimental spectra of the initial copper-bearing compounds are close (Fig. 1b).

In Fig. 2, the experimental XANES spectra of the soil samples with high rates of Pb contamination are presented together with the spectra of PbO and $Pb(NO_3)_2$. The peculiarities of the spectra in the energy region of ~13,030–13,058 eV indicate the presence of Pb^{2+} ions. The maximum value of absorbance is located in the energy region of ~13,038–13,040 eV for the samples saturated with PbO and at ~13,042 eV for the samples saturated with $Pb(NO_3)_2$, which is due to the $2p_{3/2} \rightarrow 6d$ electron transition (Fig. 2b).

Only X-ray absorption spectra of PbO-saturated soil samples and PbO contain the peak, related to the $2p_{3/2} \rightarrow 6s$ electron transition, which is indicative of the 6s and 6p hybridization for Pb and $2p_{x,y}$ for oxygen. The ability of Pb^{2+} to take part in the formation of numerous distorted complexes, originates from the variety of possible O–Pb–O valent angles [5].

4 Discussion

The experimental spectrum of soil contaminated by CuO and $Cu(NO_3)_2$ are controlled by the electron transition $1s \rightarrow 4pz$ and provides evidence for electron transfer from the metal to

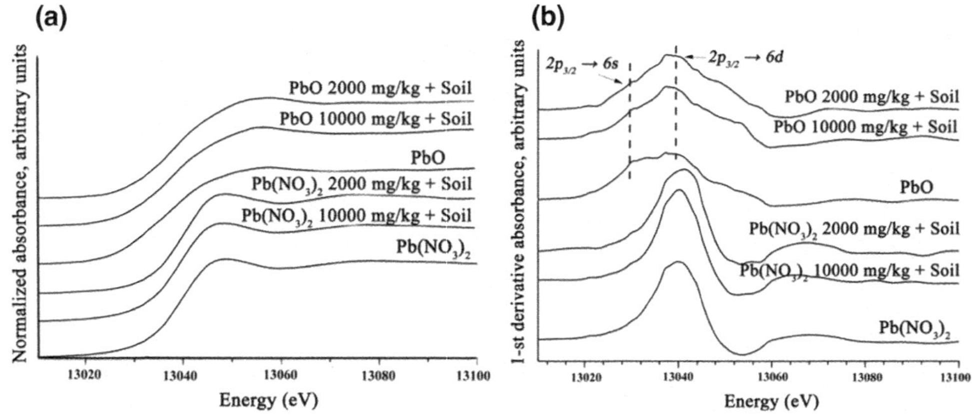

Fig. 2 Experimental spectra (**a**) and spectra of the first derivate (**b**) of X-ray absorption XANES for initial lead-bearing compounds [PbO and $Pb(NO_3)_2$] and contaminated soil samples

a ligand [6]. The interaction between Cu^{2+} ions and humic acid may result in the formation of multilateral 6-coordinated spatial structure of humate complex.

Results of EXAFS study show that the incorporation of Pb^{2+} ions into the structure of phyllosilicate minerals results in a decrease of the Pb^{2+}–O bond distances in equatorial and axial coordination positions in Pb-bearing octahedrons [5]. Divalent Pb has the $6s^2$ electronic configuration of the outer shell. A strong deformation of divalent Pb in polyhedrons is due to the lone electron pair, which is often stereochemically active. Thus, a conclusion can be made that Pb is sorbed as a bidentate inner-sphere complex at the edges of the octahedrally coordinated aluminum ions.

5 Conclusions

The application of the methods based on synchrotron radiation for soils makes it possible to obtain direct information on the kind of bonds between soil components and metals. The structural organization studied in Haplic Chernozem artificially polluted by $Cu(NO_3)_2$ and $Pb(NO_3)_2$ while applying X-ray absorption spectroscopy allowed revealing the mechanism of metals associated with soil phases. The analysis of the XANES and EXAFS results suggested that copper ions from $Cu(NO_3)_2$ are highly ordered inner-sphere sorption complexes with humic acids and incorporated in the octahedral and tetrahedral sites of minerals in soil. Lead ions from $Pb(NO_3)_2$ were incorporated in the silicates structure with bidentate inner-sphere complex formation. This led to decreasing the bond distances between Pb^{2+} ions and O atoms in equatorial and axial coordination positions in Pb-bearing octahedrons.

Acknowledgements This research was supported by the Ministry of Education and Science of the Russian Federation, project no. 5.948.2017/PP.

References

1. Synchrotron-Based Techniques in Soils and Sediments. In: Singh, B., Gräfe, M. (eds.), vol. 34, pp. 1–480. Elsevier B, The Netherlands (2010)
2. IUSS Working Group WRB. In: World reference base for soil resources 2014, update 2015. International soil classification system for naming soils and creating legends for soil maps. World Soil Resources Reports, No. 106. FAO, Rome (2015)
3. Chernyshov, A.A., Veligzhanin, A.A., Zubavichus, Y.V.: Structural materials science end-station at the Kurchatov Synchrotron radiation Source: recent instrumentation upgrades and experimental results. Nucl. Instr. Meth. Phys. Res. A **603**, 95–98 (2009)
4. Minkina, T.M., Soldatov, A.V., Motuzova, G.V., Podkovyrina, Y. S., Nevidomskaya, D.G.: Speciation of copper and zinc compounds in artificially contaminated chernozem by X-ray absorption spectroscopy and extractive fractionation. J. Geochem. Explor. **144**, 306–311 (2014)
5. Nevidomskaya, D.G., Minkina, T.M., Soldatov, A.V., Shuvaeva, V. A., Zubavichus, Y.V., Podkovyrina, Y.S.: Comprehensive study of Pb(II) speciation in soil by X-ray absorption spectroscopy (XANES and EXAFS) and sequential fractionation. J. Soils Sediments **16**(4), 1183–1192 (2016)
6. Xia, K., Bleam, W., Helmke, P.: Studies of nature of binding sites of first row transition elements bound to aquatic and soil humic substances using X-ray absorption spectroscopy. Geochim. Cosmochim. Acta **61**(11), 2223–2235 (1997)

Hydro-Geochemical Behavior of Acid Formation of Sulfide Bearing Rocks Based on Kinetic Column Tests

Guzide Kalyoncu Erguler and Zeynal Abiddin Erguler

Abstract

Acid mine drainage (AMD) is an important environmental hazard due to its serious chemical contamination to the surface and groundwater resources. To provide enough and representative data for developing restoration techniques, this time-dependent geochemical process should be investigated based on kinetic principles. Thus, the kinetic column test was generally utilized in the earlier studies without any accepted procedures related to the column dimension and properties of materials (such as particle size and mass of samples). To overcome the dissimilarities between the mass release rate of contaminants specified in the laboratory and in the field, and to upscale laboratory-based measurements to the field, kinetic column tests were performed using different columns filled with crushed coarse and fine ore samples. The fluctuations of pH values and the concentration of various constituents were determined during kinetic column tests. The effluents of columns turned to acid after lag time of 21–65 weeks depending on the column dimensions and particle size. Statistically significant predictive models for upscaling geochemical behavior of AMD processes were presented based on simple and multiple regression analyses among column dimensions and main parameters controlling the rate of acid generation.

Keywords

Acid mine drainage • Kinetic column test • Oxidation • Sulfide minerals

G. Kalyoncu Erguler
Mineral Research and Exploration General Directorate (MTA), 06800 Çankaya, Ankara, Turkey

Z. A. Erguler (✉)
Department of Geological Engineering, Dumlupinar University, 43100 Kütahya, Turkey
e-mail: zeynal.erguler@dpu.edu.tr

1 Introduction

Acid mine drainage (AMD) is thought as one of the key mining-related ecological issues that results in releasing contaminants and decreasing the pH of surface and groundwater. It is well known that this natural process is a product of the oxidation of sulfide minerals whenever oxygen dissolved in water comes into contact with these minerals.

The determination of AMD potential of any site at laboratory condition is based on two main different testing techniques, which are known as static and kinetic tests. In comparison with a static test, the kinetic tests are better in representing field conditions [1]. Up to now, different approaches considering kinetic principles have been developed and recommended for use in the prediction and/or assessment of AMD. Among these approaches, the column test is known as more representative for simulation of actual weathering processes [1]. However, Strömberg and Banwart [2] verified that the behavior of acid generation is still unclear because of inconsistencies between the laboratory and field based AMD measurements. Despite extensively performing kinetic column test to predict AMD [2–4], the particle and column size dependent mass release rate of oxidation products and other constituents have not been investigated in detail. Therefore, considering the constantly increasing AMD's impact on the environment over time and the deficiencies related to upscaling results of the kinetic column test, a research program has been conducted to find suitable practical understandings by using key parameters controlling the rate of acid generation for upscaling geochemical behavior and kinetic principles of AMD generation. For this purpose, columns were manufactured in various sizes based on recommendations of previous studies. The length and diameter were kept as 16–72 cm and 2–7 cm, respectively. The collected ore samples consisting of sulfide minerals were subjected to a crushing process for assessing particle size dependent AMD generation.

A. Kallel et al. (eds.), *Recent Advances in Geo-Environmental Engineering, Geomechanics and Geotechnics, and Geohazards*, Advances in Science, Technology & Innovation, https://doi.org/10.1007/978-3-030-01665-4_5

Predictive models thought to be useful for upscaling kinetic principles based on the geochemical behavior of AMD processes were obtained by performing simple and multiple regression analyses among column dimensions and main parameters controlling the rate of acid generation.

2 Materials and Methods

The samples were collected from an open pit quarry located in Murgul, Artvin City in North-eastern Turkey. The samples were crushed to obtain relatively coarse (≤ 3.35 mm) and fine (≤ 0.625 mm) particles to investigate the effect of particle size distribution on the AMD generation. Samples consisting of two different particle size distribution curves were placed in columns for kinetic tests. It was found in X-ray diffraction (XRD) measurements which, in addition to quartz, chalcopyrite and pyrite known as the main reason for acid formation, constitute a significant proportion of samples' mineralogical composition. Mineral liberation analyzer (MLA) was also used to achieve mineral percentage of fresh samples. These analyses indicated that fresh ore samples were composed mainly of chalcopyrite (32–38%), quartz (32.5–37.5%), pyrite (19.5–23%), biotite (4.8–6.5%) and ankerite (2.3–3%) with less than 0.5% of other minerals that are unimportant for the tests kinetic behavior. Samples with relatively coarse and fine particle sizes were placed in twenty-four columns set up in different dimensions to investigate the effect of particle size and column dimensions on the AMD generation rate. The amount of distilled water required for initiation of kinetic column tests was defined based on fifty-year meteorological records. Kinetic column tests were performed by dropping water inside columns. After water had flown through the column, contaminated solutions were collected in glass jars and then analyzed for monitoring variations in pH and chemical composition at specific time intervals.

3 Results and Discussion

In previous studies [2–4], investigators generally evaluated the weathering kinetics of samples by plotting changes in key indicator parameters (such as pH values, concentration of Ca^{2+}, Mg^{2+}, Mn^{2+}, Mn^{2+}, SO_4^{2-}) as a function of time, except for a detailed research on time-dependent variations in cumulative oxidation and neutralization products. In addition to the above given relationships between cumulative concentrations of Ca^{2+}+Mg^{2+}+Mn^{2+} and SO_4^{2-}, it is thought that evaluations including the time dependent variations in the cumulative values of pH values, concentration of Ca^{2+}, Mg^{2+}, Mn^{2+}, Mn^{2+}, SO_4^{2-} would be useful for a better understanding of the process of acid generation throughout column tests. To find the effect of time on the cumulative values of $Ca^{2+} + Mg^{2+} + Mn^{2+}$ and SO_4^{2-}, and the cumulative pH measurements during chemical oxidation processes, further analyses were performed by taking these parameters as dependent and time as independent variables. Statistically significant simple correlations were obtained between time and cumulative values of $Ca^{2+} + Mg^{2+} + Mn^{2+}$, SO_4^{2-} and pH. By symbolizing the constants of these correlations as "a_1", "a_2" and "a_3", the general equations given below are obtained for the prediction of time-dependent variations in cumulative values of $Ca^{2+} + Mg^{2+} + Mn^{2+}$, SO_4^{2-} and pH:

$$\text{Cumulative } SO_4^{2-} = a_1 x \quad (1)$$

$$\text{Cumulative } Ca^{2+} + Mg^{2+} + Mn^{2+} = a_2 x \quad (2)$$

$$\text{Cumulative pH} = a_3 - b * \ln(x) \quad (3)$$

where x is time in week, a_1, a_2, a_3 and b are dimensionless constants. The variations in the constants of the above given equations are thought to depend on column dimensions. Therefore, to predict these constants for practical engineering application, statistically significant regression models were developed between the constants of the above equations and column dimensions based on simple regression analyses (Table 1). To incorporate the effect of time and pH, multiple regression analyses were performed by taking cumulative SO_4^{2-} and cumulative $Ca^{2+} + Mg^{2+} + Mn^{2+}$ as dependent variables (y) and cumulative pH and time as independent variables (x). The statistically significant empirical approaches were obtained as results of these multiple regression analyses. The R-squared values of these multiple regression analyses reveal that these predictive empirical models explain 90–99% of the variability in cumulative SO_4^{2-} and cumulative $Ca^{2+} + Mg^{2+} + Mn^{2+}$ released from coarse samples.

4 Conclusions

Contrary to the fluctuation in the instantaneous values of $Ca^{2+} + Mg^{2+} + Mn^{2+}$, SO_4^{2-} and pH measurements over time, the cumulative values of these parameters show significant correlations with time during the column tests. Statistically significant relationships were obtained between time and cumulative values of these measurements. Furthermore, these empirical models were simplified to produce three general equations to be used for the prediction of time-dependent variations in cumulative values of $Ca^{2+} + Mg^{2+} + Mn^{2+}$, SO_4^{2-} and pH. In addition, to predict the constants of these empirical models and determine

Table 1 Summary of the statistical analysis for prediction a_1, a_2, a_3 and b from column dimensions

Dependent variable (y)	Coarse particle size (−3.35 mm)		
	Independent variable		
	Length (cm) (x)	Diameter (cm) (x)	Volume (cm^3) (x)
a_1	$y = 0.031x^{2.3167}$ ($R^2 = 0.88$)	$y = 4.991x^{2.4599}$ ($R^2 = 0.94$)	$y = 0.82x^{0.8403}$ ($R^2 = 0.96$)
a_2	$y = 0.025x^{2.1181}$ ($R^2 = 0.90$)	$y = 2.634x^{2.2376}$ ($R^2 = 0.95$)	$y = 0.505x^{0.7657}$ ($R^2 = 0.98$)
a_3	$y = 7.118x^{-0.026}$ ($R^2 = 0.423$)	$y = 6.756x^{-0.031}$ ($R^2 = 0.58$)	$y = 6.895x^{-0.01}$ ($R^2 = 0.55$)
b	$y = 0.156e^{0.0053x}$ ($R^2 = 0.53$)	$y = 0.152e^{0.0584x}$ ($R^2 = 0.62$)	$y = 0.176e^{0.0001x}$ ($R^2 = 0.64$)
Dependent variable (y)	Fine particle size (−0.625 mm)		
	Independent variable		
	Length (cm) (x)	Diameter (cm) (x)	Volume (cm^3) (x)
a_1	$y = 0.035x^{2.22}$ ($R^2 = 0.98$)	$y = 3.118x^{2.4512}$ ($R^2 = 0.86$)	$y = 0.672x^{0.8101}$ ($R^2 = 0.93$)
a_2	$y = 0.154x^{1.7262}$ ($R^2 = 0.83$)	$y = 5.386x^{1.8661}$ ($R^2 = 0.71$)	$y = 1.622x^{0.6218}$ ($R^2 = 0.77$)
a_3	$y = 8.072e^{-0.002x}$ ($R^2 = 0.76$)	$y = 8.080e^{-0.023x}$ ($R^2 = 0.55$)	$y = 8.852x^{-0.031}$ ($R^2 = 0.62$)
b	$y = 0.002x + 0.002$ ($R^2 = 0.73$)	$y = 0.026e^{0.2469x}$ ($R^2 = 0.51$)	$y = 0.010x^{0.339}$ ($R^2 = 0.58$)

cumulative values of SO_4^{2-} and $Ca^{2+} + Mg^{2+} + Mn^{2+}$ for any desired length, diameter and volume of column, some practical approaches were suggested based on simple regression analyses.

Predictive empirical relationships were obtained based on multiple regression analyses by taking cumulative SO_4^{2-} and $Ca^{2+} + Mg^{2+} + Mn^{2+}$ values as dependent variables and cumulative pH and time as independent variables to incorporate combine effect of time and pH conditions, which are thought to be key parameters controlling the rate of acid generation.

Acknowledgements The authors gratefully acknowledge the financial support of the Mineral Research and Exploration General Directorate (MTA Project no: 2012-37-14-01-9).

References

1. Bradham, W.S., Caruccio, F.T.: A comparative study of tailings analysis using acid/base accounting, cells, columns and Soxhlets. In: 2nd Int Conf Abatement of Acidic Drainage, Montre´al, vol. 1, CANMET, Ottawa, Canada, pp. 157–173 (1991)
2. Strömberg, B., Banwart, S.: weathering kinetics of waste rock from the aitik copper mine, Sweden: scale dependent rate factors and pH controls in large column experiments. J. Contam. Hydrol. **39**, 59–89 (1999)
3. Benzaazoua, M., Bussière, B., Dagenais, A.M., Archambault, M.: Kinetic tests comparison and interpretation for prediction of the Joutel tailings acid generation potential. Environ. Geol. **46**, 1086–1101 (2004)
4. García, C., Ballester, A., González, F., Blázquez, M.L.: Pyrite behaviour in a tailings pond. Hydrometallurgy **76**, 25–36 (2005)

Interaction of CuO Nanoparticles with *Hordeum Sativum* Distichum in an Aquatic Medium and in the Soil

Vishnu D. Rajput, Tatiana Minkina, Alexey Fedorenko, Grigoriy Fedorenko, Saglara Mandzhieva, Svetlana Sushkova, Natalya Chernikova, Nadezhda Duplii, Anatoly Azarov, and Alexsandr Usatov

Abstract

Copper nanoparticles arise in soil and water from different industries, and well known for their adverse effects on aquatic and terrestrial ecosystems. The soil and hydroponic experiments were conducted to assess the toxicity of excessive amounts of CuO nanoparticles spiked in the soil and in hydroponic conditions on the root and leaf cells of spring barley. The microscopic study indicated changes in vascular bundles, stele, accumulation of electronically dense material in epidermis and vacuoles. The root hairs reduced by excess CuO nanoparticles even disappeared in hydroponic condition. Moreover, there were no significant changes visualized in semithin sections of leaf cells. The descriptive study on ultrastructure also indicated several changes in the chloroplast, thylakoids, plastoglobules, starch granules, peroxisomes and mitochondrial cristae. The negative effects of CuO nanoparticles on barley were more significant in the aquatic medium than in the soil.

Keywords

Spring barley • Nanoparticles • Soil spiking • Toxicity • Ultrastructure

1 Introduction

The involvement of Cu in human life is well known, and that's why the toxicity of CuO nanomaterials attracted global scientific attention. The rapid increases in Cu manufacturing uses and disposal inevitably increased the level of Cu concentration in the soil and aquatic ecosystem. When Cu ends up in the soil, it strongly adheres to organic matter and soil minerals and interrupts the functionality of soil ecosystem. Plants are sensitive to Cu toxicity with the value approximately 20–30 mg/kg on dry biomass basis [1]. Impact on root growth and damage to cell membranes, vascular bundles could be an immediate toxic response to the excess concentration of CuO nanoparticles. Copper is not only toxic to the plant but also dangerous to other organisms i.e., algae, mussels, crustaceans, fishes [2].

In recent years, CuO nanoparticles have received notable attention due to their unique physical and chemical properties, and beneficiary applications in a variety of ways, for instance, antimicrobial activity, skin products, semiconductors, catalysis, boat antifouling paints, electronics and optics. It is assumed that the CuO nanoparticles are the most bioreactive metal oxide particles, and highly toxic compared to other carbon or metal oxide nanomaterials [3]. Studies focused on the environmental toxicity of Cu-based nanoparticles due to their increasing application in consumer products [4]. These studies are warning an increasing application of CuO nanoparticles to impose a threat to the ecosystem. The accumulation of nanoparticles in soil and groundwater may potentially increase the accumulation in plant tissues. Several studies indicated that CuO nanoparticles affected crop plants growth by reducing germination rate, photosynthesis, transpiration rate, quality of yields, damage or corroded roots, decreased root and shoot length, and biomass [5]. These studies are only limited to morphometric parameters and did not examine the structural and ultrastructural changes. Therefore, the two analogous experiments conducted in the soil and hydroponic condition to observe effects of excess CuO nanoparticles on spring barley (*Hordeum sativum distichum* cv. Travnik), one of the most important food grain crops since ancient times, and widely grown in this region. It is also recommended to use this species as a bioindicator for evaluating the effects of heavy metals.

V. D. Rajput · T. Minkina (✉) · A. Fedorenko · G. Fedorenko
S. Mandzhieva · S. Sushkova · N. Chernikova · N. Duplii
A. Azarov · A. Usatov
Southern Federal University, Rostov-on-Don, Russia
e-mail: tminkina@mail.ru

A. Fedorenko · G. Fedorenko
Southern Scientific Center of Russian Academy of Sciences, Rostov-on-Don, Russia

A. Kallel et al. (eds.), *Recent Advances in Geo-Environmental Engineering, Geomechanics and Geotechnics, and Geohazards*, Advances in Science, Technology & Innovation, https://doi.org/10.1007/978-3-030-01665-4_6

2 Materials and Methods

The commercial grade copper (II) oxide (CuO) nanoparticles powder used in soil and hydroponic experiments conducted together. The excess CuO nanoparticles (10,000 mg/kg) were spiked with soil to make it possible to reveal the mechanisms of metal transformation. Soil without CuO nanoparticles addition served as a control. The high concentration was selected by considering the existing levels of Cu in soils adjacent to industrial enterprises [6]. Several other studies also indicated a similar concentration of Cu in Kola Peninsula soils of the Russian territory. After a month of soil spiking, seeds were sown in respective pots. Similarly, the hydroponic experiment was conducted in the lab using same concentration of CuO nanoparticles (10,000 mg/L) with control (0 mg/L). Surface sterilized seeds were germinated in Petri plates and well-grown seeds transferred to hydroponic vessels. Samples from the soil and hydroponic experiments were examined after 30 days of growth. Ultrathin sections were prepared by the microtome (Leica EM UC6, Germany) and examined under the light-optical microscope (Mikmed-6 St. Petersburg, Russia) and transmission emission microscopy (TEM; Tecnai G2 Spirit Bio TWIN, Netherlands). Images were processed by Image_J and Adobe Photoshop software.

3 Results and Discussion

Since the root is the first target tissue exposed to pollutants, it seems that the functional and structural disorders appear more often in root than in aerial tissues. Ultrathin sections show the endoderm and stele of root cells were disorganized (Fig. 1). The number and lengths of root hairs is reduced compared with the control, even disappeared in hydroponic conditions. The internal structure of the leaves suffered less than the root, and changes were not noticeable. Previous studies also indicated that the nanoparticle can uptake by roots and transport through vascular bundles to above-ground parts of plants depending on plant types, anatomy and composition, shape and size of nanoparticles [5, 7]. Accumulation of Cu in the parenchyma cells was observed in the vacuoles and cell walls.

Several ultrastructure modifications were also observed in the plastid, mitochondria, peroxisomes, plastoglobules, starch granule, protoplasm (figure not shown). It is confirmed, that the metal-tolerant plant prevents the uptake of toxic pollutants into aboveground tissues by restricting the transport across the epidermis, endodermis and storing them in cortex and hyper-accumulating plants accumulate excess amounts of toxic elements in the harvestable tissues [8]. Typically, the plant limits the translocation of metal in

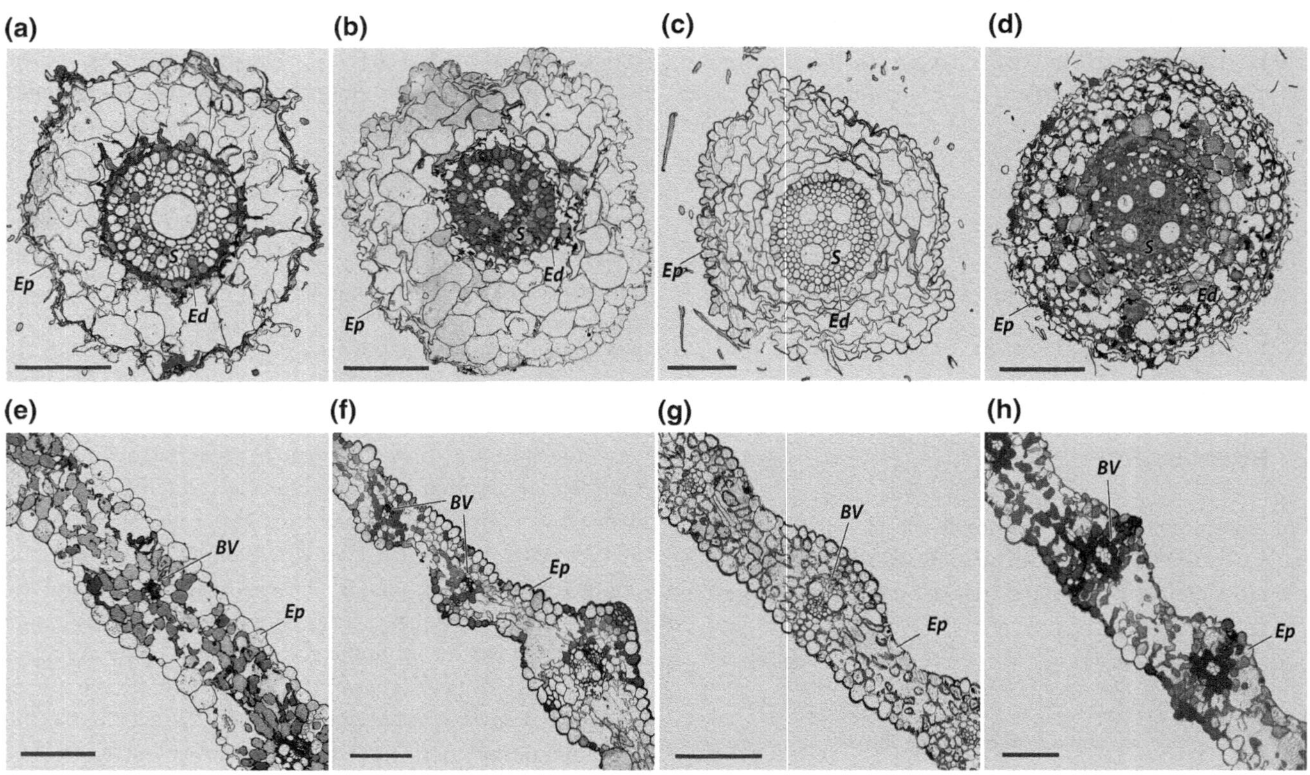

Fig. 1 Effects of CuO nanoparticles on root cells: **a**—control, **b**—in pollution (in hydroponic); **c**—control, **d**—in pollution (in soil) and leaf cells: **e**—control, **f**—in pollution (in hydroponic); **g**—control, **h**—in pollution (in soil). Ep—epidermis, Ed—endodermis, BV—vascular bundle, S—stele. Scale bar—100 μm

shoots to avoid toxicity [9]. It is expected that the identified structural changes of the epidermis and mesoderm affect the radial transport of liquid to the root from the peripheral areas and may be one of the causes for a decline in both sequestration and translocation of mineral nutrients from the roots to other parts of the plant. It is well known that the Cu accumulates in vacuoles and cell walls [10]. Copper deposition and long-term toxicity decreased photosynthetic activity, disturbed the architecture of chloroplasts, particularly thylakoids membranes [11].

4 Conclusion

Copper oxide nanoparticles affected vascular bundles of the leaf and root cells, cell membrane, plastid, thylakoids, plastoglobules and starch granules and mitochondrial cristae. The findings indicated that the CuO nanoparticles in hydroponic conditions were more toxic than in the plants grown in the soil. This could be due to the dissolution of Cu ions. However, the mechanism is not well understood. Therefore, the descriptive studies are needed to understand the mechanism of CuO nanoparticles ion dissolution and translocation through cell organelles. These changes may be considered as an important characteristic to evaluate the CuO nanoparticles toxicity to spring barley and protective function of the soil in reducing the toxic effect for plants.

Acknowledgements The work was supported by the Russian Science Foundation, project no. 16-14-10217.

References

1. Marscner, H.: Mineral nutrition of higher plants. Academic, London (1995)
2. Katsumiti, A., Thorley, A.J., Arostegui, I., Reip, P., Valsami-Jones, E., Tetley, T.D., Cajaraville, M.P.: Cytotoxicity and cellular mechanisms of toxicity of CuO NPs in mussel cells in vitro and comparative sensitivity with human cells. Toxicol. Vitro **48**, 146–158 (2018)
3. Wang, Z., Li, N., Zhao, J., White, J.C., Qu, O., Xing, B.: CuO nanoparticle interaction with human epithelial cells: cellular uptake, location, export, and genotoxicity. Chem. Res. Toxicol. **25**, 1512–1521 (2012)
4. Adeleye, A.S., Pokhrel, S., Madler, L., Keller, A.A.: Influence of nanoparticle doping on the colloidal stability and toxicity of copper oxide nanoparticles in synthetic and natural waters. Water Res. **132**, 12–22 (2018)
5. Rajput, V.D., Minkina, T., Suskova, S., Mandzhieva, S., Tsitsuashvili, V., Chapligin, V., Fedorenko, A.: Effects of copper nanoparticles (CuO NPs) on crop plants: a mini review. BioNanoScience **8**(1), 36–42 (2018)
6. Minkina, T.M., Linnik, V.G., Nevidomskaya, D.G., Bauer, T.V., Mandzhieva, S.S., Khoroshavin, V.: Forms of Cu (II), Zn (II), and Pb (II) compounds in technogenically transformed soils adjacent to the Karabashmed copper smelter. J. Soils Sediments **18**, 2217–2228 (2018)
7. Rico, C.M., Majumdar, S., Duarte-Gardea, M., Peralta-Videa, J.R., Gardea-Torresdey, J.L.: Interaction of nanoparticles with edible plants and their possible implications in the food chain. J. Agric. Food Chem. **59**, 3485–3498 (2011)
8. Kopittke, P.M., Menzies, N.W.: Effect of Cu toxicity on growth of cowpea (*Vigna unguiculata*). Plant Soil **279**, 287–296 (2006)
9. Manceau, A., Nagy, K.L., Marcus, M.A., Lanson, M., Geoffroy, N., Jacquet, T., Kirpichtchikova, T.: Formation of metallic copper nanoparticles at the soil-root interface. Environ. Sci. Technol. **42**, 1766–1772 (2008)
10. Vesk, P.A., Nockolds, C.E., Allaway, W.G.: Metal localization in water hyacinth roots from an urban wetland. Plant, Cell Environ. **22**, 149–158 (1999)
11. Olchowik, J., Bzdyk, R.M., Studnicki, M., Bederska-Błaszczyk, M., Urban, A., Aleksandrowicz, T.M.: The effect of silver and copper nanoparticles on the condition of english oak (*Quercus robur* L.) seedlings in a container nursery experiment. Forests **8**, 310 (2017)

Behavioural Responses of *Armadillidium Granulatum* (Crustacea, Oniscidea) to Zinc Contaminated Soil

Raja Jelassi, Wafa Hammami, Chedliya Ghemari, and Karima Nasri-Ammar

Abstract

Avoidance and locomotor behaviours of *Armadillidium granulatum* Brandt, 1833 collected from sebkha El Ouafi were studied. Avoidance test was conducted in individuals exposed to zinc contaminated soils. Globally, the animals did not show an avoidance behaviour after 48 h of exposure. The determined concentration effect of this metal was equal to 1043 mg/L. The locomotor activity rhythm was studied in the control animals as well as those exposed to Zn contaminated soils. The results revealed the existence of nocturnal behaviour whatever the experimental conditions. Furthermore, the Periodogram analysis showed the existence of ultradian and circadian periods. This latter was longer for contaminated specimens than the control ones under entraining conditions as well as under constant darkness. In addition, exposed animals showed the most important activity time whatever the regimen imposed and were characterized by a phase advance contrary to the control animals which showed a phase delay.

Keywords

Sebkha • Zinc • Soil • Terrestrial isopod • Avoidance test • Locomotor behaviour

1 Introduction

Wetlands are a major landscape in the Mediterranean region. They are threatened by human activity. Heavy metal pollution was considered one of the major threats that can lead to the degradation of these ecosystems [4]. To assess the quality of soil contaminated by heavy metals, biological indicators were used. In fact, soil fauna, especially invertebrates, were considered as good bio-indicators of the soil quality [10]. These organisms were known to maintain good soil structure [3]. Among soil invertebrates, terrestrial Isopods were considered as important players in soil ecology. In the present study, we chose the terrestrial isopod *Armadillidium granulatum* living in the supralittoral zone of sebkha El Ouafi and we proposed to study, firstly, its avoidance behaviour after Zn exposure, and secondly, the effect of this metal on its locomotor behavior.

2 Materials and Methods

The present study was carried out on the banks of sebkha El Ouafi (N37°09′21.7″ E010°13′36.2″) connected to Ghar El Melh lagoon. Adult individuals of *A. granulatum* were manually collected in the morning under stones and between the plant roots. Then, they were transported to the laboratory in perspex boxes containing soil from the collection site.

2.1 Soil Preparation and Avoidance Test

The soil sampled from the collection site was dried in an oven during 24 h in 40 °C. Then, it was contaminated with a zinc chloride ($ZnCl_2$) salt solution according to Köhler et al. [7]. Four concentrations (C_1, C_2, C_3, C_4) were prepared with three replicates for each concentration, and with 20 individuals per replicate. A control was also prepared. After its contamination, the soil was acclimated during 72 h in 25 °C before the experiment. The avoidance test consisted in maintaining *A. granulatum* individuals in rectangular plastic boxes which were divided into two equal compartments by a removable plastic wall. The test period was limited to 48 h. After this period, the plastic wall was reintroduced and individuals at each compartment were counted. A Probit

R. Jelassi (✉)
National Institute of Marine Sciences and Technologies, Salammbô, Tunisia
e-mail: djelassi.raja@gmail.com

R. Jelassi · W. Hammami · C. Ghemari · K. Nasri-Ammar
Tunis El Manar University, Faculty of Sciences of Tunis, Research Unit of Bio-Ecology and Evolutionary Systematics, El Manar II, Tunis, Tunisia

A. Kallel et al. (eds.), *Recent Advances in Geo-Environmental Engineering, Geomechanics and Geotechnics, and Geohazards*, Advances in Science, Technology & Innovation, https://doi.org/10.1007/978-3-030-01665-4_7

regression was used to calculate the EC_{50} conducted by *Xlstat* software. The avoidance test results were represented in terms of average net response (NR) expressed in percentage and calculated according to Amorim et al. [1] as follows: NR = ((C-T)/N) * 100 (*C = individuals observed in the control litter; T = individuals observed in the tested litter; N = total number of individuals per replicate*).

2.2 Locomotor Behaviour

To study the locomotor behaviour of *A. granulatum*, experiments were performed under two successive regimens: the light/dark cycle (nLD) and the constant darkness (DD). In the two experiments, the adult control specimens (N = 21) and specimens exposed to Zn were individually transferred to separate annular actograph containing soil. Then they were kept in a controlled environment cabinet maintained at a constant temperature of 18 ± 1 °C. The recording apparatus and software were constructed in the workshop of the School of Biosciences, University of Birmingham, U.K.

The results were initially presented for analysis in the form of double-plotted actograms using the Chart software package version 35 (D.D. Green, University of Birmingham, U.K.). Periodogram analysis was performed using the program based on the method of Dorscheidt and Beck [6]. Each time series was scanned for periods between 10 and 33h20' and was accepted as rhythmic if the periodogram peaks exceeded the 99% confidence level. This same technique also gave a good measure of the rhythm definition, in the form of the Signal to Noise Ratio (SNR), measured as the difference between the periodogram peak value and its corresponding 95% confidence limit. The waveform of activity was further examined using the astronomic time under entraining conditions and circadian time under free-running conditions. Phase values were calculated using a technique in which the median point of the main activity bloc was correlated to the onset time of the dark phase, expressed in degrees. The percentage of survival, number of animals showing periodicity, mean periods, activity time, phase shift, and SNR (a measure of rhythm definition) were calculated for each individual and for the whole population. Differences between percentages were analyzed using the $\chi 2$ test, and differences between all mean values were calculated using the Wilcoxon test.

3 Results

3.1 Avoidance Behaviour

A low mortality in *A. granulatum* individuals was shown after Zn exposure. The estimated EC_{50} was equal to 1043 mg L^{-1}. Considering the net response, mean values were summarized in Table 1; they fluctuated between concentrations. Our results showed a positive net response (0 ± 24.74) after ZnC_2 exposure. Concerning the other concentrations, a negative response was observed (Table 1).

3.2 Locomotor Behaviour

Our data showed that at the end of these experiments, all control individuals and 42.9% of those exposed to Zn survived in the recording chambers. The χ^2 test revealed a highly significant difference ($\chi^2 = 7.83$; df = 1; $p = 0.0051$).

The most representative actograms were shown in Fig. 1. Concerning the control, the activity was strictly concentrated during the experimental night. Individuals started their activity with the dusk. Under constant darkness, specimens delayed their activity showing thus the drift to the right. When exposed to Zn, *A. granulatum* concentrated its activity in the experimental night during the two first days. From the 3rd day until the 5th day, an anticipation of the activity was highlighted. From the 6th day, the animal began to delay its activity slightly day after day but globally a resynchronization of the activity was maintained. Furthermore, different locomotor patterns were obtained. In the control, the locomotor patterns were in most of the cases unimodal whatever the experimental conditions. Exposure to Zn, locomotor patterns became in majority multimodal. Under nLD cycle, a shift of the most important activity peak from the dusk in the control (70.6%) towards the middle of the experimental night (55.6%) for individuals exposed to Zn was observed. Under continuous darkness, this peak of activity was well marked around the subjective dusk (61.5%).

Through the Periodogram analysis, ultradian and circadian components were determined. The first one was observed only in the control under constant darkness with a mean period equal to τ_{12hDD} = 13h02 ± 1h39. Concerning the circadian component, whatever the experimental conditions, the mean period was shorter in the control than in the

Table 1 Results (average values ± standard error) of the net response of *A. granulatum* individuals exposed to Zn contaminated soil

$ZnCl_2$ nominal concentrations	Net response
C_1 (400 mg L^{-1})	−52 ± 12.59
C_2 (500 mg L^{-1})	0 ± 24.74
C3 (600 mg L^{-1})	−24 ± 12.59
C4 (700 mg L^{-1})	−57 ± 8.24

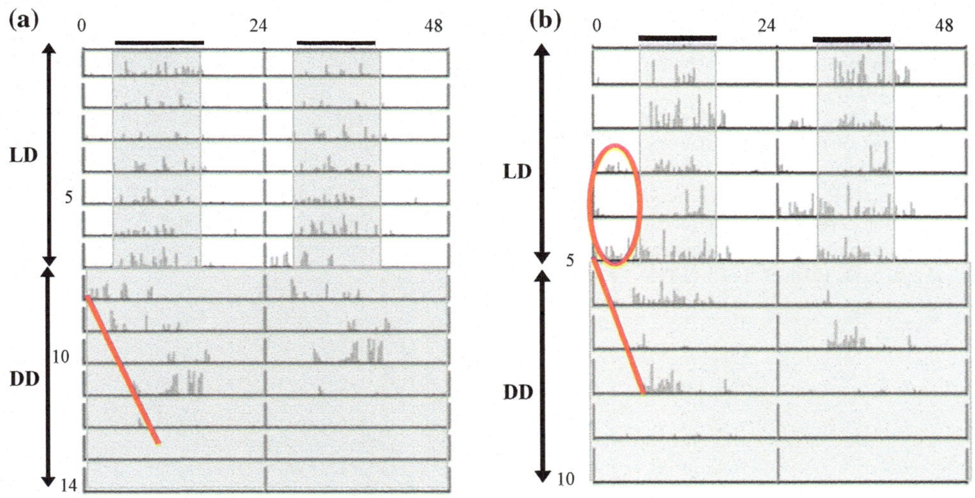

Fig. 1 Double plotted actograms of the locomotor activity rhythm of *Armadillidium granulatum* under LD-DD regimens obtained for the control individuals (a) and those exposed to zinc (b). Grey area corresponds to dark condition

contaminated animals which exhibited a high activity time. Furthermore, a phase advance equal to $\Delta\varphi = 24° \pm 47°$ was calculated for the individuals exposed to Zn contrary to the control ones which were characterized by a phase delay ($\Delta\varphi = -3° \pm 31°$).

4 Discussion

The study of avoidance behaviour highlighted a low mortality showing thus the tolerance of *A. granulatum* to Zn, as essential metal, like the majority of terrestrial isopods [9] explaining thus the obtained EC_{50} estimated to 1043 mg L^{-1}. Regarding the net response, our results were in accordance with those of Ghemari et al. [5] who showed a negative net response with the lowest concentration in *P. laevis* individuals. These results suggested an adaptive strategy of individuals which may use their antenna as chemoreceptive to avoid metal contamination [8]. Furthermore, exposed and unexposed *A. granulatum* individuals were characterized by the nocturnal behaviour. This behaviour was typical for the majority of terrestrial isopods [2]. The unimodal pattern of the locomotor activity observed in the control animals became generally multimodal in individuals exposed to Zn. This variation in locomotor profiles could be explained by the effect of contamination. In addition, the multiplication of locomotor activity peaks corresponds to the various activities of the species in its biotope such as burrowing.

5 Conclusion

In the present study, the majority of *A. granulatum* individuals did not avoid the contaminated substrate explaining thus the highest EC_{50} values. The study of the locomotor rhythm showed that this species was characterized by a nocturnal behaviour with the presence of multimodal patterns when exposed to Zn. Other experiments will be needed to confirm our results.

References

1. Amorim, M.J., Novais, S., Römbke, J., Soares, A.M.: *Enchytraeus albidus* (Enchytraeidae): a test organism in a standardized avoidance test? Effects of different chemical substances. Environ. Int. **34**, 363–371 (2008)
2. Bohli-Abderrazak, D., Ayari, A., Morgan, E., Nasri-Ammar, K.: Towards a characterization of the locomotor activity rhythm of the supralittoral Isopod Tylos europaeus. Chronobiol. Int. **29**, 166–174 (2012)
3. Cortet, J., Gomot-De Vauflery, A., Poinsot-Balaguer, N., Gomot, L., Texier, C., Cluzeau, D.: The use of invertebrate soil fauna in monitoring pollutant effects. Eur. J. Soil Biol. **35**(3), 115–134 (1999)
4. Davis, J.A., Froend, R.: Loss and degradation of wetlands in southwestern Australia: underlying causes, consequences and solutions. Wetlands Ecol. Manage. **7**(1–2), 13–23 (1999)
5. Ghemari, C., Ouni, A., Ayari, A., Nasri-Ammar, K.: The use of *Porcellio laevis* (Crustacea, Isopoda) as organism for the

avoidance test in response to a metal contaminated litter. Proceeding in the 3rd International Conference on Integrated Environmental Management for Sustainable Development, 296–298 (2018)
6. Harris, J.G., Morgan, E.: Estimates of significance in periodogram analyses of damped oscillations in a biological time series. Anal. Lett. **3**, 221–230 (1983)
7. Köhler, H.R., Hüttenrauch, K., Berkus, M., Grăff, S., Alberti, G.: Cellular hepatopancreatic reactions in *Porcellio scaber* (Isopoda) as biomarkers for the evaluation of heavy metal toxicity in soils. Appl. Soil. Ecol. **3**, 1–15 (1996)
8. Lukkari, T., Haimi, J.: Avoidance of Cu- and Zn-contaminated soil by three ecologically different earthworm species. Ecotoxicol. Environ. Saf. **62**, 35–41 (2005)
9. Mazzei, V., Longo, G., Brundo, M.V., Copat, C., Oliveri Conti, G., Ferrante, M.: Bioaccumulation of cadmium and lead and its effects on hepatopancreas morphology in three terrestrial isopod Crustacean species. Ecotoxicol. Environ. Saf. **110**, 269–279 (2014)
10. Wolters, V.: Biodiversity of soil animals and its function. Eur. J. Soil Biol. **37**(4), 221–227 (2001)

Heavy Metals Distribution in Soils of an Agricultural Area Impacted by Former Mining Activities: Case of Trozza Mine, Tunisia

Intissar Elmayel, Pablo L. Higueras, Jalel Bouzid, Eva M. Garcia Noguero, and Zouhair Elouaer

Abstract

Mining waste management is currently a major environmental concern due to the presence of high concentrations of potentially toxic metals (Pb, Zn, Cd, among others) in the surroundings of old mining areas. The large quantities of tailings are a direct cause of serious metal contamination, which is a threat to the environment as well as human beings. The decommissioned Jbal Trozza mine in Kairuan (Tunisia) is considered a serious problem, and in order to evaluate the distribution of metals contents in the mine area, tailings and soils samples were collected in the vicinity of the mine. The mine tailings are characterized by high levels of Pb (1830–5950 mg.kg^{-1}), Zn (7590–12480 mg.kg^{-1}) and Cd (85.95–123.25 mg.kg^{-1}). The adjacent soils were also highly contaminated with metals. The results show that the average concentrations of Pb, Zn and Cd in the soil of the mine area clearly exceed the Tunisian standard values in agricultural areas (100 mg. kg^{-1} for Pb, mg.kg^{-1} for Zn and 3 mg.kg^{-1} for Cd).

Keywords

Soil mining • Contamination • Heavy metals mobility

1 Introduction

Today there is a recent awareness in North Africa about the risks associated with former mining activities. This awareness is already widespread in the world and many environmental studies showed interest in the assessment of contamination related to mining activities, the dynamics of trace elements in the short and long runs (mobility and bioavailability) and risks due to the release and dissemination of heavy metals, which may favor their incorporation into food chains [2]. The present work main objective was the study of the potential impacts on the environment (soil, water, plants) of potentially toxic metals, mainly from the dumps, in the former mining site of Jebel Trozza, Tunisia (Kairouan region).

The old Pb/Zn mine of Trozza is located approximately 18 km north of Hajeb El Ayoun in central northern Tunisia, approximately 150 km SSW Tunis. The tailings present in the area are a result of mining activities between 1907 and 1937. Its excavations, plants, and tailings have been totally abandoned since then with no reclamation measures.

The Trozza mine area lies along the trend of the Sbiba Fault, a major NE–SW striking thrust structure approximately coincident with the northern margin of the central Tunisian uplift, also known as the Serdj-Ressas line, which forms part of the Atlas thrust front [3].

2 Materials and Methods

A sampling campaign was conducted in December 2015, during which we achieved samplings of tailings and soils from the Jebel Trozza mine site. The location of each sample was recorded using GPS (Global Positioning System) technology as shown on Fig. 1.

The wastes sampling was carried out from two types of tailing residues: four samples from the fine fraction tailings DF (from the top and the base of the tailings, referred to as DF1 to DF4); and five samples from the coarse-fraction tailing DG (DG1 to DG5). In order to evaluate the effect of the abandoned mine residues and estimate the lateral dispersion, 16 samples were collected from near and throughout the wadi after removing the first layer of surface soil (2 cm); samples were taken from the top 20 cm, [4] considering that the most important part of the soil profile for the control of degradation is the topsoil. The topsoil analysis is also the

I. Elmayel (✉) · J. Bouzid · Z. Elouaer
Ecole Nationale D'ingenieurs de Sfax, 3029 Sfax, Tunisia
e-mail: ma-intissar@hotmail.com

P. L. Higueras · E. M. Garcia Noguero
Instituto de Geología Aplicada, University of Castilla-La Mancha, Ciudad Real, Spain

A. Kallel et al. (eds.), *Recent Advances in Geo-Environmental Engineering, Geomechanics and Geotechnics, and Geohazards*, Advances in Science, Technology & Innovation, https://doi.org/10.1007/978-3-030-01665-4_8

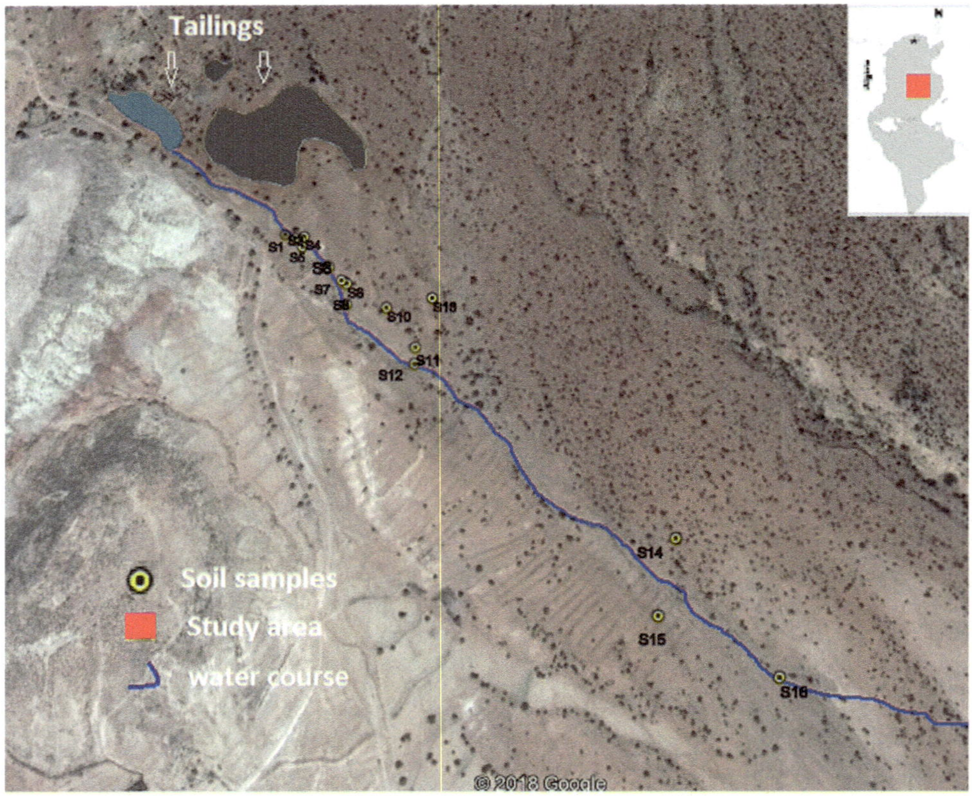

Fig. 1 Site of studied area and sampling points of soils

best way to evaluate the incorporation of trace elements into soil profiles.

In the laboratory, the waste and soil samples were air-dried, transferred to clean, polyethylene bags and subsequently sieved to a <2 mm and <63μm. The fraction <63μm was reserved for the geochemical analysis and small quantities from this fraction were further ground with an agate mortar to 0.10–0.15 mm, whereas the soil fraction <2 mm was used to determine the physicochemical parameters including pH, electrical conductivity (EC), organic matter content (SOM), and carbonate (%$CaCO_3$).

For the physicochemical analysis, the pH of the soil was measured in a solid/solution ratio of 1: 2.5 with distilled water [5]. The level of carbonate calcium was determined by means of the Bernard calcimeter method; and organic matter content (SOM) was determined according to loss on ignition method.

For the chemical analysis, Multi-elemental X-ray fluorescence (XRF) analyses using a Panalytical Epsilon 1 device were performed.

3 Results and Discussion

3.1 Mine Tailings Characterization

Table 1 shows the heavy metals concentrations in tailing samples. The results display high concentrations in Pb (1830–5950 mg.kg^{-1}), Zn (7590–12,480 mg.kg^{-1}), Cd (85.95–123.25 mg.kg^{-1}) and Fe (25.29–42.55%). Furthermore, they also show that the highest concentrations in Pb and Zn are associated to samples DF3 and DG3 (5.95 and 12.48%, respectively); those richer in Cd are associated to DF4 (123.25 mg.kg^{-1}) and DF3 (115.9 mg.kg^{-1}) and those richer in Fe are associated to DF2 (42.55%).

These high concentrations of heavy metals in mining waste should be the cause of soil pollution in the vicinity of the mine, in combination with meteorological factors favoring the environmental dispersion of the concerned elements.

Table 1 Major elements and heavy metals contents in mine tailings

	Heavy metals contents (mg kg^{-1})			Major elements concentration (%)					
	Pb	Zn	Cd	SiO_2	Al_2O_3	Fe_2O_3	CaO	MgO	P_2O_5
DF1	4000	8480	85.95	18.66	3.61	32.80	10.5	4.55	0.19
DF2	5160	7600	92.50	20.97	3.79	42.55	6.2	2.47	0.18
DF3	5950	7630	115.90	20.43	5.47	32.98	10.1	4.22	0.25
DF4	3250	12440	123.25	18.09	3.70	30.22	12.9	6.98	0.16
DG1	2860	11130	102.30	18.05	4.05	26.87	16.2	7.50	0.14
DG2	2900	11870	113.00	16.90	3.55	28.08	15.3	7.63	0.15
DG3	2980	12480	113.25	20.15	3.58	29.91	12.9	6.75	0.16
DG4	2080	10460	105.35	16.06	2.74	26.63	18.2	9.88	0.15
DG5	1890	9822	109.95	16.97	3.01	25.29	19.5	10.50	0.14

3.2 Physical–Chemical Properties and Total Heavy Metal Levels in Soils

Physiochemical properties and the XRF results are presented in Table 2. The results show that the reactivity (pH) of the soils in the vicinity of the mine varies from 7.6 to 8.8 with a mean value of 7.9, indicating the absence of soil acidification, which, together with the alkalinity detected in the soil samples, should be associated to the high concentration of carbonates (varying between 4 and 26.01%). In Tunisia, the soils that are developed on sedimentary rocks are mainly calcareous in the North and gypseous in the South [1]. The analyzed soil samples were characterized by relatively high organic matter content: values vary between 2.35 and 6.18% (Table 2); for the background samples the values are similar and this is due to the soil nature (arable and cultivated land).

The soil samples had high concentrations of heavy metals, with values ranging from 190 to 9500 mg kg^{-1} for Pb, 500 to 40,280 mg kg^{-1} for Zn and 0.5 to 39 mg kg^{-1} for Cd. The results are shown in Table 2.

The highest values of Cd and Pb were found in S2 and S4 and the highest values of Zn were observed in S4 and S6.

Table 2 Physiochemical and total heavy metal contents in soils in the vicinity of the abandoned mine area

Sample type	pH	$CaCO_3$ (%)	SOM (%)	Cd (mg kg^{-1})	Pb (mg kg^{-1})	Zn (mg kg^{-1})
Polluted soils						
S1	8.6	19.20	3.73	12.50	6740	8760
S2	7.6	15.34	4.89	34.25	8010	17540
S3	7.7	18.75	3.19	25.80	6400	20520
S4	7.7	20.88	3.18	34.00	9580	40280
S5	7.7	25.43	3.51	3.75	1000	2190
S6	7.6	25.68	2.75	32.75	4960	28770
S7	7.6	21.17	2.87	0.50	230	500
S8	7.6	20.33	3.70	1.35	190	500
S9	7.9	23.86	3.31	5.10	1740	2880
S10	8.1	26.01	5.05	1.50	820	1610
S11	8.2	25.90	5.18	3.00	490	1270
S12	7.7	18.96	2.35	39.00	7150	37770
S13	8.4	4.00	2.87	2.85	510	2020
S14	7.7	4.00	6.18	1.10	250	900
S15	8.2	21.47	4.30	6.05	1010	4450
S16	7.8	21.59	3.30	16.75	7310	11030
Control soil						
T1	8.8	4.63	2.36	0.55	25.25	67.3
Reference levels CEE	–	–	–	3	100	300

The heavy metals contents of the soil samples nearest to the mine area clearly exceed the Tunisian standards in agricultural areas (100 mg kg^{-1} for Pb, 300 mg kg^{-1} for Zn and 3 mg kg^{-1} for Cd) by a factor of 37, 39 and 4, respectively. Besides, it is worth noting that the heavy metal contents in the control soil with respect to the Tunisian standards are of a value of 0.55 mg kg^{-1} for Cd, 25.25 mg kg^{-1} for Pb and 67.3 mg kg^{-1} for Zn.

4 Conclusions

The main Conclusions of the present study are the following:

- The studied wastes and soils from the mining areas show that, in general, all the samples presented a neutral to alkaline pH, similar to the background sample, the alkaline pH could be attributed to the presence of carbonates.
- Soils from the area show relatively high organic matter contents, which is evidence that the influence on the soil productivity is lower than expected.
- Based on the Tunisian standards for agricultural areas, we conclude that the majority of tailing samples and soils surrounding the mine of Jebel Trozza had high concentrations of Zn, Pb and Cd, and so most of the wastes and soils studied here represent a potential source of contamination for the surrounding area and its environmental compartments.
- Further studies are still needed in order to properly assess the biogeochemical risks involved in the pollution of this region and similar areas.

References

1. Elouear, Z., Bouhamed, F., Boujelben, N., Bouzid, J.: Assessment of toxic metals dispersed from improperly disposed tailing, Jebel Ressas mine, NE Tunisia. Environ. Earth Sci. (2016)
2. Higueras, P., Esbrí, J.M., García-Ordiales, E., González-Corrochano, B., López-Berdonces, M.A., García-Noguero, E.M., Alonso-Azcárate, J., Martínez-Coronado, A.: Potentially harmful elements in soils and holm-oak trees (Quercus ilex L.) growing in mining sites at the Valle de Alcudia Pb-Zn district (Spain)—some clues on plant metal uptake. J. Geochem. Explor. **182,** Part B: 166–179 (2017)
3. Sainfeld, P.: Les gîtes plombozincifères de la Tunisie. Annales des mines et de la géologie **9** (1952)
4. Spaargaren, O., Nachtergaele, F.: Topsoil characterization for sustainable land management. Food and Agriculture Organization of the United Nations. ftp://fao.org/agl/agll/docs/topsoil.pdf (1998). Accessed 04 May 2015
5. Thomas, G.W.: Soil pH and soil acidity. In: Sparks, D.L., et al. (eds.) Methods of soil analysis part 3. Chemical method, pp. 475–490. American Society of Agronomy, Madison, WI (1996)

Statistical Distribution of Geochemical Elements in Stream Sediments and the Influence of Flood Phosphate Mud on the Mining Area, Metlaoui, Southwest Tunisia

Feyda Srarfi, Raouen Rachdi, Roland Bol, Nadhem Brahim, and Najet Slim Shimi

Abstract

In this study we elucidated the relationships between geochemical composition of the stream sediment samples and industrial pollution (phosphate mud flood). We offered an approach based on multivariate statistics: the factor analysis with a series of factor score of the investigated area. This type of numerical analysis, applied to 21 samples of surveyed Metlaoui stream, should also offer an additional insight into the possible anthropogenic impacts that might be present in the stream sediment geochemistry. The model is presented in the form of a variance factor matrix in which the four factors explain almost 90% of the total system variability. As can be seen from the statistics results, the first factor (F1) predominates and accounts for almost twice as much of the total explained percentage variance as the second (F2). The other factors show a decline in magnitude (F3 and F4). Three main metal associations were individualised in the stream sediment of Metlaoui. Ti–Zr; Fe–Mn and Zn–Ni–Cu–Mo. The increase of all concentration values (Ti, Zr, Fe, Mn, Zn, Ni, Cu and Mo) reflects that the mining activity (phosphate mud flood, or airborne deposition) could be associated with Fe–Mn oxyhydroxide that play a significant role in precipitating and thus fixing heavy metals in the proximity of mine contamination source.

F. Srarfi (✉) · R. Rachdi · N. Brahim · N. S. Shimi
Department of Geology, Faculty of Sciences of Tunis, University of Tunis El Manar, 2092 Tunis, Tunisia
e-mail: feydasrarfi@gmail.com

R. Rachdi
e-mail: raouenrachdi@hotmail.fr

N. Brahim
e-mail: nadhem.brahim@gmail.com

N. S. Shimi
e-mail: najetshimi@yahoo.fr

R. Bol
Agrosphere Institute (IBG-3), Forschungszentrum Jülich GmbH, 52428 Jülich, Germany
e-mail: r.bol@fz-juelich.de

Keywords

Multivariate statistics • Stream sediment • Metal Associations • Phosphate mud • Metlaoui • Tunisia

1 Introduction

During Phosphate manufacturing, ground phosphate rock is firstly extracted from the mines and sent to recovery units to separate sand and clay and remove impurities. Most of the processes are wetted to minimize the emission of dust and facilitate transport. During this course, starting from a mine or quarry, washing is a means of enriching the crude phosphates by releasing the phosphate grains trapped in the clay and limestone clumps by mixing water with the removal of coarse and fine particles. The enrichment of the phosphates by a humid process can be accompanied with rejects of mud poured in hydrographic network of the region of Gafsa-Metlaoui [2]. So the activity of phosphate ore generates a significant discharge of polluted wastewater called "phosphate mud". These releases are extremely heterogeneous. They are loaded with various pollutants that can cause environmental degradation and harm human health. According to Galfati et al. [1], these releases are rich in phosphate and trace elements; the average is around these values: Zn: 260.91 mg/Kg, Cu: 16 mg/kg, and Ni: 26 mg/kg.

2 Materials and Methods

2.1 Sampling

Samples were collected according to an irregular network. The first sample S_0 (control sample) was designed in a way that the sample site is kept away from inhabited areas. The samples were collected in four stations, the first one was

A. Kallel et al. (eds.), *Recent Advances in Geo-Environmental Engineering, Geomechanics and Geotechnics, and Geohazards*, Advances in Science, Technology & Innovation, https://doi.org/10.1007/978-3-030-01665-4_9

200 m from the washing site discharge point of phosphate. At the sampling site, at least five samples of sediment were collected from different places along a 100 m stream section. Sampling was carried out during the dry season.

2.2 Evaluation of Basic Statistical Parameters and Multivariate Statistics

The evaluation of basic statistical parameters for all elements was performed on the analytical data from the sample database. Factor analysis was performed for the data set of each element with the aim to observe similarities of geochemical behaviour of the analysed samples. All statistical analyses were performed with the STATISTICA software.

3 Results

As can be seen from Table 1, although the first factor (F1) predominates and accounts for almost twice as much of the total explained percentage variance as the second (F2), the other factors show a slow decline in magnitude which is a sure indicator of weakness. Interdependence among the observed properties with the first factor F1 accounts for 42.25% of the total system variability. It is a monopolar factor (or slightly bipolar as far as pH is concerned) which is essentially composed of variables positively correlating elements such as Zn, Ni, Cu and Mo. This indicates that they may originate in the same source. The second factor F2 explains a further 19.60% of the total information in the data matrix. It is largely concerned with negative associations within Ti, Zr and Mn. A notable absence of other elements was remarked. The third factor F3, explaining a further 15.50% of the total variance, is of a bipolar nature on the basis of the negative relationship that unites Fe and Mn with pH, standing for a metallic-hydroxide factor. The fourth factor F4, accounts for 10.15% of the total variance examined. As can be seen, it is dominated by only one variable, conductivity. The reason for this is the considerably reduced number of variables loading significantly on each factor.

4 Discussion

The geochemical variability within the stream sediment of Metlaoui river can be described by a four factor model, which accounts for 87.54% of the data variability. Although there is some similarity in the element associations, there are also some significant differences of the element associations. The bedrock lithology and the hydroxide of Fe and Mn seem to be an important supply of material for corresponding suite of elements including Ni, Cu, Zn, and Mo (Factor 1, Table 1). This factor accounts for 42.25% of the data variability. The Ni, Cu, Zn, and Mo association seems to have as principal origin the dispersion from mining activities (flood of the phosphate mud). The factor 2 has high negative loadings for Ti and Zr considered as conservative elements in alluvial soils especially Ti [3]. Mn has a moderate contribution on this factor axis. Factor 4 is concerned essentially with conductivity. The correlation matrix (Table 1) shows that conductivity slightly influences Cu, Mo and Zr dispersal in the stream sediments.

5 Conclusion

Three main heavy metal associations are highlighted by statistical results. Mo–Ni–Cu–Zn; Ti–Zr and Fe–Mn. The poor role of pH may indicate the physical phenomenon that involves interactions of electrostatic type: adsorption/desorption of these element by different material surfaces (hydroxides, clay, organic molecules). When the values were compared with the local geochemical background station S_0, it is possible to identify an increase of Mo, Zn, Fe, Zr, Ti,

Table 1 The distribution of both physical parameters and trace elements in the analyzed sediment stream is given on factor plots

Variable	Factor 1	Factor 2	Factor 3	Factor 4	Factor 5	Factor 6
Mo	**0.924**	−0.019	0.267	−0.079	0.113	−0.142
Zn	**0.859**	0.215	0.378	0.127	0.033	−0.085
Ni	**0.971**	0.122	0.021	−0.081	0.0004	−0.112
Mn	0.096	**−0.611**	**−0.665**	0.158	−0.016	−0.311
Fe	**0.541**	−0.343	**−0.660**	−0.188	−0.042	0.126
Cu	**0.843**	−0.045	0.062	−0.012	−0.497	0.059
Ti	−0.081	**−0.850**	0.366	0.023	−0.152	0.237
Zr	0.227	**−0.723**	0.359	−0.420	0.283	−0.033
pH	**−0.720**	−0.256	0.423	0.041	−0.261	−0.323
CE	0.309	−0.301	0.094	**0.865**	0.146	0.067

The bold value shows important correlations, generally identified with a value > 0.5

a slight increase in Ni, Cu and Mn contents in the samples collected. The increase of some concentration values of Mo, Zn, Ni, Cu either reflects mining activity (phosphate mud flood, or airborne deposition), or could be associated with Fe-Mnoxyhydroxide, organic matter (humic acid) and clay minerals. The pollution effects caused by human activities can be clearly deduced especially in the floodplain regions that were flooded by waters that drained phosphate mining storage area in the past. Protection measures of phosphate mining are not sufficient to prevent the loss and release of heavy metals from phosphate mud.

References

1. Galfati, I., Bilal, E., Sassi, A.B., Abdallah, H., Zaier, A.: Accumulation of heavy metals in native plants growing near the phosphate treatment industry, Tunisia. Carpathian J. Earth Environ. Sci. **6**(2), 85–100 (2011)
2. Marzougui S., Sdiri A., Rkhiss F.: Heavy metals mobility from phosphate washing effluents discharged in the Gafsa area (South-western Tunisia). Arab. J. Geosci. (2016)
3. Miko, S., Halamic, J., Peh, Z., Galovi, L.: Geochemical baseline mapping of soils developed on diverse bedrock from two regions in Croatia. Geol. Croat. **54**(1), 53–118 (2001)

Assessment of Heavy Metal Contamination in the Sediment of the Bizerte Lagoon in Northern Tunisia

Ibtihel Saidi, Olfa Ben Said, Jamel Ben Abdelmalek, Luis Chicharo, and Hamouda Beyrem

Abstract

Metals concentrations, Geoaccumulation index (Igeo), Factor contamination (CF) and degree of contamination (DC) were used to assess heavy metals contamination in Bizerte lagoon sediments. The highest metal concentrations were detected in stations located close the industrial sewages discharge points 1089.96 ± 28.5 mg/kg of dry weight (mg/kg of dry wt). The DC was 15.65 reflecting a high degree of contamination for this station located close to the industrial swage.

Keywords

Marine sediments • Geo-accumulation index • Heavy metals • Sediment contamination

1 Introduction

Heavy metals are micro-pollutants, which can affect the marine environment, since they do not undergo biological or chemical degradation. Thus, an index of environmental pollution may be found not only in the presence of metals, but when their environmental concentrations exceed the geochemical background that was determined by geological and geochemical processes [1]. The contamination of aquatic systems by heavy metals, especially in sediments, has become one of the most challenging pollution issues. As is the case for many Mediterranean coastal lagoons, Bizerte lagoon is a polluted ecosystem subject to both urbanization and industrialization [2]. Thus, direct and indirect discharge of waste and runoff has resulted in a chemical contamination of the lagoon by various toxic compounds such as heavy metals [3]. The objectives of this study were to examine the spatial variations of the heavy metal concentration in the surface sediment of Bizerte, to explore the degree of heavy metal contamination using different pollution indices.

2 Materials and Methods

Sediment and fauna were collected at the Bizerte lagoon, a coastal Mediterranean lagoon located in the Northern part of Tunisia [4]. Three sampling sites were selected based upon cement and steel industrial plants wastewater discharge points into Bizerte lagoon (Fig. 1). Sediment granulometry [5], sediment metal contents [arsenic (As), barium (Ba), boron (B), chromium (Cr), copper (Cu), lead (Pb), manganese (Mn), nickel (Ni), vanadium(V), and zinc (Zn)] [6] were determined using atomic emission spectrometry with inductively coupled ICP_AES plasma.

Different indices such as Contamination factor CF [7], Contamination Degree (CD) [8] and Geo-accumulation index Igeo [9] were calculated.

3 Results

The granulometric analysis (Table 1) showed the particle size indicating that the soil consists principally of mud and sand. The highest mud values were observed at S1, followed by S2, while the lowest ones were revealed at S3.

S1 showed the highest score for total metals contents 872.13 ± 4.37 mg/kg of dry wt. The metal concentrations in collected surface sediments were ranked in a decreasing order

I. Saidi (✉) · O. B. Said · H. Beyrem
Laboratory of Environment Biomonitoring, Faculty of Sciences of Bizerte (FSB), University of Carthage, 7021 Zarzouna, Bizerte, Tunisia
e-mail: saidi_ibtihel@hotmail.fr

J. B. Abdelmalek
Laboratory Quality Control, Cement Bizerta Bay Sabra, 7018 Bizerte, Tunisia

L. Chicharo
Faculty of Sciences and Technology, Marine and Environmental Research Centre (CIMA), University of Algarve, Portugal Campus de Gambelas, 8005-139 Faro, Portugal

A. Kallel et al. (eds.), *Recent Advances in Geo-Environmental Engineering, Geomechanics and Geotechnics, and Geohazards*, Advances in Science, Technology & Innovation, https://doi.org/10.1007/978-3-030-01665-4_10

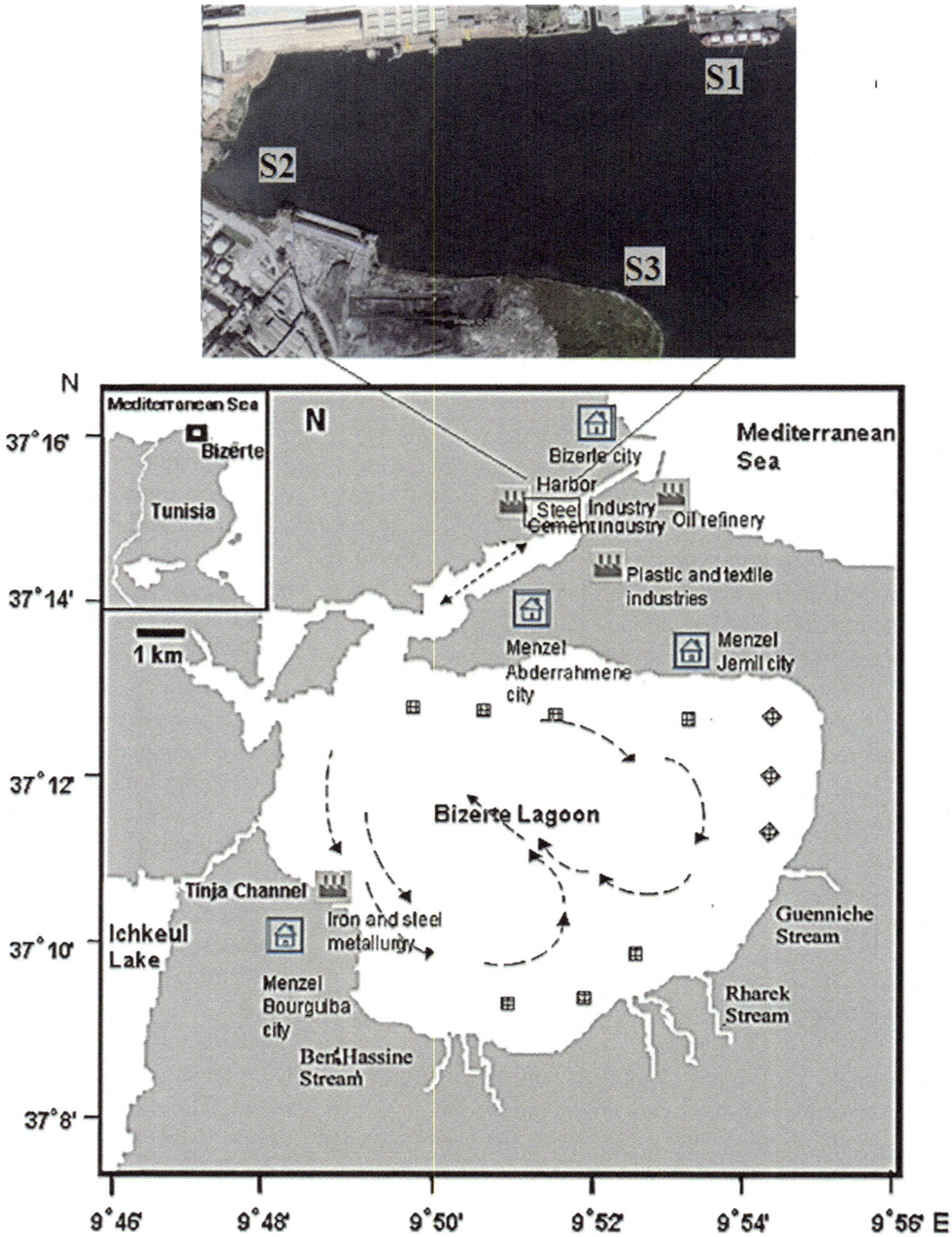

Fig. 1 The Bizerte lagoon and the three selected sampling stations, S1 and S2: station upstream of the discharge; S3: downstream station of the discharge

Table 1 Percentage of grain size distribution in coastal superficial sediments

	S1 (%)	S3 (%)	S5 (%)
Gravel	0.0	0.0	0.0
Sand	76.6	89.4	99.8
Mud	23.4	10.6	0.2

Mn > Zn > Ba > B > V > Cr > Pb > Cu > Ni (Fig. 2). Higher concentrations of As, Zn, B and Pb were found in the S1 located in the upstream of raw effluent, showing the highest score for metal content, mainly contaminated with Mn, Zn and Cr (respectively 114, 364 and 55 mg/kg of dry wt). The highest zinc content (364.7 mg/kg of dry wt) was recorded at S1. The contamination factors calculations showed low contamination factors of Cr and Ni observed in almost all the three stations (Table 2). Cd reaches a moderate contamination level for the three sediment samples from S1. Low contamination factors exist in the sediments of site S3 for all metals.

Fig. 2 Levels of metals in the sediment of different locations

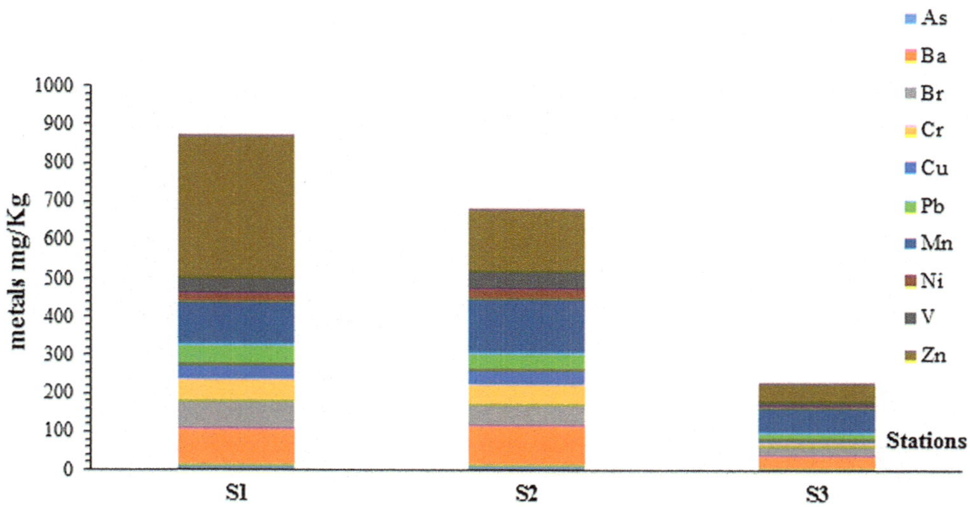

Table 2 Contamination factors (CF), Degree of contamination (Dc) and geo-accumulation index (Igeo) of sediments of Bizerte lagoon

Stations	S1		S2		S3	
Metals	(CF)	Igeo	(CF)	Igeo	(CF)	Igeo
Cd	2.91	0.54	2.9	0.59	1	0
Cr	0.66	0.14	0.61	0.12	0.12	0.02
Cu	1.62	0.35	1.56	0.31	0.32	0.06
Pb	2.93	0.38	2.54	0.5	0.94	0.19
Ni	0.45	0.11	0.58	0.11	0.1	0.02
Zn	5.13	0.47	2.23	0.44	0.63	0.12
Mn	1.95	0.4	2.47	0.5	1.14	0.23
Contamination degree (CD)	15.65		12.89		4.25	

The Igeo for Cr, Cu, Pb and Ni indicated unpolluted to moderately polluted sediment quality (class 1) for all the studied sites. Zn showed a slight spatial variation in the station close to steel industry S1 (class 1 to 2) moderately polluted. The present study showed a low contamination in sediments of site 3 for all metals. The degree of contamination was 15.65 for S1 and 12, 89 for S2 and 4 at S3, $12 < Cd < 24$ reflecting the considerable degree of contamination for the stations located close the industrial swage.

4 Discussion

Fine sediments containing only mud primarily characterized the grain size distribution in the sampled stations and sand consequently accumulated more pollutants and organic matter [10]. It has long been recognized that sediment characteristics greatly influence the structure and diversity of benthic faunal communities [11]. Comparing the Igeo results from this study with previous works suggests that there is a progressive increase in metal contamination in surface sediment of Bizerte lagoon [12]. Ben Mna et al. [13] revealed Igeo values for Cd, Pb, Ni and Zn above 1, suggesting a widespread pollution and showing that these heavy metals originate from anthropogenic sources mainly urban runoffs and industrial effluent discharges. Zaaboub et al. [14] mentioned that the most accumulated trace elements in surface sediments are Zn and Pb in Bizerte lagoon. The dominant taxon was represented by nematodes in surface sediments station located in upstream effluent are moderately polluted sediment with Cd, Cr, Pb, Ni and Zn. Van der Wurff et al. [15] showed that nematodes can accumulate and tolerate higher levels of copper, cadmium, zinc and lead than other meiofaunal organisms. The observed high densities of nematodes is an indicator of a disturbed environment, rich muddy substrates relatively higher densities of copepods in rural coastal areas, which could be an indicator of undisturbed environment, were associated with coarser sediments and low in organic matter content.

5 Conclusion

The contamination factor, the geo-accumulation index, and the degree of contamination show that, currently, the lagoon of Bizerte, is unpolluted to moderately polluted. Most of the superficial sediments of Sabra Bay of Bizerte lagoon are moderately contaminated, by mainly manganese, lead, zinc, chromium, barium and arsenic with a slighter degree nickel. The results show that the Bizerte lagoon is relatively polluted based on the ratio of nematode and copepod densities.

References

1. Zhang, L.J., Wang, G., Yao, D., Duan, G.Z.: Environmental significance and research of heavy metals in offshore sediments. Mar. Ecol. Lett. **19**, 6–9 (2016)
2. Béjaouia, B., Harzallaha, A., Moussab, M.: Annie Chapellec and Cosimo Solidorod. Analysis of hydrobiological pattern in the Bizerte lagoon (Tunisia). Estuar. Coast. Shelf Sci. **1**, 121–129 (2008)
3. Yoshida, M., Hamadi, K., Ghrabi, A.: Solid waste landfills and soil/sediment contamination around Bizerte lagoon: possible pollution sources. In: Ghrabi, A., Yoshida, M. (eds.) Study on Environmental Pollution of Bizerte Lagoon, vol, 55 (2002)
4. Blott, S.J., Pye, K.: GRADISTAT: a grain size distribution and statistics package for the analysis of unconsolidated sediments. Earth Surf. Process. Land. **26**, 1237–1248 (2001)
5. Loring, D.H.: Regional mineralogical variability in Eastern Canadian Marine Sediments. Nantes, France, ICES WG/MS (1993)
6. Varol, M.: Assessment of heavy metal contamination in sediments of the Tigris river (Turkey) using pollution indices and multivariate statistical techniques. J. Hazard. Mater. **195**, 355–364 (2011)
7. Muller, G.: Die Schwermetallbelastung der Sedimenten des Neckars und Seiner Nebenflusse. Chem. Ztg **6**, 157–164 (1981)
8. Hakanson, L.: An ecological risk index for aquatic pollution control: a sedimentological approach. Water Res. **14**, 975–1001 (1980)
9. Carvalho, S., Gaspar, M.B., Moura, A., Vale, C., Antunes, P., Gil, O., Cancela da Fonseca, L., Falcao, M.: The use of the marine biotic index AMBI in the assessment of the ecological status of the O′ bidos lagoon (Portugal). Mar. Pollut. Bull. **52**, 1414–1424 (2006)
10. Gray, J.S.: Animal-sediment relationships. Oceanogr. Mar. Biol. Annu. Rev. **12**, 223–261 (1974)
11. Coblentz, K.E., Henkel, J.R., Sigel, B.J., Taylor, C.M.: Influence of sediment characteristics on the composition of soft-sediment intertidal communities in the northern Gulf of Mexico. PeerJ (2015). https://doi.org/10.7717/peerj.1014
12. Barhoumi, B., Elbarhoumi, A., Clérandeau, C., Al-Rawabdeh, A. M., Atyaoui, A., Touil, S., Driss, M.R., Cachot, J.: Using an integrated approach to assess the sediment quality of an Mediterranean lagoon, the Bizerte Lagoon (Tunisia). Ecotoxicology **25**(6), 1082–1104 (2016)
13. Ben Mna, H., Oueslati, W., Helali, M.A., Zaaboub, N., Added, A., Aleya, L.: Distribution and assessment of heavy metal toxicity in sediment cores from Bizerte Lagoon, Tunisia. Environ. Monit. Assess. **7**, 189–356 (2017)
14. Zaaboub, N., Martins, M.V.A., Dhib, A., Béchir, B., Galgani, F., El Bour, M., Aleya, L.: Accumulation of trace metals in sediments in a Mediterranean lagoon: usefulness of metal sediment fractionation and elutriate. Environ. Pollut. **207**, 226–237 (2015)
15. Van der Wurff, A.W.G., Kools, S.A.E., Boivin, M.E.Y., Van den Brink, P.J., van Megen, H.H.M.: Type of disturbance and ecological history determine structural stability. Ecol. Appl. **17**, 190–192 (2007)

Assessment of Heavy Metals Along a Contamination Gradient in Soils Collected from Industrial Areas in Northern Tunisia

Ahmed Ouni, Chedliya Ghemari, Amina Ben Said, Christelle Pruvot, Francis Douay, and Karima Nasri-Ammar

Abstract

Many studies were interested in the assessment and the monitoring of soil contamination with heavy metals. This study aimed to evaluate the effects of emissions from the Tunisian Steel Company El Fouladh in Bizerte (Tunisia). In this context, we measured Zinc (Zn), Copper (Cu), Lead (Pb) and Cadmium (Cd) soil's concentrations from 8 sites along a contamination gradient. Besides, physicochemical parameters, pH, organic carbon, CEC, and $CaCO_3$ were measured. Heavy metals were determined using an atomic absorption spectrometry. The results revealed a high concentration of heavy metals in the sampled soils. The contamination factor (CF) was determined to assess the soil contamination of the sampling sites, and revealed that the level of contamination varied along a contamination gradient, which may be due to the mobility of metals.

Keywords

Assessment • Pollution • Soils • Heavy metals • Sampling sites

1 Introduction

Metallic contamination has become a worldwide threat due to the environmental persistence of heavy metals issued from the surrounding industrial activities [1–4]. Consequently, many studies were carried out to assess soil contamination by heavy metals [5–7]. Furthermore, soil quality depends on the nature of the physicochemical parameters within the soil and is influenced by the heavy metals contents [8].

Bizerte, situated in Northern Tunisia, was characterized by a significant industrial activity generating high contamination. Contrariwise, few studies were conducted to qualify this anthropic activity except those of Ghemari et al. [9] using the terrestrial isopod *Porcellio laevis* and those of Ghannem et al. [10] using beetles. Thus, the main objectives of this study were (i) to measure the concentrations of Zn, Cd, Cu and Pb and the physico-chemical parameters in the soils to assess the contamination around the sampling sites (ii) and to follow the mobility of the heavy metals throughout a contamination gradient.

2 Materials and Methods

Samples were taken near the Tunisian Steel company (El Fouladh), located in Bizerte Northern Tunisia, in the region of Menzel Bourguiba, the south side of the Bizerte's lake and where soil samples were taken at a depth of 10–15 cm (Fig. 1).

In the laboratory, the collected soil samples for each station were dried at 40 °C in an oven aiming to determine the pH, $CaCO_3$, organic carbon, and metal concentrations in the soils. Carbonate and organic carbon contents were determined according to the NF ISO 10693 and NF ISO 10694 standards respectively. Soil pH was measured according to the NF ISO 10390 standard. The Cation Exchange Capacity (CEC) was determined according to the NF X 31-130 standard. The determination of heavy metals concentrations requires the mineralization of the samples. The digestion consists in adding 4.5 mL of 37% hydrochloric acid (HCl) and 1.5 mL of 70% nitric acid (HNO_3) using a Hot Block system-assisted digestion (Environmental Express® SC100, Charleston, SC, USA) during 1 h 30 at 120 °C. After that, the analytical

A. Ouni (✉) · C. Ghemari · A. Ben Said · K. Nasri-Ammar
Faculty of Science of Tunis, Research Unit of Bio-Ecology and Evolutionary Systematic, University of Tunis El Manar, 2092 El Manar II, Tunis, Tunisia
e-mail: ahmedouni2@yahoo.fr

C. Pruvot · F. Douay
Institut Supérieur d'Agriculture (ISA), Laboratoire Génie Civil et géo-Environnement (LGCgE), Yncréa Hauts-de-France, 48 boulevard Vauban, 59046 Lille Cedex, France

A. Kallel et al. (eds.), *Recent Advances in Geo-Environmental Engineering, Geomechanics and Geotechnics, and Geohazards*, Advances in Science, Technology & Innovation, https://doi.org/10.1007/978-3-030-01665-4_11

determination was conducted using an atomic absorption spectrometry (AAS) by flame (FAAS-6800, Shimadzu).

The contamination factor (CF) corresponds to the ratio between the metal concentration in soil (in mg/kg dry weight) and its baseline value (in mg/kg dry weight). According to Hakanson [11], CF values were interpreted as follows:

- $CF < 1$: means that low contamination exists.
- $1 < CF < 3$: means that moderate contamination exists.
- $3 < CF < 6$: means that very high contamination exists.
- $CF > 6$: means that very high contamination exists.

3 Results

The physicochemical properties and heavy metals concentrations in the soils were determined and summarized in Table 1. The results showed that the organic carbon values oscillated between 5.05 and 35.61 (g/kg). Considering the $CaCO_3$, the values ranged from 34.90 (g/kg) to 297.02 (g/kg) reflecting a high variability in the sampling sites. The lowest value of CEC was observed in S3 (3.49 mg/kg), while the highest was found in S6 (22.28 mg/kg). Furthermore, the soil pH values ranged between 8.04 and 9.84 in S7 and S4, respectively reflecting that all the sampling sites were characterized by an alkaline pH (>7).

Considering the metals concentrations in soils, Zn concentration ranged from 50.60 to 3222.33 mg kg^{-1} dw, where the highest concentration was observed in S8. Similarly, the highest Cu concentration was obtained in S8 (440.91 mg kg^{-1} dw). Besides, Cd concentration oscillated from 0.43 in S3 and S7, to 2.24 in S5. However, compared to the rest of the areas, soils Pb concentration was remarkably high in S2 with 1672.23 mg kg^{-1} dw.

The results of the contamination factor showed an important variability between the sampling sites (Table 2). Thus, for S6 and S7, whatever the metals, low contamination exists ($CF < 1$). Regarding S1, it exhibited moderate contamination with the measured metals. Also, very high contamination was obtained in S2, S5 and S8 for Zn and Pb. Regarding Cd, moderate and high contamination were reported in (S1, S2, S8) and in (S5) respectively. Considering Cu, only high contamination was observed for S8.

To better understand the soil distribution along the studied physicochemical parameters (pH, C.org, CEC, and $CaCO_3$), and the heavy metals concentrations (Cd, Pb, Zn, and Cu), a principal correspondence analysis (PCA) was performed (Fig. 2). The first and the second axis of PCA revealed a cumulative percentage of 77.00%. Except for pH, all quantitative data were projected on the first axis 'F1'. Considering the second axis, the metals were positively projected and correlated with Org. C. The sampling sites S2, S5 and S8 were positively correlated with heavy metals, and are considered as the most contaminated sites. However, CEC and $CaCO_3$ values characterized the two sampling sites S7 and S6, and were negatively correlated with pH (Fig. 2).

4 Discussion

The soils were sampled from 8 sites and belonged to the industrial area of Bizerte (Northern Tunisia). The first site S1, considered as the source, showed moderate contamination. However, we observed low contamination in the more distant sites (S6 and S7). Also, a decrease in soil contamination was observed away from the source of pollution. In this context, Van Straalen et al. [12] showed a comparable gradient of contamination, where the closest site to the pollution was not the most contaminated.

The relationship between heavy metals concentrations and soil physicochemical parameters was evaluated using PCA analysis. A strong correlation was obtained between the studied metals and organic carbon. Similarly, a study was carried out in Nanjing area (China), where a positive

Table 1 Heavy metals concentrations and physicochemical parameters of the sampling soils

	Zn	Cu	Pb	Cd	Org. C	$CaCO_3$	CEC	pH
S1	895.25	107.21	165.37	0.91	27.08	97.41	7.17	8.69
S2	2497.26	36.11	1672.23	1.99	32.30	96.52	5.53	8.69
S3	429.70	42.73	219.57	0.43	7.46	71.48	3.49	9.09
S4	436.32	154.38	67.16	0.66	14.00	155.88	4.38	9.84
S5	2503.71	178.43	596.15	2.24	34.14	34.90	8.98	9.19
S6	50.60	14.09	0.42	0.55	5.05	297.02	22.28	8.24
S7	149.34	44.34	54.78	0.43	16.32	245.30	13.19	8.04
S8	3222.33	440.91	390.29	1.50	35.61	63.78	7.70	8.32

Org. C organic carbon content in g kg^{-1}; **$CaCO_3$** total carbonates contents in g kg^{-1}; **CEC** cation exchange capacity (Cmol kg^{-1}); **Zn, Cu, Pb and Cd** Cadmium, Lead, Zinc and Copper soil's concentrations in mg kg^{-1} dw

Table 2 Contamination factor (CF) of Zn, Cu, Pb and Cd in soils from the different sampling sites

	Zn	Cu	Pb	Cd
S1	2.98	1.07	1.65	1.31
S2	8.32	0.36	16.72	2.84
S3	1.43	0.43	2.20	0.61
S4	1.45	1.54	0.67	0.94
S5	8.35	1.78	5.96	3.20
S6	0.17	0.14	0.01	0.79
S7	0.50	0.44	0.55	0.61
S8	10.74	4.41	3.90	2.15

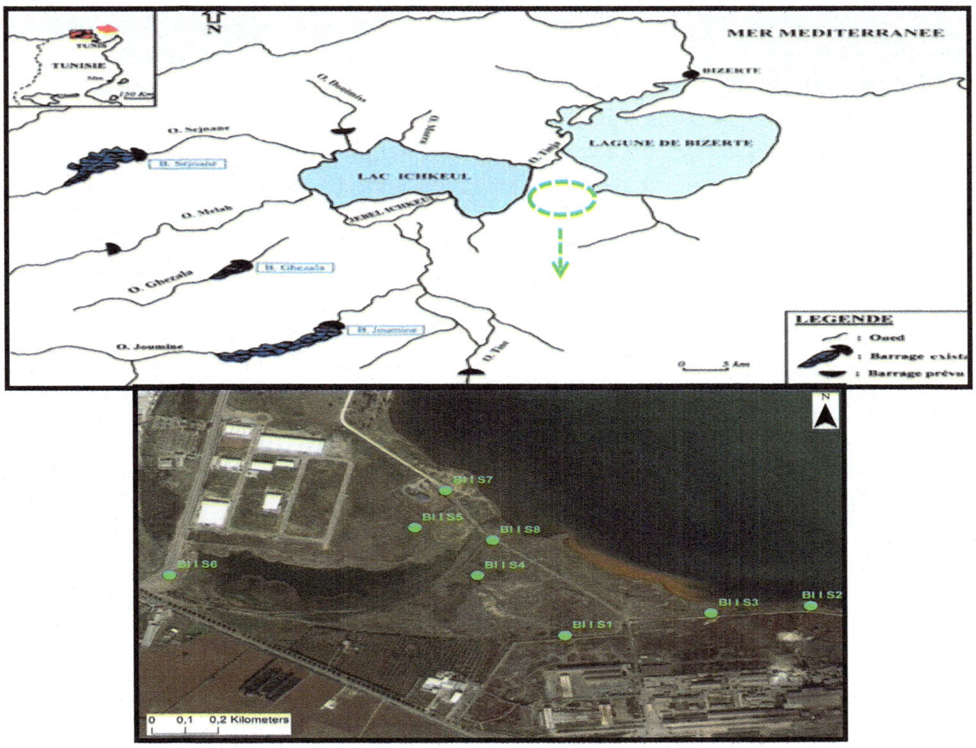

Fig. 1 Localization of the sampling sites in the region of Bizerte

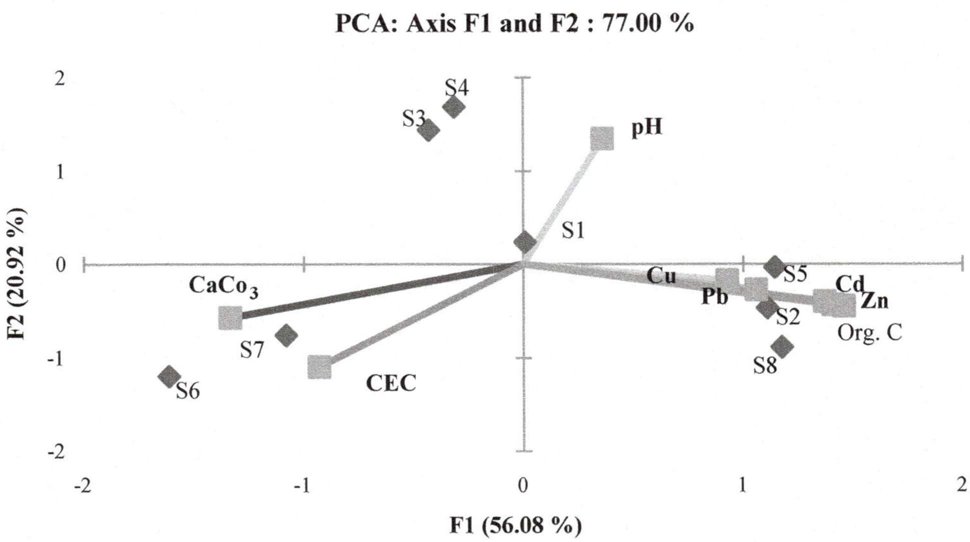

Fig. 2 Plot of the two principal components from PCA of the measured soil properties: physicochemical parameters (pH, $CaCO_3$, CEC, Org. C) and heavy metals concentrations (Cd, Pb, Cu and Zn), and the situation of sampled soils along the two axes

correlation was obtained among Cu, Zn and Pb [7]. Also, a negative correlation was obtained between metals and pH, confirmed the decrease of the solubility in soils with the increasing pH [13].

5 Conclusion

Although they are near to the pollution source, metal concentrations in the soil were sometimes low but globally, they indicated that the soils were polluted with Cd, Pb, Zn and Cu. The present study confirmed the findings of Ghemari et al. [9] which revealed the impact of soil contamination on terrestrial organisms.

Thus, the data obtained here in, provided values that could be useful in future studies to assess soil pollution by heavy metals in Tunisian industrial areas and will be helpful to set up monitoring programs to reduce the pollution impact on terrestrial ecosystems.

References

1. Boldina-Cosqueric, I., Amiard, J.C., Amiard-Triquet, C., Dedourge-Geffard, O., Métais, I., Mouneyrac, C., Moutela, B., Berthet, B.: Biochemical physiological and behavioural markers in the endobenthic bivalve Scrobicularia plana as tools for the assessment of estuarine sediment quality. Ecotoxicol. Environ. Saf. **73**(7), 1733–1741 (2010)
2. Clemens, S.: Toxic metal accumulation, responses to exposure and mechanisms of tolerance in plants. Biochimie **88**, 1707–1719 (2006)
3. Thompson, J., Bannigan, J.: Cadmium: toxic effects on the reproductive system and the embryo. Reprod. Toxicol. **25**, 304–315 (2008)
4. Zhuang, S.S., Zhang, H.M., Qin, R., Jiang, W.S., Liu, D.H.: Cadmium induction of lipid peroxidation and effects on root tip cells and antioxidant enzyme activities in Vicia faba L. Ecotoxicol. **18**, 814–823 (2009)
5. Chlopecka, A., Bacon, J.R., Wilson, M.J., Kay, J.: Forms of cadmium, lead and zinc in contaminated soils from southwest Poland. J. Environ. Qual. **25**, 69–79 (1996)
6. Culbard, E.B., Thornton, T., Watt, J., Wheatley, M., Moorcroft, S., Thompson, M.: Metal contamination in British urban dusts and soils. J. Environ. Qual. **17**, 226–234 (1998)
7. Lu, Y., Gong, Z., Zhang, G., Burghardt, W.: Concentrations and chemical speciations of Cu, Zn, Pb and Cr of urban soils in Nanjing China. Geoderma **115**, 110–111 (2003)
8. Selim, H.M., Sparks, D.L.: Heavy Metals Release in Soils. CRC Press, Boca Raton (2001)
9. Ghemari, C., Waterlot, C., Ayari, A., Leclerq, J., Douay, F., Nasri-Ammar, K.: Assessment of heavy metals in soil and terrestrial isopod Porcellio laevis in Tunisian industrialized areas. Environ. Earth Sci. **76**, 223 (2017)
10. Ghannem, S., Khazri, A., Sellami, B., Boumaiza, M.: Assessment of heavy metal contamination in soil and Chlaenius (Chlaeniellus) olivieri (Coleoptera, Carabidae) in the vicinity of a textile factory near Ras Jbel (Bizerte, Tunisia). Environ. Earth Sci. **75**, 442 (2016)
11. Hakanson, L.: An ecological risk index for aquatic pollution control. A sedimentological approach. Water Res. **14**(8), 975–1001 (1980)
12. Van Straalen, N.M., Butovsky, R.O., Pokarzhevskii, A.D., Zaitsev, A.S., Verhoef, S.C.: Metal concentrations in soil and invertebrates in the vicinity of a metallurgical factory near Tula (Russia). Pedobiologia **45**, 451–466 (2001)
13. Krásný, J., Sharp, J.M.: Groundwater in Fractured Rocks: IAH Selected Paper Series, vol. 9. ISBN: 9780415414425 (2007)

PAHs Monitoring in Soils Affected by Electric Power Station

Svetlana Sushkova, Abdulmalik Batukaev, Tatiana Minkina, Elena Antonenko, Irina Deryabkina, Jana Popileshko, and Tamara Dudnikova

Abstract

The environmental pollution in soils affected by emissions from the Novocherkasskaya Electric Power Station (NEPS) was monitored according the monitoring data of two years (2016–2017). The monitored sites are located on fallow lands 20 km around NEPS. PAHs extraction from the soil samples was performed using ecologically clean newly developed method of subcritical water extraction. The 5-km zone located in the northwestern direction from the power station, in direction of predominant winds was the most subjected to PAHs contamination, mostly high at a distance up to 5.0 km from the contamination source. The level of high-molecular PAHs exceeding the level low-molecular PAHs form the monitoring sites situated in direction of predominant winds from the energy plant. The vise-versa dependence was established for sites around NEPS. According the 16 priority PAHs content in the studied soils. The most subjected to environmental pollution of soils were the monitored sites located in direction of predominant winds from NEPS.

Keywords

PAHs • Environmental pollution • Monitoring • Electric power station

1 Introduction

Polycyclic aromatic hydrocarbons (PAHs) are persistent hydrophobic compounds and one of the most dangerous toxicants of the environment affecting all living organisms mutagenicity and carcinogenity. From 16 up to 32 PAHs compounds have been subjected to control in countries that have a legislative regulation that depends on toxicants carcinogenicity, mutagenicity and toxicity [1]. Monitoring of soil technogenic contamination with PAHs has been carried out over many years [2]. The PAHs presence in soils is caused by increased level of the environmental pollution. Energy power industry is one of the main PAHs contamination sources especially thermal power plants. Novocherkasskaya Electric Power Station (NEPS) has become the greatest power plant in the European part of Russian Federation as well as in the whole Europe for years. NEPS produces electricity using the low-quality coal as the main energy source. After coal burning emissions fall through 185 and 250 m high pipes, they directly spread to the nearby environment. NEPS territory reaches the city Novocherkassk 300–400 thousand people and has a significant contribution to the air pollution of the entire region. The purpose of this investigation was to develop PAHs monitoring in soils affected by the NEPS.

2 Materials and Methods

The soils within the NEPS impact zone were the main object of our study. Figure 1 shows satellite images of the impact zone and the monitored sites location. The major part of the territory in the impact zone of the NEPS has mostly ordinary calcareous chernozems (Haplic Chernozems) (Cho). In addition, the meadow-chernozemic soils [Haplic Chernozems (Stagnic)] (MChS) (plot 3sw) and alluvial soils (Al) (Fluvisols) are distinguished within the

S. Sushkova (✉) · T. Minkina · E. Antonenko · I. Deryabkina
J. Popileshko · T. Dudnikova
Southern Federal University, B. Sadovaya str. 105, 344006 Rostov-on-Don, Russia
e-mail: snsushkova@sfedu.ru

A. Batukaev
Chechen State University, Dudaeva blvd. 17, 366007 Grozny, Russia

A. Kallel et al. (eds.), *Recent Advances in Geo-Environmental Engineering, Geomechanics and Geotechnics, and Geohazards*, Advances in Science, Technology & Innovation, https://doi.org/10.1007/978-3-030-01665-4_12

Fig. 1 Map of monitored sites of the environmentally polluted energy producing plants zone

Tuzlov River floodplain [3]. Chernozems and meadow-chernozemic soils have deep humus horizons (70–100 cm), the middle levels for the region organic matter (2.4–2.9%), cation exchange capacity (CEC) (31.2–47.6 cmol (+)/kg). The content of exchangeable calcium was 76–90% of the sum of exchangeable cations, pH of water extract was neutral or weakly alkaline (7.4–7.7). By their particle-size distribution, the soils are classified as heavy loamy and light clayey varieties formed on calcareous loess-like deposits with physical clay content 51–55% and clay content 27–30%. The alluvial soils are specified by a coarser texture, less thick humus horizon (40–60 cm) which were characterized by the lower organic matter content (up to 1.8%); CEC (10.6 cmol(+)/kg) and high content of exchangeable calcium.

The soils were sampled annually in June during the 2 years 2016–2017. The newly developed method of subcritical water extraction was used for PAHs determination in the soil samples [4]. The content of PAHs in the extracts was quantified by the HPLC method with ultraviolet and fluorescence detection simultaneously according ISO 13877 requirements [5]. The solvents and reagents used in the research had HPLC grade mark and were represented by ethanol, n-hexane, potassium hydrate, acetonitrile, NaOH, and anhydrous Na_2SO_4. The PAHs standards in acetonitrile in a concentration of 200 $\mu g\ cm^{-3}$ (Sigma-Aldrich/Merch) was used as standard solution prepared during HPLC analyses.

3 Results

The study of PAHs content in 20 cm soil layer of monitored sites located in the NEPS impact zone showed an intensive accumulation of polyarenes in 2016. The PAHs content in all studied soils increased in 2017 compared to 2016 data. It was found that the average value of PAHs total concentration in 20 cm soil layer of the monitored sites in direction of predominant winds from NEPS was 1196.9 ± 17.0 µg/kg in 2016 and 1514.1 ± 12.1 µg/kg in 2017 (Fig. 2a). The average value of PAHs total concentration in 20 cm soil layer of monitored sites around NEPS was 580.8 ± 6.6 µg/kg in 2016 and 946.3 ± 7.4 µg/kg in 2017 (Fig. 2b). Accordingly, the dynamic of PAHs total concentration in 20 cm soil layer of the monitored sites has been found to be on the increase.

4 Discussion

The PAHs accumulation in the monitored sites soils was shown to be in direction of predominant winds depending mostly on distance from NEPS. The most affected monitored soil was site no. 4 (1.6 km sw) with maximum PAHs concentration in 2017—2352.9 ± 15.9 µg/kg, and in 2016—1958.61 ± 23.0 µg/kg. A sharp increase of total PAHs content was noticed in the soil of the monitored site no. 8 (5.0 km nw) rising from 1080.8 ± 12.1 µg/kg in 2016 up to 1484.9 ± 16.9 µg/kg in

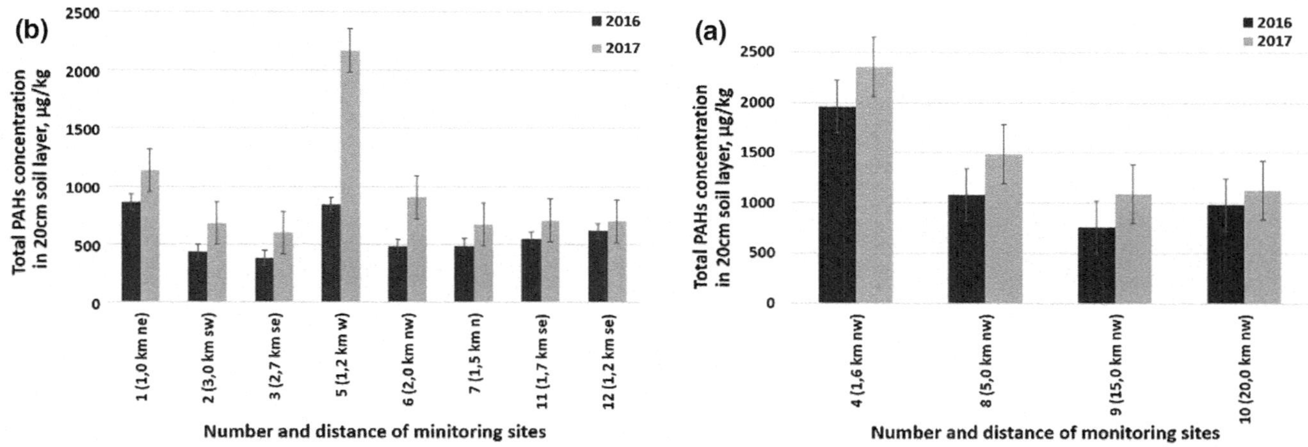

Fig. 2 Dynamic of PAHs total concentration in 20 cm soil layer of monitoring sites (**a**) in direction of predominant winds from NEPS (**b**) around NEPS in 2016–2017

2017 proving that the territory of environmental pollution from NEPS emissions reached 5.0 km in the predominant wind direction. Dynamics of PAHs total concentration in 20 cm soil layer of monitored sites showed the PAHs content decreased in site no. 9 (15.0 km nw) 760.7 ± 8.6 µg/kg in 2016 and 1091.0 ± 10.4 µg/kg in 2017. A relatively close contamination level was found for site no 10 (20.0 km nw) 987.6 ± 11.7 µg/kg in 2016 and 1127.8 ± 16.2 µg/kg in 2017 (Fig. 2). Thus, the most affected sites were no. 4 (1.6 km nw) and 8 (15.0 km nw) located at the distance 1.6–5.0 km from NEPS, respectively. This data are in agreement with previous research on NEPS emission zone and show that NEPS still have the prevailing effect on close territories and cause a high level of environmental pollution [6].

The data on PAHs concentrations in 20 cm soil layer of monitoring sites around NEPS in 2016 and 2017 showed that the level of total PAHs varied from 383.3 ± 11.6 µg/kg up to 842.5 ± 8.4 µg/kg in 2016 and from 600.3 ± 11.0 µg/kg up to 1135.2 ± 13.4 µg/kg. The total PAHs content in the monitored site no. 5 (1.2 km w) was the highest of all sites around NEPS from 863.5 ± 8.3 µg/kg up to 2168.0 ± 11.6 µg/kg due mostly to its connection to the site location.

5 Conclusion

The aerotechnogenic emissions of power plant NEPS were prevailing source of environmental pollution and PAHs accumulation in soils of the studied region. For the first time during 2 monitoring years, a total of sixteen PAHs content was studied for soils affected by energy-producing plants. The total PAHs content was revealed to increase gradually in soils of the studied territories over the period 2016–2017 explained by the increase of contaminants emission. The most polluted area by PAHs was the soil situated in the distance 5 km in the north-west direction, and most of the NEPS deposition has been emitted to a distance of about 1.6 km and decreases at the distance of 15 km from 62% up to 54% in 2016 and 2017, respectively.

Acknowledgements The investigation has been supported by the projects of RFBR 16-35-60051, Grant of the President of Russia MK-3476.2017.5, Ministry of Education and Science of Russia 5.948.2017/PCh.

References

1. Wenzl, T., Simon, R., Kleiner, J., et al.: Analytical methods for polycyclic aromatic hydrocarbons (PAHs) in food and the environment needed for new food legislation in the European Union. Trend. Anal. Chem. **25**(7), 716–725 (2006). https://doi.org/10.1016/j.trac.2006.05.010
2. Pereira, T., Laiana, S., Rocha, J., Broto, F., Comellas, L., Salvadori, D., Vargas, V.: Toxicogenetic monitoring in urban cities exposed to different airborne contaminants. Ecotoxicol. Environ. Saf. **90**, 174–182 (2013)
3. IUSS Working Group WRB: World reference base for soil resources 2014, update 2015 International soil classification system for naming soils and creating legends for soil maps. World Soil Resources Reports. No. 106. FAO, Rome (2015)
4. Sushkova, S.N., Minkina, T.M., Mandzhieva, S., et al.: New alternative method of benzo[a]pyrene extraction from soils and its approbation in soil under technogenic pressure. J. Soils Sed. **16**(4), 1323–1329 (2016). https://doi.org/10.1007/s11368-015-1104-8
5. ISO 13877-2005: Soil quality-determination of polynuclear aromatic hydrocarbons—Method using high—performance liquid chromatography (2005)
6. Sushkova, S., Minkina, T., Turina, I., et al.: Monitoring of benzo[a]pyrene content in soils under the effect of long-term technogenic poluttion. J. Geochem. Explor. **174**, 100–106 (2017). https://doi.org/10.1016/j.gexplo.2016.02.009

Kinetics of Crude Oil Desorption from Contaminated Soil

Rahal Soufiane, Hadidi Noureddine, and Moulai Mostefa Nadji

Abstract

The objective of this work was to study the power of surfactant solutions on the remediation of soils contaminated by crude oil during a batch washing process. We used as surfactants Sodium dodecyl sulfate (SDS), Sodium dodecyl benzenesulfonate (SDBS). In this study, we examined the effect of various parameters (level of contamination, type of surfactant and contamination age) on the surfactant elimination of oil from soil. The results obtained in this study showed the potential of surfactants in the cleanup of soils contaminated by hydrocarbons, and that the surfactants structure has a crucial impact on the elimination kinetic.

Keywords

Crude oil • Contamination • Surfactant • Desorption kinetics • Remediation

1 Introduction

Hazardous and Toxic pollutants become a widespread environmental problem that can lead to soil contamination. This contamination can be due to industrial accidents or voluntarism (spills of industrial waste, leaking underground storage tanks and pipelines…) [1–3]. The elimination of these hazardous compounds from soils has become a major concern, and presents a challenge to scientists and engineers. In general, the organic pollutants have a high interfacial tension and low solubility, which makes their disposal of contaminated soil is difficult [3–7]. Problems associated with crude oil-contaminated site in environmental media have received increasing attention. To solve such problems, an innovative technology for the cleaning up process is a must. Solubilization of oil by the surfactants is a key factor in the remediation of soils contaminated by hydrocarbons [6–10]. The use of surfactants can improve the solubility of hydrcarbones by distributing them into the hydrophobic cores of the surfactant micelles [7, 10, 11]. The objective of the present work was to study the effects of surfactants solutions and even their structure as well as the sand nature on the remediation of crude oil-contaminated soils by a batch washing process.

2 Materials and Methods

2.1 Materials

We used an Algerian crude oil, which was selected as a representative organic pollutant model. It was obtained from a collection center located in HaoudhElhamra (HassiMessaoud, South of Algeria). The anionic surfactants were obtained via a dispenser and were used without further purification (Table 1). In this study, we used three sand types (Table 2):

- Sand from Adrar town about 1500 km from Algiers.
- Sand from Ouargla, located 820 km south of Algiers.
- Sand from operating station hydrocarbons Hadjret annas, located in Ouargla.

2.2 Methods

In this experiment we chose 25 and 50 g of oil contaminated in a 1 kg sand for Periods of one and 10 months to see the effect of age contamination on the washing process by surfactants solutions. For washing, the samples were taken as a

R. Soufiane (✉) · M. M. Nadji
Materials and Environmental Laboratory, University of Medea, Medea, Algeria
e-mail: rahalsoufiane@yahoo.fr

H. Noureddine
Department of Process Engineering, University of Medea, Medea, Algeria

A. Kallel et al. (eds.), *Recent Advances in Geo-Environmental Engineering, Geomechanics and Geotechnics, and Geohazards*, Advances in Science, Technology & Innovation, https://doi.org/10.1007/978-3-030-01665-4_13

Table 1 Characteristics of the selected surfactants

Surfactant	Molecular formula	MW	CMC (mM)
SDS	$C_{12}H_{25}SO_4Na$	288.38	8.3
SDBS	$C_{12}H_{25}C_6H_4SO_3Na$	348.5	3–2.8

Table 2 Characteristics of the selected sand

Sand	pH	Humidity	Density
Ouargla	6.87	0.138	2.5
Station	7.13	0.109	2.667
Adrar	7.12	0.280	2.64

function of time. We put individual samples consisting of 100 ml of surfactant solution containing 10 g of contaminated sand in 250 ml Erlenmeyer flasks. These samples were sealed with a cap to prevent loss of solution. They were then stirred for a period of 2, 4, 6, 8, 18 h at 200 rpm using a shaker at room temperature to ensure maximum solubility. After stirring, the mixture was filtered using filter paper and diluted by (1/50) for the COD analyses. The concentration of desorbed oil was determined by COD method. The surfactant concentration was kept constant for the blank and the samples in order to eliminate the effect of the surfactant. The COD value was determined by a standard method and the concentration of desorbed oil was deduced from the standard curve of crude oil established in water solution.

3 Results and Discussion

Crude oil desorption profiles versus time in anionic surfactants solutions in the washing processes carried out with three kinds of sand and two surfactants are shown in Figs. 1, 2 and 3, respectively. Figures 1, 2 and 3 show that the curves display an increase in the recovered oil content followed by stabilization. According to these figures, we note that for the three sands curves are similar. The mass of oil in the liquid phase increases during the washing until it reaches a maximum.

From Figs. 1, 2 and 3, and at the same concentrations of surfactants, the highest concentration of the oil solution was found in the presence of SDBS. However, the lowest was found with SDS, for different oil contents and for all three sands. These figures also indicate that the nature of sand and surfactant have a very significant effect on the evolution of the mass of oil in the liquid phase. It was found that the maximum of the mass was reached after 20 h of washing. According the Figs. 2 and 3 the age of contamination has a remarkable effect in the desorption process. The evolution of the crude oil desorption in function of time represented on Figs. 1, 2 and 3 show that the desorption kinetics follow a pseudo-first order mass transfer model (Eq. 1), that describes the crude oil concentration in all surfactants solution:

$$\frac{dS_C}{dt} = k\left(S_C^* - S_C\right) \qquad (1)$$

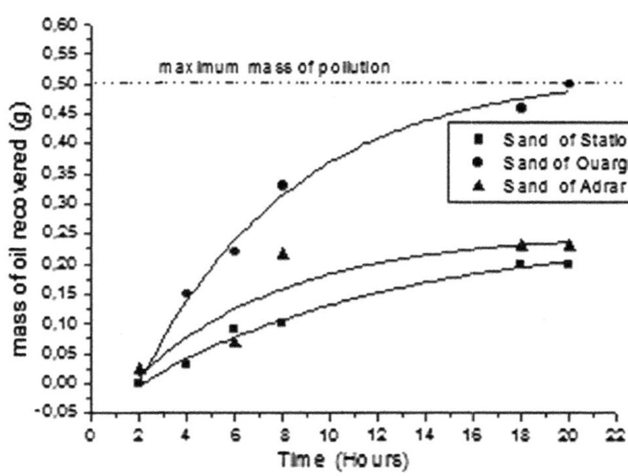

Fig. 1 Oil desorption evolution in the presence of SDBS for 10 months of contamination (50 g/1 kg)

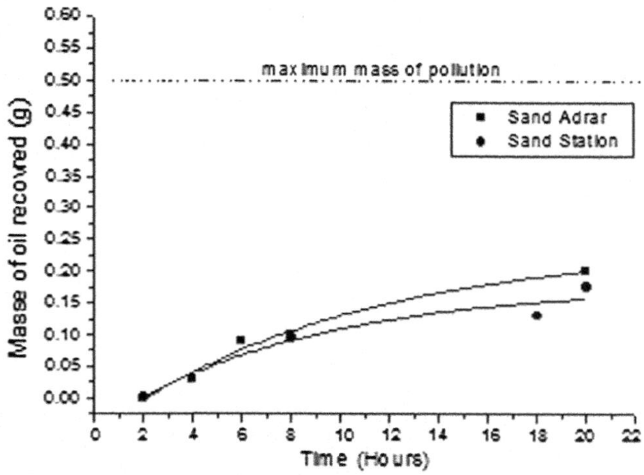

Fig. 2 Oil desorption evolution in the presence of SDS for 10 months of contamination (50 g/1 kg)

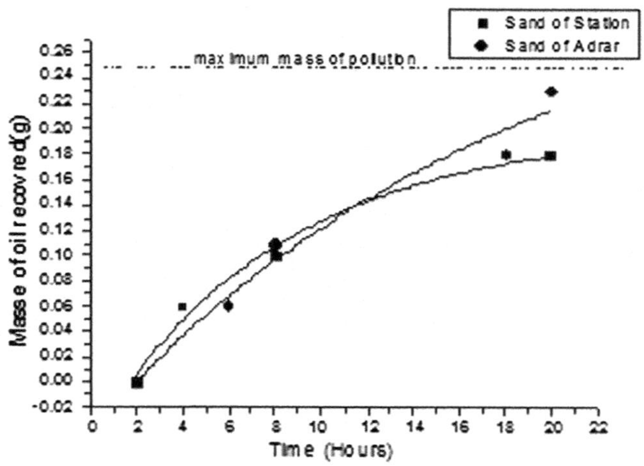

Fig. 3 Oil desorption evolution in the presence of SDS for 1 month of contamination (25 g/1 kg)

Table 3 Mass transfer coefficients, correlation coefficients of crude oil

	10 months of contamination				1 months of contamination	
Sand	SDBS		SDS		SDS	
	R^2	k (h^{-1})	R^2	k (h^{-1})	R^2	k (h^{-1})
Ouargla	0.93	0.065	–	–	–	–
Station	0.87	0.0003	0.88	0.038	0.926	0.048
Adrar	0.88	0.018	0.92	0.0247	0.945	0.002

where S_C^* represents the mass of oil saturation (g), S_C is the mass of oil in the surfactant solution at time t(g) and k is the mass transfer coefficient (h^{-1}). Integrating (Eq. 1) yields the following equation:

$$S_C = S_C^*\left(1 - e^{-kt}\right) \quad (2)$$

The fitting of Eq. (2) to the experimental data (Figs. 1, 2 and 3) was used to find the parameter k for each surfactant given in Table 3. Indeed, these results indicate that the desorption kinetics of crude oil depend on the surfactants structure [7, 10], sand and age of contamination. Comparing the correlation coefficients (R^2) to the experimental data of each surfactant, it was remarked to vary between 0.87 and 0.945. The results obtained show the structure and the CMC of the surfactants (Table 1) at a remarkable effect on the decontamination process. The mass of oil recovered by the surfactants clearly show the effectiveness of the anionic surfactant in the decontamination processes. According to Pennell et al., anionic surfactants are still qualified by their solubilizing power, which allows a moderate solubilization of apolar substances, even at low concentrations [3].

4 Conclusion

The results obtained in this study, showing that the anionic surfactants are very effective in cleaning contaminated soil by hydrocarbons. Indeed, the effectiveness of these surfactants in a batch washing process is always meaningful. The results obtained in this study show that SDBS presents the best results by SDS in terms of performance, and the desorption kinetics depend on the sand, contamination age and the surfactants nature.

References

1. Pearl, M., Pruijn, M., Bovendeur, J.: The application of soil washing to the remediation of contaminated soils. Land Contam. Reclam. **14**(3), 713–726 (2006)
2. Raton, F.L., Riser-Roberts, E.: Remediation of Petroleum Contaminated Soil: Biological, Physical, and Chemical Processes. Lewis Publishers, Boca Raton, FL (1998)
3. Pennell, K.D., Abriola, L.M., Weber, W.J.: Surfactant-enhancedsolubilization of residual dodecane in soil columns: 1. Experimental Investigation. Environ. Sci. Technol. **27**, 2332–2340 (1993)
4. Chen, J., Wang, X.J., Hu, J.D., Xu, F.L., Tao, S.: Numerical simulation of PAHs sorption/desorption on soil with the influence of Tween80. J. Environ. Sci-China **18**(4), 716–720 (2006)
5. Khalladi, R., Benhabiles, O., Bentahar, F., Moulai-Mostefa, N.: Surfactant remediation of diesel fuel polluted soil. J. Hazard. Mater. **164**, 1179–1184 (2009)
6. West, C.C., Harwell, J.H.: Surfactants and subsurface remediation. Environ. Sci. Technol. **36**, 2324–2330 (1992)
7. Rosen, M.J.: Surfactants and Interfacial Phenomena, 3rd edn. Wiley, New Jersey (2004)
8. ASTI Homepage, http://www.springer.com/ASTI, Last accessed 21 Nov 2016
9. Salager, J.L., Forgiarini, A.M., Marquez, L., Manchego, L., Bullon, J.: How to attain an ultralow interfacial tension and a three-phase behavior with a surfactant formulation for enhanced oil recovery: A review. Part 2. J. Surfact. Deterg. **16**, 631–663 (2013)
10. Rahal, S., Khalladi, R., Moulai-Mostefa, N.: Solubilization of crude oil by extended and otheranionic surfactants. Arab. J. Sci. Eng. **41**(1), 111–117 (2016)
11. Bourrel, M., Shechter, R.S.: Microemulsions and related systems: formulation, solvency, and physical properties. In: Surfactant Science Series, vol. 30, Marcel Dekker Inc., New York and Basel (1988)

Determination of Groundwater Quality Near a Non-engineered Landfill Using Electrical Resistivity Tomography

Arindam Saha, Debaprakash Parida, and Ashim Kanti Dey

Abstract

Groundwater is one of the major sources of drinking water. Groundwater often gets contaminated which makes very dangerous when consumed. Groundwater sources present in an immediate environment of a non-engineered landfill are more prone to get contaminated. Prediction of groundwater contamination using the platitudinous method using a few test wells is time consuming and cannot accurately illustrate the contamination of the whole site. Electrical resistivity tomography (ERT), an advanced non-invasive technique, can effectively and economically be used to delineate and predict the contaminated zoned of groundwater. This study developed a correlation with which the quality of groundwater near a non-engineered landfill can be known without making any borehole and laboratory analyses. For this study a model non-engineered landfill of dimension 4 m × 2.4 m × 0.6 m depth was prepared to replicate the original site condition. Electrical resistivity test was performed on and around the landfill to visualize the groundwater contamination. Groundwater was collected regularly near the landfill for five months and analyzed for contaminant concentrations. Water quality index (WQI) was calculated to determine the quality of water for each sample and correlated with the resistivity data. Design chats were prepared by correlating water quality with resistivity value. The study evidenced the efficiency of resistivity surveys in predicting the contaminated zone of groundwater.

Keywords

Groundwater contamination • Non-engineered landfill • Electrical resistivity tomography (ERT) • Water quality index (WQI)

1 Introduction

Landfills are the systems which facilitate the containment of wastes keeping the surrounding environment safe. But if these landfills are not properly designed or well-maintained, they cause serious environmental threats to its immediate environment. Nowadays, major drinking water sources are mostly affected by the detrimental effects of non-engineered landfills. Leachate is a toxic chemical solution which gets emanated from the landfill when water flows through the landfill wastes [1]. Leachate percolates through the soil and ultimately contaminates the groundwater source below the landfill. The prediction of this contamination using conventional methods is difficult and time consuming. Moreover, it fails to provide full information on the whole site contamination. Geophysical methods are found to be more effective and efficient in predicting the spread of groundwater contamination. Among all the geophysical methods, electrical resistivity tomography (ERT) technique is found to be very much adaptable for this concern [2].

Electrical resistivity tomography technique is a non-invasive geophysical technique which depends upon the property of the material in carrying the electrical current. This technique is performed on the surface with the help of a resistivity measuring instrument and some electrodes to get the subsurface information [3]. Though this technique is very old, its effective applications in the study of emerging environmental issues are yet to be explored extensively. In this method, Current is passed through the subsoil from current electrodes and the potential difference between two points is measured by the potential measuring electrodes. Depth of observation depends on the spacing between the

A. Saha (✉) · D. Parida · A. K. Dey
National Institute of Technology Silchar, Cachar, 788010, Assam, India
e-mail: 15-3-01-107@student.nits.ac.in

A. Kallel et al. (eds.), *Recent Advances in Geo-Environmental Engineering, Geomechanics and Geotechnics, and Geohazards*, Advances in Science, Technology & Innovation, https://doi.org/10.1007/978-3-030-01665-4_14

electrodes which is in turn responsible for the resolutions of the data quality. This technique has higher control over the telluric noise which tends to distort the observed data. Resistivity value obtained at the site is nothing but the apparent resistivity value of the subsurface material which is later processed and inverted using RES2DINV and RES3DINV software to get the true resistivity value in 2D and 3D resistivity survey, respectively. Determining the quality of water from the resistivity value is one of its kinds in the field of environmental geo-technique.

The quality of water at any stage is defined by the Water Quality Index (WQI) which is determined from the physico-chemical analyses of the given water sample. WQI is an effective tool which facilitates the information about the quality of water to the concerned body. WQI reflects the combined effects of various parameters on the water quality. The quality of water is categorized as excellent, good, poor, very poor and extremely harmful depending upon the WQI values.

The objective of this pilot study was to correlate the groundwater quality, calculated using WQI, with the resistivity values obtained from the ERT survey over a period of time at a non-engineered landfill site.

2 Materials and Methods

A non-engineered landfill was made at a site within the campus area of NIT Silchar, with the latitude 24° 45′ and longitude 92° 47′. The site is chosen because the water table is not very deep, so that leachate from the landfill could easily contaminate the ground water within a short time span. The landfill was prepared in the month of November. The study area receives an average annual perception of around 300 cm. The water table is around 0.7 m below the ground surface. The site had mainly silty clayey soil with permeability of 3.174×10^{-6} cm/s.

A trench of dimension 4 m × 2.4 m × 0.6 m was prepared. The trench was prepared on the lowest side of the slope and near a pond. Waste comprising food waste from a hostel, paper waste, plastics and cow dung were dumped in the trench and made levelled with the ground surface. It was intended to replicate the original landfill site present at Silchar in Assam, India through this pilot study. The model landfill was prepared in the month of November.

Two ERT tests were conducted on downstream side of the landfill and at a perpendicular distance of 0.5 and 1.5 m from the landfill. 48 numbers of electrodes were used at a spacing of 0.5 m and performed at every month. Two observatory wells, each located in the middle of the ERT profile, were prepared to monitor contamination of groundwater and collect the samples. The depth of the observatory well was kept to 1.2 m. ERT tests were conducted every month along these two profiles and subsequently water from these two boreholes was drawn and tested in laboratory.

The laboratory test of the water samples includes pH, turbidity, alkalinity, chloride, iron, sulphate, nitrate, dissolved oxygen and hardness. To represent the quality of groundwater simply, a water quality index (WQI) was used [4]. To calculate the water quality index, firstly each parameter of the nine considered, was given some weight, ranging from 1 to 5, depending upon the impact of each parameter on human health. Secondly, the relative weight of each parameter was deduced by dividing its weight prescribed by the sum of total weights of all the parameters. Thirdly a quality rating scale for each parameter was prepared by dividing the concentration of each parameter by the standard value of drinking water prescribed in BIS guideline, and multiply the value by 100 [5]. Finally, the water quality index was obtained by the summation of the product of each relative weights and their quality rating scale.

3 Results

3.1 Resistivity Survey

The resistivity survey has shown variations in the resistivity values below the landfill before and after the filling of waste. The resistivity profiles before and after four months of dumping are shown in Figs. 1 and 2, respectively. According to the observed depth of contaminant flow, resistivity values were interpreted from the true resistivity profiles for a depth range between 0.7 and 1.2 m. The representative resistivity values were taken from the groundwater table to the contaminated depth. The range of resistivity values observed at the site for boreholes 1 and 2 up to 0.7–1.2 m depth is presented in Table 1. It is observed that the resistivity values near both boreholes decrease with time; which indicates that the contaminant concentration in soil is increasing.

3.2 Water Quality Data

For each successive month, the quality of the contaminated water was determined according to the physicochemical analyses of the water samples. Water quality index (WQI) was calculated for the samples from each borehole as per the obtained physicochemical concentrations for each of the observed months. The water quality was categorized as per the WQI values and was presented for corresponding boreholes in Table 2.

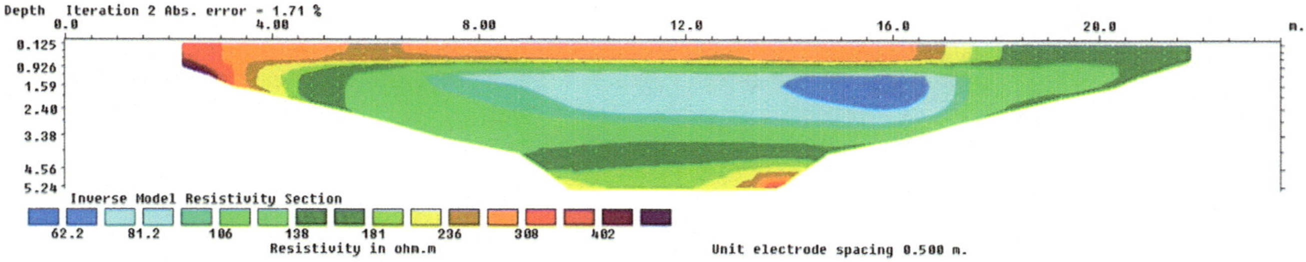

Fig. 1 ERT profile before making landfill on Nov-2017

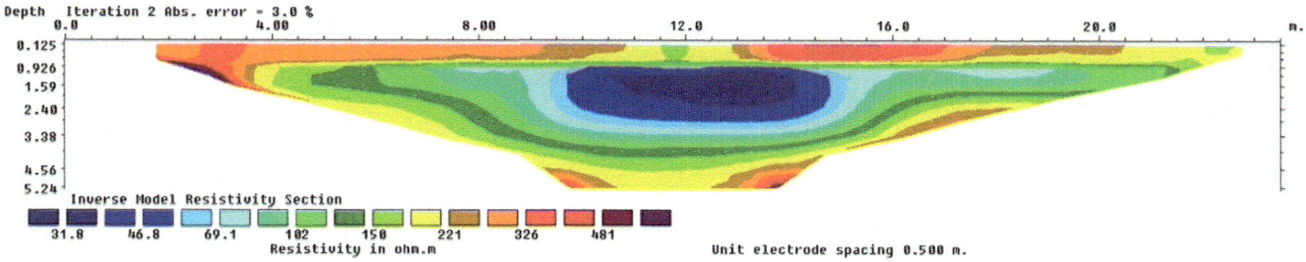

Fig. 2 ERT profile after four months of dumping waste in Mar-2018

Table 1 Resistivity values corresponding to the time of observation for borehole 1 and 2

Sl. No.	Month	Resistivity (Ohm-m)	
		Near borehole 1	Near borehole 2
1	November-2017	70–75	75–80
2	December-2017	55–60	40–50
3	January-2018	50–55	30–40
4	February-2018	40–45	20–30
5	March-2018	45–50	30–35

Table 2 WQI and corresponding water quality for borehole 1 and 2

Month	Borehole 1		Borehole 2	
	WQI	Water quality	WQI	Water quality
November-2017	89.19	Good	89.19	Good
December-2017	111.36	Poor	121.72	Poor
January-2018	119.76	Poor	240.35	Very poor
February-2018	114.63	Poor	264.09	Very poor
March-2018	109.55	Poor	238.72	Very poor

3.3 Correlations Between Resistivity and Water Quality

The variation of the resistivity values with the quality of water for corresponding distances from the landfill are also presented in Table 3. A good regression coefficient for the correlation was obtained from the analysis. The obtained resistivity data and the water quality index over the observed periods are correlated and the resulting graph is presented in Fig. 3. The figure helps in determining the water quality once resistivity of the ground is known.

4 Discussion

From this pilot study, it was observed that the resistivity values decrease over a period of time indicating the contamination of groundwater at corresponding depths. The contaminant concentration increases with time, however, in the month of March, the resistivity values increased suggesting a decrease in contaminant concentration. The only reason for this anomaly was a sudden downpour in the early days of March. This heavy downpour decreased the

Table 3 Variation of resistivity with quality of water

Month	Borehole 1		Borehole 2	
	Resistivity range	Water quality	Resistivity range	Water quality
November-2017	70–75	Good	75–80	Good
December-2017	55–60	Poor	40–50	Poor
January-2018	50–55	Poor	30–40	Very poor
February-2018	40–45	Poor	20–30	Very poor
March-2018	45–50	Poor	30–35	Very poor

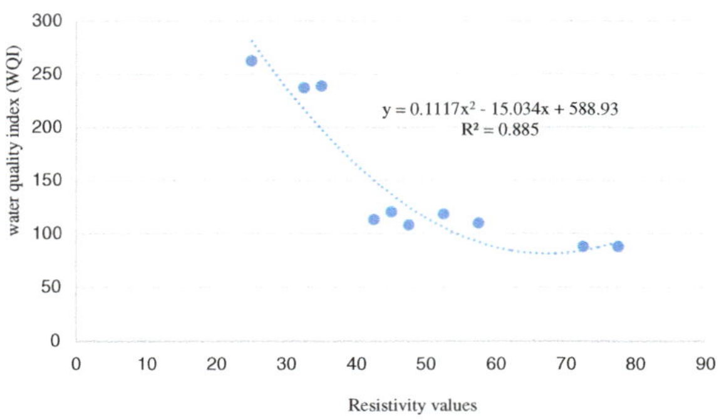

Fig. 3 Correlation between resistivity value and water quality index over a period of five months

contaminant concentration, thereby increasing the resistivity. The physico-chemical analyses of the collected water samples confirmed the decreased value of the contaminant concentration after the rainfall. The resistivity of the area below the landfill was about 80 ohm-m before contamination but at the same place the resistivity value was reduced to about 20 ohm-m after four months of contamination. The spread of contamination was clearly visible in the resistivity profiles of the subsurface below the landfill. The spread of contaminant flow, indicated by the decreased resistivity value was about 6 m long across the landfill and more than 2.5 m deep in the month of March-2018 as shown in Fig. 2.

5 Conclusion

The experiment showed that the variation in the degree of contamination of groundwater over a period of time can effectively be predicted from the resistivity survey. The following conclusions can be drawn.

- ERT is effective in mapping the flow of contaminated groundwater.
- The changes in contaminants concentration in groundwater can be easily captured through ERT.
- Quality of ground water in terms of Water Quality Index near non-engineered landfill can be predicted through resistivity survey.
- ERT is successful in providing subsurface information within a short period of time.

References

1. Mor, S., Ravindra, K., Dahiya, R.P., Chandra, A.: Leachate characterization and assessment of groundwater pollution near municipal solid waste landfill site. Environ. Monit. Assess. **118**, 435–456 (2006)
2. Koda, E., Thaczyk, A., Lech, M., Osinski, P.: Application of electrical resistivity data sets for the evaluation of the pollution concentration level within landfill subsoil. Appl. sci. **7**, 262 (2017)
3. Loke, M.H., Chambers, J.E., Rucker, D.F., Kuras, O., Wilkinson, P. B.: Recent developments in direct current geoelectrical imaging method. J. Appl. Geophys. **95**, 135–156 (2013)
4. Ramakrishnaiah, C.R., Sadashivaiah, C., Ranganna, G.: Assessment of water quality index for the groundwater in Tumkur Taluk, Karnataka State, India. E-J. Chem. **6**(2), 523–530 (2009)
5. BIS (Bureau of Indian Standard) 10500, Indian Standard drinking water specification, 2nd Revision (2012)

Physical and Chemical Characteristics of Municipal Solid Waste in Gabes

Oumaima Chamem and Moncef Zairi

Abstract

The most common way of solid waste disposal in many countries is landfilling. In Tunisia 65% of the solid waste is landfilled. However, although landfilling is cheap in comparison to other waste management options, it requires significant funds for the construction, the operation phase and the post-closure phase. The mechanical and biological pretreatment processes, applied for decades in developed countries, reduce the economic and environmental costs of landfilling by reducing the volumes of waste to be buried and stabilize them. In addition, recovering valuable materials upstream of the preprocessing can generate a product with high added value. A survey was conducted in 2017 to characterize the municipal solid waste (MSW) composition. Thereto the waste was analyzed during January–February (dry season) and August–September (wet season). Municipal solid waste (MSW) in Gabes includes of a wide range of heterogeneous materials. The organic waste was the main fraction, followed by textiles, fines, paper, plastics, leather, rubber, glass, ceramic, and metal. The average MSW moisture content in Gabes landfill amounts to 69%.

Keywords

Municipal solid waste • Developing countries • Arid climate • Landfill • Gabes • Management

1 Introduction

The rapid increase in population, booming economy, rapid urbanization and the rise in the community living standards is coupled with an accelerated municipal solid waste generation rate in developing countries [1, 2]. In general, the operation and management of municipal solid waste (MSW) collection services in many developing countries are the responsibility of the Municipal Corporation and some private companies; both are often rudimentary, poorly developed and incomplete. This was revealed by the lack of information about the quantities and types of MSW collected, the amounts recovered and recycled, the insufficient infrastructure and weak strategic planning [3, 4].

Accurate data on generation and composition of MSW is pivotal in order to decide towards the appropriate waste management system.

The production and composition of MSW depends on many factors, such as economic status; climatic and geographical conditions, and collection frequency [5, 6]. The moisture content, organic matter and heavy metals are all parameters which are of importance for waste biodegradation in landfills, as well as for the most economically feasible choice and design for transport and treatment of waste materials [7].

In this paper, MSW characteristics from Gabes Municipality were investigated in order to evaluate the current methods of treatment and propose an alternative waste management plan.

2 Materials and Methods

2.1 Samples Collection

Two Samples were collected from Gabes landfill in two different seasons, one during the wet season (January 17, 2017)

O. Chamem (✉) · M. Zairi
Water, Energy and Environment Laboratory LR3E, The National School of Engineering of Sfax (ENIS), Sfax, Tunisia
e-mail: oumaima.chamem@gmail.com

M. Zairi
e-mail: moncef.zairi@enis.rnu.tn

A. Kallel et al. (eds.), *Recent Advances in Geo-Environmental Engineering, Geomechanics and Geotechnics, and Geohazards*, Advances in Science, Technology & Innovation, https://doi.org/10.1007/978-3-030-01665-4_15

and the other in the dry season (August 28, 2017) in order to examine the seasonal variations in MSW composition.

According to the MODECOM (Methodology for municipal solid waste characterization), the MSW sampling was performed and classified manually into seven fractions: (1) organics, (2) cardboard paper, (3) plastics, leather and rubber, (4) textiles and sanitary textiles, (5) metals, (6) glass and ceramic and (7) fines waste [2].

2.2 Laboratory Analyses

In this study, according to the standard NF M03-002, the moisture content of typical MSW was determined, by drying 2 kg of each fraction using hot air oven set at 105 °C for, 24 h. The moisture content is expressed in percentages, which is loss in the weight from the original wet weight of each sample [7].

The organic matter content (OM) was determined according to procedure 2540E in Standard Methods [8] by igniting for at least 4 h at 550 °C of about 50 g of MSW until a constant weight was achieved [9]. The organic content is the percent weight lost on ignition. According to the standard NF EN ISO 11885(1998), the determination of heavy metal contents (Ni, Zn, Pb, Cd, Cr, and Cu) by ICP–OES (Optima 2100DV, Perkin-Elmer, USA) in each fraction of MSW required the following steps: First, about 0.5 g of each fraction was transferred to a polytetrafluorethylene digestion capsule.

Next, all the samples were dissolved in 5 mL of nitric acid, 5 mL of hydrofluoric acid and 3 mL of perchloric acid. and then heated on an electric heating plate. After evaporation, each capsule was rinsed by 10 mL of hydrochloric acid and demineralized water.

Finally, the digestion solution was filtered through a 0.45-m membrane and the filtrate was diluted to 100 mL in a flask [10].

3 Results and Discussion

3.1 Average Composition of MSW in Gabes City

The waste composition analysis indicated that organic matter is the largest waste fraction in MSW of Gabes city, which consists mainly of food residue, slaughterhouse waste, decayed vegetables and yard waste (31% kg/kg-dry and 59% kg/kg-dry of the total MSW collected during the winter and summer respectively), followed by Textiles which accounts for 12% kg/kg-dry in winter and 16% kg/kg-dry in summer, fine waste (10% kg/kg-dry in summer and 14% kg/kg-dry in winter), papers-cardboard (9% kg/kg-dry in dry season and 10% kg/kg-dry in wet season), plastics, leather and rubber (7% kg/kg-dry in wet season and 11% kg/kg-dry in wet season), glass and ceramics and metals each accounted for 3% or less (see Fig. 1).

3.2 Physicochemical Characteristics of MSW

Moisture content

In Fig. 2 the moisture content of each fraction of MSW in Gabes city is shown. The far largest water content (77%) was observable for the organic fraction (indicated as fermentable).

Previous studies have reported that in developing countries high water contents of MSW is considered normal,

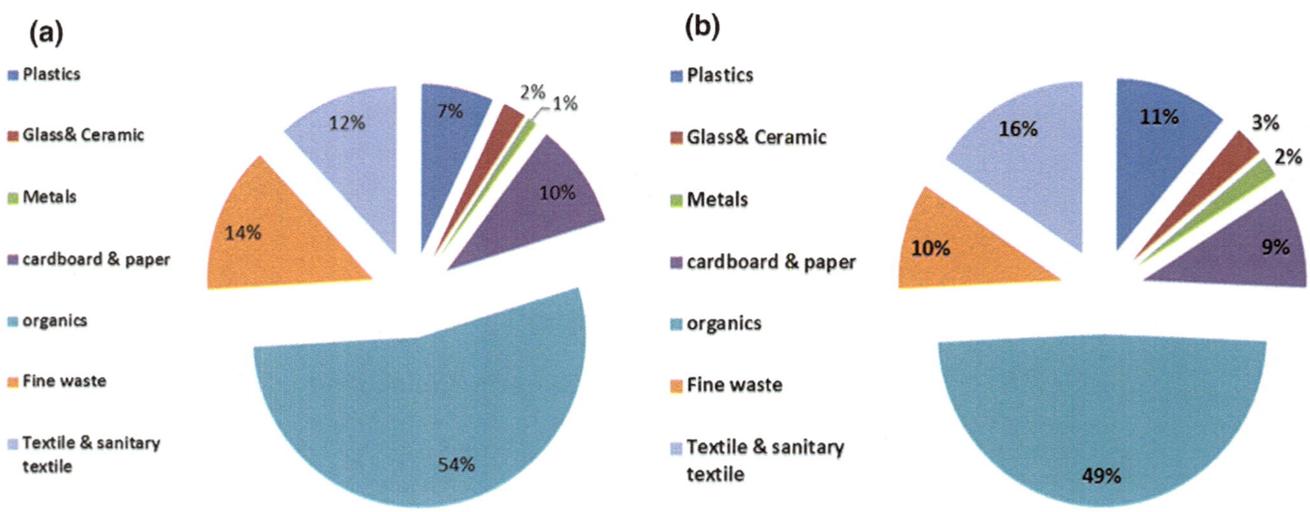

Fig. 1 MSW composition in wet season (**a**) and dry season (**b**)

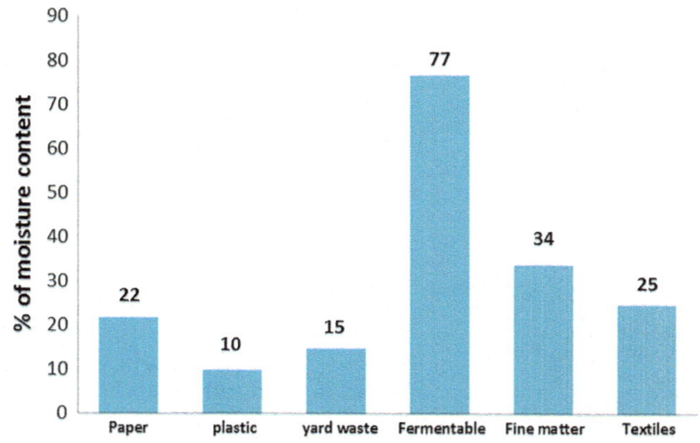

Fig. 2 Moisture content of each waste fraction (given in %)

Table 1 Heavy metal content mg/kg dry weight) of each waste fraction

Element	Fe	Cd	Zn	Cu	Cr	Ni	Pb
Paper	447	8	24	25	2	5	23
Plastic	189	7	45	23	1	3	25
Textile	349	8	25	9	0	5	32
Fermentable	318	10	59	19	1	9	23
Yard	654	6	24	10	1	9	20
Fine matter	4258	12	53	21	5	8	34
Ranges of household waste [9, 16]	–	03–41	380–2677	77–1048	21–426	16–200	100–800

while in industrialized countries such high contents are not observable [7].

Indeed, the given value of moisture content in MSW of developing country can be explained by the specific food consumption which largely consists of fresh fruits and vegetables [7].

Organic matter

The organic matter content was estimated to about 69% by dry matter. This high value is explained by the large share of food waste [7]. The observations fit the data of most developing countries, which have tremendous organic content of MSW such as, Mauritius 85% [11], Tanzania 80% [12], Morocco 68% [13] and Algeria 64.6% [14].

This result indicates that composting has a potential for waste management in Gabes [7, 15].

Heavy metal contents

The total metal contents in MSW of Gabes landfill are summarized in Table 1. It showed that compared to the concentrations reported by François [9] and Aloueimine [16], the concentrations of all heavy metal ions were relatively low and below the reference limits.

In fact, this result indicate the possibility of using composted organic waste as fertilizer and soil amendment [7].

3.3 MSW Valorisation

In addition to the aforementioned results, a supplementary valorisation of the MSW was evaluated applying the Tanner graphic method [17, 18], which uses three parameters: Moisture (% M), Ashes (% Ce) and Volatile Matter (% VM).

The results of this analysis show that the waste is suitable for composting, because of its high contents of moisture and volatile matter (% M: 69%, VM: 69%, Ce: 31%) (see Fig. 3).

4 Conclusion

This paper focused on determining the physical composition of Gabes MSW, and then determining its optimum physiochemical parameters. These parameters proved that the high amount of organic waste in Gabes landfill can be effectively used as organic manure through composting which would be an appropriate method of MSW management in Gabes city.

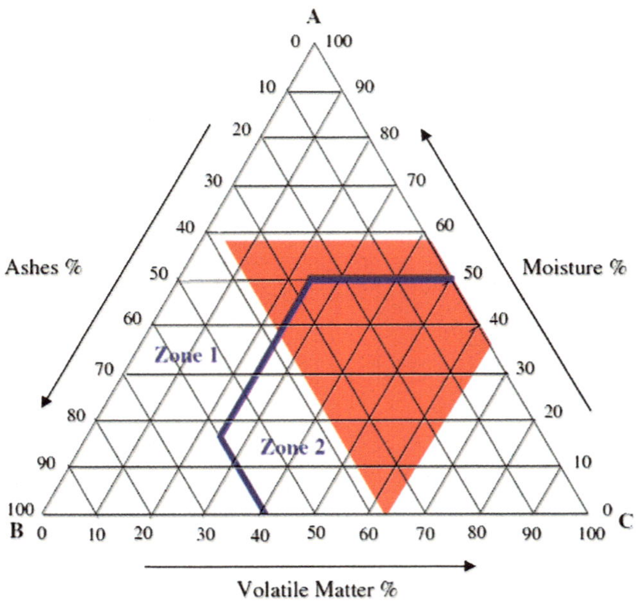

Fig. 3 Tanner ternary diagram

References

1. Minghua, Z.X., Xiumin, F., Rovetta, A., Qichang, H., Vicentini, F., Bingkai, L., Giusti, A., Yi, L.: Municipal solid waste management in Pudong New Area, China. J. Waste Manag. **29**, 1227–1233 (2009)
2. Srivastava, V., Ismail, S.A., Singh, P., Singh, R.P.: Urban solid waste management in the developing world with emphasis on India: challenges and opportunities. Rev. Environ. Sci. Biotechnol. **14**(2), 317–337 (2015)
3. Ayininuola, G.M., Muibi, M.A.: An engineering approach to solid waste collection system: Ibadan North as case study. Waste Manage. **28**, 1681–1687 (2008)
4. Buenrostro, O., Bocco, G.: Solid waste management in municipalities in Mexico: goals and perspectives. Resour. Conserv. Recycl. **39**, 251–263 (2003)
5. Tchobanoglous, G., Hilary, T., Vagil, S.A.: Integrated solid waste management: engineering principles and management issues. Mc-Graw Hill, New York (1993)
6. Collivignarelli C., Sorlini S., Vaccari M.: Solid wastes management in developing countries. CD-ROM of ISWA World Congress, Rome (2004)
7. Zairi, M., Aydi, A., Dhia, H.B.: Leachate generation and biogas energy recovery in the Jebel Chakir municipal solid waste landfill, Tunisia. Mater. Cycles Waste Manag. **16**, 141–150 (2014)
8. Eaton, A.D., Franson, M.A.H.: Standard Methods for the Examination of Water & Wastewater. American Public Health Association (2005)
9. Francois, V.: Détermination d'indicateurs d'accélération et de stabilisation de déchets ménagers enfouis. Etude de l'impact de la recirculation de lixiviats sur colonnes de déchets, thèse de doctorat No 8-2004. Université de Limoges, pp. 158. Yousuf TB, R. M. (2007). Monitoring quantity and characteristics of municipal solid waste in Dhaka City. Environ. Monit. Assess. **135**, 3–11 (2004)
10. Xiaoli, C., Shimaoka, T., Xianyan, C., Qiang, G., Youcai, Z.: Characteristics and mobility of heavy metals in an MSW landfill: implications in risk assessment and reclamation. J. Hazard. Mater. **144**, 485–491 (2007)
11. Mohee, R.: Assessing the recovery potential of solid waste in Mauritus. Resour. Conserv. Recycl. **36**, 33–43 (2002)
12. Mbuligwe, S.E., Kassenga, G.: Feasibility and strategies for anaerobic digestion of solid waste for energy production in Dar EsSalaam city, Tanzania. Resour. Conserv. Recycl. **42**, 183–203 (2004)
13. Zahrani, F.: Contribution à l'élaboration et validation d'un protocole d'audit destiné à comprendre les dysfonctionnements des centres de stockages des déchets (CSD) dans les pays en développement. Application à deux CSD: Nkolfoulou (Came roun) et Essaouira (Maroc). de Lyon, France: The`se de doctorat, Institut National des Sciences Appliquées (2006)
14. Guermoud, N., Ouadjnia, F., Abdelmalek, F., Taleb, F., Addou, A.: Municipal solid waste in Mostaganem city (Western Algeria). Waste Manag. **29**, 896–902 (2009)
15. Gajalakshmi, S., Abbasi, A.S.: Solid waste management by composting: state of the art. Crit. Rev. Environ. Sci. Technol. **38** (5), 311–400 (2008)
16. Aloueimine, S.: MSW Characterization Methodology in Nouakchott, Mauriatnia (2006)
17. Antonopoulos, I.S., Karagiannidis, A., Kalogirou, E.: Estimation of municipal solid waste heating value in Greece in the frame of formulating appropriate scenarios on waste treatment (2010)
18. Gillet, R.: MSW Management and its Application in Developing Countries, vol. 1. OMS Publisher, PUND, Copenhagen, Denmark (1985)
19. MODECOM (Methode de caractérisation des ordures ménagères): Methodology for municipal solid waste characterization. ADEME Report, 2nd edn., pp. 1601–2766 (1993)

Portland Cement Exhaust Characterization and Its Potential Use in Mineral Carbon Sequestration

Freeman E. D. Senzani and Antoine F. Mulaba-Bafubiandi

Abstract

Portland cement manufacturing produces, in its exhausts, carbon dioxide, carbon monoxide, sulphur dioxide, oxygen, nitrogen, water vapour, and argon. Significant amounts of carbon dioxide and sulphur dioxide are therefore released into the atmosphere. For the purpose of minimizing the input of greenhouse and acid rain gases, ways of capturing the carbon dioxide by, for instance, mineral carbonation, are being investigated around the world. This study is a review of the masses of gases, their volumes, temperatures, and the heat energy they release. It forms the first step in determining ways for on-site carbon dioxide and sulphur dioxide sequestration. The study shows that for a tonne of clinker, the masses of the gases produced, in tonnes, are 1.2 CO_2 (63%), 0.004 SO_2 (0.2%), 0.008 CO (0.4%), 0.0549 O_2 (2.82%), 0.5 H_2O (25.8%), 0.16 N_2 (0.1%) and Ar 0.0028 (0.096%). The total volume of the gases leaving the stack, if the exit temperature is 300 °C, is 2.975 m^3, but reaches 4.532 m^3 at 600 °C, while the heat energy potentially available for downstream use, is 1272 and 1938 MJ, again at 300 and 600 °C, respectively. This is wasted heat energy that cement companies should use to capture the gases that have negative environmental consequences.

Keywords

Cement • Clinker • Sequestration • Mineral carbonation

1 Introduction

During the manufacture of Portland cement, calcium oxide (CaO), obtained from limestone calcining ($CaCO_3$) is fused at 1450 °C with silica (SiO_2), alumina (Al_2O_3) and iron oxide (Fe_2O_3 or Fe_3O_4) (Fig. 1). The fused product is cooled rapidly to solidify as a set of new minerals, known as Portland cement clinker, or, just clinker (Fig. 2, Table 1). The main raw material which supplies the calcium oxide is the sedimentary rock limestone or, alternatively, its igneous and metamorphic equivalents, namely carbonatite and marble, respectively. They need to contain at least 42% CaO, or 75% $CaCO_3$ for this purpose. The other inputs are quartz rich sand to provide silica, bauxite for alumina and haematite or magnetite for iron-oxide. Clay may also supply a combination of the silica, alumina and iron oxide. The raw materials are blended to satisfy the Lime Saturation Factor (LSF), such that,

$$LSF = (CaO)/(\{2.8\,x\,SiO_2\} + \{1.18\,x\,Al_2O_3\} + \{0.65\,x\,Fe_2O_3\}) \approx 1. \quad (1)$$

The most common fuels used are coal and petroleum, but utilisation of various waste materials is on the increase. The conversion of the raw materials and burning of the fuels lead to the production of exhaust gases consisting of different proportions of carbon dioxide, carbon monoxide, sulphur dioxide, oxygen, nitrogen, water vapour, and argon (Table 2). These are expelled at temperatures between 300 and 600 °C [1].

This study involves a determination of the heat energy carried and dispersed into the atmosphere during the fusing process [1]. The aim is to find ways of utilising the energy to speed up the process of the carbonation of olivine and pyroxene in order to contribute to the reduction of the amount of greenhouse gases released to the atmosphere. It is part of a broader investigation focusing on developing methods that exploit the earth raw materials on an artisanal to small scale, which is expected, in turn, to support physical

F. E. D. Senzani (✉) · A. F. Mulaba-Bafubiandi
Mineral Processing Technology Research Centre, Department of Metallurgy, School of Mining Metallurgy and Chemical Engineering, Faculty of Engineering and the Built Environment, University of Johannesburg, P. O. Box 17911 Doornfontein, 2028, South Africa
e-mail: fsenzani@uj.ac.za; freeman_senzani@hotmail.com

A. Kallel et al. (eds.), *Recent Advances in Geo-Environmental Engineering, Geomechanics and Geotechnics, and Geohazards*, Advances in Science, Technology & Innovation, https://doi.org/10.1007/978-3-030-01665-4_16

Fig. 1 Simplified flow chart of Portland cement manufacture

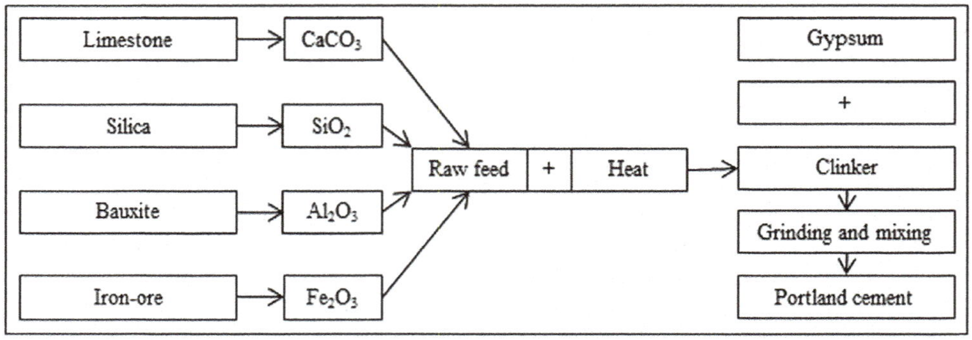

Fig. 2 Portland clinker (**a**) nodules (**b**) micrograph blue = alite; purple = belite; white = calcium aluminoferrite (ferrite); grey = tricalcium aluminate (aluminate); field of view = 300 μm

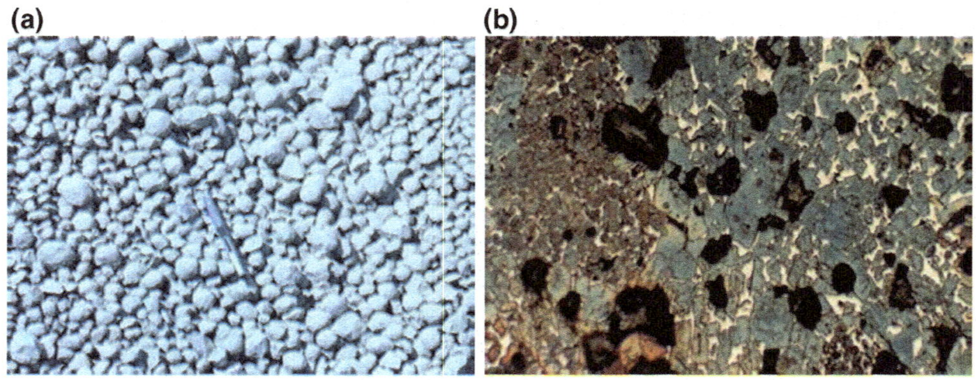

Table 1 Mineral and chemical composition of Portland cement clinker

Mineral	Chemical formula	Typical % in Portland cement
Tricalcium silicate (alite)	Ca_3SiO_5	45
Dicalcium silicate (belite)	Ca_2SiO_4	27
Tricalcium aluminate (aluminate)	$Ca_3Al_2O_6$	11
Tetracalcium aluminoferrite (ferrite)	$Ca_4Al_2Fe_2O_{10}$	8

See Refs. [1, 3–6]

Table 2 Heat energy generated by flue gases during manufacture of 1-tonne Portland cement

Gas		Heat capacity [J/(kg/K)]	Heat energy available (MJ)	
			at 300 °C (573 K)	at 600 °C (873 K)
Carbon dioxide (calcination)		844	340.46	518.72
Combustion gases	Sulphur dioxide	640	1.47	2.23
	Carbon monoxide	1020	4.68	7.12
	Oxygen	919	6.85	10.43
	Carbon dioxide	844	248.09	377.98
Water vapour from coal and raw meal		1930	552.95	842.45
Sucked in air	Nitrogen	1040	93.07	141.80
	Oxygen	919	22.06	33.62
	Argon	520	0.55	0.84
	Carbon dioxide	844	0.04	0.06
	Water vapour	1930	2.21	2.22
Total			1272.43	1937.47

See Ref. [1]

development and increase employment opportunities in rural areas [2].

2 Methods

The study was carried out by reviewing from different sources the exhaust gases quantities of the cement manufacturing process, and their physical conditions, namely the masses, their volumes, temperatures, and the heat energy they potentially contain at 300 and 600 °C. The two temperatures represent the range at which the gases are emitted [1]. There are chemical thermodynamic tables with information that allows the calculation of the heat energy contained in the gases as they are expelled at these temperatures, as performed in this study. Based on the masses and heat capacities, this investigation determined the thermal energy contained in the gases, as 1 tonne of clinker is produced.

3 Results

For a tonne of clinker, the masses of the gases produced, in tonnes, are estimated to be 1.2 CO_2 (63%), 0.004 SO_2 (0.2%), 0.008 CO (0.4%), 0.0549 O_2 (2.82%), 0.5 H_2O (25.8%), 0.16 N_2 (0.1%) and Ar 0.0028 (0.096%). The total volume of the gases leaving the stack, if the exit temperature is 300 °C, is 2.975 m^3, but reaches 4.532 m^3 at 600 °C, while the heat energy available for downstream use, is 1272 and 1938 MJ, again at 300 and 600 °C, respectively (Table 2) [1–6].

4 Discussion

Because 1272–1938 MJ heat energy is released to the atmosphere, mostly as high-temperature greenhouse and acid rain gases, there is potential to tap the thermal power and utilize it in the process of carbonation of the contained CO_2. Such chemical reactions are speeded up to viable rates if the water in which they are conducted is at a high temperature [7–9]. The heating up of such aqueous solutions and suspensions forms the first possible use for the energy which is otherwise wasted. For instance, if 3 tonne (3000 kg) of water are required to form the carbonation solution/suspension of exhausts from production of 1 tonne of cement, it will need 1005 MJ for its temperature to rise from 20 to 100 °C. The remainder of the energy would be available to stir the mixture, and maintain the high temperature until acceptable levels of carbonation have been achieved. A similar review of exhausts from fossil fuel based-power generation plants would achieve a similar purpose. Parallel studies should also be conducted for SO_2, which contributes to acid rain.

5 Conclusions

This study reviewed the flue gases that are known to be produced and discharged at high temperatures of between 300 and 600 °C into the atmosphere during the manufacture of cement. Published information from a variety of sources made available the masses, volumes, as well as heat capacities. These were used to determine how much heat energy might potentially be made available from the exhausts. The total flue gas mass was 1.94 tonne. At a temperature of 300 °C, the volume was 14.87 m^3, while the heat energy available was 1272.43 MJ for every tonne of Portland cement produced. At 600 °C, the volume was 22.66 m^3, while the heat energy was 1937.47 MJ. The energy that is lost to the atmosphere could be utilized in other processes such as mineral carbonation. It is recommended that future designs of cement plants be required to utilize the wasted energy for this process, among others.

References

1. Peray, K.E.: The Rotary Cement Kiln. Chemical Publishing Company Inc., New York NY (1982)
2. Tonnegu, E.L.: Limestonnee Report. Council for Geoscience, Pretoria, South Africa (2007)
3. British Geological Survey (BGS), Mineral Profile, Cement Raw Materials. Office of the Deputy Prime Minister (2005)
4. Eckel, E.C.: Cement Materials of the United States. United States Geological Survey, United States Department of the Interior (1905)
5. Taylor, H.F.W.: Cement Chemistry. Academic Press, London (1990)
6. Wright, J.B.: Constructional and Other Bulk Materials. The Open University Press. Miltonne-Keynes, UK. 88 p (1974)
7. Sanna, A., Uibu, M., Carmanna, G., Kussik, R., Maroto-Valer, M. M.: A review of Carbonation Technologies to Sequester CO_2. Chem. Soc. Rev. **48**, 8049–8080 (2014)
8. Zevenhoven, R., Fagerlund, J.: Mineralisation of carbon dioxide (CO_2). In: Maroto-Valer, M.M. (ed.) Developments and Innovation in Carbon Dioxide (CO_2) Capture and Storage Technology, pp. 433–462. Woodhead Publishing Ltd, Oxford (2010)
9. Hogan, C.M.: Abiotic factor. In: Monosson, E., Cleveland, C. (eds.) Encyclopedia of Earth. National Council for Science and the Environment. Washingtonne DC (2010)

Evaluation of Air Pollutants and Dispersion Patterns for the Adjacent Areas of Mellitah Gas Complex, Libya

Abdulhamid B. M. Danna, Amjad Kallel, and Mohamed Jamel Rouis

Abstract

This study aimed to assess the dispense of concentrations and patterns of dispersion of gases and particulate matter emitted from the Mellitah gas complex in Libya, specifically the adjacent areas. Concentrations of these pollutants and dispersion patterns are very variable with the wind direction. Therefore, it was found that the concentrations of pollutants in the areas that were in the wind direction were higher than those that were to the opposite of the wind direction. The highest concentrations were recorded in the late night and early morning, specifically between 10 p.m. and 7 a.m. It was noted that during these times, the air may be stable or static, which confirms that the wind has an influence on the monitoring processes either negatively or positively.

Keywords

Air pollution • Oil installations • Gas installation • Gas emission • Risk sources

1 Introduction

Libya, situated on the Mediterranean coast experiences temperate humid, cold, and hot climate. The local meteorological conditions most of the year are mild humidity, temperate calm north east wind. This paper focused mainly on the follow-up of major air pollutants near the Mellitah gas complex shown in Fig. 1. Libya is among the major oil exporting countries in Africa and its economy is heavily dependent on oil and Gas production. It has several oil and Gas fields all over Libya. Flaring is the common methods used in almost all oil and Gas installation in Libya for safe disposal of undesired products of produced hydrocarbons. By burning these products of hydrocarbons and converting them largely to carbon dioxide and water their environmental impacts are greatly reduced [1]. Air pollution is one of the most critical issues. EPA has identified nitrogen oxides, sulfur oxides, carbon monoxide, ozone (O_3), particulates (PM10) as an indicator of air quality. They are well known as sources emitted from combustion of hydrocarbons of oil and Gas installations. Air pollutants are divided into two types: primary pollutants emitted from a specific source. Among these, the most important are SO_2, CO, NO_x, SO_x, hydrocarbons (VOCs) and minerals. Secondary contaminants are those formed in the atmosphere by chemical reaction of primary pollutants including O_3, oxidized hydrocarbons and other photochemical oxidants [2]. To know the quality of the atmosphere for the cities required an extensive analysis. Monitoring programs should be performed on a daily basis and in particular in the surrounding area of the industrial plants in order to protect the environment. The air pollution is not yet a major problem in almost all the Libyan Cities due to the fact of the low emission for the cities being open which makes dilution more efficient [3].

A. B. M. Danna (✉)
College of Engineering Technology, Janzour, Libya
e-mail: danna1970@hotmail.com

A. B. M. Danna · A. Kallel · M. J. Rouis
Sfax National School of Engineers, University of Sfax, Sfax, Tunisia

2 Materials and Methods

In cooperation with the National Environment Agency, the research was carried out for a specific period of time for some areas adjacent to the Mellitah gas complex in Libya. The self-measuring devices used to measure the pollutants were installed onboard of a mobile car unit. The system absorbs air, analyzes and obtains the quantity and type of recorded contaminants. The most important compounds measured in the air were nitrogen oxides (NO_x–NO–NO_2), SO_2, H_2S, CO, O_3 (PM1–PM2–PM10), weather conditions also were measured (humidity, temperature, wind direction and speed).

A. Kallel et al. (eds.), *Recent Advances in Geo-Environmental Engineering, Geomechanics and Geotechnics, and Geohazards*, Advances in Science, Technology & Innovation, https://doi.org/10.1007/978-3-030-01665-4_17

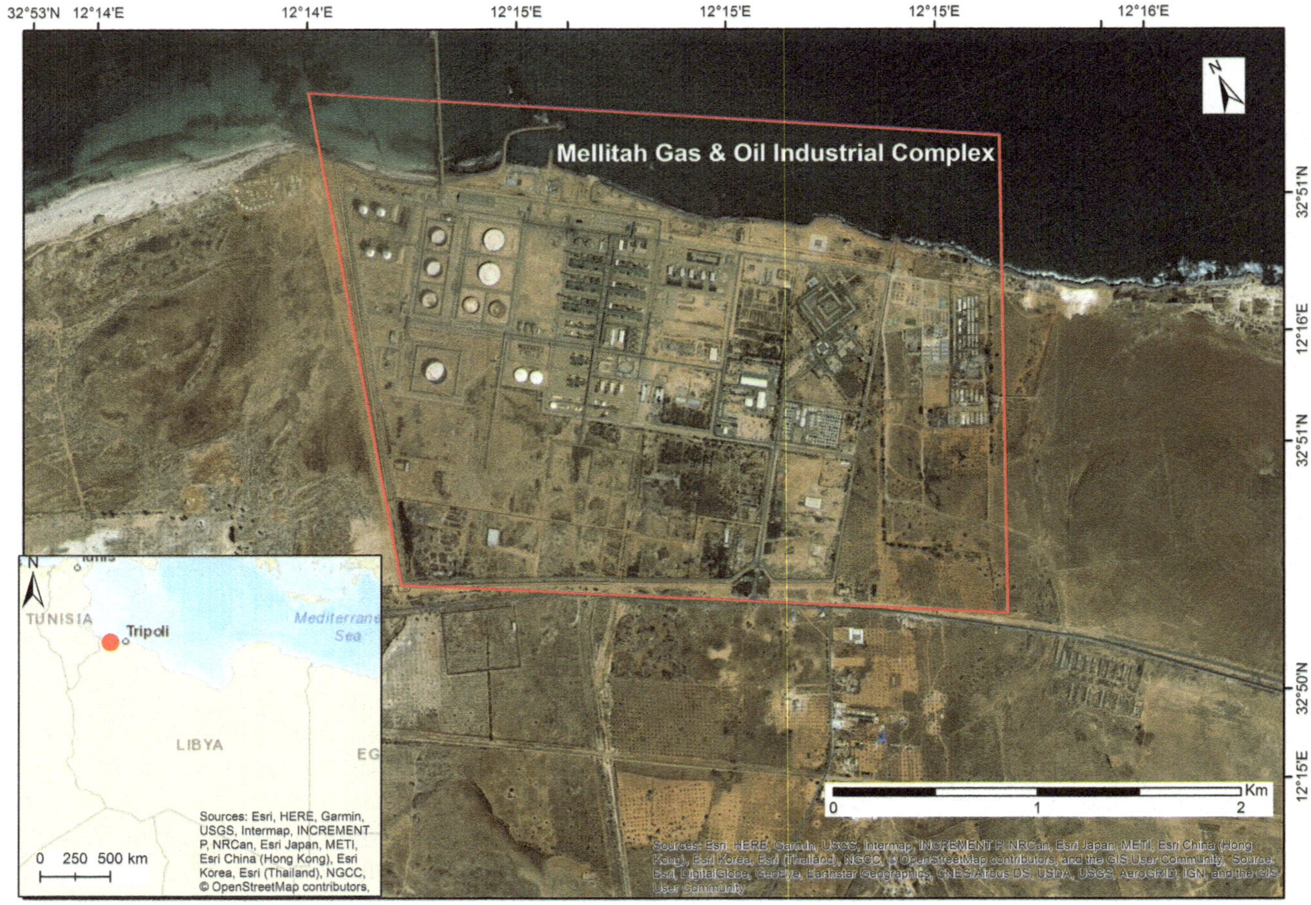

Fig. 1 Mellitah gas complex site location

3 Results and Discussion

The most important compounds measured in the air were nitrogen oxides (NO_x–NO–NO_2) as shown in Table 1. Table 2, presents data of particulates and SO_2, H_2S, O_3, CO, shown in Table 3. The daily average of the readings obtained, as well as the highest and minimum readings, were presented to facilitate the possibility of tracking these results. Figure 2 presents changes in concentration of (NO, NO_x, NO_2), whilst Fig. 3 displays particulates amounts changes and Fig. 4 presents shows the recorded amounts of SO_2, H_2S, O_3, CO.

The presented measurements are part of the results obtained, displaying the readings (every half hour) over 24 h, equivalent to (48) readings per day. When comparing the results obtained with the international and Saudi standards [4, 5], they may be somewhat acceptable, but this does not confirm the absence of pollution in the region due to several conditions, including the location of the monitoring site, the determination of distances and directions and the monitoring periods. The wind speed was relatively high on the first days of the monitoring process. The wind speed was 6–9 m/s, and the direction was as measured EN. The location to measure the discharged pollutants is situated north of the observation point. After that, the speed of the wind decreased but only at night. The direction was also northeasterly, but during the daytime it changed to the north, and it was accompanied by an increase in its velocity ranging from 6 to 8 m/s and continued on to the third day. 80% of the highest rates for each gas recorded in the time period were between 11 p.m. and 7 a.m., when the wind speed is between (0.5–2 m/s), which confirms that the wind is the main factor of the monitoring process, whether negative or positive.

4 Conclusion

Air pollution in areas adjacent to Mellitah gas complex, Libya was assessed by measuring the pollutants NO_X, NO, NO_2, SO_2, H_2S, CO, O_3, PM1, PM2 and PM10. Although

Table 1 The daily average of nitrogen oxides levels (ppb)

	NO	NO_x	NO_2
1	5.2	8.2	3
2	15.8	23	7.2
3	0.5	1.8	1.2
4	0.7	3.6	2.9
5	1.7	6.2	4.5
6	0.9	5.1	4.2
7	0.8	4.6	3.8
8	1	7	6
9	1.1	4.7	3.6

Table 2 The daily average of particulates levels (μg/m^3)

	PM1	PM2	PM10
1	6.6	15.7	22.5
2	5.9	14.4	37.8
3	6.3	12.4	28
4	7.6	11.8	18.1
5	13.3	20	29.1
6	18.1	25.9	35.1
7	15.1	21.4	30.7
8	12.3	20.7	50.7
9	10.4	17.1	32.2

Table 3 Concentration levels of SO_2, H_2S, O_3 and CO (ppb)

SO_2	H_2S	O_3	CO
4.6	1.6	32.1	3
3.9	1.1	38.1	2
2.5	1.2	46.5	1
8.5	1.8	41.1	2
7.9	2.2	39.2	2
4.8	1.5	38.4	3
6	1.5	37.6	2
6.4	1.6	40.8	3
6.3	1.4	35	2

the obtained results are in general in the range of international regulations, they do not confirm the absence of pollution in the region due to many factors. From this point of view, it can be said that these monitoring studies may require longer periods during which all the variables are taken into account and a better selection of control sites and control techniques should be selected to cover such a large area to be effective and without any restrictions.

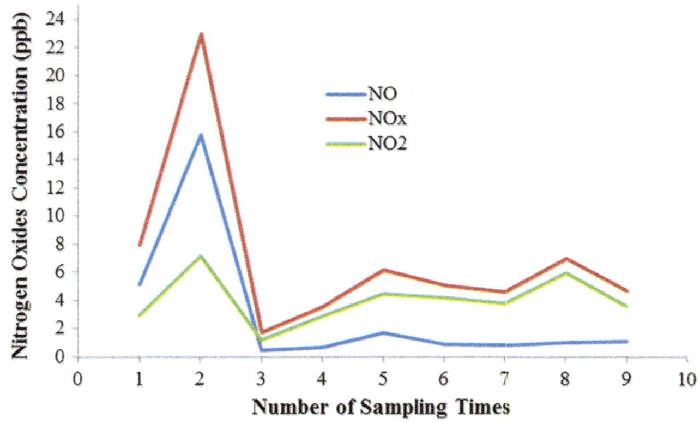

Fig. 2 Mean nitrogen oxides concentration

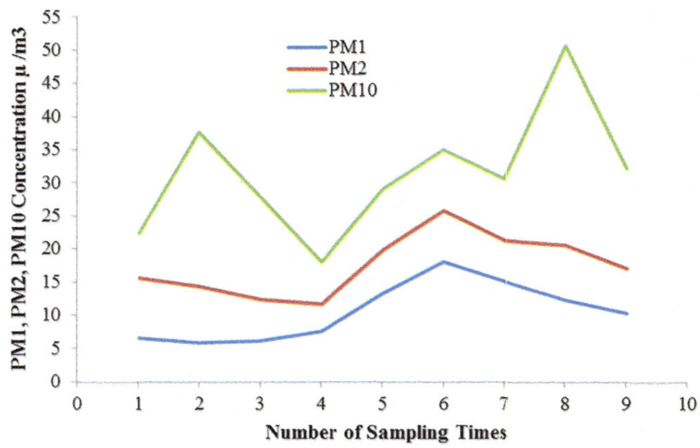

Fig. 3 Mean particulates values

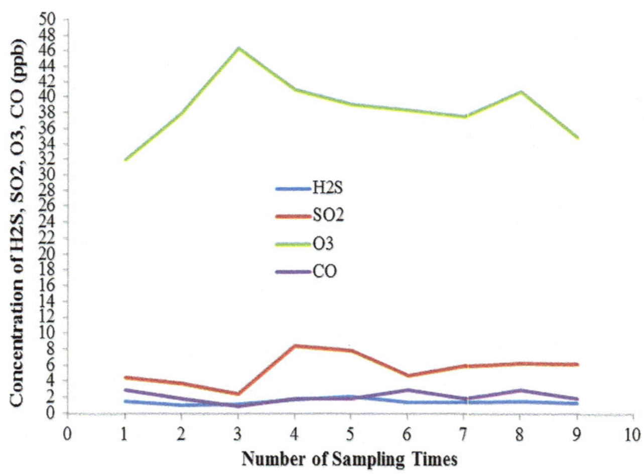

Fig. 4 Mean H_2S, SO_2, O_3, CO concentration

References

1. ENPI-SEIS Country report—LIBYA, Mar 2015
2. Abogrean, E.M., Elssaidi, M.A., Almathnani, A.M., Alansari, M.H.: Seasonal behavior of gaseous, PM10 and VOCs pollutants of Tripoli Ambient Air, Libya (ICCEBS' 2015)
3. Danna, A.B.M., Jamel, R.M.: Environmental Risk Modeling of Power Plants & Oil Installations in Libya, https://doi.org/10.1109/SM2C.2017.8071829
4. World Health Organization: Ambient (outdoor) Air Quality and Health, http://www.who.int/mediacentre/factsheets/fs313/en. Accessed 13 Apr 2015
5. Royal Commission Environmental Regulations: Royal Commission for Jubail and Yanbu. Yanbu, Royal Commission Environmental Regulations, Environmental Control Department, Saudi Arabia (2004)

Study of Chemical Composition in Wet Atmospheric Precipitation in Karachi, Pakistan

Saiyada Masood, Sumayya Saied, Azhar Siddique, Shaikh Mohiuddin, Mirza Hussain, Muhammad Khan, and Haider Khwaja

Abstract

Anthropogenic and natural emissions in atmosphere directly affect the rainwater chemistry as its chemical speciation is representative of emission status in the surrounding area. A comprehensive study on the chemical composition of rainwater was carried out in Karachi, a mega-city of Southeast Asia to delineate the urbanization impact on the local environment. Rainwater samples were analyzed for conductivity, pH, HCO_3^-, F^-, Cl^-, NO_2^-, NO_3^-, SO_4^{2-}, Na^+, NH_4^+, K^+, Mg^{2+}, Ca^{2+}, $HCOO^-$, CH_3COO^-, $C_2O_4^{2-}$, pyruvate, malonate, propionate, glyoxylate, and total organic carbon (TOC) levels. The ionic load in rainwater samples was found to be high in the densely populated sampling sites experiencing heavy traffic activity and located adjacent to industrial zones. Na^+ was the most dominant ionic species, followed by SO_4^{2-}, Ca^{2+}, Cl^-, HCO_3^-, K^+, NH_4^+, Mg^{2+}, NO_3^-, CH_3COO^- and $HCOO^-$ in sequence. $HCOO^-$ and CH_3COO^- were found to be the predominant carboxylic acids and their mean concentrations were 4.9 ± 7.3 and 9.4 ± 16.0 μeq/L, respectively. These organic acids together contributed with 7% to the TOC in precipitation. Source apportionment of the contaminants was accomplished by statistical techniques, which indicated strong impact of anthropogenic pollution on the rainwater chemistry in Karachi.

Keywords

Rainwater • Anthropogenic pollution • Organic acids • Sources • Karachi

1 Introduction

The study of chemical species in wet precipitation is of great importance because of their adverse environmental and human health effects [1–3]. The chemical composition of precipitation plays a significant role in scavenging soluble components from the atmosphere and helps to understand the atmospheric pollutants contribution [4]. These scavenged pollutants affect the natural chemical composition and pH of precipitation, which leads to the damage of ecosystem, human health, wild life, and the built environment [5, 6]. Relatively few systematic studies of rainwater chemistry have been carried out in developing countries with arid climates in comparison with widespread studies in Europe and North America [7–15]. In Pakistan, the rapid economic growth, industrial expansion, increase in population, and urbanization have led to an increase in pollutant emissions and deteriorartion of air quality [16, 17]. Therefore, information gathering of the current status of precipitation chemistry in urban centers of Pakistan through monitoring has become essential.

Karachi, the fastest growing megacity of the world, representing the urban and coastal area of Pakistan, has a unique geography and demography that greatly contributes to changes in atmospheric chemistry. The major objectives of this paper were to: (1) provide a detailed assessment of the chemical composition of precipitation; (2) compare the results of this study with other studies from various places in the world; and (3) investigate possible sources of various constituents of precipitation in the study area.

S. Masood · S. Saied · S. Mohiuddin · M. Khan
University of Karachi, Karachi, Pakistan

A. Siddique
Hamad Bin Khalifa University, Qatar Foundation, Doha, Qatar

M. Hussain · H. Khwaja (✉)
Wadsworth Center, New York State Department of Health, Albany, NY, USA
e-mail: hkhwaja@albany.edu

M. Hussain · H. Khwaja
University at Albany, Albany, NY, USA

A. Kallel et al. (eds.), *Recent Advances in Geo-Environmental Engineering, Geomechanics and Geotechnics, and Geohazards*, Advances in Science, Technology & Innovation, https://doi.org/10.1007/978-3-030-01665-4_18

2 Materials and Methods

The megacity Karachi (Fig. 1) is the most urbanized, industrialized, and affluent city in Pakistan. There are various industrial areas located close to or even merged with residential areas in Karachi. Twenty one rainwater samples from such areas, considered to be representative of these sites, were collected using polyethylene buckets (Fig. 1).

The pH and electrical conductivity (EC) of rainwater samples were measured immediately with conductivity (4320 JENWAY) and pH (3310 JENWAY) meters. The Dionex IC-3000 system was used to determine the major anions (F^-, Cl^-, NO_2^-, NO_3^-, SO_4^{2-}, $C_2O_4^{2-}$, $HCOO^-$, CH_3CHOO^-, pyruvate, malonate, propionate, glyoxylate) and cations (Na^+, NH_4^+, K^+, Mg^{2+}, Ca^{2+}) concentrations. Total Organic Carbon (TOC) measurement was performed by Shimadzu 5000 TOC analyzer.

A statistical analysis was conducted using the software SPSS 20 to identify the sources of observed species. Descriptive statistics, correlation matrix and factor analyses (using PCA) were also performed on the precipitation data. A stringent quality control protocol including calibration of the instruments, analyses of duplicate samples, spike recoveries were used to control data quality.

3 Results and Discussion

The samples pH profile indicate marked variations ranging from 3.31 to 7.9 with an average of 6.84 (Table 1). The lowest pH was observed at AFOHS site along with a high concentration of SO_4^{2-}, reflecting strong impact of anthropogenic pollution on the rainwater chemistry. The AFOHS site is located adjacent to a dumping yard facing an industrial zone. The high EC values at FBA, NN, and NZ reflect the worst atmospheric environmental quality in the study area.

The ionic species mean concentration in precipitation follows the order of $Na^+ > SO_4^{2-} > Ca^{2+} > Cl^- > HCO_3^- > K^+ > NH_4^+ > Mg^{2+} > NO_3^- > CH_3COO^- > HCOO^-$. The dominance of SO_4^{2-} in anions suggests that a SO_2 emission from fossil-fuel combustion was a major reason for rainwater acidification. The concentration of SO_4^{2-} was found to be much higher than those of major cities in North America, Asia, Mexico, South America, and the Middle East, indicating strong influence of anthropogenic pollution in Karachi. The high NH_4^+ concentrations in Karachi precipitation could possibly be attributed to neutralization by such key factors as overuse of fertilizers, biomass burning, livestock breeding, and agricultural activities.

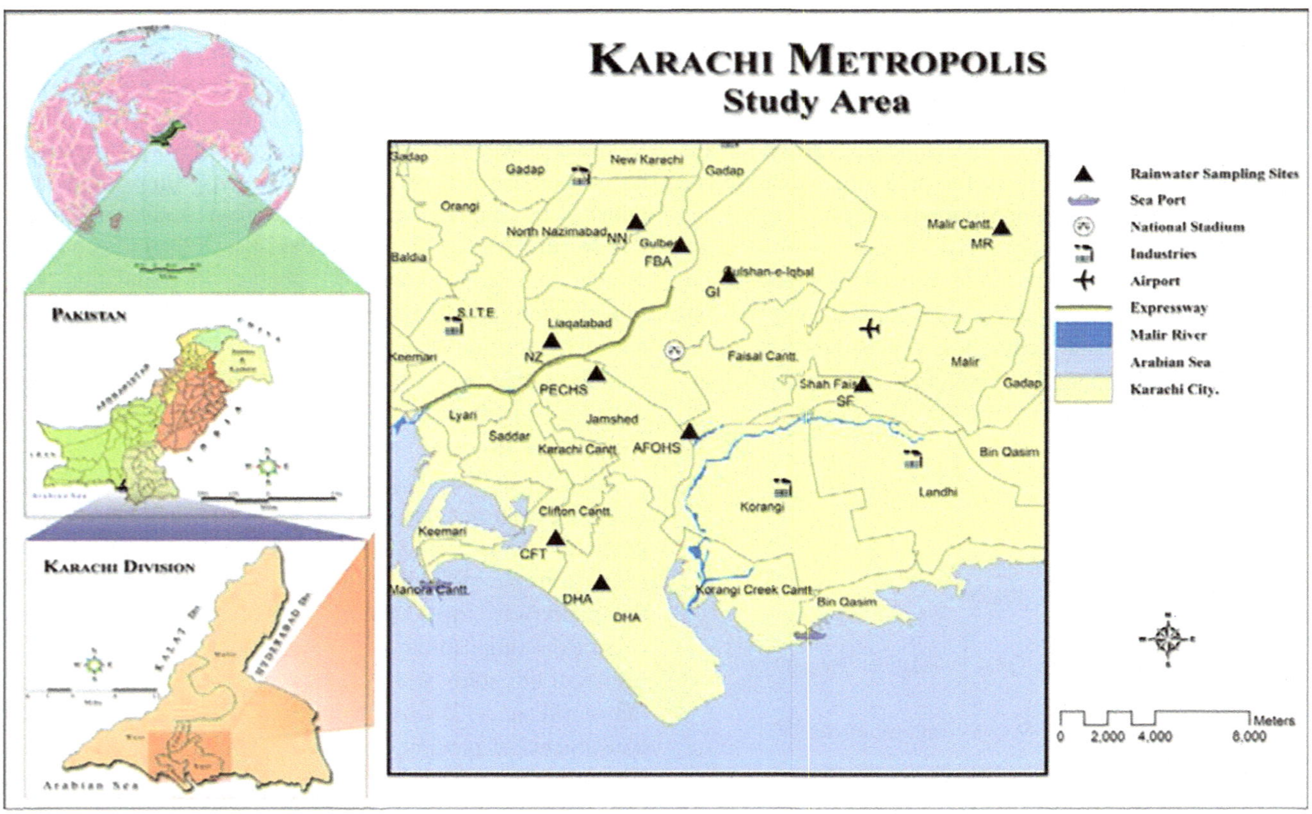

Fig. 1 Map of Karachi showing the sampling sites

Table 1 Descriptive statistics of chemical compositions in precipitation in Karachi

Parameters	Minimum	Maximum	Annual arithmetic mean
Precipitation (mm)	0.25	10.9	8.80 ± 3.10
TOC (mg/L)	0.05	16.2	3.90 ± 4.60
EC (μS/cm)	13.7	476	115 ± 121
pH	3.31	7.9	6.84 ± 0.93
H^+ (μeq/L)	0.01	490	23.5 ± 107
HCO_3^- (μeq/L)	0.02	708	164 ± 211
Na^+ (μeq/L)	61.8	3510	484 ± 759
NH_4^+ (μeq/L)	7.0	359	123 ± 104
K^+ (μeq/L)	3.6	454	132 ± 120
Mg^{2+} (μeq/L)	4.1	362	97.2 ± 108
Ca^{2+} (μeq/L)	7.0	2210	333 ± 531
Cl^- (μeq/L)	10.4	1090	242 ± 289
NO_3^- (μeq/L)	0.36	109	27.1 ± 31.0
SO_4^{2-} (μeq/L)	21.6	1690	397 ± 467
$HCOO^-$ (μeq/L)	0.44	27.1	4.90 ± 7.30
CH_3COO^- (μeq/L)	0.34	66.9	9.4 ± 16.0

The TOC concentrations spanned two orders of magnitude; the mean value, 3.9 ± 4.6 mg/L (Table 1), indicated that Karachi's rainwater is relatively loaded with organic compounds. Concentrations of $HCOO^-$ and CH_3COO^- in Karachi's rainwater were comparable to urban areas, implying the importance of anthropogenic sources of carboxylic acids. A positive correlation was found between the concentrations of $HCOO^-$ and CH_3COO^-, suggesting that their sources are closely related to, and/or controlled by a common reaction.

The SO_4^{2-}/NO_3^- ratios are characteristic of the site and reveal the anthropogenic influence of its immediate environment. Compared to the SO_4^{2-}/NO_3^- ratios reported for other cities in the world, Karachi had one of the highest at 14.6. The dominance of SO_4^{2-} can be attributed to the influence of the surrounding anthropogenic sources (combustion, vehicular traffic, and garbage burning).

Correlation coefficients between the individual organic ions and various inorganic parameters were lower than the organic-organic correlation coefficient. A high correlation was seen between H^+ and SO_4^{2-} and between SO_4^{2-} and NO_3^-. It is interesting to note that Ca^{2+} and NH_4^+ also had a high correlation coefficient with SO_4^{2-} (r = 0.50 and 0.60, respectively). NH_4^+ also showed strong correlation with NO_3^- (r = 0.60). The most likely explanation of these results is that in the atmosphere of Karachi, a major portion of SO_4^{2-} is associated with the H^+ ion. They also occurred as salts associated with other cations e.g., $CaSO_4$, $(NH_4)_2SO_4$, NH_4HSO_4, and NH_4NO_3.

The four factors corresponding to four major sources explained 81.2% of the total variance with a factor loading greater than 0.5. Factor 1, which explains 35.6% of the total variance, has high loading of Cl^-, Mg^{2+}, NH_4^+, K^+, SO_4^{2-}, NO_3^-, and Na^+ in a decreasing order, implying a mixed source of anthropogenic (indicated by SO_4^{2-}, NO_3^-, NH_4^+ and K^+) and marine (indicated by Cl^-, Mg^{2+}, and Na^+) pollution. Factor 2, explaining 16.4% of the total variance has large loading in $HCOO^-$, CH_3COO^- and HCO_3^-. This factor represents the direct emissions of these low-molecular weight carboxylic acids from anthropogenic sources. Factor 3, with 16.4% of the total variance, shows high loadings in Ca^{2+} and Na^+ and moderate value in CH_3COO^-. This factor appears to be linked to the terrestrial source. Factor 4, explaining 12.8% of the total variance, has large loading in H^+ and SO_4^{2-}, suggesting the influence of anthropogenic source.

The metropolitan Karachi has an arid climate and is bordered by Balochistan and Iran to the northwest, desert of Sindh and India to the east and southeast, and Arabian Sea to the west southwest. Seasonal differences in air mass trajectories were observed. The random air mass trajectories was noted in winter precipitation due to the thunderstorms passing the study area, whereas a common pattern of air mass trajectories from the Arabian Sea were observed in the summer monsoon rainfall. The most distinctive effect of the air masses on the rainwater composition of the Karachi metropolitan area was observed in the terrestrial component Ca^{2+}.

4 Conclusion

A detailed study on precipitation chemistry was performed in a mega-city Karachi. The results revealed that SO_4^{2-} and Cl^- are the dominant anions, while Na^+ and Ca^{2+} are the major cations, respectively. $HCOO^-$ and CH_3CHOO^- were found to be the predominant carboxylic acids. We believe this is the first report of the precipitation concentration levels of carboxylic acids in Karachi. These organic acids together contributed with 7% to the TOC in precipitation. SO_4^{2-}/NO_3^- ratio showed that the contribution of SO_4^{2-} to acidity was 14.6 times higher as that of NO_3^-. The high enrichment factor of seawater for Ca^{2+}, K^+, and SO_4^{2-} suggested that these ions mainly originated from non-marine sources. The use of principal component analysis allowed the interpretation of rainwater chemistry characterization. Four components that accounted for 82.1% of the total variance were extracted. The coupling between the chemical content and air masses back trajectories highlighted the importance of the origin of air masses in the acquisition of rainwater chemistry. The chemical composition of precipitation in this study was influenced by natural (sea-salt and soil components) and anthropogenic (combustion, traffic, biomass burning, refuse combustion) sources.

References

1. Tsakovski, S.L., Simeonov, V.D.: Chemometric study of atmospheric wet and dry precipitates from an urban region. Toxicol. Environ. Chem. **65**(1–4), 203–216 (1998)
2. Al-Khashman, O.A.: Ionic composition of wet precipitation in the Petra Region, Jordan. Atmos. Res. **78**, 1–12 (2005)
3. Báez, A., Belmont, R., García, R., Padilla, H., Torres, M.C.: Chemical composition of rainwater collected at a southwest site of Mexico City, Mexico. Atmos. Res. **86**(1), 61–75 (2007)
4. Kulshrestha, U., Kulshrestha, M.J., Sekar, R., Sastry, G., Vairamani, M.: Chemical characteristics of rainwater at an urban site of South-Central India. Atmos. Environ. **37**(21), 3019–3026 (2003)
5. Senthilnathan, T.: Measurements of urban ambient air quality of Chennai city. Indian J. Air Pollut. Control **8**(1), 35–47 (2008)
6. Al-Khashman, O.A.: Chemical characteristics of rainwater collected at western site of Jordan. Atmos. Res. **91**, 53–61 (2009)
7. Galloway, J.N., Likens, G.E.: Calibration of collection procedures for the determination of precipitation chemistry. Water Air Soil Pollut. **6**(2–4), 241–258 (1976)
8. Munger, J.W., Eisenreich, S.J.: Continental-scale variations in precipitation chemistry: ion concentrations are dominated by land use and proximity to man-made emissions. Environ. Sci. Technol. **17**(1), 32–42 (1983)
9. Smirnioudi, V., Siskos, P.: Chemical composition of wet and dust deposition in Athens, Greece. Atmos. Environ. Part B. Urban Atmos. **26**(4), 483–490 (1992)
10. Downing, C., Vincent, K., Campbell, G., Fowler, D., Smith, R.: Trends in wet and dry deposition of sulphur in the United Kingdom. Water Air Soil Pollut. **85**(2), 659–664 (1995)
11. Nilles, M.A., Conley, B.E.: Changes in the chemistry of precipitation in the United States, 1981–1998. Water Air Soil Pollut. **130** (1–4), 409–414 (2001)
12. Ito, M., Mitchell, M.J., Driscoll, C.T.: Spatial patterns of precipitation quantity and chemistry and air temperature in the Adirondack region of New York. Atmos. Environ. **36**(6), 1051–1062 (2002)
13. Hontoria, C., Saa, A., Almorox, J., Cuadra, L., Sánchez, A., Gascó, J.M.: The chemical composition of precipitation in Madrid. Water Air Soil Pollut. **146**(1–4), 35–54 (2003)
14. Menz, F.C., Seip, H.M.: Acid rain in Europe and the United States: an update. Environ. Sci. Policy **7**(4), 253–265 (2004)
15. Celle-Jeanton, H., Travi, Y., Loÿe-Pilot, M.-D., Huneau, F., Bertrand, G.: Rainwater chemistry at a Mediterranean inland station (Avignon, France): local contribution versus long-range supply. Atmos. Res. **91**(1), 118–126 (2009)
16. Parekh, P.P., Khwaja, H.A., Khan, A.R., Naqvi, R.R., Malik, A., Shah, S.A., Khan, K., Hussain, G.: Ambient air quality of two metropolitan cities of Pakistan and its health implications. Atmos. Environ. **35**(34), 5971–5978 (2001)
17. Biswas, K.F., Ghauri, B.M., Husain, L.: Gaseous and aerosol pollutants during fog and clear episodes in South Asian urban atmosphere. Atmos. Environ. **42**(33), 7775–7785 (2008)

Part III

Geo-Environmental Engineering: Remediation for Geo-Environmental Problems

Remediation Treatments and Economic Assessment of Oil Residual Sludge from the Bottom of Tunisian Refinery Crude Oil Storage Tanks

Olfa Ben Said, Rihab Belgacem, Boudour Ben Gaffar, Hamouda Beyrem, and James R Kahn

Abstract

The residual sludge from the bottom of refinery crude oil storage tanks causes major environmental and societal problems in Tunisia. Due to its high viscosity, it must be removed from the storage tanks in order to utilize the crude oil. However, its disposal in the environment leads to the infiltration of toxic metals into groundwater and eventual contamination of surface water bodies. The aim of this work was to characterize and treat sludge from Tunisian refinery by two different approaches (chemical and biological) and evaluate the economic efficiency of sludge remediation. Our results showed that both approaches are useful in terms of reducing the metal content of the sludge. However, the biological approach is more economically efficient and leads to better ecological outcomes.

Keywords

Sludge • Petroleum • Biological treatment • Chemical treatment • Metals

1 Introduction

With the accelerated development of the industrial sector, humans are increasingly responsible for the pollution of the environment to such an extent that scientists are now labelling the current period as the Anthropocene [1]. The diversity, the scale and range of production of the industrial outputs have led to a considerable increase in the quantity of waste. Petroleum activity, in particular, produces an increasing amount of sludge in oil storage tanks which threatens the environment and human health by the infiltration of toxic substances into the ground and consequently into the water table [2]. This situation poses a real ecological and social problem, due to the bad management of storage and disposal. The spread of untreated crude oil residual on agricultural land creates an economically and environmentally irreversible problem. The strong persistence of the metal contamination makes it environmentally irreversible, requiring hundreds of years for the metals to break down. The dispersal of the pollution makes the cost of post-spreading treatment too high to remediate [3]. Hence the economic and environmental need for adequate treatment of sludge before spreading or landfilling. In order to reduce the volumes of residues to be stored several treatment methods have been applied including chemical and physical treatment. The biological treatment alternative represents a reliable, low cost and environmentally friendly technology compared to the chemical and physical methods [4]. Sludge treatment is one of the most complicated pollution abatement processes with necessities and requirements of environmental and public health concerns. The economic magnitude of this problem is illustrated by the high cost of both investment and operation. The present work aimed to characterize the residual sludge from the bottom of an actual crude oil storage tank of the Tunisian refinery "la Société Tunisienne des Industries de Raffinage STIR", to perform remediation treatments to reduce metal and to evaluate the sludge remediation economically.

2 Materials and Methods

2.1 Sludge Sampling, Characterization and Remediation Treatments

STIR uses cylindrical tanks with floating covers to stabilize stored oil. The sludge sample was taken from the bottom of

O. Ben Said (✉) · R. Belgacem · B. Ben Gaffar · H. Beyrem
Bizerte Faculty of Sciences, Environment Biomonitoring Laboratory LBE, Bizerte, Tunisia
e-mail: nourelimen@yahoo.fr

J. R. Kahn
Department of Economics, Washington and Lee University, Lexington, VA 24450, USA

A. Kallel et al. (eds.), *Recent Advances in Geo-Environmental Engineering, Geomechanics and Geotechnics, and Geohazards*, Advances in Science, Technology & Innovation, https://doi.org/10.1007/978-3-030-01665-4_19

the RS9 crude oil storage tank; this sludge represents the result of crude oil (AZERI) decanting.

Several chemical analyzes were conducted both before and after remediation treatments such as moisture content [5], organic carbon content [6], metal content [7].

Chemical treatments were performed with Stabilization-Leaching. Various types of hydraulic binders were used (Portland cement, lime, gypsum and clay) [8].

The biological treatment involved sludge inoculation with enriched consortia of microbial inoculants of oily sludge [4].

2.2 The Economics of Sludge Remediation

The economics of sludge remediation can be better understood by looking at three basic economic concepts (1) Pollution abatement is generated by paying high initial capital costs and the receiving benefits for a long period of time (2) In the presence of irreversible environmental change, a precautionary approach should be implemented (3) One of the keys to sustainability is to use waste outputs as economic inputs.

Pollution remediation is associated with high initial costs, but gives continuous benefits over a long term. Economists compare costs and benefits in different periods by calculating the present value of the stream of net benefits (benefits minus costs) in each time period. Equation (1) represents a typical calculation of the net present value NPV, where B(t) represents the benefits in year t, C(t) represents the costs in year t, T represents the number of periods in which the project generates costs or benefits, and r represents the rate of time preference, or the discount factor.

$$NPV = \int_0^T (B(t) - C(t))e^{-rt}dt \quad (1)$$

3 Results

3.1 Characterization and Remediation Treatments

The sludge is very rich in organic matter (60.4%) and contains a high percentage of dry matter (98.7%). Metal analysis showed the presence of iron Fe, magnesium Mg, sodium Na, barium Ba, zinc Zn, copper Cu, Manganese Mn, nickel Ni, lead Pb, vanadium V and chromium Cr (Table 1). Fe, Mg, Na and Cu were the most dominant.

Almost with all hydraulic binders used, metal sludge content has decreased. For Cu, Pb and Zn, the results clearly show a reduction percentage that exceeded 90% (Table 2). However, for Mg the highest percentage was obtained with lime and then cement. For Na, a reduction was found only with cement and then limes.

Following the biological treatment, a significant decrease of Cu, Pb, Mg, and Na was observed with reduction percentages greater than 97% (Table 2).

3.2 Economics of Sludge Remediation

A Table 3 examines a hypothetical project to remediate sludge which has an initial cost of 30 million Euros, and yields benefits in terms of reduced cancer rates of 2 million Euros per year for 100 years. The project looks like a winner, with 30 million in costs and 400 million in benefits over the years, but the calculation of NPV generates a far different calculation of net benefits [9]. Table 3 presents the NPV of this project calculated with a typical discount rate of 10, 2.5% (a good approximation of the long term real rate of growth of GDP) and a lower rate of 1%, consistent with Weitzman's admonition to discount the future at the lowest possible rate. It also shows the present value of net benefits that occur during year 10, year 20, year 30 and year 100. The project would have a negative NPV of 10.9 million Euros (10%), 72 million (2.5%) and 126 million (1%).

4 Discussion

The sludge used in present study has a high concentration of dry matter and could be classified as dry sludge. Fe, Mg, Na and Cu were the most predominant elements; this could be associated with the origin of crude oil stored in STIR and the tank bottom corrosion phenomenon [4].

With almost all of the hydraulic binders used, metal levels decreased mainly for Cu, Pb and Zn. Similar results were reached with the biological treatment.

Table 1 Metal content (ppm) in residual sludge from RS9 tank before treatment

Fe	Mg	Na	Ba	Zn	Cu	Mn	Ni	Pb	V	Cr
50	28.4	15.4	0.7	5.4	12	1.5	1.3	2.3	0.4	0.3

Table 2 Percentage of metal reduction in sludge after remediation treatments

		Mg	Na	Zn	Cu	Pb
Chemical treatment	Cement	93	60	99	96	91
	Lim	99	42	99	99	95
	Clay	62	0	99	99	99
	Gypsum	12	0	99	99	99
Biological treatment		99	97	0	100	99

Table 3 Present value of net benefits of a hypothetical project with €30 million initial cost and benefits of 2 million/year for 100 years

	Discount rates		
	10%	2.5%	1%
Present value of net benefits of the entire project	−€10.900,000	€43,000,000	€96,000,000
Present value of the 2 million benefit in year 10	€740,000	€1,500,000	€1,800,000
Present value of the 2 million benefit in year 20	€270,000	€1,200,000	€1,64,000
Present value of the 2 million benefit in year 30	€100,000	€940,000	€1,480,000
Present value of the 2 million benefit in year 100	€91	€164,000	€740,000

The NPV shows the importance of evaluating projects with long time horizons with a low discount rate, otherwise the distant future is essentially discounted to zero.

5 Conclusion

Both chemical and biological treatments have shown efficiency in immobilizing metals in sludge, while the biological one was more selective. This process has the advantage of being ecological, which will allow an alternative for the elimination of sludge and their valorization as an agricultural input. Although we have not conducted a cost-benefit analysis, economic arguments associated with irreveribilities and sustainability suggest that bioremediation should be aggressively pursued in order to protect water resources, ecosystems, and public health.

References

1. Steffen, W., Grinevald, J., Crutzen, P., McNeill, J.: The Anthropocene: conceptual and historical perspectives. Philos. Trans. Roy. Soc. London Math. Phys. Eng. Sci. **369**, 842–867 (2011)
2. Liu, W., Luo, Y., Teng, Y., Li, Z., Ma, L.Q.: Biodegradation of oily sludge contaminated soil by stimulating indigenous microbes. Environ. Geochem. Health **32**, 23–29 (2010)
3. Kahn, J., O'Neil, R.: Ecological interaction as a source of economic irreversibility. South. Econ. J. **66**(2), 381–402 (1999)
4. Gallego, J.L.R., García-Martínez, M.J., Llamas, J.F., Belloch, C., Pelaez, A.I., Sanchez, J.: Biodegradation of oil tank bottom sludge using microbial consortia. Biodegradation **18**, 269–281 (2007)
5. Black, C.A.: Methods of Soil Analysis, Part 1, Physical and Mineralogical Properties. American Society of Agronomy, Madison, WI, USA (1965)
6. Schnitzer, M.: Organic matter characterization. In: Methods of Soil Analysis, Part 2. Chemical and Microbiological Properties, pp. 581–594. American Society of Agronomy, Madison, WI, USA (1982)
7. Marchandise, P., Olió, J.L., Robbe, D., Legret, M.: Trace metal determination in river sediments and sewage sludge. Inter laboratory comparison of extraction techniques. Environ. Technol. Lett. **3**, 157–166 (1982)
8. Pereira, C.F., Pinero, M.R., Vale, J.: solidification/stabilisation des poussières de four à arc électrique utilisant les analyses des cendres volantes du processus de stabilisation. Hazard. Mater. **82**, 183–195 (2001)
9. Weitzman, M.L.: Why the far-distant future should be discounted at its lowest possible rate. J. Environ. Econ. Manag. **36**(3), 201–208 (1998)

Proposing Rehabilitation Scenarios for Limestone Quarries with 3D Modeling and 3D Print: Case of Jbel El Oust (Tunisian Atlas)

Soumaya Ben Fredj, Fetheddine Melki, and Kamel Jridi

Abstract

The production of building materials in Tunisia has recently undergone a notable development thanks to several factors. Among these, we can mention the growth of property sector, the increase of the construction sites of public works (roads, motorways and bridges) and the use of the calcareous materials in cement production. All these factors led to the exploitation of new quarries of different types and ages, such as quarries of carbonated rocks of the Jurassic, Cretaceous and Eocene age. The exploitation of these quarries affects the environment and the landscape. Although the Tunisian legislation imposed an impact study (law no. 89-20 of 22 February 1989) carried out by the operating company to leave the quarry in a good state quarry being exploited, many quarries are abandoned without being redeveloped and integrated into their natural environment. Thus, good rehabilitation management performed by proposing redevelopment scenarios using three-dimensional modeling developed by specific software can be efficient to overcome the negative impacts of the quarries exploitation on the environment.

Keywords

Quarries • Environment • Rehabilitation • Limestone • Jbel EL oust • Modeling 3D • 3D print

S. B. Fredj (✉) · F. Melki
Geomatic Research Unit, Structural and Applied Geology (UR11ES13), Faculty of Sciences of Tunis, Université de Tunis El Manar, 1060 Tunis, Tunisia
e-mail: soumaya.benfredj@fst.utm.tn

F. Melki
e-mail: fetheddinemelki@yahoo.fr

K. Jridi
SOROUBAT Group 34, 36 Avenue de la Gare, Mégrine Riadh, Ben Arous, Tunisia
e-mail: Jeridi.kamel@yahoo.fr

1 Introduction

Jbel El Oust is located in delegation of Zaghouan, about thirty kilometers south of Tunis, near the village of Cheylus (delegation of Zaghouan). The outcrop of Jebel El Oust is a vast complex anticline with a folded and faulted heart, formed by Lias limestone. The sides of the outcrop consist of Cretaceous age and Tertiary grounds. The study area is characterized by semi-arid climate of an average annual temperature equal to 18.6 °C and average precipitation varying between 300 and 400 mm/year. The area has a particular dense hydrographic network, and the Meliane River also known as Oued Miliane is a main component of this network. The hot thermal waters of Jbel El Oust have been exploited since 1976 in order to feed a modern thermal establishment where the modern cryotherapy is practiced [1]. The establishment is known today by the thermal station of Jbel El Oust. The site is characterized by a huge diversity of plants and animals. The plants' richness of the site is mainly represented by the forest trees such as pines, holm oaks, eucalyptus, subshrubs like acacias and many different types of scrub plants on the southern flanks of the mountain. The environment site contains a diverse fauna including mammals such as wild boars, foxes, wild rabbits, hedgehogs which are among the smallest mammals in the world, Lizards and birds as well as partridges. The natural conditions and benefits favor the presence of habitats, explaining the existence of an ancient Roman city that occupies the eastern slope of Jbel El Oust. During the last decades, the northern flank of Jbel El Oust has been exploited by different companies mainly the SOROUBAT one. This company has token advantage from an industrial quarry. The actions of the exploiting companies were to collect and produce limestones for the building projects and the public works sector. The recent exploitation has been a great negative impact on the environment. The current study aimed to propose rehabilitation scenarios for the SOROUBAT quarry using 3 Dimensional (3D) simulation software and 3D

A. Kallel et al. (eds.), *Recent Advances in Geo-Environmental Engineering, Geomechanics and Geotechnics, and Geohazards*, Advances in Science, Technology & Innovation, https://doi.org/10.1007/978-3-030-01665-4_20

printing. These scenarios have a purpose to protect the environment by integrating the quarry in its natural environment after the exploitation's cessation.

2 Materials and Methods

The quarry which is the subject of this study, operated by the SOROUBAT Company, located in Jebel El Oust, in the northwest of Zaghouan and north-south of the city of BirMcherga, is divided into two compartments; northern and southern compartments. The northern compartment dates to the Jurassic age, represented by the lower and the middle lias (Fig. 1) with a fault system characterized by the dominance of the North-South direction and North 45 direction (N45). The N45 are responsible for the appearance of extrusive Lias. The Second compartment is represented by the lower Lias limestones crossed by an N60–70 fault system intersected by N130 to N160 systems. The proposed scenarios were introduced to put into perspective the site landscape in its natural environment. Then, a geotechnical study was performed on the northern compartment to assess the stability of the different fronts. In a second part, according to the heights of the fronts and slope angles, a 3D simulation of two scenarios was carried out using two software programs. The first software is "Autodesk 3dsMax" designed for 3D modeling, animation, rendering and visualization was used. The second software is the "SketchUp" used for the drawing the landscape and 3D modeling.

The representing model of the quarry after the final exploitation was printed in a 3D model using the "Ultimaker 2+ 3D printer", based on the side plan of the quarry. These 3D representations could potentially provide a better understanding of the proposed rehabilitation scenarios.

3 Results

3.1 Geotechnical Site Characterization

A geo-technical study was carried out in the studied quarry; the limestones of Jbel El Oust have great mechanical quality.

The main geotechnical characteristics of these limestones are: Los Angles 22, Duval humide 3.9, Duval sec: 11.4 and Adhesion to bitumen: 99%. The mechanical resistance of their blocks can be approximated by the following values: Compressive strength: 50–8 MPa, tensile strength 8 MPa, internal friction angle: 45 and cohesion: 10 MPa.

From these characteristics and according to the HoeK-Bray chart, the results obtained by calculating the maximum angle of repose for the northern compartment are the following: Front F1: $\alpha = 80.5°$; Front F2 $\alpha = 81.55°$ Front F3 $\alpha = 83.6°$ (Fig. 1) and the edge angle calculation result $\beta = 46.65$.

4 Discussion

In order to prepare the rehabilitation step of, we designed the quarry shape after the end of the exploitation and a 3D printing was obtained from the plan side of the quarry employing SketchUp (Fig. 2).

Two scenarios were purposed to integrate the site in its natural environment. The first scenario consists of re-implanting plants and trees in the quarry's fronts, (Fig. 3). Reforestation of the rehabilitated quarry can be achieved by direct sowing or planting [2]. It favors the stabilization of slope and restoring the site to its original environment.

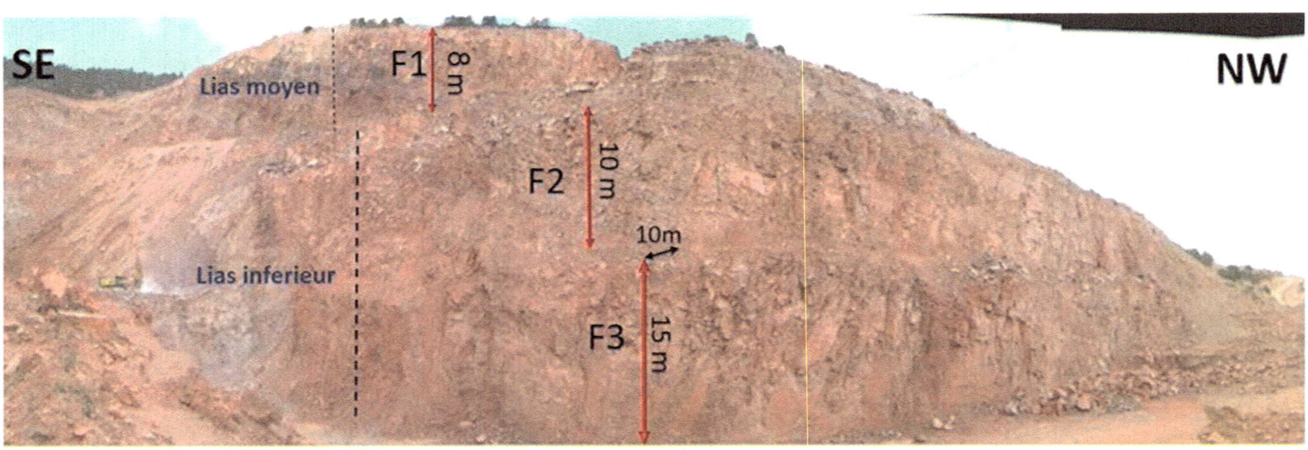

Fig. 1 Panoramic view of the northern compartment of the Jbel El Oust quarry

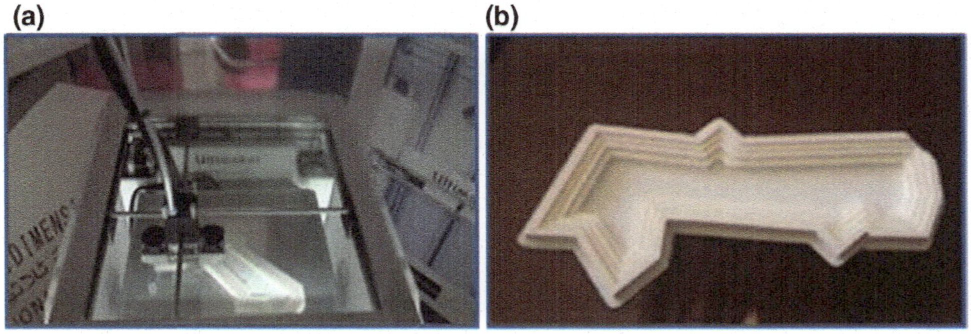

Fig. 2 3D models of the quarry shape after the end of exploitation. **a** the step of 3D print with "Ultimaker 2+ 3D printer"; **b** the final quarry model

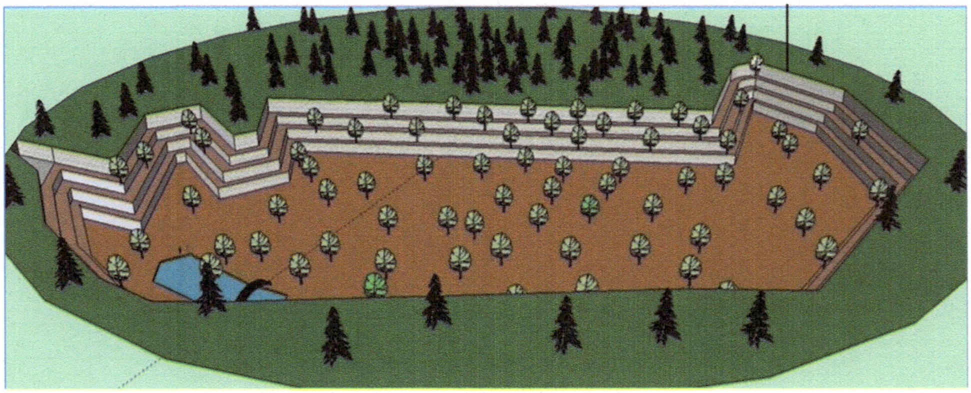

Fig. 3 First 3D rehabilitation scenarios using SketchUp software

Fig. 4 Second 3D rehabilitation scenario obtained using 3dsMax software

The second scenario consists in redeveloping a tourist area by the construction of natural wood guest houses as an annex to the thermal station of Jbel El Oust. In fact, the tourist development of the quarry helps to highlight the historical and natural values of the outcrop represented by the archaeological site of Jbel El Oust (Fig. 4).

5 Conclusions

In order to protect the environment, we propose two rehabilitation scenarios in this research work using 3D modeling software and 3D printing.

These scenarios presented in 3D models showed the final state of the quarry after the rehabilitation and facilitated the decision making process by the quarrying companies.

In our future works, we are planning to create a new web application for conception of rehabilitation numerical model.

References

1. Curie, J.: Les travertins anthropiques, entre histoire, archéologie et environnement: étude geoarchéologique du site antique de Jebel Oust (Tunisie). Géographie, Université de Bourgogne (2013)
2. Vanpeene-Bruhier, S., Piedallu, C., Delory, I.: Réaménagement forestier des carrières de granulats. Cemagref editions (2002)

Potential Remobilization of Heavy Metals by Wave Friction Case of Algiers Bay

Atroune Farid, Hemdane Yacine, and Bouhmadouche Mohamed

Abstract

Marine pollution is one of the majors problems of concern to coastal populations. This concern resulted principally from multiple forms of marine pollution. Algeria is one of the most affected countries by this pollution, resulting in an obstacle to economic and social development. However, the uncontrolled urbanization of watersheds and the presence of high-risk industrial sites, close to urban concentrations, have generated considerable pollution and degradation, which has endangered the environment and public health. The pollution by heavy metals of superficial sediments, carried by the El Harrach river, is spread over Algiers Bay in a diversified way. These sediments generally have a high affinity for fine-grained sediments. Maouche (Thèse de 3ème cycle. Océanologie. Univ. Perpignan, 225 p + annexes, 1987 [1]) specified that the fraction below 40 μm occupies more than 75% of the superficial sediments of the Bay of Algiers. Heavy metals are usually associated with the fine fraction. Then, numerical modeling of waves and sediment transport were applied in order to observe the potential dynamic of marine pollution in Algiers Bay. The results show that heavy metals can be remobilized by moderate waves (frequent). This work allowed to highlight the possible role of waves in the resuspension of contamined sediments deposited on the shallow water of the bay of Algiers. Therefore, it is highly recommended to monitor the water quality of Algiers Bay in order to protect both the human health and marine ecology of the region (including the coastal waters of Algiers Bay).

Keywords

Marine pollution • Superficial sediments • Heavy metals • Algiers Bay

1 Introduction

The watershed of Oued El Harrach belongs to the large Algiers coastal watershed. It covers an area of 1270 km^2 and stretches over 51 km from North to South for and 31 km from East to West. The concentration of heavy metals along the Oued continues to increase especially for mercury (Hg), lead (Pb), manganese (Mn) and zinc (Zn). Consequently, the coastline of Algiers Bay is very concerned with heavy metal pollution which affects the bathing water quality. Although these pollutants deposited at relatively large depths, these sediments are nevertheless susceptible to be eroded by the waves combined with the general current. The aim of this work was to study the possible remobilization of heavy metals during the frequent waves of NE characterizing the bay of Algiers.

2 Methods

Surficial nearshore sediment samples were taken using a van Veen. The positioning plan was based on the mission carried out by the Marine Geology Laboratory.

The particle size analysis was carried out by dry mechanical sieving on a series of AFNOR type sieves on the coarse fraction (>40 μm) for 15 min. Carbonate levels were determined by calcimetry on the total fraction of the sample.

Metals (Cu, Zn, Ni, Fe, Pb and Hg) were determined using an atomic absorption spectrometer). Statistical treatment by A.C.P (Principal Component Analysis) was applied to all the geochemical data obtained by analysis. *Simulating Waves Nearshore Model* [2] was used to simulate waves of the NE sector that are frequent in the Bay. Then, these

A. Farid (✉) · H. Yacine · B. Mohamed
Laboratoire Géo-Environnement, Faculté des Sciences de la Terre, Géographie et Aménagement du Territoire, Université Houari Boumediene, BP 32, El-alia, Algiers, Algeria
e-mail: faridatroune@gmail.com

A. Kallel et al. (eds.), *Recent Advances in Geo-Environmental Engineering, Geomechanics and Geotechnics, and Geohazards*, Advances in Science, Technology & Innovation, https://doi.org/10.1007/978-3-030-01665-4_21

Table 1 Evolution of heavy metals in Algiers Bay

Authors/µg/g	Cu	Zn	Cr	Mn	Fe	Pb
Maouche [1]	24–51	123–231		250–316	3.8–4.8	20–161
Chouikhi et al. [4]		72–222	11–84	109–350		18–117
Boudjellal et al. [5]	23.4–72.7	60.25–255				16.19–93
Yoshida et al. [6]	28–86	98–520	36.2–95	630–1100	18.84	37–127
Atroune et al. [7]	22–98	128–375	38–105	258–398	3.9–4.5	33–154

results were inserted in the model Sedtrans05 [3] to know if these waves cause a transport on sedimentary bottom. For the implementation of this model it was assumed that the speed of the general current is equal to 2 m s^{-1}. These applications concerned the surrounding area of Algiers Port located at a depth of approximately 25 m.

3 Results

From the spatial distribution of these metallic elements it can already be seen that, the high contents of heavy metals present in the superficial sediments of the Bay of Algiers are located in the western sector of the bay between the port (discharges of hydrocarbons, chemicals) and in the immediate vicinity of the mouth (discharges from domestic and industrial tributaries). The metal concentrations (Table 1) vary between 128 and 375 µg/g for zinc, 33 and 154 µg/g for lead, 38 and 105 µg/g for chromium, 22 and 98 µg/g for copper and between 0.04 µg/g. and 8.74 µg/g for mercury. The results show that frequent NE waves with heights of 1 m could mobilize sediments on the 25 m deep coastal bottom in presence of current of 0.3 m s^{-1}. The simulation of the effect of bottom on the waves shows that the waves are modified by this effect at the surrounding study area (See Fig. 1).

4 Discussion

High concentrations of heavy metals are usually associated with fine sediments of a silty nature and rich in organic matter. Low concentrations are located in the bay area, where relatively coarse sediments predominate over the eastern part of the bay. The results show that frequent NE waves with heights of 1 m could mobilize sediments near the Algiers Harbour (depth about 25 m). These results were obtained using a general current of 0.3 m s^{-1}. It is important to emphasize that this current can have higher speeds. As a result, sediment transport would be more important. In addition, the simulation of shoaling coefficient (Ks) shows that the waves are significantly modified by shallow waters of the harbour area. This modification was induced by the friction of waves on the bottom. Consequently, sedimented heavy metals could be remobilized pollution that could spread to adjacent areas of the Harbour.

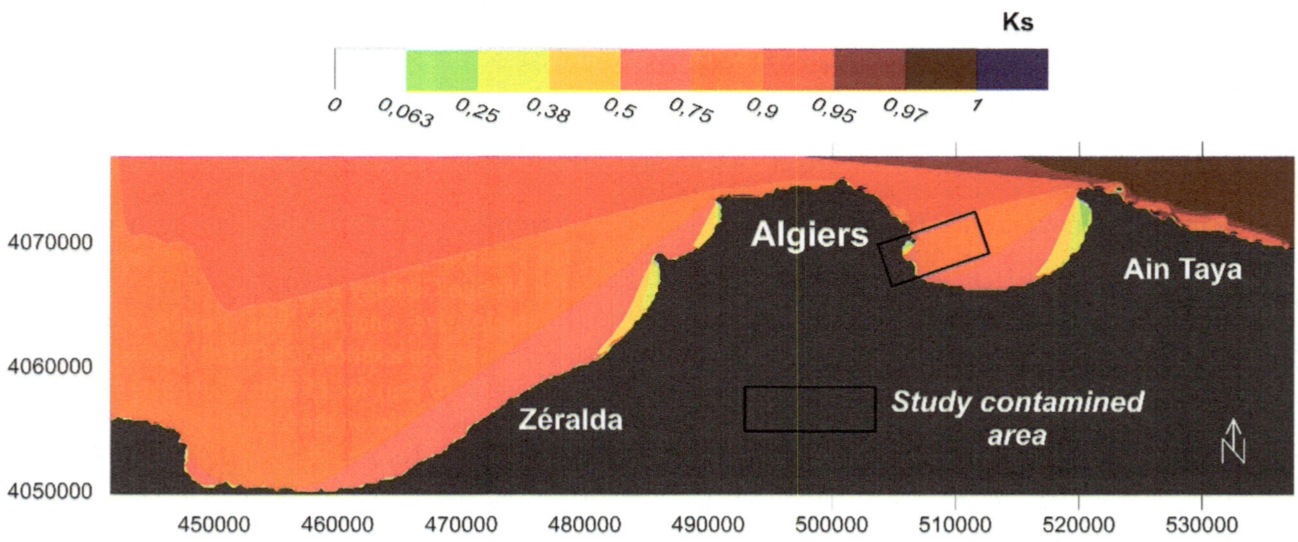

Fig. 1 Simulation of shoaling coefficient (Ks); effect of bottom on the waves around Algiers harbour. SWAN model. Hs: 1.5 m; Tp: 6 s and D°: N45°

5 Conclusion

The spatial distribution of heavy metals shows that superficial sediments are in perpetual motion and hydrodynamics can change the distribution of fine sediments at any time.

Moreover, the simulation of waves and sediment transport evidenced that this pollution can possibly be reactivated by bottom friction in the bay of Algiers. Hence, given the state of critical pollution levels of the waters of the sea and rivers, the need for the establishment of wastewater treatment plants is of major importance. This solution would improve the quality of fluvial waters and contribute to improving the state of the coastal marine environment of the Bay of Algiers.

References

1. Maouche, S.: Mécanismes hydrosédimentaires en baie d'Alger (Algérie): Approche sédimentologique, géochimique et traitement statistique. Thèse de 3ème cycle. Océanologie. Univ. Perpignan, 225 p + annexes (1987)
2. Booij, N., Ris, R.C., Holthuijsen, L.H.: A third-generation wave model for coastal regions. 1. Model description and validation. J. Geophys. Res. **104**(C4), 7649–7666 (1999)
3. Neumeier, U., Ferrarin, C., Amos, C.L., Umgiesser, G., Li, M.Z.: Sedtrans05: an improved sediment-transport model for continental shelves and coastal waters with a new algorithm for cohesive sediments. Comput. Geosci. (2008) https://doi.org/10.1016/j.cageo.2008.02.007
4. Chouikhi, A., Boulahdid, M., Sellali, B., Boudjellal, Y., et Azzouz, M.: Distribution des sels nutritifs des eaux interstitielles et de métaux lourds dans les sédiments superficiels du golfe d'Arzew et de la baie d'Alger. In: Symposium international sur la pollution des eaux marines. Casablanca, p. 10 (1991)
5. Boudjellal, B., Sellali, B., Benoud, D., Mallem, M.T.: Métaux lourds dans les sédiments superficiels de la baie d'Alger. In: Résultats du workshop sur la circulation des eaux et pollution des côtes méditerranéennes du Maghreb. Rabat, pp. 153–156 (1993)
6. Yoshida, M., Moali, M., Houas, O., Lakhdari, M., Nechaoui, L., Guerrida, D., Chatal, A., Oussalem, S., Makour,F., Khelifi, F., laleg, A.: Environmental pollution in oued El Harrach area, Alger. A preliminary report on mercury and heavy metals contaminations. In: Compte rendu du séminaire sur la pollution et la protection de l'environnement en Algérie, Alger, pp 19–37 (2005)
7. Atroune, F., Boutaleb, A.: Les métaux lourds dans les sédiments superficiels de la baie d'Alger: Influence de l'Oued El Harrach sur leur distribution. Bulletin du service géologique National **23**(2), 169–178 (2012)

Mitigation of Salinity Hazard from Low Permeable Soil by Electrochemical Treatment: A Laboratory Based Investigation

Mohammed Mustapha Bessaim, Hanifi Missoum, Karim Bendani, and Mohamed Said Bekkouche

Abstract

Soil salinization has become a serious problem and an environmental issue of worldwide significance, affecting the ecosystem biodiversity, human health and civil engineering infrastructures. This research investigated the efficiency of the electrochemical treatment (ECT) for remediation and mitigation of salinity hazard. The salinity of soil samples, including magnesium (Mg^{2+}), sulfate (SO_4^{2-}) and chloride (Cl^-) contents, was investigated using different duration time of 72 and 120 h. The originality of this research is to investigate the effect of the processing time on the migration and removal of harmful salt ions. During the treatment time, Mg^{2+} ions were transferred toward the cathode, whereas SO_4^{2-} and Cl^- ions were transported toward the anode by electro-migration. The optimal removal efficiency for Mg^{2+} was 90 and 93% after 72 and 120 h respectively, while the extraction rate of SO_4^{2-} ions was 5 and 15%. The removal of magnesium was substantially higher than that of sulfate and chloride ions, owing to the combined action of electro-osmosis and electro-migration mechanisms. On the other hand, increasing in the processing time allowed better transfer of ions species, enhancing thereby salts extraction. This study demonstrated that the harmful effect of salinity can be readily mitigated by ECT treatment. Therefore, ECT treatment could be successfully used for mitigation of salinity hazard and restoration of salt affected soils for environment protection and civil engineering constructions management.

Keywords

Salinization • Hazard • Remediation • Electrochemical • Saline soil

1 Introduction

Soil salinization is a real threat and the main reason of land degradation in many areas around the world [1–3].

This phenomenon has compromised approximately 7% of the earth's land surface, 20% of agricultural lands, and 50% of the cultivated land in the world [1]. This scourge is namely related to climate changes. The high temperatures and low rainfall rates in arid and semi-arid areas led to excessive salt accumulation in soil sub-surface. The excessive amount of salt within the soil matrix can have adverse effects on plant and agriculture productivity, including reduced water availability, osmotic adjustment, nutrient uptake and tissue injury [3, 4].

Furthermore, the high rate of salt can have detrimental effects on human health such as birth malformation, methemoglobinemia (blue baby syndrome), infant mortality, goiter, hypertension and epidemiological gastric cancer [5, 6]. In terms of constructions, excessive concentration of salts leads to adverse impacts affecting roads and bridges, buildings and steel structures. In addition, it alters soil properties, causing surface crusting, permeability deterioration, erosion, water logging, and decrease in soil bearing capacity [7]. This creates an urgent need to solve these problems. Generally, soil salinity can be reduced by several techniques such as irrigation, soil substitution, chemical amendment and bio-remediation.

However, these methods were found to be expensive and mostly ineffective, namely when dealing with soft soils, due to their low permeability and mineralogical complexity. The electrochemical treatment (ECT) can be an innovative and low-cost technique that can solve those deficiencies [8]. The principles of this process involve application of an

M. M. Bessaim (✉) · H. Missoum · K. Bendani
M. S. Bekkouche
Abdelhamid Ibn Badis University of Mostaganem, Architecture and Civil Engineering Department, Mostaganem, Algeria
e-mail: mohammed.bessaim@univ-mosta.dz

M. M. Bessaim · H. Missoum · K. Bendani
Construction, Transport and Protection of Environment Laboratory (LCTPE), Mostaganem, Algeria

A. Kallel et al. (eds.), *Recent Advances in Geo-Environmental Engineering, Geomechanics and Geotechnics, and Geohazards*, Advances in Science, Technology & Innovation, https://doi.org/10.1007/978-3-030-01665-4_22

electrical potential via inserted electrodes within the polluted media.

Salts, which have high water solubility and exist as charged ions, are transported and removed by electro-migration, electro-osmosis and electrophoresis from soil surface [1, 9]. The combined effect of these processes alters the chemistry of the medium, changing thereby the physico-chemical properties of the soil in question.

This research investigated the viability of ECT treatment on mitigation of salinity hazard for enhancing agriculture and constructions uses.

2 Materials and Methods

2.1 Saline Soil

The saline soil used in this study was collected from Mostaganem, Algeria. The soil was mixed with deionized water, so that the water content was adjusted to ensure sufficiently the ECT process. Table 1 shows the physico-chemical properties of the saline soil.

2.2 Electrochemical Cell

The diagram of the electrochemical cell used in these experiments is shown in Fig. 1. The bench scale was consisted of three compartments: soil box, anolyte and catholyte reservoirs. Soil samples were placed in the central box. Filter paper was placed between the specimen and electrolytes reservoirs to avoid loss of soil. Anolyte and catholyte were filled with deionized water solution and circulated continuously over the processing time. A measuring cylinder was connected to the cathodic chamber to collect the electro-osmosis flow (EOF). Gas vents were provided in the electrolyte chambers to allow gases resulting from the electrochemical treatment to escape.

Table 1 Physico-chemical properties and composition of saline soil

Parameters	Values
pH	8.10
EC (dS/m)	18.32
Organic content (%)	5.30
Particle size analysis (%)	
Sand	12
Silt	48
Clay	40
Initial concentration (mg/kg)	
Mg^{2+}	297
SO_4^{2-}	490
Cl^-	142

3 Results and Discussion

3.1 Removal Efficiency of Harmful Salts

The removal efficiency of salts present in soil sections at the end of the treatment is shown in Fig. 2. The highest concentration of Mg^{2+} was observed near the cathodic section, during the 120 h. The increase of the processing time, allow the development of the acidic front at the anode, which enhance the desorption of these species. Thus, their transport toward the cathode, via electro-migration and electro-osmotic flow.

In case of chloride ions, they moved and accumulated at the anodic area by electro-migration, explaining the high rate recorded in soil sections near this region during the 72 h. The extraction rate of chloride was 62 and 63% for 72 and 120 h respectively.

However, it was found, that the optimal removal was achieved for magensium and sulfate ions, reaching an efficiency of 93 and 15% during 120 h, while, only 90 and 5% were eliminated during 72 h. Increasing in treatment time leads to better removal for both magnesium and sulfate ions, because longer periods allowed more electrical transport through the soil matrix [1]. Among all investigated salts, Mg^{2+} showed the highest removal rate. This is related to the combined action of electro-osmosis and electro-migration mechanisms. Moreover, sulfate and chloride ions form complexes with other cations present within the media, reducing significantly their migration and solubility, hindering thereby their extraction [1, 4]. Due to the removal of those harmful salts, a significant decrease of soil salinity was achieved, mitigating hence the harmful hazards of this phenomenon. As a result, a direct influence on the enhancement of agronomic productivity as well as management of civil engineering structures can be achieved.

4 Conclusions

An ECT process was effectively used to mitigate salinity hazard and reclaim saline affected soil. The proposed technique achieved 93, 63 and 15% removal rates for magnesium, chloride and sulfate ions, respectively. Increasing in process time allows better electrical transport, leading an optimal transport of ionic species. The removal of magnesium ions was greater than those of anionic salts, despite their fast ionic mobility. This is related to the chemical form of sulfate and chloride ions, forming complexes with cations present in interstitial pore fluid, decreasing their

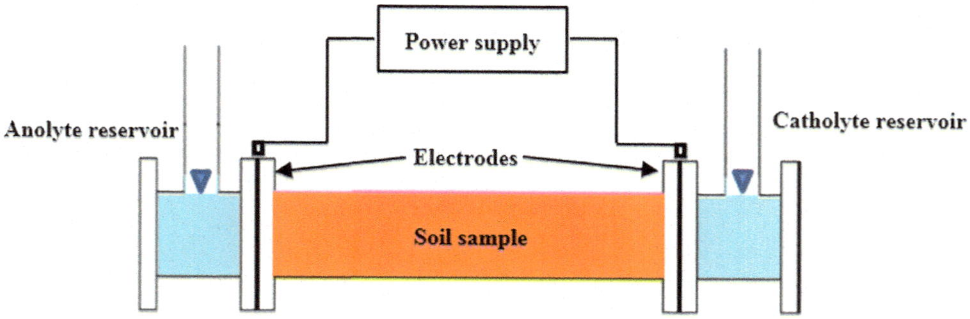

Fig. 1 Schematic diagram of the electrochemical cell

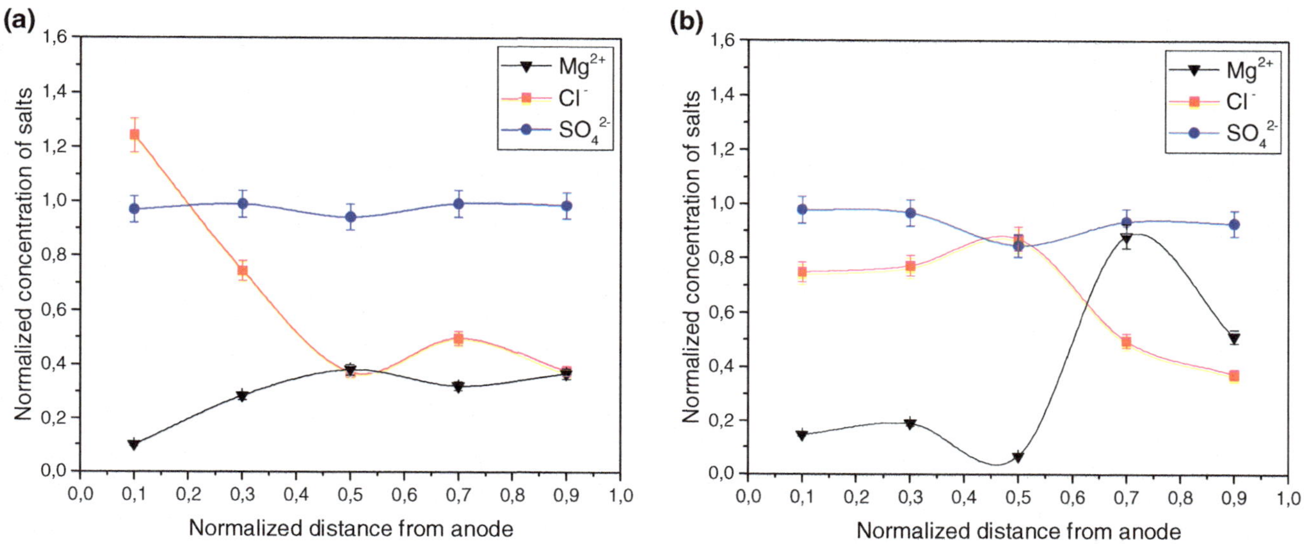

Fig. 2 Salts distribution at the end of the ECT. **a** 72 h, **b** 120 h

mobility, solubility and hindering their removal. These results lead to a net reduction in soil salinity, thus reducing the real hazard of soil salinity.

References

1. Cho, J.M., Park, S.Y., Baek, K.: Electrokinetic restoration of saline agricultural lands. J. Appl. Electrochem. **40**(6), 1085–1093 (2010)
2. Lukman, S., Mu'azu, N.D., Essa, M.H., Usman, A.: Optimal removal of cadmium from heavily contaminated saline-sodic soil using integrated electrokinetic adsorption technique. Arab. J. Sci. Eng. **40**, 1289–1297 (2015)
3. Khanamani, A., Fathizad, H., Karimi, H., Shojaei, S.: Assessing desertification by using soil indices. Arab. J. Geosci. **10**, 287 (2017)
4. Kim, K.J., Cho, J.M., Baek, K., Yang, J.S., Ko, S.H.: Electrokinetic removal of chloride and sodium from tidelands. J. Appl. Electrochem. **40**(6), 1139–1144 (2010)
5. Reddy, K., Cameselle, C.: Electrochemical Remediation Technologies for Polluted Soils, Sediments and Groundwater. Wiley, New York (2009)
6. Choi, J.H., Lee, Y.J., Lee, H.G., Ha, T.H., Bae, J.H.: Removal characteristics of salts of greenhouse in field test by in situ electrokinetic process. Electrochim. Acta **86**, 63–71 (2012)
7. Jayasekera, S.: Electrokinetics to modify strength characteristics of soft clayey soils: a laboratory based investigation. Electrochim. Acta **181**, 39–47 (2015)
8. Mosavat, N., Oh, E., Chai, G.: Laboratory assessment of kaolinite and bentonite under chemical electrokinetic treatment. J. Civil Environ. Eng. **3**, 1–7 (2013)
9. Acar, Y.B., Alshawabkeh, A.N.: Principles of electrokinetic remediation. Environ. Sci. Technol. **27**(13), 2638–2647 (1993)

Heated Blends of Phosphate Sludge: Thermal Transformation and Microstructure Characterization

Hajer Baccour and Samir Baklouti

Abstract

The mining industry faces many environmental challenges resulting from the huge amounts of waste generated by mines such as phosphate sludge. This waste deposited in the mine site is a potential source of pollution. The aim of this study was to valorize Tunisian phosphates sludge in ceramic manufacturing. To this end, the microstructure of heated blends was characterized using X-ray diffraction (XRD), inductively coupled plasma and atomic emission spectrometry (ICP-AES), scanning electron microscope (SEM), thermal analyses (DTA-TG) and dilatometry. Their ceramic properties: shrinkage, water absorption, density and compressive strength were followed as a function of heating temperature and kaolin content. The DTA curve shows an exothermic peak at 900 °C, which corresponds to the neoformation of the gehlenite phase. Moreover, fluor-apatite phase remains stable up to about 1100 °C. Based on their mineralogical and ceramic properties, heated blends of phosphate sludge may be a used in the ceramic industry. The addition of kaolin at different proportions with phosphate sludge improved the performance of samples sintered at 1100 °C and proved that the content of phosphate sludge in the mixture remained limited to 30% wt against 70% wt of kaolin.

Keywords

Ceramic • Phosphate • Sludge • Dilatometry • Pollution • Kaolin

H. Baccour (✉) · S. Baklouti
Laboratoire de Chimie Industrielle, Ecole Nationale D'Ingénieurs de Sfax, BP W 3000 Sfax, Tunisia
e-mail: hajerbaccour@yahoo.fr

1 Introduction

The washing of phosphate in the laundry of Kef Eddour (Metlaoui) generates a large amount of phosphate sludge, which contains, essentially apatite minerals associated with carbonate and silicate minerals. Moreover, the accumulation of this waste constitutes a potential source of pollution [1] and changes the landscape [2]. The awareness about the protection of the environment and waste management urges us to value the latter in building materials, particularly in ceramic products. Unfortunately, the reuse of phosphate sludge has been limited in Tunisia. Phosphate sludge consists of fluorapatite, quartz, carbonate and clay mineral [3]. Because the low contents of alumina and silica, phosphate sludge only is unable to afford a stable ceramic material [4]. It should be enhanced with alumino-silicates such as clay [5]. The physical properties of ceramics are related to the microstructure, which depends on the mineralogical and the chemical composition of the raw materials [3, 4]. Furthermore, the effects of the clay addition and heating temperatures on some physical ceramic properties were investigated.

2 Materials and Methods

The phosphate sludge (without flocculant) used in this work was taken from the storage ponds in Kef Eddour area located in the region of Metlaoui. The analytical methods were used during this study in order to characterize these samples. The ceramic samples were prepared in the form of pellets. Then they were shaped by uniaxial pressing using a hydraulic press with a pressure of 100 MPa. The pellets obtained with a mass of about 4 g are dried at 105 °C for 24 h to evaporate the residual moisture and then are heated at 900, 1000 and 1100 °C for up to 2 h on programmable electric furnace. The density of fired samples was determined with **AccuPyc II 1340 pycnometer**. The compressive strength was achieved with **LLOYD Instruments LR50 K**.

A. Kallel et al. (eds.), *Recent Advances in Geo-Environmental Engineering, Geomechanics and Geotechnics, and Geohazards*, Advances in Science, Technology & Innovation, https://doi.org/10.1007/978-3-030-01665-4_23

3 Results

3.1 Chemical Composition

For the preparation of the ceramic samples, a suspension in well-defined proportions of the phosphate sludge powder (BL) and/or kaolin and distilled water was prepared. Sodium tripolyphosphate (STPP) was added as a deflocculant. The composition of the studied mixtures of raw materials is given in Table 1.

The chemical analysis showed that phosphate sludge consisted essentially of SiO_2 (29.44wt%), CaO (25wt%), P_2O_5 (9wt%) and Al_2O_3 (6.21wt%). Also, it was revealed that the percentage of SiO_2 was relatively large, probably indicating the presence of free silica in the wash sludge. In addition, the high content of CaO suggests the presence of large quantities of carbonates and calcium minerals. This hypothesis was supported by the relatively high value of loss on ignition for phosphate sludge due to its elevated organic matter content [6]. Also, the significant weight percentage of phosphorus pentoxide (P_2O_5 = 9%) may reflect the existence of phosphate residues in the form of apatites such as fluorapatite. Furthermore, it was noted that the content of potassium alkalis (K_2O = 4.58%) was important and the existence of magnesium oxide (MgO = 1.77%) probably indicated the presence of dolomite ($MgCa(CO_3)_2$) [7] (Table 2).

3.2 X-Ray Diffraction

The diffractogramme of the phosphate sludge obtained by X-ray diffraction showed several crystalline mineralogical phases. Calcite ($CaCO_3$), carbonate-fluorapatite ($Ca_{9.35}$ $(PO_4)_{4.72}F_{1.98}$ $(CO_3)_{1.483}$) as major minerals. Also, the diffractogramme revealed the presence of quartz (SiO_2), zeolite heulandite type and dolomite (Ca Mg $(CO_3)_2$). These results are in agreement with those of the chemical analyses.

Table 1 Composition of raw studied mixtures blends, BL: phosphate sludge

	Mixtures
M0	100% Kaolin
M1	90% Kaolin +10% BL
M2	80% Kaolin + 20% BL
M3	70% Kaolin + 30% BL
M4	60% Kaolin + 40% BL
M5	50% Kaolin + 50% BL
M6	100% BL

3.3 Thermal Analysis

The DTA analysis indicated that the curve of sludge sample showed endothermic peaks at 622, 739 and 824 °C which are attributed to the decomposition of $CaCO_3$, $CaMg(CO_3)_2$ and $Ca_6Al_{2,40}Si_{6,60}O_{18}(H_2O)_{4,77}$, respectively [3]. The DTA curve also presented a pic exothermic at 900 °C, which corresponds to the neoformation process, particularly of the gehlenite [5]. The results of the dilatometric analysis of the phosphate sludge are shown in Fig. 1. These results show that: at 200 °C a slight expansion appeared, probably due to the elimination of residual water. From 200C up to 700 °C, the curve showed a slow and stable shrinkage probably related to the dehydroxylation of heulandite in addition to the expansion of grains. Between 750 and 900 °C a shrinkage due to an important phenomenon the decarbonation of the phosphate sludge powder accompanied by the gaseous releases of CO_2. Above 1000 °C, a start of shrinkage was observed which continued even at a temperature of around 1200 °C, probably due to the sintering phenomenon of the phosphate sludge. Thus this material sintered only at a temperature above 1100 °C [6].

X-ray diffraction patterns of raw and heated sludge (Fig. 1) indicated a gradual disappearance of carbonates as a function of temperature and achieved at 900 °C. The persistence of fluorapatite, this phase is therefore relatively stable at this temperature range. Fluorapatite persisted up to 1100 °C [5]. The appearance of a new phase crystallized gehlenite at the temperature 900 °C. Thermal analysis showed that carbonate was removed, but the amount of quartz and fluorapatite remained almost stable to about 1000 °C, then decreased with further heating. It was believed that ghlenite occured from the breakdown products of kaolin and carbonates derived lime.

Density increased as a function of temperature and was generally followed by a decrease in the total porosity of the samples (Fig. 2). From these results, it is noted that the compressive strength and the Young's modulus of all the prepared samples increased with the firing temperature. This increase was more remarkable from 1000 to 1100 °C. Thus, ceramic samples sintered at 1100 °C have the most important mechanical properties. This improvement of the mechanical properties is explained by the increase of the density and the decrease of the total porosity. For samples containing phosphate sludge content between 30% wt and 100% wt, the mechanical properties remained almost constant or decreased slightly.

Table 2 Chemical composition of phosphate sludge and pharmaceutic kaolin

	CaO	P_2O_5	SiO_2	C.org	Al_2O_3	Fe_2O_3	MgO	Na_2O	K_2O	LOI
BL	25	9	29.44	2.05	6.21	1.13	1.77	0.55	4.58	17.35
Kaolin	1.81	–	44.17	–	36.55	1.69	0.27	1.83	0.73	12

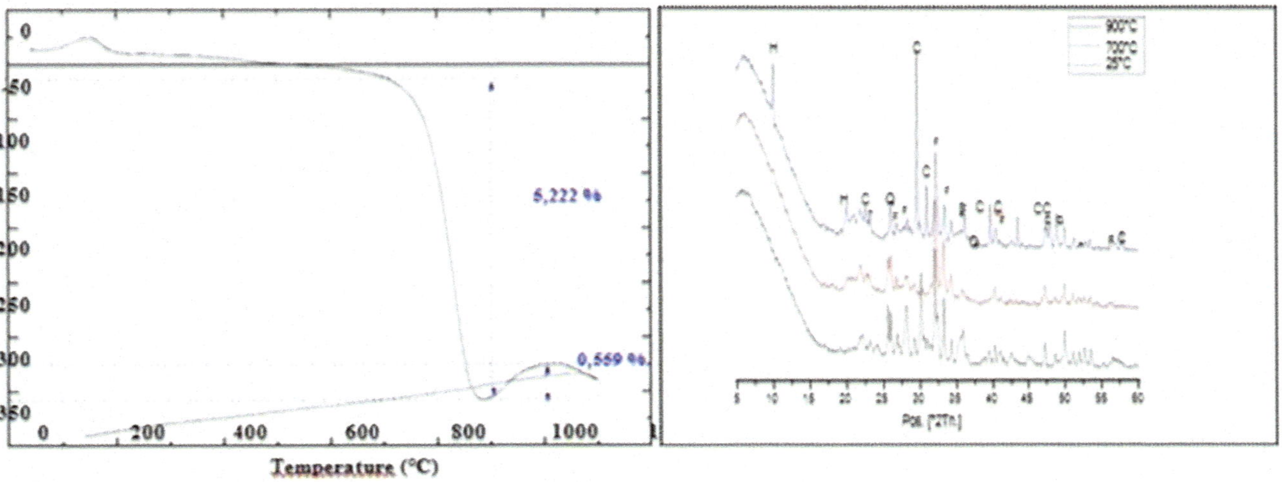

Fig. 1 Dilatometric curves of the phosphate sludge and X-ray diffraction patterns of heated phosphate sludge

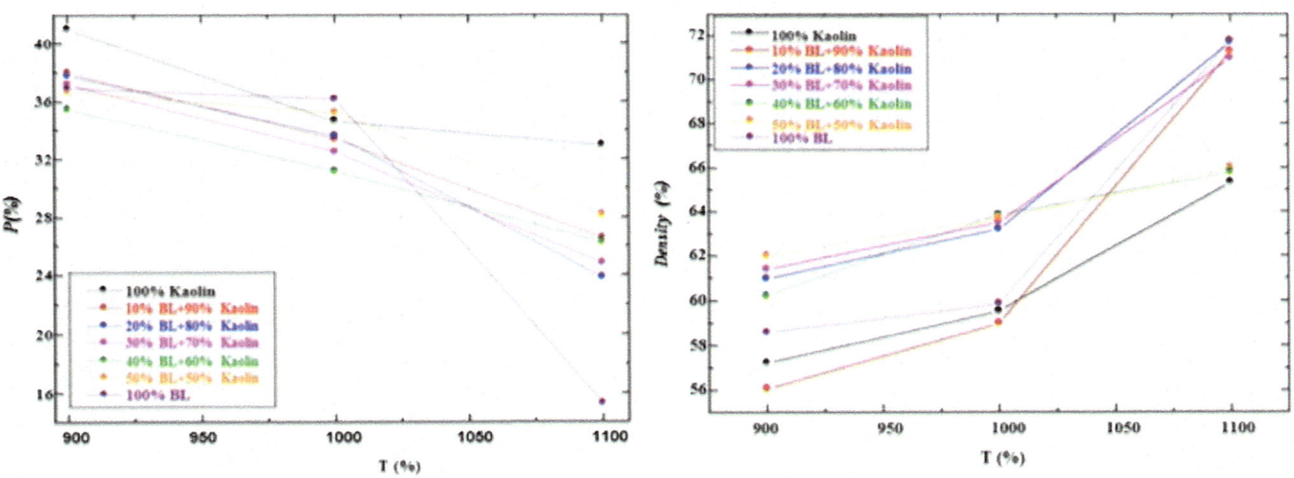

Fig. 2 Evolution of density and porosity as a function of heating temperature and kaolin addition

4 Conclusion

Phosphate sludge constitute a huge amount of pollutants, their valorization in building materials has become a necessity. The results show that the phosphate sludge alone can be used in ceramics since it allows obtaining samples sintered at 1100 °C with acceptable mechanical properties except for shrinkage which remains high. The use of kaolin with phosphate sludge improves the performance of samples sintered at 1100 °C provided that the content of BL in the mixture remains limited to 30% wt.

References

1. Sis, H.: Reagents used in the flotation of phosphate ores: a critical review. Miner. Eng. **16**, 577–585 (2003)
2. González, C.: Microstructure and mineralogy of lightweight aggregates manufactured from mining and industrial wastes. Constr. Build. Mater. **25**, 3591–3602 (2011)
3. Loutou, M.: Heated blends of phosphate waste: microstructure characterization, effects of processing factors and use as a phosphorus source for alfalfa growth. J. Environ. Manag. **177**, 169–176 (2016)
4. Fakhfakh, E.: Effects of sand addition on production of lightweight aggregates from tunisian smectite-rich clayey rocks. Appl. Clay Sci. **35**, 228–237 (2007)
5. Loutou, M.: Phosphate sludge: thermal transformation and use as lightweight aggregate material. J. Environ. Manag. **130**, 354–360 (2013)
6. Rachid, H.: Valorization of phosphate waste rocks and sludge from the Moroccan phosphate mines: challenges and perspectives. Procedia Eng. **138**, 110–118 (2016)
7. Wissem, G.: Beneficiation of phosphate solid coarse waste from Redayef (Gafsa Mining Basin) by grinding and flotation techniques. Procedia Eng. **138**, 85–94 (2016)

Geochemical of Fossil Diatoms and Its Utilization as Adsorbent in Water Treatment

Touina Amel, Chernai Safia, and Hamdi Boualem

Abstract

Diatomite is characterized by exceptional physicochemical properties and its many industrial applications. Using a variety of techniques in this study, we characterized the chemical, physicochemical and structural properties of a diatomaceous earth sampled in the deposit of sig Algeria. Natural diatomite was tested as a potential sorbent for the removal of heavy metal ions. The results show that the high sorption ability of diatomite makes the material a suitable candidate in the removal of heavy metals (Pb(II), Zn(II) and Cd(II)) from water.

Keywords

Diatomite • Geochemistry • Adsorbent • Waster • Treatment

1 Introduction

Diatomite is a sedimentary rock composed of the skeletal remains of single-cell aquatic plants called diatoms. The remarkably complex skeletons are unusual in that they are composed of natural silica with various amounts of impurities and a high degree of unique structure [1]. In Algeria, the principal deposits of diatomite belong to the category of the deposits of continental context, formed in lake media with lagoons of Miocene age (-10 AM). It is the acid volcanicity of this period which brought the silica stock favorable to the development of the diatoms: these deposits are located in the tertiary basins of miocene of the west of Algeria.

Diatomite's highly porous structure, good sorption ability, chemical inertness, low density and high surface area result in a number of industrial applications as filtration media for various organic and inorganic chemicals as well as an adsorbent [2]. It is insoluble in water, and extremely stable in the ambient conditions. A number of laboratory-scale studies were performed to investigate the potential application of diatomite as adsorbent in water treatment [3].

The present work reports on a variable geochemical and mineralogical composition of samples of diatomite deposit from Sig region in western of Algeria. The local diatomaceous earth (78.41% SiO_2) was tested for its potential use as a natural adsorbent for the removal of heavy metals as Pb (II), Zn(II) and Cd(II) ions in the water treatment.

2 Materials and Methods

The natural sample was collected from Sig deposit (Algeria). After dry state crushing and grinding of the material in a toothed roll crusher, a representative sample, designated ND of granulometry 40–100 μm was selected. This sample has already been described in previous papers [4]. Several techniques have been used to characterize diatomite (X-ray diffractometry, X-ray florescence, scanning electron microscopy, etc.). The adsorption of Pb(II), Zn(II) and Cd(II) on diatomite was investigated using batch technique at 25 °C.

3 Results

3.1 Deposit Site

In this research, samples of diatomite were obtained from the location of Sig (Mascara-Algeria) deposits for microscopic, mineralogical studies and/or chemical and physical analyses. The internal and external views of this deposit are shown in Figs. 1 and 2.

T. Amel · C. Safia · H. Boualem (✉)
LCVRM, ENSSMal, BP 19 Bois Des Cars, Dely Ibrahim, 16320 Alger, Algérie
e-mail: bhamdi_99@yahoo.fr

A. Kallel et al. (eds.), *Recent Advances in Geo-Environmental Engineering, Geomechanics and Geotechnics, and Geohazards*, Advances in Science, Technology & Innovation, https://doi.org/10.1007/978-3-030-01665-4_24

Fig. 1 An external view of Sig deposit

Fig. 2 An internal view of Sig deposit

3.2 Geochemical Composition

The geochemical composition analysis showed that the diatomite samples contain on average SiO_2 (78.41%), CaO (11.09%), Al_2O_3 (2.48), Fe_2O_3 (1.2), MgO (1.34), Na_2O (0.62), K_2O (0.59), TiO_2 (0.21) with trace metal element as Cu, Zn and Pb.

3.3 Mineralogy Phases

The X ray pattern revealed that the raw diatomite contains, in addition to amorphous, SiO_2 (quartz form and Tridimite), $CaCO_3$ in the calcite and Ankerite. It is important to note that diatomite contains more than 56% amorphous silica (Table 1) and does not contain any clay phase, contrary to many other diatomite layers from other origin.

3.4 SEM Microscopy

The observation by Scan Electronic Microscopy showed that the samples of the taken sediments were obviously very rich in diatoms. Their forms are often broken but their structures remain preserved very well. The centric form is the prevalent. The SEM images show that the material is characterized by a high porosity [5]. The key to the exceptional properties of diatomite lies more in the microstructure—each diatom is peppered with thousands of holes, usually of three distinct sizes, ranging from a few microns in diameter down to sub-micron diameters. The number and sizes of holes vary with the species. The combination of the natural silica composition, the overall structure of the diatom particles, and the network of holes in the structure are responsible for the unique properties of these multi-functional mineral products (Fig. 3).

3.5 Physicochemical Properties

The physicochemical properties of natural diatomite are summarized in Table 2. We notice that the natural product is characterized by a high porosity, a mesoporous surface, a basic pH related to the presence of calcium carbonate.

3.6 Adsorption Capacities

Figure 4 shows the adsorption kinetics (Fig. 4a) and adsorption isotherms (Fig. 4b) of the metal ions onto natural diatomite at 25 ± 1 °C and at pH 5.0. The results show that diatomite is characterized by a high capacity to eliminate metal pollutants.

Table 1 Mineralogy phases for diatomite sample

Phases	Formula	%
Amorphous silica	SiO_2	56
Crystalline silica	SiO_2	8
Calcite	$CaCO_3$	24
Ankerite	$Ca(Fe, Mg)(CO_3)_2$	7

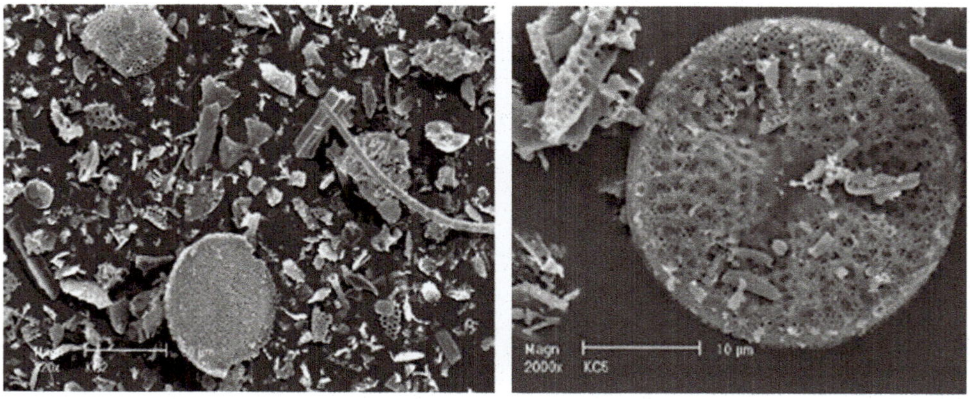

Fig. 3 SEM images of natural diatomite

Table 2 Physicochemical properties of natural diatomite

Parameters	Value
pH	8.25
Real density (g/cm^3)	2.24
Apparent density (g/cm^3)	0.57
Porosity (%)	74.55
Surface area (m^2/g)	23.27

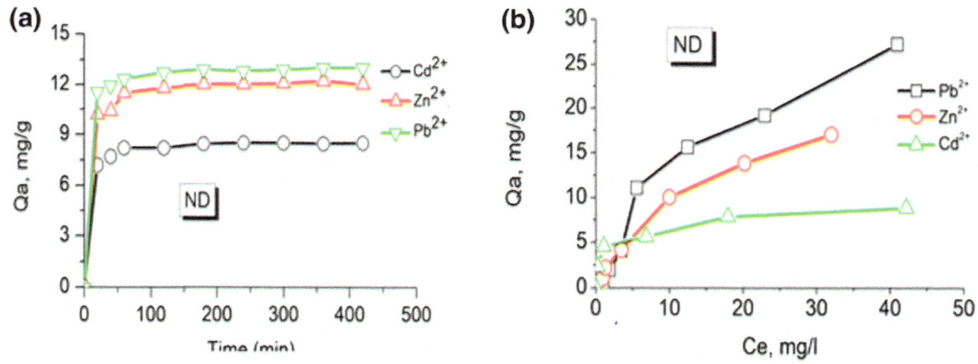

Fig. 4 Adsorption capacity of natural diatomite. **a** Adsorption kinetics. **b** Adsorption isotherms

4 Conclusions

In conclusion, the use of the diatomite samples as an adsorbent for the removal of Pb(II), Zn(II) and Cd(II) from heavy metal-contaminated water under various conditions gave excellent performances. The results demonstrated that the adsorption of heavy metals was strongly dependent of contact time, initial concentration, temperature and pH of the medium.

References

1. Yuan, B., Ding, S., Wang, D., Wang, G., Li, H.: Heat insulation properties of silica aerogel/glass fiber composites fabricated by press forming, Mater. Lett. **75**, 204–206 (2012)
2. Ivanov, S.É., Belyakov, A.V.: Diatomite and its applications. Glass Ceram. **65**(1–2), 18–21 (2008)
3. Sheng, G., Wang, S., Hua, J., Lua, Y., Li, j., Dong, Y., Wang, X.: Adsorption of Pb(II) on diatomite as affected via aqueous solution chemistry and temperature. Colloids Surf. A Physicochem. Eng. Aspects **339**, 159–166 (2009)
4. Hamdi, B., Kessaissia, Z., Donnet, J.B., Wang, T.K.: Analytical study of physico-chemical, textural and structural of Kieselguhr. Analusis J. **26**, 8–12 (1998)
5. Hadjar, H., Hamdi, B., Jaber, M., Brendlé, J., Kessaïssia, Z., Balard, H., Donnet, J.B.: Elaboration and characterisation of new mesoporous materials from diatomite and charcoal. Microporous Mesoporous Mater. **107**, 219–226 (2008)

Elaboration and Characterization of New Adsorbent Using Oil Shale Ash for Dyes Removal from Aqueous Solutions

Youssef Miyah, Anissa Lahrichi, Fatiha Mejbar, Anis Khalil, Meryem Idrissi, and Farid Zerrouq

Abstract

The aim of this work was the elaboration and characterization of a zeolite (OSZ) in an alkaline medium from the oil shale ash for the removal of the two dyes Crystal Violet (CV) and Red Congo (RC) in aqueous milieu. The adsorbent material quality (OSZ) prepared was studied by X-ray diffraction, scanning electron microscopy (SEM), transmission electron microscopy (TEM) coupled by EDX, Fourier transform infrared spectroscopy (FTIR) and fluorescence of X-rays spectroscopy (FRX). In addition, the adsorption in batch system of the two dyes CV and RC on the OSZ material was studied. Therefore, we accomplished a parametric study of the adsorption by studying the effect of several important parameters on the decolorizing power of the used material, namely, the contact time, the adsorbent dose, the pH solution and the temperature. The decolorization percentage was 97 and 81% for CV and RC dye respectively. Kinetic studies of dyes carried out the pseudo-second order at different dye concentration values. The Langmuir model describes satisfactory adsorption on the OSZ. Thermodynamic studies indicated that dye adsorption process by OSZ was spontaneous, physisorption and endothermic in nature.

Keywords

Zeolite • Oil shale ash • Isotherm • Kinetic • Thermodynamic • Adsorption

Y. Miyah (✉) · F. Mejbar · A. Khalil · M. Idrissi · F. Zerrouq
Laboratory of Catalysis, Materials and Environment, School of Technology, University Sidi Mohammed Ben Abdellah Fez, Fez, Morocco
e-mail: youssef.miyah@gmail.com

F. Zerrouq
e-mail: zerrouq1959@gmail.com

A. Lahrichi
Laboratory of Biochemistry, Faculty of Medicine and Pharmacy, University Sidi Mohammed Ben Abdellah Fez, Fez, Morocco

1 Introduction

The work described herein was undertaken in order to respond to a major environmental challenge that is water pollution and its impacts on receiving media, especially those arising from pollution caused by textile industry is the major concern for the developing countries [1]. The small and large scale operations in this textile sector cause hazardous effect to environment [2]. The textile manufacturing processes involves large consumption of water and various chemicals which will generate waste [3]. Wastewater from the textile industry is rated as the most polluting among all industrial sectors [4]. Dyes present even in low concentrations are highly visible [5]. They affect the photosynthetic activity of the aquatic life due to the reduced light penetration [6]. Dyes are known for their stability and recalcitrant ability, making them very dangerous to human health, because they have a carcinogenic, mutagenic, teratogenic, and toxic power [7]. Additionally it may also cause severe damage to human beings such as kidney dysfunction, reproductive system, liver brain and central nervous system damage [8].

Consequently, the development of an efficient and sustainable process to manage the pollution resulting from the dye industry has long been a considerable undertaking for environmental protection [9]. In this respect, the classical physico-chemical processes for dye removal from wastewater include oxidation, photochemical degradation, reverse osmosis, adsorption, membrane separation and coagulation/flocculation [10–12]. The removal of dyes in an economic way remains an important issue for researchers and environmentalists. The adsorption is an excellent alternative especially using inexpensive and a high adsorption capacity adsorbent without requiring any additional pre-treatment step before application. Adsorption is superior to other wastewater treatment techniques in terms of its initial cost, flexibility and simplicity of design, ease of operation and insensitivity to toxic pollutants [13]. Currently, several

A. Kallel et al. (eds.), *Recent Advances in Geo-Environmental Engineering, Geomechanics and Geotechnics, and Geohazards*, Advances in Science, Technology & Innovation, https://doi.org/10.1007/978-3-030-01665-4_25

studies have been performed on the possibility of adsorbents such as corn cob, sawdust, clay, fly ash, activated carbon, chitin and others [13, 14]. However, these adsorbents need to be improved in their adsorption capacities while other adsorbents are still under development [15].

In this work, we were interested in the study of the adsorption of the dyes on the oil shales taken of a layer of Timahdit, located in the mountains of the Average Atlas (Morocco) and spread out over a surface of 1000 km^2. The schist layer thickness varies from 100 to 150 m. Oil shale is be one of the most interesting energy and chemical sources in the world after the exhaustion of oil deposits [16]. Morocco is one of the world leading countries in terms of oil shale resources [17]. They are very promising and estimated at 86 billion tons in Tarfaya deposit, located in the south western part of Morocco within a 2000 km^2 area and in Timahdit deposit, located about 250 km southeast of the capital Rabat, at 18 billion tons within a 196 km^2 area. Mining and processing of the oil shale will significantly disturb the environment, as a result of pollution by dust particles and ash derived from the oil shale. Valorisation of these valuable by products will reduce the environmental impact and will make oil shale development economically feasible for energy production in the future [18].

The target of this study was, then, to develop adsorbent systems which are effective in the treatment of water and particularly for the removal of textile dyes. To this end, we were interested in the adsorption of dyes on a very abundant material such as the oil shale ash. The batch technique was used under a variety of conditions to produce quantitative adsorption, namely amount of adsorbent, dye concentration, contact time, pH solution and temperature. In addition, kinetic models were applied to experimental results such as the pseudo-first order and pseudo-second order model. The adsorption isotherm data were well described by the Langmuir and Freundlich models. Furthermore, the thermodynamic analysis suggests that the adsorption is a spontaneous process and endothermic in nature.

2 Materials and Methods

The sample of oil shale studied comes from the region of Timahdit (Morocco). It was washed with ultra pure water, centrifuged at 3000 rpm for 1 h and sieved at 63 μm. The obtained material was noted OSB. It was calcined at 950 °C for 5 h in the oven to remove organic matter and the obtained material is noted OSC. Adsorption measurements were carried out by mixing various oil shale ash quantities for eliminating two dyes Crystal Violet (CV) and Red Congo (RC) dyes potential in a series of beakers containing 200 mL of solution colored by different pHs using heating with numerical control and a magnetic stirrer (Stuart). Dye solutions were prepared using ultrapure water (MILLIPORE, direct-Q, UV3 with Pump) to prevent and minimize any possible interferences. Specimens were accomplished using a syringe filter of 0.45 μm diameter (Minisart, sartorium stedim biotech). The effect of several variables such as the adsorbent amount (0.5–2 g/L), the contact time (0–120 min), and the initial concentration while coloring (50–100 mg/L), the pH (4–12) and the temperature of the solution (20–50 °C) were all studied. After 20 min, the balance was established. At the end of adsorption experiments, the dye concentration was given by measuring the solution absorbance using a UV–visible spectrophotometer (Jasco V530). All experiments were carried-out in triplicate and the medium values were presented.

Hence, the dye removal percentage was calculated using this equation:

$$\%\ dye\ removal = \frac{C_0 - C_e}{C_0} \times 100 \quad (1)$$

where C_0 and C_e are the initial and final concentrations of mixture dye respectively (mg/L) in the solution.

3 Results and Discussion

The study of the kinetics of the removal of Crystal violet and Crystal Violet (CV) and Red Congo (RC) by the ashes of zeolitized oil shales in an aqueous medium shows that equilibrium time was very fast and can be in less than 10 min, it was also independent of the initial concentration.

The amount adsorbed by the OSZ adsorbent increases as a function of the initial CV and RC concentration (Table 1). This can be explained by the presence of a large number of molecules that diffuse to the sites on the surface of the adsorbent and therefore the partial adsorption depends on the initial concentration of the dye.

The temperature of the material is a parameter that significantly influences the amount of adsorption. The results obtained show that the amount of adsorption increases with the temperature in the studied range. This means that the interaction between the OSZ adsorbent and the two CV and RC adsorbents was of an endothermic nature.

To optimize the amount of OSZ added to the CV and RC solution, experiments were performed using 200 ml of the 100 mg L^{-1} dye solution and a dose of OSZ adsorbent, which results show that the adsorption rate increases as a function of the concentration of the adsorbent used. A concentration of 2 g L^{-1} of the ashes of zeolitized oil shales makes it possible to reach the maximum adsorption with a percentage of 97% and afterwards stabilization was observed, indicating a saturation of the pores of the adsorbent.

Table 1 Percentage by OSZ at different initial concentrations of the two CV and RC dyes

$[CV\ and\ RC]_0$ (mg L^{-1})		100	80	70	60	50
Percentage (%)	CV	97	94.2	90.1	86.3	76.8
	RC	81	70.5	50.1	41.3	38.1

pH is among the main variables that affect the adsorption process, influencing not only the adsorbent surface charge, the degree of ionization of the material present in the solution and the dissociation of functional groups at the active sites of the adsorbent. Adsorbent, but also the chemical properties of the dye solution. In this case, the pH must control the adsorption of the dyes on the particles in the suspension, because the adsorbed molecules and the adsorbent particles can be affected at the level of their functional groups by the concentration of H^+ ions involved in the molecular adsorption process. The corresponding parameters which also present the correlation coefficients R^2. These calculated coefficients are closer to the kinetic model of the pseudo-second order. This indicates that this model describes well the experimental results of the adsorption of the two dyes CV and RC.

4 Conclusion

The present study proved that CV can be removed from dye bearing effluent in an eco-friendly way using OSZ. In this respect, the adsorption was rapid and could be considered to fit pseudo-second order kinetics model in the case of CV and RC was increased with the increasing initial concentration of dye solution. The removal percentage of the cationic CV dye was greater than that of the anionic dye RC which indicates that our materials OSZ and are effective for the removal of cationic dyes. The Langmuir model described satisfactorily the adsorption on OSZ. Finally, the thermodynamic studies indicated that the dye adsorption process by oil shale ash was physisorption and endothermic in nature.

References

1. Pal, A., He, Y., Jekel, M., Reinhard, M., Gin, K.Y.H.: Emerging contaminants of public health significance as water quality indicator compounds in the urban water cycle. Environ. Int. **71**, 46–62 (2014)
2. Wojciech, K., Cendrowski, K., Bazarko, G., Mijowska, E.: Study on efficient removal of anionic, cationic and nonionic dyes from aqueous solutions by means of mesoporous carbon nanospheres with empty cavity. Chem. Eng. Res. Des. **94**, 242–253 (2015)
3. Haik, Y., Qadri, S., Ganoe, A., Ashraf, S., Sawafta, R.: Phase change material for efficient removal of crystal violet dye. J. Hazard. Mater. **176**, 1110–1112 (2010)
4. Heibati, B., Rodriguez-Couto, S., Al-Ghouti, M.A., Asif, M., Tyagi, I., Agarwal, S., Gupta, V.K.: Kinetics and thermodynamics of enhanced adsorption of the dye AR 18 using activated carbons prepared from walnut and poplar woods. J. Mol. Liq. **208**, 99–105 (2015)
5. Mittal, A., Mittal, J., Malviya, A., Kaur, D., Gupta, V.K.: Adsorption of hazardous dye crystal violet from wastewater by waste materials. J. Colloid Interface Sci. **343**, 463–473 (2010)
6. Miyah, Y., Lahrichi, A., Idrissi, M., Boujraf, S., Taouda, H., Zerrouq, F.: Assessment of adsorption kinetics for removal potential of Crystal Violet dye from aqueous solutions using Moroccan pyrophyllite. J. Assoc. Arab Univ. Basic Appl. Sci. **23**, 20–28 (2017)
7. Mohamad Amran, M.S., Dalia, K.M., Wan, A.W.A.K., Azni, I.: Cationic and anionic dye adsorption by agricultural solid wastes: a comprehensive review. Desalination **280**, 1–13 (2011)
8. Miyah, Y., Idrissi, M., Zerrouq, F.: Study and modeling of the kinetics Methylene blue adsorption on the clay adsorbents (Pyrophillite, Calcite). J. Mater. Environ. Sci. **6**(3), 699–712 (2015)
9. Miyah, Y., Lahrichi, A., Idrissi, M.: Removal of cationic dye – Methylene blue– from aqueous solution by adsorption onto corn cob powder calcined. J. Mater. Environ. Sci. **7**(1), 96–104 (2016)
10. Duta, A., Visa, M.: Simultaneous removal of two industrial dyes by adsorption and photocatalysis on a fly-ash–TiO_2 composite. J. Photochem. Photobiol. A Chem. **306**, 21–30 (2015)
11. Gupta, V.K., Suhas: Application of low-cost adsorbents for dye removal—a review. J. Environ. Manage. **90**, 2313–2342 (2009)
12. Mohamad Amran, M.S., Dalia, K.M., Wan, A.W.A.K., Azni, I.: Cationic and anionic dye adsorption by agricultural solid wastes: a comprehensive review. Desalination **280**, 1–13 (2011)
13. El Haddad, M., Slimani, R., Mamouni, R., ElAntri, S., Lazar, S.: Removal of two textile dyes from aqueous solutions onto calcined bones. J. Assoc. Arab Univ. Basic Appl. Sci. **14**, 51–59 (2013)
14. Idrissi, M., Miyah, Y., Chaouch, M., El Ouali Lalami, A., Lairini, S., Nenov, V., Zerrouq, F.: CWPO of phenol using manganese-based catalysts. J. Mater. Environ. Sci. **5**(S1), 2309–2313 (2014)
15. Cao, J.S., Lin, J.X., Fang, F., Zhang, M.T., Hu, Z.R.: A new absorbent by modifying walnut shell for the removal of anionic dye: kinetic and thermodynamic studies. Bioresour. Technol. **163**, 199–205 (2014)
16. Dahri, M.K., Kooh, M.R.R., Lim, B.L.: Water remediation using low cost adsorbent walnut shell for removal of malachite green: equilibrium, kinetics, thermodynamic and regeneration studies. J. Environ. Chem. Eng. **2**, 1434–1444 (2014)
17. Saoiabi, A., Doukkali, A., Hamad, M., Zrineh, A., Ferhat, M., Debyser, Y.: Timahdit (Morocco) oil shales: composition and physicochemical properties. Comptes Rendus de l'Académie des Sciences-series IIC-chemistry; Paris, Chimie/Chemistry **5**(4), 351–360 (2001)
18. Miyah, Y., Idrissi, M., Lahrichi, A., Zerrouq, F.: Removal of a cationic dye—Méthylène Bleu—from aqueous solution by adsorption onto oil shale ash of Timahdit (Morocco). Int. J. Innov. Res. Sci. Eng. Technol. **8**(3), 15600–15613 (2014)

Biochars Induced Changes in the Physicochemical Characteristics of Technosols: Effects of Feedstock and Pyrolysis Temperature

Manel Kammoun-Rigane, Hajer Hlel, and Khaled Medhioub

Abstract

Soils in urban areas are anthropized and unfavorable for agricultural practices. Constructed Technosols (TS) are mainly manmade and easily manipulated by mixing different proportions of anthropogenic materials. To achieve this study, twelve types of biochars were produced at 350 and 450 °C, using olives husks (OH), almond shells (AS) and coffee husk (CH) to improve Sandy soil (SS) quality during ten weeks. The main aim of this work was to assess the short-term impact of biochar addition on soil properties. An open-air pot experiment was designed in a pilot scale adding biochars to SS at the rates of 5%. The results showed that the addition of biochars, in the beginning, improved soil Organic matter (OM) and N contents and there were no significant effects on the pH and EC values in all Technosols compared to control (SS). The OM contents and total N remained stable during the incubation for Technosol treated with biochars produced at 350 °C indicating high recalcitrant indices. However, the OM and N contents decreased for the lower recalcitrance materials prepared at 450 °C. The pH values remained stable and EC increased in soil with all biochars. Thus, the use of biochar contributed to soil carbon sequestration and improved OM content in Technosols.

Keywords

Constructed technosols • Urban area • Soil quality • Biochars • C recalcitrance

M. Kammoun-Rigane (✉) · H. Hlel
Faculty of Sciences, Sfax University, 3000 Sfax, Tunisia
e-mail: manelriganekamoun@yahoo.fr

K. Medhioub
IPEIS, Sfax University, 3000 Sfax, Tunisia

1 Introduction

Human activities such as industrialization, urbanization and intensive agriculture, cause soil degradation. Construction of Technosols, an innovative approach, consists of mixing soils with recycled waste and industrial by-products. This process presents quadruple benefit as it (i) treats degraded lands, (ii) valorizes by-products and wastes, (iii) stocks organic carbon in soil and (iv) creates a substrate with selected physical, chemical and biological properties. The addition of composts and biochar to reconstituted soils was proposed as a potential method for improving soil quality [1, 2]. The beneficial effects of biochar are associated to their impacts on the physical, chemical and biological characteristics of soils [3]. Considering all these effects, biochar addition to the soil can enhance plant productivity and crops performance [4]. The objectives of this study were to compare changes in the physicochemical characteristics of technosols prepared by mixing biochars (provided from different feedstocks and prepared at tow pyrolysis temperatures) with Sandy soil. This kind of soil is characterized by low OM and nutrients contents, very extended in the south of Tunisia and considered without agronomic values.

2 Materials and Methods

Twelve types of biochars (B) were produced at 350 and 450 °C using four types of olives husks (OH), almonds shells (AS) and coffee husk (CH) to improve Sandy soil (SS) quality during ten weeks. OH originated from different kind of olive oil industries (OHI, OH1, OH2, OH3). Pyrolysis was achieved using four production techniques under oxygen-limited conditions. Characterizations of biochars indicated that pH was neutral for those produced at 350 °C and basic ranging from 8.3 to 10.3 for biochars prepared at 450 °C. All biochars were characterized by higher EC values than 4.5 mS/cm and may be a source of N

A. Kallel et al. (eds.), *Recent Advances in Geo-Environmental Engineering, Geomechanics and Geotechnics, and Geohazards*, Advances in Science, Technology & Innovation, https://doi.org/10.1007/978-3-030-01665-4_26

Table 1 Abbreviation used for all mixtures

SS + B 350 °C	SS + B 450 °C
TS1 (i, f): SS + OHI	TS7 (i, f): SS + OHI
TS2 (i, f): SS + OH1	TS8 (i, f): SS + OH1
TS3 (i, f): SS + OH2	TS9 (i, f): SS + OH2
TS4 (i, f): SS + OH3	TS10 (i, f): SS + OH3
TS5 (i, f): SS + CH	TS11 (i, f): SS + CH
TS5 (i, f): SS + AS	TS12 (i, f): SS + AS

Sandy Soil (SS): Control, in the beginning of the experiment (i), At the end of the experiment (f)

and recalcitrant carbon. An open-air pot experiment was conducted during ten weeks in order to study the impact of biochar addition on soil properties. Table 1 presents the abbreviation used for all mixtures. Constructed Technosols prepared at the ratio of B/SS = 5/95% (V/V) were placed in pots in open-air condition and wetted twice a week. All mixtures were performed in three replications and SS was used as control.

The physicochemical analyses were conducted according to NFT 01-013 NFT-012 for pH and NFT-90-031 for electric conductivity (EC mS/cm). The OM and TN contents were evaluated following the methods JIS K0102.14.4, (1995) and ISO 11261 (1995), respectively. The total organic carbon was calculated using the formula:

TOC% = OM%/1724.

All the analysis were conducted in triplicate and the mean values are presented in Table 2.

3 Results and Discussion

The characteristics of the SS used as control and TS in the beginning and at the end of the experiment are presented in Table 2. We evaluated the effects of different kinds of biochar on soil quality during ten weeks. The pH of SS was neutral (7.3), the EC value was 1.1 mS/cm, TN, OM and C content were low. The biochar application had no significant effect in terms of pH values and EC in all mixtures in the beginning of the trial (ranging from 7.2 to 7.5 and 1.1 to 1.53 mS/cm respectively) despite the alkaline pH and the high EC values of biochars. At the end of the experiment, pH values were reduced slightly and ranged from 7.7 to 6.8. The highest reduction was observed for TS1 and TS10 (0.6 and 0.5 unit pH, respectively). These values led to improve the bioavailability of important nutrients such as phosphorus. Addition of 5% biochar, produced at 450 °C to the Technosol increased the EC values in all mixtures and the highest was observed in TS7, 9, 10, 11 and 12 (less than 1.5 mS/cm in the beginning and ranged from 4.65 and 4.95 mS/cm at the end).

Various salts were found in the ash fraction of biochar causing a deterioration or loss of one or more soil functions. Chintala et al. [5] attributed the increment of EC to the release of weakly bounded nutrients (cations and anions) of biochars into the soil solution. However, higher soluble salts related to the high EC values should cause soil salinization. No significant effect was observed for the TN compared to control plots. Biochar additions increased TS organic matter (SOM) and carbon (SOC) contents. An increment of SOM contents was recorded for all TS by 100–700% comparing to SS. The highest SOC contents were observed in TS9 and TS10 (4.24 and 3.02% respectively) prepared using biochar of OH pyrolysis at 450 °C. At the end of our experiments, we observed a decrease of SOM mainly for TS5, TS7 TS9, TS10 (from 3.03 to 1.99; 3.24 to 1.94; 7.4 to 2.54; and 2.64 to 2.1 respectively) as observed by Liang et al. [6] which is associated to the mineralization of organic matter released from biochar. Inversely, a SOM conservation was observed for TS1, TS3, TS6, TS8 and TS12. Zimmerman et al. [7] observed similar results and explained the phenomenon by adsorption due to the affinity of biochar surfaces and pores for SOM and protect them from being decomposed. The TS2, TS4 and TS11 showed a SOM stabilization as observed in Cross and Sohi [8] who found an inhibition of SOM decomposition during biochar incubation experiments with two different soils.

4 Conclusion

In our study, we used biochars prepared by pyrolysis at different temperatures of olive oil industries by-products, almonds shells and coffee husk to improve sandy soil quality during a short-term experiment. The prepared Technosols were placed in pots in open air conditions. In conclusion, the incorporation of these highly carbonaceous biochar materials can induce ameliorations mainly in the SOM contents of Sandy soil and stabilize pH at neutral value. Besides, biochars had no significant effect on TN contents. The liming potential of biochars produced at 450 °C can be attributed to the high EC causing soil salinization. The creation of Technosols with biochar is an effective strategy to sequester

Table 2 The influence of biochars on Technosol properties

		pH	EC (mS/cm)	TN (%)	OM (%)	TOC (%)
SS		7.3	1.13	0.03	0.99	0.57
TS1	i	7.4	1.10	0.07	1.96	1.13
	f	6.8	1.93	0	4.24	2.55
TS2	i	7.2	1.10	0.04	2.3	1.32
	f	7	1.57	0.05	1.99	1.15
TS3	i	7.2	1.16	0	2.10	1.20
	f	7.1	2.02	0.02	3.13	1.82
TS4	i	7.2	1.27	0.05	2.84	1.63
	f	7.2	1.90	0.03	2.54	1.47
TS5	i	7.2	1.10	0.02	3.03	1.74
	f	7.2	2.06	0.05	1.99	1.15
TS6	i	7.5	1.30	0.02	2.29	1.31
	f	7.2	1.99	0.03	3.38	1.96
TS7	i	7.4	1.27	0.03	3.24	1.86
	f	7.4	4.65	0.04	1.94	1.12
TS8	i	7.3	1.53	0.02	2.04	1.17
	f	7.7	1.80	0.02	4.74	2.75
TS9	i	7.4	1.29	0.03	7.4	4.24
	f	7.4	4.84	0	2.54	1.47
TS10	i	7.5	1.26	0.03	5.26	3.02
	f	7	4.71	0	2.64	1.53
TS11	i	7.4	1.31	0.08	2.10	1.20
	f	7.3	4.89	0	2.44	1.41
TS12	i	7.5	1.19	0.1	1.70	0.98
	f	7.4	4.95	0	3.38	1.96

carbon in soil and improve soil structure. This study should be completed by increasing the ratio of biochars mainly for Technosols prepared for use in urban areas and needing high OM and nutrient contents.

References

1. Kammoun Rigane, M., Medhioub, K.: Assessment of properties of Tunisian agricultural waste composts: application as components in reconstituted anthropic soils and their effects on tomato yield and quality. Resour. Conserv. Recycl. **55**(8), 785–792 (2011)
2. Jones, D.L., Rousk, J., Edwards-Jones, G., DeLuca, T.H.: Murphy, biochar-mediated changes in soil quality and plant growth in a three year field tria. l. Soil Biol. Biochem. **45**, 113–124 (2012)
3. Jeffery, S., Verheijen, F.G.A., van der Velde, M., Bastos, A.C.: A quantitative review of the effects of biochar application to soils on crop productivity using meta-analysis. Agric. Ecosyst. Environ. **144**, 175–187 (2011)
4. Graber, E.R., Harel, Y.M., Kolton, M., et al.: Biochar impact on development and productivity of pepper and tomato grown in fertigated soilless media. Plant Soil **337**, 481–496 (2010)
5. Chintala, R., Schumacher, T.E., McDonald, L.M., et al.: Phosphorus sorption and availability from biochars and soil biochar mixtures. CLEAN Soil Air Water **41**, 1–9 (2013)
6. Liang, B., Lehmann, J., Sohi, S.P., Thies, J.E., O'Neill, B., et al.: Black carbon affects the cycling of non-black carbon in soil. Org. Geochem. **41**, 206–213 (2010)
7. Zimmerman, A.R., Gao, M., Ahn, M.Y.: Positive and negative carbon mineralization priming effects among a variety of biochar-amended soils. Soil Bio. Biogeochem. **43**, 1169–1179 (2011)
8. Cross, A., Sohi, S.P.: The priming potential of biochar products in relation to labile carbon contents and soil organic matter status. Soil Biol. Biochem. **43**, 2127–2134 (2011)

Short-Term Effects of Sewage Sludge Compost Application on Some Chemical Properties of Sandy Soil

Houda Oueriemmi, Kaouther Ardhaoui, and Mohamed Moussa

Abstract

The application of sewage sludge compost in agricultural soils has a dual interest. This allows preserving nature and improving the soil quality by supplying plant nutrients. A field experiment was carried out to evaluate the effect of different amounts of sewage sludge compost (0, 20, 40 and 60 t ha^{-1}) on chemical soil properties. The considered soil depths are (0–20; 20–40 and 40–60 cm). OM, N, P, K, pH, CE determination followed amendment with sewage sludge compost. The results indicated that the evaluated chemical soil parameters were changed by the application of this biosolid. In addition, sludge application increased the organic status of the soil and nutrients.

Keywords

Reuse • Sewage sludge compost • Soil fertility • pH • Electrical conductivity

1 Introduction

Sewage sludge is a concentrated suspension of solids that has been produced during the treatment of urban waste water. Increasing costs of commercial fertilizers, shortages in organic matter content and the generation of high volumes of sewage sludge have made the re-use of this biowaste in agriculture an alternative solution. Sewage sludge compost is a beneficial soil amendment, it contains mainly organic matter and macronutrients (N, P, K) that are essential for plant growth. The application of this biosolid can contribute to the improvement of the soil chemical (EC, pH, CEC), biological (microbial biomass, nitrogen cycle) and physical (water retention, aggregate stability, bulk density) properties [7, 8]. In this present work, we assessed the land application of sewage sludge compost on the soil's chemical properties. The experiment was conducted using a randomized complete block design with three replications. Therefore, three levels of sewage sludge compost (20, 40 and 60 t ha^{-1}) have been applied to the field.

2 Materials and Methods

This research was gegun in September 2017 at the field experiment of the Institute of Arid Regions of Medenine at El Fje, located at 33° 16′ 21″ North latitude and 10° 19′ 30″ East longitude. The site is characterized by a semi-arid to arid climate with a mean annual precipitation of 150 mm and an average temperature of 18–20 °C.

The used soil had a sandy texture. The sewage sludge compost was prepared at the pilot composting station of Medenine. It was produced under aerobic conditions by a mixture of garden waste and an activated sewage sludge collected from the wastewater treatment plant of Medenine. The main characteristics of sewage sludge compost used in this study are shown in Table 1.

The parcel design was a randomized complete block, consisting of an untreated plot (control) and three rates of sewage sludge compost (20, 40 and 60 t ha^{-1}) and three blocks. Each block consisted of four plots, and each plot had a total area of 1×1 m^2. After being air-dried, the soil sample was crushed and passed through a 2-mm sieve.

Soil pH was analyzed with a pH electrode in the saturated past extract (EC). The electrical conductivity was measured using a conductimeter. Soil organic matter was measured by rapid oxidation and subsequent titration with ferrous ammonium sulphate [9]. Total nitrogen (TN) in all samples

H. Oueriemmi (✉) · K. Ardhaoui · M. Moussa
IRA, Medenine, Tunisia
e-mail: houdaoueriemmi@outlook.com

K. Ardhaoui
e-mail: ardhaouikaouther@gmail.com

M. Moussa
e-mail: mohamed.moussa@ira.rnrt.tn

K. Ardhaoui
ISBAM, Gabes University, Gabes, Tunisia

A. Kallel et al. (eds.), *Recent Advances in Geo-Environmental Engineering, Geomechanics and Geotechnics, and Geohazards*, Advances in Science, Technology & Innovation, https://doi.org/10.1007/978-3-030-01665-4_27

Table 1 Main characteristics of sewage sludge compost used in this study

Characteristics	Sewage sludge compost
pH	6.75
CE (mS cm^{-1})	11.46
Sodium (%)	0.45
Potassium (%)	0.47
Calcium (%)	1.60
Magnesium (%)	0.59
Copper (%)	0.008
Zinc (%)	0.02
Iron (%)	0.30
Manganese (%)	0.006

was determined using the Kjeldahl's method [1] by attacking the matter at high temperatures using concentrated sulphuric acid in the presence of a catalyser. The available phosphorus (P) was measured by Olsen method [6].

3 Results

3.1 Effect of Sewage Sludge Compost Application on the Percentage of Organic Matter

The percentage of organic matter showed a decrease with soil depth (Fig. 1). The results indicate that the addition of sewage sludge compost (40 t ha^{-1}) causes an increase in the content of OM in soil surface, it is about 0.80%.

In addition, the decrease in OM content with depth reflects its decomposition and rapid mineralization by the microorganisms.

3.2 Effect of Sewage Sludge Compost Application on Macronutrient Concentrations in Soil

The macronutrient contents (N, P, and K) increased by adding biosolids such as sewage sludge compost (Fig. 2).

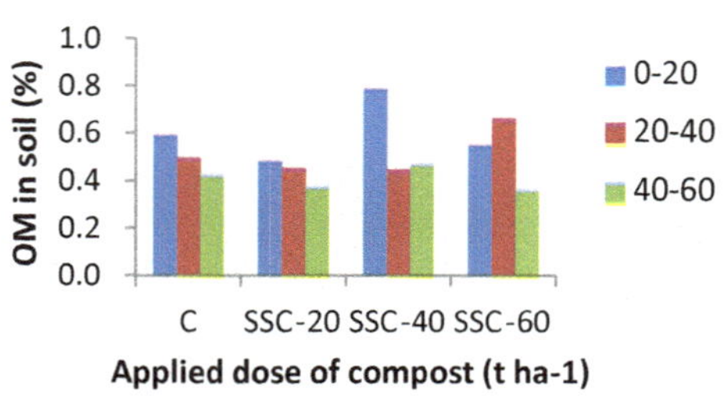

Fig. 1 Influence of sewage sludge compost application on organic matter content

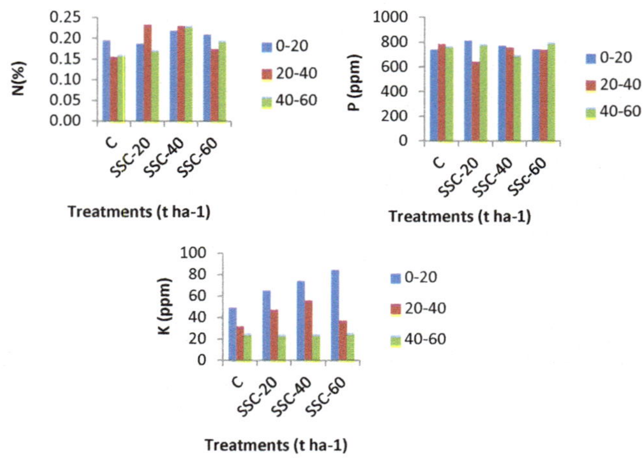

Fig. 2 Effect of sewage sludge compost on N, P and K contents in soil

3.3 Effect of Sewage Sludge Compost Application on pH and Soil Salinity

As first observation, the pH profiles showed that soil is generally neutral. In fact, pH values at 0–20 depth were varying from 7.50 to 7.38 for CSS-20 and SSC-60, respectively. Furthermore, the soil electrical conductivity increased with sewage sludge application. The highest (EC) value was found with the medium sewage sludge compost application dose (40 t ha^{-1}). In addition, these results showed that the measured EC increased with depth. For instance, EC reached 5.38 mS/cm and 5.31 mS/cm in the treatments SSC-20 and CSS-60, respectively at 40 and 60 cm depths.

4 Discussion

Our study showed that soil fertility parameters such as OM, N, P and K were improved after short-term application of sewage sludge compost. This increase can be explained by the fact that sewage sludge compost generally contains a high amount of nutrients. Our investigation is in agreement with the previous works obtained by Latare and Walter [4, 10]. In fact, they stated that sewage sludge application had a positive impact on soil fertility.

However, it was noted that organic matter and total nitrogen in soil did not increase significantly with sewage sludge compost treatments, possibly because of the lower biosolid rate [3].

Considering the soil salinity reflected by (EC), we remarked a significant increase of this salinity in the treated soil. This is in line with the results of Mendoza et al. [5]. In addition, Courtney and Mullen [2] suggested that potassium and sodium are the main contributors to this increase of EC.

5 Conclusions

Land application of sewage sludge compost has become a widespread practice in Tunisia. The use of this compost could improve soil fertility mainly by increasing pH, total nitrogen, organic matter and phosphorus amounts in the sandy soil situared in an arid area.

References

1. Bremner, J.M., Mulvaney, C.S.: Nitrogen—total. In: Page, A.L. (ed.) Methods of soil analysis, Part 2, pp. 595–624. American Society of Agronomy and Soil Science Society of America, Madison (1982)
2. Courtney, R.G., Mullen, G.J.: Soil quality and barley growth as influenced by the land application of two compost types. Bioresour. Technol. **99**(8), 2913–2918 (2008)
3. Garrido, S., Del Campo, G.M., Esteller, M.V., Vaca, R., Lugo, J.: Heavy metals in soil treated with sewage sludge composting, their effect on yield and uptake of broad bean seeds (*Vicia faba* L.). Water Air Soil Pollut. **166**, 303 (2005)
4. Latare, A.M., Omkar, K., Singh, S.K., Gupta, A.: Direct and residual effect of sewage sludge on yield, heavy metals content and soil fertility under rice–wheat system. Ecol. Eng. **69**, 17–24 (2014)
5. Mendoza, J., Tatiana, G., Gabriela, C., Nilsa, S.M.: Metal availability and uptake by sorghum plants grown in soils amended with sludge from different treatments. Chemosphere **65**, 2304–2312 (2006)
6. Olsen, S.R., Cole, C.V., Watanabe, F.S., Dean, L.A.: Estimation of available phosphorus in soils by extraction with sodium bicarbonate. U. S. Department of Agriculture Circular No. 939. In: Banderis, A.D., Barter, D.H., Anderson, K. (eds.) Agricultural and Advisor (1954)
7. Poulsen, P.H.B., Magid, J., Luxhoi, J., De Neergaard, A.: Effects of fertilization with urban and agricultural organic wastes in a field trial-waste imprint on soil microbial activity. Soil Biol. Biochem. **57**, 794–802 (2013)
8. Roig, N., Nadal, M., Sierra, J., Ginebreda, A., Schuhmacher, M., Domingo, J.L.: Novel approach for assessing heavy metal pollution and ecotoxicological status of rivers by means of passive sampling method. Environ. Int. **37**, 671–677 (2011)
9. Walkley, A., Black, I.A.: An examination method for determination for role organic matter and proposed modification of the chromic acid titration method. Soil Sci. **37**, 29–38 (1934)
10. Walter, I., Cuevas, G., Garcia, S., Martinez, F.: Biosolid effects on soil and native plant production in a degraded semiarid ecosystem in central Spain. Waste Manag. Res. **18**, 259 (2000)

Part IV

Geomechanics and Geotechnics: Characterization of Geomaterials

Generic Classification of Hoars in the Northeastern Part of Bengal Basin, Bangladesh

Mohammed Masum and Mohammad Omer Faruk Khan

Abstract

The northeastern part of Bangladesh which is also the northeastern part of the Bengal Basin is an area of interest for many disciplines for its unique geologic, tectonic, structural, ecologic, environmental and socio-economic aspects. This part of the country is important for its richness in wetland resources locally termed as haors. The current study analyzed the wetlands from a generic aspect and attempted to classify them according to the dominant tectonic and structural influence. It is expected that such a study would reveal new insights about these haors.

Keywords

Haor • Basin • Tectonic and generic

1 Introduction

The northeastern part of Bangladesh which is under the jurisdiction of the Sylhet Division (Fig. 1) is well known for its richness in wetland resources locally termed as haors. The areas underlined by haors are generally discussed under a unique framework such that the term 'haor basin' is widely used for this part of the country. However, while there is a very little visible difference among the haors being enormous sweet water bodies, geological factors such as tectonic settings and structural conditions reveal that the haor basin can be broadly subdivided into two categories and their mode of origin may vary accordingly. In Bangladesh, the local term haor is applied to a bowl-shaped depression that goes under water for about half a year during the wet period. The term haor is a corrupted form of its Bengali word 'Sagar' or Arabic word 'Bahar' meaning 'Sea' [4]. However, during the rest of the year i.e. the dry period, the haors get dried up but only some localized places still retain some water which is termed as 'beels' (shallow water body).The haor areas are important from socio-bio-geo-economic point of views. There are more than four hundred haors in the area which occupy more than 6000 km^2 of land [6]. The haors span over the districts of Sunamganj, Kishoreganj, Netrokona, Sylhet, Habiganj, Moulvibazar and B. Baria [1]. Most of the people that live by the haor area are poor. The people in the haor areas have no or little access to the basic services compared to that of the people in the mainland [3]. Poverty is mainly caused by natural disasters such as flood, flash flood, drought etc. The biodiversity of haor areas are ample and many rare and endangered species of birds, reptiles and mammals thrive in the areas. Swamp forests are also very important natural resource that contributes significantly to the uniqueness of haor areas. The landscapes of haor areas are also very prospective for tourism development [5]. A generic classification is a geoscientific approach to characterize the areas dominated by haors depending on their mode of formation. This approach should give more insight as to the generation of the lowlands.

2 Structure and Tectonics

Tectonically the Sylhet Division is situated on the Sylhet Trough which is in the northern end of the Central Deep Basin geo-tectonic province of the Bengal Basin. The trough terminates against the Shillong Massif in the north truncated by the Dauki Fault. On the western part of the trough is the Hinge Zone and on the further west the Stable Shelf geo-tectonic province of the Bengal Basin. The trough is bounded by the Chandpur High on the southwest and on the further south by the Faridpur Trough. The trough is closed on the south and east by the westward migrating Indo-Burman Range (Alam et al. [2]). The important tectonic elements are shown in Fig. 2.

Structurally the trough is a basin shaped depression filled with thick sediments. The sediments are folded into a series of anticlines and synclines in the southern and eastern parts

M. Masum (✉) · M. O. F. Khan
Geological Survey of Bangladesh (GSB), 153, Pioneer Road, Shegunbagicha, 1000 Dhaka, Bangladesh
e-mail: masum613@yahoo.com

A. Kallel et al. (eds.), *Recent Advances in Geo-Environmental Engineering, Geomechanics and Geotechnics, and Geohazards*, Advances in Science, Technology & Innovation, https://doi.org/10.1007/978-3-030-01665-4_28

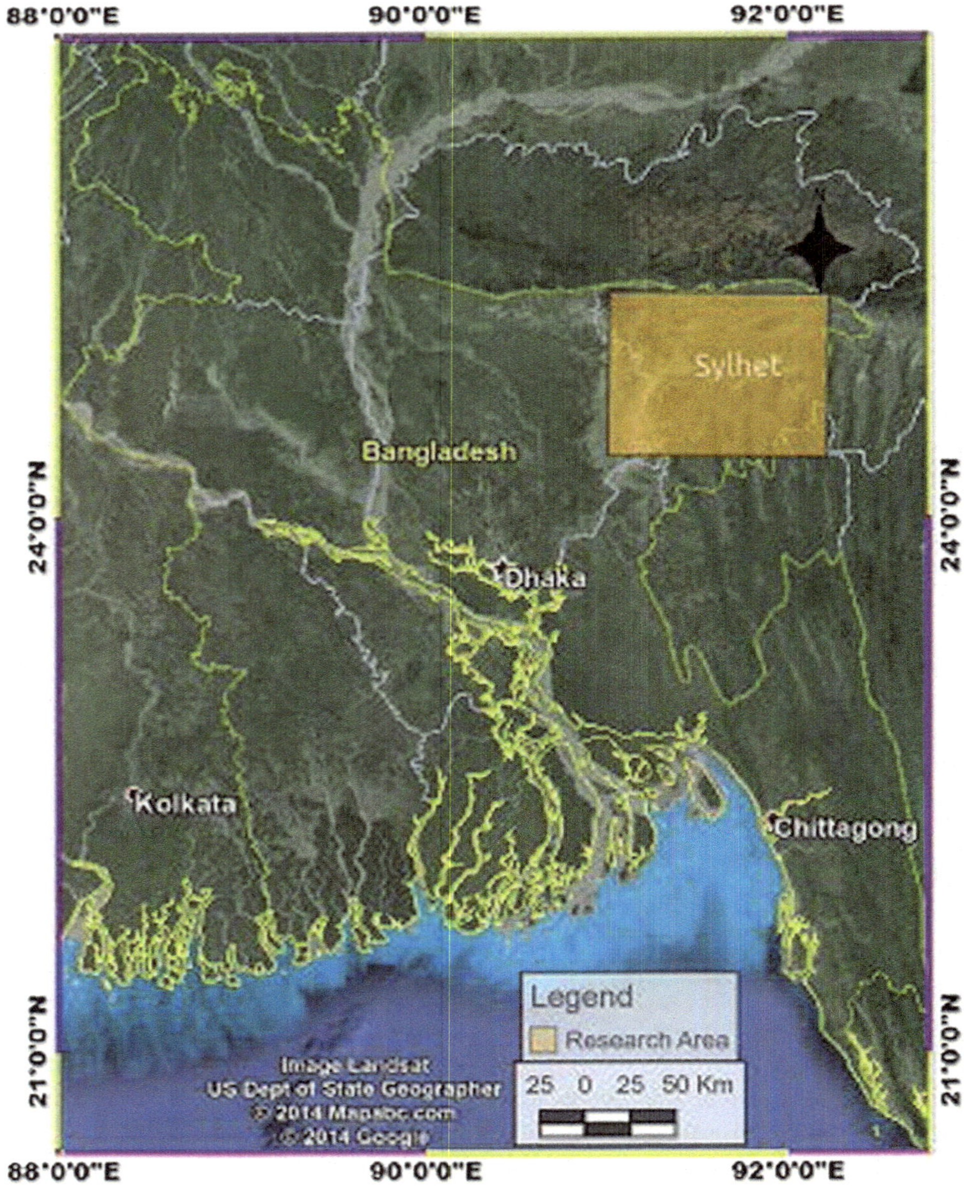

Fig. 1 Location map of the Northeastern Haor dominated area

which are mere northern extensions of the western part of the Indo-Burman Range. The Dauki Fault is a reverse fault (Alam et al. [2]) along which the Shillong Massif pops-up and acts as a major source of sediments and water for the northern part of the Haor Basin. The sediments of the Haor Basin are folded to form an isolated anticline in the middle of the Sylhet Trough. The Sylhet Trough in general and its northern half in particular are mostly influenced by subsidence as the trough is under thrusting against the uplifting Shillong Massif (Alam et al. [2]).

3 Methodology

Ideally an observational approach was implemented in categorizing the haor areas. Aerial imagery from Google Earth has been used for their ease of access and clarity. Maps were generated using the Free and Open Source (FLOSS) GIS software QGIS. The authors experience from working in haor areas helped tremendously in the development of concepts and ideas approached in the current study.

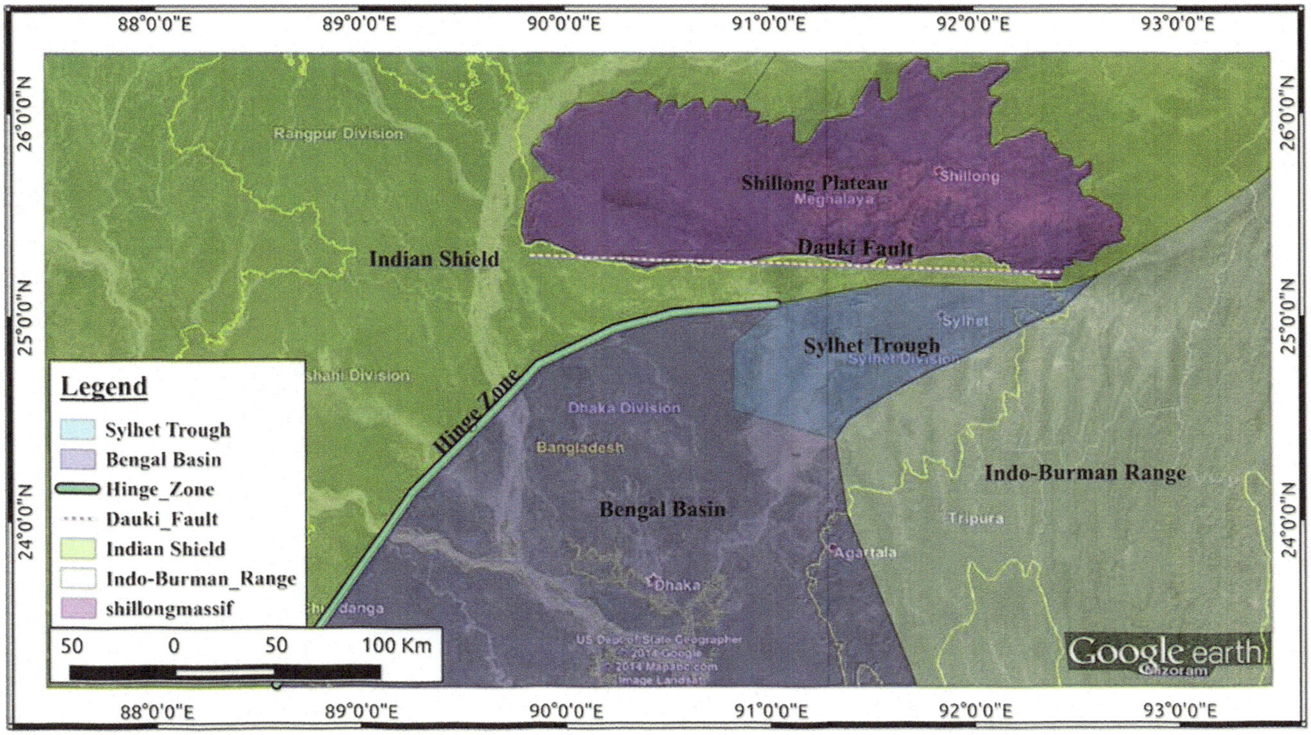

Fig. 2 Important tectonic elements surrounding Sylhet Trough

4 The Classification and Its Basis

The Haor Basin of the northeastern Sylhet Division can be broadly subdivided into two depression areas as (i) Northern Depression and (ii) Southern Depression. Most of the haors occur in the Northern Depression and are represented by many numbers of closely spaced; often interconnected haors, while the Southern Depression is represented by only several haors that are separated from each other for considerable distances by intervening hill ranges. The haors of the Northern Depression are clustered while in the Southern Depressions they are distributed in a sequential manner. The haors in the former area are distributed following an east-west direction while the haors of the latter area are distributed following a loosely northeast south direction.

The above mentioned characteristics of those two areas are closely related to their tectono-structural setting. The important tectonic elements around the Sylhet Trough are shown in (Fig. 2) and structural features are shown in (Fig. 3). The Northern Depression is located to the immediate south of the Dauki Fault. As the Sylhet Trough is a slowly subsiding area against the uplifted Shillong Massif, the Northern Depression is dominated by subsidence. As a result, the haors occupying the areas i.e. the Northern Depression can be classified as “Subsidence Haors” owing to the subsidence of the Sylhet Trough which greatly influences their development.

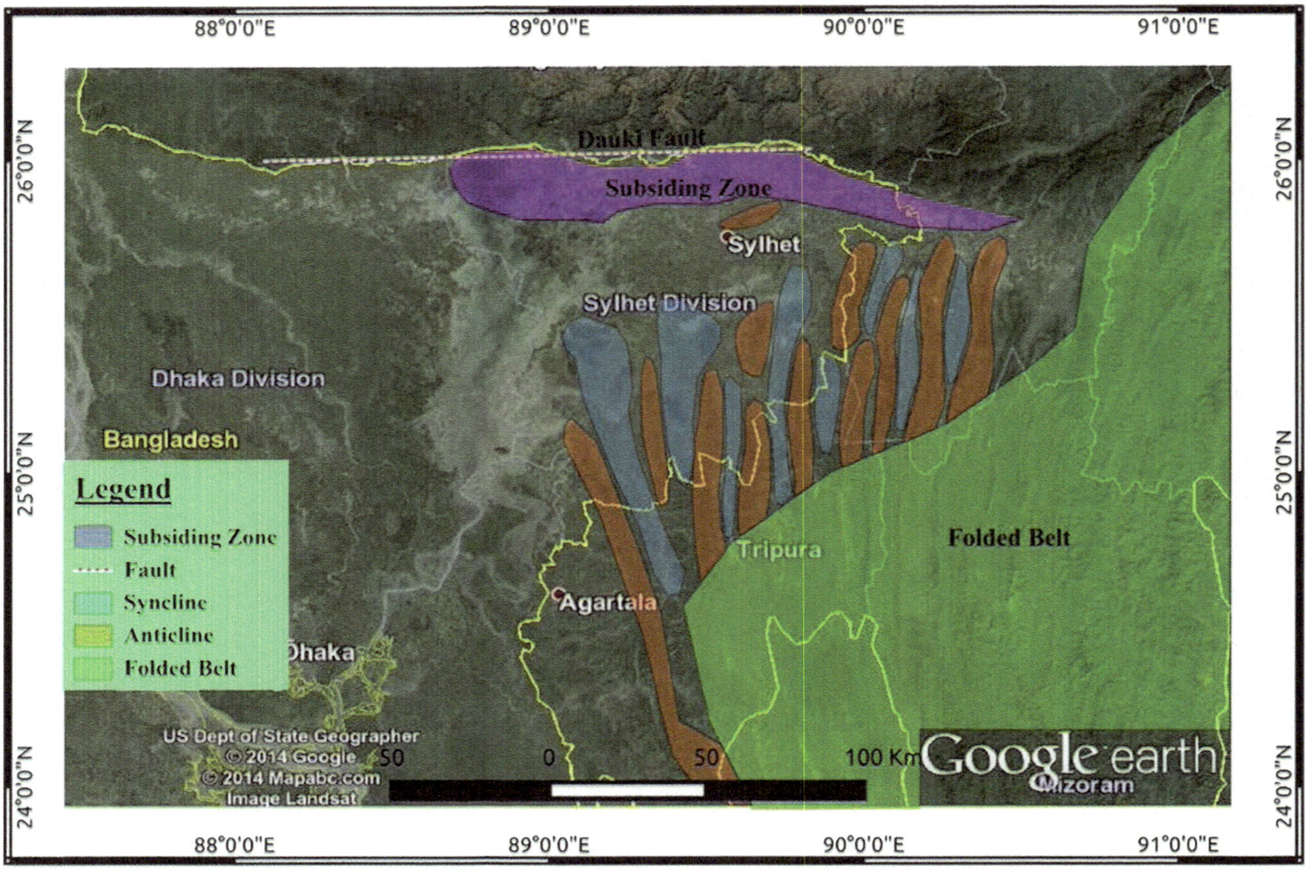

Fig. 3 Major structural features influencing the study area (The N-S ridges are formed due to E-W compression)

In contrast, the haors of the Southern Depression are located largely in the synclines between the anticlines of the Indo-Burman Range. These anticlines are expressed as north-south elongated hills and plunge in the north into the southern and eastern margin of Sylhet Trough. Thus, the depressions created by these anticline-syncline systems are occupied by the characteristic haor of the Southern Depression which can be classified as "Synclinal Haors" (Fig. 4).

5 Conclusion

The Subsidence induced Haors are mainly tectonic controlled while the Synclinal Haors are mainly structure controlled. These facts may subtly influence the geologic, environmental and socio-economic aspects of the areas and special consideration should be given to any future studies with these facts in mind.

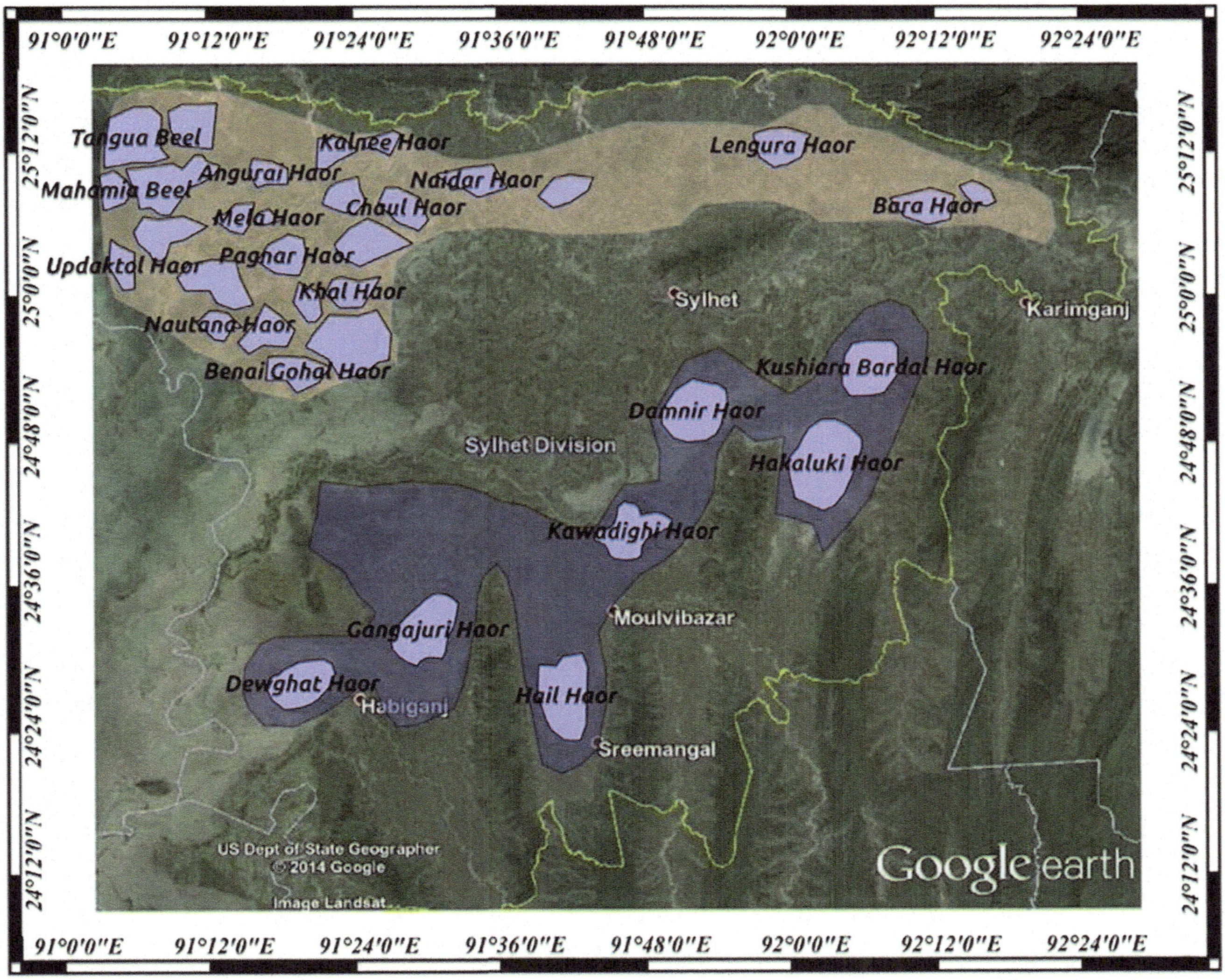

Fig. 4 Classification of the Haors based on dominant tectonic and structural signatures

References

1. Alam, M.S., Quayum, M.A., Islam, M.A.: Crop production in the Haor areas of Bangladesh: insights from farm level survey. The Agriculturists **8**(2), 88–97. Sci. J. Krishi Foundation. ISSN-1729-5211 (2010)
2. Alam, et al. Geological Map of Bangladesh, GSB. vol. 14, p. 14 (1990)
3. BHWDB. Brief Outline for Development of Haor Master Plan. Bangladesh Haor and Wetland Development Board (BHWDB). Ministry of Water Resources (2014)
4. Choudhury, G.A., Nishat, A.: Hydro-Meteorological Characteristic of Hakaluki Haor. Union Bangladesh Country Office, Dhaka, Bangladesh, x + 91p (2005)
5. IUCN-Bangladesh, CNRS. Natural Resource Economic Evaluation of Hakaluki Haor. BGD/99/G31-Coastal and Wetland Biodiversity Management Project: Hakaluki Haor Component (Draft Final. Ministry of Environment and Forests, Government of Bangladesh (2006)
6. Salauddin, M., Islam, A.K.M.S.: Identification of land cover changes of the Haor area of Bangladesh using MODIS images. In: 3rd International Congress on Water & Flood Management (2011)

Geotechnical Characterization of the Batoufam Lateritic Gravels (West Cameroon) for Road Construction Purpose

Takala Boris Honore and Mbessa Michel

Abstract

This study involves the geotechnical characterization of lateritic soils from Batoufam (West Region of Cameroon). The methodology involved the collection of samples from the field and laboratory analyses which involved identification and characterization tests at the National Advanced School of Engineering of Yaoundé. The results obtained show that: In the quarry, two stratigraphic levels are observed, a nodular and a superior sandy clay level on which vegetation is present; The material studied is a gravelly sandy clay of the A-2-7 (0) class according to the HRB classification; The gravelly materials have an optimum dry density of 2, 23 and an optimum water content of 13%. The direct bearing capacity values of CBR and by immersion are 42, 5 and 31 respectively, classifying them of the S5 class. The values of resistance to compression and traction are 15.44 bars; 13.78 bars and 8.82 bars and of 1.7 bars; 1.5 bars and 1.4 bars respectively after exposure for 4.7 and 28 days in open air. According to the practical guide involving road construction in tropical countries, the material studied is useful as platform and base layer for all traffic classes (T1–T5), as well as foundation base for low traffic (T1–T3).

Keywords

Batoufam • West region • Basalts • Geotechnical properties • Cameroon

1 Introduction

Batoufam is a locality of the Western Cameroon, located between latitude 5°14′ and 5°18′ North, and between longitude 10°26′ and 10°31′ East. According to Guiffo [1], the general morphology of the Batoufam relief is that of the High plateau. Soils are mostly lateritic. This work dealt with the Batoufam soils and devoted their quintessence to the determination of their geotechnical characteristics in relation to their use in road construction.

2 Methods

In the case of this study, the methodology is based on field works, on the one hand, and on laboratory analysis which involves an identification test and a physical characterization, on the other. On the field, the extraction reveals a visible pedologic profile that defines the main horizons. This quarry situated at an altitude 1435 m matches with the coordinates latitude 05°17′20.1″N, and longitude 10°28′14.6″E.

The identification test involved the particle-size analysis (Wet sieving obtained in accordance with the NF P 94-056 norm, and Sedimentation with the NF P 94-057 norm) and Atterberg limits (carried out in conformation to the NF P 94-051 norm). For the characterization tests, the Proctor tests (in conformity with the NF P 94-093 norm) and CBR (precisely by the NF P 94-078 norm), as well as simple compression tests (in accordance with the NF P 94-420 norm) and indirect traction (carried out according to the NF P 94-422 norm) were performed.

T. B. Honore (✉)
University of Yaoundé 1, Yaoundé, Cameroon
e-mail: honoretakala@uy1.uninet.cm

M. Michel
National Advanced School of Public Works, Yaoundé, Cameroon

A. Kallel et al. (eds.), *Recent Advances in Geo-Environmental Engineering, Geomechanics and Geotechnics, and Geohazards*, Advances in Science, Technology & Innovation, https://doi.org/10.1007/978-3-030-01665-4_29

Table 1 Percentages of the different size fractions

Classes	Gravel	Coarse sand	Fine sand	Silt	Clay
Percentage (%)	74	9.3	4.5	2.2	10

Table 2 Summary of the obtained results from the Atterberg limit

Parameter	*wl*	*wp*	*Ip*	*IC*	*Il*	*f X Ip*
Value	42	22	20	1.75	−0.75	280

3 Results

3.1 Physical Characteristics

For the particle-size analysis and Atterberg Limits, the results are presented respectively in Tables 1 and 2. The numeric results obtained led to the establishment of the granulometric curve in Fig. 1a, and the curve of the variation in water content following the number of drop in Fig. 1b.

3.2 Behaviour Parameters

The immediate CBR index and immersion values were determined from the dry density variation curves following these indices (respectively at Fig. 2a, b). We thus found 42.5 and 31, respectively.

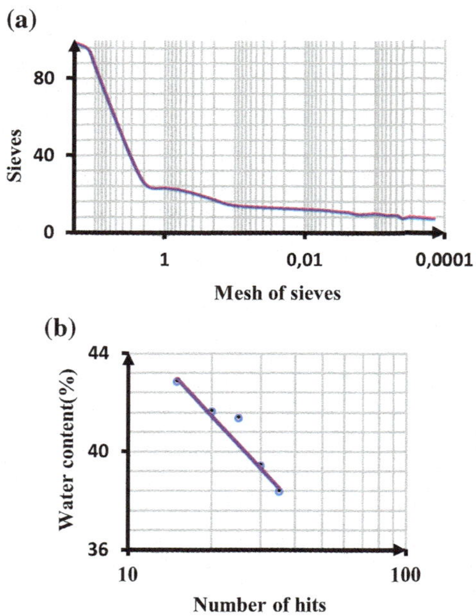

Fig. 1 **a** Granulometric curve; **b** variations of the water content according to the number of hits (Atterberg limits)

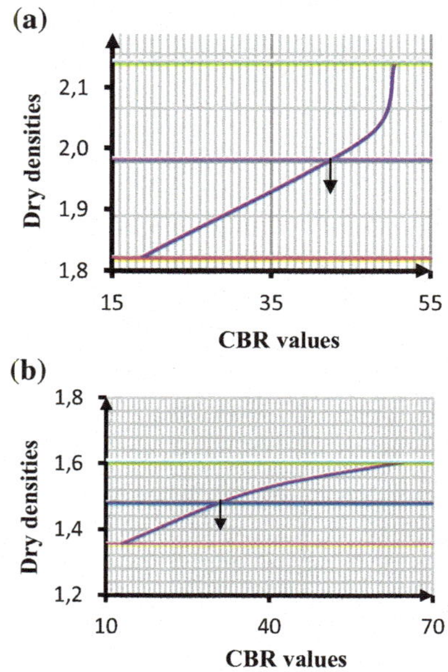

Fig. 2 **a** Variation in densities following the CBR index (CBR by immersion); **b** variation in densities following the CBR (immediate CBR)

3.3 Compressive Strengths Simple and the Indirect Traction

The values of the resistance to traction and compression are shown in Tables 3 and 4.

4 Discussion

The materials have a low percentage in fine particles, either 14% of the material, lower than (34.20 and 16.20%) in Ebolowa by Nanga Bineli [2], but greater than those obtained in 2012 by Djuickouo (12%) in Dschang. Liquidity and plasticity limits are lower than 63 and 43% obtained by Djuickouo [3], and 48.6% obtained by Ananfouet Djeufack [4] on hardened basalt soils, both in Dschang locality. However, compared to the three authors mentioned, the plasticity index is respectively equal, greater and lower than theirs (20; 18.3; 27.4 and 33.5% respectively).

Table 3 Resistance to traction

Days	4	7	28
P max (N)	5000	4000	1000
Rit	0.14	0.11	0.03
Rt (Mpa)	0.11	0.09	0.02
Rt (bars)	1.10	0.88	0.22

Table 4 Resistance to compression

Days	4	7	28
F max (N)	28,000	25,000	16,000
S (mm^2)	18136.6	18136.6	18136.64
Rc (Mpa)	1.54	1.38	0.88
Rc (bars)	15.44	13.78	8.82
Cu	30.88	27.57	17.64

For Autret [5], the Proctor values (2.23), accommodate with those recommended for foundation layers; $\gamma d > 1.90\ g/cm^3$ and in base layer, $\gamma d > 2\ g/cm^3$. According to the CEBTP (1984), specifications, the lateritic gravels of Batoufam belong to the bearing class of S5 (CBR > 30). Resistance to compression and traction decreases with the number of days spent in open air. Thus, with time there is the creation and the appearance of fissures on the material which gradually disintegrates losing its resistance and later becomes friable. This is as well illustrated by the values of cohesion of the material that follows the same evolution as that of the resistance to compression.

5 Conclusion

The studied material is classified as a gravelly sandy-clay with a well graded grain size particle, average plasticity, of the class A-2-7(0) according to the HRB classification and as such being part of gravels and sandy silt or clayey.

In accordance with the practical guide for the dimensioning pavements for tropical countries, it is realised that the lateritic gravels of Batoufam are usable for pavement layer notably in platform and in form layers for all classes of traffic and as foundation layers for low traffic; T1–T3. They, however, require the optimization of their characteristics especially for their use in a foundation layer for important traffics (T4 and T5) as well as in base layers.

References

1. Guiffo, J.P.: Les Bamiléké de l'intérieur et leurs problèmes. éditions de l'Essoah, 163–175 (2009)
2. Nanga Bineli, M.T.: Caractérisation géologique et géotechnique des graveleux latéritiques sur granites de la région d'Ebolowa (Sud-Cameroun). Master thesis, University of Yaoundé I, Yaoundé, 73p (2016)
3. Djuickouo, C.N.: Caractérisation géotechnique des graveleux latéritiques sur basaltes de la zone de Maka (Dschang, ouest-Cameroun). D.I.P.E.S II thesis, Yaoundé, (2012), 70p + annex
4. Ananfouet Djeufack, C.Y.: Cartographie et caractérisation géotechnique des sols développés sur roches volcaniques de la ville de Dschang, Master thesis, Falculty of sciences, University of Dschang, Dschang, 89p (2012)
5. Autret, P.: Latérites et graveleux latéritiques, Document of the Institute of Science and Technology Equipment and Environment for Development, France, 38p (1983)

Identification of Geotechnical Properties of Weak Rock Masses and Stockpiles at Tunçbilek Open Pit Mining and the Related Slope Stability Analyses

Zeynal Abiddin Erguler, Huseyin Karakus, I. Goktay Ediz, and Cem Sensogut

Abstract

Tunçbilek lignite basin situated in the western part of Turkey has been increasingly excavated by open-pit and underground mining for a very long time to meet Turkey's energy demand. The blasted and then the excavated stratified rock masses in this basin consist mainly of clay-bearing rocks such as claystone, mudstone, siltstone and marl. It was found that the strength and deformation properties of these clay-bearing rocks decrease with the increase in water content. In addition, time-dependent attenuations were also recorded in these mechanical properties due to disintegration caused by physical weathering processes. Therefore, several time-dependent instabilities were observed in the slopes left after open pit mining activities. This study aimed to determine the design parameters of weak rock masses and stockpiles at Tunçbilek basin and then to recommend safe slope geometry for open pit mining activities. For this purpose, the geotechnical properties of rock masses and stockpiles were obtained based on in situ tests, characterizations and back analyses of instabilities.

Keywords

Instability • Open-pit mining • Slope stability • Stockpile • Weak rock mass

Z. A. Erguler (✉) · H. Karakus
Geological Engineering Department, Dumlupinar University, 43270 Kütahya, Turkey
e-mail: zeynal.erguler@dpu.edu.tr

I. G. Ediz · C. Sensogut
Mining Engineering Department, Dumlupinar University, 43270 Kütahya, Turkey

A. Kallel et al. (eds.), *Recent Advances in Geo-Environmental Engineering, Geomechanics and Geotechnics, and Geohazards*, Advances in Science, Technology & Innovation, https://doi.org/10.1007/978-3-030-01665-4_30

1 Introduction

The Neogene aged Tunçbilek basin is generally known for weak, thinly bedded, stratified and fractured rock masses. Thus, a considerable time-dependent slope instabilities were recorded at this basin due to low shear strength parameters as well as decreases in these parameter with water content increase and progressive disintegration behavior. The Neogene deposits in the Tunçbilek basin begin with the Beke formation reaching a thickness of about 1000 m. The Beke formation consists of pebbles, sandstones, mudstones and thin lignite levels. Tunçbilek formation involving lignite layers, which can be economically produced by open-cast and underground mining techniques, conformably overlies Beke formation. This lignite containing formation mainly consists of clay-bearing rocks such as marl and claystone, which indicates very fast disintegration and decomposition behavior against physical and chemical weathering processes, sandstone, conglomerate and limestone levels. The lignite bearing levels are found inside this basin extending to NE and SW directions.

Due to the water sensitive and disintegration characteristics of rock masses studied, identifying representative engineering parameters for these geomaterials requires comprehensive field-based evaluations. Thus, in order to collect the required data to incorporate with slope stability analyses of rock masses and stockpiles in Tunçbilek basin, a research involving in situ tests, measurements and observations, back analyses of previous instabilities and data mining on obtained variables was carried out. In addition to these field investigations, rock mass classification systems and empirical failure criteria were also used for this particular purpose. The results of all of these geotechnical studies and associated stability analyses were discussed in the following sections.

2 Geotechnical Properties of Rock Masses and Stockpiles

The slopes of open-pit mining activities were excavated within weak rocks masses involving claystone, mudstone, siltstone, marl, clayey limestone, limestone, conglomerate and sandstone. Uniaxial compressive strength (UCS) was utilized as an input parameter in rock mass classification system. UCS of these weak geological units were predicted by performing the needle penetrometer test [3]. In addition to the strength of rock materials, degree of weathering, geomechanical properties of discontinuities (type, dip, fracture frequency, roughness, aperture, and infilling) were also very important for understanding the failure behavior of rock masses. These properties were measured during field investigation. Furthermore, for the fractures quantitative description, Rock Quality Designation (RQD) values of rock masses were also determined as 20–93% with a mean value of 63.1% by using core boxes of 20 different core runs of a drilling program in Tunçbilek basin.

Rock Mass Rating (RMR) system recommended by Bieniawski et al. [1] was also used for the engineering classification of the rock masses encountered within this project. By utilizing charts and graphs proposed by Bieniawski et al. [1] and using the data obtained from field and laboratory studies, the RMR values for clay-bearing rock levels (e.g., marl, claystone etc.) and other geological units involving conglomerate-sandstone were approximated as 42 and 38, respectively. The shear strength parameters (cohesion and internal friction angle) were predicted from RMR values of associated geological materials. When the approaches recommended by Bieniawski et al. [1] were applied to rock masses, the cohesion of clay-bearing rocks changes was found to be between 200 and 300 kPa and the internal friction angle of the same rock units to range between 25° and 35°. For geomaterials mainly involving conglomerate and sandstone, the cohesion and internal friction angle are 100–200 kPa and 15° and 25°, respectively. In addition to the RMR system [1], Hoek-Brown failure criteria [2] were also used to understate the failure behavior of rock masses in this study site. The results obtained by using the RMR system [1] and Hoek-Brown failure criteria [2] are summarized in Table 1.

Beside the rock masses, the geotechnical properties of stockpiles produced after open-pit and underground mining activities were also significant to achieve safety condition for slopes excavated in the Tunçbilek basin. Therefore, the wet unit weight of stockpiles was measured at field conditions by digging large-sized trenches. After in situ volume and collected samples mass basic measurements, the unit weight of stockpiles changes was determined to be between 17.9 and 21.3 kN/m^3 with a mean value of 20.1 kN/m^3. Since it is very difficult to accurately determine the shear strength parameters of stockpiles using small-size samples in laboratory, it was decided to use the average of previously published cohesion and internal friction angle values given in Table 1 in slope stability analyses of these geomaterials.

3 Limit Equilibrium Analyses of Slopes and Stockpiles

Based on the methods of Bishop simplified, Janbu simplified and Fellenius, limit equilibrium analyses were performed using the data given in Table 1 and considering the earthquake effect. In order to be used in the limit equilibrium analysis, geological cross-sections with 19 different directions were taken to provide a reliable representation of the slopes within the study site. In addition, three more topographical cross-sections were prepared from stockpiles for associated analyses in this geomaterial. The dynamic limit equilibrium analyses were performed on slopes in different geometries having overall slopes angles ranging between 25° and 45° for all the geological cross-sections by fixing the bench height to 9 m. As a result of these analyses, the proposed design geometry for stable slopes excavated in undisturbed rock masses involving both sequences of clay-bearing rock and conglomerate-sandstone unit is illustrated in Fig. 1. Limit equilibrium analyses involving investigation on the slope stability of stockpiles were performed by taking average shear strength parameters given in Table 1 and considering various slope geometries. The results of these analyses on three different topographical cross-sections of deposited stockpiles indicate that any decrease in factor of safety is mainly controlled by an increase in bench height of stockpiles. When all the analyses

Table 1 Input parameters used for slope stability analyses of rock masses and stockpiles

Parameters	Stockpile	Rock mass I	Rock mass II
c (kPa)	20.5 (5–35.9)	200–300	100–200
ϕ (°)	32.3 (23.6–41)	25–35	15–25
c′(kPa)	–	80–237	65 (27–94)
ϕ′(°)	–	24.6–40.6	17 (12.8–26.1)
γ (kN/m^3)	20.1 (17.9–21.3)	22.6	22.6

c: cohesion; ϕ: internal friction angle, c′: effective cohesion; ϕ′: effective internal friction angle; γ: unit weight; Rock mass I: clay-bearing rocks; Rock mass II: conglomerate-sandstone

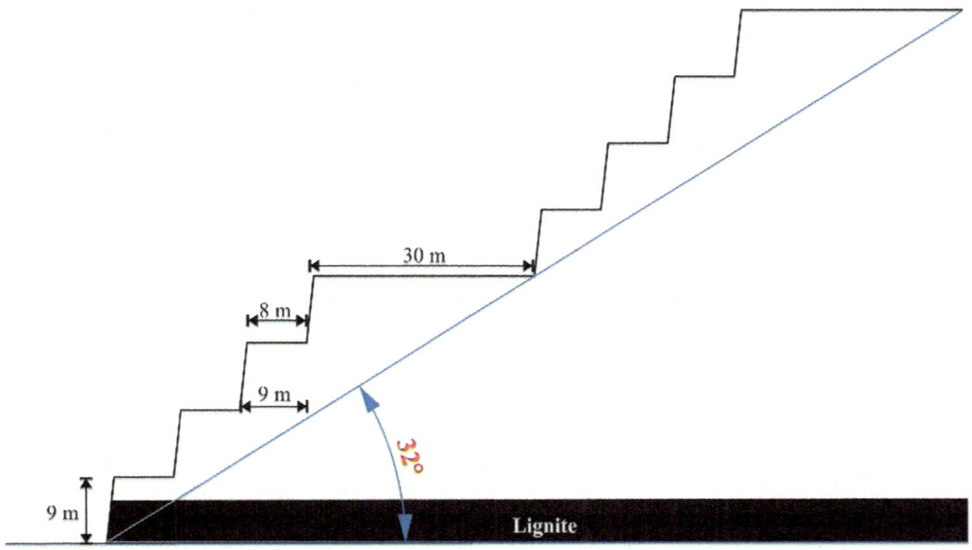

Fig. 1 The geometry proposed for the slopes excavated in undisturbed rock masses

are considered, the stable condition for the deposition of stockpiles can be proposed as 30 m, 40 m and 36° for the bench height, width and angle, respectively.

4 Conclusion

The representative cohesion and internal friction angle of rock masses consisting of clay-bearing rocks were specified as 200 kPa and 30°, respectively. However, relatively lower shear strength parameters (150 kPa and 25°) were obtained for the conglomerate and sandstone-rich sequences. Based on the given shear strength parameters, it was found that the overall slope angle might be 32° with a haul road width of 30 m and bench width of 8 m. The unit weight of stockpiles ranges from 17.9 to 21.3 kN/m^3 with a mean value of 20.1 kN/m^3 based on the in situ measurements. The dynamic limit equilibrium analyses results indicate that the stockpiles can be safely deposited if average bench height, width and angles are kept as 30 m, 40 m and 36°, respectively.

Acknowledgements The authors gratefully acknowledge the financial support of the Western Lignite Corporation (WLC).

References

1. Bieniawski, Z.T.: Engineering Rock Mass Classifications. Wiley-Interscience, New York, 264 p (1989)
2. Hoek, E., Carranza-Torres, C.T., Corkum, B.: Hoek–Brown failure criterion—2002 edition. In: Hammah, R., Bawden, W., Curran, J., Telesnicki, M. (eds.) Proceedings of the Fifth North American Rock Mechanics Symposium (NARMS-TAC), University of Toronto Press, Toronto, pp. 267–273 (2002)
3. Ulusay, R., Erguler, Z.A.: Needle penetration test: evaluation of its performance and possible uses in predicting strength of weak and soft rocks. Eng. Geol. **149–150**, 47–56 (2012)

Review on the Mechanical Behavior of Soil-Structure Interface

Xue-Ying Yang, Li Yuan, and Zhen-Dong Cui

Abstract

At the soil-structure interface, there are some frontier problems, such as non-linearity, large deformation and local discontinuity, which is one of the core topics in the study of soil and structure interaction. This paper introduced the progress and present situation of the mechanical behavior of soil-structure interface in laboratory test, constitutive model and numerical simulation, analyses the problems that have not been studied in depth and pointed out that the determination of thickness of the contact surface, the sensitivity analysis of the mechanical parameters of the contact surface and the dynamic contact questions would be the main research trend in the future.

Keywords

Soil-structure interface • Mechanical behavior • Shear test

1 Introduction

In underground engineering, soil-structure interaction problems are widespread. However, under the load action, the mechanical response at the interface between soil and structure is neither different from soil nor structural material. It involves nonlinear, large deformation, local discontinuity and other mechanical frontiers, which makes contact surface research extremely difficult.

At present, the research thoughts on the soil-structure interface study are as follows: Through the laboratory test method, research focuses on the study of the failure and deformation mechanism, draw the stress-strain relationship curve, obtain the rule of stress distribution and interface displacement at the soil-structure interface, propose relevant constitutive models, and finally commit to the needs of numerical simulation and practical engineering.

This paper introduced the progress and presented situation of the mechanical behavior of soil-structure interface in a laboratory test, constitutive model and numerical simulation, analyzed the problems that have not been studied in depth, and predicted the future research trends.

2 Soil-Structure Interface Shear Tests

Nowadays, the shear apparatus mainly includes direct shear apparatus and single shear apparatus. However, the direct shear apparatus suffers from many defects such as fixed shear plane, the decrease of the shear area and so on. Though the single shear apparatus has overcome these defects, it was also subject to the uneven distribution of shear stress and strain in process, and failed to accurately reflect the interface behavior. Therefore, many scholars began to adopt a torsional shear apparatus or a resonant column apparatus to improve the test process. In order to reduce the size effect, the test instrument is also developing towards large-scale and smartness. For example, Zhang et al. have recently designed a new automated large-scale apparatus (3DMAS) for 3-D cyclic testing, photographic and schematic views of the 3DMAS are presented in Figs. 1 and 2 [1].

In addition, more attention has been paid to the related macro and micro testing technology and the intellectualization of data acquisition systems. Hoping to reproduce the characteristics of the true three-dimensional, discontinuous, large deformation and diversity of soil-structure interface.

X.-Y. Yang · L. Yuan (✉) · Z.-D. Cui
State Key Laboratory for Geomechanics and Deep Underground Engineering, School of Mechanics and Civil Engineering, China University of Mining and Technology, Xuzhou, 221116, Jiangsu, People's Republic of China
e-mail: yuanli-1213@163.com

A. Kallel et al. (eds.), *Recent Advances in Geo-Environmental Engineering, Geomechanics and Geotechnics, and Geohazards*, Advances in Science, Technology & Innovation, https://doi.org/10.1007/978-3-030-01665-4_31

Fig. 1 Photograph of the 3DMAS

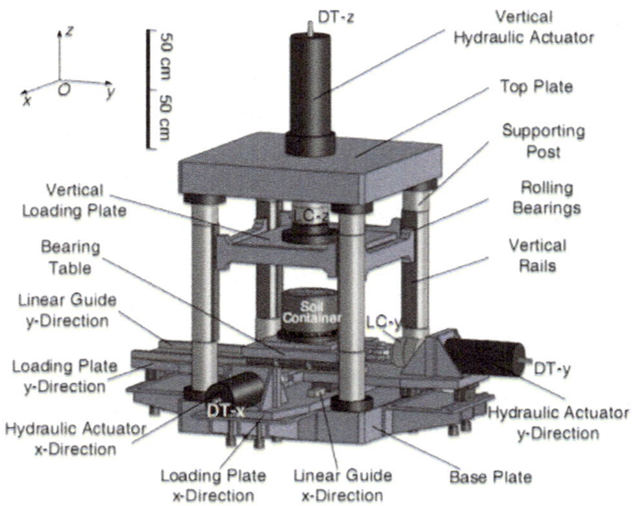

Fig. 2 3-D schematic diagram of the 3DMAS

3 Constitutive Model of Soil-Structure Interface

The constitutive model of soil-structure interface is the comprehensive reflection of macroscopic mechanical properties of soil materials or structures. At present, it mainly consists of nonlinear elastic, elastic-plastic, rigid-plastic, coupling, damage and three-parameter models.

As early as 1970s, Clough and Duncan put forward a hyperbolic relationship based on the direct shear test results, which represented the nonlinear elastic model [2]. Subsequently, Brandt introduced a simplified elastic-plastic model, which was described by 2 polygons [3]. Then, Yin et al. thought that the interface failure form should be rigid-plastic [4]. However, both them could not reflect the nonlinear, hardening, strain softening and dilatancy characteristics during the interface deformation. For the first time, Desai et al. applied the basic theory of damage mechanics to the interface constitutive relation [5]. The elastic-plastic damage model of coarse-grained soil-structure interface was been established by Zhang [6]. It can be used to describe the interface mechanical response under monotonous and cyclic loading. Wang derived the 3 parameter-model of the interface stress and strain relationship from the microcosmic angle, which has been proved by mathematical principles and experimental results [7].

Although there are various interface constitutive models, the Clough-Duncan hyperbolic model and the ideal rigid-plastic model are widely used in numerical simulation. The main reason is that the model is simple and the parameters are easy to determine. The elastic-plastic model develops quickly, but a simple and practical model is not widely accepted. The interface damage model has just been developed in recent years, and has been applied in some large projects; however, it is still concentrated in a few scientific research colleges and universities and needs further promotion.

4 Numerical Simulation of Soil-Structure Interface

Under load, the soil or structure will slip or crack at the contact surface. The traditional method of continuous deformation analysis will no longer apply to the study of this problem. And as commonly known there is a shear zone between the different materials, which has a given thickness. Here is an example for gravelly soil-structure interface (see Fig. 3) [8]. To simulate these engineering and physical problems, the existing methods mainly include finite and discrete element methods.

The current finite element method to solve contact problems divides the contact area into several elements with simple shape, and through the specified constitutive relation of contact elements simulates the interface mechanical behavior. These interface elements mainly contain two spring node elements, zero-thickness element, thickness element and contact friction element. Zero-thickness Goodman element and thin-layer Desai element are the most common.

The discrete element method is used from micro aspect to solve the discontinuous media problem. The process

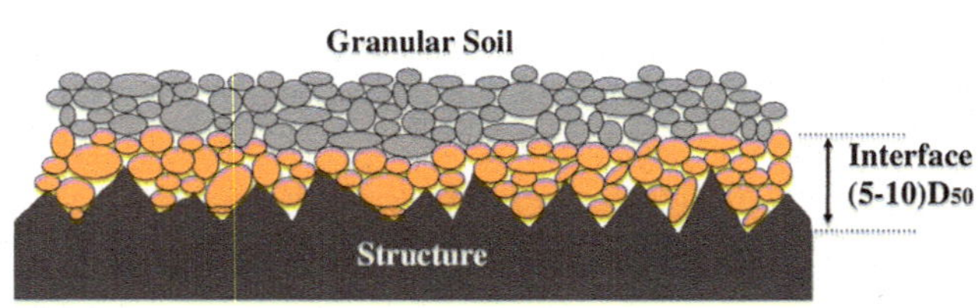

Fig. 3 Schematic of an interface layer in soil-structure interaction system

includes three steps: generating particles and making them compact, defining contact type and material properties, defining the boundary and initial conditions. Besides, it is complex and difficult to match the mechanical parameters of mesoscopic granular materials with macroscopic samples. Nowadays, discrete element software, represented by Particle Flow Code (PFC), has been widely applied to the study of soil-structure interaction problems.

5 Prospect

Although many achievements have been made in the study of soil-structure interface at home and abroad, there are still many problems to be further studied:

(1) In the shear testing system, the research on larger size, complex stress path, high stress state, special soil, precise macro and micro measurement technology and micro failure mechanism should be further strengthened.
(2) The determination of the interface thickness should be further studied. With different soil-structure interfaces, the interface thickness is different. Therefore, more experiments should be carried out to establish corresponding databases and summarize the regularities.
(3) Carrying out more sensitivity analysis of the interface mechanical parameters. Standardizing the influence factors and quantifying the influence degree, so as to define the interface parameters and ranges more comprehensively and precisely.
(4) To strengthen the study on dynamic contact problem. In the present numerical calculation, there are few dynamic constitutive models of soil-structure interface that can be selected, and the interface dynamic mechanical behaviors are not sufficient.

Acknowledgements This work was supported by National Key Research and Development Program (Grant No. 2017YFC1500702).

References

1. Zhang, J.M., et al.: An automated large-scale apparatus for 3-D cyclic testing of soil-structure interfaces. Geotech. Test. J. **41**(3), 459–472 (2018)
2. Clough, G.W., Duncan, J.M.: Finite element analyses of retaining wall behavior. J. Soil. Mech. Found. Div. ASCE **97**(12), 1657–1673 (1971)
3. Brandt, J.R.T.: Behavior of Soil-concrete Interface. University of Alberta, Edmonton (1985)
4. Yin, Z.Z., et al.: Deformation and mathematical simulation of contact surfaces between soil and structural materials. Chin. J. Geotech. Eng. **16**(3), 14–22 (1994)
5. Desai, C.S., et al.: Cyclic testing and modeling interfaces. J. Geotech. Eng. Div. ASCE **111**(6), 793–815 (1985)
6. Zhang, G.: Study on static and dynamic characteristics and elastoplastic damage theory of interface between coarse-grained soil and structure. Tsinghua University, Beijing (2002)
7. Wang, W.: Study on soil-structure interface model based on potential energy dissipating principle and its application. Hohai University, Nanjing (2006)
8. Saberi, M., et al.: Constitutive modeling of gravelly soil-structure interface considering particle breakage. J. Eng. Mech. **143**(8), 04017044 3 (2017)

Influence of Aggressive Groundwater Stream in Substrate for Lateral Loaded Piles

Janusz Kozubal and Marek Wyjadłowski

Abstract

This article described and defined the sulphate corrosion effects. This type of corrosion is one of the most important factors that limit the lifespan of concrete construction. Harsh environmental conditions have a large impact on the operational costs of concrete piles or columns. The presented phenomena are non-deterministic; therefore, the reliability analysis was used for the assessment of the impact. The strength characteristics of the soil around the construction modeled as one-dimensional random field, and corrosion defined by a set of random variables. Fick's second law described the penetration of sulphate ingress into the concrete material with explicit numerical solutions for boundary conditions and an increase in the transition factor under the progress of corrosion. This process was solved in an analytical and numerical way to calculate corrosion, generate random fields and determine the reliability. A numerical example was provided to illustrate the proposed method to prevent unexpected structural failures during pile service life. The proposed methodology can assist designers and the management of the existing piles in the decision making concerning interventions that ensure safe and serviceable operations of a geotechnical construction.

Keywords

Concrete piles • Sulphate corrosion • Safety level • Reliability • Geotechnics • Horizontal load

J. Kozubal (✉) · M. Wyjadłowski
Wroclaw University of Science and Technology, Wyb. St. Wyspiańskiego 27, 50–370, Wrocław, Poland
e-mail: janusz.kozubal@pwr.edu.pl

1 Introduction

The approach adopted in this work was dedicated to situations where the horizontal displacements of the pile caps are critical. A description of the material by a reduction in its compression strength replaced in this study with an analysis of the crack propagation more appropriate for concrete structures. Direct studies of corrosion involving concrete sampling and chemical groundwater analysis or the performance of non-destructive testing in the case of corrosive burns are rare and difficult to carry out [1]. The motivation to present the reliability assessment of concrete piles originated from the evaluation of real foundation cases. Problematic foundations were located in areas containing substrates consisting of non-bearing sediments and seasonal fluctuations in groundwater levels. The propagation of cracks and material deterioration caused a chemical aggression [2, 3]. Reliability problems were defined in many previous studies, i.e., [4], and in the precursor [5] approach to geotechnical problems. The one-dimensional random field was associated with the analyzed substrate trough non-linear stiffness in the general homogenic layer.

The task was modelled numerically using the finite element method with elastic material for pile subjected to elasto-plastic material stretched on one dimensional random field with solution of Bernoulli's deflection of beam from the *p-y* method. The numerical variant of the task was previously presented by [6, 7].

2 Sulphate Corrosion in Time-Dependent Formula as Reliability Process

In the present article the percolation model was used to describe the micromechanical sulphate corrosion process, the progress measure is a concrete crack density correlated with concentration of sulphate ions. It is assumed that the high density of fissures allows the concept of percolation.

A. Kallel et al. (eds.), *Recent Advances in Geo-Environmental Engineering, Geomechanics and Geotechnics, and Geohazards*, Advances in Science, Technology & Innovation, https://doi.org/10.1007/978-3-030-01665-4_32

The diffusion coefficient and the mechanical characteristics of the concrete can be described as a set of threshold concentrations like conductivity percolation threshold, in which micro cracks are connected to form continuous channels that provide fluid flow. Concentration less than k_{th} without wasted water transport; k_{dg} rigidity percolation threshold, in which mutual influence of connected channels of microcraks causes a sharp loss of material stiffness. In the presented method, deterioration ingress was described with the assumption of the same values as K_{th} and K_{dg}. The factor of diffusion in concrete has got direct dependency on porosity. The model of the ingress of sulphate in concrete is function of time and depth under Fick's second diffusion law. The authors created an axial symmetrical model, to describe the degradation process of the concrete pile. The boundary conditions and penetration process into concrete were provided to the numerical model. The model was written and solved in the FlexPDE program where α and κs coefficients were calibrated to the results of laboratory experiments results from literature data for two ions concentrations 10 and 20% [8].

2.1 The Mechanical Model Pile—Soil

For a pile with a varied diameter into substrate with random stiffness, the *p-y* model was used. The soil stiffness was assumed to be dependent on depth z under terrain as non-linear function. The stiffness of the pile determined the elementary stiffness for each of its segments. Taking into account a depth of elements in the later part of the calculation, the beam segments were connected to a zone covered by corrosion. A solution was obtained iteratively due to the non-linear calculations in Mathematica in Wrocław Centre of Server Supercomputing.

2.2 Corrosion as Non-deterministic Process

The calculations performed using the Monte-Carlo method based on 10^7 drawing number for progressive corrosion with two selected random variables: C_0 (ion concentration at pile skin) and λ (time of corrosion attack per year). All the random variables were defined as symmetrical probability beta distributions, they were scaled to domain divided by mean value of random parameters [9].

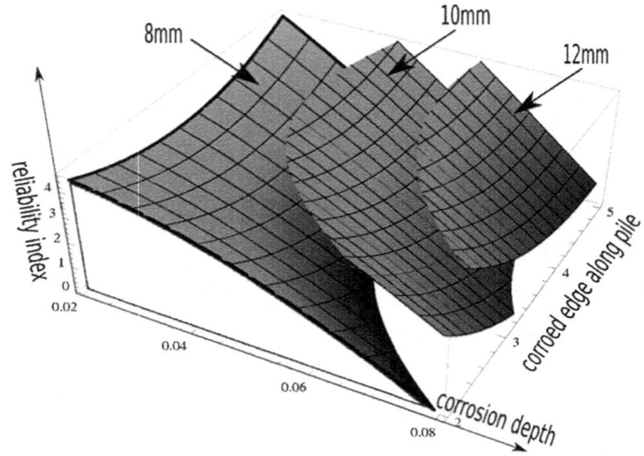

Fig. 1 The value of reliability index β as surfaces depending on corrosion depth, corroded edge along pile and the assumed limitations of a displacement pile cap {8, 10 and 12 mm}

3 Results

Continuous corrosion reduced the thickness of the pile column over time as probability process. The function of the log-normal probability distributions was fit to the results of corrosion for continuous description of its progress in time. This is a reliability process solved independently from deformation calculation.

An impact of the destructed zone depth and length for the reliability index is shown in Fig. 1.

We can observe two important phenomena: a decrease in the limit restriction leads to higher values of the safety factor and an increase of the depth of corrosion into material leads to lower values of the safety factor. The derivative effect is the occurrence of the falloff associated with the length of the corrosion (in middle parts of each safety surface), and this phenomenon has origins in pile mechanics coupled with the soil. The two presented random elementary processes, the corrosion and the deflection of structure, are connected by the corrosion depth variable.

4 Discussion

This work was devised in response to the lack of general guidelines beyond design and classification rules for piles and Controlled Modulus Columns. The complexity of the problem for horizontally loaded piles was described by the following: an environment with progressive corrosion; inhomogeneous material under destruction process; random substrate material as a mechanical substrate.

A simple procedure has been presented in this work and can be used for evaluating the reliability of piles in the face of ongoing material degradation. Moreover, we made some simplifying assumptions to emphasize pre-selected effects: random fields that modelled the soil properties are one dimensional; the model of soil-column interaction needs further studies for full representation as a FEM 3D model; the corrosion process is only restricted to pure concrete material; investigators focus on controlling the rust creation mechanism [10].

The results are consistent with engineering intuition. The presented processes during the life span of the construction consider time as next operating factor for engineers or it involves reducing the level of confidence for existing objects. Three main conclusions can be drawn from this study: Corrosion provides positive results of reliability corresponding to experimental results. We observed that a decrease in the reliability index is connected to both dimensions of the corroded zone and time for its parameters that are typical and common in field testing; the algorithm for generating random field GRF is fast and convenient for solving the one-dimensional problems. It established the assumed spatial correlation of a random variable; the modified response surface method in this same level of quality produces low values of probability with lower time consumption in the calculation process and combines the advantages of the direct methods from the Monte Carlo family, discarding problems connected with first and second order methods.

The use of ground improvement as piles provide cost and time savings and a more sustainable solution for the construction of foundations on sites with poor quality soils compared to more traditional solutions. The presented approach to measuring a sensitive complex system reliability with time dependent processes is an answer to the main question designers have about the dependency between limitations of head displacement conditions and time and safety.

References

1. Mori, Y., Ellingwood, B.R.: Reliability based service life assessment of aging concrete structures. J. Struc. Eng. **119**, 1600–1621 (1993)
2. Luping, T., Andersen, A.: Chloride ingress data from five years field exposed in a Swedish marine environment. In: 2nd International Workshop on Testing and Modelling the Chloride Ingress into Concrete (pp. 1–15), Paris (2000)
3. Mahmoodian, M., Alani, A.M.: Multi-failure mode assessment of buried concrete pipes subjected to time-dependent deterioration, using system reliability analysis. J. Fail. Anal. Prev. **13**, 634–642 (2013)
4. Ekici, A., Huvaj, N.: Validation of 3D finite element solution for laterally loaded passive piles. 8th European Conference on Numerical Methods in Geotechnical Engineering, June 2014. https://doi.org/10.1201/b17017-117
5. Biernatowski, K., Pula, W.: Probabilistic analysis of the stability of massive bridge abutments using simulation methods. Struct. Saf. **5**, 1–15 (1988)
6. Bauer, J., Kozubal, J., Puła, W., Wyjadłowski, M.: Application of HDMR method to reliability assessment of a single pile subjected to lateral load. Stud. Geotech. et Mech. **34**, 37–51 (2012)
7. Muszyński, Z., Rybak J.: Horizontal displacement control in course of lateral loading of a pile in a slope. In: IOP Publishing Ltd IOP Conference Series: Materials Science and Engineering, vol. 245, pp. 1–8 (art. 032002) (2017)
8. Ming, F., Deng, Y.S., Li, D.Q.: Mechanical and durability evaluation of concrete with sulphate solution corrosion. Adv. Mater. Sci. Eng. (2016). https://doi.org/10.1155/2016/6523878 (Article ID 6523878)
9. Hong, H.: Assessment of reliability of aging reinforced concrete structures. J. Struct. Eng. ASCE **126**, 1458–1465 (2000)
10. Stewart, M., Rosowsky, D.: Structural safety and serviceability of concrete bridges subject to corrosion. ASCE J. Infrastruct. Syst. 41, 146–155 (1998)

Experimental Investigation of the Effect of Internal Erosion on the Behavior of Collapsible Soils

Med Salah Laouar, Adel Djellali, and Abdelkader Houam

Abstract

This is an experimental study on an actual topic of collapsible soils, which has offered a range of diversified applications. These soils are characterized by the unsaturation and the abrupt collapse after wetting, located in many parts of the world, especially when they are arid or semi-arid. These types of soils are located in a significant number of countries in particular those of the northern hemisphere between the 30th and 55th parallels as well as countries of South America. The cycles of prolonged dryness which occurred in the past years on several occasions and in several regions of the world modified the parameters governing the behavior of the soil and gave rise to new collapsible soils zones. The results of the tests of this experimental study, conducted on reconstituted soils in a large number of tests, indicated the influence of the parameters retained on the collapse potential, the migration of fines and the ultrasonic speeds. Similarly, the fine particles played an important role in the formation of macroporous loose structures.

Keywords

Collapsible soil • Migration • Fine particles • Hydraulic gradient

1 Introduction

The prevention of natural or industrial risks is an integral part of the concerns of the research teams, in this sense, the dimensioning of works able to repose on collapsible soils is a major problem of civil engineering. The unpredictable nature of the collapse, which can get dangerous ampleness, does not allow the use of purely deterministic approaches. Indeed, the comprehension of the collapse phenomenon, of its causes and mechanisms is essential. Various theoretical approaches and experimental methods for predicting this type of soils are in progress, offering some remedies for geotechnical problems identified with these soils [1]. Shows that the most vulnerable soils to the suffusion are well-graded, and that the internal erosion is one of the main causes of failure of hydraulic structures. According to [2], 71 cases of internal erosion counted in France between 1970 and 1997. The results of this experimental work conducted on reconstituted soils, composed of sand and kaolin in different proportions, indicate the influence of the initial moisture content, the state of compactness and the hydraulic gradient over the collapse potential and fines migration. They reveal a certain number of aspects of the hydro-mechanical behavior of collapsible soils and show the prediction tool of the migration of fine particles.

2 Properties of Materials and Experimental Program

2.1 The Sand

The sand used for soils reconstitution was washed river sand extracted from the river of Mellegue in the Tebessa region. Given the low percentage of its fine content, it was used for making concrete. The grain size of the sand was between 2 and 0.08 mm and 2.83% of the particles' weight was less than 80 μm. The chemical composition and the geotechnical characteristics of this sand are represented in Tables 1 and 2, respectively.

M. S. Laouar (✉) · A. Djellali
Civil Engineering Department, Larbi Tebessi University, 12002 Tebessa, Algeria
e-mail: C_laouar@yahoo.fr

A. Houam
Civil Engineering, Department College of Engineering, King Khalid University, Abha, Kingdom of Saudi Arabia

A. Kallel et al. (eds.), *Recent Advances in Geo-Environmental Engineering, Geomechanics and Geotechnics, and Geohazards*, Advances in Science, Technology & Innovation, https://doi.org/10.1007/978-3-030-01665-4_33

Table 1 Chemical composition of the adopted sand

Composition	SiO_2	FeO_3	Al_2O_3	CaO	MgO	NaCl	P.F
Content (%)	14.28	3.44	2.81	43.40	3.62	0.29	31.20

Table 2 Geotechnical characteristics of the tested sand

γs (g/cm^3)	γd_{min} (g/cm^3)	γd_{max} (g/cm^3)	e_{min}	e_{max}	Cu	Cc	Es (%)	Grain shape
2.67	1.34	1.73	0.54	0.85	2.29	0.96	69.25	Rounded

2.2 The Kaolin

The used Kaolin in the soils reconstitution (<80 μm) was taken from Tamazert deposit in the region of Elmilia-Algeria. It is generally used in the manufacture of fine porcelain used as raw materials for pottery and ceramic products. The chemical composition and the kaolin geotechnical characteristics of are presented in Tables 3 and 4.

2.3 Reconstituted Soils

The tests were performed on five reconstituted soils consisting of fine particles of kaolin and sand in various proportions. The geotechnical properties of the same soils are presented in Table 5.

Experimental Program

The experimental program included several numbers of oedometer, internal erosion and ultrasonic tests. The retained parameters in each test series and the number of tests are detailed in Table 6.

3 Results and Discussion

3.1 Oedometer Tests

The collapse potentials obtained from various tested soils vary as follows: for soil 1, from 0.88 to 14%; for the soil 2, from 0.95 to 15.9%; for the soil 3, from 3.45 to 17.2%; for the soil 4, from 3.95 to 18.65%; for the soil 5, from 4.5 to 19.61%; The collapse criteria reported by [3] show that the tested soils are susceptible to collapse.

3.2 Influence of Fine Particles on the Potential of Collapse

It is clear that whatever the values of compaction energy (E) and of the hydraulic gradient (i), the soil5, containing the highest quantity of fine particles, exhibits the greatest collapse susceptibility, Fig. 1. The compaction energy and the hydraulic gradient influences the behavior of the tested soils

Table 3 Tamazert kaolin chemical composition

Composition	SiO_2	Al_2O_3	Fe_2O_3	TiO_2	CaO	MgO	K_2O	Na_2O
Content (%)	71.51	18.48	0.58	0.15	0.22	0.26	3.15	0.10

Table 4 Kaolin geotechnical characteristics

Specific surface	Ø < 2 μm (%)	w_L (%)	w_p (%)	IP (%)	Gs	Ac
15 m^2/g	42	65.85	39.34	26.42	2.45	0.63

Table 5 Reconstituted soils geotechnical properties

	Soil 1	Soil 2	Soil 3	Soil 4	Soil 5
Kaolin proportion (%)	10	20	30	40	50
Sand proportion (%)	90	80	70	60	50
G_S	2.46	2.56	2.59		2.65
w_L (%)	16.47	18.47	26.63	28.97	35.37
w_p (%)	11.03	11.95	14.77	15.37	20.87
d_{max} (g/cm^3)	1.84	1.96	1.82	1.95	2.04
w_{opt} (%)	13.88	12.82	14.67	9.43	8.62
Ø < 2 μm (%)	4.91	7.03	9.84	14.12	16.74

Table 6 Tests program and parameters

Test type	Retained parameters	Number of tests
Oedometer	Moisture contents: 2, 4, 6 and 8% Compaction degrees: 20, 40 and 60 blows/layer Hydraulic gradient: 0, 15, 30, 45 and 60	60
Internal erosion	Moisture contents: 4% Compaction degrees: 20, 40 and 60 blows/layer Hydraulic gradient: 0, 7.5, 10, 12.5 and 15	15
Ultrasonic	Moisture contents: 2, 4, 6 and 8% Compaction degrees: 20, 40 and 60 blows/layer Hydraulic gradient: 10	12

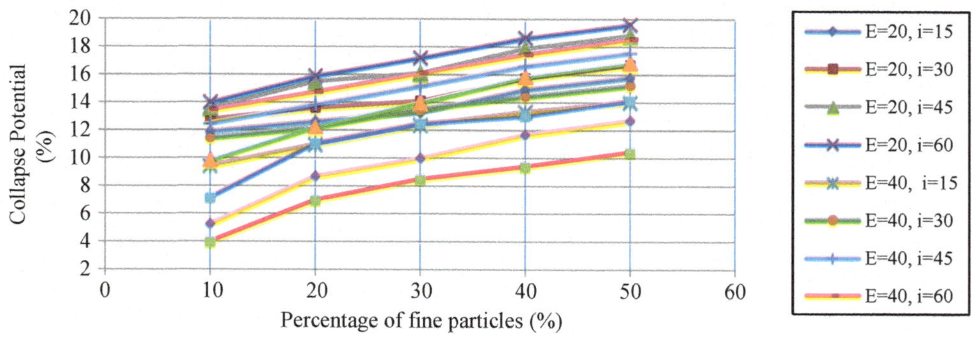

Fig. 1 Variation of collapse potential with percentage of fine particles

differently: for the same hydraulic gradient value, the increase of the compaction energy promotes the decrease of void volumes, the stability of the soil structure and the settlements' reduction. Given that the increase of moisture content by saturation decreases suction, at the same time it favors the breaking of inter-granular bonds. The result is a soil structure devoid of fines, rich in macrospores, susceptible to the collapse.

3.3 Influence of Moisture Content and Compaction Energy on the Ultrasonic Speed

In view of the results obtained in the series of ultrasonic tests Fig. 2, it should be noted that for all the tested soils, the ultrasonic speed is proportional with the initial moisture content and the compaction energy. The findings mentioned

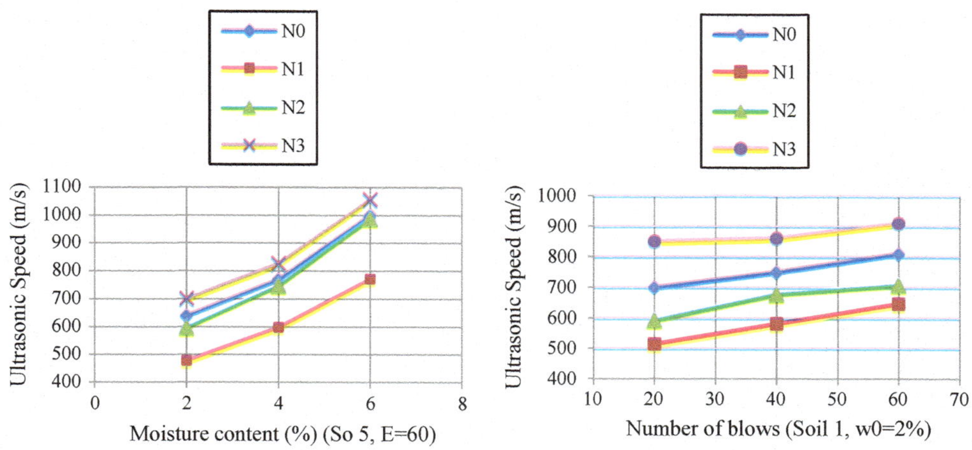

Fig. 2 Variation of ultrasonic speed with moisture content and number of blows

above were confirmed by representing in the same benchmark the variation of ultrasonic speeds depending on the number of blows for all soils.

4 Conclusion

The importance of this experimental study was the synchronization between the tests series. The internal erosion tests show that the extraction of the fine particles under the effect of hydraulic flow followed of the fines' migration after eliminating the cohesion is the main cause of the soils' collapse. During the flood of collapsible soils the fine particles move, in the direction of flow, through their structures from one level to another. The ultrasonic auscultation is a new method for predicting the soil's collapse. This is a very relevant approach, that can be used to highlight the heterogeneity of the soil, locate the voids volumes, pores and the state of compactness, on the one hand, and quantify and monitor the process of internal erosion of the fine particles, on the other hand.

References

1. Chetti, A., Hazzab, A., Korichi, K.: Modélisation d'un écoulement diphasique érodant. Larhyss **J. 9**, ISSN 1112–3680 (2011)
2. Fry, J.J., Degoutte, G.: L'érosion interne: typologie, détection et réparation, Barrages et réservoirs, n°. 6, 26 p. Comité Français des Grands Barrages (1997)
3. Ayadat, T., Ouali, S.: Identification des sols affaissables basée sur les limites d'Atterberg. Note Technique. Revue française de géotechnique (1999)

Study on the Deformation Properties of Functionally Gradient Metro Tunnel Lining Structure

Tong-Tong Zhang and Zhen-Dong Cui

Abstract

In this article, a literature review about FGMs and a new idea were presented. In the past few years, the concept of FGMs has been introduced in the geotechnical engineering; some scholars have done some research about it. On this basis, this paper provided a new idea about the future application of FGMs in geotechnical engineering. The results show that the functionally gradient lining structure can meet the deformation requirement when the elastic modulus is reduced in some positions. In other words, the functionally gradient lining structure can reduce the material cost greatly.

Keywords

FGMs • Functionally gradient lining structure • Deformation • Structure design

1 Introduction

Great performance achievements are already well at hand for the class of materials called composites, in which one type of hybrid material is functionally gradient materials (FGMs) [1]. It is a new functional material whose composition and structure change continuously from one direction to another, making the properties of materials change continuously [2, 3]. The variations of composition and properties in conventional composite materials and FGMs are shown as Fig. 1.

The concept of functionally gradient material can introduced into the design of metro tunnel lining structure to reduce its cost. On this basis, the concrete mark can be reduced in some position of the lining structure. If the deformation can meet the requirement, the construction of the lining structure will greatly reduce the cost.

T.-T. Zhang · Z.-D. Cui (✉)
China Skate Key Laboratory for Geomechanics and Deep Underground Engineering, School of Mechanics and Civil Engineering, China University of Mining and Technology, Xuzhou, 221008, Jiangsu, People's Republic of China
e-mail: cuizhendong@cumt.edu.cn

2 The Research Status on Mechanical Characteristics in the Functionally Gradient Cylinder

Some scholars used different forms of elastic modulus functions to analyze the internal force distribution of functionally graded cylinder. In general, the elastic modulus is considered as a function of radius, and the Poisson's ratio is assumed to be a constant. By summarizing the literature, the functions of elastic modulus along the radius can be roughly divided into the following laws: linear law [4], power law [5], exponential law [6] and other laws [7]. The specific forms are shown as follows.

$$E(r) = Ar + B \tag{1}$$

$$E(r) = E_0 r^{\beta} \text{ or } E(r) = E_0\left(\frac{r-a}{b-a}\right)^{\beta} \text{ or } E(r) = E_0\left(\frac{r}{a}\right)^{\beta} \tag{2}$$

$$E(r) = E_0 e^{\left(\frac{\beta(r-a)}{b-a}\right)} \text{ or } E(r) = E_0 e^{\left(\frac{\beta r}{b}\right)} \tag{3}$$

$$E(r) = E_0(1 + \beta s + \gamma s^2), \quad s = \frac{r-a}{r-b} \tag{4}$$

where a is the inner radius, b is the outer radius.

To improve it, the elastic modulus function was set changing with three parameters, and the Poisson's ratio is considered a constant, the specific form is shown as follows [8].

$$E(r) = E_0[1 - n(\frac{r}{b})^{\mathrm{k}}] \tag{5}$$

A. Kallel et al. (eds.), *Recent Advances in Geo-Environmental Engineering, Geomechanics and Geotechnics, and Geohazards*, Advances in Science, Technology & Innovation, https://doi.org/10.1007/978-3-030-01665-4_34

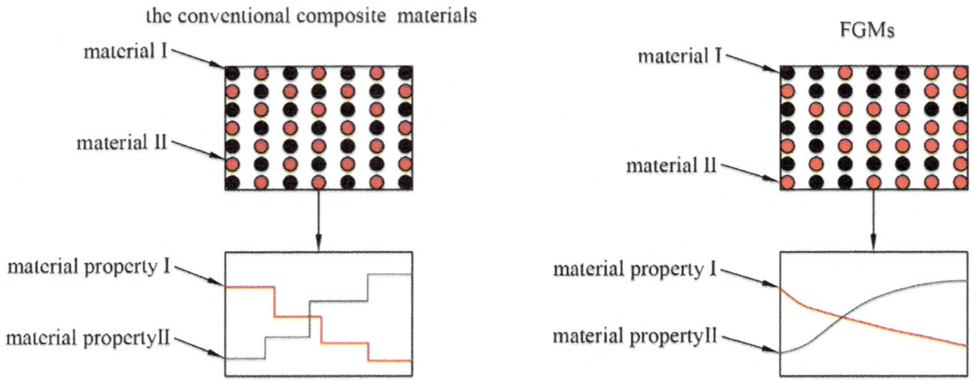

Fig. 1 Variations of composition and properties in conventional composite materials and FGMs

To find the optimal elastic modulus function along the radius in the functionally gradient cylinder, back analysis method was used based on a specific failure criterion [9].

3 The Research Status of FGMs in Geotechnical Engineering

The radial function of elastic modulus was derived in the functionally gradient vertical shaft based on the equal stress difference $(\sigma_\theta - \sigma_r)$ under uniform surrounding rock pressure [9]. The result is as follows.

$$E(r) = B\left[\frac{1-2\mu}{1-\mu}\ln(r/R_1)+1\right]^{\frac{2(1-\mu)}{1-2\mu}} \tag{6}$$

In addition, a function/structure integration model of functionally gradient concrete segment was developed where there is no interface between the functional layers [10]. The schematic diagram of the model is as shown in Fig. 2.

4 Discussion

From the above literature review, we can remark that a huge number of research efforts has been made in the internal force analysis of the FGMs under axisymmetric loads. In addition, the research of functionally gradient concrete metro tunnel segments has made some achievements. However, no one has analyzed the internal force and deformation of the functionally gradient metro tunnel lining structure under non-axisymmetric loads.

Above all, we can do some research in the deformation properties of the functionally gradient metro tunnel lining structure whose elastic modulus or other material properties change along a certain direction. In this point, the elastic modulus was set as changing with the angle, then, the deformation of functionally graded lining structures is derived based on structural mechanics and material mechanics. The deformation of homogeneous lining structure was compared to the functionally gradient structure; the result is shown as Fig. 3.

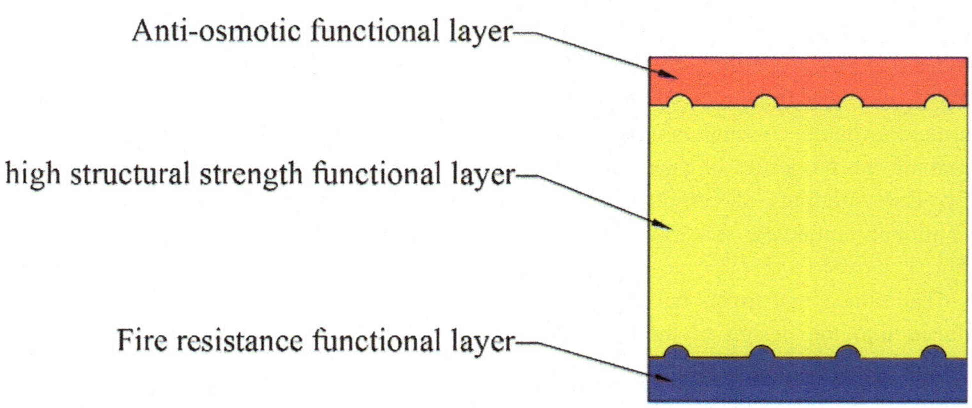

Fig. 2 Schematic diagram of integrated design of function/structure

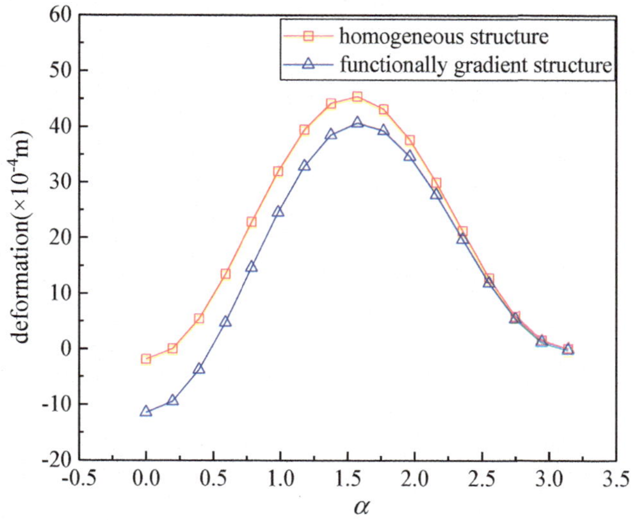

Fig. 3 The deformation of homogeneous lining structure and functionally gradient lining structure

5 Conclusion

A critical review and a new idea was given in the present paper, in which the research status has been discussed and some research results about the idea have been presented. The general conclusions are as follows.

(1) The study of functionally gradient structures is mainly focused on mechanical properties while the mechanical analysis of the functionally gradient metro tunnels lining structure has not started yet.
(2) The functionally gradient lining structure can meet the deformation requirement when the elastic modulus is reduced in some positions (Fig. 3); therefore, the functionally gradient lining structure can reduce the material cost while satisfying the normal use.

Acknowledgements This work was funded by National Key R&D Program of China (2016YFC0600904).

References

1. Gupta, A., Talha, M.: Recent development in modeling and analysis of functionally graded materials and structures. Prog. Aerosp. Sci. **79**, 1–14 (2015)
2. Koizumi, M.: FGM activities in Japan. Compos. Part B **28**, 1–4 (1997)
3. Wang, S.S.: Fracture mechanics for delamination problems in composite materials. Stud. Appl. Mech. **6**(3), 369–383 (1983)
4. Shi, Z., Zhang, T., Xiang, H.: Exact solutions of heterogeneous elastic hollow cylinders. Compos. Struct. **79**(1), 140–147 (2007)
5. Dai, H.L., Fu, Y.M., Dong, Z.: Exact solutions for functionally graded pressure vessels in a uniform magnetic field. Int. J. Solids Struct. **43**(18), 5570–5580 (2006)
6. Tutuncu, N.: Stresses in thick-walled FGM cylinders with exponentially-varying properties. Eng. Struct. **29**(9), 2032–2035 (2007)
7. Shokrolahi-Zadeh, B., Shodja, H.M.: Spectral equivalent inclusion method: Anisotropic cylindrical multi-inhomogeneities. J. Mech. Phys. Solids **56**(12), 3565–3575 (2008)
8. Eraslan, N.A., Akis, T.: Elastoplastic response of a long functionally graded tube subjected to internal pressure. Turk. J. Eng. Environ. Sci. **29**(6), 361–368 (2005)
9. Zhang, N.: Study on Bearing Behavior of Functionally Gradient Concrete Shaft Lining. North China Electric Power university (2012)
10. Wang, X.G.: Research and Application of Functionally Gradient Concrete Segment Used in River-crossing or Sea-crossing Tunnels. Wuhan University of Technology (2007)

Geological Context and Fracturing State of the Rock Massifs of the Jijelian Ledge (Northeast Algeria)

Chahra Yellas and Riad Benzaid

Abstract

The national road n°43 connecting the cities of Jijel and Bejaia (North-East of Algeria) runs along the Mediterranean and crosses over more than 50 km a volcano-sedimentary rock mass. The road development of this ledge is constantly apprehended for the needs of road traffic. Several topographic and geotechnical constraints permanently threaten the rocky cliffs overlooking the sea. In this article, we studied the fracturing of the rock by estimating its occurrence on the massif by statistical methods. The large number of discontinuities observed in the study area by the azimuthal orientation measurements of the dip of the rupture plane of the various joints recorded, the spacing and the distribution of the discontinuities served as a basis for our work.

Keywords

Fracturing • Rock mass • Discontinuity • Stereographic projection

1 Introduction

The behavior of a rock mass depends not only on the mechanical properties of the rock itself but also on the discontinuities it contains [1–3]. The existence of structural planes induces discontinuity and anisotropy of the rock masses, which influences the mechanical properties and infiltrations to a large extent [4].

The road section concerned by our study is part of the national road n°43 (RN43) linking Jijel to Béjaia (Northeast Algeria). Several instability points have been identified along this section of the road, whose morphology is dominated by very high reliefs and very steep and jagged cliffs. The study of the state of fracturing of a rock mass of this ledge by statistical and numerical methods was addressed in this article to define the families of discontinuities and their influence on the stability of the studied massif. Based on geometric parameters proposed by the ISRM [5, 6], we evaluated the state of fracturing of a rocky massif on a sensitive section of the RN43 on the jijelian ledge.

2 Geological Context

Oriental Babors are part of the internal tellian domain. It is the geographical entity extending from West to East, and from North to South, from the Mediterranean to the Babors Mountains [7]. Several distinct tectonic units have been recognized [8]. These are characterized by their stratigraphic composition and metamorphic evolution, each unit being subdivided into subunits (Fig. 1).

Geographically, Brek's unit presents a long limestone chine extending from south to north from the Adrar El-Alem to the Tazeguezaout Mountain.

This limestone chine is interrupted in its middle by a western basin of Beni Zegoual and an oriental basin drained by the river of Dar El-Oued and separated by a line of crest joining the southern chine to the chains [7] (Fig. 2). The Brek-Gouraya tectonic unit located north-east of the Baborien massif is crossed by the road section, subject our study. It consists of a set of carbonate scales often cut into vertical strips.

3 Description and Morphology of the Massif

Our study focused on a mountainous massif with steep slopes often cut by cliffs. The discontinuities resulting from fragmentation in blocks were diffused and favor the instability of the slope. Among the discontinuities found on the

C. Yellas (✉) · R. Benzaid
Laboratoire de Génie Géologique (LGG), Université Mohammed Seddik Benyahia- Jijel, BP. 98, Ouled Aissa, 18000 Jijel, Algérie
e-mail: chahra.yellas@gmail.com

A. Kallel et al. (eds.), *Recent Advances in Geo-Environmental Engineering, Geomechanics and Geotechnics, and Geohazards*, Advances in Science, Technology & Innovation, https://doi.org/10.1007/978-3-030-01665-4_35

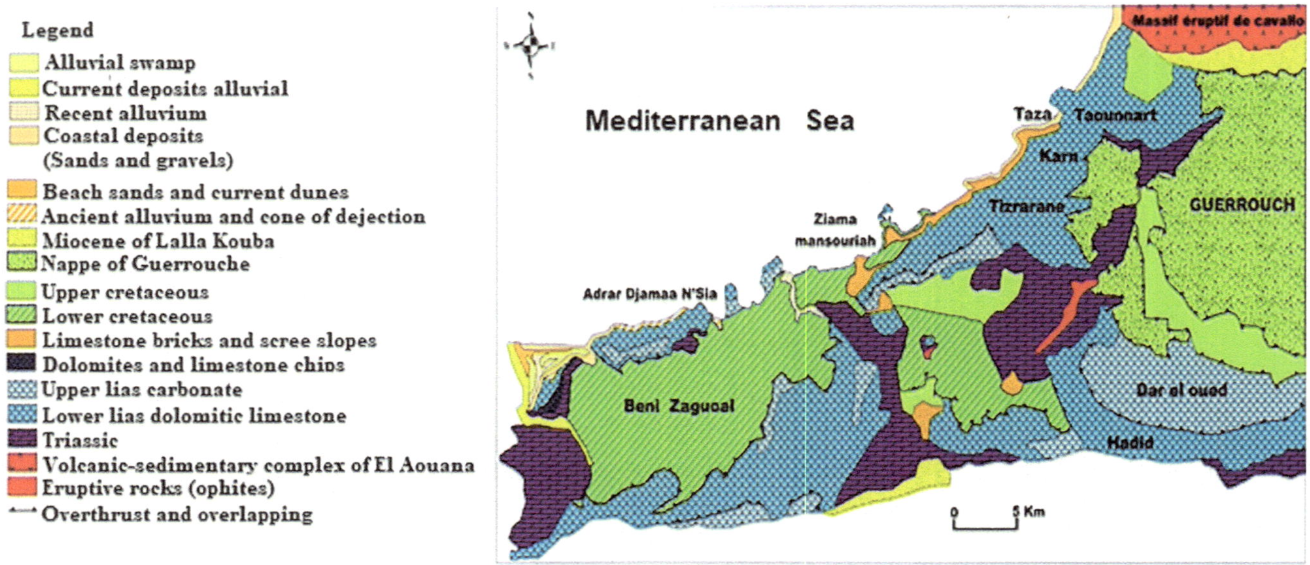

Fig. 1 Geological sketch of the Brek-Gouraya unit Oriental Babord, (from Obert [7], slightly modified)

Fig. 2 Interpretative geological section of the Baborian tectonic units, Jijel- Algeria

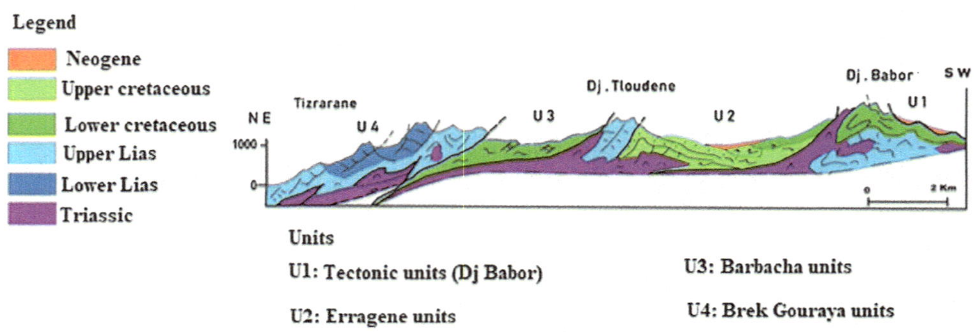

Fig. 3 Different lines of measurement of the rock mass

magma massif of El-Aouana, faults, cracks, and joints (Fig. 3). It is noteworthy to remark that within the Baborian carbonates, the karst flow network adds to the series of discontinuities and further complicates the analysis of fracturing inside the massif because of the lack of information on the extension of these cracks in depth.

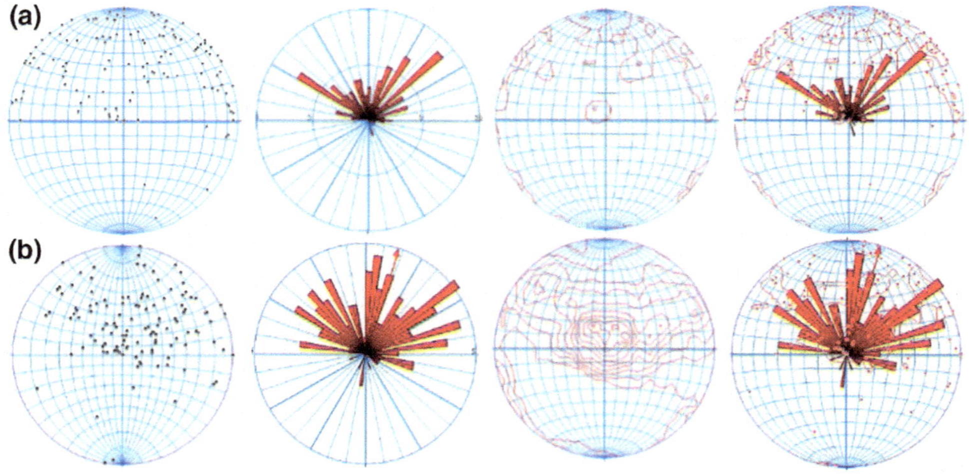

Fig. 4 Stereographic projection results. **a** Measures taken on the Aftis site, **b** measures taken on the El-Aouana site

Petrographic and mineralogical studies on samples from the El-Aouana and Aftis sites have revealed the existence of volcanic (dacite, andesite) and subvolcanic (micro diorite, micro grano-diorite and diorite) adakitic facies [9].

4 Results and Discussion

The study of the natural discontinuities of the rock mass was conducted according to several classes observed in the field. Each discontinuity class is characterized by its azimuth, dip, cliff spacing, infill and extension, although the latter is locally difficult to follow on the cliff overlooking the road. Field data were processed by stereographic projection on the lower hemisphere of the Schmidt Canvas, the results of which are shown in the stereograms of Fig. 4. The stereographic projection applied to the magma massif of El-Aouana clearly clarified the state of fragmentation of the front of the massif. The compilation of data using the stereonet revealed three dominant families of vertical and oblique fracturing: the first family N130 to N150, the second family N70 to N80 of tectogenetic order related to the structuring phases of the chain and a third N10 to N20 fracture family appears to be of lesser extent, probably due to the decompression of the ascending magmatic mass, and presents one of the subhorizontal dips.

The high density of the fractures indicates the proximity of the regional shear zones linked to tertiary tectogenesis of the massif. Our work also allowed highlighting two different fracturing states depending on the magmatic (El-Aouana site) or sedimentary (Aftis site) nature of the massifs.

5 Conclusion

The characterization of the network of discontinuities of a rock mass crossed by a road layout requires a capital importance for the durability of this infrastructure and the safety of the traffic.

As part of this work, we noticed that the intensity of the fracturing cuts the studied massif into nested blocks. The different families of identified discontinuities interfered in the rock mass according to their respective directions. This conditioned both the mechanical and hydraulic properties of the massif as well as its stability.

The stereographic projection applied to the magma massif of El Aouana allowed us to evaluate the state of fragmentation of the front of the massif. Compiling the data and using the stereonet revealed three dominant families of vertical and oblique fractures namely, the N130 to N150 family, the N70 to N80 family of tectogenetic order related to the structuring phases of the chain and a last family of fracture N10 to N20 which seems to be of less extension and presents one of the subhorizontal embeddings.

References

1. Hoek, E., Martin, C.D.: Fracture initiation and propagation in intact rock e A review. J. Rock Mech. Geotech. Eng. **6**, 287–300 (2014)
2. Li, X., Zhu, W.: The damage fracture analysis of ajointed rock mass and its application in engineering. Eng. Fract. Mech. **43**(2), 165–I 70 (1992)

3. Singhal, B.B.S., Gupta, R.P.: Fractures and discontinuities. In: Applied Hydrogeology of Fractured Rocks, 2nd edn. Springer, Heidelberg (2010)
4. Chen, W.Z., Yang, J.P., Tan, X.J., Yu, H.D.: Study on mechanical parameters of fractured rock masses. Sci. China Tech. Sci. **54**(1), 140–146 (2011)
5. ISRM: International Society for Rock Mechanics. Suggested methods for the quantitative description of discontinuities in rock masses. Int. J. Rock Mech. Min. Sci. Geomech. Abstr. **15**, 319–368 (1997)
6. Kimour, M., Serradj, T.: Characterization of geological rock mass case of the Socar Heliopolis – Guelma, Algeria Aggregate Quarry. Proc. Earth Planet. Sci. **15**, 205–212 (2015)
7. Obert, D.: Etude géologique des babors orientaux (Domaine tellien, Algérie) Ph.D. thesis, University of Paris 6, Paris, France (1981)
8. Leikine, M., Obert, D., Bellier J.-P.: Integration des Babors aux nappes telliennes; existence d'un metamorphisme ante-nappe. Bulletin de la Société Géologique de France **S7-XVII**(5), 764–772 (1975)
9. Benali, H., Semroud, B., Belanteur, O.: Sur la présence d'Adakites à El Aouana (Algérie). Première Conférence Internationale de Géologie Africaine, Assiut, Egypte (1999)

Engineering Geological Assessment Using Geochemical, Mineralogical and Petrographic Analysis Along the Riyadh Metro Line 3 (Saudi Arabia)

Manuel Cueto, Carlos López-Fernández, Luis Pando, and Daniel Arias

Abstract

Anticipating the geotechnical constraints for construction is necessary to ensure the viability of engineering and building projects, especially in challenging geological contexts such as Riyadh. This paper shows the use of geological identification tests to detect and assess hazards along the Line 3 of the Riyadh Metro Project, which is the largest metro project ever built from scratch. A total of 475 samples of soil and rock were chemically, mineralogically and petrographically analyzed prior to construction. The obtained results enabled the successful assessment of potential geotechnical hazards such as collapse, swelling, stickiness and abrasiveness along Line 3. These results were essential to consequently implement corrective measures during construction.

Keywords

Engineering geology • Geotechnical hazard • Riyadh • Metro line 3

1 Introduction

Geological constraints are one of the main causes of construction delay and excessive costs in large infrastructure projects. In engineering geological studies, it is crucial to understand the soils and rocks' chemical and mineralogical composition. These materials can produce processes of swelling, aggressiveness, etc., resulting in bad ground conditions. This article aimed to examine the main geological hazards identified within Line 3 (L3), which is the longest line under construction in the KSA capital (41 km) in the framework of the largest metro project ever built from scratch in the world. The Riyadh Metro Project totals 176 km, with 85 stations and 7 depots. Completion is expected to be by the end of 2018. In addition to the 342 boreholes and geophysical surveys (39 km of integrated seismic refraction, electrical resistivity and ground penetration radar), geochemical X-ray fluorescence (XRF), mineralogical X-ray diffraction (XRD) and petrographic analysis (PA) were also carried out to assess the potential hazards identified along L3. Consequently, preventive and corrective measures were implemented in order to avoid future geotechnical problems. The number of geological identification tests carried out was almost certainly higher than ever in engineering, due to the project importance and the complex geological settings of the site.

2 Geological Setting

Riyadh is located in the center of the Arabian Peninsula belonging to the Interior Homocline of the stable Arabian Shelf that crops out in a great curved belt flanking the eastern margin of the Arabian Shield [1]. Metro Line 3 runs through Late Jurassic to Lower Cretaceous carbonate rocks, which from bottom to top are: Jubaila Formation, Arab Formation (A, B, C and D members), Hith Formation and Sulaiy Formation (Fig. 1). In L3, even though karst is the most threatening geo-hazard [2], other potential dangers were identified prior to construction such us swelling, stickily-clay, mixed-face conditions, etc.

3 Methodology

A number of 475 rock and soil samples were collected from borehole cores (466) and open outcrops (9) along the entire L3 with special emphasis on the underground section through Riyadh's downtown. The chemical composition was determined by 458 XRF spectrometry tests. A Philips

M. Cueto (✉)
IDOM, Avda. Monasterio de El Escorial 4, 28049 Madrid, Spain
e-mail: mcueto@idom.com

C. López-Fernández · L. Pando · D. Arias
Department of Geology, University of Oviedo, Jesús Arias de Velasco s/n, 33005 Oviedo, Spain

A. Kallel et al. (eds.), *Recent Advances in Geo-Environmental Engineering, Geomechanics and Geotechnics, and Geohazards*, Advances in Science, Technology & Innovation, https://doi.org/10.1007/978-3-030-01665-4_36

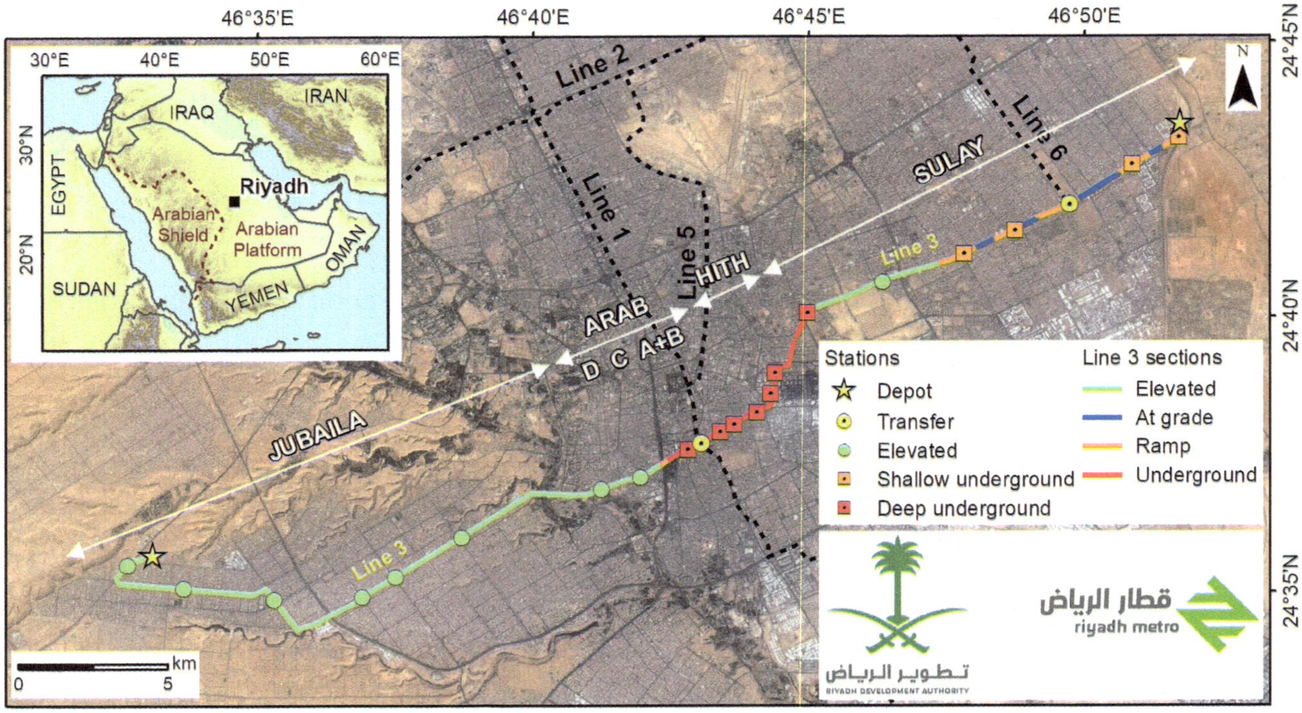

Fig. 1 General layout of the Riyadh Metro Project highlighting the different types of sections and stations along the Line 3. The main geological units are shown in white (RDA-ANM-IDOM)

PW2404 spectrometer equipped with a rhodium anode tube allowed the examination of major and trace elements according to ASTM C1271 [3]. The identification of total sulfur content based on ASTM D1619 [4] was assessed through a Leco sulfur analyzer in 458 samples. The mineralogical composition was identified by 453 XRD using a Philips X'Pert Pro diffractometer equipped with a copper anode tube. The analysis was performed according to the procedures of the International Centre for Diffraction Data. The quantitative phase study was based on the Rietveld method. Finally, the petrographic analysis was carried out by examining thin sections with a Zeiss-made microscope to identify textures and mineralogical components.

4 Results and Discussion

4.1 Geochemical Analysis

Most of the rock samples showed high calcium carbonate and very low sulphate content, corresponding to pure limestone, with a small percentage of argillaceous limestone and the dolomite presence limited to Arab D and Jubaila (Fig. 2). The linear trend is indicative of geochemical affinity.

Using the geochemical analysis, the total dissolution of the anhydrite levels in Arab and Hith formations was confirmed, as previously described by Sharief et al. [5]. This result discards the risk of structural damage by floor heaving in tunnels (swelling in anhydrite) or fracturing due to settlements (dehydration of gypsum). The soil layers showed high content of silica, calcium oxide, loss pack ignition elements and calcium carbonate. However, only the anthropic soils showed a significant presence of calcium sulphate minerals (gypsum, bassanite and anhydrite). The sulphate content was restricted to the top three meters of the ground in relation to the anthropic waters (8.2%). These results made essential the use of sulphate-resistant cement.

4.2 Mineralogical Analysis

Through XRD in the infilling soils of large sinkholes up to 12% of paligorskite and 8% of montmorillonite content was identified. These are potentially expansive minerals. It was concluded that in general their presence was generally low and this risk was finally discarded using specific geotechnical tests in the most hazardous areas (swell pressure and free swelling test). All the rock units showed similar mineralogical composition with low clay minerals presence and absence of expansive minerals. The presence of quartz is low (1.0 to 6.8%) which results in a low abrasive ground that would otherwise severely damage the cutting tool of the

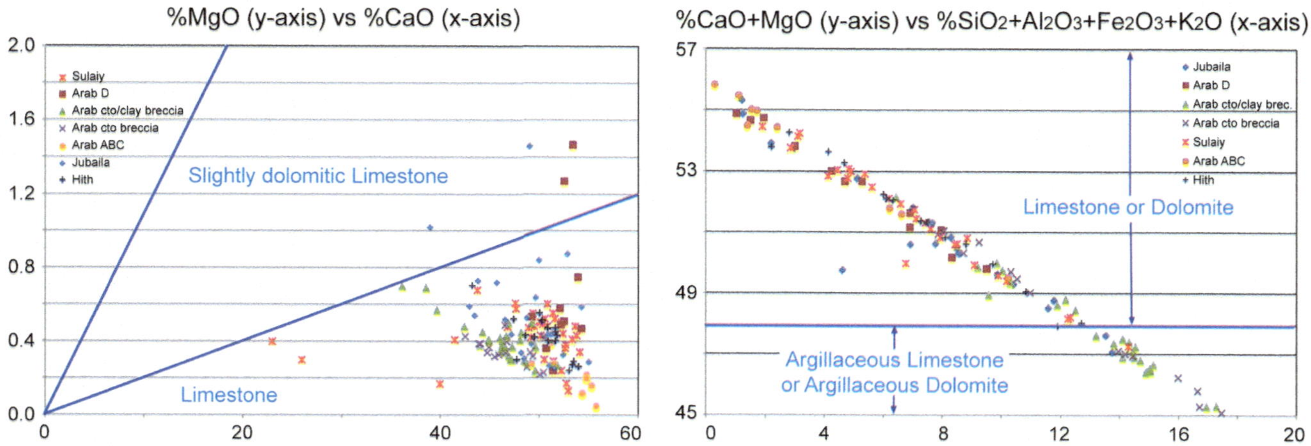

Fig. 2 Left: trend evolution between dolomite (%MgO) and limestone (%CaO) in rocks of L3 showing limestone predominance. Right: carbonate (%CaO & MgO) versus siliciclastic elements (%SiO_2, Al_2O_3, Fe_2O_3 & K_2O) with differentiation of limestone (Jubaila, Arab D & ABC and Sulaiy) from argillaceous limestone (Arab carbonate & clay breccia) (RDA-ANM-IDOM)

TBM. This fact was later confirmed using 11 Cerchar tests (CAI 0.32 to 0.95), which corresponds to low abrasiveness.

4.3 Petrographic Study

The petrographic study enabled the identification of different engineering geological units, which are listed in Fig. 2. Furthermore, in highly adhesive clayey units, stickiness causes difficulties during the use of TBM. Sticky-clay can affect the production rate by clogging moving parts of the TBM and sticking to the exposed steel [6]. Based on the test results, potential sticky ground conditions were predicted mainly in the Arab carbonate and clay breccia and Hith. On the other hand, the friction process and the water fluids can also adversely affect the TBM progress under mixed-face conditions, such as brecciated units with potential clayey matrix washing.

5 Conclusions

This study showed how the exhaustive multi-technique geological testing (X-ray fluorescence, X-ray diffraction and petrographic analysis) was crucial to shed light on potential geotechnical hazards, anticipating engineering problems which would otherwise have led to delays and excessive costs during construction. The systematic geochemical, mineralogical and petrographical tests were found as a highly successful and economic tool to predict geotechnical properties and bad ground conditions for the Line 3 of the Riyadh Metro. Further analysis will seek to show correlations between the aforementioned identification geological analysis and common geotechnical tests used in civil engineering projects, such as Cerchar, swell pressure, free swelling tests, etc.

Acknowledgements We would like to thank IDOM Consulting, Engineering, Architecture, S.A.U. for supporting this research and especially all the staff involved in the design of the Line 3. Also, the authors gratefully acknowledge Saudi Arabia's Riyadh Development Authority (RDA) promoting the Riyadh Metro Project and especially ArRiyadh New Mobility Consortium (ANM) selected to deliver the Package 2- Line 3.

References

1. Powers, R.W., Ramirez, L.F., Redmond, C.D., Elberg Jr., E.L.: Geology of the Arabian Peninsula: sedimentary geology of Saudi Arabia. U.S. Geological Survey Paper (1966)
2. Cueto, M., Olona, J., Fernández-Viejo, G., Pando, L., López-Fernández, C.: Karst-induced sinkhole detection using an integrated geophysical survey: a case study along the Riyadh Metro Line 3 (Saudi Arabia). Near Surf. Geophys. **16**, 270–281 (2018)
3. ASTM C1271-99: Standard Test Method for X-ray Spectrometric Analysis of Lime and Limestone. ASTM International, West Conshohocken, PA (2012)
4. ASTM D1619-16a: Standard Test Methods for Carbon Black—Sulfur Content. ASTM International, West Conshohocken, PA (2016)
5. Sharief, F.A., Khan, M.S., Magara, K.: Outcrop-subcrop sequence and diagenesis of Upper Jurassic Arab-Hith Formations, Central Saudi Arabia. J. King Abdulaziz Univ.-Earth Sci. **4**, 105–136 (1991)
6. Khademi, H.J., Shahriar, K., Rostami, J.: Double shield TBM in challenging difficult ground conditions—a case study from Nosoud water transfer tunnel, Iran. In: Rapid Excavation and Tunneling Conference (RETC), Nevada, USA (2009)

Strength Estimation of Evaporitic Rocks Using Different Testing Methods

Hasan Arman, Osman Abdelghany, Ala Aldahan, Mahmoud Abu Saima, Bahaa Mahmoud, Saber Hussein, Abdel-Rahman Fowler, and Saaed AlRashdi

Abstract

Rock strength is recognized as the ability of a rock to resist stress or deformation without breaking down. Testing methods are suggested by ISRM (International Society of Rock Mechanics) and ASTM (American Standards Testing Material), and include Unconfined Compressive Strength (UCS), Point Load Index (PLI), Indirect Tensile Strength (ITS), Schmidt Hammer Rebound (SHR) and Sonic Velocity (SV). In the investigation presented here a comparison of these methods was performed using evaporitic rocks. These sedimentary rocks are common in the Arabian Peninsula as exposures or in the subsurface and they may constitute the foundations of buildings. We have chosen samples of the Lower Miocene evaporitic rocks from Al Ain city, United Arab Emirates (UAE). 35 UCS, 51 PLI, and ITS, 640 SHR and 86 SV tests were carried out on either core samples or rock blocks according to ASTM Standards. The results revealed moderate to weak correlations between the UCS and V_p, PLI, ITS and SHR_{Rcor} for the evaporitic rocks. This behavior may relate to textural and mineralogical heterogeneity within the samples, despite overall homogenous hand specimen characteristics, and calls for caution in applying a single rock strength value in engineering applications.

Keywords

Rock strength • Evaporitic rocks • Unconfined compressive strength • Schmidt Hammer • Point load index

1 Introduction

Rock strength measurement and characterization of discontinuities (mainly fractures) are important tasks in engineering applications. Overall, rock strength is defined as the ability of rocks to resist stress or deformation without breaking down [1]. Inhomogeneity and rapid changes in the physical properties of rock masses cause unexpected rock failure, particularly when these factors are combined with the presence of water. Therefore, it is common to measure rock strength using several testing methods and approaches [2]. A number of methods are suggested for the testing of rock strength. These include Unconfined Compressive Strength (UCS), Indirect Tensile Strength (ITS), Point Load Index (PLI) and Schmidt Hammer Rebound (SHR) and Sonic Velocity (SV) tests. In this study we applied these tests on a suite of evaporitic rocks, with the aim of providing data that can be used to better constrain their rock strength properties. These sedimentary rocks are common in the Arabian Peninsula as exposures or in the subsurface and they may represent the foundations of buildings and infrastructure constructions.

The evaporitic rocks used here are found in the Al Ain region of the United Arab Emirates. The rocks belong to the Lower Miocene (Lower Fars Formation) and outcrop on the eastern limb of the Hafit Mountain (Fig. 1).

H. Arman (✉) · O. Abdelghany · A. Aldahan · M. A. Saima · B. Mahmoud · S. Hussein · A.-R. Fowler
United Arab Emirates University, P.O. Box: 15551 Al Ain, UAE
e-mail: harman@uaeu.ac.ae

O. Abdelghany
e-mail: osman.abdelghany@uaeu.ac.ae

A. Aldahan
e-mail: aaldahan@uaeu.ac.ae

M. A. Saima
e-mail: m.abusaima@uaeu.ac.ae

B. Mahmoud
e-mail: bahaa.mahmoud@uaeu.ac.ae

S. Hussein
e-mail: s_hussein@uaeu.ac.ae

A.-R. Fowler
e-mail: afowler@uaeu.ac.ae

S. AlRashdi
University of the Western Cape, P/Bag X17, Cape Town, 7535, South Africa
e-mail: saeed.alrashedi@milmail.ae

A. Kallel et al. (eds.), *Recent Advances in Geo-Environmental Engineering, Geomechanics and Geotechnics, and Geohazards*, Advances in Science, Technology & Innovation, https://doi.org/10.1007/978-3-030-01665-4_37

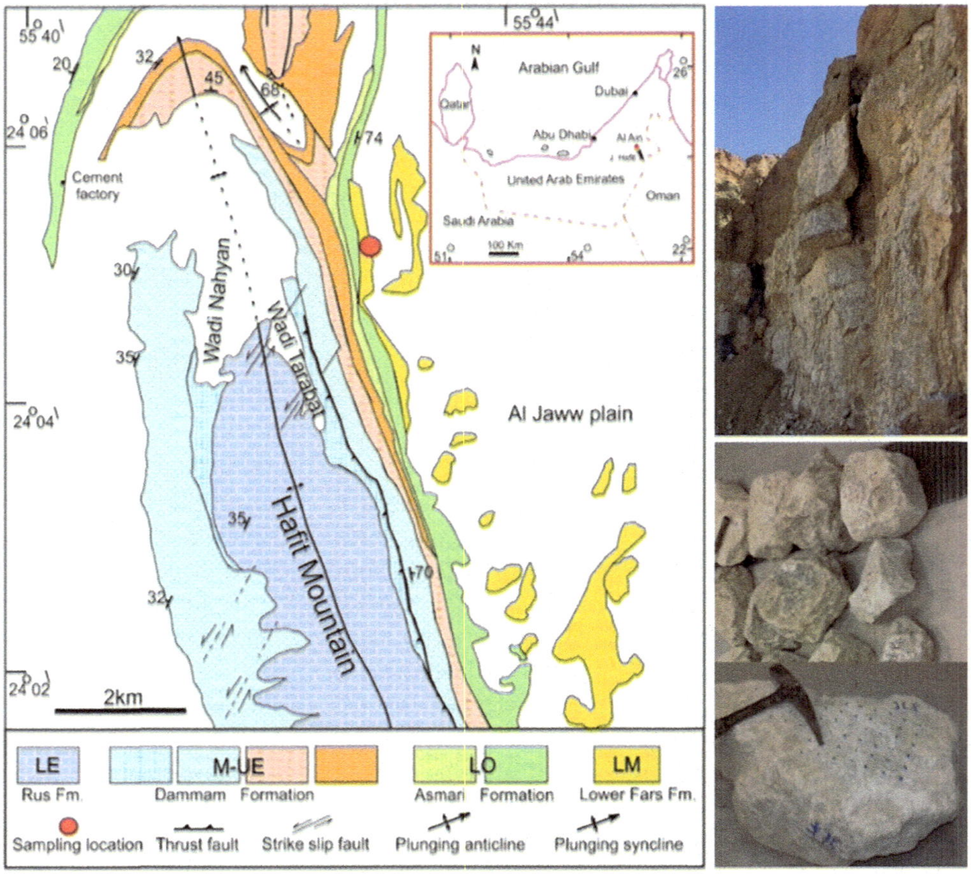

Fig. 1 Geological map and sampling location with exposure and rock block samples. LE = Lower Eocene, 55-49 Myr; M-UE = Middle to Upper Eocene, 49-34 Myr; LO = Lower Oligocene, 34-29 Myr and LM = Lower Miocene, 23-16 Myr. These formations are dominated by different types of carbonate rocks, with the exception of evaporates, which dominate the Lower Fars Formation. From top right to bottom right: photographs of quarry sample site (looking E); boulder sized rock samples; rock sample showing pattern of test points for SHR

2 Samples and Rocks Strength Testing

The evaporitic rocks used in this investigation lie on the western limb of Hafit Mountain (a major NNE trending asymmetric anticline) (Fig. 1). These rocks are part of the Lower Fars Formation (Lower Miocene 23-16 Myr) [3]. The evaporitic rocks are dominated by nodular, granular and crystalline gypsum as massive gypsiferous layers and veins that are interbedded with friable marls and mudstones. Within the exposed area of the Lower Fars Formation, a quarry site was located and used for the collection of 26 representative samples, over a formation thickness of about 10–20 m (Fig. 1). To reduce effects of anisotropy, samples showing any bedding signs were avoided. A total of 640 laboratory SHR were carried out on the rock block samples, before preparing the coring for other tests. In each rock block a minimum of 20 surface SHR tests were performed with a vertically downward orientation for the rock hammer. The SHR_{Rcor} (corrected value of SHR) value of each test was determined and reported according to ASTM Standard. A total of 97 core samples (NX, 54 mm in diameter) were prepared from the rock blocks for the 35 UCS, 51 PLI and ITS and 86 SV (for V_p and V_s) tests.

3 Results and Discussion

Table 1 presents the statistical results of the performed tests. According to the intact rock classification [4] the studied evaporitic rocks can be classified as medium strong using the UCS and PLI mean values. The relations between the mean PLI for the 50 mm in diameter size rock sample, ITS, SHR_{Rcor}, V_p and the mean UCS (Fig. 2) indicate medium to weak linear correlations. These mean values and relationships between the tests for evaporitic rocks are expected due to the relatively fragile crystalline structure of gypsum and the aggregated properties of the rocks in general. Furthermore, the interlayering occurrence of marl and mudstones on macro- and micro-scales affects the strength properties of the rocks. The rock strength test values of the evaporitic rocks are also lower than those for carbonate rocks in the same area [5] and thus provide a complimentary data set for engineering evaluation of this terrain. The strength data for evaporitic rocks presented here also draws attention to the variability in rock strength despite the texturally rather comparable rock samples that were used. This is an important issue for evaporitic rocks locally and in general, considering the broad range of textures and mineralogical compositions for these rocks.

Table 1 The statistical results for the tested samples

	UCS (MPa)	PLI (MPa)	ITS (MPa)	SHR_{Rcor} (N) blocks	SV (km/s)	
					V_p	V_s
Number of samples	35	51	51	26	86	86
Minimum	17.46	0.53	1.47	19.55	4.19	1.65
Maximum	40.03	2.19	4.39	27.55	6.15	3.43
Average	27.86	1.40	2.80	23.22	5.35	2.47
Standard deviation	5.36	0.49	0.67	2.29	0.50	0.44

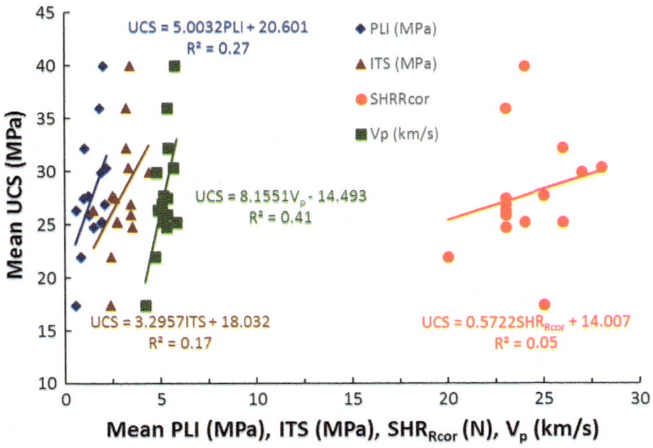

Fig. 2 Mean PLI, ITS, SHR_{Rcor}, V_p versus mean UCS

4 Conclusion

The results of this study indicate scattered values for the UCS, PLI, and ITS, SHR_{Rcor} and SV tests in the evaporitic rocks. This feature, in addition to the moderate to weak relationships between the different test parameters, are within expected values and trends for evaporitic rocks. The data also suggest that further caution should be taken during engineering applications dealing with evaporites, due to the sensitivity of their strength properties to textural variations.

References

1. Hawkins, A.B.: Aspects of rock strength. Bull. Eng. Geol. Environ. **57**, 17–30 (1998)
2. Arman, H., Hashem, W., Abdelghany, O., Aldahan, A.: Effects of lithofacies and environment on in situ and laboratory Schmidt hammer tests: a case study of carbonate rocks. Q. J. Eng. Geol. Hydrogeol. **50**, 179–186 (2017)
3. Boukhary, M., Abdelghany, O., Bahr, S.: *Nummulites alsharhani* n. sp. (Late Lutetian) from Jabal Hafit and Al Faiyah: Western side of the Northern Oman Mountains, United Arab Emirates. Revue Paleobiol Geneve **21**, 575–585 (2002)
4. Marinos, P., Hoek, E.: Estimating the geotechnical properties of heterogeneous rock masses such as Flysch. Bull. Eng. Geol. Environ. **60**, 85–92 (2001)
5. Arman, H., Hashem, W., Tokhi, E.M., Abdelghany, O., Saiy, E.A.: Petrographical and geomechanical properties of the lower oligocene limestones from Al Ain city, United Arab Emirates. Arab. J. Sci. Eng. **39**(1), 261–271 (2014)

Characterization of Soil Stability to Withstand Erection of High-Rise Structure Using Electrical Resistivity Tomography

Theophilus A. Adagunodo, Lukman A. Sunmonu, Olagoke P. Oladejo, and Anuoluwapo M. Olanrewaju

Abstract

In this paper, we used the Electrical Resistivity Tomography (ERT) technique to examine the suitability of the subsurface for its ability to withstand erection of a proposed high-rise structure in Emmanuel Alayande College of Education, Oyo, Nigeria. The Wenner array was used for the ERT survey, with the varying electrode separations of 1.0, 3.0, 6.0 and 8.0 m respectively, and the electrode increment of 5.0 m across the three (3) traverses that were established in the study area. The traverses were of distance 100 m each, with W-E orientation that would enable the subsurface imaging of the study area. The subsurface features experienced in the study area were topsoil/laterites, weathered layer, clayey zone, and bedrock. The inverse model along traverses 1 and 2 revealed that the clayey zones beneath these traverses are very thick, which showed that the study area was unsuitable for construction of high-rise building without the certified building engineers' advice.

Keywords

Soil characterization • Soil stability • High-rise structure • Electrical resistivity tomography • Pre-foundation investigation

T. A. Adagunodo (✉)
Department of Physics, Covenant University, Ota, Nigeria
e-mail: taadagunodo@yahoo.com

L. A. Sunmonu
Department of Pure and Applied Physics, Ladoke Akintola University of Technology, Ogbomoso, Nigeria

O. P. Oladejo
Department of Physics, Emmanuel Alayande College of Education, Oyo, Nigeria

A. M. Olanrewaju
Department of Mathematics, Covenant University, Ota, Nigeria

1 Introduction

The required time and financial implications of geotechnical investigation for civil engineering activities can discourage the structural engineers and developers [5]. In contrast, geostatistical investigation is inexpensive and fast [10]. Besides, the geotechnical test results are meant for point-to-point information [6]; it is unsuitable to produce volumetric information about the subsurface structure.

In this study, the characterization of soil stability for superstructure construction was achieved by carrying out Electrical Resistivity Tomography (ERT) measurements adopting Wenner array in order to relate the images generated from electrical signatures to the subsurface stability information. Wenner array measures the subsurface lateral variation, while the electrode spacing corresponds to the depth of investigation. Applications of Electrical Resistivity (ER) technique have been reported from appreciable number of publications, among which are Samouelian et al. [9], Sudha et al. [10], Adagunodo et al. [4], Adagunodo et al. [3], Adagunodo et al. [2], Wahab and Saibi [11], Adagunodo et al. [1], and Oyeyemi et al. [8].

The management of Emmanuel Alayande College of Education, Oyo has recently approved of the construction of a mega structure auditorium for academic purposes. The auditorium is expected to be erected between the Fine and Applied Art and Sandwich Degree Building within the institution. The site is resting on the basement rocks of SW Nigeria (1) as revealed in Fig. 1. This work was expected to provide information to the professional builders prior to construction.

2 Materials and Method

A digital PZ-02 earth resistivity meter was used to carry out the ERT survey in the study area. The Wenner array was used for the ERT survey, with the varying electrode

A. Kallel et al. (eds.), *Recent Advances in Geo-Environmental Engineering, Geomechanics and Geotechnics, and Geohazards*, Advances in Science, Technology & Innovation, https://doi.org/10.1007/978-3-030-01665-4_38

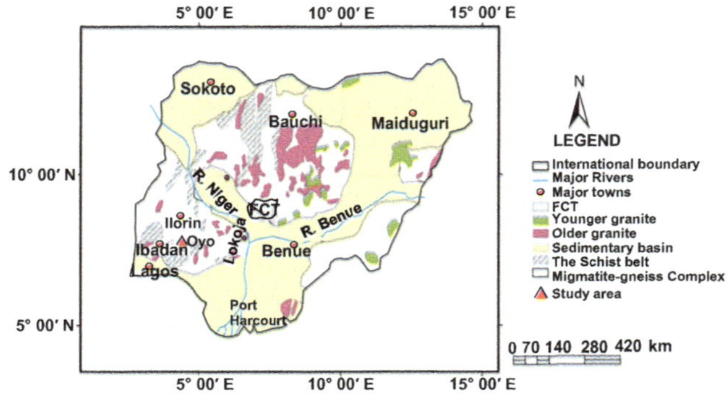

Fig. 1 Geological domains of Nigeria (Adapted and modified from [2])

separations of 1.0, 3.0, 6.0 and 8.0 m respectively, and the electrode increment of 5.0 m across the three (3) traverses that were established in the study area. The traverses are of distance 100 m each, with W-E orientation that will enable the subsurface imaging of the study area. The apparent resistivity (ρ_a) data was computed and processed based on the literature from Adagunodo et al. [2]. The obtained data was processed using RES2DINV software [7], which is capable of producing inverted data through a nonlinear Least-square technique.

3 Results and Discussion

The ρ_a pseudosection obtained for the three (3) traverses on the proposed site for superstructure erection is presented in Fig. 2. The image depicts the subsurface geological features in the study area. From the inverse model section of traverse 1, the first layer in the structure has a resistivity ranging from 46.4 to 168 Ωm at the depth ranging from 0.25 to 1.27 m. This layer is considered as topsoil. The weathered layer (mixture of sand and clay) was observed under the topsoil. The next observed zone towards the western part of traverse 1 is the clayey zone, which ranged from 17.7 to 33.0 Ωm at the depth of 2.49–5.73 m. The dynamic movement of this zone could cause differential settlement to the column of the walls from that axis if not properly taken care of from the designing stage. The bedrock (fresh/fractured) under the clayey and weathered zones in traverse 1 showed that the fractured bedrock covered a wider range than the fresh bedrock. This could be disastrous for construction of high-rise foundation without a proper design.

The inverse model of traverse 2 revealed that the topsoil ranged from 0.25 to 2.3 m. The weathered layer fell directly under the topsoil, but was experienced as sandy clay/silty sand towards the base of the traverse. The clayey zone

Fig. 2 Imaging of the ERT traverses in the study area

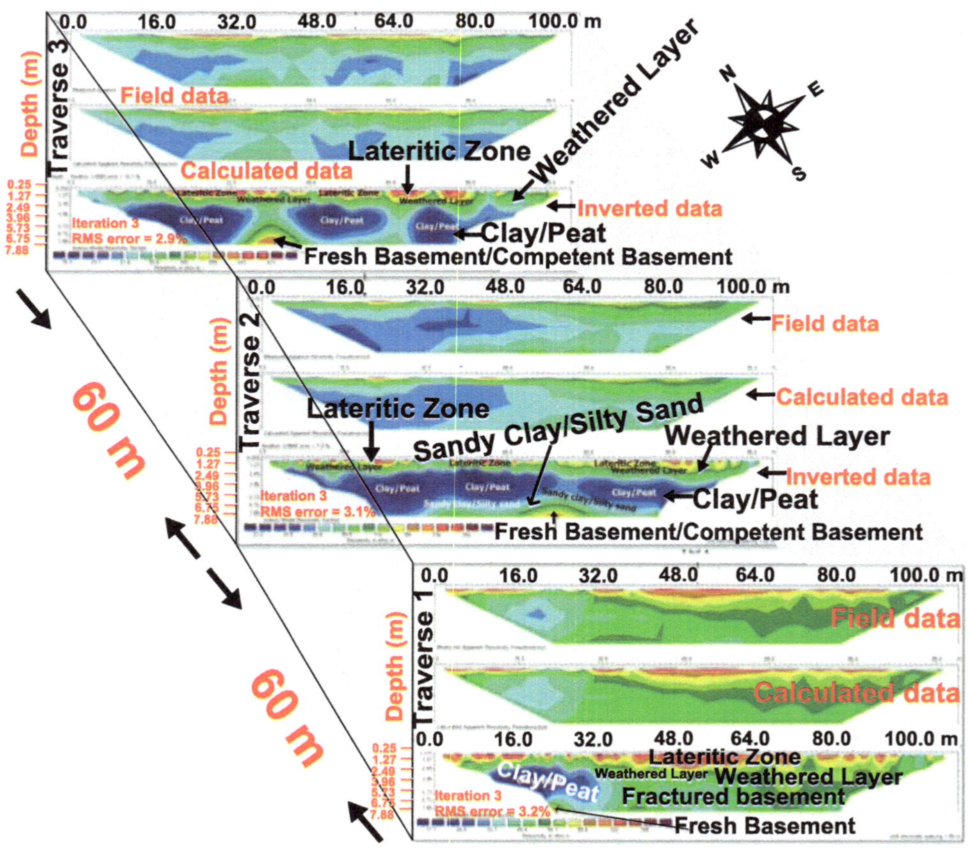

(a highly conductive region) was entrapped by the weathered layer and mostly spanned through the entire profile with the depth ranging from 2.3 to 7.2 m. This clayey zone is unfavourable for civil engineering constructions, because it could set the foundation of the overlying structure into motion during the seasonal variation.

The same subsurface trend was observed in traverses 2 and 3. In traverse 3, qualitative interpretation of this profile showed that the conductive zone is also too thick and could cause foundation problem in the future if proper channels (building engineers) were not consulted before the construction processes commence. The bedrock beneath the clayey/weathered zone was observed along 36.0–45.0 m, at the depth of about 6.0 m. This showed that the bedrock of the study is not a shallow type.

4 Conclusion

The suitability of the proposed site for construction of a mega auditorium in Emmanuel Alayande College of Education, Oyo was investigated via the ERT technique. The inverse model along traverses 1 and 2 revealed a very thick clayey zone, indicative that the study area is unsuitable for construction of high-rise building without proper consultations from the building engineers.

Acknowledgements We appreciate the conference support received from Covenant University, Nigeria.

References

1. Adagunodo, T.A., Adeniji, A.A., Erinle, A.V., Akinwumi, S.A., Adewoyin, O.O., Joel, E.S., Kayode, O.T.: Geophysical investigation into the integrity of a reclaimed open dumpsite for civil engineering purpose. Interciencia J. **42**(11), 324–339 (2017)
2. Adagunodo, T.A., Sunmonu, L.A., Adeniji, A.A.: Effect of dynamic pattern of the saprolitic zone and its basement on building stability: a case study of a high-rise building in Ogbomoso. J. Appl. Phys. Sci. Int. **3**(3), 106–115 (2015)
3. Adagunodo, T.A., Sunmonu, L.A., Oladejo, O.P.: Effect of constructing high-rise buildings without a geophysical survey. Niger. J. Phys. 91–100 (Special Edition September 2014)
4. Adagunodo, T.A., Sunmonu, L.A., Oladejo, O.P., Ojoawo, I.A.: Vertical electrical sounding to determine fracture distribution at Adumasun area, Oniye, Southwestern Nigeria. IOSR J. Appl. Geol. Geophys. **1**(3), 10–22 (2013)
5. Adewoyin, O.O., Joshua, E.O., Akinwumi, I.I., Omeje, M., Joel, E.S.: Evaluation of geotechnical parameters using geophysical data. J. Eng. Technol. Sci. **49**(1), 95–113 (2017)
6. Adewoyin, O.O., Joshua, E.O., Akinyemi, M.L., Omeje, M., Joel, E.S.: Investigation to determine the vulnerability of reclaimed land to building collapse using near surface geophysical method. J. Phys: Conf. Ser. **852**(1), 012030 (2017)
7. Loke, M., Barker, R.D.: Practical techniques for 3D resistivity surveys and data inversion. Geophys. Prospect. **44**, 499–523 (1996)
8. Oyeyemi, K.D., Aizebeokhai, A.P., Adagunodo, T.A., Olofinnade, O.M., Sanuade, O.A., Olaojo, A.A.: Subsoil characterization using geoelectrical and geotechnical investigations: implications for foundation studies. Int. J. Civ. Eng. Technol. **8**(10), 302–314 (2017)
9. Samouelian, A., Cousm, I., Tabbagh, A., Bruand, A., Richard, G.: Electrical resistivity survey in soil science: a review. Soil Tillage Res. **83**, 173–193 (2005)
10. Sudha, K., Israil, M., Mittal, S., Rai, J.: Soil characterization using electrical resistivity tomography and geotechnical investigations. J. Appl. Geophys. **67**, 74–79 (2009)
11. Wahab, S., Saibi, H.: 2-D inversion of electrical resistivity data around sayanokami spring in Ito Campus (Kyushu University, Fukuoka, Japan). In: International Workshop on Earth Resources Technology, Kyushu University, Fukuoka, Japan, 8–9 Dec 2016

Correlation Between Uniaxial Compressive and Shear Strength Data of Limestone Rocks by Regression Analysis and ANFIS Model

Masoud Rashidi, Adel Asadi, and Biltayib Misbah Biltayib

Abstract

To determine rock mechanical properties like uniaxial compressive strength and shear strength accurately, it is required to put considerable time to find and collect suitable samples for laboratory testing. To improve the time and cost efficiency, many empirical relationships have been proposed in literature. The purpose of this study is to develop a model to correlate uniaxial compressive strength and shear strength data of intact rocks. In this study, two mathematical methods, adaptive neuro-fuzzy inference systems (ANFIS) and regression analysis, were used to correlate the uniaxial compressive and shear strength. A new approach based on artificial intelligence techniques is considered to develop and train UCS-τ data. A total of 40 sets of data were used to correlate UCS and τ data of limestone rocks. The resulted regression equation shows that the relationship between uniaxial compressive and shear strength has an acceptable determination coefficients of R^2. Results of this research study has also indicated that, because of their acceptable accuracy in development of an efficient correlation between UCS and τ data, adaptive neuro-fuzzy inference systems are appropriate tools to correlate UCS-τ data, in addition to the regression model proposed in this paper.

Keywords

Uniaxial compressive strength • Shear strength • Limestone rocks • Adaptive neuro-fuzzy inference systems

M. Rashidi (✉) · B. M. Biltayib
Petroleum Engineering Department, College of Engineering, Australian College of Kuwait, Kuwait City, Kuwait
e-mail: rashidimasoud@yahoo.com

A. Asadi
Department of Petroleum Engineering, Science and Research Branch, Islamic Azad University, Tehran, Iran

1 Introduction

Unconfined Compressive Strength (UCS) represents one the most common parameters for describing intact rock strength in the field of engineering geology and rock engineering. The uniaxial compressive strength of an intact rock is calculated by dividing the compressive force at failure, applied in one direction, by the sample's area of cross section. Laboratory tests are the most accurate approach for prediction of UCS. There are also more indicators of rock strength available in scientific and professional literature including triaxial, point load, Schmidt rebound hammer, scratch, indentation, thick wall cylinder (TWC), etc.

In case of failure of rock mass, intact rock blocks forming the mass may slide, translate, rotate, shear or split. Consequently, the shear behavior of the mass is substantially different from that of a single discontinuity. At low normal stress level, shearing results in dilation of the mass associated with rotation of blocks. The friction angle of mass at low normal stress is high, whereas the cohesion is low. As the normal stress increases, the dilation is suppressed and shearing of the intact rock material commences. This results in a relatively higher cohesion and lower friction angle. "Due to continuous change in mechanism of failure, the shear strength envelope of rock mass is highly curvilinear, especially in low normal stress range. It is not feasible to prepare and test specimens of rock mass in the field" [3].

Recently, artificial intelligence (AI) based models are successfully employed by some researchers to solve difficult non-linear problems in geomechanical models and geotechnical projects. In this regard, in order to correlate UCS and τ data, as two major geotechnical parameters, an artificial intelligence (AI) method named adaptive neuro-fuzzy inference system (ANFIS) is developed. In addition, a strong correlation based on a regression model is achieved in this study, and the results of regression and ANFIS models are compared to have a better understanding of both methods and their applications in making predictions in complex areas.

A. Kallel et al. (eds.), *Recent Advances in Geo-Environmental Engineering, Geomechanics and Geotechnics, and Geohazards*, Advances in Science, Technology & Innovation, https://doi.org/10.1007/978-3-030-01665-4_39

2 ANFIS

Artificial inference techniques like artificial neural networks, genetic algorithms and fuzzy logic have been employed vastly in recent years. Artificial neural network can efficiently recognize patterns and to adapt the model to deal with unstable environment. Fuzzy logic has the applicability to simulate the use of human knowledge and expertise for uncertainty analysis. Therefore, ANFIS, as an integration of both ANNs and Fuzzy Logic approaches, has recently become a popular tool to solve geomechanical and geotechnical engineering as well as petroleum and engineering geology problems. ANFIS is an artificial neural networks model that functions equivalently with the fuzzy inference model. It can be trained to develop If-Then fuzzy rules and determine membership functions for input and output variables of the system. ANFIS is a fuzzy Sugeno model which has the advantageous ability of learning and adaptation. This framework makes adaptive neuro-fuzzy inference systems automatically expert and less dependent on human knowledge and expertise. The applicability of ANFIS model has been reviewed and reported to solve complex geomechanical problems in the paper by Asadi [1].

3 Data Analysis

According to the study conducted by Khanlari and Abdilor [2], 40 sets of data reached from experimental tests on limestone rocks were published as summarized in Table 1. The UCS-τ data from their paper was analyzed and hired to correlate uniaxial compressive strength and shear strength data of limestone rocks by the application of two different methods named regression model and adaptive neuro-fuzzy inference systems (ANFIS).

3.1 Regression Model

Regression models are generally considered traditional predictive models to correlate rock mechanical properties. Many simple models to estimate UCS & shear strength has been developed before using the dry density value, the P-wave velocity, the point load index and other properties. The aim of this study is to correlate data of UCS and τ. The outcome of regression analysis using MATLAB software Curve Fitting App is shown in Eq. 1.

$$\tau = (0.243 \times \text{UCS}) - 6.893 \quad (1)$$

The value of R^2 is equal to 0.8344, and the linear plot is shown in Fig. 1. Another set of data including 13 samples was used to validate this correlation, and the coefficient of 0.7963 was reached which confirmed the accuracy of Eq. 1.

3.2 ANFIS Performance

The data sets extracted from limestone core specimens including 40 pairs of UCS-τ data are used as training, checking and testing data of ANFIS in MATLAB.

Table 1 Data statistics

Variable	UCS (σ)[a]	Shear Strength (τ)[a]
Min	56.15	5.41
Max	103.99	18.39
Mean	87.0542	14.0387
Median	89.98	15.07

[a]All units are MPa

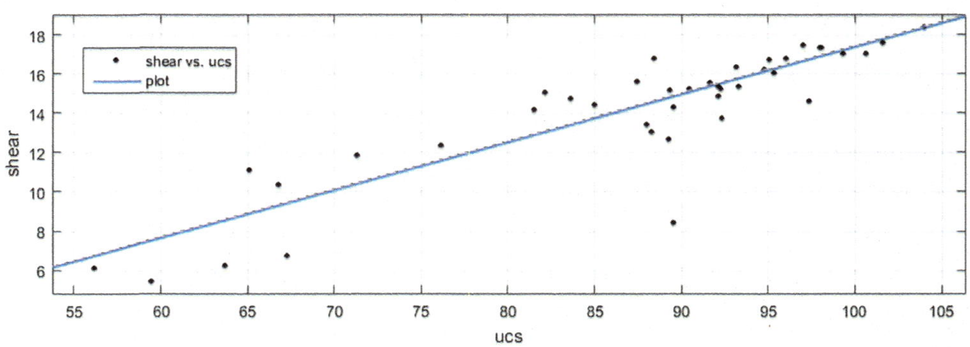

Fig. 1 UCS versus Shear Stress (τ) plot

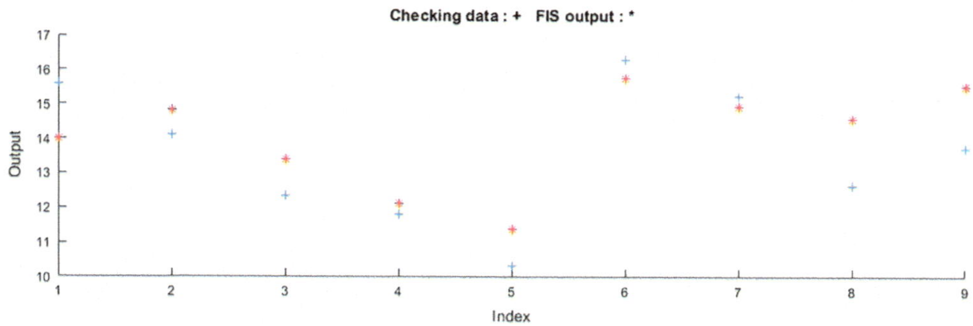

Fig. 2 Checking prediction data by ANFIS

The dataset were divided into 3 parts of Training, Testing and Checking pairs of UCS-τ data. After testing and changing different parameters for ANFIS, the best one with four Triangular-shaped built-in membership function and linear output MF was derived, and the fuzzy inference system (FIS) was optimized with hybrid method after 50 epochs during ANFIS training stage. Because of the fact that error of the ANFIS model was low, and output data were near the targets, the proposed method is applicable for UCS assessment using shear strength data. The error of the ANFIS model was reached the value of 0.671 (Fig. 2).

4 Conclusions

Our results showed that although the excellence of artificial intelligence techniques over other regression models should not be considered as guaranteed, these highly applicable and practical methods are definitely a magnificent development in data science. Indeed, AI methods including adaptive neuro-fuzzy inference systems, can have a great impact on our studies to produce prediction models as a requirement of development in the fields of geomechanics and geotechnical engineering. The results of this study have showed us that the adaptive neuro-fuzzy inference systems are powerful tools to estimate shear strength using uniaxial compressive strength data for limestone rocks, and could be significantly useful and practical in relevant industries. We suggest that researchers and industry experts could study the applicability of ANFIS model regarding our proposed scope, and try to find and apply bigger data sets belonging to limestone formations. Another proposal would be to test this method on different rock types in different locations and geological conditions.

References

1. Asadi, A.: Application of adaptive neuro-fuzzy inference system for the assessment of excavation damaged zone using uniaxial compressive strength data. In: Schubert, W., Kluckner, A. (eds.) Proceedings of the ISRM Regional Symposium EUROCK 2015 & 64th Geomechanics Colloquium—Future Development of Rock Mechanics, pp. 291–296. Austrian Society for Geomechanics, Salzburg, 7–10 Oct 2015
2. Khanlari, G., Abdilor, Y.: Estimation of strength parameters of limestone using artificial neural networks and regression analysis. Aust. J. Basic Appl. Sci. **5**(11), 1049–1053. ISSN 1991-8178 (2011)
3. Singh, M.: A document on: assessment of engineering properties of rock mass in civil engineering applications. In: IGS Working group on Rock Mechanics and Tunneling

Numerical Investigation of the Interface Shear Behaviors Between Double Soil Layers Using PFC2D

Zhong-Liang Zhang, Zhen-Dong Cui, and Ling-Zi Zhao

Abstract

The influence of the interface morphology and stiffness ratio of the deformation modulus of upper and lower layers on the stress-shear displacement curves and shear strength parameters were investigated with numerical shear tests in PFC2D. The shear strength of the specimen increases with the increase of the normal stress. The lager the normal stress, the later the peak shear stress would appear. In case of saw teeth at the interface, the peak values of the shear stress appear in the end of the shear tests. The increase of the stiffness ratio can cause the connection fracture in the lower layer particles with smaller stiffness and reduce the thickness of the shear band. The smaller saw teeth size and stiffness ratio, the larger the shear stress.

Keywords

Double soil layers • Interface • Shear test • Saw teeth • Stiffness ratio

1 Introduction

In the soft subsoil with upper crust, the soft subsoil will squeeze around and generate relative larger horizontal shear force within a certain range. Therefore, the study of shear behavior of the horizontal interface between double soil layers is an important topic in geomechanics.

The contact and friction effects of the interface between soil layers have been studied by many scholars. Hu and Pu [1] studied the mechanical characteristics of soil-structure interface through the direct shear tests. Based on the simple shear tests, the interface behavior and the non-linear failure process of interfaces between two granular soils were analyzed by Hu et al. [2] with the perfect-plastic model. Lashkari [3] proposed a simple semi-hyperbolic state-dependent constitutive model for sand-structure interfaces, considering the influence of normal stiffness on volume change and stress path. Kock et al. [4] established the soil shear model with PFC2D to study the influence of different particle shapes on the shear behaviors of sedimentary soil. Gu et al. [5] analyzed the details of the formation process of the shear band with a discrete element model of the particle material. Zhu et al. [6] and Lei et al. [7] investigated the deformation process and pattern of shear band through plane shear tests and numerical simulations, considering the influences of vertical pressure and shear rate on the properties of the shear band.

In this study, the interface shear test model of the double-layer soil was developed with PFC2D and the influence of the interface morphology and soil parameters on shear force-displacement behaviors was analyzed.

2 Set-up of the Shear Test Model in PFC2D

Before the numerical shear test with PFD2D, the parameter of the materials should be calibrated. In the shear model, sandy and clayey soils were selected as the upper crust in the upper shear box and soft subsoil in the lower shear box, respectively.

Based on the calibration of the porosity, the cohesive strength and the internal friction angle, the minimum particle radius and the maximum particle size ratio were selected as 1.5 mm and 1.5, respectively. The friction coefficient $\mu = 0.39$. The calibration of normal stiffness was accomplished by simulating the natural stacking process and the stiffness and contact strength of the particles were determined by the plane biaxial compression test. The normal stiffness of the upper sand and lower soft clay were selected as 1.66×10^5 and 3.27×10^4 kPa/m, respectively. The ratio of the normal stiffness to the shear stiffness was

Z.-L. Zhang · Z.-D. Cui (✉) · L.-Z. Zhao
State Key Laboratory for Geomechanics and Deep Underground Engineering, School of Mechanics and Civil Engineering, China University of Mining and Technology, Xuzhou, China
e-mail: czdjiaozuo@163.com

A. Kallel et al. (eds.), *Recent Advances in Geo-Environmental Engineering, Geomechanics and Geotechnics, and Geohazards*, Advances in Science, Technology & Innovation, https://doi.org/10.1007/978-3-030-01665-4_40

selected as 2.5. In addition, the normal bond strength of the upper sand and lower soft clay were selected as 0 and 1.02 kPa, respectively.

As shown in Fig. 1, the shear model with size of 400 mm × 200 mm was established. The boundary effect can be eliminated owing to the length to height ratio of the model being 2. Moreover, the shear stress can reach its peak within the shear displacement of 35 mm. Four different normal stresses (50, 100, 150 and 200 kPa), two saw teeth of different sizes (50 and 25 mm) at the interface and two stiffness ratios of deformation modulus of upper and lower layers (3.98 and over 10) were selected as the variables in the numerical shear tests.

3 Results of Numerical Shear Tests

The variation of the shear stress-displacement curves are illustrated in Fig. 2. It can be seen that the shear strength of the specimen increases with the increase of the normal stress. There is a hardening phenomenon in the early stage of shearing and the normal stress increases the hardening process. The stress experiences the peak value in the shear displacement range of 5 ~ 10 mm. The lager the normal stress, the later the peak shear stress would appear.

4 Discussion

When there are saw teeth at the interface between the double soil layers, the deformation of the saw teeth makes the shear stress of the interface more complicated. In the shear process, the sand upper crust is lifted up and the shear stress experiences the peak value at the end of the shear. Since the small and dense saw teeth can enhance the integrity of the shear strength of the interface, the shear stress of the specimen with 25 mm saw teeth is larger than that of the specimen with 50 mm saw teeth. The shear stress decreases with the increase of the stiffness ratio of the deformation modulus of upper and lower layers. When the stiffness ratio was over 10, there was only hardening, but no softening process occurred.

The shear stress and normal stress were fitted with the linear functions after the shear tests. The strength parameters including the cohesive strength and the internal friction angle of the shear specimen under different variables were summarized in Table 1. The internal friction angle decreases with the increase of the stiffness ratio. In the numerical shear process, the increase of the stiffness ratio will cause the connection fracture in the lower layer particles with smaller stiffness and reduce the thickness of the shear band.

5 Conclusion

In this study, the numerical investigation of the interface shear behaviors between the double soil layers was established with PFC2D. The effects of the interface morphology and soil parameters on shear force-displacement behaviors were analyzed. The main deduced conclusions can be stated as follows.

(1) The shear strength of the specimen increases with the increase of the normal stress. The lager the normal stress, the later the peak shear stress would appear.
(2) The interface morphology and stiffness ratio have influences the shear behaviors of the shear interface. The smaller saw teeth size and stiffness ratio, the larger the shear stress.

Fig. 1 Preparation of shear specimen

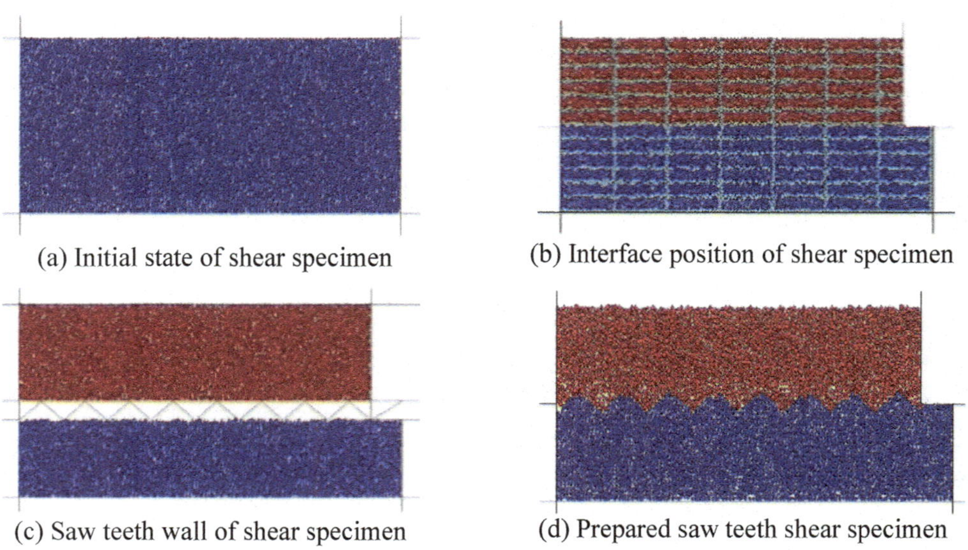

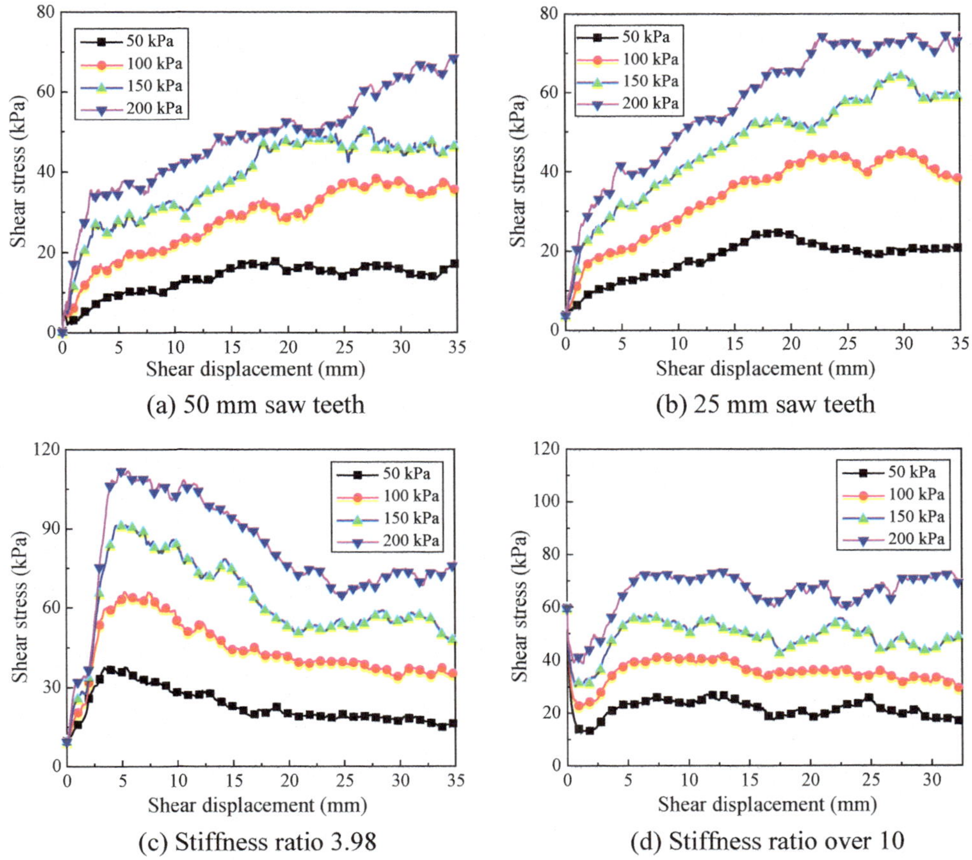

Fig. 2 Shear stress-displacement relationship of samples with different conditions

Table 1 Calculation of shear strength parameters

Variables	Cohesive strength (kPa)	Internal friction angle (°)
50 mm saw teeth	2.52	18.47
25 mm saw teeth	9.78	18.93
Stiffness ratio 3.98	14.61	26.53
Stiffness ratio over 10	9.91	17.64

Acknowledgements This work was funded by the National Key R&D Program of China (2016YFC0600903).

References

1. Hu, L., Pu, J.: Testing and modeling of soil-structure interfaces. J. Geotech. Geoenviron. Eng. **130**(8), 851–860 (2004)
2. Hu, L., Zhang, B., Ma, J.: Mechanical characteristics for interface between granular materials. Mech. Res. Commun. **37**(1), 42–46 (2010)
3. Lashkari, A.: A simple critical state interface model and its application in prediction of shaft resistance of non-displacement piles in sand. Comput. Geotech. **88**, 95–110 (2017)
4. Kock, I., Huhn, K.: Numerical investigation of localization and micromechanics in a stratified soil specimen. Struct. Geol. **29**(10), 1679–1694 (2007)
5. Gu, X.Q., Huang, M.S., et al.: Discrete element modeling of shear band in granular materials. Theoret. Appl. Fract. Mech. **72**, 37–49 (2014)
6. Zhu, H., Zhou, W.H., et al.: Numerical study of the formation of shear bands in soil under interface shearing. Procedia Eng. **175**, 102–109 (2017)
7. Lei, D., Huang, J., et al.: Deformation analysis of shear band in granular materials via a robust plane shear test and numerical simulation. Powder Technol. **323**, 285–392 (2018)

Comparison of Soil Strength Parameters in a Small and Large Scale Direct Shear Test

S. Farid F. Mojtahedi, Saeed Rezvani, and Ali Nazari

Abstract

According to the recommended criteria in the instruction, the maximum grain diameter of the samples should not exceed a certain size to meet the standard requirements for the direct shear test. It can be said that there is a tendency to use a small-scale shear test instead of a large-scale one, and the reason is that problems such as expensive equipment among others arise when using a large-scale direct shear test. Since small-scale direct shear test is recommended only for fine-grained soils, coarse soil should be tested on a large-scale, or they should be adjusted to be tested in a small direct shear by modifying grain distribution. In this study, two different methods were investigated for grading modification and using a small-scale shear test. The advantage of these methods is that they assess the change in the internal friction angle. To this end, six soil samples containing coarse particles were collected from Ardabil city. All the samples were tested for the determination of soil resistance parameters in a large direct shear test. Samples were tested on a small-scale direct shear test after applying the modified method mentioned above. Finally, soil strength parameters were compared with two different scales. The results clearly indicate that the strength parameters obtained from the replacement method lead to a closer approximation to the resistance parameters obtained by the large-scale direct shear test.

Keywords

Direct shear test • Scale effect • Replacement method • Elimination method

1 Introduction

The direct shear test is a useful and practical geotechnical test to obtain strength properties of soil. This test is very practical due to its speed and simplicity. It is performed on undisturbed soils. According to the recommended criteria in the instruction [1], the maximum grain diameter of the samples should not exceed a certain size to meet the standard requirements for the for the direct shear test. Many geotechnical laboratories currently use direct shear tests to determine the soil characteristics, including the angle of friction or the shear strength of the sand. However, today, there are different sizes of shear boxes and the question arises, as how would the internal friction angle changes be in terms of sample size. Parsons [2] compared the results of direct shear tests on the crushed quartz and a clean uniform sand. In this study, the angle of internal friction decreases by rising the direct shear box. Parson evaluated three different parts of direct cutting boxes. In another study by Skempton and Hutchison [3], the effect of the direct shear box size was investigated on the undrained shear strength of the clay. Those tests were conducted on specimens with the following dimensions: 300 600 mm. The authors indicated that the shear strength of the samples in these dimensions is 66% of the standard shear strength. The standard shear strength is the shear strength of the tests on samples with 38 mm in diameter. Lo [4] specified that the undrained shear strength of clay in samples with 300 mm in diameter is about 65% of the shear strength in samples of 38 mm in diameter. The overall result of this study is that the shear strength decreases by increasing the sample size. In total, there are 4 methods presented to modify grading of the soil, with two methods common for the coarse particles: parallel method [5], screening method [6], tetragonal method [7] and the replacement method [8].

S. F. F. Mojtahedi (✉) · S. Rezvani · A. Nazari
Sharif University of Technology, Tehran, Iran
e-mail: Farid.Fazel71@student.sharif.edu

A. Kallel et al. (eds.), *Recent Advances in Geo-Environmental Engineering, Geomechanics and Geotechnics, and Geohazards*, Advances in Science, Technology & Innovation, https://doi.org/10.1007/978-3-030-01665-4_41

2 General Laboratory Plans

Large-scale direct shear test apparatus (a square box with a cross-section of 300 × 300 mm) was used to determine the strength parameters of coarse-grained soils, and samples were prepared by removing seeds larger than the standard value to determine the soil strength parameters by small-scale direct shear test apparatus (box A square cross section of 100 × 100 mm). Two modification methods of gradation were used in this paper. In the first method, called the elimination method, aggregates larger than the No. 4 sieve were removed from the mixture. In the second method, called the replacement method, after removing aggregates larger than No. 4 sieve, the removed weight of the aggregates were replaced with aggregates between No. 4 and No. 16 sieves, and the resulting samples of this method are named with the prefix R. Then, the results of Large-scale and small-scale direct shear tests were compared to each other.

3 Materials and Methods

The soil samples used in this study were natural materials of Ardabil city. Ardabil plain is located at the geographic center of the province, and between 38, 3 to 38, and 27 degrees on Northern latitude against the equator axis and on 4 and 55 to 48 and 20 degrees on eastern longitude from Greenwich meridian (Fig. 1).

4 Direct Shear Tests

Direct shear tests were conducted on the soils according to ASTM D3080-98. The results of direct shear and index tests are shown in the Table 1.

5 Interpretation of Results

5.1 The Effects of Scale on Internal Friction Angle

Tests have shown the difference of 1 to 4 degrees in the internal friction angle of soils, and often, internal friction angle values of small-scale apparatus are larger than the values of large-scale apparatus. However, the results are influenced by some other factors such as the moisture content of the sample, the range of strain rate, the value of the vertical load, the cross section of the shear box (circular or square) and so on. The results indicate a one-degree difference in the friction angle as shown in Table 2.

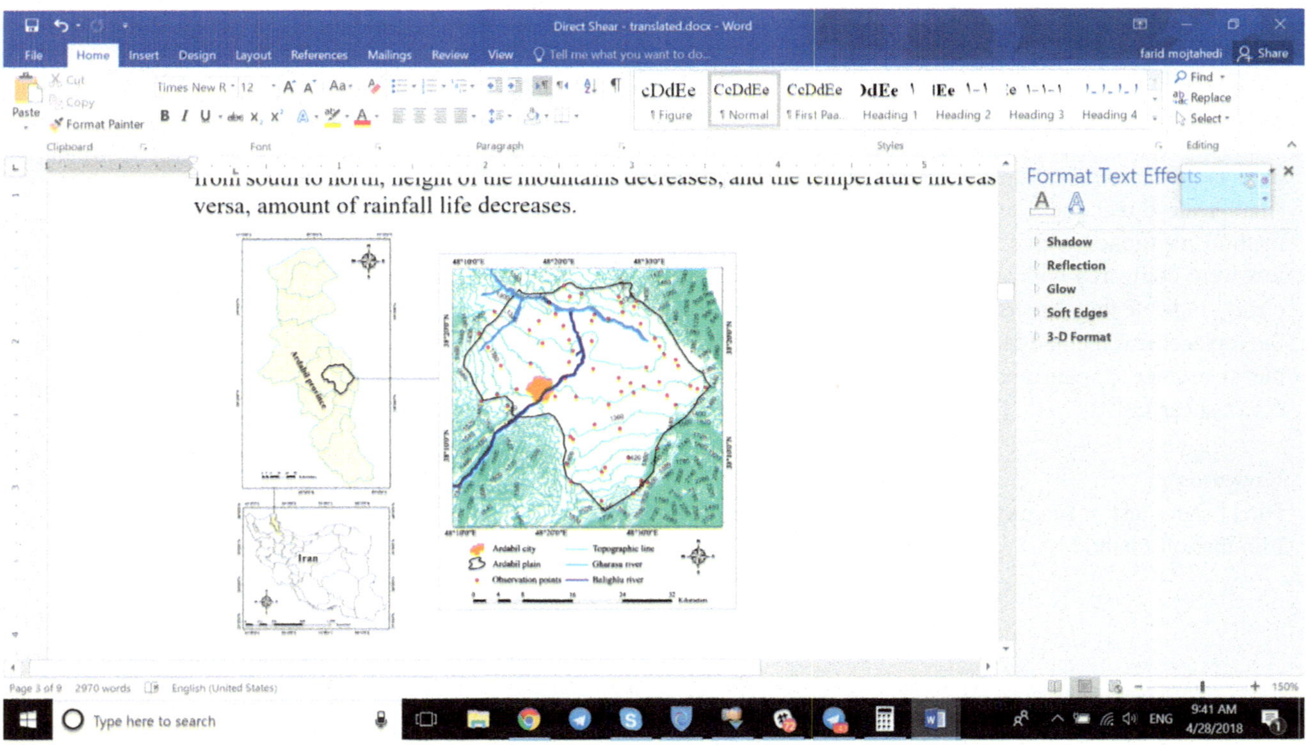

Fig. 1 The approximate location of Ardabil plain

Table 1 The results of direct shear and index tests performed on collected samples

Sample No.	c (kPa)		Gs	e_{max}	e_{min}	$g_{d,max}$	w_{opt}
1	6.67	41	2.688	0.9	0.4	19.6	12
2	0.06	44	2.689	0.82	0.36	20.4	11
3	3.18	42	2.69	0.92	0.41	19.8	12
4	5.26	40	2.675	0.85	0.38	19.9	11
5	1.10	39	2.69	0.87	0.39	19.8	12
6	0.30	46	2.683	0.83	0.33	20.6	10

Table 2 The results of standard sand in large and small shear boxes

Sample No.	Friction angle (degree)		Dry unit weight (kN/m^3)
	Small box	Large box	
1	34.0	33.0	17.6
2	35.5	35.0	18.3

5.2 The Effects of Removing Coarse Grains on Internal Friction Angle

The internal friction angle of six sandy soil samples of which out-of-range aggregates (larger than No. 4 sieve) were removed, was measured by small-scale direct shear tests. For saturated soils, due to neglecting the effect of cohesion parameter (c), the failure envelope passes through zero on the shear axis, as shown effective stress versus shear stress space. In some results, the failure envelope passes automatically through zero, but there is a little cohesion in some others which can be assumed as the resistance of particles' bonding. The ultimate friction angle was measured at 10 percent strain. However, the sand has continued to expand and never reached a constant volume in a few samples. Due to the high density of the samples (90% density), dilation was observed, and the shear stress was still increasing after the strain of 10 percent, but the increase was not remarkable. As expected, the results show the larger values of internal friction angle compared to the results of large-scale direct shear tests. Briefly, the results show a difference between 3 and 6 degrees between the internal friction angle measured by large-scale and small-scale direct shear apparatus, and the difference increased with the increase of out-of-range particles (larger than No. 4 sieve).

5.3 The Impact of Placement Method on Internal Friction Angle

By removing particles larger than No. 4 sieve and replacing them with aggregates between No. 4 and No. 16 sieve, a soil was obtained which meets the requirements of maximum particle size in small-scale apparatus. The results are shown in Fig. 2. The results show that the values of internal friction angle of the small-scale test were greater than those of the large-scale test.

5.4 Comparison of Grading Correction Methods

The results of direct shear tests are shown in Fig. 3. As seen in this figure, the internal friction angle of the replacement method is much closer than the results obtained from eliminating method. So, it can be expected that the replacement method provides better and more reliable results for a sand with a low amount of fines.

6 Conclusion

The results of the direct shear tests are as follows:

- According to the previous studies, the scale affects the internal friction angle up to 3 degrees; however, in this study, the observed scale effect was 1.5 degrees of difference between elimination and large-scale direct sheer method.
- The internal friction angle obtained by the elimination method showed a 4-to-6-degree difference with the results of the large-scale direct shear test on initial samples, and this difference increases with the increase of out-of-range particles.
- The replacement method results had a 2-to-3-degree difference compared to the results of the large-scale direct shear test on initial samples.
- Generally, the results of the replacement method present closer values to the initial samples results, compared to the elimination method. But, there is no certain comment about soils which more than 40% of their weight is

(a)

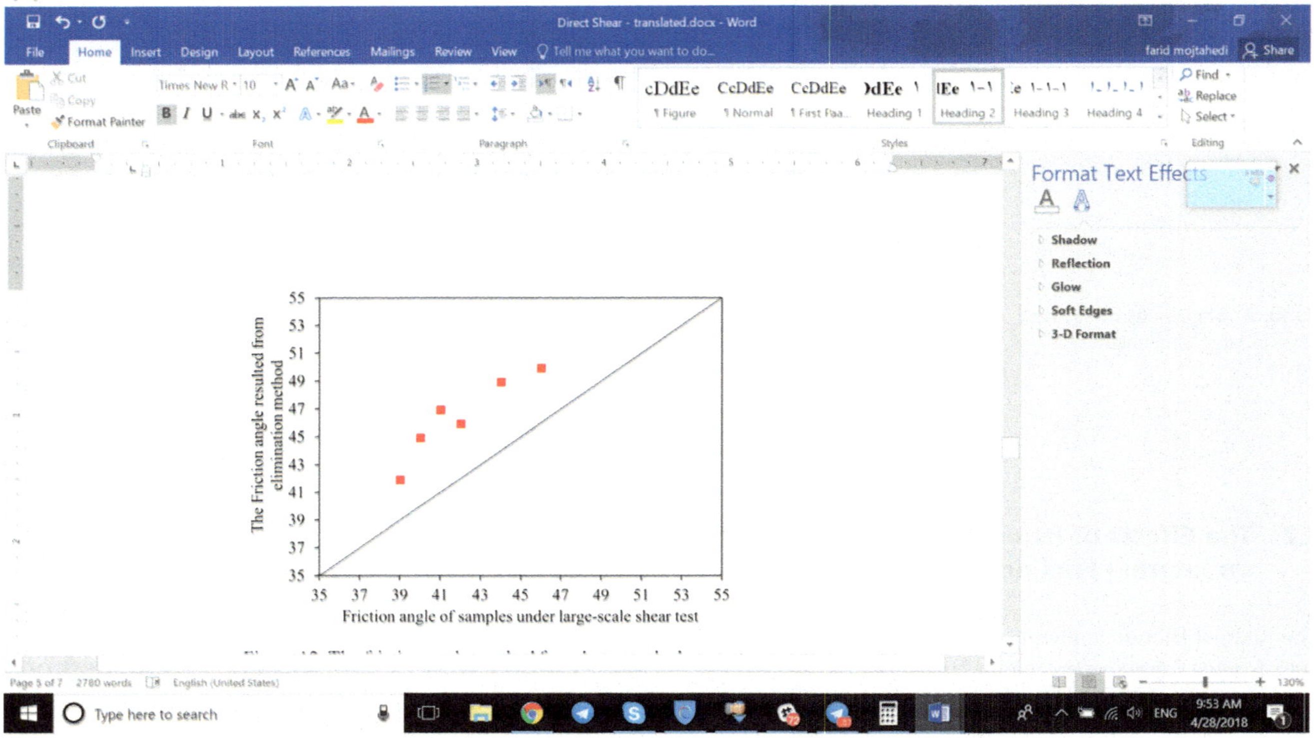

(b)

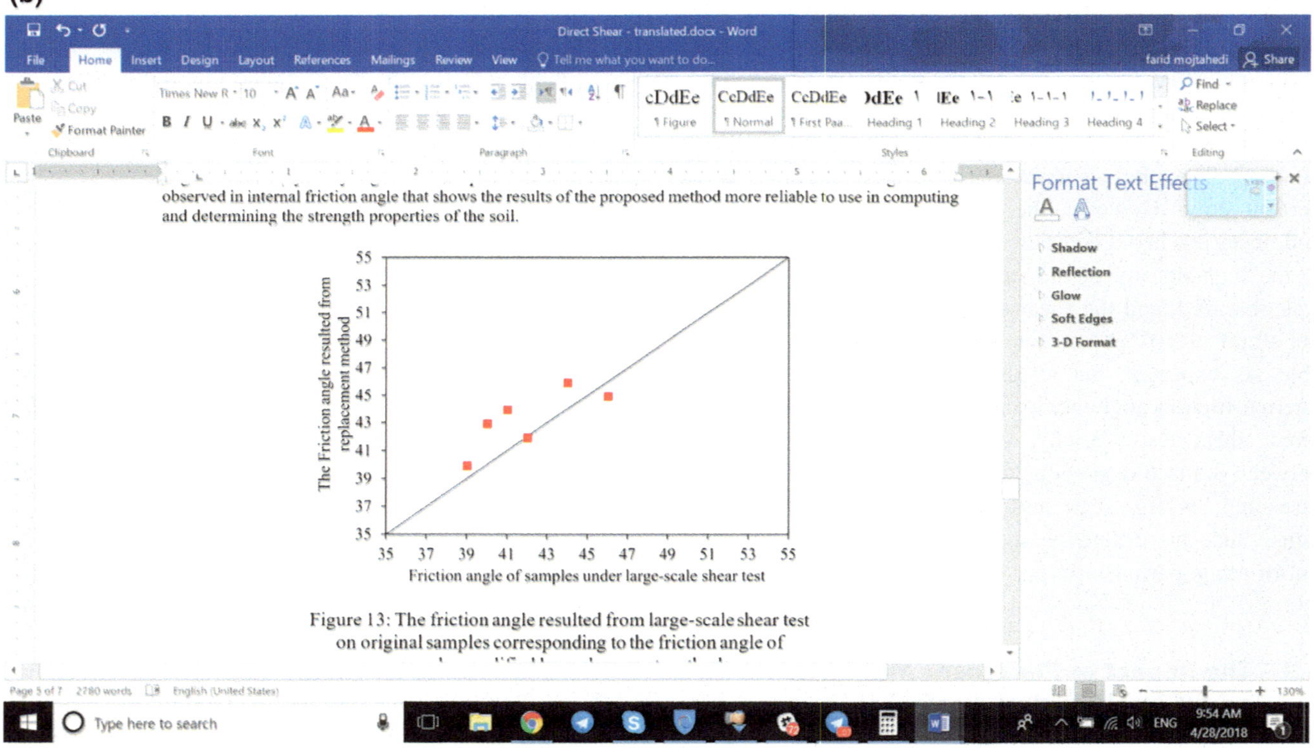

Fig. 2 The friction angle resulting from large-scale shear test on original samples corresponding to the friction angle of samples modified by **a** elimination method, **b** replacement method

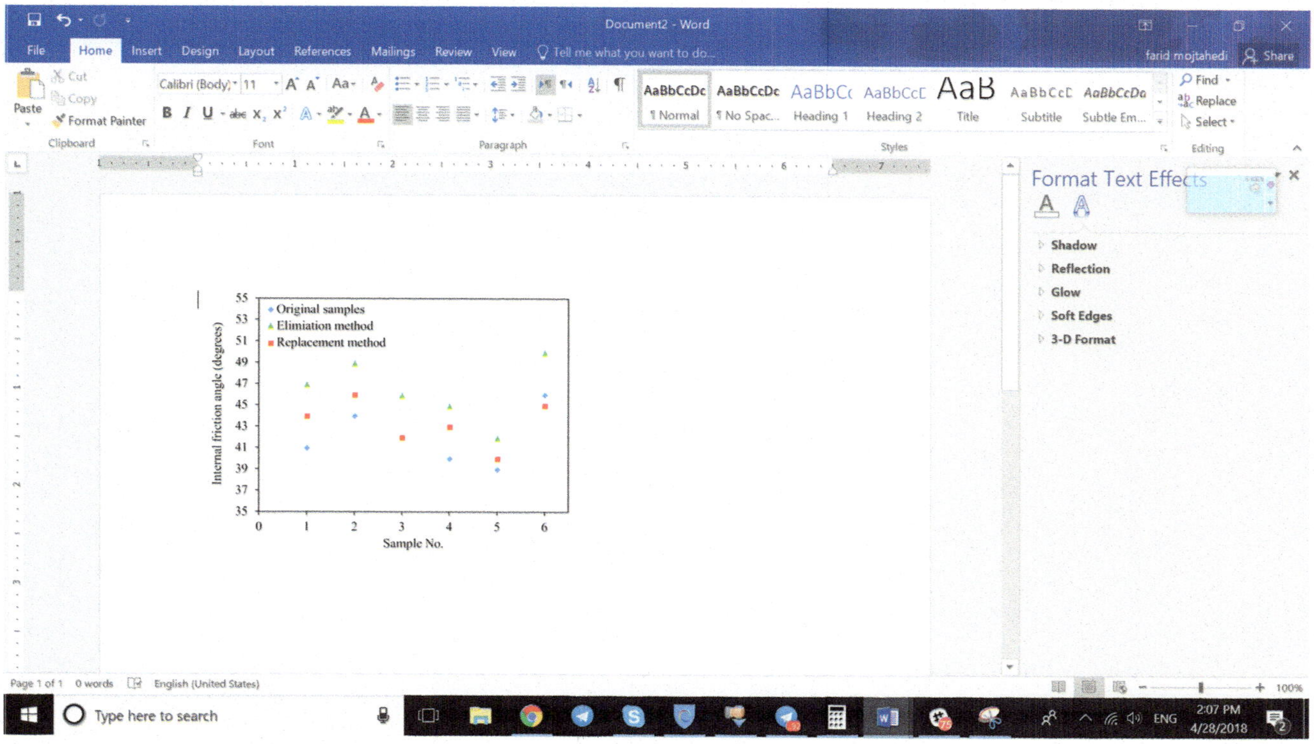

Fig. 3 The comparison of all friction angles of shear tests

comprised of out-of-range particles, so it is recommended that large-scale direct shear test be performed on these soils.

References

1. ASTM D3080-98: Standard Test Method for Direct Shear Test of Soils Under Consolidated Drained Conditions, ASTM International, West Conshochocken, PA (1998)
2. Parsons, J.D.: Progress report on an investigation of the shearing resistance of cohesionless soils. In: Proceedings of the 1st International Conference on Soil Mechanics and Foundation Engineering, vol. 2, pp. 133–138 (1936)
3. Skempton, A.W., Hutchinson, J.: Stability of natural slopes and embankment foundations. In: Soil Mech & Fdn Eng Conf Proc/Mexico/ (1969)
4. Lo, K.Y.: The operational strength of fissured clays. Geotechnique **20**(1), 57–74 (1970)
5. Lowe, J.: Shear strength of coarse embankment dam materials. In: Proceedings of 8th International Congress on Large Dams, vol. 3, pp. 745–761. International Commission on Large Dams, Paris (1964)
6. Zeller, J., Wullimann, R.: The shear strength of the shell materials for the Göschenenalp Dam. Butterworths Scientific Publication, Switzerland (1957)
7. Fumagalli, E.: Tests on cohesionless materials for rockfill dams. J. Soil Mech. Found. Div. 92(SM5, Proc Paper 490) (1969)
8. Frost, R.J.: Some testing experiences and characteristics of boulder-gravel fill in earth dams. In: Evaluation of Relative Density and its Role in Geotechnical Projects Involving Cohesionless Soils. ASTM International (1973)

Effect of Organic Matter on Geotechnical Behavior of Soils

Rinku Varghese, S. Chandrakaran, and K. Rangaswamy

Abstract

The unpredictable behaviour of organic soil makes the soil complicated to handle for geotechnical engineers. In this paper an attempt was made to study the effect of organic content on the geotechnical behaviour by studying a natural organic soil and comparing the results with the same soil after removing the organic content. The organic content from the soil was removed by hydrogen peroxide method. Significant changes in the values were noticed with and without the organic content. From the results it was observed that the presence of organic content decreases the strength and increases compressibility of the soil.

Keywords

Organic soils • Geotechnical behavior • Shear strength • Compressibility

1 Introduction

In the recent years considerable research has been conducted on the effect of organic content on the geotechnical behaviour due to its unpredictable behaviour. Organic soils show high variation in the plasticity characteristics with slight change in the water content. This is also one of the main concerns of geotechnical engineers. Studies have been conducted on the plasticity [1], compaction [1, 2], consolidation [3, 4] and shear strength characteristics [5, 6] to understand the effect of organic content on the soil properties. The more number of studies were conducted on high organic content natural soils like peat and seldom studies on the low to medium amorphous type of organic content. In this paper an attempt was made to study the effect of low organic content by conducting tests on the natural organic soil and after treating the same soil with hydrogen peroxide to remove the organic content.

2 Materials Used for the Study

The natural organic soils were collected at a depth of 3 m from local site near mills at Calicut city, Kerala. The organic matter in the soil mainly consists of decomposed wood. The collected samples were crushed and sieved through 425 micron sieve and stored in plastic bags. Before the test, the organic content in the soils were found out by loss of ignition method. From the test results it was found that a 12% of organic matter is present in the soil. The soil showed a low specific gravity of about 1.81 which also indicated the presence of organic content.

In order to study the geotechnical behavior, the natural organic soil was subjected to experiments like Compaction, unconfined compressive strength and consolidation. These test results were compared to an organic soil after removing organic content.

The organic content from the natural organic soil was removed by hydrogen peroxide method. Schmidt [7] said in his paper that the hydrogen peroxide method have no significant effect on Atterberg limits, clay mineralogy, consolidation and shear strength characteristics. Hydrogen peroxide method for organic content removal was conducted as per the ASTM D2974. After treating the soil, it was subjected to Loss of ignition tests to confirm the removal of organic content. After the confirmation the soil was air dried and the various geotechnical tests were conducted.

3 Results and Discussion

The test results by comparing the natural organic soil and the same soil after removal of organic content by hydrogen peroxide method is given below.

R. Varghese (✉) · S. Chandrakaran · K. Rangaswamy
Department of Civil Engineering, National Institute of Technology Calicut, NIT Campus P.O., 673601 Kerala, India
e-mail: rinku_p150028ce@nitc.ac.in

A. Kallel et al. (eds.), *Recent Advances in Geo-Environmental Engineering, Geomechanics and Geotechnics, and Geohazards*, Advances in Science, Technology & Innovation, https://doi.org/10.1007/978-3-030-01665-4_42

3.1 Compaction Characteristics

Figure 1 shows the compaction curve for natural organic soil and treated soil. From the compaction curve it is noticed that the organic matter reduces the maximum dry density while it increases the optimum moisture content of the soil. That is maximum dry density and optimum moisture content for organic soil is 1.69 kN/m^2 and 39%, respectively. While maximum dry density and optimum moisture content for treated soil is 1.81 kN/m^2 and 25%, respectively. The trend shown in the figure goes on well with the tendency of organic matter to adsorb water and the reduction of specific gravity with increasing organic content and also matches with the literature [4, 8]. It was found that organic matter reduces the soil's compactability by increasing the stability of the soil and also by retaining greater water content which cushions the soils against compaction. The organic particles are generally stiff when compressed, but when they absorb water they become soft and sponge-like. This indicates that at relatively low water contents much of the water added to the soil becomes in effect part of the organic matter which results in reducing the efficiency of adjusting the water content to optimise the compaction behaviour [1].

3.2 Unconfined Compressive Strength

The results of unconfined compression tests on organic and treated soil are shown in Fig. 2. The treated soil showed higher strength compared to the organic soil. The decrement in strength is due to the presence of organic matter. The organic matter has high water holding capacity which increases the liquid limit of the soil and thereby decreasing strength. It is also noticed from the graph that the strain value at which failure occurs is higher in case of organic soil compared to the treated soil. This shows that with the presence of organic matter the nature of the soil is also changing. That is the soil is trying to behave more like ductile soils.

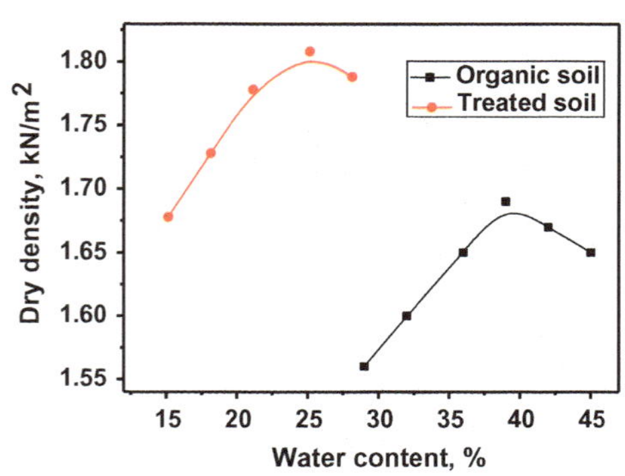

Fig. 1 Compaction curve for organic and treated soil

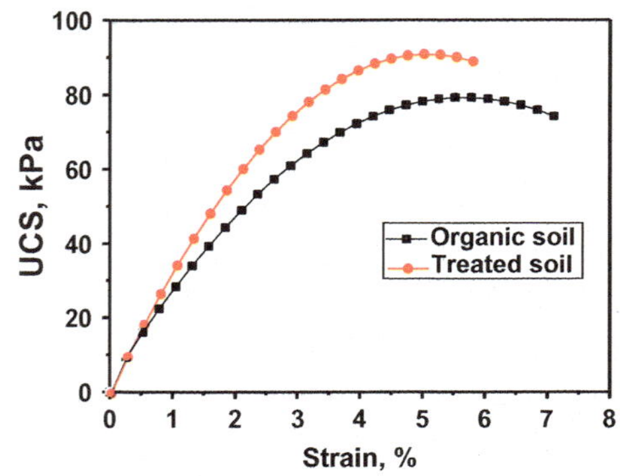

Fig. 2 Unconfined compressive strength versus strain for organic and treated soil

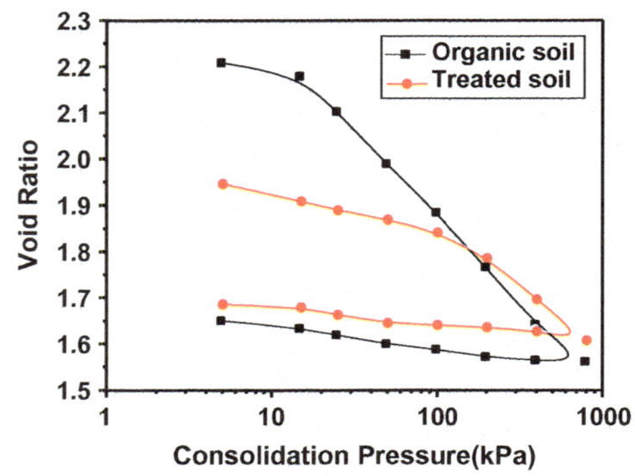

Fig. 3 Variation of void ratio with different consolidation pressure for organic and treated soil

3.3 Consolidation Characteristics

Figure 3 shows the variation of void ratio with respect to consolidation pressure for organic and treated soil. It is noticed that the organic soil has high void ratio compared to treated soil. The high void ratio in organic soils is associated with the low specific gravity of the organic matter and high water content of the soils [3, 9]. From the graphs, it can also be noted that the reduction in void ratio with respect to time and consolidation pressure is less or gradually decreasing in the case of a treated soil while in the case of organic soils it is high or the trend is steeper.

4 Conclusions

From comparing the test results on the natural organic soil and the same soil after hydrogen peroxide treatment the conclusions drawn are listed below.

- Significant changes are observed in the geotechnical behaviour with the presence of organic matter in soils, even though the percentage of organic content was less.

- Maximum dry density decreased while optimum moisture content increased with the presence of organic matter.
- The unconfined compressive strength was higher for a treated soil compared to the organic soil. This indicates that the presence of organic matter decreases the strength of the soil. The failure strain value also showed prominent changes with the presence of organic matter
- High void ratio is seen for organic soils

References

1. Husein Malkawi, A.I., Alawneh, A.S., Abu-Safaqah, O.T.: Effects of organic matter on the physical and the physicochemical properties of an illitic soil. Appl. Clay Sci. **14**(5), 257–278 (1999)
2. Thiyyakkandi, S., Annex, S.: Effect of organic content on geotechnical properties of Kuttanad clay. Electron. J. Geotech. Eng. **16 U**, 1653–1663 (no. Ayyar 1971) (2011)
3. Reddy, B.K., Sahu, R.B., Ghosh, S.: Consolidation behavior of organic soil in normal Kolkata deposit. Indian Geotech. J. **44**(3), 341–350 (2014)
4. Puppala, A.J., Pokala, S.P., Intharasombat, N., Williammee, R.: Effects of organic matter on physical, strength, and volume change properties of compost amended expansive clay. J. Geotech. Geoenvironmental Eng. **133**, 1449–1461 (2007)
5. Devi, K.R., Sahu, R.B., Mukherjee, S.: Shear strength of organic clay in Kolkata region. Indian Geotech. J. **45**(1), 25–34 (2015)
6. Cola, S., Cortellazzo, G.: The Shear Strength Behavior of Two Peaty Soils, pp. 679–695 (2005)
7. Schmidt, N.O.: A Study of the Isolation of Organic Matter as a Variable Affecting Engineering Properties of a Soil, Urbana, Illinois (1965)
8. Sen, P., Dixit, M., Chitra, R., Ratnam, M.: Effect of organic content on the index properties and compaction parameters of soil, **4**(4), 354–359 (2014)
9. Santagata, M., Bobet, A., Johnston, C.T., Hwang, J.: One-dimensional compression behavior of a soil with high organic matter content. J. Geotech. Geoenvironmental Eng. **134**, 1–13 (2008)

Sand Failure: Effect of Biocide on the Geomechanical Properties of Outcrop Carbonate Rock Under Static Conditions

Elizabeth Wuyep, Gbenga Oluyemi, Kyari Yates, and Alfred Akisanya

Abstract

The effects of chemical interaction of a biocide with formation rocks on the rock geomechanical strength were examined. A combination of analytical tests (Scanning Electron Microscopy/Energy-Dispersive X-Ray Analysis —SEM/EDX, X-Ray Powder Diffraction—XRPD and Particle Size Distribution) and uniaxial compressive test was used in this study. The particle size distribution showed an increase in D_{50} with poor sorting for the chemically treated carbonate core sample. The XRPD shows evidence of altered minerals in the chemically treated samples. It was observed that the interaction led to precipitation of new materials that clogged the pore space of the rock samples leading to about 150% increase in compressive strength of the carbonate following treatment with the biocide. The results give more insight into the limitations of the existing sand production prediction models with respect to the effect of oilfield chemicals on the strength of the reservoir rocks.

Keywords

Chemical interaction • Glutaraldehyde • Carbonate • Uniaxial compressive strength • Reservoir rock

E. Wuyep (✉) · G. Oluyemi
School of Engineering, Robert Gordon University, Aberdeen, AB10 7GJ, UK
e-mail: e.o.wuyep@rgu.ac.uk

K. Yates
School of Pharmacy and Life Science, Robert Gordon University, Aberdeen, AB10 7GJ, UK

A. Akisanya
School of Engineering, University of Aberdeen, Aberdeen, AB24 3UE, UK

1 Introduction

Biocides are typically used in the oil and gas industry to control the activities of undesirable bacteria and microorganisms that can cause corrosion of the pipelines and also produce some substances such as H_2S, organic acids, etc. that affect the yield and quality of oil and gas negatively. Unfortunately, the deleterious effect of the interaction of the biocide with the reservoir rock has not always been accounted for. Seto et al. [4] examined the effect of polyethylene oxide (PEO), aluminium chloride ($AlCl_3$) and dodecyltrimethyl ammonium bromide (DTAB) on sandstone under static conditions and found that these chemicals did not cause any significant change in the compressive strength of the sandstone. Nevertheless, the chemistries and composition of the chemicals used in the work are different from those of biocide and other commonly used oilfield chemicals.

The current work investigated the effects of biocide on the geomechanical strength of reservoir rock and sand production potentials under static condition. The results give more insight into the limitations of the existing sand production prediction models with respect to the effect of oilfield chemicals on the reservoir strength.

2 Materials and Experimental Implementation

Glutaraldehyde ($C_5H_8O_2$), a biocide was obtained from REDA oilfield UK Ltd; whilst a synthetic brine was used as control. Four (4) cylindrical outcrop core samples with a diameter of 38 mm and length 51 mm obtained from Texas, USA through Kocurek Industries Ltd were the substrates.

A combination of saturation test, analytical tests, particle size distribution analysis and mechanical tests were carried out on untreated and chemically treated (biocide) core samples (carbonate). X-ray Powder Diffraction (XRPD) and

A. Kallel et al. (eds.), *Recent Advances in Geo-Environmental Engineering, Geomechanics and Geotechnics, and Geohazards*, Advances in Science, Technology & Innovation, https://doi.org/10.1007/978-3-030-01665-4_43

Energy Dispersive X-Ray Analyser (EDXA) techniques were used to determine mineral and element compositions respectively. Scanning Electron Microscope (SEM) was used to analyse the mineral textural characteristic. While the particle size distribution was analysed with the use of Malvern Mastersizer. Compressive strength was investigated under uniaxial compression test. The results of the tests were then analysed using an integrated approach.

3 Results

3.1 Evaluation of Elemental and Mineralogical Composition of the Rock Sample

A cross-section of untreated carbonate (Edward brown) under SEM revealed dolomitized limestone with moderately sorted euhedral-subhedral dolomite ($CaMg(CO_3)_2$) mineral with sucrosic texture [6]. Observed in the SEM micrograph are cloudy centres and light rims which are typical of dolomites. Several interconnected pore spaces were observed in the sample. The EDX spectrum indicates the presence of Mg, C, O, Ca, Al, K, Cl, Si and Fe.

On the other hand, the SEM micrograph shows the glutaraldehyde treated carbonate to have about 90% of the pore spaces filled with the precipitated materials which are believed to have originated from the dissolution and precipitation. EDX scan revealed 58 and 11% increase in Ca and Mg content respectively in the glutaraldehyde and brine treated samples. According to existing literature [3], three simultaneous reaction mechanisms are identified with calcite dissolution process. Thus, in this work, the likely reactions involving the calcite dissolution and dolomite precipitation when the rock is exposed to brine and glutaraldehyde are as presented below:

$$CaCO_3 \overset{H^+}{\leftrightarrow} Ca^{2+} + HCO_3^- \quad (1)$$

$$CaCO_3 + H_2CO_3^0 \leftrightarrow 2HCO_3^- \quad (2)$$

$$CaCO_3 + H_2O \leftrightarrow Ca^{2+}HCO_3^- + OH^- \quad (3)$$

$$Ca^{2+} + Mg^{2+} + 2C^{+}7O^{-} \overset{2H^+}{\leftrightarrow} CaMg(CO_3)_2 + H_2O \quad (4)$$

3.2 Grain Size Distribution Analysis

The D_{10}, D_{50} and D_{90} values for the glutaraldehyde effluents from carbonate increased remarkably (a factor of 10) relative to those of brine (Table 1).

3.3 Response of Strength and Young's Modulus to Chemical Treatment

Figure 1a shows an increase in mean strength of untreated carbonates (Edward brown) from 5.6 MPa to about 8.4, and 15 MPa owing to treatment with brine and glutaraldehyde (1 and 2) respectively. The 50% and approximately 150% increase in strength of carbonate post chemical treatment with brine and glutaraldehyde confirms the pore space filling by the precipitated materials made possible by the dissolution/precipitation reaction.

The measured Young's modulus, E, of carbonate from the test results shows an increase from 4 to 12 GPa following treatment with glutaraldehyde and no change (5 GPa) with brine treated sample (Fig. 1b).

4 Discussion

The 89–92% decrease in calcite content indicated by XRPD suggests a dissolution reaction; whilst the 25–29% increase in dolomite indicates a precipitation reaction. The clay fraction shows little or no change with the clay minerals content (kaolinite, illite and mixed-layered illite-smectite) when exposed to brine and glutaraldehyde.

The interaction between glutaraldehyde and the carbonates as well as the interaction between the brine and glutaraldehyde can lead to precipitation of new materials that fill the pore spaces (pore clogging).

The increment of the particle size distribution on exposure to glutaraldehyde suggests some dissolution/precipitation reaction orchestrated by diffusion and adsorption might have taken place between the chemical and the rock samples that led to deterioration of the grain to grain binding resulting in grains detachment and release into the fluid stream. The result is consistent with earlier works [2, 5].

Table 1 Grain size distribution with sorting of the original brine and the effluents

Effluents	Particle size (μm)			Folks [1]	
Carbonate	D10	D50	D90	Sorting (σ_1)	Description
Original brine	3	7	30	0.01	Very well sorted
Carb-Brine	2	15	41	0.54	Moderately sorted
Carb-Glut 1	7	70	407	0.84	Moderately sorted
Carb-Glut 2	7	71	408	0.85	Moderately sorted

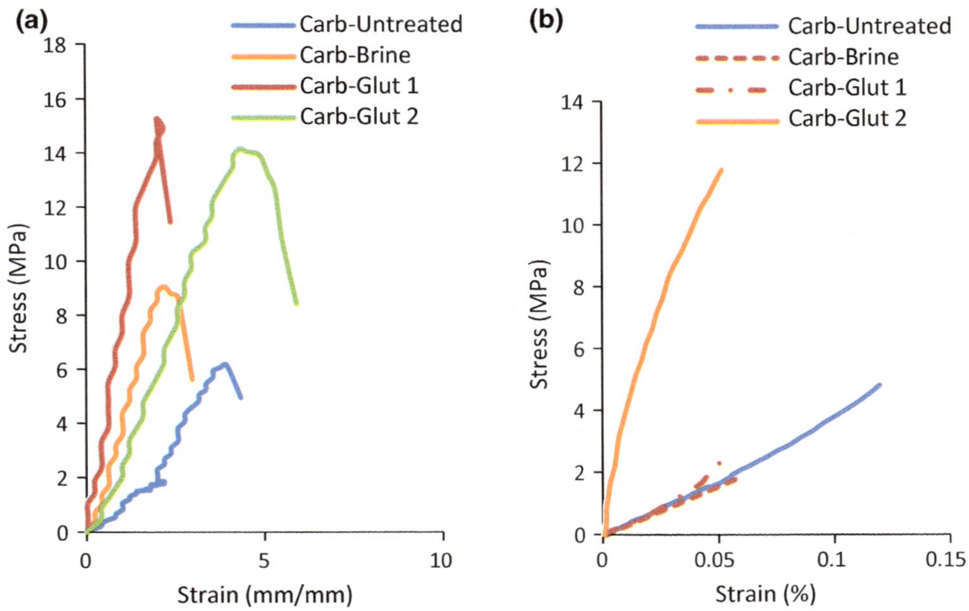

Fig. 1 Stress-strain relationship in plastic regions of untreated and chemically treated Edward brown carbonate: **a** complete response to failure; **b** stress-small strain response

5 Conclusion

The results show that chemical reactions, namely, diffusion, adsorption, dissolution and precipitation took place. The interaction between biocide and carbonate rock led to precipitation of new materials that filled the pore spaces giving rise to approximately 150% uniaxial compressive strength increase, a phenomenon that potentially leads to formation damage and hydrocarbon production impediment.

References

1. Folk, R.: A review of grain-size parameter. Sedimentology **6**(2), 73–93 (1966)
2. Oluyemi, G.: Conceptual physicochemical models for scale inhibitor-formation rock interaction. Pet. Sci. Tech. **32**(3), 253–260 (2014)
3. Plummer, L.N., Wigley, T.M.L., Parkhurst, D.L.: The kinetics of calcite dissolution in CO_2-water systems at 5 degrees to 60 degrees C and 0.0 to 1.0 atm CO_2. Am. J. Sci. **278**(2), 179–216 (1978)
4. Seto, M., Nag, D.K., Vutukuri, V.S., Katsuyyama, K.: Effect of chemical additives on the strength of sandstone. Int. J. Rock Mech. Min. Sci. **34**, 3–4 (1997)
5. Wuyep, E., Oluyemi, G., Yates, K., Akisanya, A.: Geochemical effects of oilfield chemicals on sand failure in reservoir rocks. J. Pet. Sci. Eng. **165**, 347–357 (2018)
6. Wuyep, E., Oluyemi, G., Yates, K., Akisanya, A.: Evaluation of the interactions between oilfield chemicals and reservoir rocks. J. Pet. Sci. Eng. (2018b)

Statistical Analysis of Non-destructive Evaluation of Concrete Strength in Several Case Studies of Literature: Effect of the Number of Cores on the Assessment Predictive Capacity

Faiza Neggaz, Khoudja Ali-Benyahia, and Mohamed Ghrici

Abstract

Strength assessment of existing buildings is a challenge for structural engineers who need to feed structural computations with material data. Such assessment is required in various contexts like the prediction of their seismic capacity, change of use or evaluation after a seismic damage. Destructive testing (DT) (coring) in conjunction with non-destructive testing (NDT) offer an alternative of concrete strength assessment in existing constructions. The accurate quality of this strength assessment is a fundamental point. The Rebound Hammer (RH) and Ultrasonic Pulse Velocity (UPV) techniques which appear to be complementary are often used in combination with the objective of improving the strength of the estimations. The present study is based on on-site data of several case-studies selected from literature. The thorough statistical analysis of the data and the performance of the models allowed showing the effect of the number of cores used for the model identification on the assessment quality. In several situations, the conclusions aimed to quantify the number of cores needed to offer a better effectiveness of the assessment process.

Keywords

Cases-study • Non-destructive evaluation • RH • UPV • Number of cores • Assessment quality

F. Neggaz · K. Ali-Benyahia · M. Ghrici (✉)
Department of Civil Engineering, University Hassiba Benbouali, 02000 Chlef, Algeria
e-mail: m.ghrici@univ-chlef.dz

K. Ali-Benyahia
Department of Technology, University Djillali Bounaama, 44225 Khemis-Miliana, Algeria

1 Introduction

There may be a need to assess in situ concrete strength when evaluating the structural seismic capacity of existing constructions, designed according to previous seismic criteria of which the seismic safety factors are currently considered low [10]. For instance, in North Algeria, thousands of public buildings (hospitals, schools …) have to comply with the recent seismic criteria of the local code. The fundamental questions which are nowadays of concern for the managers are the estimate of residual life span and the re-categorization of the structures [5].

This estimation is usually performed by compression tests of cores that are drilled from structural elements. The coring method is expensive, invasive, time consuming and in some cases difficult or impossible to be performed. In any case, NDT offers an important approach, since it gives access to information of material properties with moderate cost and rapid execution [6]. The DT in conjunction with NDT offer an alternative method of concrete strength assessment in existing buildings [9, 11–13]. The non-destructive assessment requires a relationship establishment between the results of non-destructive tests and the compressive strength of cores [2, 8, 11]. The accuracy of the strength assessment depends mainly on the number of cores used for calibration [3, 4, 7]. In general, when the number of cores increases, a good quality of assessment is observed. However, the possible negative effects on the structural stability (during and after coring) and the budget requirements limit the number of cores [10]. The European standard EN 13791 [8] recommends at least 18 pairs of data (core strengths and NDT test results). ACI standard [1] recommends also six to nine areas of measurement with two cores in each. The real challenge aims to reduce the number of calibration cores without reducing the assessment quality. Some research works in recent literature [3, 4] tackled the question: how many cores are needed to offer a better effectiveness of assessment process? These conclusions depend especially on

A. Kallel et al. (eds.), *Recent Advances in Geo-Environmental Engineering, Geomechanics and Geotechnics, and Geohazards*, Advances in Science, Technology & Innovation, https://doi.org/10.1007/978-3-030-01665-4_44

the case-study analyzed. The present paper focused on several real case-studies of existing constructions with the objective of analyzing the accuracy of this strength assessment depending of the number of calibration cores and providing some recommendations for more reliable assessment using the root mean square error "RMSE" and the coefficient of determination "r^2" as statistical indicators.

2 Methodology of Strength Assessment

An assessment methodology (Fig. 1) is proposed in order to analyze the effect of the number of cores on the errors convergence of the predictive capacity. All the data of each case-study selected from literature is considered as the same population "N". The data collected on site is in form of triplet (core strength "F", Rebound "R" and Ultrasonic Pulse Velocity "V"). The cores which are used for calibration are randomly selected among all the data "N". The number NC varies between 2 and 20. For each NC, a statistical regression leads to a specific model $F_m = f(NDT_m)$. The model shapes were tested here: the power, exponential and double power laws for RH, UPV and combined method respectively. Only the models of physical sense are accepted among the identified models. RMSE and r^2 values are thus calculated at the identification (on sample NC) and prediction (on sample N-NC) stages. All this procedure is repeated 500 iterations for each NC. Statistical properties (average, median, 95% percentile and variability) are quantified for the results of 500 iterations with the objective of analyzing the robustness of the assessment quality.

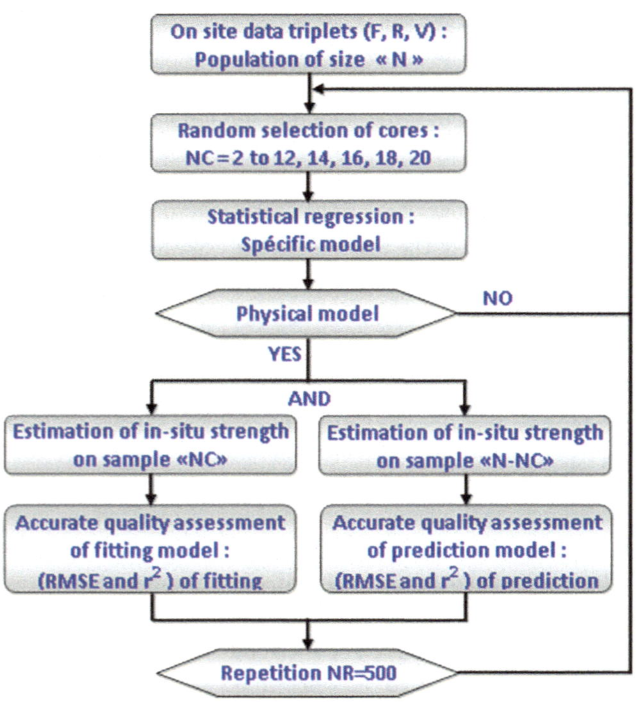

Fig. 1 Strength assessment process at fitting and prediction stages

3 Analysis of the Effect of Cores Number on the Assessment Quality

The common engineering approaches are to test the model precision on the sample where it is identified. The real challenge here is to test the assessment accuracy of the conversion model on new data (prediction stage). The difference between fitting and prediction errors is usually confused. This model can be used to evaluate strengths on the other measurement points (N-NC). The indicators RMSE and r^2 are thus used in order to estimate the distance between measured and estimated strengths on the (N-NC). The difference in precision between the two stages is very important, especially if NC is low, but the real practice is often based on only the precision of fitting model [3]. A fitting model can present a good precision quality but a very low prediction capacity [4]. The fitting and prediction errors present contradictory behaviors in accuracy when NC increases (Fig. 2). This study aimed to show the adequate number of cores at which the fitting and prediction errors are stabilized.

4 Conclusion

A several real case-studies of literature were used as a basis for the analysis of the effect limit of the number of cores on concrete strength assessment using the RMSE and coefficient r^2 as statistical indicator. The fitting and prediction errors present contradictory behaviors in precision quality if NC increases. This study allowed quantifying the number of cores needed to obtain a better effectiveness of the assessment process.

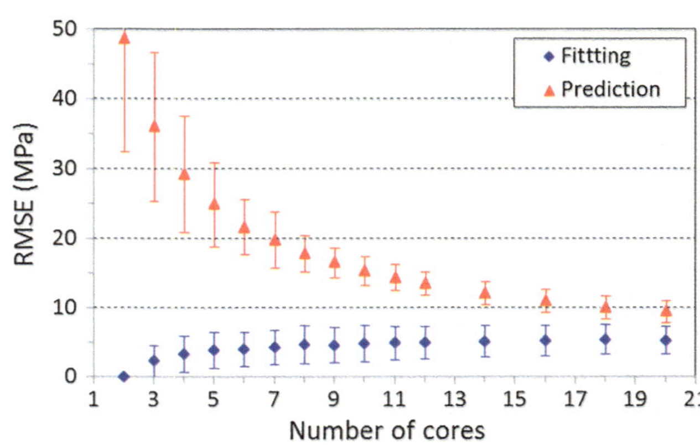

Fig. 2 Fitting and prediction RMSE against the number of cores

References

1. ACI 228 1R.: In-Place Methods to Estimate Concrete Strength. American Concrete Institute, USA (2003)
2. Ali-Benyahia, K., Ghrici, M., Kenai, S., Breysse, D., Sbartaï, Z. M.: Analysis of the relationship between nondestructive and destructive testing of low concrete strength in new structures. Asian J. Civ. Eng. **18**(2), 191–205 (2017)
3. Ali-Benyahia, K., Sbartaï, Z.M., Breysse, D., Kenai, S., Ghrici, M.: Analysis of the single and combined non-destructive test approaches for on-site concrete strength assessment: general statements based on a real case-study. Case Stud. Constr. Mater. **6**, 109–119 (2017)
4. Alwash, M., Breysse, D., Sbartaï, Z.M.: Non-destructive strength evaluation of concrete: analysis of some key factors using synthetic simulations. Constr. Build. Mater. **99**, 235–245 (2015)
5. Balayssac, J.P., Laurens, S., Breysse, D., Garnier, V.: Evaluation of Concrete Properties by Combining NDT Method. RILEM, pp. 187–192 (2013)
6. Breysse, D.: Nondestructive evaluation of concrete strength: an historical review and a new perspective by combining NDT methods. Constr. Build. Mater. **33**, 139–163 (2012)
7. Breysse, D., Martinez-Fernandez, J.L.: Assessing concrete strength with rebound hammer: review of key issues and ideas for more reliable conclusions. Mater. Struct. **47**, 1589–1604 (2014)
8. EN 13791.: Assessment of In-Situ Compressive Strength in Structures and Precast Concrete Components. CEN (2007)
9. Fiore, A., Porco, F., Uva, G., Mezzina, M.: On the dispersion of data collected by in situ diagnostic of the existing concrete. Constr. Build. Mater. **47**, 208–217 (2013)
10. Masi, A., Chiauzzi, L.: An experimental study on the within-member variability of in situ concrete strength in RC building structures. Constr. Build. Mater. **47**, 951–961 (2013)
11. Pucinotti, R.: Reinforced concrete structure: non-destructive in situ strength assessment of concrete. Constr. Build. Mater. **75**, 331–341 (2015)
12. Uva, G., Porco, F., Fiore, A., Mezzina, M.: Proposal of a methodology for assessing the reliability of in situ concrete tests and improving the estimate of the compressive strength. Constr. Build. Mater. **38**, 72–83 (2013)
13. Vona, M., Nigro, D.: Evaluation of the predictive ability of the in situ concrete strength through core drilling and its effects on the capacity of the RC columns. Mater. Struct. **39**, 149–160 (2013)

Correlation of Electrical Resistivity Test with the Geotechnical Parameters of Sandy Soil

Ahsan Naseem, Fazal-e Jalal, Hans De Backer, Ken Schotte, and Muhammad Kashif

Abstract

Non-destructive tests (NDTs) are the most economical and easy to use techniques to determine different soil properties. They expedite the process of determining sub-surface characteristics. These include Ground Penetrating Radars (GPRs), Seismographs, Shear Wave Velocity (SWV) and Electrical Resistivity Test (ERT) etc. ERT is nowadays getting worldwide popularity for determining the sub-surface geology in geotechnical engineering, as it does not require extensive testing. This research aimed at developing empirical correlations of ERT with different soil parameters by performing extensive conventional laboratory tests in order to get all the required soil parameters by just performing ERT in the future, which otherwise require great time and effort to be determined by the conventional laboratory tests. The developed correlations include relationship of resistivity value obtained in the field with the internal friction angle (r^2 = 0.964), cohesion (r^2 = 0.946) and the bearing capacity of shallow foundation (r^2 = 0.90) for depth to width ratios of 0.5 and 1.0 respectively. The regression coefficients obtained ensured the development of quite good correlation for the sandy-clayey soil.

Keywords

Non destructive tests • Electrical resistivity test • Soil parameters

1 Introduction

ERT is an in situ NDT which can be used to have a good knowledge about sub-surface profile. The major benefit of this test is that it does not require extensive testing and material transportation to the laboratory thus not only does it save a lot of money but it can enable us to get rid of the extensive laboratory testing too. The other benefits include less expertise, less operational costs, faster operation and less personnel required [1, 2]. These benefits make ERT very popular among the investigators to use for reconnaissance survey of any site, to determine different sinkholes, study crack propagation in soils [3], problematic soil seams, settlement issues in an already constructed building, factor of safety (F.S) of any landslide etc. [4, 5]. But, as it is still a new technique in the geotechnical engineering, it lacks reliable and sufficient amount of research which can be readily used. Despite being an expedite and easy to use technique, one has to conduct extensive laboratory tests after performing ERT in order to determine different geotechnical properties of the soil for different design procedures. The currently existing correlations include a relationship of resistivity with cone penetration and moisture content [2], resistivity with SPT blows [1], hydraulic conductivity of compacted clays [4, 6], Atterberg's limits and dry density etc. [7]. But no research has been able to develop comprehensive relationship with a high confidence value between ERT and geotechnical soil parameters in such a way that performing ERT alone would be enough to use the developed correlations to determine almost each soil parameter, which otherwise would require a lot of testing, time and money to be determined.

The correlations which were developed based upon this research included the relationship of resistivity with shear strength parameters (internal friction angle (ø) and cohesion (c)) and allowable bearing capacity for shallow foundations with depth to width ratio (D/B) of 0.5 and 1.0.

A. Naseem (✉) · H. De Backer · M. Kashif
University of Ghent, 9052 Ghent, Belgium
e-mail: ahsan.naseem@ugent.be

Fazal-eJalal · K. Schotte
COMSATS University Islamabad, Abbottabad, 22060, Pakistan

A. Kallel et al. (eds.), *Recent Advances in Geo-Environmental Engineering, Geomechanics and Geotechnics, and Geohazards*, Advances in Science, Technology & Innovation, https://doi.org/10.1007/978-3-030-01665-4_45

2 Methodology

The research included performing the ERT using Wenner-Schlumberger array at 7 different locations on the site, each of 1 km long and then the bore holes were drilled at the test location to recover the soil samples. A total of 7 boreholes were drilled using motorized auger. 28 probes were erected at 1 m c/c spacing and contours were generated. The soil samples were recovered using shelby tubes from each 1.5 m depth up to a maximum depth of 7.5 m. These soil samples were then waxed and transported to the laboratory, where the conventional laboratory tests including soil gradation, direct shear tests and triaxial tests were performed in accordance with the ASTM (2014) standard procedures.

Finally, the resistivity values obtained at each depth were plotted against the obtained laboratory soil parameters of the same depth and the empirical correlations were developed using the regression analysis.

3 Results

3.1 Correlation with Shear Strength Parameters

In order to develop the correlations of resistivity with the strength parameters, direct shear tests were performed on the soil samples recovered from different depths. The values of internal friction angle (ø) and cohesion (c) were then plotted against the resistivity (R) obtained at the same depth as shown in Fig. 1. The developed equations are given by Eqs. 1 and 2.

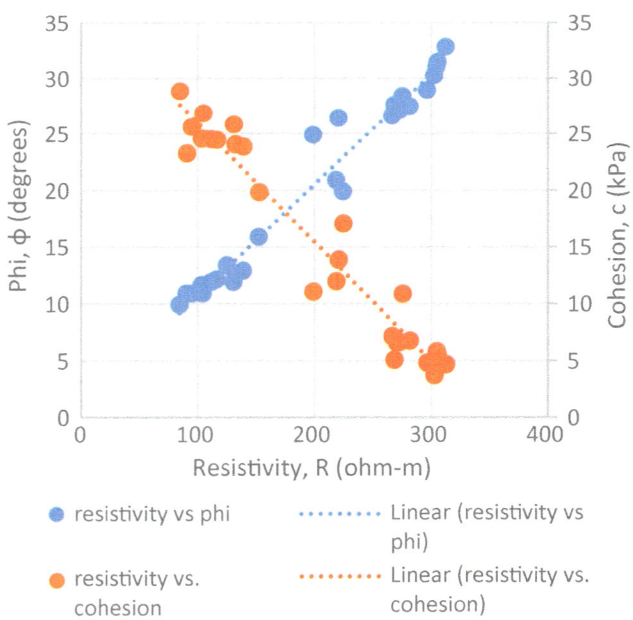

Fig. 1 Correlation of resistivity with the shear strength parameters

$$\phi(^\circ) = 0.0985R + 0.973 \quad (r^2 = 0.964) \tag{1}$$

$$c(\text{kPa}) = 36.569 - 0.1052R \quad (r^2 = 0.964) \tag{2}$$

3.2 Correlation with Bearing Capacity

Bearing Capacity analysis at any depth requires detailed calculations of overburden stresses, cohesion, friction angle, bearing capacity factors, relative density etc., if conventional equations are to be used. So, in the development of this correlation, the allowable bearing capacity (q_a) of soil was calculated with D/B ratios of 0.5 and 1.0 using conventional Terzaghi's bearing capacity equation by keeping F. S = 4.0 and then plotting these values for each depth against the obtained resistivity at that corresponding depth as shown in Fig. 2. Based on the developed plot, Eq. 3 was suggested for the calculation of allowable bearing capacity. From Fig. 2, it is also evident that D/B ratio has no major influence on the empirical correlations developed.

$$q_a(\text{kPa}) = 48.446e^{0.0083R} \quad (r^2 = 0.903) \tag{3}$$

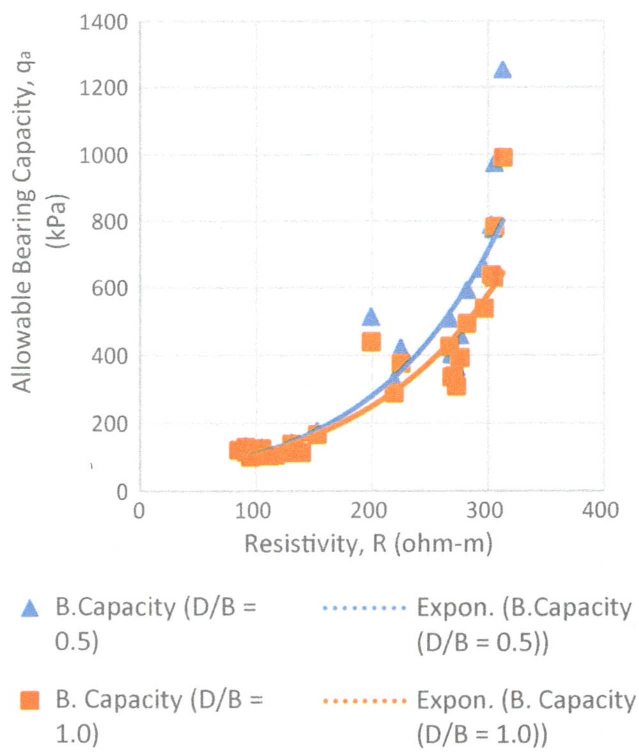

Fig. 2 Correlation of resistivity with allowable bearing capacity (D/B = 0.5, 1.0)

4 Discussion

Figure 1 shows that the trend lines developed have high regression values of $r^2 = 0.964$ and $r^2 = 0.946$ for friction angle and cohesion, respectively. It is also evident from the plot that the greater the density and internal friction angle, the greater the resistivity would be, while the greater the cohesion, the smaller the resistivity value would be. It means that as the particle size increases, its resistivity value also increases and vice versa.

Figure 2 shows that the trend lines developed from the plots yielded high regression values of about $r^2 = 0.912$ and $r^2 = 0.903$, respectively. The benefit of this correlation is, that just using the resistivity value now, one can obtain the allowable bearing capacity value instead of using the existing cumbersome conventional equations which require many inputs calculated using laborious tests to yield this value.

5 Conclusion

From this research, it is concluded that: firstly, ERT can be well correlated with the different geotechnical soil parameters. Secondly now ERT alone can be used to determine the soil parameters for the design procedures by using the developed correlations, which previously had to be determined using extensive laboratory testing.

References

1. Sudha, K., Israil, M., Mittal, S., Rai, J.: Soil characterization using electrical resistivity tomography and geotechnical investigations. J. Appl. Geophys. **67**(1), 74–79 (2009)
2. Cosenza, P., Marmet, E., Rejiba, F.: Correlations between geotechnical and electrical data: a case study at Garchy in France. J. Appl. Geophys. **60**(3), 165–178 (2006)
3. Hassan, A., Toll, D.G.: Electrical resistivity tomography for characterizing cracking of soils. In: Geo-congress: stability and performance of slopes and embankments III, ASCE, pp. 818–827 (2013)
4. Abu-Hassanein, Z.S., Benson, C.H., Blotz, L.R.: Electrical resistivity of compacted clays. J. Geotech. Eng. **122**(5), 397–406 (1996)
5. Joab, M.J., Andrewsb, M.: Investigating slope failures using electrical resistivity: case studies. Assoc. Prof. Eng. Trinidad Tobago **38**(1), 66–75 (2009)
6. Shevnin, V., Delgado-Rodríguez, O., Mousatov, A.: Estimation of hydraulic conductivity on clay content in soil determined from resistivity data. Geofísica internacional **45**(3), 195–207 (2006)
7. Bryson, L.S.: Evaluation of geotechnical parameters using electrical resistivity measurements. Earthq. Eng. Soil Dyn. ASCE (2005)

Evaluation of the Physical and Mechanical Properties and Structure of Geological Materials by Broadband Ultrasonic Structuroscopy

Ivan A. Shibaev, Oleg D. Belov, Alexander Kravcov, and Svetlana P. Mesyats

Abstract

This paper presented the results of a laboratory study of pyroxenite and carbonate core samples taken at different depths in an open pit mine. Longitudinal and shear wave velocities in these samples were measured with an error of 0.1%. The local porosity of the samples was evaluated with the use of broadband ultrasonic structuroscopy.

Keywords

Geophysics • Structuroscopy • Local moduli of elasticity • Fractures in rocks • Porosity

1 Introduction

The evaluation of the physical and mechanical properties of geological materials at different scales is very important for the mineral mining and processing industry. At the macro-level, these properties are studied by such geophysical methods as seismic tomography [1], electrical prospecting, etc. At the micro-level, the porosity, fractures, and other structural features of rock samples are chiefly characterized by electron microscopy [2] and X-ray tomography [3].

However, these methods are rather laborious: they require that samples be specially prepared. That is why broadband ultrasonic structuroscopy has recently become increasingly popular as a means for evaluating the physical and mechanical properties and internal structure [4] of geological materials at the micro-level.

This method involves using a laser to generate high-power short pulses and employing a broadband piezoelectric transducer to record signals travelling through the sample and scattered by its internal heterogeneities; the method measures velocities of elastic waves with an error of 0.1%, attaining a depth resolution of 30 μm and a lateral resolution of 0.5 mm. In [5, 6], such resolutions are achieved through the use of specially designed opto-acoustic generators combined with a piezoelectric transducer. We used seismic tomography to study the structure and properties of a rock sequence. Pyroxenite and carbonate samples taken from the rock sequence were studied using broadband ultrasonic structuroscopy.

2 Methods and Materials

Geophysical monitoring was first carried out to determine the initial state of certain layers of the rock sequence. The seismic survey involved evaluating longitudinal and shear wave velocities in the rock sequence, based on which the range of Poisson's ratio ν was estimated.

Rock samples for physical and mechanical testing and for laboratory studies by broadband ultrasonic structuroscopy were taken from different depths in the open pit mine.

The core samples were 60–120 mm long on average.

Longitudinal and shear wave velocities in pyroxenite and carbonate samples were calculated from the difference in the arrival times of the signals reflected from the front surface and bottom of the samples.

Poisson's ratio and Young's modulus were determined from these velocities.

I. A. Shibaev (✉) · O. D. Belov
The National University of Science and Technology "MISiS", Leninsky Prospect 4, Moscow, 119991, Russia
e-mail: sh1baev@yandex.ru

O. D. Belov
e-mail: below2idknet@inbox.ru

A. Kravcov
Department of Construction Technology, Faculty of Civil Engineering in Prague, Thákurova 7/2077, 166 29 Prague 6—Dejvice, Czech Republic
e-mail: kravtale@fsv.cvut.cz

S. P. Mesyats
The Establishment of RAS Mining Institute, 14, Fersman Str, Apatity, Murmansk reg., 184209, Russia
e-mail: mesyats@goi.kolasc.net.ru

A. Kallel et al. (eds.), *Recent Advances in Geo-Environmental Engineering, Geomechanics and Geotechnics, and Geohazards*, Advances in Science, Technology & Innovation, https://doi.org/10.1007/978-3-030-01665-4_46

Table 1 Acoustic and elastic properties of rocks of the Kovdorskoye iron ore deposit

Seismic monitoring data				
Rock	Longitudinal wave velocity, C_p (m/s)	Shear wave velocity, C_s (m/s)	Poisson's ratio	
1. Pyroxenite	2860–6820	1018–4034	0.17–0.32	
2. Carbonatite	2640–5810	1500–4080	0.01–0.33	
Laboratory data				
Rock	Average value of longitudinal wave, C_p (m/s)	Average value of shear wave, C_s (m/s)	Poisson's ratio (v)	Young's modulus (GPa)
1. Pyroxenite	6893	4390	0.16	138
2. Carbonatite	4337	2625	0.23	53

3 Results

The results of the examination of the rock samples from the Kovdorskoye complex ore deposit (iron ore deposit in the Murmansk region of Russia) by the ultrasonic method for laboratory determination of elastic and strength properties are given in Table 1.

Longitudinal wave velocities measured in the northern side of the open pit mine vary from 2600 to 6800 m/s, which is in good agreement with the laboratory data.

The seismic monitoring revealed that longitudinal wave velocities vary over a wide range; this means that there is a high degree of variability of rock properties, which is primarily due to fractures in the rocks.

If we introduce coefficient B_e for estimating the degree of fracturing, which value is calculated by the below formula (1):

$$B_e = \left(\frac{C_{p0}}{C_{pi}}\right)^2 - 1 \qquad (1)$$

where c_{p0} is the velocity of longitudinal waves in a solid rock mass and c_{pi} is the velocity of longitudinal waves in a fractured rock mass, then it is possible to evaluate the degree of fracturing of the rock mass, which determines its stability.

Also, using the well-known expression [7], the porosity of the rock samples was recalculated by the following formula (2):

$$P = \left[1 - \left(\frac{C}{C_0}\right)^2\right]^{\frac{3}{2}} \qquad (2)$$

where C_0 is the maximum velocity of longitudinal waves in the sample and C is the local velocity of longitudinal waves.

Additional measurements of local shear wave velocity made it possible to determine the elastic characteristics of the sample.

Figure 1 shows a porosity map of a pyroxenite sample.

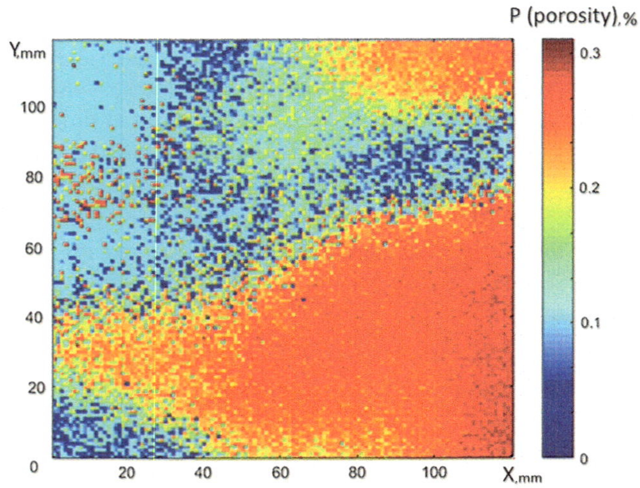

Fig. 1 Porosity map of pyroxenite sample

4 Discussion

The results of the ultrasonic study allowed the mechanical characteristics of rocks to be described at different scales without resorting to mechanical and destructive testing [8]; also, the energy consumption of an ore concentration process can be predicted.

Accordingly, the inefficiency of beneficiation of certain minerals can be justified.

5 Conclusion

Based on the aforesaid results, it can be stated that differences in the elastic characteristics of pyroxenite were successfully studied by broadband ultrasonic structuroscopy, compared to seismic monitoring data.

Broadband ultrasonic structuroscopy allows minor changes in the internal structure of rock samples to be revealed; the method makes it possible to determine the local elastic moduli and porosity of rock samples with high accuracy without resorting to X-ray tomography.

Acknowledgements This work was carried out with the financial support from the Russian Science Foundation (grant no. 16-17-10181).

References

1. Melnikov, N.N., Mesyats, S.P., Ostapenko, S.P., Cherepetskaya, E. B., Shibaev, I.A., Morozov, N.A., Kravcov, A.N., Konvalinka, A.: Investigation of disturbed rock zones in open-pit mine walls by seismic tomography. Key Eng. Mater. **755**, 147–152 (2017)
2. Ming, Z., Ning, J.G., Zhang, H.B., Liu, X.S.: Study on microscopic structure and mineral composition of shallow rock using SEM. Key Eng. Mater. **303–306**, 2552–2558 (2013)
3. Kravcov, A., Svoboda, P., Konvalinka, A., Cherepetskaya, E.B., Sas, I.E., Morozov, N.A., Zatloukal, J., Koťátková, J.: Evaluation of crack formation in concrete and basalt specimens under cyclic uniaxial load using acoustic emission and computed X-Ray tomography. Key Eng. Mater. **722**, 247–253 (2017)
4. Cherepetskaya, E.B., Karabutov, A.A., Makarov, V.A., Mironova, E.A., Shibaev, I.A., Vysotin, N.G., Morozov, D.V.: Internal structure research of Shungite by broadband ultrasonic spectroscopy. Key Eng. Mater. **755**, 242–247 (2017)
5. Kravcov, A., Svoboda, P., Konvalinka, A., Cherepetskaya, E.B., Karabutov, A.A., Morozov, D.V., Shibaev, I.A.: Laser-ultrasonic testing of the structure and properties of concrete and carbon fiber-reinforced plastics. Key Eng. Mater. **722**, 267–272 (2017)
6. Podymova, N.B., Karabutov, A.A., Cherepetskaya, E.B.: Laser optoacoustic method for quantitative nondestructive evaluation of the subsurface damage depth in ground silicon wafers. Laser Phys. **24**(8), 086003 (2014)
7. Podymova, N.B., Karabutov, A.A., Cherepetskaya, E.B.: Measuring the dependence of the local Young's modulus on the porosity of isotropic composite materials by a pulsed acoustic method using a laser source of ultrasound. J. Appl. Mech. Tech. Phys. **54**(3), 500–507 (2013)
8. Shibaev, I.A., Cherepetskaya, E.B., Bychkov, A.S., Zarubin, V.P., Ivanov, P.N.: Evaluation of the internal structure of dolerite specimens using X-ray and laser ultrasonic tomography. Int. J. Civil Eng. Tech. **9**(9), 84–92 (2018)

Part V

Geomechanics and Geotechnics: The Behavior of Soils Under Dynamic Loading

Mechanical Instability of Sandy Soils Under Seismic Effect (Algeria)

Mohammed Bousmaha, Mohamed Bensoula, Renaud Toussaint, Hanifi Missoum, and Karim Bendani

Abstract

The aim of this work was to study the saturated sand behavior under seismic effects which are stressed with several seismic vibrations and a reproduced seismic shaking frequency. The external excitation was operated by regular horizontal oscillations with controlled frequency and amplitude. The liquefaction had then a direct impact on the vertical displacement of intruders. We concluded that the movement of the intruder depends on the imposed acceleration and the relative density.

Keywords

Saturated sand • Seismic shaking • Frequency • Vertical displacement • Acceleration

1 Introduction

Liquefaction is an important phenomenon in the mechanical instability of a geotechnical study. Its factors have been constantly modified to have an approach avoiding a catastrophic failure of the ground and controlling the mechanical instability under seismic vibration. An accurate determination of critical conditions has been a major challenge for geotechnical engineering. This experimental study allowed us to know the mechanical behavior in seismic conditions of a saturated reconstituted sample medium and loose density taken at the port of Oran. In a dynamic condition, the aim was to provoke the sand sample under seismic solicitations corresponding to the earthquake recorded in Boumerdes in 2003 (Algeria).

M. Bousmaha (✉)
University USTO MB of Oran, Oran, Algeria
e-mail: Bousmed2000@gmail.com

M. Bensoula · H. Missoum · K. Bendani
University of Mostaganem, Mostaganem, Algeria

R. Toussaint
University of Strasbourg, Strasbourg, France

2 Settings

2.1 Experimental Study

The soil is located in the city of Oran in northwestern Algeria known for its considerable seismic activity. The sea front of the city of Oran is at important risk of collapse by the prolonged flow of the groundwater. The sample consists mainly of sand with a low plasticity index. A series of laboratory tests were conducted to obtain the geotechnical characteristics of the sample [1]; the sample is sand and graded (SP). From the results obtained from the characterization tests, we have:

γ_s 2.663: specific gravity of soil grain,
e_{min} 0.672: minimum void ratio (corresponding to dense state),
e_{max} 0.967: maximum void ratio (corresponding to loose state).

2.2 Materials

The reconstituted loose state preparation method remains an important subject in the experimental field of geotechnics [3]. There are two sample preparation methods for both dry and saturated conditions for laboratory use such as sedimentation with or without dry or wet tamping or pluviation through a funnel. To do this, the correct choice method must reproduce the conditions in situ of the soil. The aim was to mimic the evolution of the vertical displacement of an intruder in a Plexiglas box. The experimental studies consist in applying a seismic shaking to granular medium. We used two different densities in order to cover a wider band of data (medium and loose state). The free end of the rod was

A. Kallel et al. (eds.), *Recent Advances in Geo-Environmental Engineering, Geomechanics and Geotechnics, and Geohazards*, Advances in Science, Technology & Innovation, https://doi.org/10.1007/978-3-030-01665-4_47

attached to one side of this box and the other was attached to the engine. When the motor is run horizontally, the Plexiglas box is affected by a sinusoidal movement.

3 Results

At different vibration types, saturated sand samples are sensitive to liquefying. On the other situation, it can be observed that the samples show the same behavior for the same initial relative density and at the different vibrations. The first phase of the vibratory stress tests was performed to achieve a displacement reaching a peak which corresponds to the maximum of depression of the studied sample followed by a horizontal line signifying the end of the test by a loss of resistance [2, 5]. In the vibratory test using a seismic frequency of the Boumerdes earthquake in 2003 (Fig. 1), the use of the same frequency allows us to get an idea of the soil behavior over a short time span (Fig. 2).

The aim was to have a sand behavior idea at different modes of vibratory stress for saturated sand corresponding to a minimum relative density. In the saturated cases, it is the

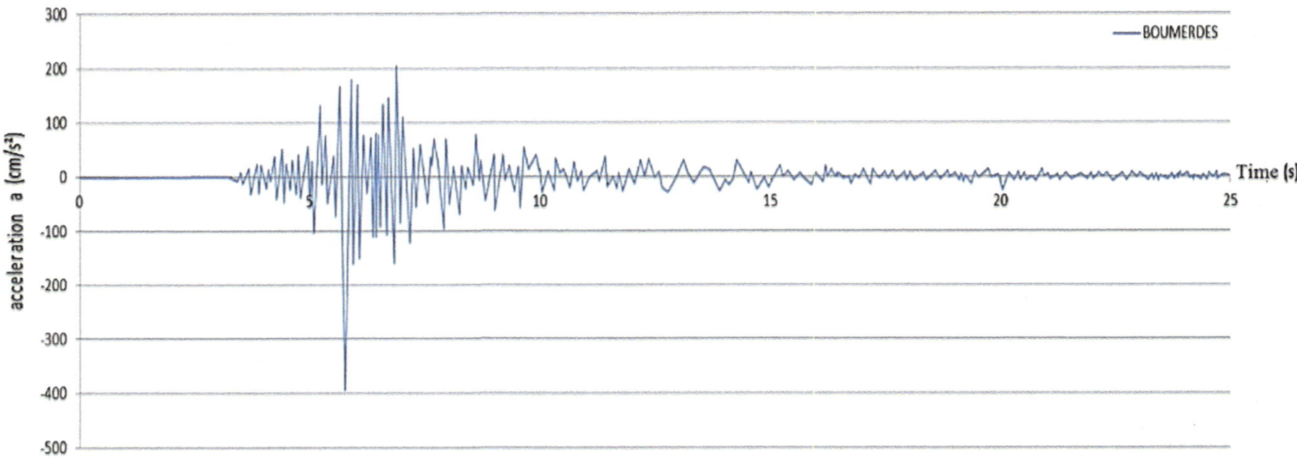

Fig. 1 Earthquake of Boumerdes 2003

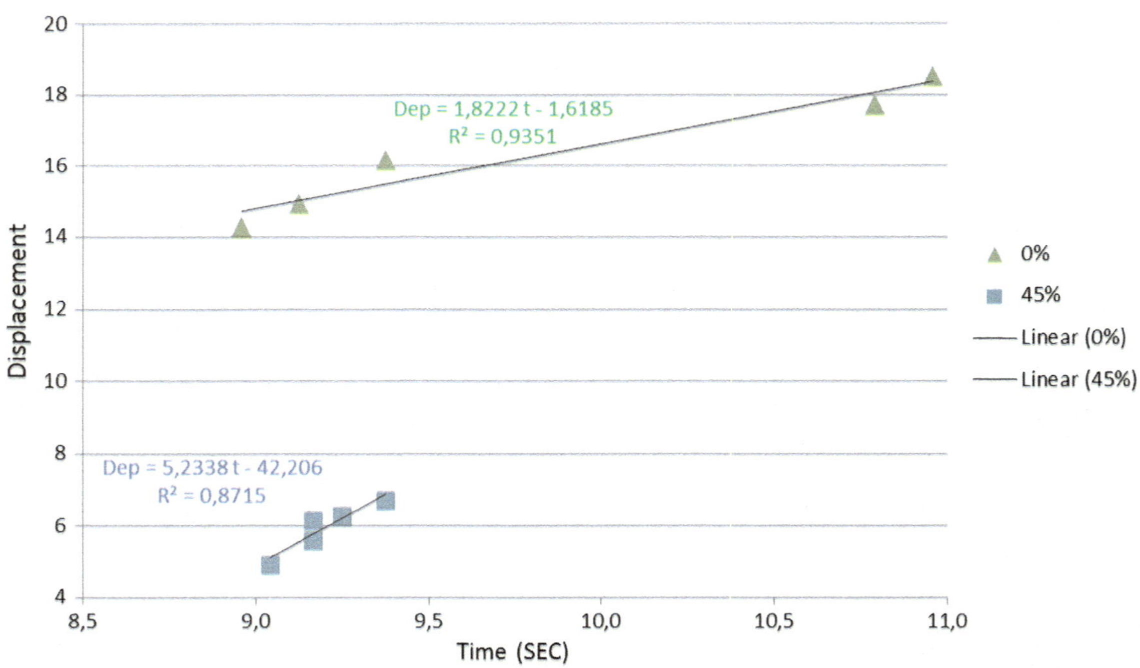

Fig. 2 Representation of vertical displacement max with both relative densities

Table 1 Recap of maximum displacements results with two different relative densities (D_r)

D_r = 0%	Start 312	Start 317	Start 319	Start 320	Start 321
Time (s)	9375	9125	1,079,167	1,095,833	8,958,333
Displacement max (mm)	1,616,667	1,491,803	1,770,492	1,852,459	142,623
D_r = 45%	Start 297	Start 299	Start 300	Start 301	Start 302
Time (s)	9,166,667	925	9,166,667	9,041,667	9375
Displacement max (mm)	6,164,384	630,137	5,616,438	4,931,507	6,712,329

presence of the Archimedes condition which is the often responsible for the penetration of the intruder [4]. From the two states test studied previously, it is possible to assume the results based on the following Table 1 and the corresponding results are shown in Fig. 2.

4 Discussion

The test consisted in following the penetration of an intruder in saturated sand with distilled water since the liquefaction of the environment occurs with the water under seismic shaking [5]. Because, the Archimedes force is present only in a saturated situation, its presence is often responsible for the intruder penetration [4]. The results analysis led to establish a relation between the intruder max displacement and the corresponding time, see the following equations:

Situation 1:	$D_{epmax} = 1.82\ t - 1.61$	With $D_r = 0\%$ and if $t \in [9, 11]$ s
Situation 2:	$D_{epmax} = 5.23\ t - 42.2$	With $D_r = 45\%$ and if $t \in [9, 9.5]$ s

5 Conclusion

The principal objective of this work was to study the phenomenon of liquefaction obtained by horizontal shaking in a saturated medium. Sand taken from the port of Oran. Seismic tests were carried out in two densities (loose and medium) for saturated sand in the case of a seismic vibration. After conducting the experiments in the case where the penetration height of the intruder exceeds 10%, the results show that the sand may be subject to a potential risk of liquefaction.

Beyond 30% of this height, liquefaction is more accentuated and this can be adequately interpreted by the relative density.

Finally, void ratio, relative density and vibratory frequency of the seism are to be considered as essential parameters for the liquefaction phenomena. With great acceleration and saturation, the sandy soil is affected by this phenomenon.

References

1. Bousmaha, M., Missoum, H., Bendani, K., Derkaoui, M., Belhouari, F.: Experimental study on mechanical instability of sand-silt mixtures. Int. J. Appl. Eng. Res **11**(3), 2149–2156. ISSN 0973-4562 (2016)
2. Clément, C., Toussaint, R., Stojanova, M., Aharonov, E.: Sinking during earthquakes: critical acceleration criteria control drained soil liquefaction. Phys. Rev. E **97**, 022905. https://doi.org/10.1103/physreve.97.022905 (2018)
3. Ladd, R.S.: Specimen preparation and liquefaction of sands. J. Geotech. Eng. ASCE **100**(GT10), 1180–1184 (1974)
4. Sanchez-Colina, G., Alonso-Llanes, L., Martinez, E., Batista-Leyva, A.J., Clément, C., Fliedner, C., Toussaint, R., Altshuler, E.: Note: «lock-in accelerometry» to follow sink dynamics in shaken granular matter. Rev. Sci. Instrum. **85**(12), 126101 (2014)
5. Toussaint, R., Clément, C., Fliedner, C., Stojanova, M., Aharonov, E., Sanchez-Colina, G.E., Altshuler, A., Batista-Leyva, J., Flekkøy, E.G.: Sink vs. tilt penetration into shaken dry granular matter: the role foundation. Geophys. Res. Abstr. **16** EGU2014, 12539 (2014)

Seismic Response and Failure Mechanism of the Subway Station: A Literature Review

Zhi-Xiang Zhan and Zhen-Dong Cui

Abstract

Because of the surrounding soil around the subway station, it is generally believed that the aseismic behavior of the underground subway station is so good that research on the seismic response of the underground structures is of no use. However, in 1995, Daikai subway station was the first large-scale underground structure that almost collapsed in the earthquake, which aroused people's attention to the aseismic design of the underground structures. In this paper, a systematic summary of the seismic response research about the subway stations was proposed as a reference for researchers based on several documents. The effects of the earthquake, geological condition and structural characteristics on the earthquake damage of Daikai subway station were summarized. In addition, some suggestions we are put forward for the further study of the seismic response and failure mechanism of the subway station. Finally, some measures for the isolation and shock absorption were proposed.

Keywords

Subway station • Seismic response • Failure mechanism • Isolation and shock absorption

1 Introduction

The rapid development of the economy and the explosion of the population in large cities result in the traffic jam on the ground surface, which promotes the development of the subway tunnels and stations. As we know, the lateral displacement of the subway station is restrained by the surrounding soil and the seismic acceleration of the underground structure is smaller than that of the ground structure. Therefore, the aseismic design of the subway station has been ignored over the past several decades. However, the subway station would be damaged seriously under the action of the strong earthquake, such as the earthquake damage at Daikai subway station. Figures 1 and 2 [1] show the seismic response and failure mechanism of the subway station under the action of El Centro wave. Different from the ground structures, the subway station needs to bear the self-weight of the overlying soil. Under the action of strong earthquakes, the vertical bearing capacity of mid-columns and lateral walls decreases. And the self-weight of the overlying soil and the inertia force of the vertical earthquake loading will result in the large deformation of the vertical components and the roof of the subway station. The decrease of the bearing capacity of the joint between the roof and the lateral wall will increase the axial compression ratio of the mid-columns, which will accelerate their damage. What is more, the reduction of the restraint of the lateral wall to the roof accelerates the deformation of the roof. When the mid-columns are seriously damaged, the roof cannot bear the self-weight of the overlying soil and falls. Finally, the structure collapses.

Research about the seismic response and the failure mechanism was summarized systematically in this paper to give a reference to the design and construction of the subway station.

2 Analysis Methods

The methods to analyze the seismic response and the failure mechanism of the subway station mainly include the theoretical analysis, numerical simulation, model test and field measurement. The results obtained by using the first three methods are usually verified by using the last method.

Z.-X. Zhan · Z.-D. Cui (✉)
State Key Laboratory for Geomechanics and Deep Underground Engineering, School of Mechanics and Civil Engineering, China University of Mining and Technology, Xuzhou, 221116, Jiangsu, People's Republic of China
e-mail: cuizhendong@cumt.edu.cn

A. Kallel et al. (eds.), *Recent Advances in Geo-Environmental Engineering, Geomechanics and Geotechnics, and Geohazards*, Advances in Science, Technology & Innovation, https://doi.org/10.1007/978-3-030-01665-4_48

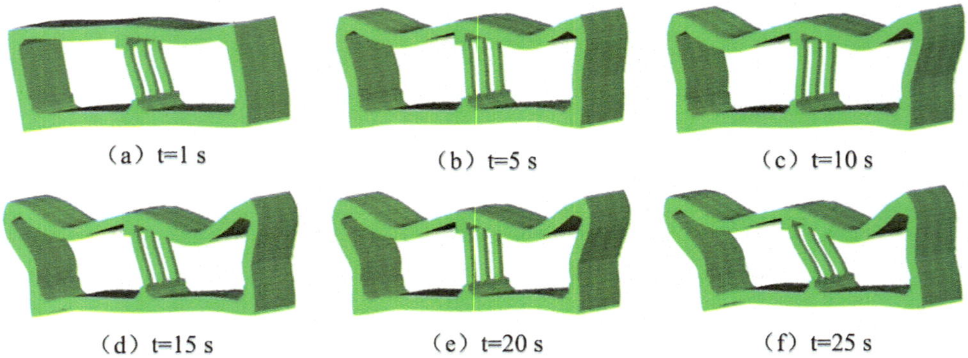

Fig. 1 The deformation evolutionary process of the subway station under the action of El Centro wave

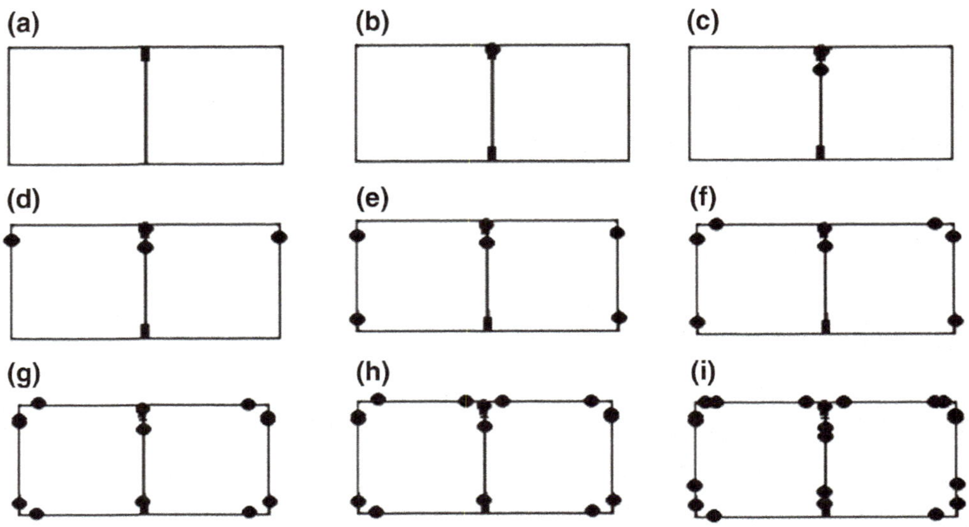

Fig. 2 The damage evolutionary process of the subway station under the action of El Centro wave

The simplified methods for seismic analysis of underground structures are summarized in Table 1 [2]. Those methods do not consider the vertical earthquake loading and are only suitable for small earthquakes. What is more, under the action of strong earthquakes, the cumulative damage of materials, the fracture, contact and collision of components, and the interaction of soil-structure should be considered.

The numerical simulation considering the interaction of soil-structure mainly includes the substructure and global analysis methods [3]. When using the method of global analysis, the non-uniform and nonlinear material properties of the generalized structure, the irregular terrain and the contact nonlinearity can be considered. Two conventional pseudo-static analyses, applying either a displacement-based or a force-based approach, were carried out to simulate the seismic response of the subway station [4]. The mid-column is the key component of the subway station. The columns with different aspect ratios have different lateral seismic capacity. So finding a reasonable aspect ratio is a significant problem. The ultimate bearing capacity analysis of isolated columns of various sizes was performed using the finite element software ABAQUS/CAE [5]. The results indicate that the ultimate bearing capacity of the isolated columns is not monotonous with the increase of the aspect ratio of columns and there is a recommended aspect ratio range. Among all the subway stations that have been built, some have irregular cross sections. Tang et al. [6] studied the damage mechanisms of a variable cross-section subway station structure in complex geological ground by constructing a three-dimensional numerical model and found that the seismic response of sub-layer of the variable cross-section subway station structure with a narrow base was stronger than that of the up-layer, which is different from that of the ordinary ones. In addition, the peak acceleration, velocity and relative lateral displacement of the variable cross-section subway station structure with a narrow base are also different from those of the ordinary ones.

The seismic damage data about the real subway stations are very scarce. So the model tests mainly including the shaking table test and centrifuge test are used to obtain some seismic damage data. Zou et al. [7] proposed an improved Finite Element method called New Pseudo-Static Analysis to predict the nonlinear behavior of underground frame structures subjected to increasing horizontal seismic excitations, which can predict the nonlinear performances of underground structures under horizontal earthquake loadings.

Table 1 Characteristics of simplified methods for seismic analysis of underground structure cross-section

Analysis methods	Key parameter	Advantage	Disadvantage	Application scope
Seismic coefficient method	The inertia force of the structure, dynamic earth pressure and the height calculation of the overlying soil	It is suitable for the ground structures	It is not suitable for the underground structures	It should be aborted
Method of free field deformation	Deformation of the free field	It has a simple form and is easy to be achieved	The interaction of soil-structure is ignored	The stiffness difference between soil and structure is small
Method of flexibility coefficient	Deformation of the free field and interaction coefficient of soil-structure	It has a simple form and the interaction of soil-structure is considered preliminarily	The burial depth of structures and the type of earthquakes are ignored	It is suitable for the conventional design
Response displacement method	Foundation spring stiffness, relative displacement of soil layers, shear force between the adjacent soil layers and the inertia force of the structure	The interaction of soil-structure is considered and the theoretical basis is enough	The foundation spring stiffness is not easy to determine	It is suitable for the conventional design
Response acceleration method	Horizontal effective acceleration	The interaction of soil-structure is considered and the theoretical basis is enough	The effective response acceleration is not accurate	It is suitable for the conventional design
Pushover analysis method	Target displacement and horizontal effective acceleration	It is the improved pushover analysis method used for the ground structures and its concept is clear	The type of the load is single, which is not easy to achieve	It is suitable for the conventional design

3 Conclusion

The seismic response and failure mechanism of the subway station were summarized in this paper. The drawn conclusions are as follows.

(1) The poor aseismic performance of the mid-column is the key factor for the damage of collapse of a subway station. Different from the ground structures, the self-weight of the overlying soil and the inertia force of the vertical earthquake loading acting on the roof of the underground structures will change the axial compression ratio of the mid-column, which has an important influence on the deformation capacity and shear strength of the mid-column.

(2) The simplified methods for the seismic analysis of the underground structures do not consider the vertical earthquake loading, so the simplified methods based on the more advanced dynamic approaches should be studied and applied.

(3) The cumulative damage of materials, the fracture, contact and collision of components, and the interaction of soil-structure should be considered in the numerical simulation.

(4) The nonlinear method for seismic response analysis under complex geological conditions should be studied.

(5) The multi-level isolation and shock absorption technology should be applied during the construction of the subway station.

Acknowledgements This work was funded by National Key R&D Program of China (2017YFC1500702)

References

1. Li, W.T.: Study on seismic responses of soil large-scale structure interaction system. Tongji University, China (2017)
2. Xu, C., Xu, Z., Du, X., et al.: Comparative study of simplified methods for seismic analysis of underground structure. Earthq. Eng. Eng. Dyn. **37**(2), 65–80 (2017)
3. Du, X.L., Li, Y., Xu, C.S., et al.: Review on damage causes and disaster mechanism of Daikai subway station during 1995 Osaka-Kobe earthquake. Chin. J. Geotech. Eng. **40**(2), 223–236 (2018)
4. Fabozzi, S., Licata, V., Autuori, S., et al.: Prediction of the seismic behavior of an underground railway station and a tunnel in Napoli (Italy). Undergr. Space **2**(2), 88–105 (2017)
5. Chen, Z.Y., Liu, Z.Q.: Effects of central column aspect ratio on seismic performances of subway station structures. Adv. Struct. Eng. **21**(1), 14–29 (2018)
6. Tang, B., Chen, S., Li, X., et al.: Numerical simulation on seismic behavior of variable cross-section subway station structure in complex geological ground. Int. Collab. Lifeline Earthq. Eng., 464–470 (2016)
7. Zou, Y., Liu, H., Jing, L., et al.: A pseudo-static method for seismic responses of underground frame structures subjected to increasing excitations. Tunneling Undergr. Space Technol. **65**, 106–120 (2017)

Dynamic Characteristics of Soft Clay Under Traffic Load

Shi-An Dai and Zhen-Dong Cui

Abstract

The widely distributed soft clay in Shanghai is particularly prone to settlement, causing many disasters. In order to predict the settlement of soft soil, an in-house dynamic triaxial test was carried out to study the influence of dynamic stress amplitude, frequency and effective average principal stress on the axial strain of soft soil. It laid the foundation for the study of settlement of soft clay in Shanghai. This article also put forward many places worth considering and investigating and hoped to help scholars.

Keywords

Cyclic loading • Part of the drainage • Interval load • Axial strain

1 Introduction

With the rapid development of the transportation industry, people have higher requirements for the durability and safety of transportation facilities. The high compressibility, high water content, low permeability, and low bearing capacity of soft soil are great challenges facing the transportation industry. Large settlements often occur in the construction of railways and highways on soft soils. Therefore, it is of great practical significance to study the dynamic response and long-term accumulation mechanism of soft soil under traffic load.

A large number of scholars have studied the long-term settlement of soft soil foundation under traffic load. Jeknins [1] summarized the contents of predecessors' studies and based on the existing research results, simulated a train dynamic load through a series of sine waves and superpositions. Later generations generally accepted this simulation method and obtained a wide range of applications. In order to study the long-term cumulative subsidence of subgrade under traffic load, the subgrade and track were numerically simulated [2]. The dynamic response of the railway subgrade was studied by changing the indoor model tests such as train running speed, axial load and train dynamic load cycle. The undrained deformation behavior of saturated soft soil under long-term cyclic loading was studied [3, 4]. The frequency of natural vibration of the foundation is about 2.3 Hz [5]. The following data is cited from the literature [6].

In this paper, the dynamic properties of saturated soft clays were studied using the intermittent and partial drainage methods. Dynamic triaxial tests were conducted to study the dynamic properties of saturated soft clays.

2 Experimental Results and Analysis

This paper studied the influence of dynamic stress amplitude, frequency and effective mean principal stress on axial strain. The mean effective principal stress is one-third of the effective axial pressure plus twice the effective confining pressure.

(1) The average effective principal stress affects the results.

In Fig. 1, the single curve in the soil is in a rising state, which increases faster after the existing turning point. There is no simple linear change between maximum dynamic strain and dynamic stress amplitude. Figure 1 is the turning point because of the existence of this test based on the undisturbed soil; when the dynamic stress amplitude is small because the soil has structural, cumulative plastic deformation, the soil mass deformation cannot take place obviously. This enhanced many scholars to put forward the law of the threshold cyclic stress of soil. The larger the average effective principal stress, the less the influence of axial strain.

S.-A. Dai · Z.-D. Cui (✉)
China State Key Laboratory for Geomechanics and Deep Underground Engineering, School of Mechanics and Civil Engineering, China University of Mining and Technology, Xuzhou, 221008, Jiangsu, People's Republic of China
e-mail: cuizhendong@cumt.edu.cn

A. Kallel et al. (eds.), *Recent Advances in Geo-Environmental Engineering, Geomechanics and Geotechnics, and Geohazards*, Advances in Science, Technology & Innovation, https://doi.org/10.1007/978-3-030-01665-4_49

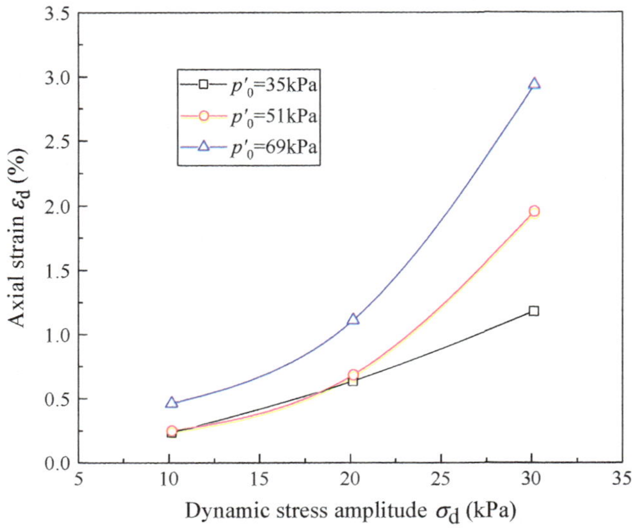

Fig. 1 Variations of axial strain with dynamic stress under different effective principal stress means

The reason is that under the impact of the main stress, the soil is squeezed, and the space that can be compressed is very small.

(2) Frequency affects the outcome

In Fig. 2, the results show that at the same frequency, the axial strain and the dynamic stress amplitude are almost linear. There is no obvious inflection point in the graph, and the higher the frequency, the smaller the effect of axial strain. The higher the frequency, the faster the vibration frequency, the easier the soil compression and the less obvious the axial strain.

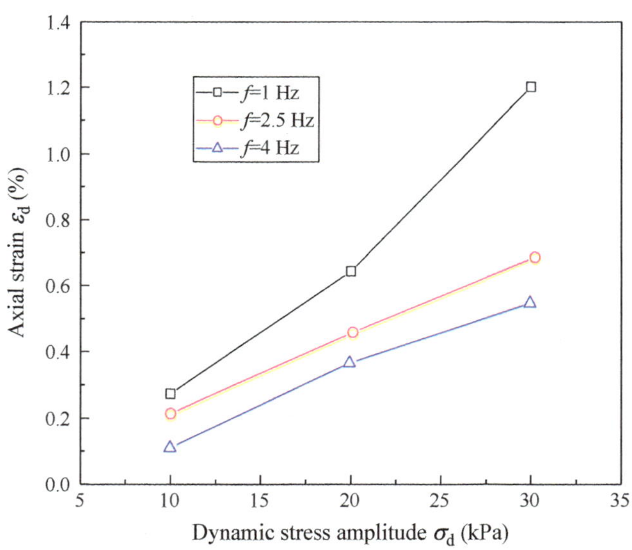

Fig. 2 Variations of axial strain with dynamic stress under different vibration frequencies

(3) The effect of stress amplitude on the test results

In Figs. 3 and 4 the soil is under different dynamic stress amplitudes, the axial strain decreases with the increase of load vibration frequency, but this is not simply a linear change between them. It shows that the relationship between the vibration frequency and dynamic strain does not simply come from the load and allows the possibility to another interpretation. There is a certain natural frequency in the foundation soil, and when the frequency of external load is close to its natural frequency, the soil will produce larger dynamic strain and smaller dynamic strength. Therefore, the

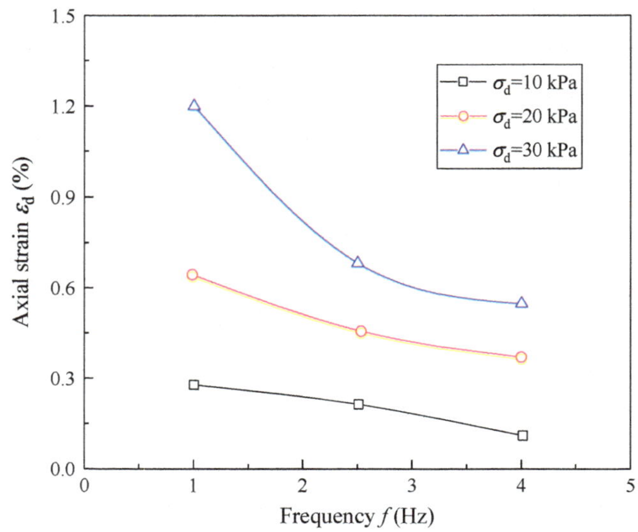

Fig. 3 Variations of maximum axial strain with vibration frequency under different dynamic stresses

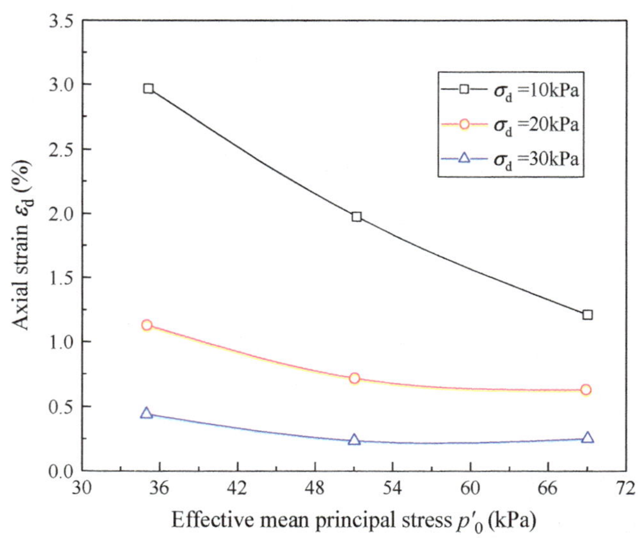

Fig. 4 Variations of maximum axial strain with effective mean principal stress under different dynamic stresses

influence of vibration frequency factor on the dynamic characteristics of saturated soft clay is not simple with the increase of vibration frequency and the decrease of dynamic strain.

3 Discussion

This paper discussed the influence of dynamic stress amplitude, frequency and effective mean principal stress on the soil axial strain. Without considering other factors, the number of test groups in this paper is relatively small, and the results may be biased. In addition, the traffic load can also cause changes in the horizontal soil stress. So it is necessary to consider the influence of confining pressure on the dynamic characteristics of the soil.

In fact, at a certain moment, there is not much ring oscillation caused by traffic load. The duration of the interval should also be considered. In addition, although the remodeled soil was relatively easy to obtain following a desired pattern, it lacks the structural properties of the original soil itself, and therefore is not practical.

4 Conclusion

This paper analyzed the influence of dynamic stress amplitude, frequency and effective average principal stress on axial strain. Its main conclusions can be stated as follows.

(1) The dynamic stress amplitude and effective mean principal stress are the main factors to determine the level of the soil dynamic strain; the soil dynamic strain increases as the dynamic stress amplitude, and decreases as the effective mean principal stress.
(2) Generally speaking, the low frequency load will cause the soil to produce larger dynamic strain, but there is a certain natural frequency for the foundation soil; when the applied load frequency is close to its natural frequency, it can cause the soil to have a more dynamic strain and its dynamic strength will be reduced.

Acknowledgements The Fundamental Research Funds for the Central Universities (2018ZZCX04) funded this work, which is greatly appreciated.

References

1. Jenkins, H.: The effect of track and vehicle parameter on wheel-rail vertical forces. Railway Eng. **3**(1), 21–27 (1974)
2. Cai, Y., Cao, X.W.: The critical dynamic stress and permanent deformation of subgrade filling soil under repeated loading. J. Southwest Univ. **31**(1), 1–5 (1996)
3. Jiang, H.G.: Research on the interaction and accumulation settlement of the high-speed railway slab track structure-subgrade dynamic interaction. Zhejiang University (2014)
4. Guo, L., Wang, J., Cai, Y., et al.: Undrained deformation behavior of saturated soft clay under long-term cyclic loading. Soil Dyn. Earthq. Eng. **50**, 28–37 (2013)
5. Bian, X.C., Chen, Y.M.: Response characteristics of layered foundation under the action of moving load. J. Rock Mech. Eng. **26**(01), 182–189 (2007)
6. Yang, J.Q.: Study on dynamic behavior and long-term cumulative settlement of saturated soft soil foundation under train load. China University of Mining and Technology, Xuzhou (2017)

Case Study of Reliability Liquefaction Analysis Based on Standard Penetration Test: Sakarya City (Turkey)

Zamila Harichane, Ayfer Erken, Mohamed Ghrici, and Alaa Chateauneuf

Abstract

In the present paper, the standard penetration test (SPT) based liquefaction trigging analysis using the first-order moment method probabilistic method (FORM) was conducted in the Sakarya (Turkey). Variability of input parameters for a specified level of risk was considered. The reliability analysis of liquefaction triggering was followed in order to take into account the uncertainties of input parameters. The SPT results were obtained in an experimental program following the 1999 Kocaeli earthquake. These results together with the empirical approach of Seed and Idriss and its updates, used in a probabilistic framework, were enough to confirm the occurrence of liquefaction in the studied area.

Keywords

Factor of safety • FORM • Uncertainty • Liquefaction • SPT

1 Introduction

During major earthquakes the occurrence of soil liquefaction is one of the most disastrous factors that may cause economic and human losses. An important aspect for earthquake risk analysis and hazard management is the evaluation of the liquefaction potential of soil [1]. Many methods have been carried out to study le liquefaction triggering but because of the difficulty and expense to preserve undisturbed samples and reproduce the stress state, empirical methods based on in situ tests such as the standard penetration test (SPT) or the cone penetration test (CPT) are rather used [2]. Deterministically, soil liquefaction is assessed using a safety factor to judge whether liquefaction would occur or not. But the inability of this kind of method to determine the liquefaction probability related to a safety factor [3], probabilistic methods provide a more reliable analysis. Several practical reliability-based methods were proposed by several researchers. Mostly, these methods are based on Seed and Idriss [4] method together with NSP- or CPT-data to assess soil liquefaction triggering.

In this paper, a probabilistic procedure for liquefaction evaluation incorporating uncertainties in the input parameters which were treated as random variables was conducted based on the popular Seed and Idriss liquefaction analysis method. The probabilistic approach was built from the steps in a deterministic liquefaction analysis and the accuracy of the procedure of analysis was studied by conducting a case study.

2 Deterministic Model for Soil Liquefaction Trigging

The deterministic evaluation of liquefaction potential of soils is frequently carried using the simplified procedure originally proposed by Seed and Idriss [4] and which, since then, has undergone several revisions and updates using the SPT. The evaluation of liquefaction resistance in the Seed and Idriss procedure involves two calculation steps: (1) the cyclic stress ratio (CSR), i.e. the cyclic loading on the soil caused by the earthquake and (2) the cyclic resistance ratio (CRR) which expresses the resistance of the soil to liquefaction.

Z. Harichane (✉) · M. Ghrici
Geomaterial Laboratory, University Hassiba Benbouali of Chlef, Chlef, Algeria
e-mail: z.harichane@univ-chlef.dz

A. Erken
Civil Engineering Faculty, Istanbul Technical University, Istanbul, Turkey

A. Chateauneuf
Institut Pascal, University of Blaise Pascal, Clermont-Ferrand, France

A. Kallel et al. (eds.), *Recent Advances in Geo-Environmental Engineering, Geomechanics and Geotechnics, and Geohazards*, Advances in Science, Technology & Innovation, https://doi.org/10.1007/978-3-030-01665-4_50

3 Reliability Analysis for Soil Liquefaction Trigging

The different variables involved in this deterministic method are prone to uncertainties. So reliability assessment of liquefaction may provide means of evaluating the combined effects of uncertainties in providing better engineering decisions [5, 6]. The liquefaction potential may be evaluated at different depths within a soil profile since both CSR and CRR vary with depth. The reliability analysis for soil liquefaction consists in defining the performance function. A specified value (or state) called the limit state of the performance function may be used to judge on the occurrence of liquefaction. A simple form of the performance function for liquefaction is expressed as:

$$Z = R - S \quad (1)$$

where S denotes the CSR and R the CRR. So, if Z = R − S < 0, the performance state is "failure", i.e., liquefaction occurs. If Z = R − S > 0, the performance state is "safe", i.e., no liquefaction occurs. If Z = R − S = 0, the performance state is on a "limit state", i.e., on the boundary between liquefaction and non-liquefaction states [3]. To account for measurements or inherent uncertainties in estimating CSR and CRR, the input parameters may be considered as random variables. The liquefaction probability P_f is defined as the probability of Z = R − S ≤:

$$p_f = P(Z \le 0) = \int_{-\infty}^{0} f_z(z)dz = F_{fz}(0) \quad (2)$$

where $f_z(Z)$ and $F_z(Z)$ denote, respectively, the probability density function (PDF) and cumulative probability function (CPF) of Z. The reliability index β is defined according to the first-order and second moment method, as

$$\beta = \frac{1}{\delta_z} = \frac{\mu_z}{\delta_z} = \frac{\mu_R - \mu_{RS}}{\sqrt{\sigma_R^2 + \sigma_S^2}} \quad (3)$$

where δ_z is the covariance coefficient, μ_R and σ_R the mean value and standard deviation, respectively, of R and μ_S and σ_S those of S. The Hasofer Lind method which is a first-order reliability method (FORM) [2] was used here.

4 Results

In an experimental program [7] led in the Sakarya-Adapazari region following the Kocaeli earthquake of Magnitude M = 7.4, SPT results are available for several places. A soil profile and index properties of a borehole under a tilted area was chosen. The safety factor against liquefaction was calculated as the ratio between the CRRand $CSR_{M=7.5}\ \sigma'_{v=1atm}$ where $CRR_{M=7.5,}\ \sigma'_{v=1atm}$ and $CSR_{M=7.5}\ \sigma'_{v=1atm}$ are, respectively, the CRR and CSR adjusted for M = 7.5 and $\sigma'_v = 1$ atm. $CRR_{M=7.5,}\ \sigma'_{v=1atm}$ is deduced from the values of the equivalent clean sand corrected SPT $(N_1)_{60cs}$ and $CSR_{M=7.5}\ \sigma'_{v=1atm}$ according to Boulanger and Idriss [8]. The deterministic data and results of the borehole BH4 are plotted in Fig. 1. Judging on factor of safety against liquefaction, the occurrence of liquefaction is expected. To conduct the reliability analysis, the input parameters (fines content (FC), SPT blow count value $(N_1)_{60}$, peak ground acceleration (a_{max}), magnitude (M_W), total pressure (σ_v), and effective pressure (σ'_v)) were considered as normally distributed random variables. The a_{max} value was assumed 0.4 g and the M_W value is 7.4. The coefficients of variation (Cv) of the parameters were estimated according to Phoon [9] (Table 1). The obtained results are shown in Fig. 2. From Fig. 2, the probability of failure is close to one (1), except for the depth z = 8.0 m because of the low value of fine content

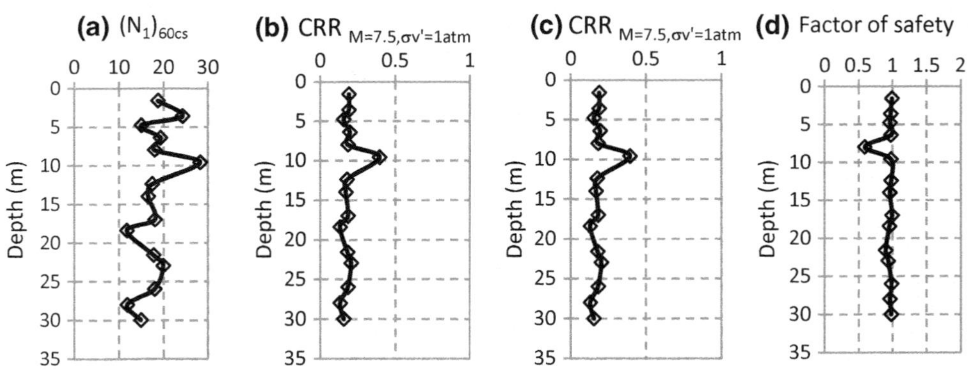

Fig. 1 Deterministic data and results of the borehole BH4: **a** $(N_1)_{60cs}$, **b** CSR $_{M=7.5,}\ \sigma'_{v=1atm}$, **c** CRR $_{M=7.5,}\ \sigma'_{v=1atm}$, **d** factor of safety (FS)

Table 1 Coefficients of variation of input parameters

Parameter	FC	$(N_1)_{60}$	a_{max}	M_W	σ_v	σ'_v
Cv	0.35	0.40	0.20	0.10	0.20	0.20

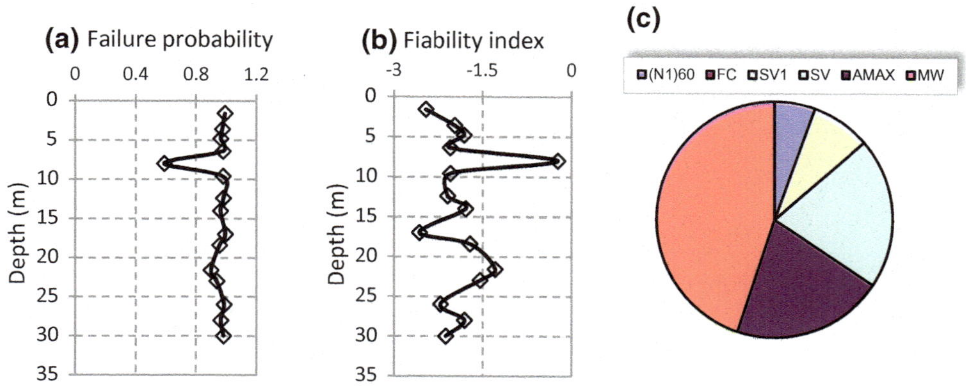

Fig. 2 **a** Probability of failure, **b** index of friability and **c** factors affecting the friability analysis against liquefaction at the borehole BH4

(FC = 15) [9]. Also, the friability index obtained according to Eq. (3) is negative because of the negative values of the limit state function (Eq. 1) at all the depths in the soil profile. The reliability method used in this study confirmed successfully the occurrence of liquefaction during the 1999 Kocaelli Earthquake.

5 Conclusion

In this study a reliability analysis of soil liquefaction trigging has been conducted based on SPT results in the framework of probabilistic methods. The reliability FORM method has shown its ability on the decision of the occurrence of liquefaction at the studied site compared to that confirmed in many studies that followed the 1999 Kocaelli Earthquake.

References

1. Huang, H.W., Zhang, J., Zhang, L.M.: Bayesian network for characterizing model uncertainty of liquefaction potential evaluation models. KSCE J. Civil Eng. **16**(5), 714–722 (2012)
2. Phoon, K.K., Ching, J.: Risk and Reliability in Geotechnical. CRC Press Taylor & Francis Group (2015)
3. Hwang, C.W., Yang, C.W., Juang, D.S.: A practical reliability-based method for assessing soil liquefaction potential. Soil Dyn. Earthq. Eng. **24**(9), 761–770 (2004)
4. Seed, H.B., Idriss, I.M.: Simplified procedure for evaluating soil liquefaction potential. J. Soil Mech. Foundations Division **97**, 1249–1273 (1971)
5. Guellil, M.E., Harichane, Z., Djilali Berkane, H., Sadouki, A.: Soil and structure uncertainty effects on the soil foundation structure dynamic response. Earthq. Struct. **12**(2), 153–163 (2017)
6. Sadouki, A., Harichane, Z., Elachachi, S.M., Erken, A.: Response of anisotropic porous layered media with uncertain soil parameters to shear body- and Love- waves. Earthq. Struct. **14**(4), 313–322 (2018)
7. Kaya, Z., Erken, A.: Cyclic and post-cyclic monotonic behavior of Adapazari soils. Soil Dyn. Earthq. Eng. **77**, 83–96 (2015)
8. Boulanger, R.W., Idriss, I.M.: CPT and SPT. Based Liquefaction Triggering Procedures. Report. UCD/CGM- 14/01, Department of Civil and Environmental. Engineering, University of California (2014)
9. Phoon, K.K.: Reliability-Based Design in Geotechnical Engineering, Computations and Applications, 1st edn. Taylor & Francis (2008)

Evaluation of Some Part of Lagos (Nigeria) Wetland for Liquefaction Vulnerability Using Integrated Approach

Hamid Oladunjoye, Kayode Oyedele, Omolara Adenuga, and Sofiat Adekoya

Abstract

The study aimed to assess the vulnerability of some parts of Lagos wetland towards the liquefaction phenomenon. Due to the population density of Lagos, the coastal area has been an alternative to the dwellers. The recent review of seismicity phenomena recorded within the recent years shows that the country might not be safe anymore from Earthquake related hazards such as liquefaction. Multi-channel Analysis of Surface waves (MASW) and Cone penetration testing (CPT) were carried out. Twenty-four (24) channels geophones were connected to the ABEM Mark 6 Seismograph to detect the generated seismic wave produced through active method. 10-Ton Dutch Cone Penetrometer was employed in acquiring the CPT sounding data. The shear wave velocity models generated showed the presence of loose sands with shear wave velocity value ranging from 140–180 m/s. The soil behaviour type chart showed that the sediments in the study area are sand mixtures. These sands have the soil behaviour type index (I_c) that is within the liquefiable boundary. The sediments are mainly dilative soil with low degree of fine particles. Simplified procedure showed that the depth of 4.50–16.00 m has factor of safety less than 1 which is an indication of probable liquefaction during sudden ground shaking.

Keywords

Liquefaction • Lagos • Wetland • MASW • Factor of safety

H. Oladunjoye (✉) · O. Adenuga · S. Adekoya
Department of Physics, Olabisi Onabanjo University, Ago Iwoye, Nigeria
e-mail: oladunjoye.titilope@oouagoiwoye.edu.ng

K. Oyedele
Department of Geosciences, University of Lagos, Akoka Lagos, Nigeria

1 Introduction

The liquefaction phenomenon is one of the geological hazards resulting from an earthquake. It occurs to young Holocene, saturated loose sands. These sands lose their stiffness and strength in response to strong ground shaking [8]. From the definition, it shows that the phenomenon is only attributed to an area with frequent earthquake occurrence in which Lagos is not included. The review of the seismicity in Nigeria showed that about 27 earth tremors have been recorded across the nation from 1933 till date. Though, the magnitudes of these tremors have not been properly recorded due to unavailability of the equipment, the magnitudes of the recent ones have been placed to loiter between 2.1 and 5.0 [1, 2]. Previous studies showed that Equatorial Fracture Zones (EFZ) such as St. Paul, Romanche, Charcot and Chain fracture zones are responsible for transporting the process via the Gulf of Guinea to the coastal part of Lagos [12]. This is represented in Fig. 1.

Liquefaction assessments have been carried out by various researchers using geophysical, geotechnical procedures or both techniques. A simplified procedure was developed using Standard Penetration Tests (SPT) blow counts [13]. The simplified procedure has been reviewed by several authors in an attempt to recalculate the parameters such stress reduction ratio [6], Magnitude scaling factors [3], etc. Robertson and Campanella [10] introduced the evaluation of liquefaction using Cone Penetration Tests (CPT) in 1985. A seismic method was used to provide an alternative procedure of evaluating liquefaction [4]. Reviews and advancements have been achieved on these techniques such as integrated approach of both CPT and SPT [9, 5]. Also the review carried out by Idriss and Boulanger gave an opportunity to evaluate liquefaction for the earthquake of magnitude that is less than 7.5 as earlier proposed through the simplified procedure of Magnitude scaling factor (MSF) [3].

A. Kallel et al. (eds.), *Recent Advances in Geo-Environmental Engineering, Geomechanics and Geotechnics, and Geohazards*, Advances in Science, Technology & Innovation, https://doi.org/10.1007/978-3-030-01665-4_51

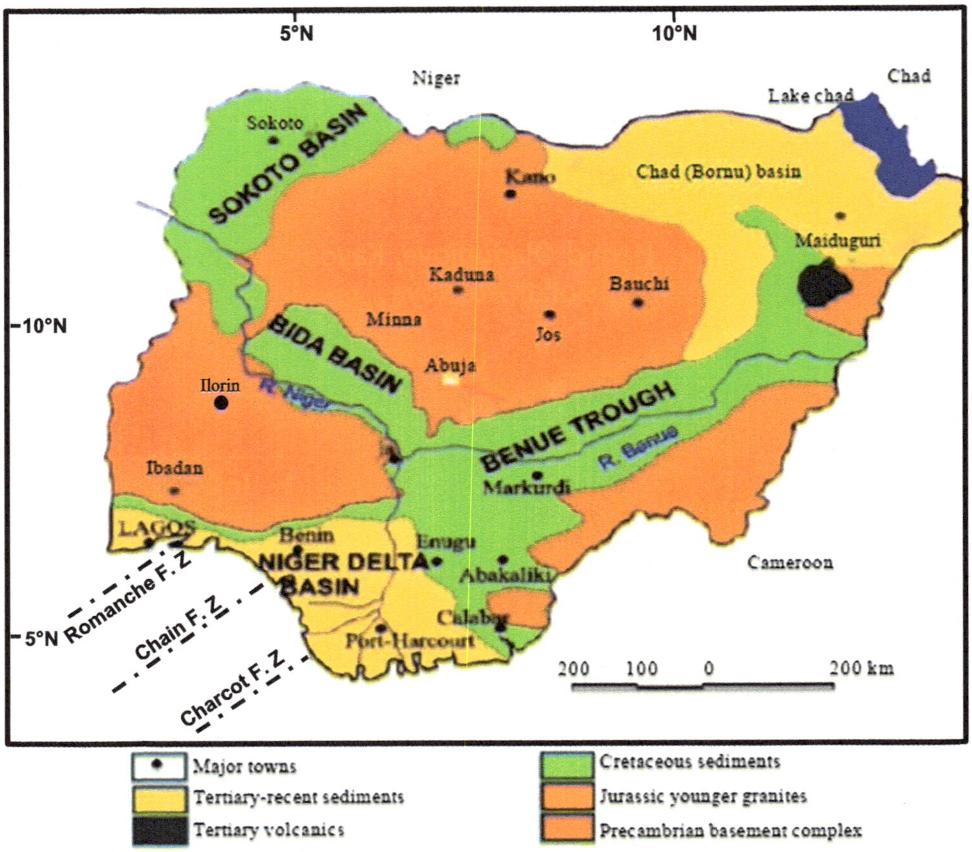

Fig. 1 Map of Nigeria showing the equatorial fracture zones (EFZ) Modified after Samaila and Likkason [12]

2 Materials and Methods

The integrated approach of geophysical and geotechnical methods were employed. Multichannel Analysis of Surface Waves (MASW) of Seismic refraction was employed to obtain the 2-D shear wave velocity model (V_s). The (V_s) obtained from the seismic line profile was used to infer the possible lithological arrangement based on the shear wave velocity values. The (V_s) is one of the basic mechanical properties of soil materials which is directly related to small-strain shear modulus G_{max} with the Eq. 1 below.

$$G_{max} = \rho V_s^2 \tag{1}$$

where ρ is the mass density of the soil; G_{max} and V_s are properties required in earthquake response and soil structure interaction analyses.

10 ton Cone of Cone Penetrating Tests (CPT) was deployed as the geotechnical measure for the purpose of this study. This provides the necessary parameters needed such as sleeve resistance for the evaluation and assessment of the Liquefaction potential of the study area. Liquefaction parameters such as Cyclic Stress Ratio (CSR) [7], Cyclic Resistance Ratio (CRR), Soil Behaviour Test Index (I_c) [11], Factor of Safety against Liquefaction (F_s) were evaluated from the acquired data using the following equations.

$$\boldsymbol{I_c} = \left[(3.47 - logQ)^2 + (logF + 1.22)^2\right]0.5 \tag{2}$$

where Q is the cone penetration resistance; F is the friction ratio.

$$\boldsymbol{CSR} = \frac{\tau_{av}}{\sigma'_{v0}} = 0.65\frac{a_{max}}{g}\cdot\frac{\sigma_{v0}}{\sigma'_{v0}}\frac{r_d}{MSF} \tag{3}$$

where τ_{av} = average shear stress; σ'_{v0} = initial vertical effective stress; a_{max} = max acceleration of the ground; σ_{v0} = total overbudden stress; r_d = stress reduction ratio; MSF = magnitude scaling factor.

$$CRR = CRR \times K_m \times K_\sigma \times K_\alpha \tag{4}$$

where K_m = magnitude correction; K_σ = overbudden stress correction; K_α = initial static shear correction.

$$\boldsymbol{F_s} = \frac{CRR}{CSR}; F_s < 1.0; Liquefaction \tag{5}$$

3 Results

See Figs. 2, 3 and 4 and Table 1.

4 Discussion

A 2D shear wave velocity was presented as one of the results of the MASW seismic refraction data acquired within the study area as shown in Fig. 2. The obtained inverted 2-D shear wave velocity (Vs) model showed the contrast in the wave velocity across the lithologic units in the study area. The Vs values across the profile 1 ranges between 140.0 and 280.0 m/s. Along the 2-D shear wave velocity (Vs) model of the profile 1, low V_s of about 162–180 m/s was delineated at the depth of about 9.0–16.0 m. This was suspected to be loose saturated sand, due to the presence of very low shear wave velocity recorded at that point. Shear wave velocity does not propagate easily in liquid medium thereby makes it have low shear wave velocity as depicted in the lithologic unit. Within the second profile, the V_s with values ranging from 130 to 170 m/s was delineated at the depth of about 12–20 m along the profile line. This area showed that the sand sediments within the lithologic unit mainly saturated loose sand which can liquefy in the presence of increased stress.

The soil Behaviour type chart obtained from the study area as shown in the Fig. 3 was obtained from the acquired CPT sounding. The chart showed that the soils are mainly cohesionless soils with formations ranging from dense sands, clean sands and sand mixtures. The friction between the granular particles of cohesionless soil was found to have low shear strength. The soils are mainly drained dilative with sand sediments of little degree of fine particles. The soil behavior type index for the sounding points in the study area ranges between 1.70 and 2.5. This means that the soil is susceptible to cyclic liquefaction. The normalized friction ratio obtained is less than 1, confirming the degree of fines of the silty sand in the study area. The friction ratio which is a function of grain characteristics increases with the increase in fine content and plasticity. This shows that the sand

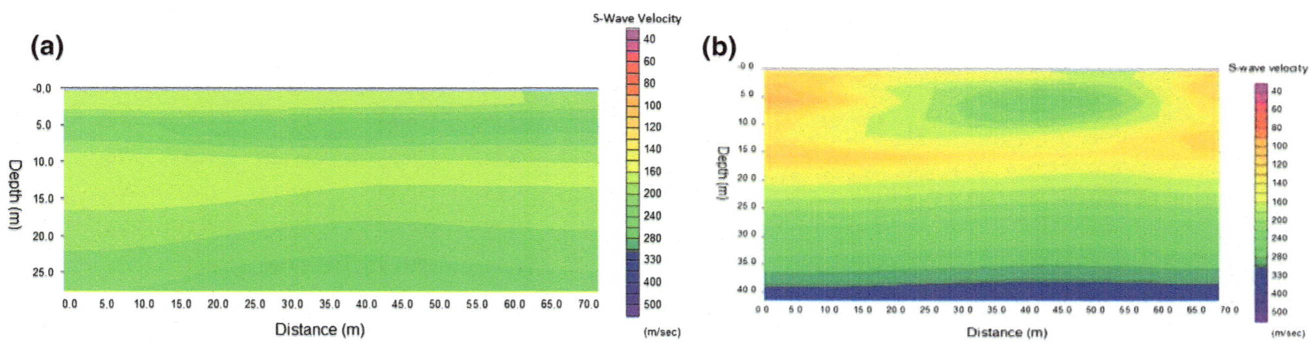

Fig. 2 **a** 2-D shear wave velocity model of the datum point 1, **b** 2-D shear wave velocity model of the datum point 2

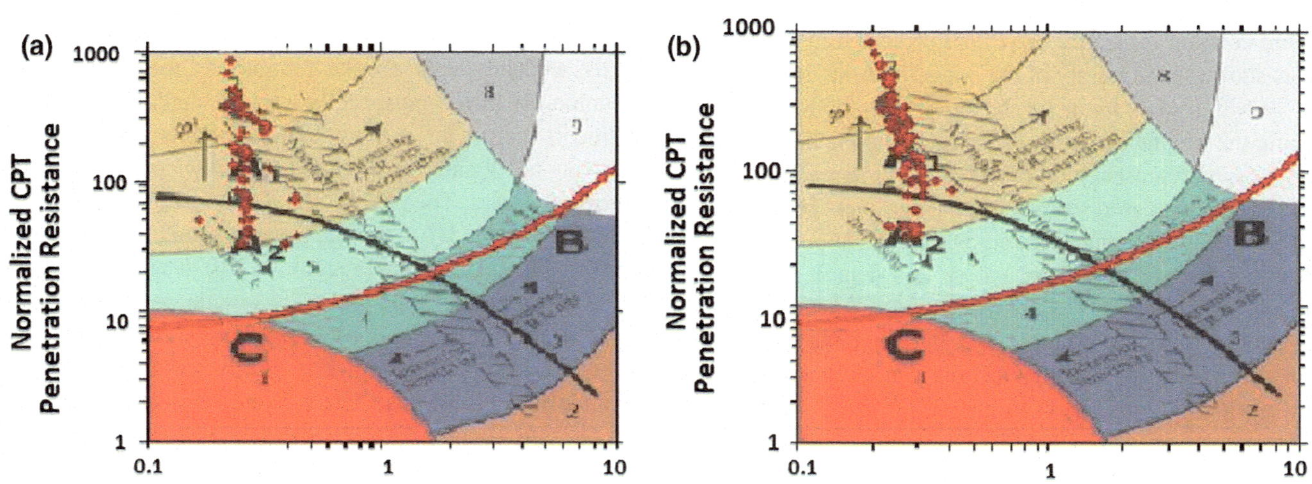

Fig. 3 **a** Soil behaviour type chart of the datum point 1, **b** soil behaviour type chart of the datum point 2

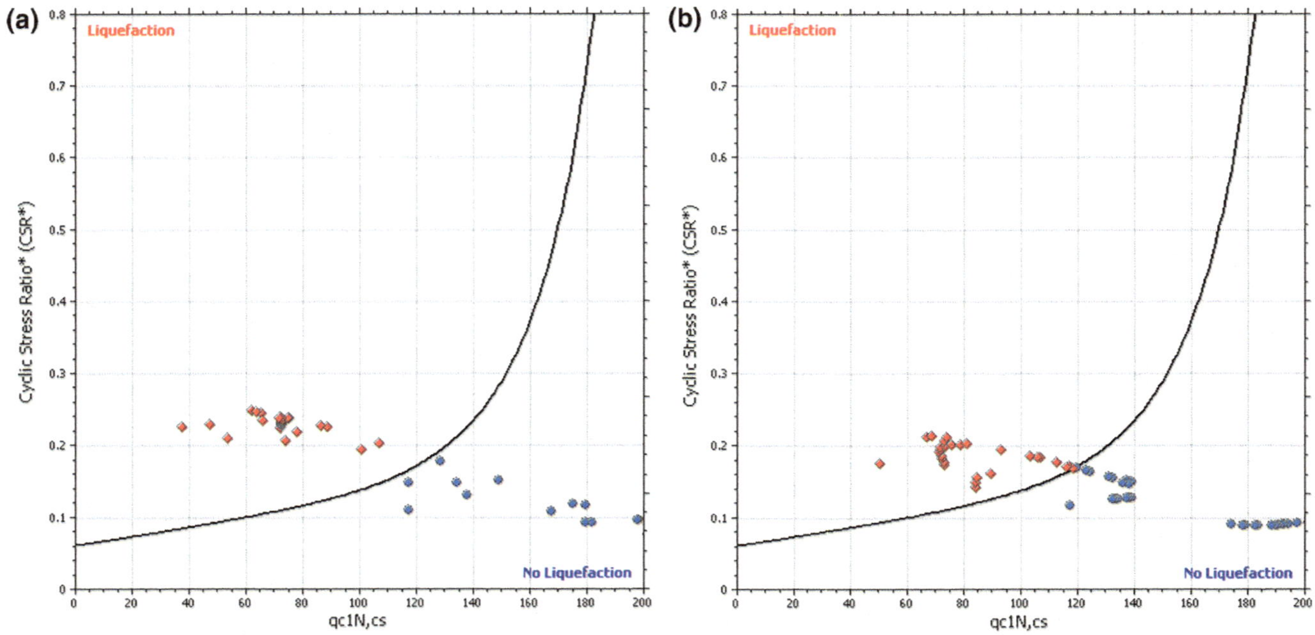

Fig. 4 **a** Liquefaction CRR curve of the CPT datum point 1, **b** liquefaction CRR curve of the CPT datum point 2

Table 1 Part of liquefaction parameters obtained from shear wave velocity from the study area

S/N	Depth (m)	V_s (m/s)	Fines %	n	V_{s1} (m/s)	CRR	F.S	Can liquefy
1	2.00	150.00	5.00	0.25	215.04	4.000	2.00	No
2	4.50	100.00	5.00	0.25	128.43	0.056	0.23	Yes
3	7.00	152.00	4.28	0.25	176.51	0.128	0.50	Yes
4	10.50	165.00	0.00	0.25	170.20	0.113	0.47	Yes
5	14.00	160.00	0.04	0.25	152.36	0.083	0.37	Yes
6	16.00	180.00	0.00	0.25	190.42	0.090	0.57	Yes
7	19.00	284.00	0.00	0.25	249.42	0.500	2.51	No

sediments in this area have a low plasticity index as depicted in the normalized friction ratio.

The plots of Cyclic Stress Ratio (CSR) against the corrected cone tip resistance (q_{c1N}) are presented in Fig. 4. The plots show the variability of the corrected cone tip resistance (q_{c1N}) with respect to the Cyclic stress ratio (CSR) generated during the ground shaking. The corrected cone tip resistance (q_{c1N}) is the representation of the resistant degree of the soil towards the increased stress within the ground. Therefore the red ink points are characterized as liquefiable points; which means points with factor of safety less than 1. The liquefaction parameters evaluated with the acquired shear wave velocity was presented in Table 1. The table shows the factor of safety against liquefaction in accordance to shear wave velocity with respect to their depth.

5 Conclusion

The study showed that the coastal sand of some parts of Lagos wetland areas are susceptible to the liquefaction phenomenon in the presence of ground shaking. The sand sediments consisting of loose saturated sands which can be easily agitated and behave like liquid when there is an increase in stress. The plot of SBTn Index showed that their indices is within 1.7–2.5 which was absolutely within liquefiable region (Liquefiable region > 2.6). The sediments in the area are dilative soils with very low degree of fines and low plasticity index. This observation was confirmed with the plot of friction ratio which is mostly less than 1. It was commonly believed that Lagos-Nigeria is safe from the liquefaction phenomenon, which needs to be reviewed as the sediment in the studied part show high degree of vulnerability. Therefore, the liquefaction assessment should be included as part of geotechnical assessment prior to the construction of engineering structures.

References

1. Adepelumi, A.A.: Short-term probabilistic forecasting of earthquakes occurrence in South-Western Nigeria. Technical Report Submitted to the Centre for Geodesy and Geodynamics, Toro, Nigeria (2009)
2. Akpan, U.O., Yakubu, T.A.: A review of earthquake occurrences and observation in Nigeria. Earthq. Sci. **23**, 289–294. https://doi.org/10.1007/s11589-010-0725-7 (2010)
3. Boulanger, R.W., Idriss, I.M.: CPT and SPT based liquefaction triggering procedures. Rep. No. UCD/CGM-14/01, University of California, Davis, CA (2014)
4. Dobry, R., Powell, D.J., Yokel, F.Y., Ladd, R.S.: Liquefaction potential of saturated sand—The stiffness methods. In: Proceedings of the 7th World Conference on Earthquake Engineering, Istanbul, vol. 3 (1980)
5. Green, R.A., Cubrinovski, M., Cox, B., Wood, C., Wotherspoon, L., Bradley, B., Maurer, B.: Select liquefaction case histories from the 2010–2011 Canterbury earthquake sequence. Earthq. Spectra **30**(1), 131–153 (2014)
6. Idriss, I.M.: An update to the Seed-Idriss simplified procedure for evaluating liquefaction potential. In: Proceedings, TRB Workshop on New Approaches to Liquefaction, Publication No. FHWARD-99–165, Federal Highway Administration (1999)
7. Idriss, I.M., Boulanger, R.W.: Soil Liquefaction During Earthquakes. Earthquake Engineering Research Institute, Oakland, CA (2008)
8. Marcusson, W.F.: Definition of terms related to liquefaction. J. Geotech. Eng. Div. ASCE **104**(9), 1197–2000 (1978)
9. Robertson, P.K.: Interpretation of cone penetration tests—A unified approach. Can. Geotech. J. **46**, 1337–1355 (2009)
10. Robertson, P.K., Campanella, R.G.: Liquefaction potential of sands using CPT. J. Geotech. Eng. ASCE **111**(3), 384–403 (1985)
11. Robertson, P.K., Wride, C.E.: Evaluating cyclic liquefaction potential using the cone penetration test. Can. Geotech. J. **35**(3), 442–459 (1985)
12. Samaila, N.K., Likkason, O.K.: Role of equatorial fracture zones on fluid migration across the South Atlantic Margins. J. Earth Sci. Clim. Change. https://doi.org/10.4172/2157-7617-004 (2013)
13. Seed, H.B., Idriss, I.M.: Simplified procedure for evaluating soil liquefaction potential. J. Soil Mech. Found. Div. ASCE **97**(9), 1249–1273 (1971)

Sand-Steel Interface Behavior Under Cyclic Loading

Mohamed Khemissa, Naoui Tallah, and Djaâfar Barkat

Abstract

This paper is to investigate the behavior of sand-steel interfaces under cyclic loading. A series of direct shear test results performed on a dune sand in contact with a stainless steel rigid plate, as well as those of a numerical simulation carried out using the Plaxis-2D (version 8.2) software were presented and discussed. The cyclic shear experimental and numerical curves are fairly well concordant. These curves make it possible to conclude that the sand-steel interface cyclic shear modulus and the corresponding damping ratio were influenced by the sand compactness and the steel plate roughness. The results also show that the rough plate is more effective than the smooth one mainly when it is in contact with a dense sand than with loose sand.

Keywords

Sand-Steel interface • Direct shear test • Maximum dynamic shear modulus • Maximum damping ratio

1 Introduction

Numerous theoretical and laboratory experimental investigations on soil-structure interfaces under monotonic and cyclic loadings have been developed during these last decades [1–5]. However, the cyclic shear effects on the interface shear modulus and the corresponding damping ratio are still unclear.

The purpose of this paper was to focus on some of the above-mentioned aspects and conclude on the effect of the cyclic shear behavior of the sand-steel interface on its stress-strain-strength dynamic properties.

M. Khemissa (✉) · N. Tallah · D. Barkat
Geomaterials Development Laboratory, University of M'sila, 28000 M'sila, Algeria
e-mail: khemissa@univ-msila.dz

2 Experimental Approach

The direct cyclic shear tests were performed using a standard shear box consisting of a fixed lower half-box and a mobile upper half-box. The test specimens consist of a dry dune sand enclosed in the mobile upper half box in contact with a stainless steel rigid plate placed in the fixed lower half box. These test specimens were confined under a conventional normal stress $\sigma_v = 100$ kPa and then sheared alternately over a horizontal travel of ± 2 cm during 20 cycles.

Figure 1 shows the smooth and rough plates used in this study. Table 1 gives the main identification characteristics of the dune sand and the stainless steel rigid plate used in this study.

3 Numerical Simulation

Figure 2 shows the geometrical model adopted in the present numerical simulation. This is comparable to the physical model used in the experimental study presented above. The calculations made using the Plaxis-2D (version 8.2) software [6] assume that the confinement of this model is ensured by the application of a distributed load on its upper side and the shearing by the application of a prescribed displacement on its lateral sides. The shear plane is defined by the interface element whose behavior obeys the Mohr-Coulomb model like that of sand but with reduced shear characteristics (the reduction coefficient, called the shape factor, characterizes the shape of the interface) [5, 6].

4 Results and Discussion

Figures 3 and 4 show the experimental and predicted cyclic shear curves for smooth and rough sand-steel interfaces respectively. The qualitative analysis of these results makes it possible to draw the same remarks made by Khemissa

A. Kallel et al. (eds.), *Recent Advances in Geo-Environmental Engineering, Geomechanics and Geotechnics, and Geohazards*, Advances in Science, Technology & Innovation, https://doi.org/10.1007/978-3-030-01665-4_52

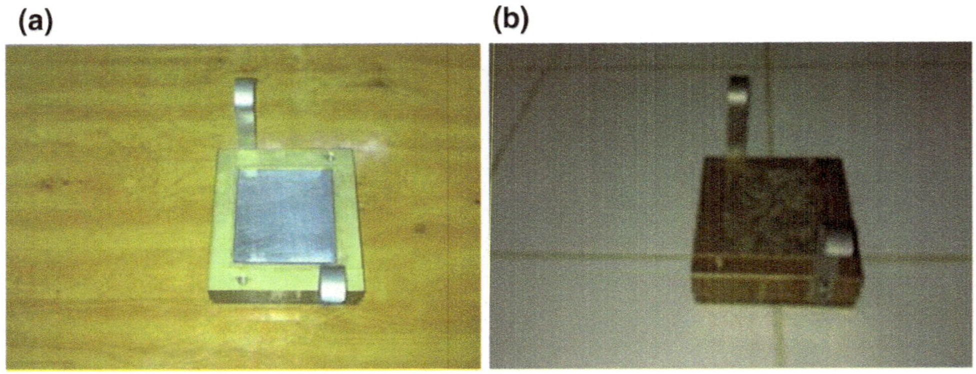

Fig. 1 Photographs of the used smooth (**a**) and rough (**b**) stainless steel rigid plates

Table 1 Main identification characteristics of the used materials

Materials	Parameters	Values
Dune sand	Specific density, γ_s	2.65
	Minimum dry density, $\gamma_{d\text{-}min}$	1.50
	Maximum dry density, $\gamma_{d\text{-}max}$	1.73
	Sand equivalent, SE (%)	76
	Coefficient of uniformity, C_u	2.33
	Coefficient of curvature, C_c	1.19
	Density index of loose sand, $I_{D\text{-}min}$ (%)	15
	Density index of dense sand, $I_{D\text{-}max}$ (%)	90
Stainless steel rigid plate	Yield stress, σ_e (MPa)	240

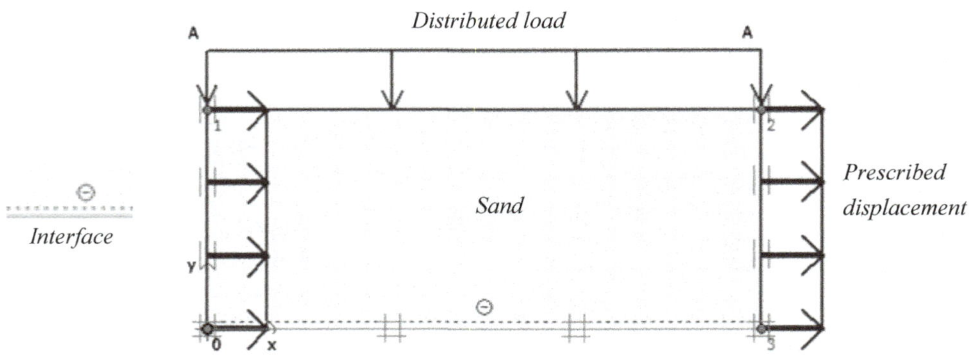

Fig. 2 Geometrical model adopted for the direct cyclic shear test

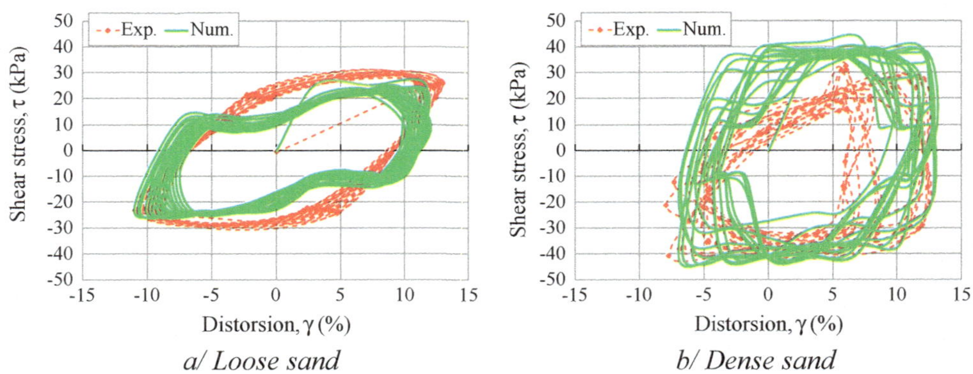

Fig. 3 Experimental and predicted cyclic shear curves for the smooth sand-steel interface

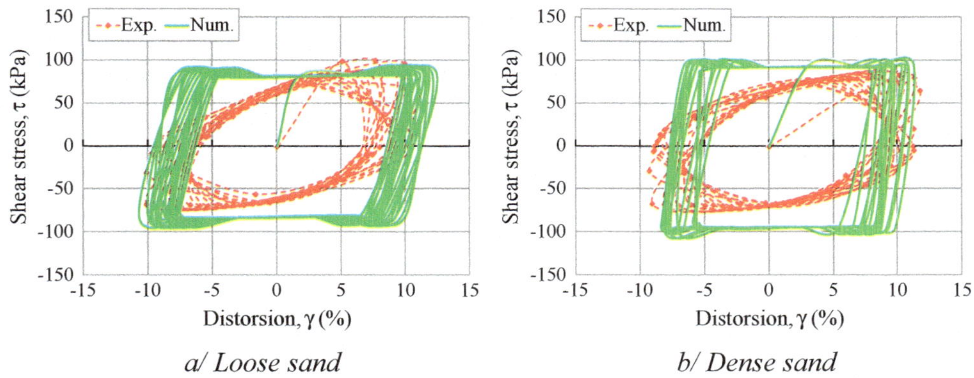

Fig. 4 Experimental and predicted cyclic shear curves for the rough sand-steel interface

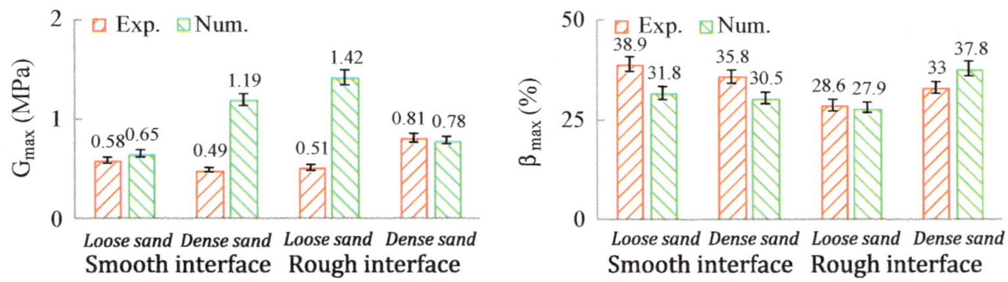

Fig. 5 Cyclic shear behavior parameters for the smooth and rough sand-steel interfaces

et al. [5] on the experimental and numerical data under monotonic loading:

- These curves have almost similar shapes for each state of sand compactness and for each form of steel roughness.
- They describe a behavior characterized by an increase of the interface shear strength (highlighted friction mobilization by adherence between the sand and the steel plate).
- The interface shear strength is higher for a rough interface than for a smooth one and with dense sand than with loose one. In addition, there may be a reasonable agreement between experimental and predicted data.

The comparative analysis of the experimental and predicted cyclic shear behavior parameters values leads to the following remarks (Fig. 5):

- The maximum shear modulus G_{max} is higher for dense sand than for smooth sand and more with smooth interface than with rough interface.
- The maximum damping ratio β_{max} is higher for smooth interface than for rough interface and more for dense sand than for rough sand.

However, a parametric analysis (not showed here) seems to indicate that these two parameters are influenced by the amplitude of the horizontal travel and the corresponding number of cycles.

5 Conclusion

The experimental and numerical results presented in this paper were intended to study the mechanical behavior of soil-structure interfaces under cyclic loading. In particular, it is dedicated to the cyclic shear direct effect on the interface shear modulus and the corresponding damping ratio.

The comparative analysis shows that the maximum shear modulus and the corresponding damping ratio are greatly affected by the dune sand compactness and the stainless steel plate roughness, as well as by the amplitude of the horizontal travel and the corresponding number of cycles.

References

1. Shahrour, I., Bencheikh, B.: Analysis of the soil-structure interaction under monotonic and cyclic loadings. In: Hirsch, C., Zienkiewicz, O.C., Onate, E. (eds.) First European Conference on Numerical Methods in Engineering, pp. 269–275. Brussels (1992)
2. Shahrour, I., Rezaie, F.: An elastoplastic constitutive relation for the soil–structure interface under cyclic loading. Comput. Geotech. J. **21**(1), 21–39 (1997)

3. Boulon, M., Ghionna, V.N., Mortara, G.: A strain-hardening elastoplastic model for sand–structure interface under monotonic and cyclic loading. Math. Comput. Model. **37**, 623–630 (2003)
4. Khemissa, M., Safer, S., Aidjouli, S.: Roughness's shapes comparative analysis of some reinforced earth elements under monotonous loading. Alexandria Eng. J. **54**, 577–582 (2015)
5. Khemissa, M., Tallah, N., Bencheikh, B.: Experimental and numerical modeling of the sand–steel interface behavior under monotonic loading. Innov. Infrastruct. Solutions **3**, 25 (2018)
6. Brinkgreve, R.B.J., Vermeer, P.A.: PLAXIS: Finite element code for soil and rock analyses, version 8.2. Balkema, Rotterdam, Brookfield (2002)

Cyclic Pressuremeter Tests Dedicated to Study the Behavior of Piles Under Cyclic Transverse Loads

Rim Baccara, Wissem Frikha, Philippe Reiffsteck, and Sébastien Burlon

Abstract

We propose a cyclic pressuremeter (CPM) test protocol deduced from the reaction of the soil-pile values derived from tests of cyclic transverse loads on pile. The relative displacements issued from CPM tests are compared to the relative displacements of the pile under cyclic transverse loads.

Keywords

Pile • Pressuremeter • Transverse • Cyclic • Load

1 Introduction

For some structures such as onshore and offshore wind turbines and oil platforms, which are supported by piles, cyclic transverse loads are very significant due to the wind and waves. Several previous studies were interested in the action of piles under cyclic transverse loads [1–3]. The pressuremeter is a widely used tool in the prediction of the behavior of a transversely loaded pile in the case of a static load [4–6]. In fact, as an application of a cavity expansion, the pressuremeter test is comparable to the soil-pile reaction under transverse loads. In the present paper, we proposed to use the results of a cyclic pressuremeter (CPM) test program to predict the behavior of a pile subjected to cyclic loading tests. The experimental site is in Plancoët (Côtes-d'Armor, France).

R. Baccara · W. Frikha (✉)
laboratoire d'ingénierie géotechnique, ENIT, Tunis, Tunisie
e-mail: frikha_wissem@yahoo.fr

R. Baccara · P. Reiffsteck
GERS-SRO, IFSTTAR, Champs sur Marne, France

S. Burlon
Terrasol, SETEC, Paris, France

2 Materials and Methods

The pressuremeter test is an in situ test consisting in the radial expansion of a cylindrical probe on the wall of a borehole. This probe was introduced into the borehole until reaching the desired depth of the test. Then, equal increments of pressure were applied with the control unit, transferred with a tubular system to the probe [7]. For *CPM* tests, the applied pressure was in loops of loading and unloading [8]. The volume variation was measured as an out-put of the test. The used material consisted of a control unit able to control electro valves by means of a computer. The borehole was drilled until 5 m of depth by hand auger and the probe was manually introduced into the soil until the required depth. The CPM tests were stopped when a stabilization of the volume variation was observed.

The CPM test program was adapted to the cyclic loading tests on a pile carried out in Plancoët [9]. The pile subjected to cyclic loading tests was an HEA28A embedded to 6.1 m. The pile was instrumented by 28 couples of strain gauges from which the moment was directly deduced. The main experimental result in this study was the reaction of the soil-pile P. P was obtained by a double differentiation of the recorded moment curve. The maximum and the minimum reactions (P_{max} and $P_{min,}$ respectively) were measured on different depths during the cyclic loads under the maximum and the minimum applied loads (H_{max} and $H_{min,}$ respectively). The pressures applied by the pressuremeter were deduced from the soil reactions P recorded during the first cycle of the loading tests on pile. The pressure of the start of the cyclic load on the CPM, also called average pressure p^*_{av}, at a given depth, was determined as follows.

$$p^*_{av} = \frac{P_{max} + P_{min}}{2 \times B} \tag{1}$$

B is the width of the pile. The pressure of the half-amplitude $p^*_{1/2}$ is given by:

A. Kallel et al. (eds.), *Recent Advances in Geo-Environmental Engineering, Geomechanics and Geotechnics, and Geohazards*, Advances in Science, Technology & Innovation, https://doi.org/10.1007/978-3-030-01665-4_53

$$p_{1/2}^{*} = \frac{\max(|P_{max}|, |P_{min}|)}{B} \quad (2)$$

In order to ensure the contact between the probe and the soil, both p_{aver} and $p_{1/2}$ were increased by the pressure of soil at rest p_0, deduced from static pressuremeter tests. Also, inversed corrections were taken into consideration, like the pressure loss p_{loss} due to the hydraulic system and to the membrane rigidity p_{mr}.

$$p_i = p_i^{*} + p_{loss} + p_{mr} \quad with \quad i = av, 1/2 \quad (3)$$

The frequencies of the CPM tests and the tests on pile were kept identical. The two CPM test programs, PCCH1 and PCCH2 are related to the average of the results of the pile tests CH1 and CH2 respectively. CH1 is the cyclic loading test with a maximum load of H_{max} = 13.3 kN. This test consisted of three series. The first and the second series are identical, with a number of cycles NC = 1000 (3.9 h). The third one is performed with NC = 10,000 (39 h). CH2 is the cyclic loading test with H_{max} = 20 kN and was conducted twice, identically as NC = 10,000 (39 h). For all the tests, a minimum load H_{min} = 6.7 kN was maintained. Table 1 sums up all undertaken parameters for the CPM tests in Plancoët.

3 Results and Interpretation

The pressure-volume curves represent the evolution of the applied pressure as a function of the observed volume change during the CPM tests (example from the PCCH2 at 2 m of depth in Fig. 1).

The cyclic loads started when the volume was about 70 ml. The accumulation of volume was relatively important in the first cycles (distant loops), when it started to stabilize in the last part of the curve (condensed loops).

Relative displacement from the results of tests performed on pile is defined as the ratio of the displacement during cyclic loads to the first value of the measured displacement derived from the first cycle y_{0i} and from the CPM tests as follows.

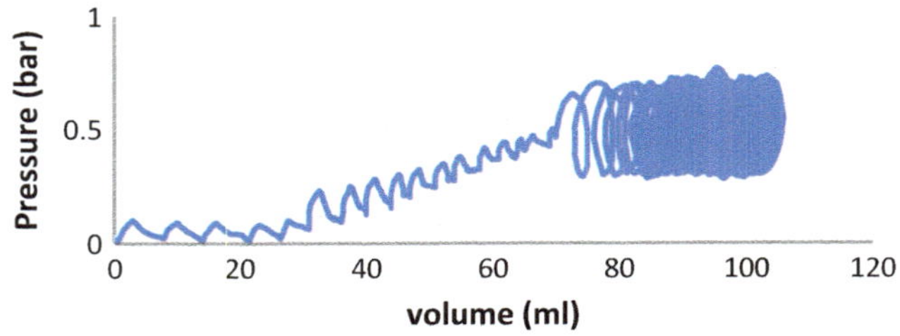

Fig. 1 Pressure-volume curve from PCCH2 at 2 m of depth

$$y_{Re,i}^{Pile} = \frac{y_i}{y_{0i}} i = max, min \quad (4)$$

$$y_{Re}^{CPM} = \frac{\Delta V}{\Delta V_0} \quad (5)$$

ΔV_0 and ΔV are the evolution of the volume measured in the first average pressure and during cyclic pressure, respectively.

Figures 2 and 3 show the results of the relative displacements for CH1 compared to PCCH1, and CH2 compared to PCCH2 at different depths.

At 1 and 2 m of depth, the accumulation of the horizontal displacement for both pile and CPM tests are less than 150%. At 3 and 4 m, the relative displacement is more important. The accumulation of the displacement under minimum loads is more important than the accumulation of the displacement under maximum loads. The area of soil between y_{min} and y_{max} is in fact the most disturbed by the cyclic loadings.

After an important number of cycles, the curves of are included between the curves of and at the first 3 m. CPM test curves and tests on pile curves have the same trend. The CPM test results are able to predict the pile displacements under cyclic loads.

Table 1 Values of CPM tests parameters (Plancoët site)

Depth (m)	PCCH1				PCCH2			
	p_{aver} (kPa)	$p_{1/2}$ (kPa)	Duration (h)	NC	p_{aver} (kPa)	$p_{1/2}$ (kPa)	Duration (h)	NC
1	37.7	13.6	1.27	287	50.5	26.6	0.85	234
2	29	9.4	1.28	332	37.5	20	0.85	220
3	19	5.2	0.93	224	30	12	0.85	225
4	17	5	1.24	317	25	11	0.85	211
5	–	–	–	–	21	10	0.88	207

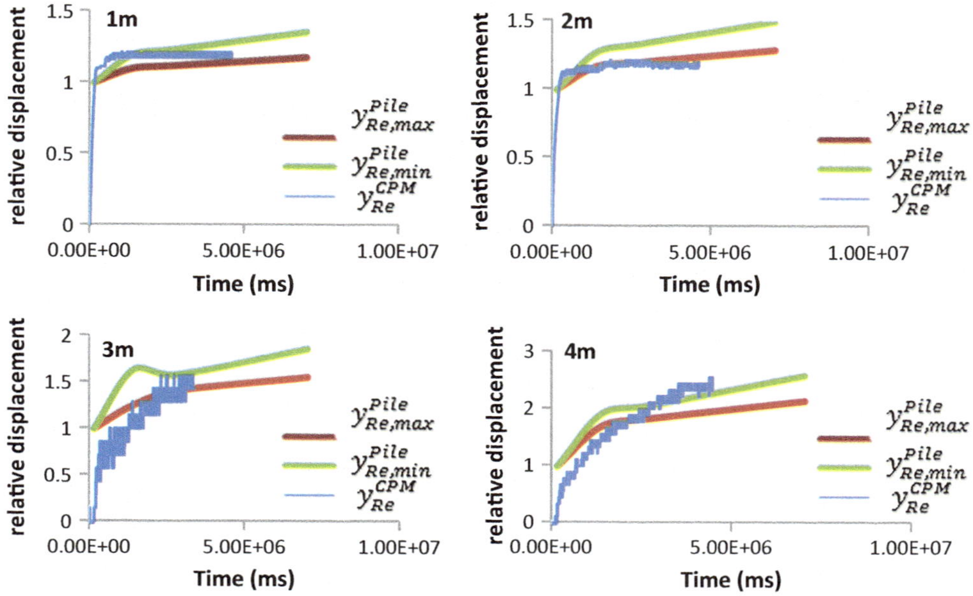

Fig. 2 Comparison of relative displacements for CH1 and PCCH1

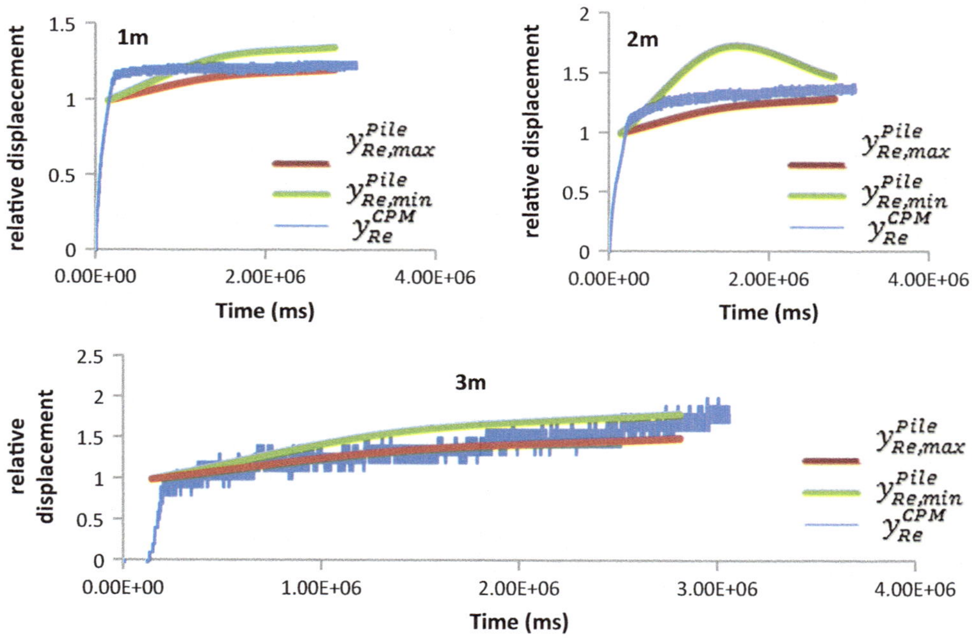

Fig. 3 Comparison of relative displacements for CH2 and PCCH2

4 Conclusion

In this paper, we presented the cyclic loading tests on a pile carried out at the site of Plancoët. Then, CPM test program deduced from the soil-pile reaction, an experimental result from the tests on pile, were developed and exerted on the same experimental site. Afterwards, a relative displacement, derived from CPM tests, was compared to the relative displacement from tests on pile. The results show that CPM tests are able to describe the behavior of the pile under cyclic transverse loads and to predict the horizontal displacement of the pile.

References

1. Little, R.L., Briaud, J.L.: A pressuremeter method for single piles subjected to cyclic lateral loads in sand. Miscellaneous paper GL-88-14 (1988)
2. Matlock, H.: Correlations for design of laterally loaded piles in soft clay. 2nd Ann. Offshore Tech. Conf., 1577–1594 Houston (1970)
3. Poulos, H.G.: Behaviour of laterally loaded piles: I—single piles. J. Soil Mech. Found. Div. ASCE **5**(97), 711–731 (1971)
4. Ménard, L., Bourdon, G., Gambin, M.: Méthode générale de calcul d'un rideau ou d'un pieu sollicité latéralement en fonction des résultats pressiométriques. Sols-Soils **VI**, n° 22–23 (1969)
5. Bigot, G., Bourges, F., Frank, R.: Etude expérimentale d'un pieu soumis aux poussés latérales du sol. Revue Française de Géotechnique N° 18, février, pp. 29–47 (1982)
6. Briaud, J.-L.: The Pressuremeter. A.A. Balkema, Rotterdam (1992)
7. Ménard, L.: Règles d'utilisation des techniques pressiométriques et d'exploitation des résultats obtenus pour le calcul des fondations. Notice générale D60 (1967, 1975)
8. Reiffsteck, P., Fanelli, S., Tacita, J.L., Dupla, J.C., Desanneaux, G.: Utilisation des essais d'expansion cyclique pour définir des modules élastiques en petites déformations. In: Proceedings of the 18th International Conference on Soil Mechanics and Geotechnical Engineering, Paris, Presses de l'ENPC(3), pp. 2383–2386 (2013)
9. Jézéquel, J.F., Saintilan, D.: Résistance latérale des pieux. Site de Plancoët. Chargement appliqué sur le pieu isolé. Rapport FAER 1.05.01.4, Laboratoire Régional de St Brieuc (1986)

Part VI

Geomechanics and Geotechnics: Recent Studies on Expansive Soils

Direct Measurements of Swell Potential of Expansive Soils with Computerized Equipment

Murat Türköz

Abstract

Many studies have mentioned that expansive soils cover wide areas in many parts of the world. Research work on this subject has been conducted and designed criteria have been proposed. The identification and classification techniques developed for expansive soils are used to qualitatively determine the possible impact of volume change on the behavior of soils. Yet, a standard measurement and identification method for determining the swell potential of the soils has not been established. The direct measurements that constitute the subject of this study are computer controlled and can be simply applied and provide very useful quantitative data for design engineers without the need for complex laboratory equipment. Within the scope of this study, first, geotechnical properties of 40 different soil samples were determined and then, compaction characteristics at the Standard Proctor energy level were obtained. The swell percentage and swell pressure values of samples prepared by compressing in optimum water content were determined using computer controlled direct methods. As a result, statistical evaluations were made for the purpose of practical use.

Keywords

Compaction • Expansive soil • Swell percentage • Swell pressure

1 Introduction

Damage caused by expansive soils is almost entirely restricted to light structures. It is also a problem outside the conventional construction practices when encountered during the construction of such structures. When the soils are compacted at high energy levels to be used in embankment, the natural structures break down and cement bonds break, water content decreases, dry densities increase and swelling indices increase. Due to this reason, swelling can reach a high value. As an example of this situation, Li and Du [1] stated that after the clay soil with plastics limits of 13–17% and liquid limits of 35–40% was kneaded and compacted, an artificially swelling soil behavior could be shown. In general, the determination of the swell potential of the soils is also made with a conventional consolidation device. ASTM D 4546 [2] describes the application of uniaxial swell tests on soils using the classical consolidation (odometer) device as defined in the standard. In these methods, the pressure required for the sample to reach the initial void ratio is defined as the swell pressure. In particular, the concentration of laboratory equipment used in determining the swell potential on a single device causes the actual indices to differ. In a study by Ali and El Turabi [3], the classical consolidation device and potential volume change (PVC) meter equipment were used to determine the swell pressures of expansive soils. Available data indicated that the swell pressures obtained from the odometer method give greater results than those obtained from PVC method and that the PVC method is more consistent [3].

In this study, for the results of the swell potential of the compacted soil samples to be comparable, experiments were performed on specimens with the same diameter and height. For this reason, the Expansion Index (EI) and Potential Volume Change (PVC) meter equipment have been converted.

2 Materials and Methods

2.1 Materials

Soil samples used in this study were taken from Eskişehir, Şanlıurfa and Afyon provinces of Turkey. To determine the

M. Türköz (✉)
Civil Engineering Department, Eskisehir Osmangazi University, 26480 Eskisehir, Turkey
e-mail: mturkoz@ogu.edu.tr

A. Kallel et al. (eds.), *Recent Advances in Geo-Environmental Engineering, Geomechanics and Geotechnics, and Geohazards*, Advances in Science, Technology & Innovation, https://doi.org/10.1007/978-3-030-01665-4_54

swelling potential of the samples, geotechnical testing program including grain size analysis (ASTM D 422-63), Atterberg limits (ASTM D 4318-00), specific gravity (ASTM D 854-00) and compaction characteristics were performed on the samples [2], and then, swell tests were performed on the compacted soil specimens using direct methods.

2.2 Methods

Swell percentage test determines the amount of vertical heave that occurs when the soil is wetted. This test was performed on specimens prepared by compressing in the desired density and water content. The swell percentage is defined as the ratio of the initial length of the sample to the final length of the sample after being soaked in water under 7 kN/m^2 [4]. The swell pressure test was carried out on the samples prepared as in the swell percentage test. The test covers the determination of the pressure resulting from the obstruction of the swelling developed after the wetting of the compressed soil sample. The pressure at the end of the test is defined as the swell pressure of the soil [4].

On the basis of the definitions given in the paragraph above, the development of the test devices to be used in determining the swell percentage and swell pressure variables was inspired by PVC Meter and EI meter equipment. The mold of the PVC meter used in the determination of the swell pressure was modified to compare the results (Fig. 1). The equipment used to determine the swell percentage was also developed by taking advantage of EI meter equipment. A weight was manufactured to place 7 kPa of pressure on the samples in 2-cm-high thin-walled rings with 7-cm diameters (Fig. 1). As soon as the samples placed in the equipment were soaked in water, the swell potential was measured at a series of time intervals using digital deformation meters. At the same time, the variation of the swell potential from graphs drawn using real-time recorded readings could also be monitored (Fig. 2).

3 Results and Discussion

The statistical evaluation of the tests results carried out in this study is presented in Table 1. The parameters from the identification tests were used in the description and classification of the swell potential of the samples. In general, higher soil plasticity indices and liquid limits imply larger swelling potentials [4]. When the parameters obtained from the identification test results were evaluated as the primary indicator of the swell potential, different swelling potentials were identified. The relationship between the swell percentage and pressure is given in Fig. 2. In this figure, the relationship between swell percentage and pressure can be observed in a high and meaningful level. Especially at the low swelling potential levels, the relationship is more pronounced. Similar evaluations could also be made based on the index values and the compaction characteristics of the soil samples.

4 Conclusion

In this study, swell percentage and pressure values of high plasticity clay soils were determined using computerized direct methods, the effect of some physical parameters on

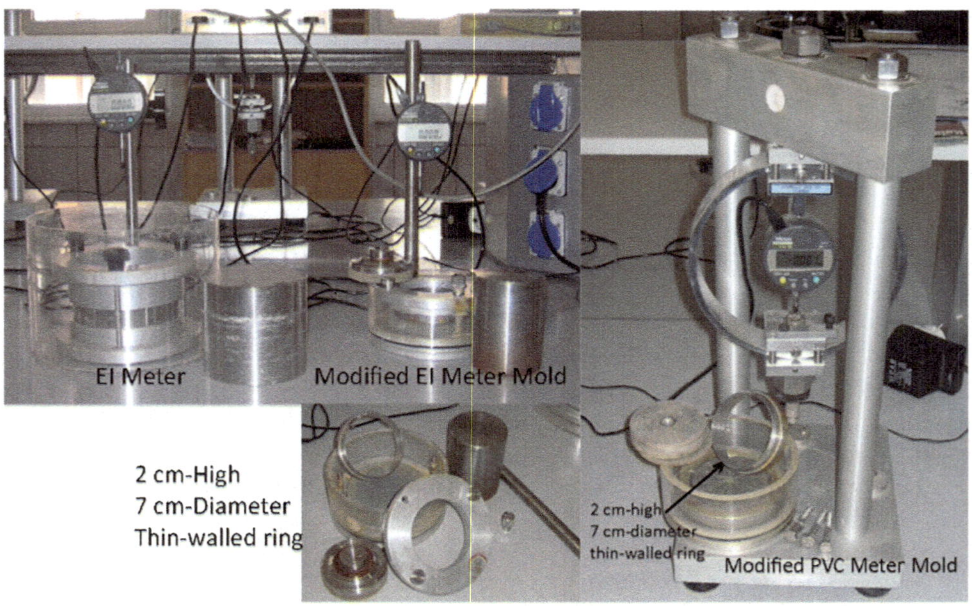

Fig. 1 Computerized equipment used for determining swell parameters of the samples

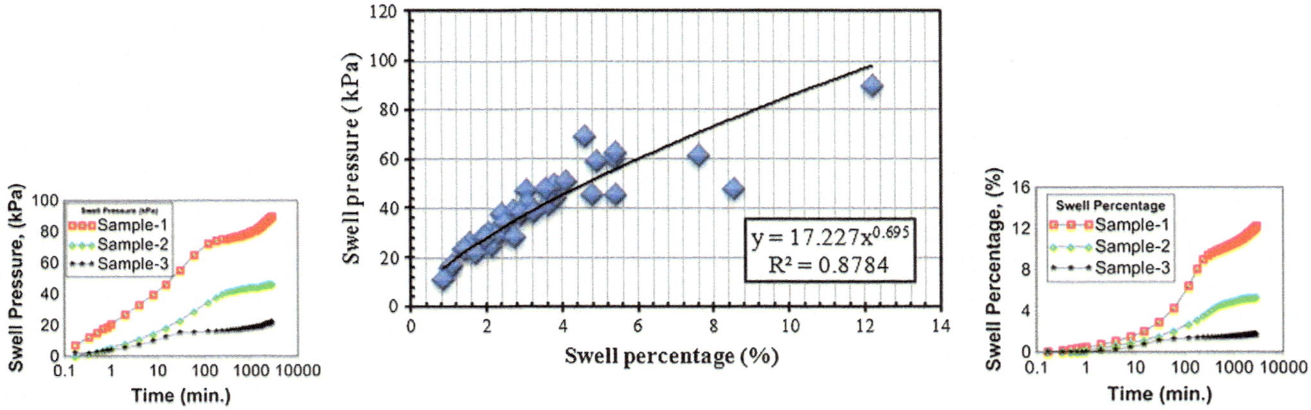

Fig. 2 The relationship between the swell percentage and pressure

Table 1 The statistical data for the considered soils

Property	Num.	Value			
		Min.	Max.	Mean	Std. deviation
Grain size					
Clay (%)	40	14.0	61.0	41.9	11.5
Silt (%)	40	32.8	60.9	45.9	8.0
Sand (%)	40	3.8	30.0	10.2	6.4
Gravel (%)	40	0.0	15.4	2.0	3.4
Atterberg limits					
Liquid limit (%)	40	40.0	75.0	60.6	7.0
Plastic limit (%)	40	19.0	36.0	28.7	3.0
Plasticity index (%)	40	20.0	42.0	31.9	5.1
Standard proctor					
w_{opt} (%)	40	18.2	29.4	23.9	2.19
ρ_{dmaks} (Mg/m^3)	40	1.400	1.653	1.534	0.05
Specific gravity (Gs)	40	2.6	2.83	2.76	0.04
Swell tests					
Swell pressure (kPa)	40	11.3	89.9	39.4	15.8
Swell (%)	40	0.84	12.2	3.44	2.19
Activity (A)	40	0.5	1.9	0.8	0.32

these values was investigated. As a result, the developed equipment is small and easy to use. Therefore, it should be considered as an advantage of these devices.

References

1. Li, S., Du, Y.: On the swelling-shrinkage properties and mechanisms of compacted expansive soils. In: Proceeding of the 30th International Geological Congress, vol. 23, pp. 253–259, China(1997)
2. ASTM: Annual Book of ASTM Standards. Soil and Rock, Vol. 04.08. American Society for Testing and Materials, Philadelphia (1994)
3. Ali, E.M., Elturabi, M.A.D.: Comparison of two methods for the measurement of swelling pressure. In: Fifth International Conference on Expansive Soils, pp. 72–74, Adelaide South Australia (1984)
4. Turkoz, M., Savas, H., Acaz, A., Tosun, H.: The effect of magnesium chloride solution on the engineering properties of clay soil with expansive and dispersive characteristics. Appl. Clay Sci. **101**, 1–9 (2014)

Temperature Effect on Alkali Contaminated Kaolinitic Clays

P. Lakshmi Sruthi and P. Hari Prasad Reddy

Abstract

It is well understood that the swelling behavior of alkali contaminated kaolinitic clays is mainly influenced by mineralogical and morphological changes. However, complete mineralogical changes occur over a long period of time due to slow reaction rates at normal temperatures or field contamination conditions. The long term effects of alkali on kaolinitic clays can be simulated by increasing the reaction rate through carrying the studies at elevated temperatures. The present study focused on understanding the effect of temperature (35, 60, 80 and 110 °C) on mineral dissolution and new mineral formations of alkali contaminated kaolinitic clays. Further, the obtained results were compared with the mineralogy changes observed in soil samples after interacting with alkali for a long period (60 days) in laboratory at normal temperature of about 27 °C, considering the field contamination condition. It is clear from the XRD and SEM studies that a complete change in morphology was observed when exposed to higher temperatures.

Keywords

Kaolinitic clays • Temperature • XRD • SEM

1 Introduction

Research studies highlighted the fact that geotechnical failures, such as differential settlements, ground heave below manufacturing units of industrial plants, occur due to alterations in the soil engineering properties. Among the various pollutants, it is the alkali contamination that has the most significant impact on soil behavior. Formation of neogenic hydrated salt crystals due to alkali contamination leads to an increase in the number of instances of structural failures [1]. Serious cracks and upliftment of floor by more than 80 mm in K-Unit plant, Runcorn and significant rise in columns and beams were also observed [2].

In this regard, intensive research has been carried out by many researchers [3, 4] to understand the mineral alterations behind soil alkali interaction. Mineralogy is a controlling factor determining the sizes and surface characteristics of the particles in soil. Thus, mineralogy is fundamental to the understanding of the geotechnical properties [5]. Kaolinite when reacting with NaOH in between 80 to 140 °C, formation of feldspathoids such as sodalite and cancrinite were noticed by [6]. From the literature, it is clear that long term effects of soil alkali contamination leads to dangerous consequences. So, in order to accelerate the reaction rate and simulate the field contamination, samples were exposed to higher temperatures.

2 Materials and Methods

2.1 Clays Used

In the present study, two kaolinitic clays with varying mineral contents were used for the experimental investigation. One is a natural red earth soil (Warangal, India) and the other is commercially available kaolin (Visakhapatnam, India). Both clays were air dried and soil passing through Indian Standard 425-μm sieve were prepared before use. The clays physical properties are presented in Table 1.

2.2 Contaminant Used

The used contaminant in this study was 4 N sodium hydroxide solution. The solution was prepared by dissolving the chemically pure salt in distilled water.

P. Lakshmi Sruthi · P. Hari Prasad Reddy (✉)
Department of Civil Engineering, National Institute of Technology, Warangal, 506004, India
e-mail: ponnapuhari@gmail.com

A. Kallel et al. (eds.), *Recent Advances in Geo-Environmental Engineering, Geomechanics and Geotechnics, and Geohazards*, Advances in Science, Technology & Innovation, https://doi.org/10.1007/978-3-030-01665-4_55

Table 1 Kaolinitic clays physical properties

Properties	Red earth	Kaolin
Specific gravity	2.62	2.70
Liquid limit (LL) (%)	38	42.9
Plastic limit (PL) (%)	22.64	29.72
Plasticity Index (PI) (%)	15.36	13.18
Free swell Index (ml/g)	1.0	1.1
Maximum dry unit weight (kN/m^3)	18.3	17
Optimum water content (%)	20.74	27.05

2.3 Experimental Investigation

The aim of the experimental investigation was to understand the effect of temperature on mineral dissolution and new mineral formations of kaolinitic clays. These studies were carried out to simulate the long-term effects of alkali on complete mineral dissolution and mineralogical changes that occur in kaolinitic clays [7], which subsequently may influence the geotechnical properties of soils. Due to the slow reaction rates at normal temperature under alkaline environment, engineering properties of soils could not be able to notify unless the scenario becomes dangerous. So, to speed up the reaction rate, and to understand the effect of varying temperature on mineral dissolution and new mineral formations, samples were exposed to different temperatures (35, 60, 80 and 110 °C). In the mineral transformation process the volume of interacting solution plays an important role [8]. Solid to liquid ratio (S/L) was set to 2.5 for red earth and S/L = 4 for kaolin, since complete transformations with 4 N NaOH occurred at that particular volume of interaction at 110 °C for one day interaction period. So, in order to maintain uniformity throughout the work, the same volume of interaction was taken for the other temperatures (35, 60 and 80 °C) also and the effect of temperature on mineral dissolution and neogenic formations was investigated.

2.4 X-Ray Diffraction and SEM Studies

After being exposed to varying temperatures, the samples were taken for X-ray diffraction and SEM analyses. XRD samples were scanned from 6° to 70° using 0.01 steps.

TESCAN VEGA 3LMU scanning electron microscope with conventional tungsten heated cathode was used to understand the morphological alterations in clays.

3 Results and Discussions

3.1 Mineralogical Changes Due to Alkali Interaction

Figures 1 and 2 show the XRD patterns of red earth and kaolin interacting with NaOH solution at different temperatures. In all the samples, complete kaolinite dissolution and new mineral formations were observed. Sodalites are the primary minerals formed in all the samples. However, the secondary mineral formations varied with temperature and clay type. Further, it can be observed that the duration of interaction for the mineralogical changes to take place varied with temperature: 1 day at 110 °C, 4 days at 80 °C, 2 weeks at 60 °C and 1 month at 35 °C for both clays. The results obtained by varying the temperature were compared with the XRD pattern of red earth (Fig. 1b) and kaolin (Fig. 2b) obtained by considering field contamination condition [7]. The mineralogical changes observed on the field contamination conditions (i.e. at 27 °C) were similar to temperature effects. However, clay-alkali reaction occurred at a very slow rate i.e. 60 days. At a 60-day interaction period, a complete dissolution of kaolinite mineral was observed in red earth only, which indicates that kaolin requires more period of interaction for a complete dissolution of kaolinite mineral. To re-confirm the formation of new minerals identified from XRD studies the samples were taken for SEM studies.

3.2 Morphological Changes Due to Alkali Interaction

Figure 3a shows the SEM photographs of red earth samples interacting with 4N NaOH solution at different temperatures. From this figure, it can be noticed that red earth exhibited a compacted and aggregated form of morphology. At 27 °C,

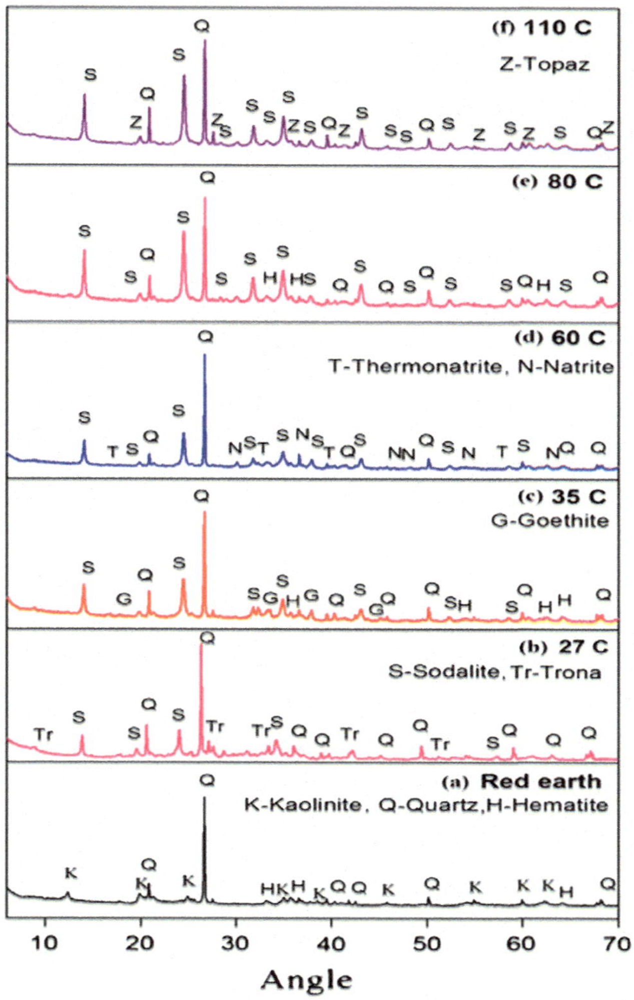

Fig. 1 XRD patterns of red earth interacted with **a** water and 4N NaOH at **b** 27 °C, **c** 35 °C, **d** 60 °C, **e** 80 °C and **f** 110 °C

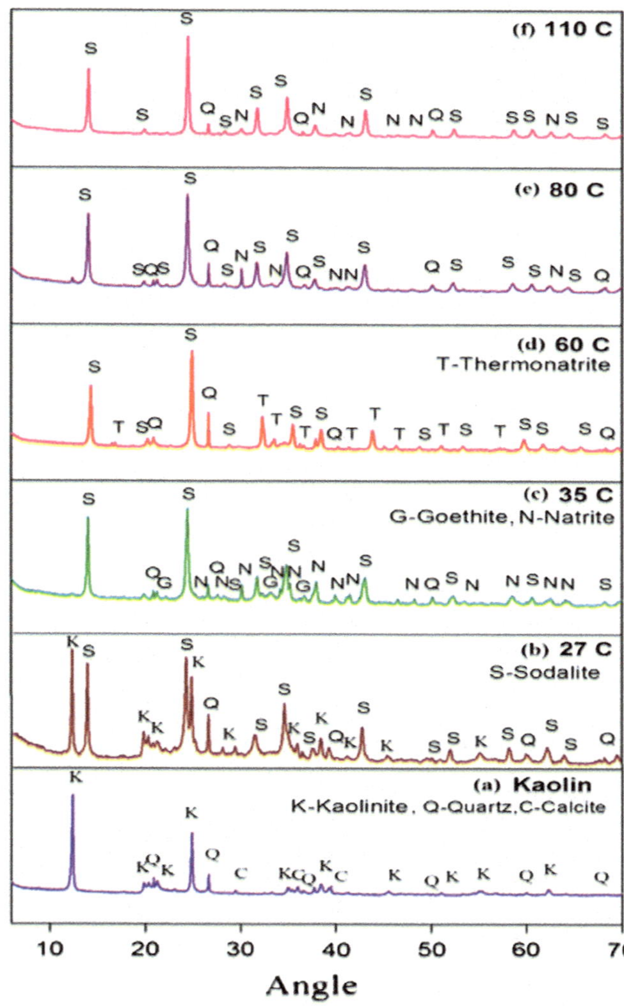

Fig. 2 XRD patterns of kaolin interacted with **a** water and 4N NaOH at **b** 27 °C, **c** 35 °C, **d** 60 °C, **e** 80 °C and **f** 110 °C

morphology shows a more disintegrated form with void spaces in between. Coalesce pellet formations with crystalline trona can be observed (Fig. 3b). At 35 °C, prismatic with more crystalline morphology was observed, which confirms the formation of goethite (Fig. 3c). Thermonatrite and natrite mineral at 60 °C represented a group of platy and flaky morphology (Fig. 3d) whereas at 80 °C, traces of cotton ball morphologies were observed (Fig. 3e). Further, at 110 °C interaction, cotton ball morphologies highlighted the formation of sodalites. Very small traces of reaction products called sodalites and crystal shaped morphology of topaz can be seen [9] (Fig. 3f). Complete transformation of morphology from aggregated form to well defined crystals indicates the dissolution of silica and alumina under alkali conditions which leads to formation of sodium based zeolites.

Figure 4a shows the microstructure of kaolin with highly loose fibrous structure [8]. Figure 4b displays a complete change in morphology at 27 °C, highlighting the formation of sodalites depicting cotton ball morphology [10]. Due to the presence of more than one mineral, proper identification of morphology cannot be observed at 35 °C (Fig. 4c). The formation of platy crystals (thermonatrite) and cotton ball morphology was clearly visible from Fig. 4d at 60 °C. Severe morphological changes have been observed with the formation of intense sodalite particles highlighting the increase in the size of sodalite particles at 80 and 110 °C. The fibrous like structures (Fig. 4a), were completely absent at all temperatures except at 27 °C (Fig. 4b). The morphological studies thus clearly support XRD studies, highlighting the new mineral formations.

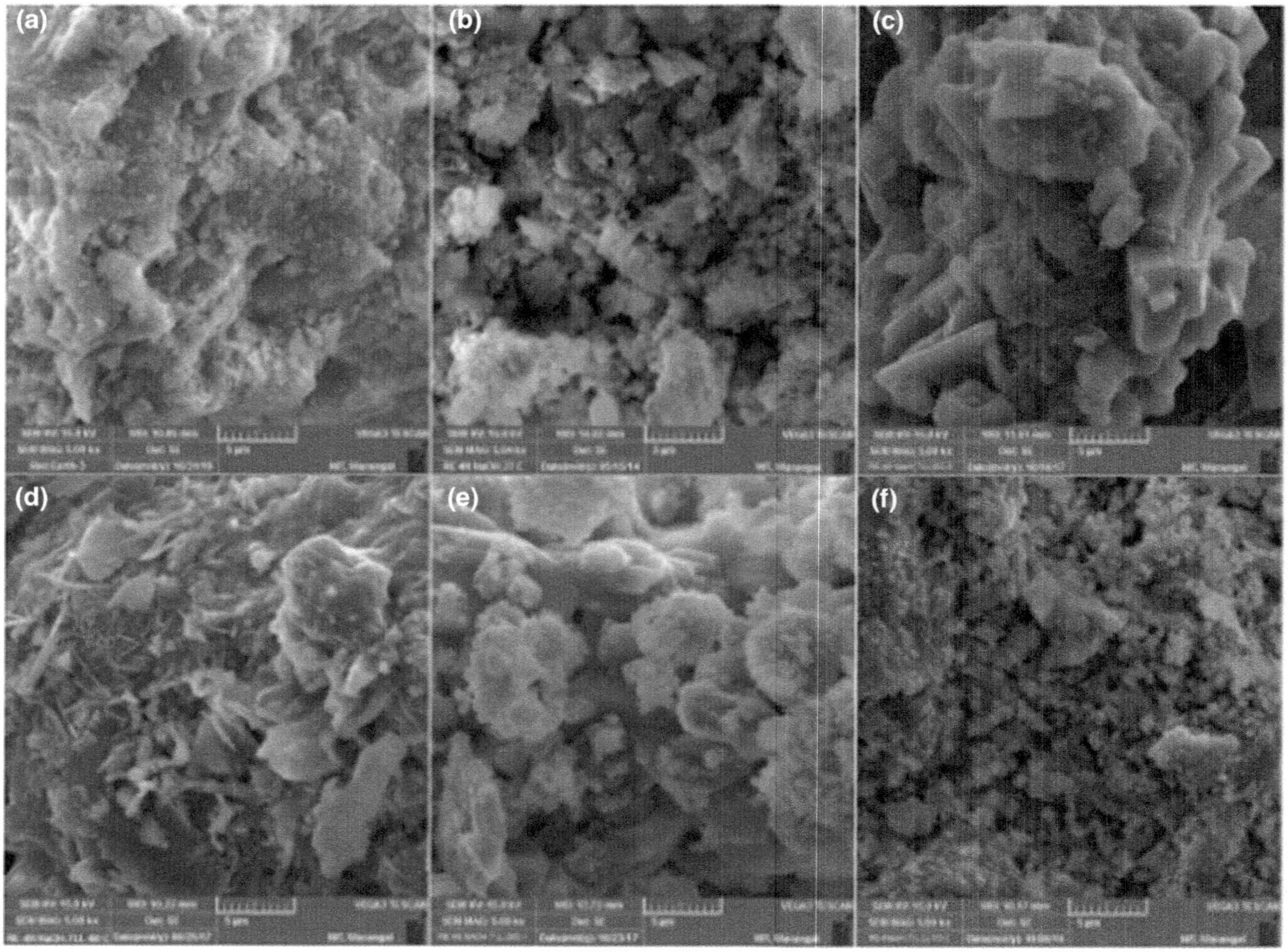

Fig. 3 SEM images of red earth interacting with **a** water and 4N NaOH at **b** 27 °C, **c** 35 °C, **d** 60 °C, **e** 80 °C and **f** 110 °C

4 Conclusion

Kaolin having more kaolinite content, volume of interacting solution required for complete dissolution of clay mineral is relatively higher when compared with red earth.

For both clays at all temperatures, the primary mineral observed was sodalite, whereas secondary minerals varied with the type of clay and temperature.

Complete dissolution of kaolinite mineral was not observed in kaolin at field contamination even after a 60-day interaction period. The same was observed with in one day at 110 °C. Thus, the long term mineralogical changes can be well understood in a short period of time at elevated temperatures.

Fig. 4 SEM images of kaolin interacting with **a** water and 4N NaOH at **b** 27 °C, **c** 35 °C, **d** 60 °C, **e** 80 °C and **f** 110 °C

References

1. Shekhtman, L.M., Baranov, V.T., Nesterenko, G.F.: Building deformations caused by the leakage of chemical reagents. Soil Mech. Found. Eng. **32**(1), 32–36 (1995)
2. Sibley, M.H., Vadgama, N.J.: Investigation of ground heave at ICI Mond division, Castner-Keller works. Runcorn. Geol. Soc. Lond. Eng. Geol. Spec. Publ. **2**, 367–373 (1986)
3. Cuadros, J., Linares, J.: Experimental kinetic study of the smectite-to-illite transformation. Geochim. Cosmochim. Acta **60**, 439–453 (1996)
4. Wang, Y., Siu, W.: Structure characteristics and mechanical properties of kaolinite soils. I. Surface charges and structural characterizations. Can. Geotech. J. **43**, 587–600 (2006)
5. Mitchell, J.K.: Fundamentals of Soil Behavior, 2nd edn. Wiley, New York, NY (1993)
6. Barrer, R.M.: Hydrothermal Chemistry of Zeolites. Academic Press, London (1982)
7. Chavali, R.V.P., Vindula, S.K., Reddy, H.P.P., Ambili, B., Rakesh Pillai, J.: Swelling behavior of kaolinitic soils contaminated with alkali solutions: a micro-level study. Appl. Clay Sci. **135**, 575–582 (2016)
8. Sruthi, P.L., Reddy, H.P.P.: Characterization of kaolinitic clays subjected to alkali contamination. Appl. Clay Sci. **146**, 535–547 (2017)
9. Russell, A.: Topaz from Cornwall, with an account of its localities. Mineral. Mag. **20**(106), 221–236 (1924)
10. Barnes, M.C., Addai-Mensah, J., Gerso, A.R.: A methodology for quantifying sodalite and cancrinite phase mixtures and the kinetics of the sodalite to cancrinite phase transformation. Microporous Mesoporous Mater. **31**, 303–319 (1999)

Study on Physical and Mechanical Properties of Clay Before and After Single Freeze-Thaw

Chong Xu and Zhen-dong Cui

Abstract

Previous research indicates that significant changes in engineering behavior of cohesive soils may occur if the soil undergoes cycles of freezing and thawing. Understanding this behavior would prove useful to engineering practice as soils are subjected to freeze-thaw cycles during construction or placement. This research study focused on the impact of a single freeze-thaw cycle on the characteristics of Shanghai quaternary Silt Soft Clay (SSSC). It was found that the changes in density and specific gravity were small, the water content and porosity ratio increased, the plastic limit stabilizes with liquid limit increasing, and the soil body was soft. The permeability coefficient increased and the permeability increased. The cohesion and internal friction angle of the clay after freezing and thawing slightly increased, the compression coefficient increased, the compression modulus decreased, the compressibility of the soil increased, and the clay had no confinement resistance after freezing and thawing. The compressive strength was only 1/3 to 1/2 of that of the undisturbed soil, and the intensity decay was obvious, and the sensitivity was reduced.

Keywords

Single freeze-thaw • Silt soft clay • Physical property • Mechanical properties

1 Introduction

The stress-strain-strength characteristic of soil is a hot spot of research. With the extensive application of artificial freezing methods in the field of reinforced formations, it is necessary to study the behavior of soils subjected to freezing and thawing cycles. This article selected the Shanghai quaternary Silt Soft Clay (SSSC) as the research object. The changes of physical and mechanical properties before and after freezing and thawing were analyzed by experiment [1].

The SSSC is a modern sediment formed in a slow-flowing environment and under the condition of the participation of microorganisms. Its project performance: high compressibility, poor permeability, low shear strength (related to loading speed and drainage conditions) and high porosity ratio [2].

2 Methods

The undisturbed soil was transported to the laboratory after being sampled on site, according to the Ministry of Communications Earthwork Test Procedure JTG E40-07, ensuring that the water content of the soil sample was exactly the same inside and outside, and the disturbance of the soil sample transport was minimized [3]. The soil sample was at −10 °C. Constant temperature freezing 72 h, to ensure uniform temperature within the sample freeze-dried, after removal, placed in a sealed desiccator 3d, until the temperature is close to the room temperature (to prevent migration of water to the soil sample), and then remove the test specimen for testing [4].

3 Results

3.1 Basic Physical Indicators

The water content increases by 12.8%. The void ratio and the liquid limit increase slightly. The density, the specific gravity and the plastic limit decrease slightly. The horizontal permeability coefficient triples these results are summarized in Table 1.

C. Xu · Z. Cui (✉)
China State Key Laboratory for Geomechanics and Deep Underground Engineering, School of Mechanics and Civil Engineering, China University of Mining and Technology, Xuzhou, 221008, Jiangsu, People's Republic of China
e-mail: cuizhendong@cumt.edu.cn

A. Kallel et al. (eds.), *Recent Advances in Geo-Environmental Engineering, Geomechanics and Geotechnics, and Geohazards*, Advances in Science, Technology & Innovation, https://doi.org/10.1007/978-3-030-01665-4_56

Table 1 Basic physical parameters of indigenous soil and frozen thaw soil at −10 °C

Soil samples	Water content (%)	Density (g/cm^3)	Specific weight	Void ratio	Liquid limit	Plastic limit	Plastic index	Vertical permeability coefficient ($\times 10^{-6}$ cm/s)	Horizontal permeability coefficient ($\times 10^{-6}$ cm/s)
Undisturbed soil	33.5	1.762	2.75	1.08	35.8	20.4	15.4	0.34	2.8
Frozen and thawed soil	37.8	1.730	2.74	1.18	36.6	20.2	16.4	0.96	8.6

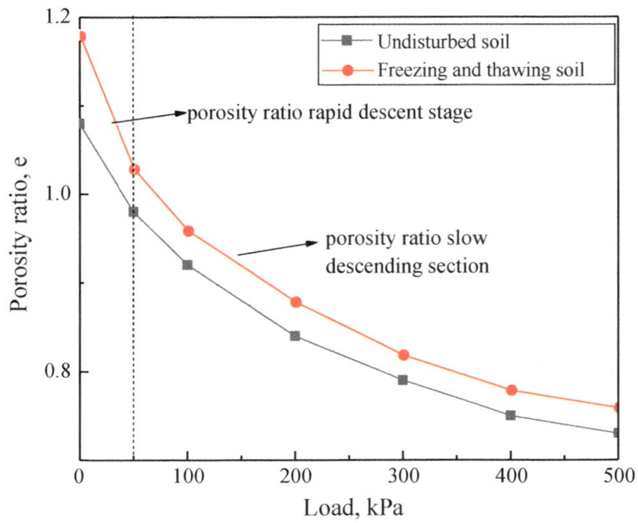

Fig. 1 Pore ratio and load curve

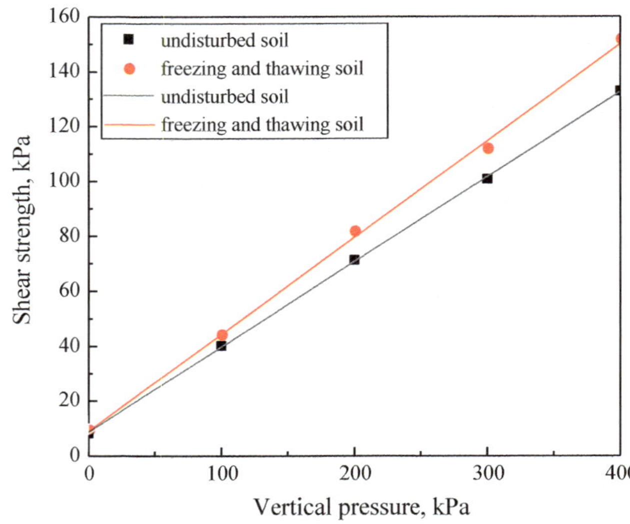

Fig. 2 Shear strength and vertical pressure curve

3.2 Mechanical Properties

The test results show that after a freeze-thaw cycle, the cohesion force and internal friction angle increase, but it is easier to compress, the compressive strength is reduced by half, and the sensitivity is reduced.

The change trend of the undisturbed soil and the frozen-thawed soil was the same [5], but it was obvious that the porosity ratio increased slightly after freezing and thawing (see Fig. 1). Under the conditions of lower positive pressure, the shear strength of both sides was close, and then with the increase of vertical pressure [6], the undisturbed soil shear strength growth rate was decreased significantly, and the results also showed that the cohesion and internal friction angle of freeze-thaw soil were both increased (see Fig. 2). The stress-strain curve of the original soil shows strain softening, and the strain that reached the peak intensity almost doubled compared to the original soil. The stress-strain curve of freeze-thaw soil shows strain

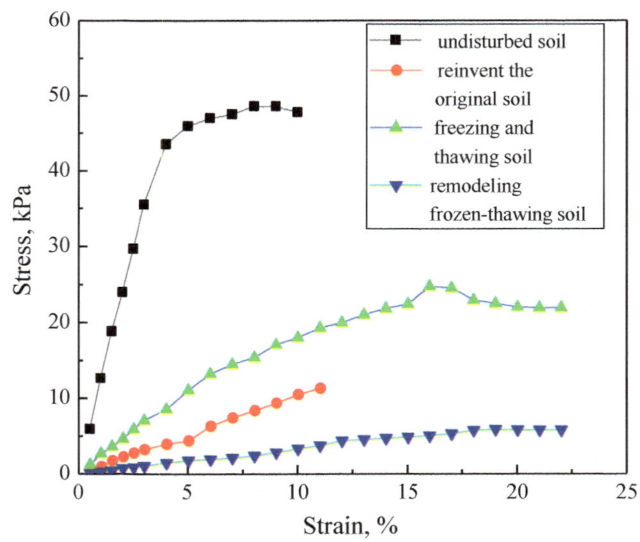

Fig. 3 Stress and strain curve

Table 2 Mechanical properties parameters of intact soil and frozen-thawed soil at −10 °C

Soil samples	Cohesive force (kPa)	Internal friction angle (°)	Compression coefficient (MPa^{-1})	Compression modulus (MPa)	Unconfined compressive strength (kPa)	Sensitivity
Undisturbed soil	8.3	17.3	0.8	2.40	48.6	4.6
Frozen and thawed soil	9.8	19.4	0.9	2.18	24.9	4.1

hardening. After a freeze-thaw cycle, the unconfined compressive strength of the soil was reduced largely, only 1/3 ~ 1/2 of the original soil, as shown in Fig. 3 and Table 2.

4 Discussion

This paper dealt with the physical properties of SSSC including water content, density, specific gravity, void ratio, liquid limit, plastic limit, plasticity index, permeability coefficient, and mechanical properties, involving compressibility, compression modulus, cohesion, and internal friction angle, unconfined compressive strength.

After a single freeze-thaw cycle, the soil properties changed. The most obvious diversification was the increased water content, increased permeability, lower compressive strength, and reduced sensitivity. The reason was the migration of external moisture during the freezing process and the reorganization of the original soil structure after freezing and thawing.

5 Conclusion

This paper systematically compared the physical and mechanical properties of the fourth layer of silty soft clay in Shanghai, which has undergone a freeze-thaw cycle, through a geotechnical test method. The present study leads to the following conclusions.

1. The changes in density and specific gravity were small, the water content and porosity ratio increased, the plastic limit stabilized with liquid limit increasing, and the soil body was soft. The permeability coefficient increased and the permeability increased, too.
2. The cohesion and internal friction angle of the clay after freezing and thawing slightly increased, the compression coefficient increased, the compression modulus decreased, the compressibility of the soil increased, and the clay had no confinement resistance after freezing and thawing. The compressive strength was only 1/3 to 1/2 of that of the undisturbed soil, and the intensity decay was obvious, and the sensitivity was reduced.

Acknowledgements This work was funded by the Fundamental Research Funds for the Central Universities (2018ZZCX04).

References

1. Huang, Y.C.: Evaluating the behavior of a cohesive soil undergoing one cycle of freeze-thaw. Geotech. Front., 643–651 (2017)
2. Hong, J.: Experimental Study on Mechanical Properties of Shanghai Saturated Soft Clay under Artificial Freezing Condition. Tongji University, Shanghai (2008)
3. Wei, R.L.: Soil extraction techniques and improvements in soft clay. Chin. J. Geotech. Eng. **8**(6), 113–125 (1986)
4. Zhang, Z.L.: Study on Dynamic Characteristics of Clay Before and After Freezing and Thawing Under Subway Vibration Load. China University of Mining and Technology, Xuzhou (2015)
5. Wang, S.F.: Microscopic pore change and fractal characteristics of artificial frozen-thawed soft clay. Chin. J. Geotech. Eng. **38**(7), 1254–1261 (2016)
6. Yin, Z.Z.: Experimental study on uniaxial unconfined compressive strength of artificial frozen clay in Shanghai. Rock Soil Mech. **33** (03), 788–792 (2012)

Geotechnical Mapping of Clayey Subgrade Soils Characteristics: A Case Study from Tebessa City (Algeria)

Adel Djellali, Med Salah Laour, and Abdelkader Houam

Abstract

Mapping of soils is essential when quantifying the relation between soils parameters and risk of swelling subgrades. Numerical mapping of soils areas affected by expansive clay minerals is essential when designing the pavements structures. In this paper, a simple method for automatic mapping of expansive clayey soils was proposed. 40 samples from subgrades of Tebessa city were subjected to various laboratory tests. The different geotechnical data were mapped to classify the subgrade soils and hazard delimitation of swelling areas using the geo-statistical method and automatic mapping technology. The results showed that soils were clayey marl. The Montmorillonite are observed abundant with small proportion of illite and kaolinite. These maps could be used to draw up the strategies of the clayey subgrade soils improvement and roads' design. Also the new soils mapping method has the advantages of being quick, objective and helpful to the engineers to localize the hazard zone in the study area.

Keywords

Soil mapping • Classification • Hazard • Swelling pressure • Subgrade

1 Introduction

The infrastructures transportation developments are principally related to earthworks, for the construction of highways [1, 2]. Numerous pavements designed on expansive soils have experienced severe asphalt cracking and early loss of serviceability [3]. As a case study, Tebessa city is located in the Northeast of Algeria (Fig. 1). The geology of the site has been studied by several researchers [4, 5]. The quaternary and plio-quaternary formations occupied the central part; they are formed by actual and recent alluvial deposits, sandstones and gravels [5].The expansive clay presents the major subgrade soil formations, where the base of principal national roads are (N10, N16 and N82). The pavements deterioration was characterized by the cracks on the top surface, and are confirmed to be due to the swelling pressure which was not introduced in design [4].

The aim of this work was to sketch a different geotechnical map to classify the subgrade soils and hazard delimitation of swelling areas, using a geo-statistical method and automatic mapping technology. The second goal was to draw up strategies of the clayey subgrade soils improvement and pavements design.

2 Materials and Methods

In this study, forty samples were selected to characterize the studied soils. Tests were carried out according to the French standards such as: particle size distribution [7], water content [8], dry density [9], liquid and plastic limits [10], methylene blue value, [11], carbonate content [12] and free swelling pressure [13]; core sample were taken along some sections of national roads of the city (N10, N16 and N82). The locations of the boreholes are positioned using GPS. They were deliberately chosen approximate to the sections where the pavements had distressed. From various boreholes, it showed that the soils were of clayey marl, clayey silts and clay. The statistical evaluation on the studied samples is given in Table 1.

A. Djellali (✉) · M. S. Laour
Civil Engineering Department, Larbi Tebessi University, 12002 Tebessa, Algeria
e-mail: adel1830@gmail.com

A. Houam
Civil Engineering Department, King Khalid University
College of Engineering, Abha, KSA, Saudi Arabia

A. Kallel et al. (eds.), *Recent Advances in Geo-Environmental Engineering, Geomechanics and Geotechnics, and Geohazards*, Advances in Science, Technology & Innovation, https://doi.org/10.1007/978-3-030-01665-4_57

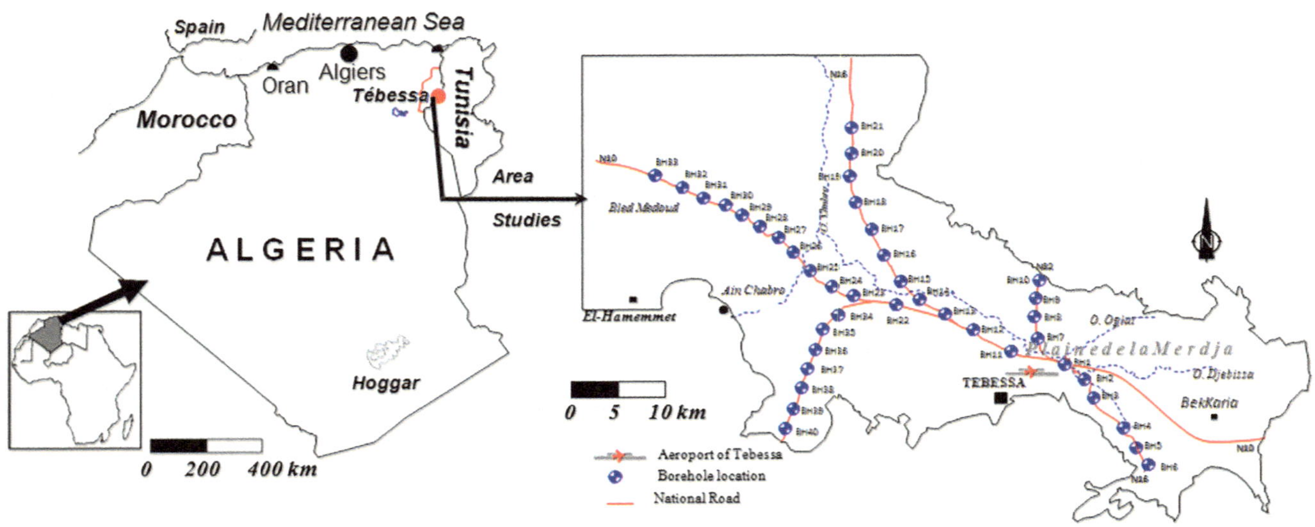

Fig. 1 Location map of the study area [6]

Table 1 Statistical data of geotechnical characteristics samples [6]

Properties	Number	Minimum	Maximum	Mean	St. deviation
Particles <0.08 mm (%) (NF P 94-056)	40	53.78	97.2	86.89	9.86
Water content, *w* (%)	40	11.71	28.55	19.14	3.85
Dry density, ρ_d (mg/m^3)	40	1.47	1.93	1.69	0.13
Liquid limit, LL (%)	40	42	77	60	9.37
Plastic limit, PL (%)	40	15	51	29.82	11.72
Plasticity index, *PI* (%)	40	16	51	30.17	8.27
Methylene blue values, *MBV* (cm^3/g)	40	2.08	10.86	5.12	1.99
Carbonate content, $CaCO_3$ (%)	40	6.56	74.20	41.16	17.10
Swelling pressure, *Ps* (kN/m^2)	40	100	260	175.50	8.89

3 Results

3.1 Soils Classification

The subgrade soils classification, according to the GTR method [14], ranges the region of the studied soils as A2, A3 and A4, as marly clays and clays with high waterproof and plasticity. According to Dakshanamurthy and Romana chart [15], the Tebessa's samples were placed in the area between montmorillonite, smectite clay and illite, with small samples containing more or less illite and kaolinite.

3.2 Numerical Classification Maps

To generate spatial distribution maps, the kriging method was employed on the geotechnical data using a linear variogram. The used estimation method is based on a set of near values points where the calculator program Surfer 11 was

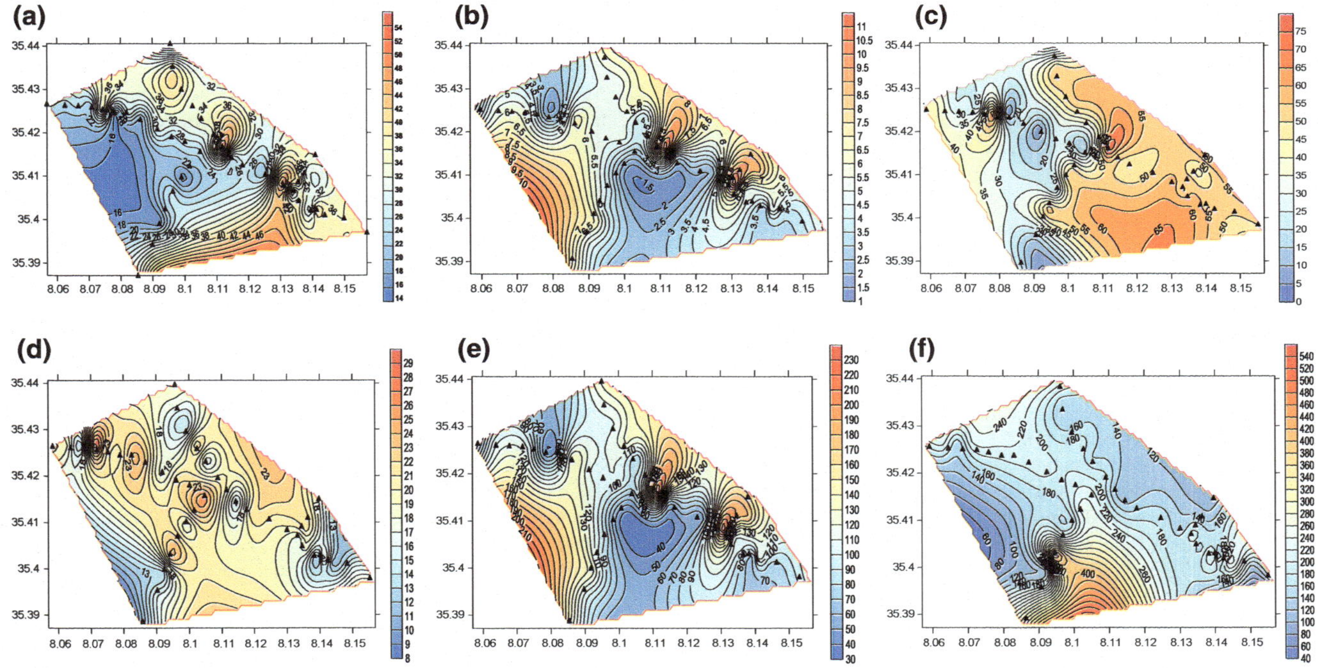

Fig. 2 Spatial distribution maps of geotechnical characteristics of subgrade soils, **a** *PI*, **b** *VBS*, **c** $CaCO_3$, **d** *w*, **e** *SSA*, **f** *Ps*

employed to recover variables maps. Each data point was weighted via its distance from the node. Based on this method, far away points from the node will have less weight in the estimated node. The Z value is calculated by [16]:

$$Z_A = \sum_{i=1}^{n} W_i Z_i \quad (1)$$

where Z_A is the calculated value of the node grid, n is the number of adjacent data values, Zi is the value at position i with weight Wi. The results of the physico-mechanical and chemical properties distribution maps of subgrade soils are shown in Fig. 2.

4 Discussion

The spatial distribution maps of geotechnical characteristics of subgrade soils represented in Fig. 2, showed that, *PI* varied from 16 to 51%, that these soils are moderately to very clayey, with swelling potential moderate to very high [17]. The, *MBV* extended from 2 to 11 cm^3/g, indicate that soils are medium plastics silts and very plastic clays. The $CaCO_3$ varied from 7 to 74%. The majority of the soils are marly clays. The *w* extended from 14 to 29%, where the site is slightly wet to wet. The Specific surface area *SSA*, varied from 44 to 228 m^2/g where the minerals clays are Illites and montmorillonites [18]. Finally the free swell pressure Ps, extended from 100 to 540 kPa which is classified as intermediate to high [18, 19].

5 Conclusion

In this paper a soil classification method was proposed, based on different authors, and a combination of geotechnical characteristics in Tebessa city in Algeria. The classification of soil is related to expansive clay minerals, determined with 40 samples around pavements sections which suffered great damage. Using kriging and stack tool from the automatic mapping software, soil maps were generated, and we provided a numerical delineation which can be used in hazard delimitation of swelling areas. The new soils mapping method has the advantages of being quick, objective and helpful to engineers to locate risky zones.

References

1. Al-Mukhtar, M., Lasledj, A., Alcover, J.F.: Behaviour and mineralogy changes in lime-treated expansive soil at 20 °C. Appl. Clay Sci. **50**, 191–198 (2010)
2. Zha, F., Liu, S., Du, Y., Cui, K.: Behavior of expansive soils stabilized with fly ash. Nat. Hazards. **47**, 509–523. https://doi.org/10.1007/s11069-008-9236-4 (2008)
3. Djellali, A., et al.: Static analysis of flexible pavements over expansive soils. Int. J. Civ. Eng. **15**(3), 391–400 (2017)

4. Vila, J.M.: La chaîne alpine de l'Algérie orientale et des confins Algéro-Tunisiens. Université Pierre et Marie curie, Paris VI, Thèse de Doctorat ès sciences (1980)
5. Fehdi, C., Rouabhia, A., Mechai, A., Debabza, M., Abla, K., Voudouris, K.: Hydrochemical and microbiological quality of groundwater in the Merdja area, Tébessa, North-East of Algeria. Appl. Water Sci. https://doi.org/10.1007/s13201-014-0209-3 (2014)
6. Djellali, A., Houam, A., Saghafi, B.: Indirect estimation of swelling pressure of clayey subgrade under pavement structures. Arab. J. Sci. Eng. (2017)
7. NF P 94-056: Sols: Reconnaissance et essais—Analyse granulométrique—Méthode par tamisage à sec après lavage. Normalisation Française (1996)
8. NF P94-050: Sols: Reconnaissance et essais—Détermination de la teneur en eau pondérale des matériaux—Méthode par étuvage. Normalisation Française (1995)
9. NF P94-053: Sols: Reconnaissance et essais—Détermination de la masse volumique des sols fins en laboratoire—Méthodes de la trousse coupante, du moule et de l'immersion dans l'eau. Normalisation Française (1991)
10. NF P94-051: Sols: Détermination des limites d'Atterberg—Limite de liquidité à la coupelle—Limite de plasticité au rouleau. Normalisation Française (1993)
11. NF P94-068: Sols: Détermination de la valeur de bleu de méthylène d'un sol ou d'un matériau rocheux par l'essai à la tache. Normalisation Française (1998)
12. NF P94-048.: Sols: Reconnaissance et essais - Détermination de la teneur en carbonate - Méthode du calcimètre. Normalisation Française (1996)
13. XP P 94-091: Sols : Essai de gonflement à l'oedomètre. Détermination des déformations par chargement de plusieurs éprouvettes. Normalisation Française (1995)
14. GTR: Guide technique pour la réalisation des remblais et des couches de forme. Edition du SETRA-LCPC, Bagneux, Fascicule I&II, Paris (2000)
15. Fredlund, D.G., Hasan, J.U., Filson, H.: The prediction of total heave. In: Proceedings 4th International Conference on Expansive Soils. Am. Soc. Civ. Eng. ASCE. **1**, 1–17 (1980)
16. Golden software Homepage. https://support.goldensoftware.com, Last Accessed 9 June 2018
17. Chen, F.H.: Foundations on expansive soils. American Elsevier Sci. Pub. Com. New York (1988)
18. Mitchell, J.K., Soga, K.: Fundamentals of Soil Behaviour. Wiley (2005)
19. Costet, J., Sanglerat, G.: Cours pratique de mécanique de sol. Troisième édition. Tome 1, plasticité et calcul de tassement. Tome 2, calcul des ouvrages, p. 464. Dunod Paris (1983)

Regression Tools to Quantify the Swelling Pressure of Expansive Soil in Tebessa Region (Algeria)

Yacine Berrah, Abderrahmane Boumezbeur, and Nouar Charef

Abstract

This study aimed to quantify the pressure generated by the clayey soil of Tebessa region in order to use it as a parameter for particular construction works. The statistical, principal component analysis and regression tools were used as simple techniques taking into account the combined effects of the soil properties on swelling behavior and dealt with the swell pressure models. In the selection of independent variables; regression analyses were carried out and the multi-colinearity problem was considered, in the multiple regression analysis. Regressors were added to obtain the final model. The obtained model is acceptable and gives the best results when compared to previous cited models in the literature. The regression analysis technique is a powerful predictive tool for assessing the swell pressure for all soil conditions.

Keywords

Clayey soil • Swell pressure • Soil parameters • Statistical tools

1 Introduction

In the N.E of Algeria, Tebessa region is characterized by an arid climate, Soil expansion was the main cause of the observed damage to different structures at the study area due to the high pressure generated within the clayey formations in a flat topography. Soils swelling hazard has become a serious problem. Several formulae have been established in order to predict the swell pressure and potential using simple geotechnical parameters [3, 4, 6–12]. Recently, the swelling behavior of soils has been studied using more sophisticated modeling techniques such as dimensional analysis [1, 2] and neural network analysis [5]. The present study relied on statistical approaches to evaluate and predict the swell pressure parameter from the correlation of physical and mechanical parameters such as the water content, the dry density, plasticity index, the clay content and the pre-consolidation pressure. The principal component analysis (PCA) and the multiple regressions were used for the studied soil data to derive the model that allows the evaluation of the swell pressure. Therefore, many aspects of statistical analyses like the type of regression, the models and results interpretations, and the determination of whether the used methods are feasible and how well the best model fits, were all included in this work.

2 Materials and Methods

As a first step, before using the principal component analysis, all the tests of descriptive statistics were carried out to screen the different variables used to derive the model. These variables are: The natural water content (w), dry density (γ_d), liquid limit (W_l), plasticity index (I_p), the % of fine fraction of less than 80 μ (F_f), the preconsolidation pressure (Pc), which only had good correlation with the swelling pressure (Ps). These parameters were chosen as mentioned on Table 1 as independent variables.

3 Results

The data used in the present study were analyzed in the first part with the principal component analysis (PCA). This technique is one of the best known multivariate analyses,

Y. Berrah (✉)
Mining Institute, Tebessa University, Tebessa, Algeria
e-mail: yacineberrah29@gmail.com

A. Boumezbeur
Department of Geology, Tebessa University, Tebessa, Algeria

N. Charef
Department of Civil Engineering, Souk-Ahras University, Souk Ahras, Algeria

A. Kallel et al. (eds.), *Recent Advances in Geo-Environmental Engineering, Geomechanics and Geotechnics, and Geohazards*, Advances in Science, Technology & Innovation, https://doi.org/10.1007/978-3-030-01665-4_58

Table 1 Descriptive statistics of the used data and variables [2]

Variable	Observations	Minimum	Maximum	Mean	Std. deviation
w (%)	86	11,710	38,240	20,433	4835
γ_d (KN/m^3)	86	11,600	19,400	16,380	1570
W_l (%)	86	36,000	160,000	65,535	22,717
IP (%)	86	19,000	85,000	42,640	14,951
F_f % < 80μ	86	44,000	97,600	88,822	10,039
Pc (KPa)	86	70,500	265,000	163,488	40,215
Ps (KPa)	86	60,000	670,000	256,622	155,683

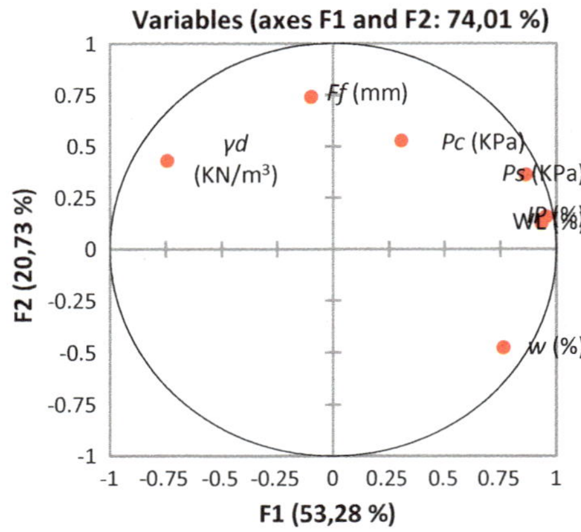

Fig. 1 The circle of correlation of variables

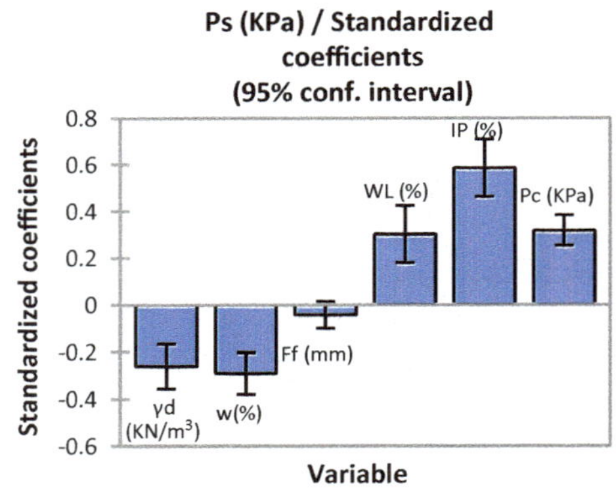

Fig. 2 Parameters contribution to the multiple regression model [2]

also known as eigenvector analysis for removing the adverse effects of collinearity. The (PC) number is based on the Kaiser's values (variances) higher than 1 [10]. The results are represented by circles of correlations which represent the projections of the variables on the first two components. A variable which is projected near the circle and close to a principal axis is well represented on that axis (Fig. 1).

In the second part of this work, two stages (learning and validating) were used; 86 data samples were applied to formulate the empirical model in the learning stage, other data were used for validation. The empirical models, were linearly fitted to the data using a single independent variable; those with negative contribution are respectively (w, γ_d, F_f), the other factors represented by (w_l, I_p, P_c) have a positive contribution to this analysis as shown by Fig. 2 and finally multiple independent variables or multiple linear regression were established for developing the empirical model to indicate a reliable assessment of swelling pressure of the studied clayey soil. These were developed using Minitab 17. The general model is presented in Table 2.

The final model in the current study consisting of six parameters showed a good correlation compared to previous models cited in the literature with coefficient of regression 0.94 (Fig. 3). The multiple linear regression models gave the best-fit for all soil conditions.

4 Conclusion

The principal component analysis results using six out of the seven soil parameters as input show strong correlation on the first principal axis and absorb about 74% of the total variance, the PCA allowed the grouping of the best correlated parameters as the first group was composed of water content, dry density and the fines fraction with a high effect on the swelling phenomenon; the second group constituted of soil indices and preconsolidation pressure had a positive contribution. The multiple regression analysis technique is a powerful and very practical method to predict the swelling pressure; it validated the obtained results by the PCA method. This methodology to derive the model is a useful and powerful tool to estimate engineering properties of swelling soils. In this investigation, there were six proprieties $\gamma_d, w, F_f, w_l, I_p$ and P_c which were strongly correlated to swelling pressure. The model proposed in the present study, showed the best results when compared to previous models cited in the literature.

Table 2 Multiple regressions to drive the model of swelling pressure Ps (KPa)

Observations	86.000
Sum of weights	86.000
DF	79.000
R^2	0.938
Adjusted R^2	0.933
RMSE	40.171
DW	2.231

Equation of the model:
$P_s = 339.43 - 25.97{*}\gamma d - 9.41{*}w - 0.68{*}F_f + 2.060{*}W_L + 6.1{*}I_P + 1.23{*}P_c$

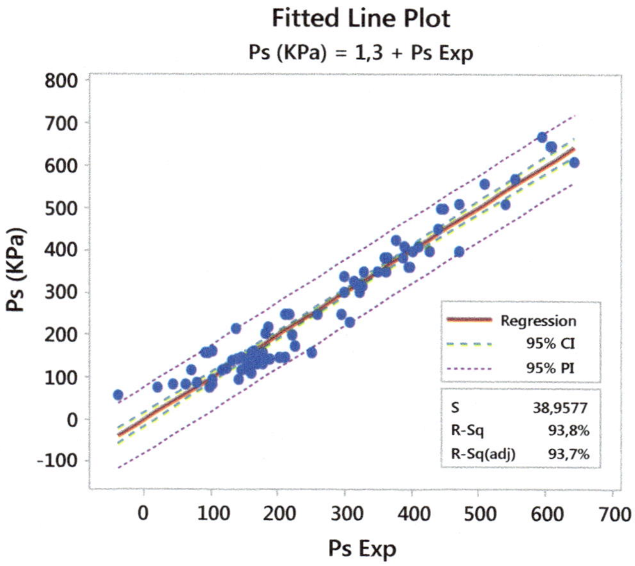

Fig. 3 Experimental (measured) versus predicted swell pressure

References

1. Buzzi, O.: On the use of dimensional analysis to predict swelling strain. Eng. Geol. **116**, 149–156 (2010)
2. Berrah, Y., Boumezbeur, A., Kherici, N.: Application of dimensional analysis and regression tools to estimate swell pressure of expansive soil in Tebessa (Algeria). Bull. Eng. Geol. Environ. (2016)
3. Daksanamurthy, V., Raman, V.: A simple method of identifying an expansive soil. Soil Found. Japan. Soc. Soil. Mech. Found. Eng. **13**(1), 97–104 (1973)
4. Erguler, Z.A., Ulusay, E.: A simple test and predictive models for assessing swell potential of Ankara (Turkey) clay. Eng. Geol. **67**, 331–352 (2003)
5. Erzin, Y.: Artificial neural networks approach for swell pressure versus soil suction behavior. Can. Geotech. J. **44**(10), 1215–1223 (2007)
6. Gray, C.W., Allbrook, R.: Relationships between shrinkage indices and soil properties in some New Zealand soils. Geoderma **108**(3–4), 287–299 (2002)
7. Kariuki, P.C., Van der Meer, F.D.: A unified swelling potential index for expansive soils. Eng. Geol. **72**, 1–2, 1–8 (2004)
8. McCormack, D.E., Wilding, L.P.: Soil properties influencing swelling in Canfield and Geeburg soils. Soil Sci. Soc. Am. J. **39**, 496–502 (1975)
9. Marquardt, D.W.: Generalized inverses, ridge regression, biased linear estimation, and nonlinear estimation. Technometrics **12**, 591–612 (1970)
10. Montgomery, D.C.: Introduction to statistical quality control, 7th edn. Wiley, New York, USA (2013)
11. Nagaraj, T.S., Srinivasa, M.B.R.: Rational approach to prediction swelling soil behavior, transportation research record (1985)
12. Agus, S.M.: Prediction and classification of expansive clay soils. Expansive Soils Recent Adv. Charact. Treat. (2006)

Dynamic Properties of Soft Clay Under Freezing-Thawing Cycle

Chen-Yu Hou and Zhen-Dong Cui

Abstract

The problems of land subsidence and deformation caused by freezing-thawing are a major problem in the construction of subway tunnels, and researchers are now focusing on them. This article summarized the research results on the dynamic characteristics of soft clay under freezing-thawing cycles, and analyzed the influencing factors to provide a reference for predicting the settlement of subway tunnels after construction by the freezing method.

Keywords

Soft clay • Freezing and thawing cycle • Dynamic characteristics • Axial strain • Pore water pressure

1 Introduction

Soft clay is widely distributed in many cities. Subway tunnels are mostly built on soft soil foundation. The soft soil foundation has low bearing capacity, low permeability, high water content, high compressibility and poor engineering properties. Therefore, the ground settlement, the open excavation and excavation instability caused by its mechanical properties are becoming more and more prominent. Freezing is often used in subway construction. After freezing and thawing cycles, the mechanical properties of the soil like frost heave and thaw will change significantly. The subway will also vibrate during the operation, which will reduce the strength of soft soil foundation and produce different degrees of settlement and deformation [1].

C.-Y. Hou · Z.-D. Cui (✉)
China State Key Laboratory for Geomechanics and Deep Underground Engineering, School of Mechanics and Civil Engineering, China University of Ming and Technology, Xuzhou, 221008, Jiangsu, People's Republic of China
e-mail: cuizhendong@cumt.edu.cn

Therefore, studying the dynamic characteristics of subway under freeze-thaw cycles will play a guiding role in the future subway construction, operation and maintenance.

2 The Review of Dynamic Properties Under Freeze-Thaw Cycles

There are many factors that may affect the dynamic characteristics of soft clay under the effect of freeze-thaw cycle. Through dynamic tri-axial tests, resonant column tests, and other test methods, the dynamic characteristics of soft clay can be obtained under different influence factors.

After applying a dynamic load, the additional deformation will be produced [2]. In the dynamic tri-axial test, the lower the loading frequency, the greater the deformation of the soil, but the plastic deformation decreases with the increase of the loading times, and the dynamic stress is reduced. There is a cyclic stress ratio [3] in saturated soft clay. When the cyclic stress is higher than this value, the soft clay will be damaged. With the increase of cyclic stress ratio and cycle times, the index of softening will gradually decrease [4]. The load and frequency will also have a significant effect on the long-term settlement of the soft substratum of the tunnel.

After freeze-thaw of soft clay, micro-cracks and pores are formed in the soil, and the structural morphology of the soil particles and the way of connection between the particles will change, which will destroy the cementation of the soil and reduce the its structural strength [5]. Whether it is loose soil or fine-grained one, the strength of the soil after the freeze-thaw cycle will be reduced [6], and the soil structure will be more seriously damaged. The elastic modulus of the soil after the freeze-thaw cycle was attenuated, and the attenuation of the elastic modulus of the soil was related to the properties of the soil [7]. The shear strength of soft clay increases first and then decreases with the increase of the number of freeze-thaw cycles [8].

A. Kallel et al. (eds.), *Recent Advances in Geo-Environmental Engineering, Geomechanics and Geotechnics, and Geohazards*, Advances in Science, Technology & Innovation, https://doi.org/10.1007/978-3-030-01665-4_59

3 Results

The following are experimental data of the axial strain and pore water pressure of soft clay under freeze-thaw cycles in some works from the literature [9, 10].

4 Discussion

From Fig. 1, we can see that the influence of frequency on the strain of soil is obvious. The lower the frequency, the time of contact between the vibration load and the soil will be longer, and more energy will be transmitted to the soil, thus the axial strain is greater.

From Fig. 2, it is obvious that the lower the freezing temperature, the greater the strain accumulation. Because the volume of water ice in pores expands after freezing and thawing, it leads to an increase in the number of large pores in the soil and a greater damage to its structural properties, which speeds up the rate of strain accumulation. Secondary freezing-thawing will exacerbate the structural destruction of the soil and produce a greater cumulative strain. Therefore, in the construction of freezing method, secondary freezing- thawing of foundation soil should be avoided as much as possible.

From Fig. 3, Freeze-thaw soil produces greater excess pore water pressure than the original soil, because the pores of freeze-thaw soils are squeezed and expanded, and the volume of water in the pores after saturation is increased. Under the same number of vibrations, the cyclic load with low frequency acts on the soil for a long time, the energy transferred to the soil structure is much greater, and the excess pore water pressure generated is greater.

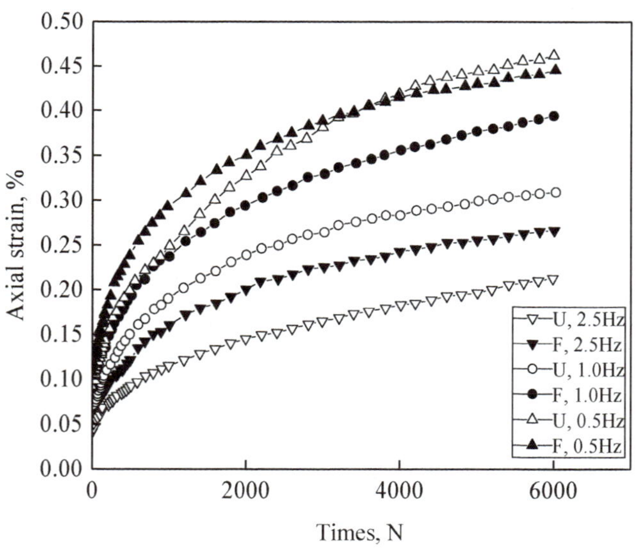

Fig. 1 Axial strain and loading times curve at different frequencies

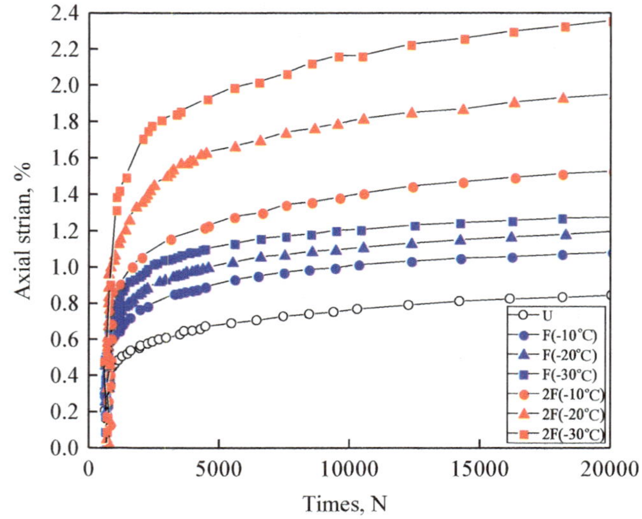

Fig. 2 Axial strain and loading times curve at different freeze-thaw cycles

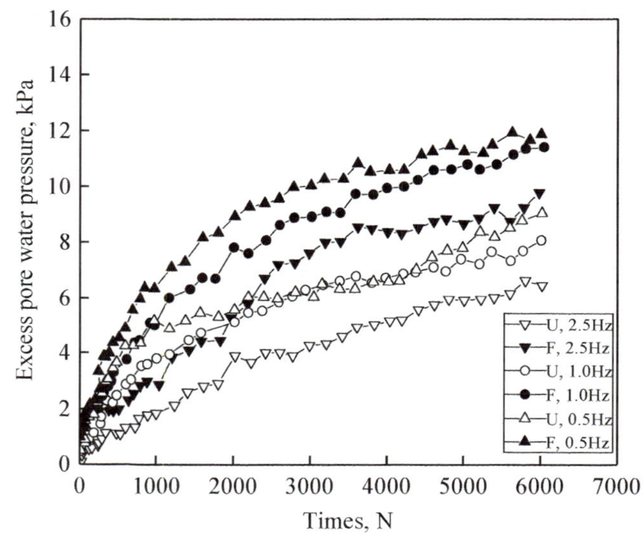

Fig. 3 Pore pressure at different frequencies

Figure 4 shows that the lower the freezing temperature is, the faster the pore pressure develops, and the pore pressure will increase when it is stable. This is because the structure of the soil is destroyed by freezing-thawing, the volume of pores increases, and the cumulative pore pressure during cyclic loading will increase as temperature decreases. The secondary freeze-thaw exacerbates the weakening of the soil structure, and has a destructive effect on the soil structure, the weakening of the structure and the increase of pores facilitate the accumulation and development of the excess pore pressure during the cyclic loading.

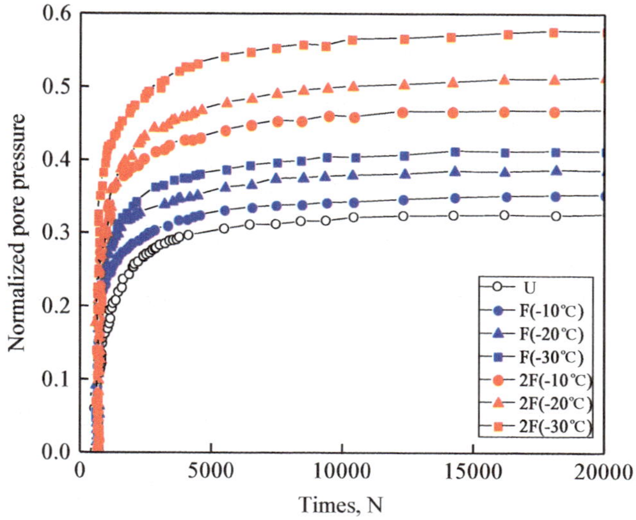

Fig. 4 Pore pressure at different freeze-thaw cycles

5 Conclusion

This paper summarizes the research results of dynamic properties of soft clay under freeze-thaw cycles, and mainly analyzes the effects of loading times, frequency, freezing temperature, and freeze-thaw cycles on the axial strain and pore water pressure of soft clay. The obtained results can be summarized as follows.

1. Under the same loading times, the lower the frequency, the greater the axial strain and excess pore water pressure of the soil, the greater the damage to the soil.
2. Freeze-thaw cycles will exacerbate the destruction of soil structure. The greater the number of freeze-thaw cycles, the more obvious the destruction will be.
3. The lower the freezing temperature, the higher the soil strain and the pore pressure accumulation rate. In addition, at lower freezing temperature, the frost heave will have more effects on the soil.

Acknowledgements This work was funded by the Fundamental Research Funds for the Central Universities (2018ZZCX04).

References

1. Wang, X.B., Yang, P.: Experimental study on effects of freezing and thawing on mechanical properties of clay. Chin. J. Geotech. Eng., 1768–1772 (2009)
2. Seed, H.B., Chan, C.K.: Effect of duration of stress application on soil deformation under repeated loading. In: Proceedings of 5th International Congress on Soil Mechanics and Foundations, Paris, pp. 341–345 (1961)
3. Larew, H.G., Leonards, G.A.: A repeated load strength criterion. In: The 41st Annual Meeting of the Highway Research Board, Washington, D.C. (1962)
4. Zhang, Y., Kong, L.W., Li, X.W.: Dynamic backbone curve model of saturated soft clay under cyclic loading. Rock Soil Mech., 1699–1705 (2010)
5. Ling, X., Li, Q., Wang, L., et al.: Stiffness and damping radio evolution of frozen clays under long-term low level repeated cyclic loading: experimental evidence and evolution model. Cold Reg. Sci. Technol., 45–54 (2013)
6. Qi, J., Vermeer, P.A., Cheng, G.: A review of the influence of freeze-thaw cycles on soil geotechnical properties. Permafrost Periglacial Process., 245–252 (2006)
7. Simonsen, E., Janoo, V.C., Isacsson, U.: Resilient properties of unbound road materials during seasonal frost conditions. J. Cold Reg. Eng., 28–50 (2002)
8. Wang, T.L., Liu, J.K., Peng, L.Y.: Study on mechanical properties of cement improved soil under freeze-thaw cycles. China Railway Sci., 7–11 (2010)
9. Zhang, Z.L.: Study on Dynamic Characteristics of Clay Before and After Freezing and Thawing Under Subway Vibration Load. China University of Mining and Technology, Xuzhou (2015)
10. Zhang, S.Y.: Experimental Study on Dynamic Characteristics and Microstructure of Frozen-Thawed Soil Under Subway Loading. Zhejiang University, Zhejiang (2016)

Stabilization of Clayey Soil Using Lime and *Prosopis* Fibers

Gopinath Rudramurthy, Poopathi Ramasamy, and Arun Rajendran

Abstract

In this paper, an experimental study was carried out to evaluate the geotechnical properties of clayey soil stabilized using lime and *prosopis* fibers. In the present study, the fibers used were at 0.0, 0.25, 0.50, 0.75 and 1.0% by weight of the soil and the amount of lime used was kept constant as 4%. From the test results, it was inferred that the optimum content of lime and *prosopis* fibers required to achieve superior geotechnical properties in clayey soil was found to be 4 and 1%. At this optimum combination, unconfined compressive strength (UCS) increases by 73.5%, California bearing ratio (CBR) resistance increases by 51.9% and linear shrinkage decreases by 69% when compared to the natural soil. Formation of cementitious compounds upon addition of lime and better interlocking of soil grains by *prosopis* fibers contributed to the significant improvement in the properties of clayey soil.

Keywords

Clayey soil • Lime • *Prosopis* • UCS • CBR

1 Introduction

An economical construction of earth embankments and pavements requires the use of the locally available soils with properties modified using appropriate stabilization methods. In India, expansive soils like black cotton soil are the second largest distributed soil covering over 20% of the total land area. Chemical additives like cement and lime prove to be an effective material for achieving the desirable properties for these types of weak soils. 4–5% lime addition can appreciably reduce the swelling and shrinkage of black cotton soils and make them stable even in wet conditions [1]. However, soils treated using lime or cement tend to exhibit high stiffness and brittle behavior [2]. The addition of natural or synthetic fibers along with these additives substantially improves the ductility and the post-peak strength of the soils [3]. *Prosopis juliflora* (PJ) is a woody weed tree that belongs to the genus *prosopis* which includes 44 species, 40 among these are native to Americas, three to Asia and one to Africa [4]. PJ is widely spread in tropical and subtropical regions of the world and grows in areas of low rainfall. In India and other regions, PJ is largely used as fuel wood by rural people and serves as food for livestock [5]. In this paper, the effect of the addition of lime and PJ fibers on geotechnical properties of clayey soil has been experimentally investigated.

2 Materials and Methods

The materials used in this work are black cotton soil, hydrated lime, and PJ fibers. The black cotton soil used in this study was brought from a site called Kutaripattu village near Tindivanam located at about 121 km from Chennai, TamilNadu, India. The soil was obtained through a pit excavated up to a depth of 1 m below the earth surface. The particle size distribution of the natural soil was determined using a hydrometer. The soil was classified as CL according to the Unified Soil Classification System (USCS) with ASTM D 2487 [6]. The plastic and liquid limits of the natural soil were 44 and 25%, respectively. After extraction by wet retting process, PJ fibers were dried in hot air oven and ground to a size of 1–3 mm.

Tests such as standard proctor compaction, atterberg limits, pH, uniaxial compressive strength (UCS), California bearing ratio (CBR), linear shrinkage and consolidation were carried out as per Indian Standards [7–13]. Six different soil compositions (untreated, treated with only lime and treated with both lime and PJ fibers) were used for the investigation. PJ fibers were used in four different percentages namely

G. Rudramurthy (✉) · P. Ramasamy · A. Rajendran
Department of Civil Engineering, University College of Engineering, Tindivanam, 604001, TamilNadu, India
e-mail: gopinathrudramurthy@gmail.com

A. Kallel et al. (eds.), *Recent Advances in Geo-Environmental Engineering, Geomechanics and Geotechnics, and Geohazards*, Advances in Science, Technology & Innovation, https://doi.org/10.1007/978-3-030-01665-4_60

0.25, 0.5, 0.75 and 1% by weight of the dry soil. In the preparation of all specimens, the method adopted differs slightly depending on the ingredients added to the soil.

3 Results and Discussion

The compaction curves obtained through proctor compaction test revealed the value of maximum dry unit weight and optimum moisture content (OMC) for the natural soil sample as 1.825 kg/m^3 and 23.72%. The optimum content of lime required for long term pozzolanic reaction can be found by performing pH test as illustrated in [14]. The pH value of the natural soil was found to be 9.36 and in lime-added soils, the pH value tends to increase with the addition of lime. At 4% lime, pH value was 9.51 and the value remained constant on further increase in the lime content. Thus, the optimum content of lime added in all the treated soil mixes was kept constant as 4%. The soil showed an immediate decrease in plasticity index (PI) upon addition of lime. The decrease in PI value indicates an improvement in the workability of the soil. The addition of 4% of lime was enough to enhance the workability of the soil by reducing the PI from 19.44 to 12.38%. The addition of lime (4% wt.) increases the optimum moisture content (OMC) from 23.72 to 25.13% and reduces the maximum dry density from 1.825 to 1.707 kg/m^3 when compared to 0% lime added. The increase in OMC upon addition of lime is mainly due to excess water required for hydration reaction and the reduction in the dry density could be due to the aggregation of cemented soil particles and the effective change in grading [15]. In soil samples treated with both lime and PJ fiber, a further increase in OMC and decrease in MDD were observed. This can be attributed to the low specific gravity of PJ fibers and the lubricating effect of absorbed water by PJ fibers which mitigate the compaction effort [16].

From UCS test results, it was inferred that the untreated soil exhibits a more ductile behavior and smaller peak strength than the lime-treated soil. Although the initial stiffness of the soil added with both lime and PJ fiber appears to be the same as that of the untreated soil, the fracture toughness of the former was found to be appreciably higher than the latter mainly due to the crack bridging ability of the fibers [17]. Also, the decrease in the failure strain of the lime-treated soil has been overcome by the fiber addition. In soils treated with both lime and fiber, the axial stress was found to increase with the increase in the axial strain until the peak value was reached, but the lime-treated soil experiences a sudden drop in stress [18]. From Fig. 1, it can be observed that addition of 4% lime increases the UCS value marginally. However, with the inclusion of PJ fibers the UCS value of soil increases sharply with the increase in the fiber content. The maximum UCS value was reported for the treated soil using 4% lime and 1% PJ fiber, about 73.5% more than that of natural soil. A similar trend was observed during CBR test wherein soil treated with 4% lime and 1% PJ fibers exhibited higher CBR resistance of about 51.93% more than the natural soil. The CBR value for soil samples under soaked condition (Fig. 2), exhibited a decreasing trend compared to soils under unsoaked condition. It is mainly because in a soaked condition, the water molecules tend to fill up the voids upon saturation [19].

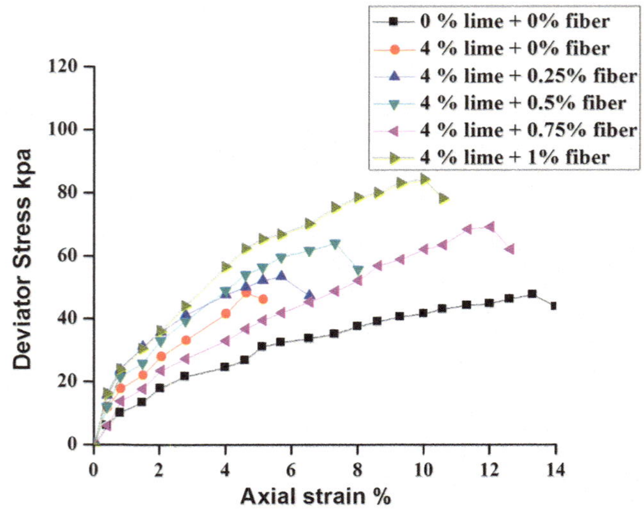

Fig. 1 Results of uniaxial compressive strength test

The soil compression index (Cc) was found to decrease with the addition of lime. This is mainly due to two basic reactions namely (a) ion exchange reaction and (b) flocculation, which transform the soil structure from plastic to granular [20].

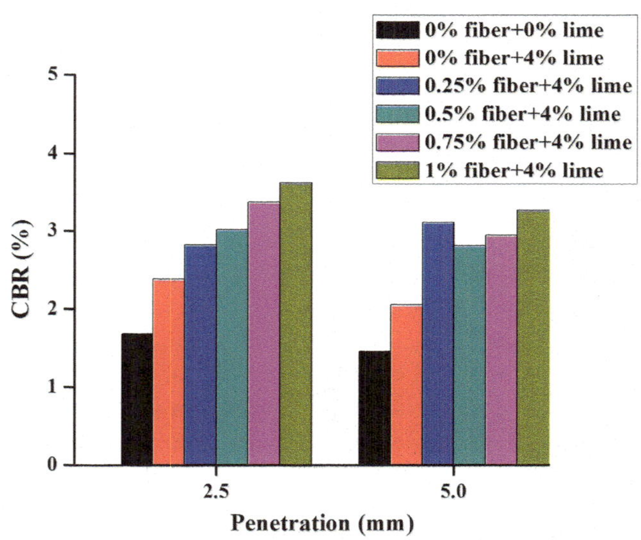

Fig. 2 Variation of CBR value at penetration 2.5 and 5 mm at soaked condition

Further, the Cc value decreases sharply with the increase in fiber content, with the maximum reduction of 52.09% observed in the soil stabilized using 4% lime and 1% fibers.

4 Conclusion

In this study, the effect of the addition of lime and PJ fibers on geotechnical properties of clayey soil was experimentally investigated. Based on the test results, it can be concluded that:

1. The optimum content of lime and PJ fibers for achieving best results was found to be 4 and 1% by weight of soil.
2. At optimum combination of lime and fibers,
   - UCS and CBR value of soil increases by 73.5 and 51.9%.
   - linear shrinkage of the soil decreases by 69%.
   - compression index decreases by 52.09%.

References

1. Uppal, H.L., Chadda, L.R.: Physico-chemical changes in the lime stabilization of black cotton soil (India). Eng. Geol. **2**(3), 179–189 (1967)
2. Ayeldeen, M., Kitazume, M.: Using fiber and liquid polymer to improve the behaviour of cement stabilized soft clay. Geotext. Geomembr., 1–11 (2017)
3. Chen, M., Shen, S.L., Arulrajah, A., Wu, H.N., Hou, D.W., Xu, Y.S.: Laboratory evaluation on the effectiveness of polypropylene fibers on the strength of fiber reinforced and cement stabilized shanghai soft clay. Geotext. Geomembr., 1–9 (2015)
4. Burkart, A.: A monograph on the genus *Prosopis* (Leguminosae subfam Mimosoideae). (Part 1 and 2). Catalogue of the recognized species of *Prosopis*. J. Arnold Arboretum **57**, 219–249, 450–525 (1976)
5. Pasiecznik, N.M., Vera-Cruz, M.T., Harris, P.J.C.: *Prosopis juliflora* withstands aridity and goat browsing in the Republic of Cap Verde. Nitrogen Fixing Tree Res. Rep. **13**, 89–91 (1995)
6. ASTM D2487-17: Standard Practice for Classification of Soils for Engineering Purposes (Unified Soil Classification System). ASTM International, West Conshohocken, PA (2017)
7. IS: 2720-5: Determination of Liquid and Plastic Limit. Bureau of Indian Standards publications, New Delhi (1985)
8. IS: 2720-5: Determination of pH Value. Bureau of Indian Standards publications, New Delhi (1987)
9. IS: 2720-26: Determination of Water Content-Dry Density Relation Using Light Compaction. Bureau of Indian Standards publications, New Delhi (1985)
10. IS: 2720-10: Determination of Unconfined Compressive Strength. Bureau of Indian Standards publications, New Delhi (1991)
11. I.S. 2720-16: Indian Standard for Laboratory Determination of CBR. Bureau of Indian Standards publications, New Delhi (1997)
12. I.S. 2720-20: Indian Standard for Laboratory Determination of Linear Shrinkage. Bureau of Indian Standards publications, New Delhi (1992)
13. I.S. 2720-15: Indian Standard for Laboratory Determination of Consolidation Properties. Bureau of Indian Standards publications, New Delhi (1986)
14. Muhmed, A., Wanatowski, D.: Effect of lime stabilisation on the strength and microstructure of clay. IOSR J. Mech. Civil Eng. **6**(3), 87–94 (2013)
15. Yadav, J.S., Tiwari, S.K.: A study on the potential utilization of crumb rubber in cement treated soft clay. J. Build. Eng. **9**, 177–191 (2017)
16. Yadav, J.S., Tiwari, S.K.: Behaviour of cement stabilized treated coir fiber reinforced clay-pond ash mixtures. J. Build. Eng. **8**, 131–140 (2016)
17. Ates, A.: Mechanical properties of sandy soils reinforced with cement and randomly distributed glass fibers (GRC). Compos. B **96**, 295–304 (2016)
18. Tang, C., Shi, B., Gao, W., Chen, F., Cai, Y.: Strength and mechanical behavior of short polypropylene fiber reinforced and cement stabilized clayey soil. Geotext. Geomembr. **25**, 194–202 (2007)
19. Yadav, J.S., Tiwari, S.K.: Effect of waste rubber fibers on the geotechnical properties of clay stabilized with cement. Appl. Clay Sci. **149**, 97–110 (2017)
20. Rao, S.M., Shivananda, P.: Compressibility behavior of lime-stabilized clay. Geotech. Geol. Eng. **23**, 309–319 (2005)

Part VII

Geomechanics and Geotechnics: Improvement of Physical and Mechanical Properties of Soils

An Experimental Study to Compare Two Soil Improvement Techniques Performance

Yasin Baskose and Candan Gokceoglu

Abstract

The main purposes of the present study were to compare in situ performances of two ground improvement techniques, which are jet grout and deep soil mixing techniques. In this study, Manisa State Hospital Project location was selected as the test site, because this location has difficult geological and geotechnical conditions. In the test site, three jet grout colons and three deep soil mix colons having same cement content were constructed. After 28 days of curing time, the core specimens were extracted for the uniaxial compressive testing, and lump specimens were collected for further tests and analyses such as X-ray diffraction, electron microscopy and determination of physical properties. The results of the uniaxial compression tests show that the uniaxial compressive strengths of the specimens extracted from the deep soil mix colons are drastically higher than those of the jet grout colons. Additionally, to understand the causes of this difference, several features of the improved soil were investigated by various mineralogical and petrographic methods. According to the results of these investigations, the fabric and some mineralogical properties were seriously affected by the improvement technique.

Keywords

Soil improvement • Jet grouting • Deep soil mixing • Mineralogy • Petrography

1 Introduction

Depending on the increase of population, construction of large engineering structures and buildings in difficult geological and geotechnical conditions is indispensable. Especially in poor and loose ground environments, depending on the feature of engineering structures, some serious foundation problems such as settlement, swelling, bearing capacity, liquefaction during earthquakes can be encountered. To overcome such types of problems, various ground improvement techniques have been applied. Ground improvement techniques have been developed and diversified, depending on the development of technology for the last five decades. Jet grout technique has been applied successfully in several projects [1–3] while deep mix technique is relatively new but successful improvement method [4, 5]. However, a comparison between the performances of these techniques applied in the same conditions has not been taken into account in the literature yet.

Manisa city center is located at an active tectonic zone and a State Hospital is planned to be constructed on weak and problematic soil conditions. To prevent settlement and liquefaction problems, a soil improvement is mandatory. Two soil improvement techniques, jet grout and deep soil mix, have been applied in situ and their performances have been compared. Three jet grout colons and three deep soil mix colons having the same cement content have been constructed. After 28 days of curing time, the core specimens have been extracted for the uniaxial compressive testing, and lump specimens have been collected for other tests and analyses (i.e. X-ray diffraction, electron microscopy and determination of physical properties).

2 Project Area and In-Situ Applications

Manisa State Hospital Construction area was selected as the test site. Manisa is located at the western part of Turkey (Fig. 1) and on the Gediz Graben. The region is an active tectonic zone and has experienced large earthquakes. For this reason, the earthquake effects on the structures and grounds should be considered.

During the investigations, a total of 13 drillholes have been performed. The depth of each varies between 20 and

Y. Baskose · C. Gokceoglu (✉)
Department of Geological Engineering, Hacettepe University, 06800 Beytepe, Ankara, Turkey
e-mail: cgokce@hacettepe.edu.tr

A. Kallel et al. (eds.), *Recent Advances in Geo-Environmental Engineering, Geomechanics and Geotechnics, and Geohazards*, Advances in Science, Technology & Innovation, https://doi.org/10.1007/978-3-030-01665-4_61

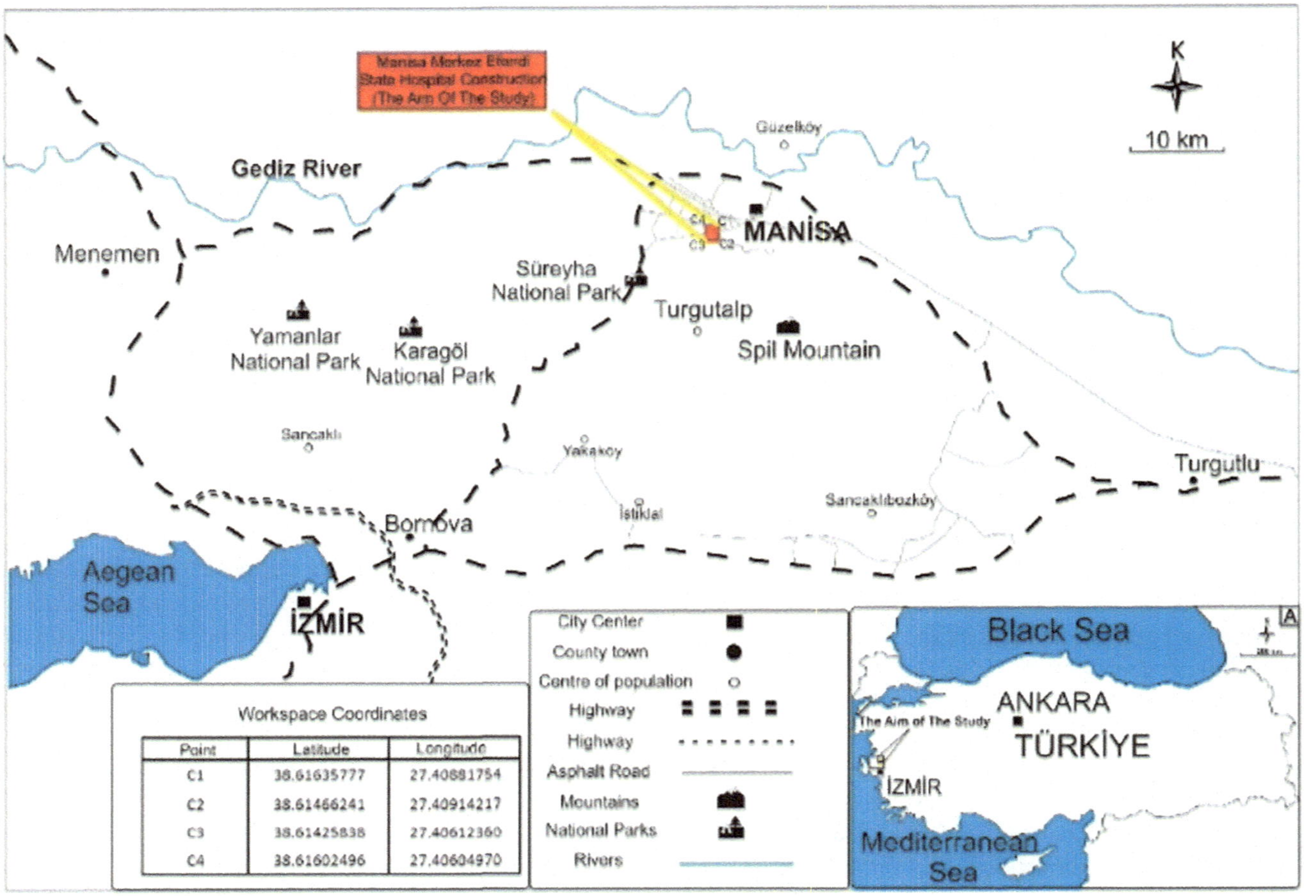

Fig. 1 Location map of the test site

25 m. Loose Quaternary sediments have been encountered during all borehole works. Quaternary sediments are formed by loose sands and silts and occasionally clay. The SPT values vary between 13 and 22. According to the results of SPT, the alluvial units can be classified as loose in general and dense occasionally. The groundwater level is about 7.5 m from the surface. The sediments were deposited by the fluvial processes but some of the materials is composed of debris transported from higher altitudes.

In the first stage of in situ applications, three jet-grout and three deep mix colons having the length of 3 m and the diameter of 0.8 m have been constructed (Fig. 2). The distances between the colons were 2 m. The ratio of water/cement was selected as 1 for both techniques.

After 28 days of curing time, the test site has been excavated and the colons have been extracted. The targeted diameter, 0.8 m, was obtained from the deep mix technique while the diameters of the colons constructed by jet-grout technique have been obtained as 0.50–0.65 m. The natural conditions and amount of cement were the same while the application procedures of the techniques followed in the

Fig. 2 The colons constructed on the test site

present study were different. This application provided a comparison between the performances of both techniques.

3 Results

To check the performances of both techniques, core specimens were extracted from the colons with a portable core drill machine. A total of 18 standard core specimens for uniaxial compressive test were obtained from the deep mix colons. The same procedure was also applied on the jet grout colons and the necessary core specimens for the tests were extracted. During the core extraction stage, extraction of standard core specimen from the jet grout colons was extremely difficult due to weak character of the colons. On these core specimens, the standard uniaxial compressive strength and unit weight tests were applied and the results are given in Table 1.

The UCS test results showed that the UCS values of the core specimens obtained from the deep mix colons are considerably higher than those from the jet grout colons. To understand the nature of this difference, the scanning electron microscope (SEM) analyses were performed on natural soil sample (Fig. 3a) and improved soil samples (Fig. 3b, c). The distribution of Ca in the improved samples is important for the strength because the increase of the strength is provided by the cement. As can be seen from Fig. 3b and 3c, the distribution of Ca in the sample taken from the deep mix colon is homogenous while that in the one taken from the jet grout colon is heterogenous. This analysis explains the strength difference clearly.

4 Conclusion

In the present study, a comparison between the performances of deep soil mix and jet grout was performed. For the purpose of the study, in situ jet and deep mix colons have been constructed. After 28 days of curing, the colons were excavated and the core specimens were extracted. The UCS and unit weight tests were applied on the core specimens. When considering the UCS test results, the core specimens obtained from the deep mix colons was found to be approximately 3–4 times higher than those obtained from the jet grout colons. To understand these drastic differences, scanning electron microscopy analyses were performed. According to the results of these analyses, the distribution of Ca in deep mix colons is rather homogenous.

Table 1 The uniaxial compressive strength (UCS) and unit weight (γ) test results

Specimen no	γ (kN/m^3)	UCS (MPa)
JET-A	16.6	2.0
JET-B	17.1	2.8
JET-C	17.2	2.9
DM-A	17.2	5.4
DM-B	16.7	6.4
DM-C	17.0	8.6

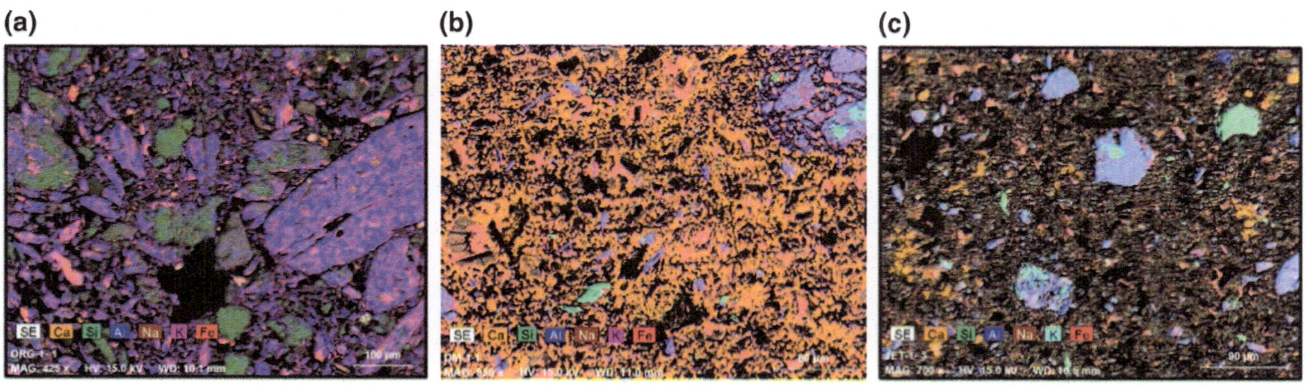

Fig. 3 Ca mappings obtained from SEM analysis: **a** the natural soil sample; **b** the sample taken from the deep mix colon; **c** the sample taken from the jet-grout colon

References

1. Wong, I.H., Poh, T.Y.: Effects of jet grouting on adjacent ground and structures. J. Geotech. Geoenviron. Eng. **126**(3), 247–256 (2000)
2. Coulter, S., Martin, C.D.: Effect of jet-grouting on surface settlements above the Aeschertunnel, Switzerland. Tunnell. Underground Space Technol. **21**(5), 542–553 (2006)
3. Shen, S.L., Wang, Z.F., Horpibulsuk, S., Kim, Y.H.: Jet grouting with a newly developed technology: the twin-jet method. Eng. Geol. **152**(1), 87–95 (2013)
4. Bruce, D.A., Bruce, M.E.C., DiMillio, A.F.: Deep mixing method: a global perspective. Civ. Eng. **68**(12), 38 (1998)
5. De Silva, M.S., O'Riordan, N.J., Parry, L.N.: Trials for the construction of a cement solidified retaining structure in a domestic landfill site using deep soil mixing. Eng. Geol. **60**(1), 49–60 (2001)

Effect of Combined Application of Mineral Fertilizer in Soil Hydraulic Properties

Nissaf Karbout, Mohamed Moussa, Nadhem Brahim, Roland Bol, and Habib Bounina

Abstract

The effects of compost, bentonite, and manure as soil amendments on soil hydraulic properties and improving water availability from saturation to oven dryness were investigated. The soil amendments were mixed with sand dune soil and compared with untreated soil (U) and reference soil (Chott Djerid soil). Two methods, hyprop and WP4 dew point potentiometer, were used to measure the whole range of soil hydraulic properties from saturation to oven dryness. The hyprop and WP4 results also exhibited that soil amendments increased the soil water content of the amended soils at low matric potentials. The results of soil water retention curves revealed that control retained less water at any matric potential compared to the amended soils. The soil water retention increased with the increase of soil amendments amount. The particle size distribution changed with the addition of bentonite. These results suggested that the soil amendments improved the soil water retentively which confirmed the appropriateness of these soil amendments for potential use in sandy soils improvements.

Keywords

Hydraulic properties • Soil amendments • Hyprop • WP4 dew point potentiometer

N. Karbout (✉) · M. Moussa
Institute of Arid Area, Mednine, Tunisia
e-mail: Nissaf.karbout@yahoo.fr

N. Karbout · H. Bounina
Institute of Agronomic of Tunisia, Tunis, Tunisia

N. Brahim
Faculty of Sciences of Tunis, University of Tunis EL Manar, 2092 Tunis, Tunisia

R. Bol
Institute of Bio-Geoscience, Agrosphere Institute (IBG-3), Juelich Research Center, 52428 Juelich, Germany

1 Introduction

Climate change accompanied by climate variability and other extreme climate events would have a direct effect on the quantity and quality of agricultural production [1], especially in semi-arid and arid regions where water scarcity is the most limiting factor for agricultural production. The quality and quantity of crop production is not only determined by total water amount and water use efficiency, but also is affected by rainfall distribution and the soil water-holding capacity, The region of Kebili south Tunisia is an arid area, where long-term intensive cultivation has led to serious soil degradation, including reduced soil water-holding capacity. In south Tunisia over half of the rural population is directly dependent on locally grown crops and date palm production. Though dates production in Tunisia hardly exceed 2% of the global production, Tunisia is considered leader in date palm export especially for Deglet Nour variety which gives importance to this sector in rural development and for the Tunisian economy [2]. Nefzaoua region (22,900 km^2) in South Tunisia covers 48% of the total area of all Tunisian oases with nearly 15,600 ha of traditional and modern palm trees plantations [3]. The Nefzaoua region contains over 4 million palm trees of which ca 1.6 million are Deglet Nour (the most valuable date variety). The date palm production in this region has decreased over the last decades from 190.6 Gg in 2011–2012 [4] to 182.5 Gg in 2014–2015. The unfavorable location of Nefzaoua oasis surrounded by Chott Jerid and Gharsa salt plains may further exacerbate soil degradation problems [5, 6] and drought and lack of water resource are the main limiting factors in the decrease of date palm production in this region. Mekki et al. [7] affirmed that the Nefzawa oasian agro-systems rely on deep fossil aquifers for their agricultural water supply. The resulting exploitation of these aquifers, increased fourfold between 1970 and 2000, and led to quantitative and qualitative degradations of water resources.

A. Kallel et al. (eds.), *Recent Advances in Geo-Environmental Engineering, Geomechanics and Geotechnics, and Geohazards*, Advances in Science, Technology & Innovation, https://doi.org/10.1007/978-3-030-01665-4_62

The aim of the present study was to examine the viability of new proposed amendment materials, consisting of mixtures of organic (farm manure or compost) and mineral (bentonite clay) materials. This combination integrated with the common management practices in Nefzaoua oases will help to improve the water holding capacity in Nefzaoua oasis system.

2 Materials and Methods

Soil samples were collected from a traditional oasis in Fantassa (33.8°N; 8.7°E), in Kebili governorate, southern Tunisia. The six treatments were (i) sandy soil + farm manure (SF), (ii) sandy soil + compost (SC), (iii) sandy soil + bentonite clay (SB), (iv) sandy soil + bentonite + farm manure (SBF), (v) sandy soil + bentonite + compost (SBC) and (vi) untreated control (U). The treatment amendments were applied on June 17, 2014, and the soil of Chatt Djred was considered as reference soil (Ch). Therefore, the soil was repacked to bulk densities measured in the field. Using disturbed soils was feasible because the natural soils did not show any structure, which could be destroyed by repacking. For the estimation of the hydraulic properties, the HYPROP® (UMS, München, Germany) method as described by Schindler et al. [8] was used in combination with the WP4® Dewpoint Potentiometer (Decagon Devices, WA, USA).

3 Results

3.1 Treatments Effects on Soil Aggregates

The soils were mostly fine texture with clay content less than 50%, the proportion of grain size distribution in different treatments were more or less similar. U soil and Ch soil have similar grain size distribution. Also, the addition of organic amendment (compost or manure) do not have a significant effect in particle size distribution, But changed with the addition of bentonite.

The clay percentage in soil increased from 36.2 to 36.5% in SBC and SBM with a decrease in sand percentage to range from 31 to 32.7%, respectively. In this context, Grandy and Robertson [9] affirmed that the addition of mineral amendment influenced in the change of particle size distribution, also affirmed that the addition of organic amendment do not have a significant influence in particle size distribution in soil.

The particle size distribution has change significantly ($p < 0.05$) in 20–40 cm of depth in SBC and SBM amendment, but we don't have significant effect with other soil amendment.

In the deep layer 40–60 cm the effect of amendment is not significant; a similar percentage of particle size distribution in all treatments was observed.

3.2 Treatments Effects on Soil Water Retention (SWR)

The water retention curve (WRC) increased significantly ($p < 0.05$) in all treatments with organic amendments (Fig. 1, Table 1). Water retention content was significantly ($p < 0.05$) lower in the untreated and Chatt Djerid soils compared to all other treatments. Corresponding to the addition of organic amendments mainly manure and compost to the soil the water retention increased, the maximum value of ðS was measured in the amended soil with bentonite and compost SBC treatment (ðS = 0.477 $cm^3.cm^3$), followed by SBF, SC and SM in which the values reached respectively (ðS = 0.462 $cm^3.cm^3$, ðS = 0.457 $cm^3.cm^3$, ðS = 0.441 $cm^3.cm^3$). The increase of water retention in the soil was related to the increase of macro and medium pores in those soils and this is confirmed with our results concerning the macroparticule size decrease with the addition of organic and mineral amendment from 33.1% for the coarse sand (ø = 600 μm) to 22% in soil amended with SBC. This result is also confirmed with the study of [10], who showed that the addition of biochar like organic amendment increases soil water retention and moisture in soil. Also, Ramos [11] specified that the addition of compost increases

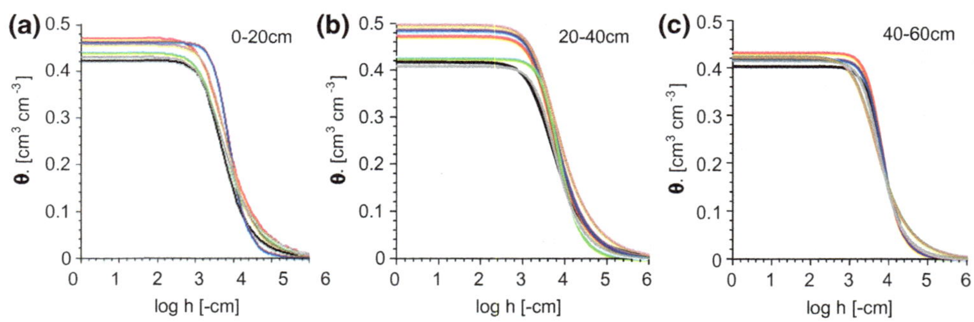

Fig. 1 Fitted retention curves for the different treatments for **a** 0–20 cm depth, **b** 20–40 cm depth, and **c** 40–60 cm depth

Table 1 Grain size distribution for the different treatments and three sampling depths

Depth	Treatment	Clay <2 µm	Fine silt 2 µm	Middle silt 6.3 µm	Course silt 20 µm	Fine sand 63 µm	Middle sand 200 µm	Course sand 600 µm
0–20 cm	U	22.1	0.0	11.7	14.6	1.0	17.5	3.1
	Ch	22.6	0.5	25.4	20.2	1.3	5.3	24.6
	SC	7.2	0.2	11.4	15.8	2.9	13.9	21.0
	SM	8.5	0.4	11.5	14.7	1.5	9.0	25.7
	SBC	36.5	0.4	16.8	38.5	2.0	13	21.9
	SBM	36.2	0.3	16.2	37.6	2.1	13.5	22.3
20–40 cm	U	23.2	0.8	11.4	5.2	3.2	21.6	34.6
	Ch	22.0	0.0	32.1	14.4	0.4	2.2	26.8
	SC	13.4	1.2	23.4	52.4	4.8	0.0	4.8
	SM	12.5	2.1	25.6	40.4	15.2	1.5	3.1
	SBC	38.3	0.0	11.3	3.2	2.4	12.7	32.1
	SBM	37.9	0.5	22.3	19.0	0.8	5.7	13.8
40–60 cm	U	22.6	0.5	45.4	6.9	1.1	7.1	17.3
	Ch	20.6	0.3	42.4	16.3	0.1	1.6	18.7
	SC	22.1	0.3	11.3	14.1	2.6	16.1	33.4
	SM	22.1	0.5	23.1	13.0	1.2	12.2	27.7
	SBC	20.5	0.5	22.5	11.2	3.2	15.4	22.5
	SBM	22.6	0.5	20.4	15.4	1.1	13.2	21.7

water retention in soil. Furthermore, Kim et al. [12] affirmed that the soil amendments may alter the pore size distribution (increase meso and micro pores) and increase the surface area which affects the water retention at low matric potentials, this increase in water retention due to the adsorptive forces between the soil solid surface and the soil solution. Therefore it is affected by specific surface area of the soil. However, at higher depths (20–40 cm), The WRC is increased, in SBM and SBC amendment because of the decrease of clay particle size distribution in the soil ($\eth S = 0.499$ and $0.487\ cm^3.cm^3$), but in 40–60 cm of depth we have a decrease and small difference in WRC in all treatments (Fig. 1, Table 1). The results was confirmed by (Mi et al. [1].

4 Conclusions

Our study provided the first experimental field-based evidence in oasis systems in the South Tunisia that adding bentonite clay (to create a mixed organic/mineral amendment) enhances the soil water retention and protects the oasis environment from degradation.

References

1. Mi, J., Gregorich, E.G., Xu, S., McLaughlin, N.B., Ma, B., Liu, J.: Effect of bentonite amendment on soil hydraulic parameters and millet crop performance in a semi-arid region. Field Crops Res. **212**(October), 107–114. Elsevier (2017)
2. Amor, R.B., Giménez, E.A., de Miguel Gómez, M.D.: The competitive advantage of the Tunisian palm date sector in the mediterranean region. Span. J. Agricult. Res. **10**(2) (2015)
3. Rurales, Alternatives: Revue de Communication et Débat Sur Les Alternatives Agricoles et Rurales (2016)
4. Anselm, D., Abdelhamid, H.: No TitleLes Oasis de Tunisie À Protéger Contre La Dégradation et Les Effets Du Changement Climatique (2012)
5. Kadri, A., Van Ranst, E.: Contraintes de La Production Oasienne et Stratégies Pour Un Développement Durable. Cas Des Oasis de Nefzaoua (Sud Tunisien). Science et Changements planétaires/Sécheresse **13**(1), 5–12 (2002)
6. Marlet, S., Bouksila, F., Bahri, A.: Water and salt balance at irrigation scheme scale: a comprehensive approach for salinity assessment in a Saharan Oasis. Agric. Water Manag. **96**(9), 1311–1322 (2009)
7. Mekki, I., Jacob, Frederic, Serge, M., Ghazouani, W.: Management of groundwater resources in relation to oasis sustainability: the case of the Nefzawa Region in Tunisia. J. Environ. Manage. **121**, 142–151 (2013)
8. Schindler, U., Durner, W., von Unold, G., Müller, L., Wieland, R.: The evaporation method– Extending the measurement range of

soil hydraulic properties using the air-entry pressure of the ceramiccup. J. Plant Nutr. Soil Sci. **173**(4), 563–572 (2010)
9. Grandy, A.S., Robertson, G.P.: Land-use intensity effects on soil organic carbon accumulation rates and mechanisms. Ecosystems **10**, 59–74 (2007)
10. Osama, M., Wolfgang D.: Effect of bentonite, hydrogel and biochar amendments on soil hydraulic properties from saturation to even dryness. Pedosphere (2017)
11. Ramos, M.C.: Effects of compost amendment on the available soil water and grape yield in vineyards planted after land levelling. Agricult. Water Manage. **191**(September), 67–76. Elsevier (2017)
12. Kim, J.S., Sparovek, S., Longo, R.M., De Melo, W.J., Crowley, D.: Bacterial diversity of terra preta and pristine forest soil from the Western Amazon. Soil Biol. Biochem. **39**, 648–690 (2007)
13. Belgacem, B.: Contribution À L'étude de Bonification Des Sols Sableux Par Un Amendement Argileux. Mémoire de 3éme Cycle, INAT (1986)
14. Fernández-Gálvez, J., Gálvez, P., Mingorance, M.D.: Soil hydrophysical properties resulting from the interaction between organic amendments and water quality in soils from Southeastern Spain—a laboratory experiment. Agricult. Water Manage. **104** (February), 104–112. Elsevier (2012)
15. Mohawesh, O., Durner, W.: Effect of bentonite, hydrogel and biochar amendments on soil hydraulic properties from saturation to oven dryness. Pedosphere, August. Elsevier (2017)

Glass Fiber Effect on the Undrained Static Response of Chlef Sand (Northern Algeria)

Leyla Bouaricha, Ahmed Djafar Henni, and Laurent Lancelot

Abstract

In this article, a laboratory study using a series of 24 consolidated undrained triaxial tests was carried out on samples consisting of two relative densities, reinforced and unreinforced by glass fibers of varying content and subjected to three confining pressures. The results showed that the presence of fibers contributed towards achieving a clear improvement on the undrained shear resistance of the tested sand. It was also noticed that the presence of randomly distributed fibers in the soil mass, provoked a considerable diminution in interstitial pressure, while contributing to the resistance improvement against liquefaction and the optimum of the resistance of fiber is 0.4% for all the performed tests.

Keywords

Triaxial test • Liquefaction • Undrained • Density • Glass fiber

1 Introduction

Soil reinforcement by glass fibers is particularly interesting considering its non-biodegradability environment, its high tensile strength, low cost and local availability [1, 2]. This paper focused on the study of the improvement and strengthening of soils with respect to liquefaction. Thus, Al-Refeai [3] noted that the inclusion of glass fibers required a higher confining pressure to prevent bond failure regardless of the type of sand. Also, he observed an increase in both the major principal stress and the secant modulus when increasing the length of the fiber. Glass fiber reinforcement has also been studied in combination with the addition of cement [4]. Triaxial compressive strength and tensile strength of glass fiber reinforced sand were sharper with higher fiber contents and longer fibers [5]. Nevertheless, there is no general consensus on this rule, but some studies have revealed few contradictions existing in this context. Bearing in mind the divergences in the published results, the literature emphasizes various combinations to know: fiber types, fiber content and length, modulus of elasticity or rigidity, as well as the physical characteristics of soil, such as the particle sizes and form, density and mode of sample preparations [6]. Several published studies in the literature have been devoted to the phenomenon of liquefaction of Chlef sand, associating triaxial tests under monotonic and cyclic loading. However, it should be noted that no study has been reported on the effect of glass fiber inclusions on the response to static liquefaction of saturated sands. The main objective of this study was to investigate the liquefaction behavior of Chlef sand, unreinforced and reinforced with glass fibers.

2 Materials and Methods

All tests in this study were carried out on samples of sand from the valley Chlef (Algeria). The glass fiber was obtained from the Company of National Products in Plastic and Rubber Chlef (Algeria). The particle size distribution curve and a scanning electron microscope (SEM) view are shown in Fig. 1. The main characteristics of this sand are summarized in Table 1 and the essential characteristics of glass fiber are summarized in Table 2. The mixtures were prepared according to the technique of Ladd [7] at two relative densities and for a fiber percentage varying from 0 to 0.6%. Thus, the average concentration of fibers included in the mixtures being determined by the following relationship:

L. Bouaricha (✉)
Laboratoire de Science Matériaux et Environnement, Université Chlef, Chlef, Algérie
e-mail: bouaricha-leyla@outlook.com

A. Djafar Henni
Laboratoire de Structures, Géotechnique et Risques, Université Chlef, Chlef, Algérie

L. Lancelot
Laboratoire de Génie Civil et Géo-Environnement, Université Lille1, Chlef, France

A. Kallel et al. (eds.), *Recent Advances in Geo-Environmental Engineering, Geomechanics and Geotechnics, and Geohazards*, Advances in Science, Technology & Innovation, https://doi.org/10.1007/978-3-030-01665-4_63

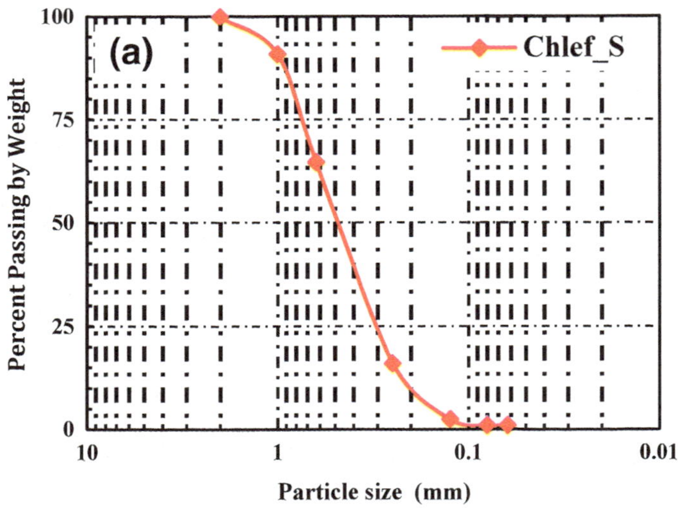

Fig. 1 Particle size distribution curve (**a**) and SEM view of Chlef sand (**b**)

Table 1 Physical characteristics of Chlef sand

D_{50} (mm)	D_{10} (mm)	Cu (−)	Cc (−)	G_s(−)	e_{min}(−)	e_{max}(−)	Shape
0.452	0.181	3.029	1.26	2.717	0.58	0.98	Rounded

Table 2 Physical and mechanical characteristics of glass fibers

Name	Unite	E-glass
Thickness (D)	(mm)	0.013
Specific density (G_s)	/	2.62
Specific weight of fibers	(g/m^2)	300
Tensile strength (T)	(N)	2500
Shear modulus (G)	(GPa)	29.2

$$w_f = (W_f/W_s) \times 100(\%) \quad (1)$$

where W_f is the weight of the fibers and W_s is the weight of the dry sand.

The fibers prepared and cut into pieces of 15 mm were manually randomly mixed with the sand until all fibers are evenly distributed in the sand. The dry deposition method was used to prepare the triaxial samples which makes it possible to pour the mixtures from a cone in the mold of 50 mm diameter and 100 mm high while respecting the state of fixed density. To ensure the saturation of our samples, the Skempton B coefficient values were all measured above 0.97. These samples were isotropically consolidated at three levels of average stress (100, 200 and 300 kPa) then subjected to undrained monotonic triaxial compression loading with a constant strain rate (500 microns/min).

3 Results and Discussions

The variations of the deviator q and the pore pressure ΔU, as well as the effective stress paths for unreinforced Chlef sand and reinforced with 0.4% of fiber, subjected to three confining pressures for loose and medium dense states are presented in Figs. 2 and 3. In the loose state (Fig. 2), the unreinforced samples can be seen to have a relaxation of the resistance beyond 1%, confirmed by the contracting behavior observed in the pore pressure curves (Fig. 2b).

However, the reinforced samples show a peak of resistance to an axial deformation of less than 0.5%. Liquefaction is observed for unreinforced samples under a confining pressure of 100 kPa. The reinforced samples show dilatant tendencies at lateral pressures of 100, 200 and 300 kPa, while the

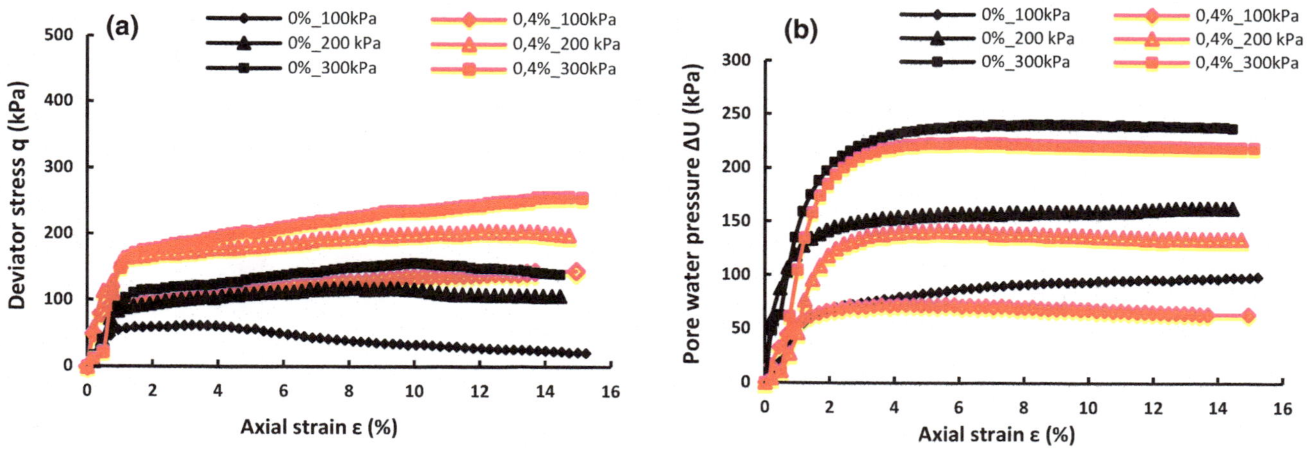

Fig. 2 Results of undrained triaxial tests for reinforced and unreinforced Chlef sand for loose state subjected to three confinements

unreinforced samples appear rather vulnerable for these same pressures and show collapse structures. For a medium density (Fig. 3), there is a marked improvement in the shear strength of the reinforced sand compared to what is observed in Fig. 2 for the loose state. The obtained results are consistent with those of Chen and Loehr [8] and Liu et al. [9].

The variations of the cohesion and the internal friction angle as a function of the fiber content are summarized in Table 3 for both relative densities. The results show a significant increase of these parameters with the increase of the fiber content, in the loose and medium dense states. Figure 4 shows the evolution of the calculated secant moduli for an axial strain of 0.15% versus fiber content for the two relative densities.

In the loose state (Fig. 4a), the secant modulus increases from 24 MPa for an unreinforced sand to 43 MPa for a sand reinforced with 0.4% fiber, rate at which the stiffness is maximum. For the medium dense state (Fig. 4b), the mean increase in secant modulus for a fiber content of 0.2% is about 50%, and up to 96% soil reinforced with 0.4–0.6% fiber.

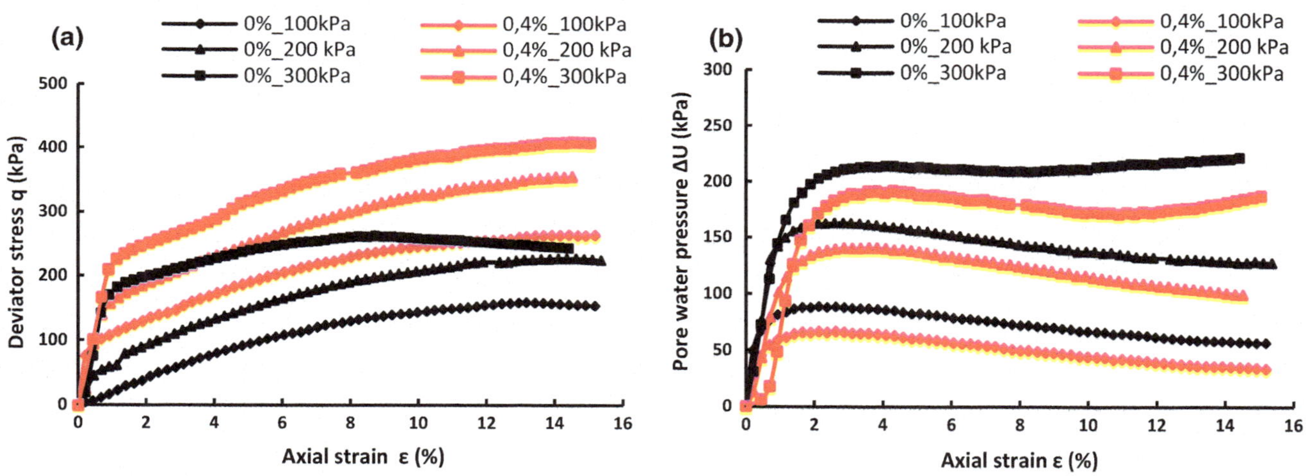

Fig. 3 Results of undrained triaxial tests for reinforced and unreinforced Chlef sand for medium and dense states subjected to three confinements

Table 3 Shear strength parameters

	Medium dense state				Loose state			
w(%)	0	0.2	0.4	0.6	0	0.2	0.4	0.6
C(kPa)	8.55	25.62	31.41	26.98	5.44	15.6	24.09	19.41
Φ(°)	25.78	29.45	34.88	32.07	17.7	23.96	27.37	26.46

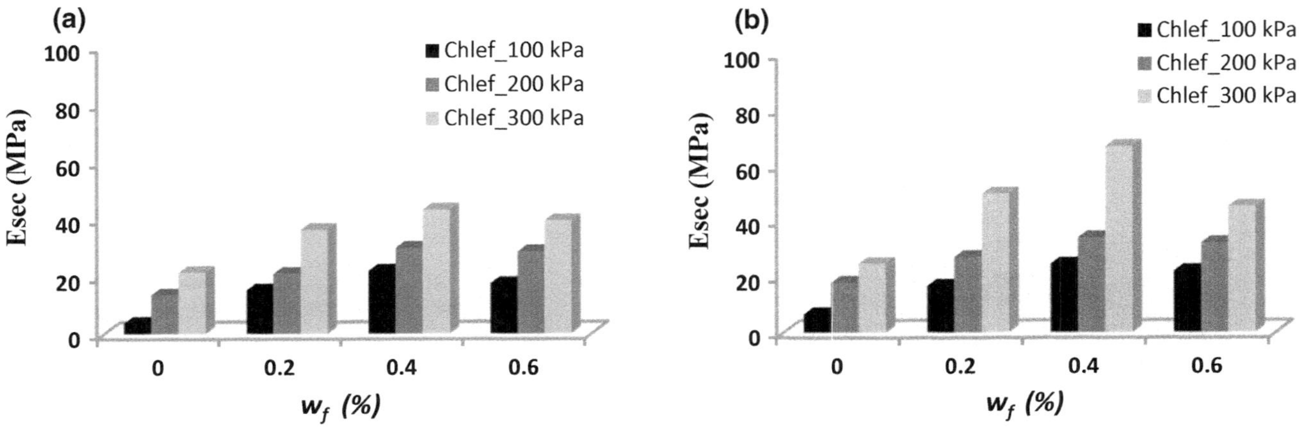

Fig. 4 Effect of fiber content on secant moduli of Chlef sand

4 Conclusion

The results of this study indicate that glass fiber reinforcement is useful for improving the resistance of sand to static liquefaction of Chlef sand. The values of the cohesion (C'), the internal friction angle (Φ) and the secant modulus (E_{sec}) increase with the increase of the percentage of fiber for a maximum rate of 0.4% fiber. Therefore, the relative density of the sand and the fiber content should be taken into account in practical applications.

References

1. Mujah, D., Ahmad, F., Hazarika, H., Safari, A.: Evaluation of the mechanical properties of recycled glass fibers-derived three dimensional geomaterial for ground improvement. J. Clean. Prod. **52**, 495–503 (2013)
2. Bouaricha, L., Djaafer Henni, A., Lancelot, L.: A laboratory investigation on shear strength behavior of sandy soil: effect of glass fiber and clinker residue content. Studia Geotechnica et Mechanica **39**(4), 462–471 (2017)
3. Al-Refeai, T.O.: Behavior of Granular Soils Reinforced with Discrete Randomly Oriented Inclusions. Geotext. Geomembr. **10** (4), 319–333 (1991)
4. Consoli, N.C., Prietto, P.D.M., Ulbrich, L.A.: Influence of fiber and cement addition on behavior of sandy soil. J. Geotech. Geoenviron. Eng. **124**(12), 1211–1214 (1998)
5. Maher, M.H., Ho, Y.C.: Behaviour of fibre-reinforced cemented sand under static and cyclic loads. Geotech. Test. J. **16**(3), 330–338 (1993)
6. Shukla, S.K.: Fundamentals of Fibre-Reinforced Soil Engineering. Springer Nature, Singapore (2017)
7. Ladd, R.S.: Specimen preparation using undercompaction. Geotech. Test. J. **1**(1), 16–23 (1978)
8. Chen, C.W., Loehr, J.E: Undrained and drained triaxial tests of fiber-reinforced sand. In: Proceedings of the 4th Asian Regional Conference on Geosynthetics, Springer, Berlin, Heidelberg, pp. 114–120 (2008)
9. Liu, J., Wang, G., Kamai, T., Zhang, F., Yang, J., Shi, B: Static liquefaction behavior of saturated fiber-reinforced sand in undrained ring-shear tests. Geotext. Geomembr. **29**(5), 462–471 (2011)

Long Term Evaluation of Wetting-Drying Cycles for Compacted Soils Treated with Lime

Maafi Nabil, Akchiche Mustapha, and Sara Rios

Abstract

This article addressed the results of an experimental study that dealt with the behavior of recycled soil treated with lime and exposed to wetting- drying cycles at the long term. The object of this research was to focus on the effect of a wide range of lime content on evolution of volume and durability of stabilized soils, when exposed to wetting- drying cycles. In this work, an experimental study was performed on soils which were treated with 8 rates (0–8%) of lime. The treated soils were conserved for three different periods of cure: 7, 28 and 180 days, and then exposed to 12 wetting-drying cycles of 48 h for each cycle. However, the behavior under wetting-drying cycles was only satisfactory for lime percentages above 6%. The formation of C-S-H and C-A-H responsible for the increased strength of stabilized soil samples are showed and the expensive ettringite was also responsible for the decreased strength of stabilized soil samples.

Keywords

Soil • Treatment • Lime • Long term

1 Introduction

Soil treatment techniques have been known for more than 60 years in the field of earthworks, in the construction of embankments and roads (highways, railways, road, airports, etc.). The treatment method by lime is an ecological and sustainable method and currently it is considered as an important method for reuse and valuation of unusable soil. Exposure to repetitive environmental conditions present one of the most destructive actions that may damage infrastructures, because their mechanical properties are severely affected and the consequence is; crack propagation and instability, occurring in geotechnical works. The most destructive actions are freezing-thawing [1] and wetting-drying cycles [2, 3]. However, the results of some research works on the effects of drying-wetting cycles on the stability and durability of soils are very disparate and depend on the type of soil, the percentage of stabilizer, test methods and cure conditions [4]. Treatment with lime has double effects, the first is to improve soil properties (in the short run), and the second is to stabilize the soil (in the medium and long runs) [5]. The immediate effect of the addition of lime causes the flocculation and the agglomeration of clay particles caused by cation exchange on the surface of soil particles. The result of this reaction in the short-run is mainly to improve workability and plasticity of soil which improves its handling and implementation by compaction to achieve the required capacity of soils [6, 7]. Over the long-run, the reactions may require time from a week to months or even a year to complete, depending on the speed of the chemical decomposition and hydration of aluminates and silicates. This process results in the formation of cementitious materials which bind the soil particles and improve the mechanical properties of the treated soil. Stabilization is a process of changing (usually in the medium or long runs), the soil characteristics to bring it to final state of stability, particularly against water action, and result in a strong soil. It retains its original form, its stability and its quality when exposed to the environment and maintains adequate residual strength at the long term. It also provides sufficient strength to resist climate conditions. The lack of knowledge about stabilization due to the reaction between lime and soil, and about the behavior of soils treated with lime in the long-term, and subject to cycles of drying-wetting are the reasons behind the lack of the use of this technique. In-depth research aiming at acquiring more knowledge should allow

M. Nabil (✉) · A. Mustapha
Department of Civil Engineering, LEEGO, University of Houari Boumedien (USTHB), Algiers, Algeria
e-mail: maafinabil@ymail.com

S. Rios
CONSTRUCT-GEO, Faculdade de Engenharia (FEUP), Universidade do Porto, Porto, Portugal

A. Kallel et al. (eds.), *Recent Advances in Geo-Environmental Engineering, Geomechanics and Geotechnics, and Geohazards*, Advances in Science, Technology & Innovation, https://doi.org/10.1007/978-3-030-01665-4_64

expanding the scope of the lime treatment in earthworks for the embankment bases in wet and floodplain zones.

2 Materials and Testing Methods

Support materials of this study are; natural soils classified according to ASTM unified classification system D 2487 (ASTM, 2000) as indicated in Table 1. The soil samples were collected from north Algeria of Sidi Aissa region of M'Sila (SAM). The untreated soil X-ray diffractogram is shown in Fig. 1. The soil mainly includes illite, kaolinite, chlorite, Smectite and quartz minerals. Tables 1 and 2 present the physical and chemical properties of soils. Lime used for the treatment of the different soils is quicklime (CaO), delivered by the ERCO company located in SAIDA, ALEGIRIAS. The physical and chemical characteristics of lime are given in the Table 3.

3 Experimental Results

The variation in the studied soils volume at different periods of cure such as 7, 28 and 180 days exposed to drying-wetting cycles are presented in Fig. 2. The increase in number of solicitation drying-wetting affects significantly the volume of treated and untreated soils. After 7 days of curing; there was a decrease of volume for all soils treated and untreated with lime and a different levels of volume change was observed. The untreated soil presented an important damage during the 1st cycle of drying-wetting. The sample with 1 and 2% of lime deteriorated during the 1st cycle and the sample with 3 and 4% of lime, had no important damage during the (2nd and 3rd), (4th and 5th) and (7th and 8th) cycles after 7, 28 and 180 days respectively. Above 4% additional lime, soils showed satisfying behavior to the last cycle. A decrease of about 78.62, 75.62 and 66.32% in volume was noted with 3% of lime after 7, 28 and 180 days respectively. However, soils with 8% lime showed a good relation between drying-wetting cycles and a loss in volume; 22.27, 13.22 and 9.23% at 7, 28 and 180 days respectively.

After 7 days of cure, the variation in water content is clear, and water content increases after each cycle for the studied soils with low lime contents (3 and 4%). Lime contents higher than 4%, water content remains almost stable after the 7th cycle for the soil. On the other hand, the water content was observed to decrease when the lime content increases. The effect of lime addition is especially visible in soils cured to 28 and 180 days, where the water content remains almost stable for all lime percentages in soil. The soil is a porous material, so it has the ability to absorb water by capillarity, and to maintain this water during the humidification. After the drying cycle, they showed a crack due to shrinkage and low tensile strength. When the water content decreases due to the repetitive exposition to drying cycles, suction forces increase until the tensile stress is equal to the cohesive forces. At this stage, the formation of cracks allows more shrinkage during visual observation of the tested samples. With the loss of additional water, primary cracks grow and allow the formation of new cracks. The external surface of soil samples was also affected due to the rapid loss of water at the surface and inside of the samples. At the beginning of the second wetting cycle, cracks are not closed or partially closed on the surface, so that cracking is not a completely reversible process (rebounding) during wetting-drying cycles. New cracks may therefore be created in soils so that during the next drying cycle, cracks occurs probably in the same spot. Crack propagation causes a loss of strength and subsequently accelerates the samples degradation with water absorption. This degradation differs according to rate of lime used during the treatment of these soils. Treated soils showed low propagation of fissures due to the formation of new hydrated C-S-H which limiting and reducing the development of cracks, when they are subjected to the drying-wetting cycle compared to untreated soils. The studied soils treatment with lime might improve the resistance when

Table 1 Geotechnical properties of the studied soils

Soils index	SAM
Parameters	Mean values
Liquid limit (ASTM D 4318 00)	61
Plastic limit (ASTM D 4318 00)	30
Plasticity index (ASTM D 4318 00)	31
Methylene blue value (ASTM C837)	6.2
Grain size (%) >0.075 mm	84
Clay content (%) (ASTM D 422 61)	22
pH (ASTM D 6276-99A)	8.46
Free swelling (ASTM D 4546) (%)	4
(UCSC) ASTM D 2487	CH

Fig. 1 XRD patterns of the soil

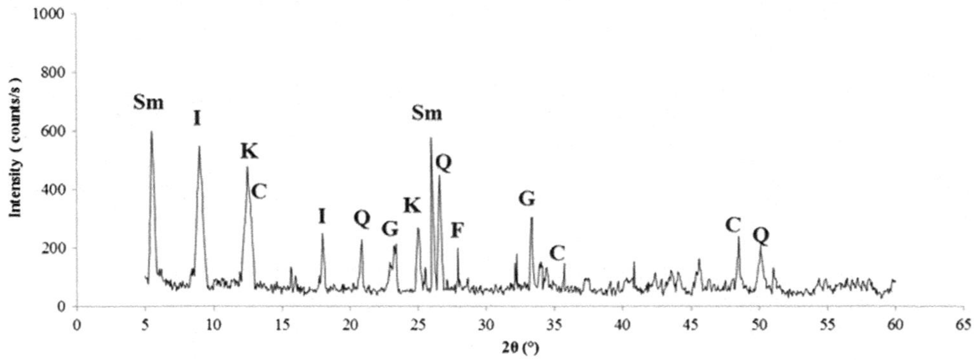

Table 2 Chemical composition of the studied soils

Constituents %	SiO_2	AlO_3	Fe_2O_3	CaO	MgO	K_2O	Na_2O	SO_3	Cl
SAM	50.01	15.81	6.76	8.39	2.42	1.57	0.5	3.31	0.03

Table 3 Physical-chemical properties of Saida's Quicklime

	Designation	(%)
Chemical composition	CaO	82.77
	MgO	1.83
	Fe_2O_3	0.87
	Al_2O_3	3.27
	SiO_2	1.35
	SO_3	0.11
	K_2O	0.151
	Na_2O	0.064
	Insoluble in HCl	<1.00
Physical properties	Bulk density Density Absorption coefficient Sensitivity to freezing b30 Volume of extinction Over 630 μm Over 90 μm	731 g/l 3.315 <5 <30 2.73 cm^3 0 <10

these soils were exposed to the drying-humidifying cycle. This effect is very remarkable especially with important periods of cure (superior than 28 days), as a consequence allowing the development of pozzoulanic reactions and formation of new hydrates C-S-H, that contribute to the reduction of porosity and consequently, the reduction of the water absorption capacity of soils treated with lime. Ultimately, stable soils were obtained as a result.

The XRD analyzes of the lime-treated soil for 180 days at 20 °C and exposed to a wetting-drying cycle are illustrated in Fig. 3. The pozzolanic reactions are the results of the combination between the silica and the alumina of the soil with lime calcium to the presence of water. As long as the residual calcium is available and the pH is high enough, the pozzolanic reaction will continue. These reactions result in the formation of calcium hydrates due to the presence of Ca^{+2}, OH^{-} and SiO_4 and AlO_6 ions [1]. The Identification revealed the formation of new calcium hydrates (CSH and CAH) in all the lime-treated samples and ettringite in samples with low lime content.

Exposure to repetitive wetting–drying cycles led to an increase in the macropores (pore diameter more than 6 μm) is illustrated in Fig. 4 due to additional ettringite formation for soil samples with 4% of lime. A slight reduction in the volume of micropores (pore diameter less than 0.3 μm) for soil samples with 8% of lime can also be observed.

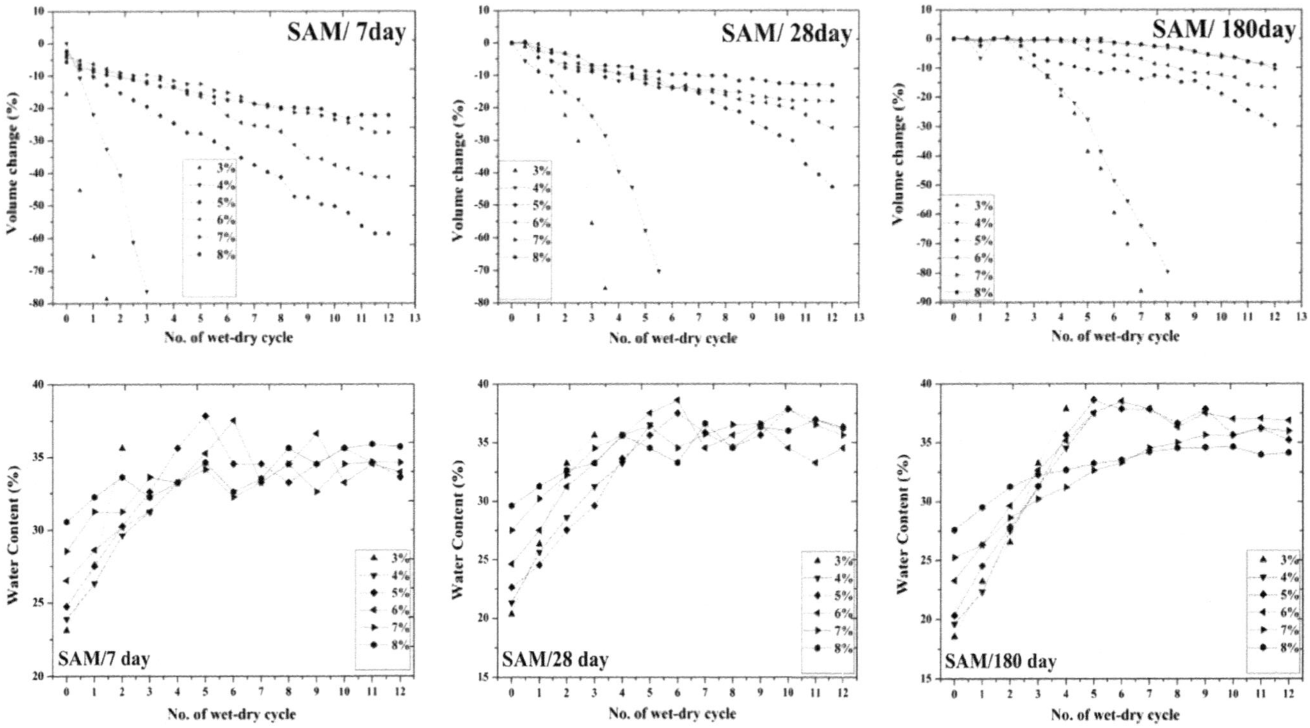

Fig. 2 The loss of volume and water content changes with wetting and drying cycles of studied soils

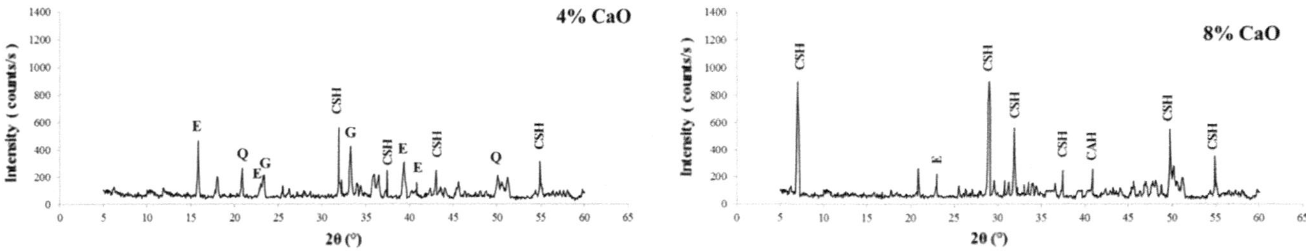

Fig. 3 XRD patterns of the soil after 180 days of curing at the end of wetting–drying cycles

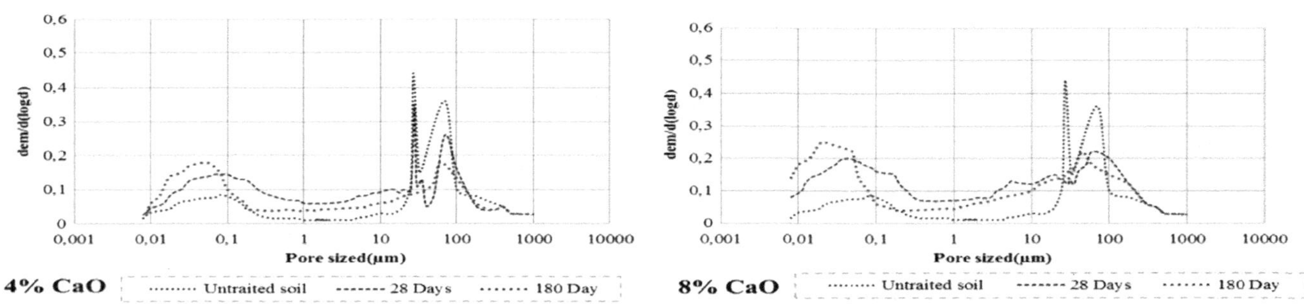

Fig. 4 Pore size distribution of the soil samples at the end of wetting–drying cycles

4 Conclusion

The cyclical drying-wetting exposure of investigated soils allows following two determining indicators of the durability (volume change and the rate of water absorbed) depending on the rate of lime and period cure of the studied soil.

Measuring the rate of water absorption collected soils at the time of the cyclic solicitations makes it possible to observe a disproportional relationship between the rate of water absorbed and lime. For superior percentage of lime; treated soils showing low water absorption rate confirm their low volume variation. The same behavior was observed for all soils with different cure periods. For inferior percentages of lime, the studied soils indicate an opposite behavior like their large volume variation.

The two distinguishing behaviors are due to the low permeability, caused by the decrease of the porosity due to the cementation following the pozzolanic reaction between mineral of sol and lime by the formation of new hydrated C-S-H and C-A-H responsible for the decrease of pores and propagation of cracks.

References

1. Aldaood, M., Bouasker, M., Al-Mukhtaret, M.: Impact of wetting–drying cycles on the microstructure and mechanical properties of lime-stabilized gypsums soils. Eng. Geol. **174**(2014), 11–21 (2014)
2. Al-Obaydi, M.A., Al-Kiki, I.M., Al-Zoubaydi, A.H.: Strength and durability of gypsums soil treated with waste lime and cement. J. Al-Rafidain Eng. **18**(1), 28–42 (2010)
3. Maafi, N., Akchiche, M., Sara, R.: Chapter 5 Wetting and Drying Compacted Soil-Lime Mixtures. Springer Nature
4. Tang, A.M., Ngoc, M., Cui, Y.J.: Effects of the maximum soil aggregates size and cyclic wetting–drying on the stiffness of a lime-treated clayey soil. J. Geotech. **61**(5), 421–429 (2011)
5. Lemaire, K., Deneele, D., Bonnet, S., Legret, M.: Effects of lime and cement treatment on the physicochemical, microstructural and mechanical characteristics of a plastic silt. Eng. Geol. **166**(8), 255–261 (2013)
6. Alrubaye, A.J., Hasan, M., Fattah, M.Y.: Improving geotechnical characteristics of kaolin soil using silica fume and lime. Spec. Top. Rev. Porous Media Int. J. **7**(1), 77–85 (2016)
7. Alrubaye, A.J., Hasan, M., Fattah, M.Y.: Stabilization of soft kaolin clay with silica fume and lime. Int. J. Geotech. Eng. **11**(1), 90–96. Taylor & Francis (2017)

Geotechnical Properties of Sandy Soil Stabilized Using Cement and *Prosopis juliflora* Fibers

Gopinath Rudramurthy, Poopathi Ramasamy, and Arun Rajendran

Abstract

In this paper, the geotechnical properties of sandy soil stabilized using cement and *prosopis juliflora* fibers have been studied. Abundant availability, superior mechanical properties and low energy consumption for processing make *prosopis juliflora* fibers an ideal choice for soil stabilization. In the present study, the fibers used were at 0.0, 0.25, 0.50, 0.75 and 1.0% by weight of the soil and the amount of cement used was kept constant as 9%. Among all the trials, the combination of 9% cement and 1% fibers was found to yield optimum results. In this combination, the shear strength of natural soil increases up to 2.36 times, the angle of internal friction increases from 47° to 68° and the CBR value increases by 68.19%. The most remarkable effect observed due to the fiber inclusion was the change in the behavior of cemented soil from brittle to ductile.

Keywords

Sand • Cement • *Prosopis juliflora* • Shear strength • CBR

1 Introduction

Stabilization of sandy soils using chemical additives and fibrous materials appreciably increases its strength and stability. Chemical additives like cement and lime proved to be an effective material for achieving the desired properties for sandy soils. However, soils treated using lime or cement tend to exhibit high stiffness and brittle behavior [1]. Addition of natural or synthetic fibers along with these additives substantially improves the ductility and post- peak strength of soils. The random distribution of fibers forms a three dimensional network and binds the soil particles together as a coherent unit and restricts the particle movement [2]. *Prosopis juliflora* (PJ) is a woody weed tree which belongs to the genus *prosopis* which includes 44 species, 40 of which are native to Americas, three to Asia and one to Africa [3]. Higher lignin content (17.11%), lower microfibril angle (10.64°) and higher tensile strength (558 MPa) makes *Prosopis juliflora* fibers, as a promising material for reinforcing weak soils [4]. In this paper, the geotechnical properties of sandy soil stabilized using cement and PJ fibers were studied.

2 Materials and Methods

The materials used in this work were sandy soil, cement and PJ fibers. The used sandy soil in this study was brought from a site called Tirukovilur town near Thenpennai river located at about 192 km from Chennai, TamilNadu, India. The particle size distribution of the raw soil samples was determined using Sieve analysis. The soil was classified as SP (poorly graded sand) according to the Unified Soil Classification System (USCS) with ASTM D 2487 [5]. Natural fibers were extracted by adopting wet retting process. As the detailed study on the stabilization of soil using natural fibers in particulate form has not been carried out yet, *prosopis* fibers used in the present work were ground to a size of 1–3 mm.

Tests such as standard proctor compaction, direct shear and, California Bearing Ratio were carried out as per Indian Standards [6–9]. The direct shear test was performed at the vertical normal stress of σ = 50, 100 and 150 kPa in order to define the shear strength parameters (c and φ). During the test, the strain rate was maintained at 0.12 mm/min. CBR test was carried out in accordance with IS 2720: part 16 (1997). Three samples from each soil mix (treated and untreated) were tested under unsoaked and soaked (in water for 96 h) conditions. The soaked condition simulates the behavior of subgrade under heavy rainfall or flooded situations.

G. Rudramurthy (✉) · P. Ramasamy · A. Rajendran
Department of Civil Engineering, University College of Engineering, Tindivanam, 604001, TamilNadu, India
e-mail: ramanaocean@gmail.com

A. Kallel et al. (eds.), *Recent Advances in Geo-Environmental Engineering, Geomechanics and Geotechnics, and Geohazards*, Advances in Science, Technology & Innovation, https://doi.org/10.1007/978-3-030-01665-4_65

Six different soil compositions (untreated, treated with only cement and treated with both cement and PJ fibers) were used for the investigation. PJ fibers were used in four different percentages namely 0.25, 0.5, 0.75 and 1% by weight of the dry soil. In a previous study [10] carried out to evaluate the effect of cement addition on sandy soils, it was inferred that the increase in shear strength was observed up to 10% addition of cement. In the present study, the amount of cement used was kept constant as 9%. In the preparation of all specimens, the method adopted differs slightly depending on the ingredients added to the soil.

3 Results and Discussion

The compaction curves obtained through proctor compaction test revealed the value of maximum dry unit weight and optimum moisture content (OMC) for the natural soil sample as 1.841 kg/m^3 and 10.06%. Stabilization of sandy soil using cement (9% wt.) increases the optimum moisture content (OMC) from 10.06 to 11.86% and reduces the maximum dry density (MDD) from 1.841 to 1.629 kg/m^3. The increase in OMC upon addition of cement is mainly due to the excess water required for the hydration reaction and the reduction in the dry density could be due to aggregation of cemented soil particles and the subsequent change in grading of soil. In soil samples treated with both cement and PJ fibers, further increase in OMC and decrease in MDD was observed. This can be attributed to the lower specific gravity and low water absorption capacity of PJ fibers [11].

From the results of the direct shear tests, it was observed that the value of peak shear strength and the failure strain of the treated soil were significantly influenced by the amount of fiber content [12]. Shear strength of soil treated using 9% cement and 1% fiber was found to be highest among all other specimens, about 2.36 times more than that of natural soil. Addition of cement appreciably increases the cohesion of soil mainly due to the formation of cementitious compounds after cement hydration. Further, with the inclusion of fibers the cohesiveness of cement stabilized soil steadily increased with the increase in the fiber content. The angle of internal friction of uncemented soil increased from 47° to 68°, when treated with 9% cement and 1% PJ fibers [13].

The variation CBR resistance for soils (untreated, treated with only cement and treated with both cement and PJ fibers) under unsoaked and soaked conditions are given in Figs. 1 and 2. Soils treated with 9% cement and 1% PJ fibers offered the highest CBR resistance of about 68.19% more than the untreated soil. The increase in CBR resistance in fiber stabilized soil could be due to the reason that randomly oriented discrete fiber inclusions in soil mass improved its load-deformation behavior by interacting with the soil particles mechanically through surface friction and by

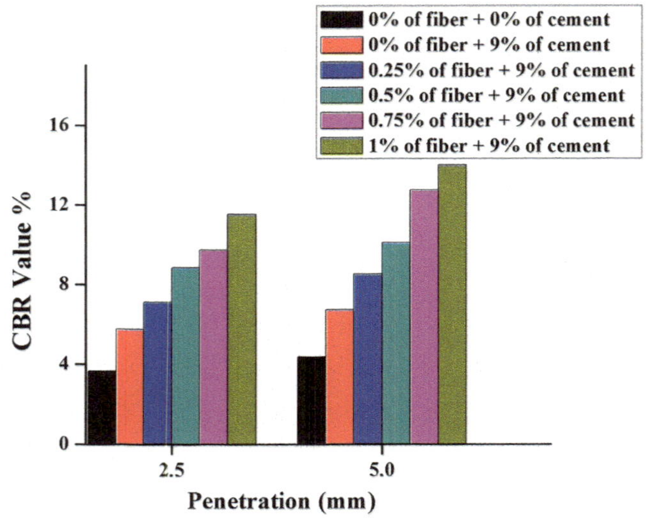

Fig. 1 CBR variation (unsoaked soil)

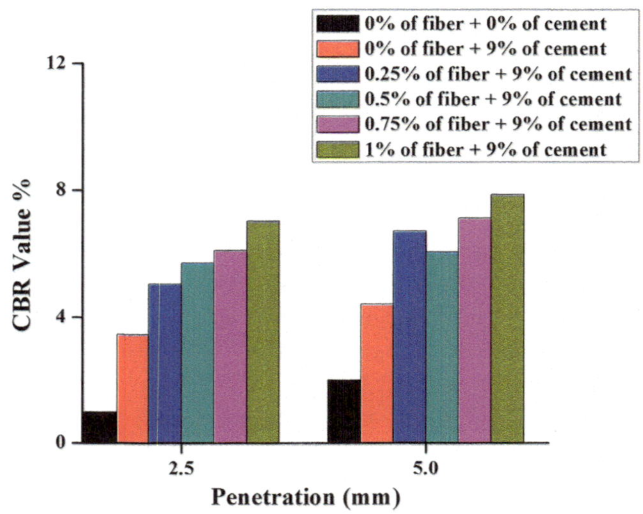

Fig. 2 CBR variation (soaked soil)

interlocking. Further, the addition of fiber bridged up the crack and prolonged the failure when compared to the untreated soil. It is also observed that the CBR value for soil samples under soaked condition exhibited a decreasing trend compared to soils under unsoaked condition. It is mainly because in soaked condition, the water molecules tend to fill up the voids upon saturation [14].

4 Conclusion

In this paper, the effect of the addition of cement and *Prosopis juliflora* fibers on the geotechnical properties of sandy soil has been studied. Based on the test results, it can be concluded that,

1. Addition of 9% cement and 1% PJ fibers could produce superior geotechnical properties in sandy soil.
2. At the optimum combination of cement and PJ fibers,
   - the shear strength of natural soil increases up to 2.36 times.
   - the angle of internal friction increases from 47° to 68°.
   - natural soil CBR resistance increases by 68.19%.

References

1. Yadav, J.S., Tiwari, S.K.: A study on the potential utilization of crumb rubber in cement treated soft clay. J. Build. Eng. **9**, 177–191 (2017)
2. Yadav, J.S., Tiwari, S.K., Shekhwat, P.: Strength behavior of clayey soil mixed with pond ash, cement and randomly distributed fibers. Transp. Infrastruct. Geotechnol., 1–19 (2018)
3. Burkart, A.: A monograph on the genus *Prosopis* (Leguminosae subfam. Mimosoideae). (Part 1 and 2). Catalogue of the recognized species of *Prosopis*. J. Arnold Arboretum **57**, 219–249, 450–525 (1976)
4. Saravanakumar, S.S., Kumaravel, A., Nagarajan, T., Sudhakard, P., Baskaran, R.: Characterization of a novel natural cellulosic fiber from *Prosopis juliflora* bark. Carbohyd. Polym. **92**, 1928–1933 (2013)
5. ASTM D2487-17: Standard Practice for Classification of Soils for Engineering Purposes (Unified Soil Classification System). ASTM International, West Conshohocken, PA (2017)
6. IS. 2720-26: Determination of Water Content-Dry Density Relation Using Light Compaction. Bureau of Indian Standards publications, New Delhi (1985)
7. IS. 2720-13: Determination of Direct Shear Strength. Bureau of Indian Standards Publications, New Delhi (1986)
8. IS. 2720-16: Indian standard for Laboratory Determination of CBR. Bureau of Indian Standards Publications, New Delhi (1997)
9. IS. 2720-15. Indian Standard for Laboratory Determination of Consolidation Properties. Bureau of Indian Standards Publications, New Delhi (1986)
10. Consoli, N.C., Vendruscolo, M.A., Fonini, A., Rosa, F.D.: Fiber reinforcement effects on sand considering a wide cementation range. Geotext. Geomembr. **27**, 196–203 (2009)
11. Yadav, J.S., Tiwari, S.K.: Behavior of cement stabilized treated coir fiber reinforced clay-pond ash mixtures. J. Build. Eng. **8**, 131–140 (2016)
12. Gray, D.H., Asce, A.M., Ohashi, H.: Mechanics of fiber reinforcement in sand. J. Geotech. Eng. **109**(3), 335–353 (1983)
13. Consoli, N.C., Prietto, P.D.M., Ulbrich, L.A.: Influence of fiber and cement addition on behavior of sandy soil. J. Geotech. Geoenviron. Eng. **124**(12), 1211–1214
14. Yadav, J.S., Tiwari, S.K.: Effect of waste rubber fibers on the geotechnical properties of clay stabilized with cement. Appl. Clay Sci. **149**, 97–110 (2017)

Mechanical Properties of Loess Treated by Calcium Lignosulfonate

Guoyu Li, Xin Hou, Wei Ma, and Fei Wang

Abstract

The chemical stabilization can consolidate the soil particles with the chemical reactions. In this paper, calcium lignosulfonate, a polymer eco-friendly material, was used for the treatment of the collapsible loess. Based on lots of tests such as Atterberg limits, uniaxial compression, loading-unloading, and X-ray diffraction, the stabilization mechanisms and strength characteristics of the calcium lignosulfonate treated loess were analyzed and compared with the NaOH and $Na_2O \cdot nSiO_2$ treated specimens, respectively. Calcium lignosulfonate can reduce the thickness of the electric double layer of particles and produce silica and carbonate cementations, thus increasing the loess strength significantly and reducing energy dissipation. The strength of the treated loess increased with the increasing amount of calcium lignosulfonate. Also, the results prove that calcium lignosulfonate is a new eco-friendly stabilizing agent.

Keywords

Calcium lignosulfonate • Eco-material • Loess • Chemical solidification • Compressive strength

1 Introduction

As a special type of soil used in engineering, the collapsible loess has poor engineering characteristics such as high sensitivity to water and large uneven settlement resulting in some engineering problems, huge economic loss and security risks [1].

For the improvement of collapsible loess foundation and subgrade, the common treatment methods include some physical and chemical solidification methods. Generally, some physical methods are easy to operate and their economy is better. So, these are most widely used in China [2]. In actual engineering, the compacted loess foundation/subgrade is often affected by wetting-drying and freezing-thawing cycles. The compression coefficient and permeability of the compacted loess is increased, and the strength decreases [1, 3, 4]. But the simple physical stabilization methods were not able to completely meet the actual needs of the loess engineering project. Many scholars focused on the study of the chemical solidification methods and developed some stabilization agents [5, 6]. These traditional stabilization agents can improve the loess structure, increase its compressive strength, reduce its compressibility, and eliminate its collapsibility in the different degrees [7–9]. These, however, have different disadvantages like their high cost. The cement soil has great shrinkage. The lime soil water and frost resistance is poor [8]. Also, the most common and biggest defect is that they are always unfriendly to the environment. Therefore, it is necessary to develop new types of the stabilizing agents for the ecological environmental protection.

In this study, the calcium lignin sulfonate was used to treat the loess, and its stabilizing effect was compared with the common stabilizers by their strength tests. The calcium lignosulfonate was produced as a by-product of lignin. It is non-toxic and environmentally friendly with high reserves and stable chemical properties [10].

2 Testing Methods

2.1 Physical Properties of the Loess Samples

The typical collapsible loess was sampled. Its physical properties such as gradation, liquid limit (multipoint liquid limit method) and plastic limit (rolling device method) and specific gravity were measured according to the related testing methods (JTG E40-2007). Its plastic limit, liquid

G. Li (✉) · X. Hou · W. Ma · F. Wang
State Key Laboratory of Frozen Soil Engineering,
Northwest Institute of Eco-Environment and Resources,
Chinese Academy of Science, Lanzhou, 730000, China
e-mail: guoyuli@lzb.ac.cn

A. Kallel et al. (eds.), *Recent Advances in Geo-Environmental Engineering, Geomechanics and Geotechnics, and Geohazards*, Advances in Science, Technology & Innovation, https://doi.org/10.1007/978-3-030-01665-4_66

limit, plasticity index and specific gravity are 18.24%, 26.29%, 8.05 and 2.70, respectively. The maximum dry density is 1.91 g/cm^3 corresponding to an optimal moisture content of 13%.

2.2 Sample Preparation

After the sample air-drying, the different types and contents of the stabilizers (calcium lignosulfonate, NaOH, and $Na_2O \cdot nSiO_2$) were added to the loess, and the mixture ratios of stabilizers are 0.5, 1.0, 2.0, 3.0, and 5.0%. The distilled water was prepared for every specimen based on the optimal water content of 13%. The calcium lignosulfonate was mixed with the loess directly, then added distilled water and stirred well. The NaOH and $Na_2O \cdot nSiO_2$ were dissolved in the distilled water to prepare the solution to be added into the loess. Finally the loess sample was stirred uniformly and compacted layer by layer to be cylindrical specimens with a diameter of 39.1 mm and a height of 80 mm according to the expected dry density and water content. These finished specimens were then placed in an open and dry room for 7 days.

2.3 Testing Methods

Atterberg limits test, uniaxial compression test, uniaxial loading and unloading test and X-ray diffraction tests were performed to study the effects of three stabilizers (calcium lignin sulfonate, NaOH, and $Na_2O \cdot nSiO_2$) on physical and mechanical properties macroscopically and microscopically. The Atterberg limits tests of loess specimens at a stabilizing agent content of 3% were carried out according to the related method (JTG E40-2007). The uniaxial compression tests were performed on an unsaturated triaxial apparatus of GDS to determine their strengths. The uniaxial cyclic loading and unloading tests were also carried out on GDS. First, they were loaded to 1 MPa (about 47.62% of the uniaxial compressive strength), and then unloaded to a stress of 0 MPa forming a loading-unloading cycle. The maximum number of cycle is six. The results of loading-unloading tests were used to analyze the stiffness and energy dissipation of the samples. The X-ray diffraction tests were carried out to qualitatively analyze the stabilization mechanisms of stabilizers by measuring phases of stabilized and unstabilized specimens using the X-ray diffractometer (XRD).

3 Results

3.1 Liquid and Plastic Limits

Following mixing NaOH at a content of 3%, the plastic and liquid limits of the soil specimens increased up to 21.30 from 18.24% and 30.10 from 26.29%, respectively. The plasticity index also increased up to 8.80 from 8.05. For the $Na_2O \cdot nSiO_2$ stabilized specimens at a same mixing content of 3%, the plastic and liquid limits also increased up to 21.50 and 29.20%, respectively. For the calcium lignosulfonate stabilized specimens, the plastic limit increased to 20.25%, and the liquid limit slightly decreased to 26.20%. Accordingly, the plasticity index decreased to 7.70.

3.2 Uniaxial Compressive (UC) Strength

Using the uniaxial compression tests, the UC strengths of the loess specimens solidified by three stabilizing agents were determined. After adding three stabilizers respectively, the UC strengths of the loess specimens have all been improved remarkably. With the increase of the mixing amount of stabilizing agents, the UC strength showed an increasing trend at different rates. When calcium lignosulfonate was mixed to reach up to 5% of the mixture, the UC strength of the solidified specimen increased to 11.8 MPa from the initial 2.12 MPa. At the same mixing ratio, the UC strengths of NaOH and $Na_2O \cdot nSiO_2$ solidified specimens were 10.2 and 7.3 MPa, respectively.

3.3 Energy Dissipation and Resilient Modulus

Based on the hysteresis loop area of cyclic loading-unloading curves, the energy dissipation could be obtained. A higher value means that more damage occurs in the interior of the soil. After one cycle of loading-unloading, energy dissipations were 33.30 J/m^3 for unsolidified specimen, 25.70 J/m^3 for calcium lignosulfonate solidified specimen, 20.95 J/m^3 for $Na_2O \cdot nSiO_2$ solidified specimen and 16.15 J/m^3 for NaOH solidified specimen, respectively. The internal damage of solidified soil under the condition of the same axial stress was smaller than that of the solidified soil. The resilient modulus of each hysteresis loop of the unsolidified loess was greater than that of the loess solidified with the calcium lignosulfonate but less than the loess solidified by NaOH and $Na_2O \cdot nSiO_2$.

3.4 Mineral Components

According to the results of the XRD tests, the main mineral components of loess are still quartz, feldspar, calcite, and dolomite before and after adding the stabilizers. The clinochlore and gaultite accounted for 9 and 11% in the untreated loess, respectively. But they were not found in the calcium lignosulfonate solidified loess. Calcium lignosulfonate produced a series of chemical reactions with clinochlore and gaultite. These reaction resulted in the increase of the contents of SiO_2 (quartz), $CaCO_3$ (calcite), and CaMg $(CO_3)_2$ (dolomite), raising the structure strength. When $Na_2O \cdot nSiO_2$ was added into the loess, a reaction occurred and the quartz content increased to 73%, also raising the strength. When NaOH was added, it reacted with SiO_2 and Al_2O_3 or aluminum silicate. And the solution of sodium silicate and sodium aluminate were formed in the loess. This reaction decreased the quartz content and increased the contents of muscovite, microcline, and other silicate minerals.

4 Conclusion

(1) The chemical reactions of calcium lignosulfonate, $Na_2O \cdot nSiO_2$, NaOH with mineral components were different. The first two decreased the loess plasticity, while the last one increased it.
(2) The calcium lignosulfonate, NaOH, and $Na_2O \cdot nSiO_2$ all increased the UC strength significantly and improved the loess engineering properties. The stabilization effect of calcium lignosulfonate was most prominent.
(3) The internal damages of the three solidified loess under the same axial stress was less than that of the unsolidified loess, and the energy dissipation of the NaOH solidified loess was the minimum one. Also, the stiffness of the NaOH and $Na_2O \cdot nSiO_2$ solidified loess was greater than that of the unsolidified loess, while the calcium lignosulfonate solidified loess was less than the pure loess.

References

1. Li, G., Ma, W., Zhao, S.: Effect of freeze-thaw cycles on mechanical behavior of compacted fine-grained soil. In: Brian, M., Guy, D. (eds.) Cold Regions Engineering 2012, pp. 72–81. American Society of Civil Engineers (2012)
2. Wu, Z.: Research on loess reinforced by modification sodium silicate. Lanzhou University, Lanzhou (2013)
3. Li, G., Ma, W., Mu, Y.: Process and mechanism of impact of freezing and thawing cycle on collapse deformation of compacted loess. China J. Highw. Transp. **24**(5), 1–5 (2011)
4. Sillanpaa, M., Webber, W.: The effect of freezing-thawing and wetting-drying cycles on soil aggregation. Can. J. Soil Sci. **41**(2), 182–187 (1961)
5. Rober, B.: Concentrated liquid stabilizers for railroad applications. In: Roadbed stabilization and ballast symposium. St. Louis, Missouri, USA (2000)
6. Tong, B., Li, Z.: Research progress in soil consolidator. J. Hefei Teach. Coll. **27**(3), 91–93 (2009)
7. Lv, Q., Wu, Z., Wang, S.: Mechanism of temperature-modification silicification grouted loess. Rock Soil Mech. **34**(5), 1293–1298 (2013)
8. Wang, Y., Yang, Z., Chen, W.: Strength characteristics and mechanism of loess solidified with new polymer material SH. Chin. J. Rock Mech. Eng. **24**(14), 2554–2559 (2005)
9. Zhang, H., Lin, C., Sheng, Y.: Experimental study of engineering properties of loess reinforced by consoled system. Chin. J. Rock Mech. Eng. **34**(supp. 1), 3574–3580 (2015)
10. Liu, S., Zhang, T., Cai, G.: Research progress of soil stabilization with lignin from bio-energy by-products. China J. Highw. Transp. **27**(8), 1–10 (2014)

Effect of Full Wrap-Around Ends of Geotextile on the Bearing Capacity of Sand

Safa Djeridi, Naima Benmebarek, and Sadok Benmebarek

Abstract

This paper presented the results of numerical model tests on the load-settlement of a strip footing on reinforced sand bed to investigate the effect of a new practical reinforcement technique using geosynthetic reinforcement to increase the load bearing capacity of shallow foundations on a sandy soil. In all previous works, the soil reinforcement method consisted of single or multiple horizontal layer(s) placed horizontally below the footing. This paper aimed to present the details of numerical analysis of strip footing resting on a reinforced sand bed with horizontal layers and two horizontal layers with full wrap-around ends of geotextile to increase the capacity of shallow foundations. Order to show more clearly the performance of the full wraparound ends of reinforcement compared to that of reinforcement placed horizontally, the effect of depth of geotextile on the increase of the bearing capacity was investigated.The results indicate that the two full wrap-around ends of the geosynthetic allowed a significant improvement in bearing capacity and space savings for building a reinforced sand bed and the proposed full wrap-around ends need less quantity of geotextile than planar geotextile placed horizontally.

Keywords

Geotextile • Reinforced sand • Strip footing • Settlement • Load bearing capacity

1 Introduction

The use of geosynthetic has revolutionized the field of ground improvement. Many researchers have studied the behavior of reinforced sandy soil beneath the foundations [1–3]. In the previous works reported in the literature, the geosynthetic reinforcement was used as a single or multi-layer laid horizontally within the soil mass underlying the footings. In order to investigate more powerful reinforcement techniques, a new construction technique using geosynthetic reinforcement with full wrap-around ends was proposed to improve the bearing capacity of shallow foundations and to provide space savings for building a reinforced sand bed. To develop a better understanding of the performance of the proposed technique, several series of numerical models were performed using FLAC code (2007) to evaluate the bearing capacity of a strip footing supported by reinforced sand with either planar layers and two fully wrap-around ends of geosynthetic.

2 Test Program and Description

The geometry of the different configurations of geosynthetic layers laid horizontally and horizontally with two full wrap-around ends in the sand bed is shown in Figs. 1 and 2, respectively. In previous studies we found the following typical parameters allowing the maximum benefit of reinforcement of strip or rectangular footing, resting on a bed of sand, reinforced with a multiple horizontal layers of geotextile/geogrid reinforcement: ($b/B = 6$ [1]); ($u/B = 0.25–0.5$ [1]); ($h/B = 0.2–0.4$ [2]).

Recently, the authors have performed numerical model tests to determine the effects of the introduction of fully wraparound ends of reinforcement on the load-settlement behavior of the reinforced soil bed [4]. These studies showed that the introduction of fully wraparound ends reinforcement requires a smaller width of the land for the construction of reinforced foundation bed, where $u'/B = 0.3$, $b'/B = 2$, $h'/B = 0.3$. In view of the earlier findings, the same parameter ratios were adopted in this study:

S. Djeridi · N. Benmebarek (✉) · S. Benmebarek
NMISSI Laboratory, Biskra University,
BP 145 07000 Biskra, Algeria
e-mail: naima_benmebarek@yahoo.com

A. Kallel et al. (eds.), *Recent Advances in Geo-Environmental Engineering, Geomechanics and Geotechnics, and Geohazards*, Advances in Science, Technology & Innovation, https://doi.org/10.1007/978-3-030-01665-4_67

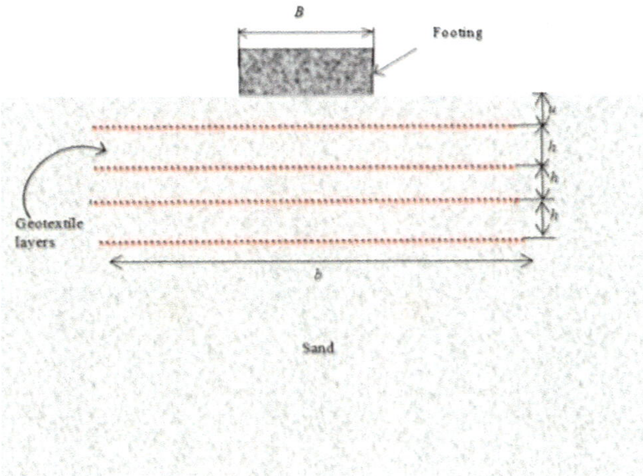

Fig. 1 Geometry of the planar reinforced foundation bed as reported in the past studies

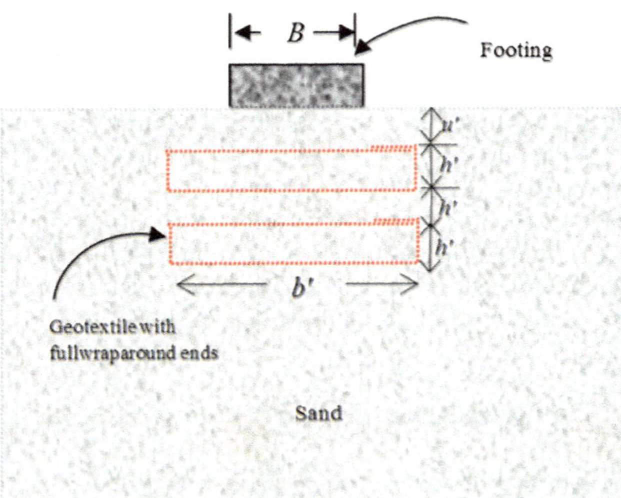

Fig. 2 Geometry of the proposed two full wrap-around ends reinforced foundation bed

- For reinforcement with planar layers: N = 4, b/B = 6, u/B = 0.3, h/B = 0.3.
- For reinforcement with fully wraparound ends: N = 2, b′/B = 2, u′/B = 0.3, h′/B = 0.3.

3 Results

To validate the FLAC numerical model, the effect of introducing four planar reinforcement layers into the soil was examined and compared with both experimental data available in literature. Figure 3 shows the present results compared to the laboratory model given by Kazi et al. [4] for unreinforced sand and N = 4, where the geotextile width and vertical spacing of reinforcement layers were kept constant as 6B and 0.3B, respectively. According to this figure, the numerical results obtained in this study seem reasonable and agree well with experimental results of Kazi et al. [4].

In order to show clearly the performance of the proposed two full wrap-around ends of reinforcement compared to that of reinforcement placed horizontally in increasing the bearing capacity, several series of numerical model were conducted using the FLAC code. The improvment of the load bearing capacity (q) with the number of horizontally reinforcement layers (N = 4) and two full wrap-around ends of geotextile is shown in Fig. 4. It can be seen that the two full wrap-around ends of geotextile provide a further improvement in the load-bearing capacity compared to the horizontally placed reinforcing layers. Therefore, it can be seen that the performance of the two full wrap-around ends of geotextile is more effective in increasing the bearing capacity of the footing than in planar geosynthetic reinforcements. The results show that the proposed full wrap-around ends of reinforcement reduces the amount of reinforcement used and saves the land space to build a reinforced sand bed system.

In order to determine the optimum depth of the full wrap-around ends of geotextile, the depth of the two full wrap-around ends of geotextile (u′) is varied between 0.1B and 0.7B, while h′/B value is kept constant at 0.3. The variation of load bearing capacity (q) versus settlement curves is plotted in Fig. 5. It can be seen that the case with u′/B = 0.2 gives highest improvement at all settlement values. Beyond u′/B = 0.2, the tests indicated that there was a slight decrease in the load bearing capacity. Therefore the maximum performance on bearing capacity was obtained at u′/B = 0.2.

4 Conclusion

From the overview of the previous work and the presented numerical analysis, the proposed two full wrap-around ends brings additional advantages than planar geosynthetic as: an additional improvement in the bearing capacity- high reduction in the needed quantity of the geotextile and space savings for building a reinforced sand bed system. The depth of the top layer of full wrap-around ends was found to be

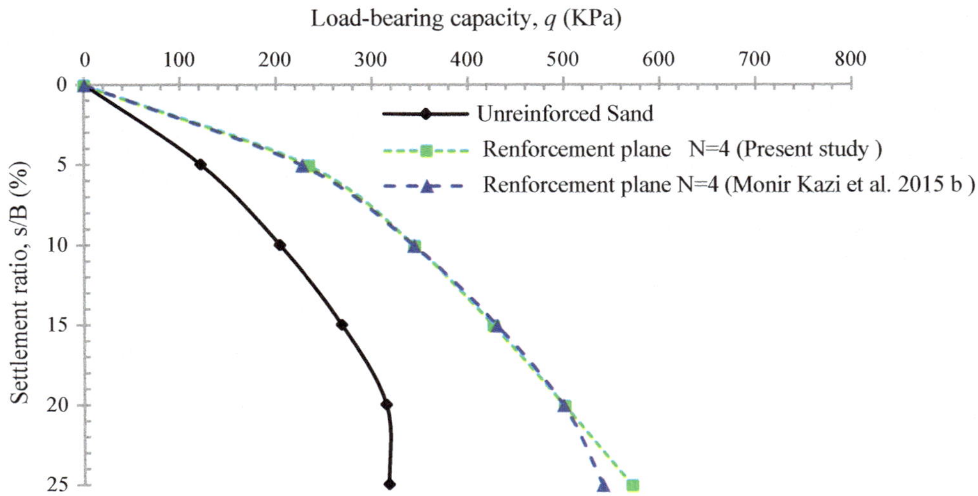

Fig. 3 Comparison of load-bearing capacity (q) versus settlement ratio (s/B) with experimental data

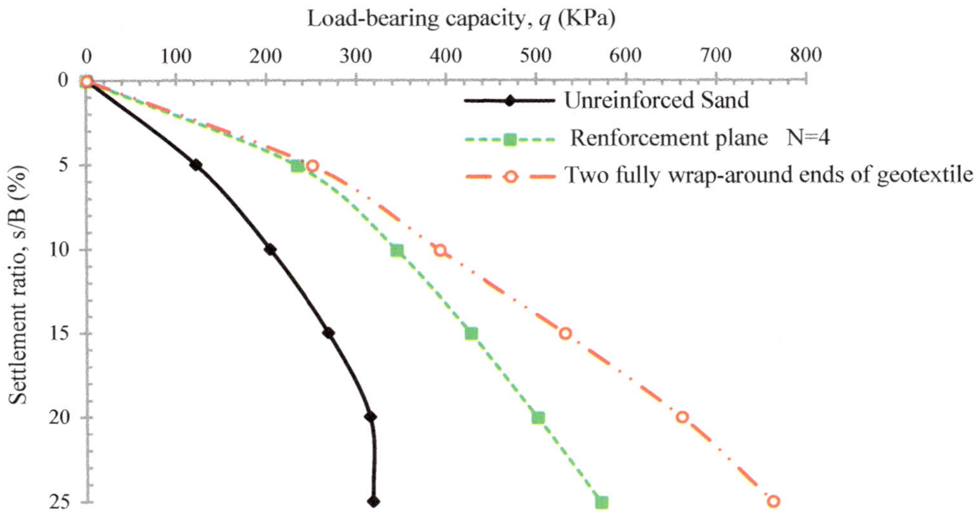

Fig. 4 Variation of load-bearing capacity (q) versus settlement ratio (s/B)—effect of two full wrap-around ends of geotextile

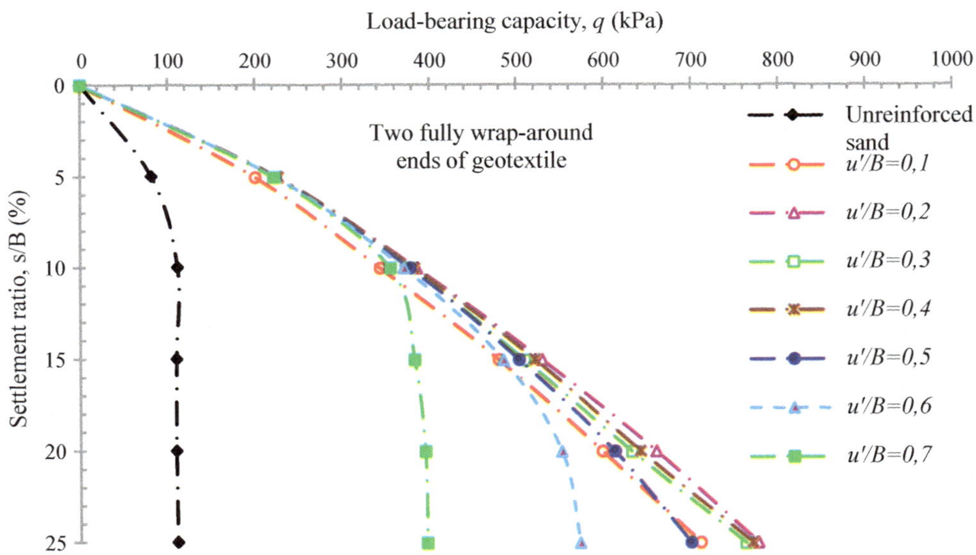

Fig. 5 Effect of the depth of the two full wrap-around ends of geotextile

about 0.2B; while for planar reinforcement may range between (0.3B–0.6B).

References

1. Khing, K.H., Das, B.M., Puri, V.K., Cook, E.E., Yen, S.C.: The bearing capacity of a strip foundation on geogrid-reinforced sand. Geotext. Geom. **12**(4), 351–361 (1993)
2. Yetimoglu, T., Wu, J.T.H., Saglamer, A.: Bearing capacity of rectangular footings on geogrid-reinforced sand. J. Geotech. Eng. **120**(12), 2083–2099 (1994)
3. Kazi, M., Shukla, S.K., Habibi, D.: Behaviour of embedded strip footing on sand bed reinforced with multilayer geotextile with wraparound ends. Int. J. Geotech. Eng. **9**(5), 437–452 (2015)
4. Benmebarek, S., Djeridi, S., Benmebarek, N., Belounar, L.: Improvement of bearing capacity of strip footing on reinforced sand. Int. J. Geotech. Eng. (1), 1–11 (2017)

A Direct Shear Investigation on the Determination of the Shearing Resistance of Reinforced Soil with Waste Rubber

Negadi Kheira and Arab Ahmed

Abstract

Disposal of Waste Rubber has an enormous impact on improvement of Environmental performance all over the World. Algeria is concerned with the problem of waste Rubber to achieve the maximum possible benefits in both of Economic and Environmental fields by solving this Issue. In recent years, there has been increasing interest in the use of recycled tires in civil engineering applications, particularly in soil improvement studies as a lightweight backfill. The aim of this Study is to get a clear vision of the effect of using Rubber grains on the shear strength behavior of Chlef River's sand. A series of direct shear test have been prepared in laboratory using mixtures of sand and rubber grains mixed randomly contained 5, 10, 15, 20 and 30% rubber grains by weight. The direct tests conducted with two relative densities of Dr = 55 and 95% and all the samples were tested under confining pressure of 100 kPa. The results show that the rubber grain content influences the shear strength characteristics of sand-rubber grain mixtures.

Keywords

Shear strength • Rubber grain • Sand • Direct shear test • Mixtures

1 Introduction

It is estimated that Algeria produces over than 26,000 tons of waste tires every year. The majority of that is presently landfill or stockpiled. This expends profitable landfill space pose a real danger to the fire and give a rearing ground to mosquitoes. Likewise, the expanding number of waste tires has represented a genuine danger to ecological security and general wellbeing endeavors lately.

Engineers and researchers are constantly endeavoring to use the waste tires as a sustainable solution for civil engineering problems as slope stability and reduce settlement risk. Still need exist to find additional and practical methods for geotechnical applications.

A variety of civil engineering applications in Algeria were based on the use of the waste tires as a lightweight fill by the technique "Pneusol". This applications ranging from slope stability and earth embankments and pavements and retaining structures and other applications [1].

Experimental results found by various researchers, conducted that using rubber pieces to the soil in geotechnical applications can increasing soil strength, improve the deformation characteristics of the backfill, and reduces the self-heating problem.

Many previous researches have worked with rubber tires as shredded or chips and they reached that we can employed as lightweight fill for embankment construction because of its high strength and low unit weight. Also can give a high shear strength and low horizontal stresses. Zornberg et al. [2] assessed the mechanical behavior of an experimental embankment fill used tire shreds and cohesive soil. The results indicate that the embankment sections built with tire shreds and cohesive soil showed satisfactory long-term performances during traffic exposure. Reference [3] have been used tire chips having size less than 20 mm to the rubber–soil mixture as a replacement. They reported improved frictional characteristics, increased resistance to dynamic loading and improved drainage characteristics of the soil.

In this research, we used a laboratory study involving direct shear test on rubber grain with different weight contents as reinforcement of Chlef sand. The objective of this study was to investigate the effect of waste rubber grains on the strength characteristics of the Clef sand; and the feasibility of utilizing waste rubber as lightweight backfill in geotechnical applications.

N. Kheira (✉) · A. Ahmed
Hassiba Benbouali University, 02000 Chlef, Algeria
e-mail: negadikh@ymail.com

A. Kallel et al. (eds.), *Recent Advances in Geo-Environmental Engineering, Geomechanics and Geotechnics, and Geohazards*, Advances in Science, Technology & Innovation, https://doi.org/10.1007/978-3-030-01665-4_68

2 Experimental Devise and Test Materials

The used experimental device in this study was a shear box with 60 × 60 cm to perform all the direct shear tests, we investigated ratios for rubber grains from 5, 10, 15, 20 and 30% by weight. The mixtures were mixed carefully and they were compacted and poured steadily into three parts on the shear box for both relative densities (55 and 95%).the confining pressure was kept at 100 kPa for all the tests.

The sand used in the mixtures of tire shred and sand namely Chlef River's sand. It is collected from the river of Chlef. Chlef city is approximately 223 km west of Algiers. Chlef sand is a medium sand with a rounded form (alluvial sand), with an average diameter of D50 = 0.612 mm. it has been classified as poorly graded sand. Rubber grains in size of 1–5 mm were used in this study. The coefficient of uniformity (Cu), coefficient of curvature (Cc), specific gravity (Gs), effective size (D10) and other engineering properties of the sand and rubber grains are presented in Table 1.

The alternative reinforcement material used in this study was rubber grains. Material tested several times with an average value of the specific gravity was found to be about 1.085, also it is to be clean and free of any Iron, Steel and rust (see Fig. 2). Other engineering properties of the two materials are presented in Table 1. The grain size distribution curves of the test materials can be seen in Fig. 1.

3 Experimental Results

3.1 Direct Shear Tests

The variation of the shear stress with horizontal displacement of rubber grains—sand mixture at 100 kPa confining pressure for the dense state (Dr = 95%) is shown in Fig. 3. It can be seen from this graphs that the effects of rubber grain contents on behavior of the mixtures is very clear in the dense mixtures, a peak shear stress is observed, demonstrating shear strength of the mixtures. The shear stress of the sand rubber grain mixtures still tends to increase slightly up to 20% rubber grain content. The mixtures at Dr = 95% show a typical dense sample behavior. The clear shear stress peak found here is similar to findings by [4], who performed laboratory triaxial tests on a mixture of 50% granular rubber mixed with 50% Ottawa sand by weight. They also obtained a clear peak for shear stress versus axial strain.

Figure 4 shows the relationship between the shear stress and horizontal displacement of sand rubber grain mixtures in the medium dense state (55%). The mixtures show typical average sand behavior with no drop in post-peak strength. It illustrates the influence of rubber grain contents on the shear strength of the reinforced sand. It clearly shows that existing more rubber grains in the sand leads to an improvement in the resistance. After that, increasing rubber grain beyond

Table 1 Materials properties used in this study

	Fc (%)	D10 (mm)	D50 (mm)	GS	Cu	Cc	e_{min}	e_{max}
Chlef sand	0	0.220	0.612	2.665	3.258	1.315	0.573	0.924
Rubber grain	100	–	–	1.085	1.136	1.0275	0.728	1.420

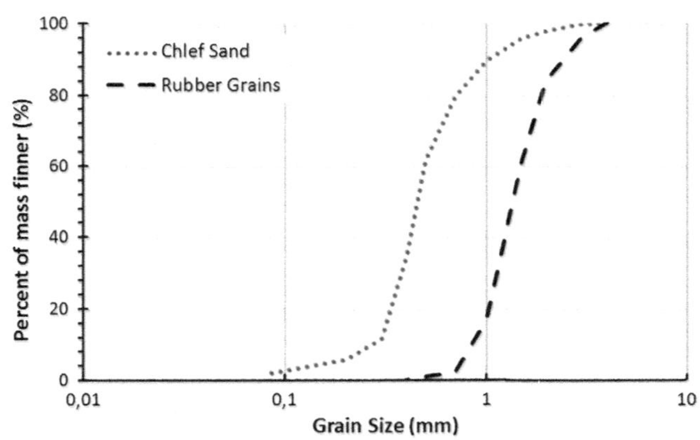

Fig. 1 Grain size distribution curves of Chlef sand and rubber grains

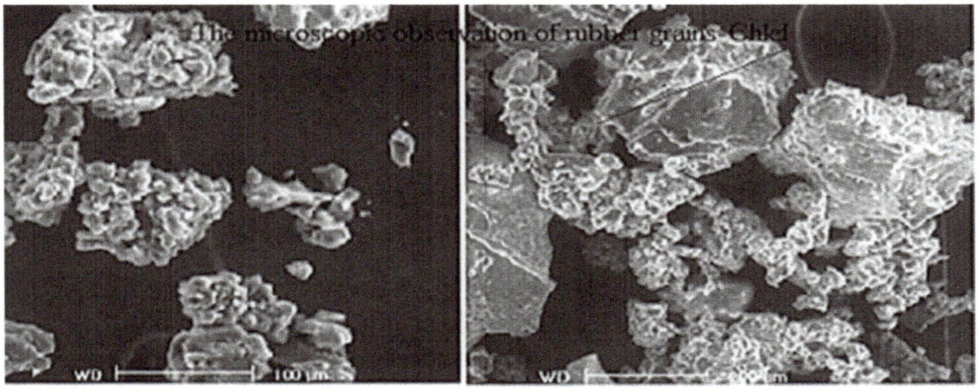

Fig. 2 The microscopic observation of rubber grains

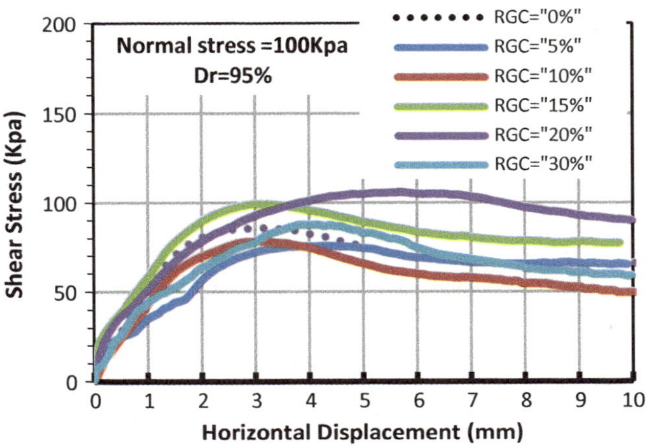

Fig. 3 Variation of shear stress versus horizontal displacement for rubber-sand mixtures at normal stress of 100 kPa

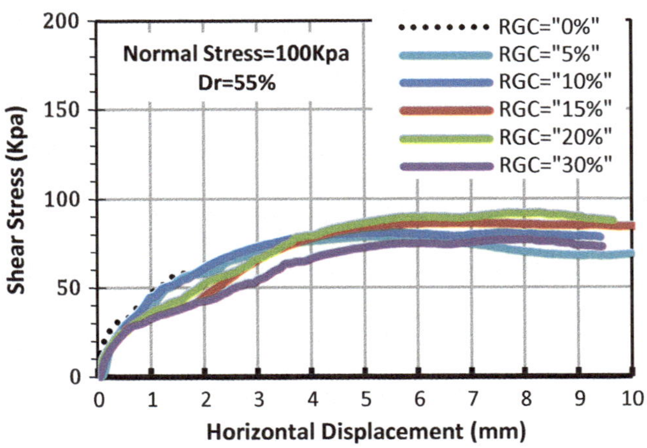

Fig. 4 Variation of shear stress versus horizontal displacement for rubber-sand mixtures at normal stress of 100 kPa

10 mm horizontal displacement led to decrease almost linearly with increasing horizontal displacement.

However, the rubber grains inclusion causes the shear strength to increase gradually. Hence, it can be concluded that the addition of rubber grains can enhance significantly the shear resistance of the mixtures. The increase in shear strength of mixtures with increasing rubber grain content can be attributed to the redistribution of the samples. Hence, the sand is pushed into the voids of rubber grains increasing the contact surface between sand particles and rubber grains. But adding further rubber grains beyond 30% to the sand is ineffective on the shear strength of rubber grain-sand mixtures. This is because, in shear zone, the rubber grains surround the sand grains and tend to roll and slide over them to make more voids. This is supported by findings of [5, 6].

3.1.1 Shear Strength Parameters

The specific gravity values of chlef-sand and rubber grains in the dense and medium dense states have been used in this study to determine the void ratios of mixtures with different relative densities using the following equation:

$$e = {}^{V_{void}}/_{V_{solid}} \tag{1}$$

$$e = \frac{V_m - \left[\frac{W_s}{G_s} + \frac{W_{RG}}{G_{RG}}\right]}{\frac{W_s}{G_s} + \frac{W_{RG}}{G_{RG}}} \tag{2}$$

where: e is the void ratio, V_{void} is the volume of void, V_{solid} is the volume of rubber grain-sand mixture, V_m is the volume of mold, W_{GR} is the weight of rubber grains, W_s is the weight of sand, G_s is the specific gravity of sand, and G_{GR} is the specific gravity of rubber grains.

As shown in Fig. 5, the void ratios at different relative densities of the mixtures of rubber grains and sand decrease linearly with increasing percentages of rubber grains. This decrease up to a point where its value is the smallest; this happens at a rubber grain content of 20% (by weight) for which the minimum and maximum void ratio reaches the minimum values of 0.442 and 0.775 respectively. This behavior can be explained based on the specific gravity G_s of the sand is larger than that of rubber grains. The specific gravity of the sand and rubber pieces is equal to 2.665 and 1.085, respectively. And then an increasing trend beyond 30% when the sand particles fill the voids between the

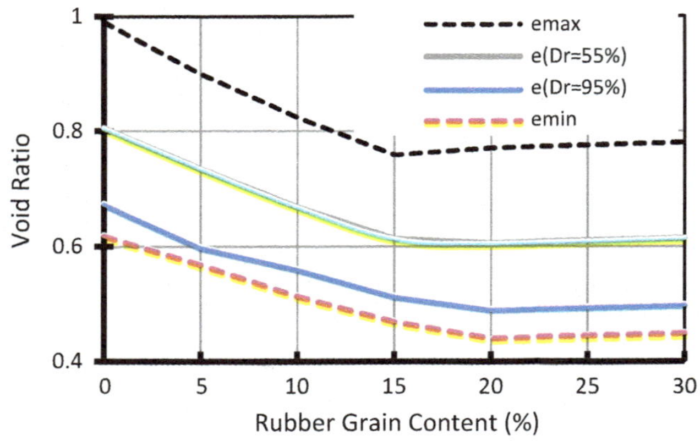

Fig. 5 Void ratios Index of the sand-rubber grain mixtures versus rubber grain content (σ_3 = 100 kPa)

rubber grains the void ratios of the mixtures increase, any further addition of rubber grains would increase the void ratio of mixtures. A similar trend was also reported by [7, 8].

Figure 6a, b show a comparison between both medium dense and dense state variations of the values of initial friction angle (φ) and apparent cohesion (c) for different sand-rubber mixtures at medium and dense state.

It was observed that by adding rubber grains, from 5 to 20% by weight, internal friction angle values were increased in both densities (55 and 95%). However, adding more rubber grains leads to a decrease in the internal friction angle.

Figure 6a, b also reveal that the cohesion at different relative densities of the mixtures follow an increasing trend from 5 to 20% rubber grain content and then a decreasing trend beyond 30%. It can be seen that the internal friction angle and the apparent cohesion of mixtures increase with increasing rubber grain contents for both relative densities, which can be attributed to the lower specific gravity of rubber pieces. Moreover, the interaction between sand particles and rubber grains leads to enhance of sand porosity. However, the fines found in sand in the dense state might penetrate into the voids between rubber grains, which could cause an increase in the apparent cohesion. Greater percentage of rubber grain is uniformly distributed in the mixtures resulting in more voids. These voids cannot be filled with sand since the amount of sand is limited. This is similar to findings of [6–11].

The obtained results indicate that the sand rubber grain mixture effectively works as a lightweight material for geotechnical applications, such as embankments, retaining wall backfill, and foundations [8–10]. Moreover, the use of scrap tires in geoengineering applications solves the problem of disposal to some extent and promotes recycling tires, which also minimizes the requirement of other construction materials providing economic benefits. The use of tire-derived material in geotechnical applications will create a sustainable future.

4 Conclusions

Direct shear tests were carried out on different sand-rubber grain mixtures in this study, and the Shear strength characteristics of mixtures have been investigated in order to use them for geoengineering applications as lightweight backfill material. The results of this research led to the following conclusions:

- The maximum and minimum void ratios decrease with increasing rubber grain contents in mixtures, which indicates better compressibility behavior. The obtained results reveal that the sand rubber grain mixtures effectively works as a lightweight material for geotechnical applications, such as embankments, retaining wall backfill, and foundations.
- The shear resistance of the mixtures increase in a regular manner with increasing rubber percentage up to 30% in the mixtures.
- The improvement in shear strength increases with increasing relative density.
- The initial friction angle of the mixtures first increases with increasing rubber grains contents and then decreases slightly with the addition of more rubber grains.
- It can be concluded that if the friction angle of the sand-rubber grains is critical, an addition of 10–20% rubber to the sand is optimal.

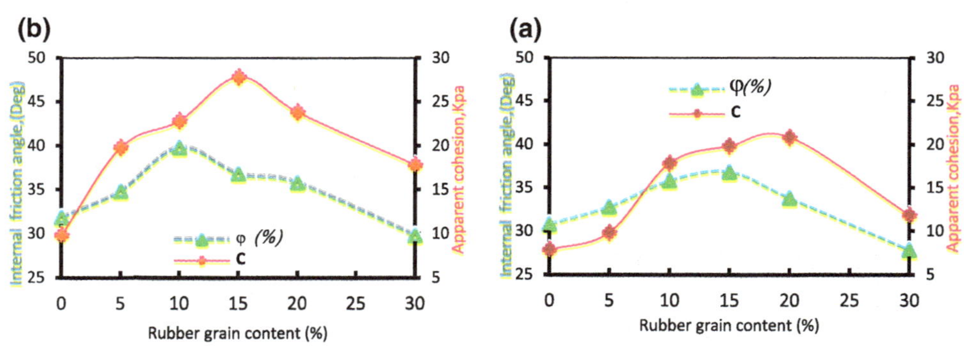

Fig. 6 **a** Variations of internal friction angle with rubber grain content at Dr = 55%, **b** Variations of internal friction angle with rubber grain content at Dr = 95%

References

1. Belabdelouahab, F., Taki, M., Djidjeli, Z., Mahiouz, M.: Large scale experimentation of soil slope stability tire. In: Inter Build CICC-Cairo, Egypt, 21–25 June (2007)
2. Zornberg, J.G., Costa, Y.D., Vollenweider, B.: Performance of prototype embankment built with tire shreds and nongranular soil. Transportation Research Record 1874, National Research Council, Transportation Research Board., Washington, D.C. pp. 70–77 (2004)
3. Bosscher, P.J., Edil, T.B., Eldin, N.: Construction and performance of shredded waste tyre test embankment, Transportation Research Record No. 1345, Transportation Research Board., Washington, D.C., pp. 44–52 (1992)
4. Edil, T.B., Bosscher, P.J.: Engineering properties of tire chips and soil mixtures. Geotech. Test. J. **17**(4), 453–464 (1994)
5. Lade, P.V., Liggio, C.D., Yamamuro, J.A.: Effects of nonplastic fines on minimum and maximum void ratios of sand. Geotech. Test. J. **21**(4), 336–347 (1998)
6. Humphrey, D.N., Katz, L.E., Blumenthal, M.: Water quality effects of tire chip fill placed above the groundwater table. In: Wasemiller, M.A., Hoddinott, K.B. (eds.) Testing Soil Mixed with Waste or Recycled Materials, ASTM STP 1275, pp. 299–313 (1997)
7. Ahmed, I., Lovell, C.: Use of rubber tires in highway construction. In: Utilization of Waste Materials in Civil Engineering Construction, ASCE, New York, N.Y., pp. 166–181 (1993)
8. Bosscher, P.J., Edill, T.B., Kuraoke, S.: Design of highway embankments using tire chips. J. Geotech. Geoenviromental. Eng. ASCE **123**(4), 297–304 (1997)
9. Humphrey, D.N., Sandford, T., Gribs, M., Manion, W.: Shear strength and compressibility of tire chips for use as retaining wall backfill. Transportation Research Record No. 1422, Transportation Research Board, Washington, D.C., pp. 29–35 (1993)
10. Masad, E., Taha, R., Ho, C., Papagiannakis, T: Engineering properties of tire/soil mixture as a lightweight material. Geotech. Test. J. **19**(3), 294e304 (1996)
11. Ahmed, I.: Laboratory study on properties of rubber-soils. Ph.D. thesis, School of Civil Engineering, Purdue University, West Lafayette (1997)

Bearing Capacity Behavior of the Clay and Sand Interface Reinforced with Geotextiles

Said Nouri, Amar Nechnech, and Maria de Lurdes Costa Lopes

Abstract

The suitability and stability of soil is usually evaluated before its use in construction of pavement. A proper analysis is necessary to ensure that public works infrastructures such as roads, rails, dams, etc. remain safe to withstand settlement and collapse. The use of geosynthetic materials to improve the bearing capacity and settlement performance of subgrade pavement has gained attention in the field of geotechnical engineering. The present study investigated the improvement in the bearing capacity of the clay and sand interface reinforced with geotextile sheets placed at different depths in the sand layer. The results show that the bearing capacity increased significantly with the increased geotextile sheets. The bearing capacity for the soil increased with an average of 53% using one geotextile sheet at interface of soils with 20% depth of sand layer and the bearing capacity increased with an average of 32% while using one geotextile in middle of the sand layer. The penetration resistance and the CBR values increased when the soils reinforced with a one geotextile sheet. At a depth of 20%, the penetration resistance is 98 kN and the CBR increased up to 3.9% compared to the unreinforced one.

Keywords

Improvement • Geosynthetics • CBR • Sand • Clay

S. Nouri (✉)
Hassiba Benbouali University of Chlef, BP 78C 02180
Ouled Fares, Chlef, Algeria
e-mail: s.nouri@univ-chlef.dz

A. Nechnech
University of Science and Technology Houari Boumediene of Algiers, BP 32 16111 El Alia, Bab Ezzouar, Algeria

M. de Lurdes Costa Lopes
University of Porto, Rua Dr. Roberto Frias, 4200-465 Porto, Portugal

1 Introduction

The bearing capacity has always been one of the most interesting topics in public works, in particular in the soil foundations of pavement. In many cases of road construction, low resistant soft clay layers may be encountered. To solve the problems associated with these low bearing capacity soil, one of the solutions is to replace some of the soil under the soil foundations with a sand reinforced with geosynthetics. The use of geosynthetics in the field of public works has developed rapidly in recent years. The use of geosynthetics enhanced the solution of a very wide range of road problems, due firstly to their ease of implementation and their low cost compared to the available conventional solutions [1, 2].

CBR tests are also carried out by introducing geotextiles and geogrids in a granular soil [3–5].

The main objective of this work was to determine the appropriate reinforcement mode for a layer of sand overlying soft clay, so as to increase the bearing capacity and reduce the settlement of a foundation of pavements. The present study investigated the bearing capacity of two layers of soil (i.e., a thin sand layer underlain by clay) with varying one geotextile sheet depth in the sand by keeping the soils properties constant. The conclusion of these results has shown that the introduction of geotextile sheets significantly increases the penetration resistance and therefore the bearing capacity and reduces the settlement of the base layer.

2 Materials and Experimental Program

Sand, clay and geotextile were used for the experimental investigations. Sand of Chlef (Algeria) is a uniform sand composed of 70% of medium sand and 30% of fine sand with an average particle diameter $d_{50} = 0.55$ mm, a uniformity coefficient $C_u = 2.52$ and a coefficient of curvature

A. Kallel et al. (eds.), *Recent Advances in Geo-Environmental Engineering, Geomechanics and Geotechnics, and Geohazards*, Advances in Science, Technology & Innovation, https://doi.org/10.1007/978-3-030-01665-4_69

Table 1 Properties of the used geotextile

Properties	Value	Units
Ultimate tensile strength	12–14	kN/m
CBR	1.75	kN
Mass per unit area	155	g/m^2
Thickness	1.50–1.70	mm
Strain at maximum strength	70–90	%

$C_c = 1.04$. The density of the solid particles is $\gamma_s = 2.6$ g/cm^3. The optimal water content $W_{op}t = 9.8\%$, the maximum dry density $\gamma_{dmax} = 1.73$ g/cm^3. The clay used in this study is a Kaolin of Jijel, its mineralogical composition is of 8% Quartz and 73% Kaolinite. It presents important features $d_{60} = 2.5$ μ, $d_{40} = 1$ μ, $d_{30} = 0.5$ μ.

A woven type polypropylene geotextile was used as reinforcement. Table 1 shows the properties of the geotextile reinforcement.

A CBR test was used to show the effect of depth of geotextile sheet on the behavior of the clay and sand interface reinforced with geotextile. A total number of 6 tests were carried out for unreinforced soils and reinforced with a single layer of geotextile. To reinforce a sample, the geotextile was placed in a single sheet at different positions: 20, 40, 60 and 80% of the specimen height from the top surface.

3 Results

3.1 Load-Penetration Behavior

The versus of the load-penetration curve for sand-clay and sand-geotextile-clay is illustrated in Fig. 1.

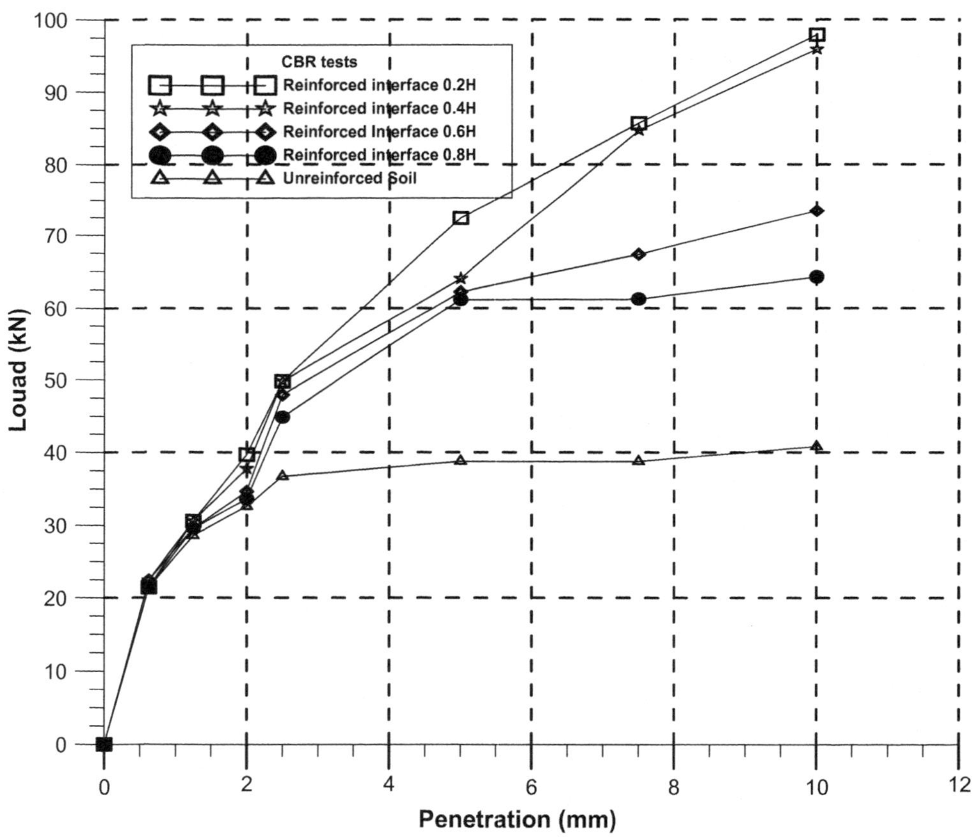

Fig. 1 Curve load-penetration unreinforced and reinforced soil systems

Table 2 CBR values for different positions of geotextiles

No test	Positions of geotextile	CBR
1	Unreinforced	2.71
2	0.2H	3.9
3	0.4H	3.57
4	0.6H	3.52
5	0.8H	3.05

3.2 Reinforcement Ratio

Based on the reinforcement ratio [6], the results of CBR values in different test conditions are shown in Table 2.

4 Discussion

From the force -penetration curve of Fig. 1, an increased resistance to penetration is clearly observed when the bilayer sand-clay soil is reinforced by the woven geotextile connected at the interface of the sand-clay. The highest value was achieved when the geotextile was placed at a depth of 20% of the height of the specimen.

The bearing capacity for the soil increased with an average of 53% using one geotextile sheet at interface of soils with 20% depth of sand layer and the bearing capacity increased with an average of 32% while using one geotextile in the middle of sand layer. The penetration resistance and the CBR values increased when the soils were reinforced with one geotextile sheet.

From Table 2, it is noted that the reinforcement ratio is higher in all the tests, which indicates that the introduction of the geotextile offers good resistance even to a lower penetration. It is clear that a considerable increase in the CBR value of the soil with a reinforcement of geotextile, for example, in the case of unreinforced soil, the CBR value is 2.71%, while with a reinforcement by geotextile the CBR value increased to 3.9%. The greatest increase in the CBR value was achieved when the geotextile sheet was placed 20% below the top of the sample.

5 Conclusion

The important results of this study are summarized below:

- The CBR value of a soil increases by 13–45% when it is reinforced with a single sheet of geotextile. The amount of improvement depends on the geotextile position.
- The force-penetration behavior of the bilayer soil under static load conditions is greatly improved when the geotextile is placed at the optimum position.
- The stress-strain behavior of soils under static load condition improved considerably when the geotextile was provided at an optimum position.

References

1. Giroud, J.P., Noiray, L.: Geotextile-reinforced unpaved road design. J. Geotech. Div. ASCE **107**(GT9), 1233–1254 (1981)
2. Basu, G., Roy, A.N., Bhattacharyya, S.K., Ghosh, S.K.: Construction of unpaved rural road using jute-synthetic blended woven geotextile—a case study. Geotext. Geomembr. **27**, 506–512 (2009)
3. Naeini, S.A., Moayed, R.Z.: Effect of plasticity index and reinforcement on the CBR value of soft clay. Int. J. Civ. Eng. **7** (2), 124–130 (2009)
4. Nair, A.M., Latha, G.M.: Bearing resistance of geosynthetic reinforced soil-aggregate systems. Ground Improv. **164**(2), 83–95 (2011)
5. Kumar, P.S., Rajkumar, R.: Effect of geotextile on CBR strength of unpaved road with soft subgrade. Electronic J. Geotech. Eng. **13**(J), 1355–1363 (2012)
6. Koerner, R.M.: Designing with Geosynthetics, 5th edn. Prentice Hall, New Jersey, pp. 184–186 (2005)

Improvement of Geotechnical Properties of Aged Municipal Solid Wastes Using Dredged Sand and Portland Composite Cement

Md. Maruful Hoque, M. Tauhid Ur Rahman, and Md. Shoriful Alam Mondal

Abstract

Closed municipal solid waste (MSW) landfill site reclamation is an opportunity for post closure and other infrastructure development in land scarce Bangladesh. But the task is challenging due to a lot of uncertainty and lack of study in the geotechnical properties of aged MSW. The test samples were prepared by mixing with a mixing ratio of 4:1 of dredged sand and Portland composite cement (PCC) type CEM II (Bangladesh standard BDS EN 197-1:2003) mixer with the 5, 10 and 15% of dry weight of aged municipal solid waste collected from *Matuail*, the largest MSW landfill site of Bangladesh. The unit weight varied from 6.5 to 13 kN/m^3. It has been showed that by increasing dredged sand-PCC mixer from 0 to 15% (0–20% dredged sand), the cohesion intercept increased from 9 to 35 kPa and the internal friction angle increased from 25° to 45°. The other two important geotechnical properties that are the compression and secondary compression indices decreased from 0.35 to 0.20 and 0.060 to 0.030, respectively. By increasing the additives of dredged sand and PCC, the shear strength, compression and mechanical properties of aged MSW samples are increased substantially.

Keywords

Dredged sand • PCC • Municipal solid waste • Ground improvement • Geotechnical properties

Md.M. Hoque (✉)
Climate Change Lab, Department of Civil Engineering, Military Institute of Science and Technology, Mirpur, Dhaka, 1216, Bangladesh
e-mail: marufpwd@gmail.com

M. T. U. Rahman
Department of Civil Engineering, Military Institute of Science and Technology, Mirpur, Dhaka, 1216, Bangladesh

Md.S. A. Mondal
Department of Civil Engineering, Bangladesh University of Engineering and Technology, Dhaka, 1205, Bangladesh

1 Introduction

Many municipal solid waste landfill sites are closed in Bangladesh. Due to scarcity of land near urban areas, it is highly demanded to use those sites for post closure and other infrastructure development. With the improvement of geotechnical properties of aged waste, the landfill site reclamation can be possible easy and economical [1]. The unpredictable heterogeneity in waste composition; presence of inorganic substances like glass, metal; unpredictable differential settlement, high moisture content and low shear strength etc. of landfill aged solid waste make closed landfill site development, a risky and challenging task [2].

The present study aimed to improve the geotechnical properties of aged municipal solid waste at landfill site using dredged river sand and Portland composite cement (PCC) in varying ratio.

2 Materials and Methods

The aged solid waste samples were collected from the first sanitary landfill of Bangladesh in the Northeast of capital Dhaka. The test samples were prepared by mixing, with a mixing ratio of 4:1 of dredged river sand, collected from the nearby the river *Buriganga* and Portland composite cement (PCC) type CEM II (Bangladesh standard BDS EN 197-1:2003) mixer, with the 5, 10 and 15% of dry weight of aged MSW. The largest MSW particle should be 1/6th of specimen used for geotechnical experiment which is smaller than 9 mm. Then the soil like MSW mixer was mixed thoroughly suing a mechanical mixer. Dry mixing method was applied due to the high moisture content of aged solid wastes and tendency of PCC to absorb water. The major geotechnical properties of MSW that were examined for the samples were waste composition, sample visual classification, sieve analysis, Atterberg's limits, compaction, direct shear test and consolidated-drained (CD) tri-axial test.

A. Kallel et al. (eds.), *Recent Advances in Geo-Environmental Engineering, Geomechanics and Geotechnics, and Geohazards*, Advances in Science, Technology & Innovation, https://doi.org/10.1007/978-3-030-01665-4_70

3 Results

The strength, compressibility and mechanical properties were found from literature summarized in Table 1. The experimental results found a substantial improvement in geotechnical properties of landfill aged waste samples. Metal, stone, polythene, rubber, fabric and waste residues were the major components of the aged MSW samples. The unit weight varied from 6.5 to 13 kN/m^3. It has been shown that by increasing dredged sand-PCC mixer from 0 to 15% (0–20% dredged sand), the cohesion intercept increased from 9 to 35 kPa and the internal friction angle increased from 25° to 45°.

By adding 15% dredged sand-PCC mixer into the waste samples, the peak strength, the brittleness index and the Young's modulus at failure increased from 500 to 1300 kPa, 0.10–0.38 and 4.5 to 30 MPa respectively, under an effective confining pressure of 300 kPa. The other two important geotechnical properties the compression and secondary compression indices decreased from 0.35 to 0.20 and 0.060 to 0.030, respectively.

4 Discussion

The absence of wood, paper, fiber and other organic matter indicate high waste residue. Compared to other landfill wastes of different countries, waste residues are much higher in Matuail Landfill site. The composition is an influential factor for determining the geotechnical properties of the aged waste. The values of unit weight of the other old landfills from literature are higher due to different waste composition, geographical locations, food habits and economic status. By increasing the additives of dredged sand and PCC, the shear strength and stiffness of MSW samples are increased. Along with the internal friction angle and cohesion intercept, the residual friction is also increased, on the contrary the residual cohesion intercept is not significantly changed. Due to increasing the cementing properties of stabilized waste samples, the brittleness increasing gradually and the peak strength also increased. As the dredged sand-PCC mixer addition increased the resistive capacity of aged wastes increased consequently the compressive properties decreased. The reduction of compression and secondary compression indices due to the cementation and pozzolanic properties of dredged sand-PCC mixer.

5 Conclusion

The chemical ground improvement technique was shown to significantly increase the compressive strength, shear strength properties, the brittleness and stiffness index but to decrease the settlement property of the aged landfill municipal solid waste samples. The chemical stabilization of landfill municipal waste using locally available dredged sand and PCC mixer will be useful and effective in the post closure development of landfill sites. The outcomes of this

Table 1 Shear strength, compressibility and mechanical properties of MSW [1, 3–13]

	Shear strength (Direct shear test)		Compressibility property	Triaxial test (CU)	
	Friction angle (°)	Cohesion (kPa)	Compression ratio (C_c)	Friction angle	Cohesion
Sowers [3]			0.1–0.41		
Landva and Clark [4]	24–42	16–23	0.17–0.24		
Gabr and Valero [5]	20–39	0–28	0.15–0.22	34	17
Wall and Zeiss [6]			0.21–0.25		
Jones et al. [7]	31	10.5			
Eid et al. [8]	35	0–50			
Pelkey et al. [9]	29	0			
Hossain [10]	24–32		0.16–0.25		
Caicedo et al. [14]	23	78			
Hossain et al. [15]			0.16–0.37		
Anderson et al. [11]			0.17–0.23		
Durmusoglu et al. [12]			0.13–0.23		
Reddy et al. [13]	26–30	31–64	0.24–0.33	12	32
Feng et al. [2]	15.7–21.9	29.1–19.3			
Pandey et al. [16]	17.5–22	27–53		7–11.5	34–46

research study may help for further foundation design calculation required in the construction of structures on closed landfill sites as well as its slope stability analysis.

References

1. Fatahi, B., Khabbaz, H.: J. Soils Sediments **13**, 1201 (2013). https://doi.org/10.1007/s11368-013-0720-4
2. Feng, S.J., Gao, K.W., Chen, Y.X., Li, Y., Zhang, L.M., Chen, X.: Geotechnical properties of municipal solid waste at Laogang Landfill, China. Waste. Manage. (2016). http://dx.doi.org/10.1016/j.wasman.2016.09.016
3. Sowers, G.F.: Settlement of waste disposal fills. In: Proceedings of 8th International Conference on Soil Mechanics Foundation Engineering, Moscow (1973)
4. Landva, A.O., Clark, J.I.: Geotechnics of waste fill. Geotechnics of waste fills theory and practice, ASTM STP 1070. In: Landva, A., Knowles, G.D. (eds.) American Society for Testing and Materials. Philadelphia, Pennsylvania, pp. 86–103 (1990)
5. Gabr, M.A., Valero, S.N.: Geotechnical properties of municipal solid waste. ASTM. Geotech. Test. J. **18**(2), 241–251 (1995)
6. Wall, D.K., Zeiss, C.: Municipal landfill biodegradation and settlement. J. Environ. Eng. **121**(3), 214–224 (1995)
7. Jones, D.R.V., Taylor, D.P., Dixon, N.: Shear strength of waste and its use in landfill stability. In: Proceedings of the Geoenvironmental Engineering Conference, Thomas Telford, pp. 343–350 (1997)
8. Eid, H.T., Stark, T.D., Evans, W.D., Sherry, P.E.: Municipal solid waste slope failure. I: Waste and foundation soil properties. J. Geotech. Geoenviron. Eng. **126**(5), 397–407 (2000)
9. Pelkey, S.A., Valsangkar, A.J., Landva, A.: Shear displacement dependent strength of municipal solid waste and its major constituents. Geotech. Test. J. **24**(4), 381–390 (2001)
10. Hossain, M.S.: Mechanics of compressibility and strength of solid waste in bioreactor landfills. Dissertation, Doctor of Philosophy, Department of Civil Engineering, North Carolina State University at Raleigh, USA (2002)
11. Anderson, E.O., Balanko, L.A., Lem J.M., Davis, D.H.: Field monitoring of the compressibility of municipal solid waste and soft alluvium. In: Proceedings 5th International Conference on Case Histories in Geotechnical Engineering, New York (2004)
12. Durmusoglu, E., Sanchez, I.M., Corapcioglu, M.Y.: Permeability and compression characteristics of municipal solid waste samples. Environ. Geol. **50**, 773–786 (2006)
13. Reddy, K.R., Hettiarachchi, H., Parakalla, N., Gangathulasi, J., Bogner, J.: Geotechnical properties of fresh municipal solid waste at Orchard Hills Landfill, USA. Waste Manage. **29**(2), 952–995 (2009)
14. Caicedo, B., Yamin, L., Giraldo, E., Coronado, O.: Geomechanical properties of municipal solid waste in Dona Juana sanitary landfill. Environ. Geotech. (4th ICEG), De Mell & Almeida (des). (2002) Swets & Zeitlinger, Lisse. ISBN 90 5809 5010
15. Hossain, M.S., Gabr, M.A., Barlaz, M.A.: Relationship of compressibility parameters to municipal solid waste decomposition. J. Geotech. Geoenvironmental Eng. **129**(12), 1151–1158 (2003)
16. Pandey, R.K.: Investigation of shear strength properties of municipal solid waste and slope stability analysis. Int. J. Res Appl. Sci. Eng. Technol. **V**(IX), 491–496 (2017)

Valorization of Foundry Green Sand in Road Construction

Hadj Bekki, M. Yacine Haouachine, and Yacine Aouci

Abstract

In foundry industry, there are two important processes used to manufacture metallic parts: in the first one, permanent molds, typically metallic, are used to produce simple form of parts; and in the second, sands are used as lost molds and cores to produce complicated forms of parts which make their use very adaptable for large scale applications. However, the latter process generates significant quantities of sand waste because only small ones are recycled in factory, that is why it is very interesting to valorize these Spent Foundry Sands (SFS) in several other industrial sectors. The objectives of the present work were to valorize/reuse the foundry sands in road construction projects. It has been observed that the addition of hydraulic binders, leads to a significant increase in the CBR (California Bearing ratio) index up to 135% with the addition of 2% of lime and up to 75% with the addition of 2% of cement.

Keywords

Spent foundry sands (SFS) • Valorization • Treatment • Road construction

1 Introduction

The world foundry sector generates about 100 million tons of spent foundry sand (SFS) each year [1]. Although the sand is reused many times in the factory before the repeated exposure to the molten metal and mechanical abrasion. This makes it undesirable for its use in the foundry process [2].

In Algerian foundries, the SFS are stored continually on the plant site, which poses a great problem for the company in the management of such waste material. On the other hand, they may cause a danger to the environment (contamination of air and ground water); in addition to the ecological risks that may occur. Therefore it is very important to valorize these spent sands in several other industrial sectors.

The largest volume of foundry sand is used in civil applications, such as embankments, landfills, concrete, clay bricks and road construction [3, 4]; yet each application requires some precautions and also contains some limitations in the reuse of these SFS, mainly in their mechanical properties.

In this paper, only the valorization (reuse) of the foundry green sand in road bases was developed.

2 Materials and Methods

The base sand used in the foundry of Tiaret in Algeria was collected from Adwan Chemicals Company located nearby Mostaganem city.

The clay used for molding is a bentonite which was collected from the deposit of Hammam Boughrara within the area of Maghnia city, Algéria. It is a white sodium bentonite.

After the use of sand in the foundry molding process, the material was collected and subjected to some identification tests in order to determine its classification.

Table 1 shows index characteristics of the studied material. The Particle size distribution curve of Green foundry sands is presented in Fig. 1.

Following these results and according to GTR guide [5], the material is a Class B sand, a sand and gravel material with fine and a subclass B6, which means that it is a sandy clay to very clayey sand. The XRD analysis reveals that the green foundry sand is mainly composed of silica (Fig. 2).

Considering that the material is a clayey soil and in order to improve its properties for its use in the road bases, it was preferred to treat the material with hydraulic binder, such as lime or cement. Seeing that there are several similarities between lime and cement as stabilizers, it was important to

H. Bekki (✉) · M. Y. Haouachine · Y. Aouci
University of Tiaret, Zaâroura, BP 78 14000 Tiaret, Algeria
e-mail: abekkihadj@yahoo.fr

A. Kallel et al. (eds.), *Recent Advances in Geo-Environmental Engineering, Geomechanics and Geotechnics, and Geohazards*, Advances in Science, Technology & Innovation, https://doi.org/10.1007/978-3-030-01665-4_71

Table 1 Main index properties of waste green sand

Identification	Green sand	Identification	Green sand
Dmax (mm)	5	$CaCO_3$ (%)	0
% passing over sieve 80 microns (%)	21.58	ES (%)	15
Plastic index (PI)	Immeasurable	Specific density (g/cm^3)	2.57
VBS (ml/g)	1.64	Density (g/cm^3)	1.26
Cu	50	Cc	22

VBS Methylene Bleu Value (Retention); *ES* Equivalent sand Test

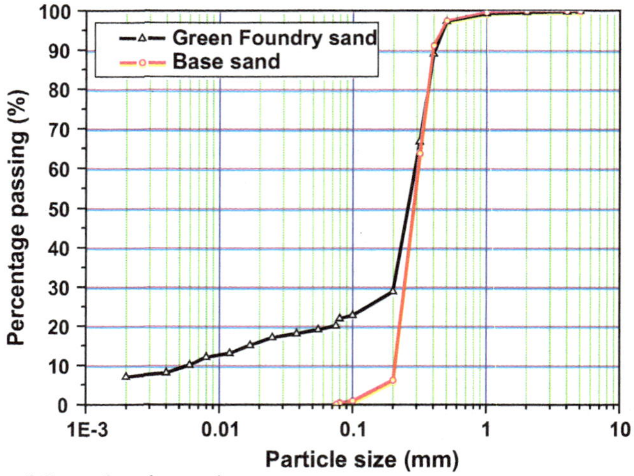

Fig. 1 Particle size distribution curve of Green foundry sands

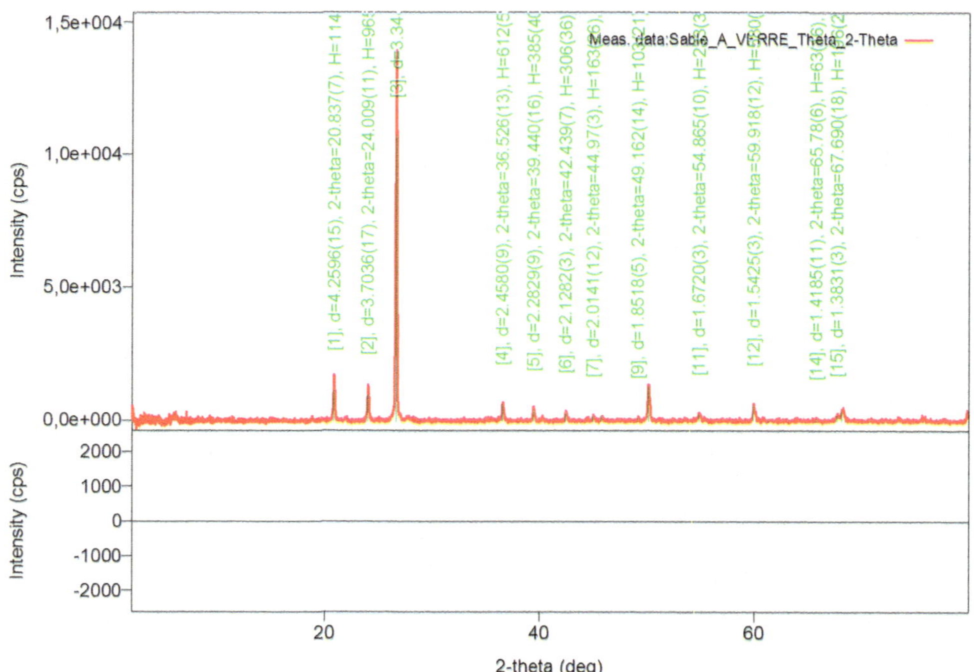

Fig. 2 XRD analysis of Green foundry sand

use both binders and compare them. In addition, the stabilization of the waste foundry sand with cement renders the material nontoxic [6].

The experimental study focused on the bearing tests, mainly Proctor test (compaction characteristics) and Californian Bearing Ratio test (CBR). The tests procedures and

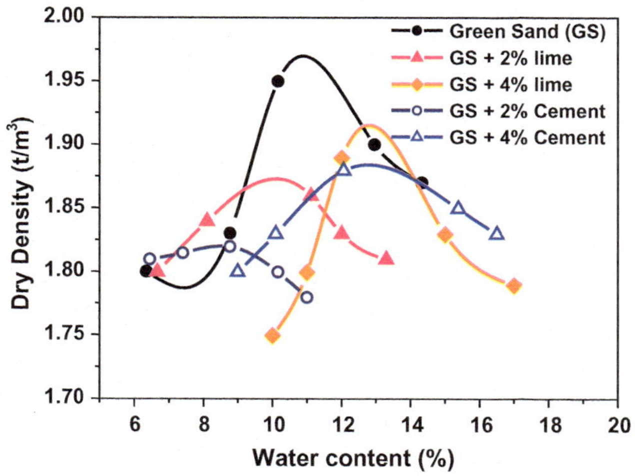

Fig. 3 Compaction curves of untreated compacted- and treated stabilized-green sand with lime/cement

the preparation of the specimens were carried out according to European and French test standards [7, 8].

3 Results

The results of the modified Proctor test conducted on different samples of SFS treated with lime/cement (at frequently used dosages) and the untreated samples are plotted in Fig. 3.

CBR-Values of green sand waste, soil-lime and soil-cement mixtures prepared at optimum water content (OMC obtained by modified Proctor Test) and at different compaction energy, are given in Fig. 4.

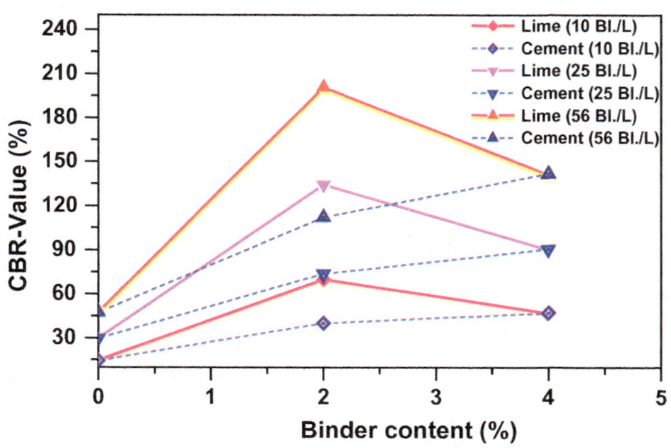

Fig. 4 CBR-value of untreated compacted- and treated stabilized-green sand with lime/cement

4 Discussion

As can be seen in Fig. 3, it may be noted that there was a decrease of the dry density for green sands treated with binders compared to the untreated green sand. This can be explained by the fact that the addition of binder increases the volume of the mixture and thus reduces the dry density after compaction. The Optimum Moisture Content (OMC) increases proportionally with the addition of the treating agent, because the binder requires an additional amount of water except for the green sand treated with 2% cement where the OMC remains low.

According to the Fig. 4, it can be noted that the addition of 2% of lime to the waste green sand, significantly affects the behavior of the material by increasing its CBR index, which confirms the concept of optimal dosage. Obviously, the dosage to be adopted for the lime treatment has a limit, beyond which soil characteristics will not change much or degrade, when increasing the dosage. Such rate of lime may produce an optimal flocculation of green sand particles, leading to a better rearrangement of soil grains by compacting at optimal water content.

The increase in CBR is due to the long-term lime effect, in which pozzolanic reactions occur by developing compounds consisting of silicates and aluminates. On the other hand, cement setting is faster than lime and the hardness occurs further.

Moreover, Nagaraj et al. [9] concluded in their study carried out on the compressed stabilized Earth Blocks (CSEBs) that the mixture with lime and cement offers a high strength in the long-term.

It can be concluded that the economic dosage to be used for the studied material is 2% of lime, then the increase in compaction intensity will be applied to reach the desired CBR index; this means that the consistency of the mixture obtained by addition of 2% of lime to the green sand is excellent. This is due to the fact that the used green sand waste is a clayey material which requires the addition of an agent that reduces its clayey character.

5 Conclusion

The results revealed that the green sand under study showed a good aptitude for its valorization in the field of road construction and should undergo a treatment with a hydraulic binder (lime or cement). It was also found that the green sand treated with a dosage of 2% lime provides an optimal treatment in which a CBR index is about 140 obtained at medium energy of compaction and 200 for intense energy compaction, which leads to an economic solution for the industrial problem.

References

1. Carnin, R.L.P., Folgueras, M.V., Luvizao, R.R., Correia, S.L., da Cunha, C.J., Zungan, R.S.: Use of an integrated approach to characterize the physicochemical properties of foundry green sands. Thermochim. Acta **543**, 150–155 (2012)
2. Dayton, E.A., Whitacre, S.D., Dungan, R.S., Basta, N.T.X.: Characterization of physical and chemical properties of spent foundry sands pertinent to beneficial use in manufactured soils. Plant Soil **329**, 27–33 (2010)
3. Salokhe, E.P., Desai, D.B.: Application of foundry waste sand in manufacture of concrete. J. Mech. Civ. Eng. (IOSR-JMCE) 43–48 (2013)
4. Alonso-Santurde, A., Andrés, A., Viguri, J.R., Raimondo, M., Guarini, G., Zanelli, C., Dondi, M.: Technological behaviour and recycling potential of SFS in clay bricks. J. Environ. Manage. **92**, 994–1002 (2011)
5. AFNOR: Classification des sols - Guide de Terrassements Routiers 'GTR' - NF P 11 300, France (1992)
6. Zirschky, J., Piznar, M.: Cement stabilization of foundry sands. J. Environ. Eng. **114**, 715–718 (1988)
7. AFNOR: Indice CBR après immersion – Indice portant immédiat, Norme NF P94-07, France (1997)
8. AFNOR: Essai Proctor normal – Essai Proctor modifié, Norme NF P 94-093, France (1999)
9. Nagaraj, H.B., Sravan, M.V., Arun, T.G.: Role of lime with cement in long-term strength of CSEBs. Int. J. Sustain. Built Environ. **3**, 54–61 (2014)

Experimental and Numerical Performance Evaluation of Cement-Calcined Kaolin-River Sand-Clay Mixture as a Highway Material

E. Arinze Emmanuel and C. Ekeoma Emmanuel

Abstract

This study evaluated the performance of cement-calcined kaolin and river sand stabilized clayey soil classified as A-7-6(3) used for a subbase of a flexible pavement in other to achieve a cost effective and environmental friendly solution to high cost construction materials. The percentages of River sand and calcined kaolin used for stabilization ranged from 0 to 20%, while the percentages of cement were kept constant at 4 and 6%. An increase in percentage of Calcined Kaolin increased the percentage of the fines, while the increase in the percentage of the River sand led to an increase in coarseness of the soil. Atterberg limits of the soil decreased with the increase in the River Sand percentage for 4 and 6% cement, but increased with the increase in the Calcined Kaolin. The soil OMC decreased from 14.2 to 11.3% at the highest addition of the River sand (OMK20RS) at 4% cement while it increased progressively with the increase in the Calcined Kaolin up to 18.2%. Displacement of the pavement was reduced by about 98.8% when used to model subbase of pavement using Plaxis, therefore cement-calcined kaolin-River sand can be used effectively for subbase and subgrade of a pavement.

Keywords

Calcined kaolin • River sand • Plaxis • Soil stabilization Deformation

1 Introduction

This is a two-facet research; the first is the experimental approach while the second is the numerical application. The geotechnical characteristics of clay mixed with cement, calcined kaolin and river sand were investigated and compared with Federal Ministry of Works FMW and AASHTO specification. A finite element program Plaxis was used to evaluate the performance of the flexible pavement whose subbase was the model with the admixture combination with the best geotechnical properties. The model was loaded with an axle load increasing from 50 to 600 kPa and the results were compared with the unstabilized soil. When poor soil is encountered in construction work, it is either excavated or stabilized. Most often it is not economical to excavate such poor bearing soil and the best option left is to stabilize such soil. Several researchers have worked on soil stabilization [1–10].

E. Arinze Emmanuel (✉) · C. Ekeoma Emmanuel
Department of Civil Engineering, Michael Okpara University of Agriculture, Umudike, Abia State, Nigeria
e-mail: emmanuel.arinze@mouau.edu.ng

C. Ekeoma Emmanuel
Research Student, Dundee University, Dundee, UK

2 Experimental Methods

2.1 Experimental Materials

The materials used for this study are as follows: Soil (Clay, River Sand), Cement and Calcined Kaolin.

2.1.1 Soil

Two different soil types were used for the study which are clay and river sand. The properties of the sand are displayed in Table 1.

2.1.2 Cement

Ordinary Portland cement (OPC) of Grade-42, was used for the experiment. The physical properties of cement are listed in Table 2.

2.1.3 Calcined Kaolin

The oxide composition can be seen in Table 4. While the physical properties are presented in Table 1. According to ASTM C618-12 requirement for pozzolans, the combination of SiO_2, Al_2O_3 and Fe_2O_3 should be greater than or equal to 70% (i.e. $SiO_2 + Al_2O_3 + Fe_2O_3$; $\geq 70\%$). From the chemical analysis result of the Calcined Kaolin in Table 3, it is remarked that the Calcined Kaolin met the specification.

A. Kallel et al. (eds.), *Recent Advances in Geo-Environmental Engineering, Geomechanics and Geotechnics, and Geohazards*, Advances in Science, Technology & Innovation, https://doi.org/10.1007/978-3-030-01665-4_72

Table 1 General properties of the materials used for the study

S/no	Properties	Clay	Calcined kaolin	River sand
1	**Color**	Reddish brown	Brown	
2	**Grain size distribution**			
i	Percentage passing sieve no 200	43.5	47.3	3.03
ii	Uniformity coefficient	–	–	2.84
iii	Coefficient of curvature	–	–	0.95
3	**Atterberg limits**			
i	Liquid limit %	41.5	38.2	–
ii	Plastic limit %	20.2	–	–
iii	Plasticity index	21.3	–	–
4	**Specific gravity**	2.62	2.34	2.71
5	**Soil classification**			
i	AASHTO classification	A-7-6(3)	–	A-3(1)
ii	Unified classification system	SC (clayey sand)	–	SP (Poor graded sand)
6	**Rating as subbase material**	Poor		Excellent
7	**Major clay mineral present**	Illite (Inorganic clay of medium plasticity)		
8	**Compaction characteristics**			
i	Maximum dry density (g/ml)	1.5	1.12	1.45
ii	Optimum moisture content (%)	14.2	28.6	8.1
iii	California bearing ratio (%)	7.5	1.2	12.2
iv	Unit weight (kN/m^3)	13.5	9.8	21.5
9	**Young's modulus** (kN/m^2)	4.5e+4		7.5e+4

Table 2 Cement properties

Properties	Results (%)
Al_2O_3	55.15
SiO_2	36.92
Fe_2O_3	1.00
TiO_2	0.70
$CaO+K_2O$	0.35
Na_2O+K_2O	1.7
Loss on ignition (LOI)	4.16

Table 3 Properties of calcined kaoline (Oxide compositions)

Fineness	Specific gravity G	Standard consistency (%)	Initial setting time (min)	Final setting time (min)	Soundness (cement expansion) (mm)
2.8	3.12	35.5	140	325	2

Table 4 Stabilizer combination scheme

S/no	Combination	Designation
1	0%MK + 20%RS	OMK20RS
2	5%MK + 15%RS	5MK15RS
3	10%MK + 10%RS	10MK10RS
4	15%MK + 5%RS	15MK5RS
5	20%MK + 0%RS	20MKORS

Note MK is Meta Kaolin (Calcined Kaolin) and RS River Sand

2.2 Testing Procedures

2.2.1 Particle Size Distribution Test

The particle size distribution PSD was carried out in accordance with BS1377: Part 2:1990. The material preparation and test procedures were followed duly in accordance with the Bs code. The PSDs were ascertained at each combination of the stabilizer.

2.2.2 Atterberg Limit Test

Atterberg Limits or consistency test were carried out on the stabilized and unstabilized soil specimen based on BS1377: Part 2:1990.

2.2.3 California Bearing Ratio CBR Test

The CBR tests on stabilized and unstabilized soil specimens were conducted in accordance with ASTM D1883-99 (2000).

2.2.4 Standard Protoctor Test

The standard protoctor test was carried out in line with BS 1377-4:1990.

3 Results and Discussion

As the percentage of Calcined Kaolin increased, the percentage of the fines increased while the increase in the percentage of the River sand lead to an increase in soil coarseness making the curve move towards the right side of the Particle Size Distribution. The increase in the coarseness of the soil particles invariable led to an improvement in the engineering properties of the soil, since is the fine percentages were reduced to 35.55, 29.78, 25.4, and 18.4% for 5MK15RS, 15MK5RS, 10MKRS10, OMK20RS, respectively, while that of 20MKORS increased to 44%.

The Atterberg limits of the soil decreased with the increase in the River Sand percentages for 4 and 6% cement but increased with the increase in the Calcined Kaolin.

The OMC of the soil decreased from 14.2 to 11.3% at the highest addition of the River sand (OMK20RS) at 4% cement while it increased progressively with the increase in the Calcined Kaolin up to 18.2%. For 6% cement OMC reduced to 12.1% from 14.4%, with an increase in the Calcined Kaolin, the OMC was reduced due to increase in the pozzolanic material. These increments in optimum moisture content with an increase in Calcined Kaolin could be attributed to the increased amount of water required in the system to adequately lubricate all the particles in the soil-cement and Calcined Kaolin mixture.

The maximum dry density of the soil increased progressively with the increase of the River sand and Calcined Kaolin. The results of the MDD are 1.56, 1.61, 1.67, 1.5, 1.42 g/ml for OMK20RS, 5MK15RS, 10MK10RS, 15MK5RS and 20MKORS combinations, respectively at 4% cement. It could be seen that the MDD dry density of the soil increased progressively from 1.5 g/ml but decreased at 20MKORS due to lower specific gravity of Calcined Kaolin [11, 12].

The CBR values at 4% cement are 12.2, 12.6, 12.5, 10.2 and 9.8% for OMK20RS, 5MK15RS, 10MK10RS, 15MK5RS and 20MKORS, respectively and 13.1, 12.8, 12.6, 11.5% and 10.3 for 6% cement. Based on IRC recommendations, the stabilized soil met the specification for a good subbase.

The stabilizer combination with the best strength properties and the unstabilizer properties of the soil were also used for finite element analysis using Plaxis to determine the deflection of the pavement under static loading condition. Various thicknesses of the layers of the flexible pavement were used for the simulation (50, 75, 100 m for the Asphalt and 100, 200, 300 for the base course and subbase. The Poisson's ratio considered for the analysis was 0.4 [13–15].

The Young's modulus values for conventional material in respective layers of haul road pavement considered were 200, 100, 50 and 50 MPa [14]. The applied pressure on the road ranged from 100 kPa to 500 kPa. 10MK10RS stabilizer combination at 4% cement would be used for the analysis since it has better properties.

4 Conclusion

The following conclusions can be drawn from this research:

- The increase in the percentages of the calcined kaolin led to the increase in the percentage fine of the clay soil while the river sand increase led to the increase in the coarseness of the clay material.
- The maximum dry density of the soil increased progressively with the increase of the River sand and Calcined Kaolin. The result of the MDD was 1.56, 1.61, 1.67, 1.5, 1.42 g/ml for 0MK20RS, 5MK15RS, 10MK10RS, 15MK5RS and 20MK0RS combination, respectively at 4% cement.
- The CBR values at 4% cement were 12.2, 12.6, 12.5, 10.2 and 9.8% for 0MK20RS, 5MK15RS, 10MK10RS, 15MK5RS and 20MK0RS respectively and 13.1, 12.8, 12.6, 11.5% and 10.3 for 6% cement. Based on IRC recommendation the stabilized soil met the specification for a good subbase.
- It was observed that the increase in the load in the unstabilized subbase pavement from 100 to 200 kPa led to an increase in the displacement of the pavement by 99.06%.

References

1. Aboubakar, M.A., Ganjian, E., Pouya, H., Akashi, A.: A study on the effect the addition of thermally treated Libyan natural pozzolan has on the mechanical properties of ordinary Portland cement mortar. Int. J. Sci. Tech. **3**(1), 79–84 (2013)
2. Alhassan, M., Mustapha, M.: Effect of rice husk ash (RHA) on cement stabilized laterite. Leonardo Electron. J. Practices Technol. (II), 47–58 (2007)
3. Amu, O.O., Salami, B.A.: Effects of common salt on some engineering properties of egg shell stabilized lateritic soil. Asian Res. Publishing Netw. (ARPN) J. Eng. Appl. Sci. **5**(9) (2010)
4. Ayininuola, G.M., Adekitan, O.A.: Compaction characteristics of lateritic soils stabilized with cement-calcined clay blends. J. Silic. Based Compos. Mater. (2017). http://doi.org/10.14382/epitoanyag-jsbcm.2017.7
5. Bhatta, N.: Engineering properties of pond ash and pond ash sand mixtures. Indian Highways 49–59 (2008)
6. Hossain, K.M.A.: Stabilized soils incorporating combinations of rice husk ash and cement kiln dust. J. Mater. Civ. Eng. **23**(9), 1320 (2011)
7. Justice, J.M.: Evaluation of calcined kaolins for use as supplementary cementitious materials, An unpublished M.Sc. thesis. School of Material Science and Engineering, Georgia Institute of Technology (2005)
8. Kolovos, K.G., Asteris, P.G., Cotsovos, D.M., Badogiannis, E., Tsivilis, S.: Mechanical properties of soilcrete mixtures modified with calcined kaolin. Constr. Build. Mater. **47**, 1026–1036 (2013). https://doi.org/10.1016/j.conbuildmat.2013.06.008
9. Olonade, K.A., Jaji, M.B., Adekitan, O.A.: Experimental comparison of selected pozzolanic materials. In: Proceedings of the 2nd International Conference on Advances in Cement and Concrete Technology in Africa, Dares Salaam, Tanzania, 27–29 Jan, pp. 309–314 (2016)
10. Das, S.K., Yudhbir, Y: Geotechnical properties of low calcium and high calcium fly ash. J. Geotech. Geol. Eng. **24**, 249–263 (2006)
11. Gupta, D., Kumar, A.: Performance evaluation of cement-stabilized pond ash-rice husk ash clay mixture as a highway construction material. J. Rock. Mech. Found. **2**(3), 345–67 (2017)
12. Das, D.: Introduction to soil stabilization. (2016). Retieved from http://www.k31.org
13. Tannant, D.D., Regensburg, B.: Guidelines for Mine Haul Road Design, 1st edn. University of Alberta, Canada (2001)
14. McCarthy, M.J., Csetenyi, L.J., Sachdeva, A.., Jones, R.: Role of fly ash in the mitigation of swelling in lime stabilized sulphate bearing soils. In: Proceedings of World Coal Ash, USA (2009)
15. Lav, A.H., Lav., M.A., Goktepe, B.A.: Analysis and design of a stabilized fly ash as pavement base material. Fuel **85**(16), 2359–2370 (2006)

Part VIII

Geomechanics and Geotechnics: Analytical and Numerical Modeling of Geo-Structures

An Analytical Model for Determining the Natural Frequency of Retaining Structures Including the Earth Pressures

Lyazid Guechi and Smain Belkacemi

Abstract

The importance of the retaining structures is crucial in geotechnical engineering and the accurate determination of earth pressures and natural frequency is important in order to study the dynamic behavior of these structures. The analytical formulae which do not consider the earth pressures behind the retaining structure are usually used. An analytical model for determining the natural frequency of retaining structures including the earth pressures by failure wedges was proposed in the present analysis. The model considers the effect of Coulomb and Mononobe Okabe earth pressures. The current model was validated using several analytical models reported in the literature of the earlier researchers.

Keywords

Earth pressures • Natural frequency • Retaining structures • Analytic model • Failure wedge

1 Introduction

An adequate determination of the natural frequency of retaining structures plays a primordial role in the study of its dynamic behavior in earthquake prone regions. Many researchers have developed several methods to determine the natural frequency of retaining structures [1–11]. The natural frequency is often calculated by the elastic wave theory, based on two parameters namely the backfill height and the soil shear wave velocity.

In this research, a simple model was proposed to determine the natural frequency of retaining structures. This simple model estimates the natural frequency of wall-backfill soil system using the Coulomb [12] and Okabe [13], Mononobe [14] to evaluate soil wedge created by active earth pressure behind retaining structures in both static and seismic cases. The validity of the proposed model was proven by using the results within the Scott [3] and Ghanbari et al. [11] models.

2 Analytical Model

This work first included the active failure wedges in the determination of natural frequency of retaining structures and proposed a new model. It is possible and common to obtain equivalent natural frequency for a backfill soil-retaining structure system from results of analytical analyses assuming the system to have a single degree of freedom. The proposed retaining structure model is illustrated schematically in Fig. 1a for static case and under seismic pseudo static conditions, and it is defined by (1) structure characteristics: weight per unit length 'ρ_{str}', modulus of elasticity 'E_{str}', thickness 'e' and height 'H' (2) backfill soil characteristics: Poisson's ratio 'υ_{soil}', soil friction angle 'φ_{soil}', Young modulus 'E_{soil}' and weight per unit mass 'ρ_{soil}' and (3) seismic excitation characteristics: horizontal and vertical seismic accelerations 'a_h, a_v'.

In view of analytical studies, this model was designed to be as simple as possible, including Coulomb and Mononobe-Okabe active failure wedges. More precisely, Fig. 1b represents the dynamic equilibrium and resultant forces acting on differential elements of the model. The following assumptions were adopted for this model: (1) The retaining wall behaves as a flexible structure [15]; (2) the static thrust was evaluated by Coulomb method [12]; (3) the seismic thrust was evaluated by Mononobe-Okabe method [13, 14]; (4) the inclined failure plane was modeled as a series of Winkler's springs [16, 17]; (5) the natural frequency of wall-backfill soil system was estimated by Rayleigh method; and (6) the backfill soil was assumed dry and cohesionless.

L. Guechi (✉) · S. Belkacemi
Ecole Nationale Polytechnique, 16200 El-Harrache, Algeria
e-mail: gue28091@gmail.com

A. Kallel et al. (eds.), *Recent Advances in Geo-Environmental Engineering, Geomechanics and Geotechnics, and Geohazards*, Advances in Science, Technology & Innovation, https://doi.org/10.1007/978-3-030-01665-4_73

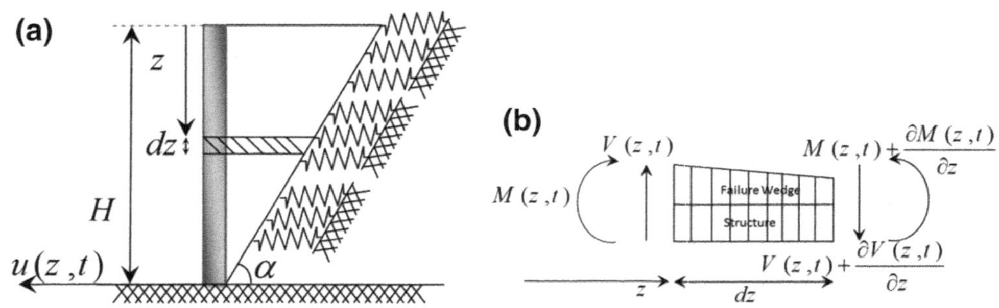

Fig. 1 A schematic system of retaining structures considering the active failure wedge and the dynamic equilibrium and resultant forces acting on differential elements of the modeled system

3 Results

3.1 The First Circular Natural Frequency (ω)

The first circular natural frequency (ω) of the system is given by an analytical equation; this equation takes into account the changes in soil wedge geometry; the first circular natural frequency (ω) of the system is function of: the retaining structure parameters, backfill soil parameters and failure wedge behind the retaining structure.

$$\omega = f(\rho_{str}, E_{str}, e, H, \rho_{soil}, E_{soil}, \upsilon_{soil}, k, \alpha, a_v, a_h) \quad (1)$$

The effects of each parameter of this equation are examined in [18].

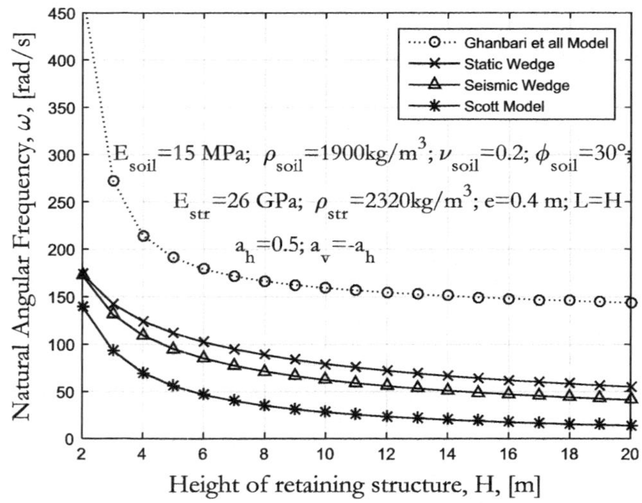

Fig. 2 Verification and comparison of proposed models with Scott and Ghanbari et al. models

3.2 Comparison and Verification

To verify the validity of the proposed model, the calculated natural frequencies were compared with those reported in the literature [3] and [11]. For different wall height values (Fig. 2), the natural angular frequency associated with Coulomb and Mononobe Okabe failure wedges models computed in the present study are located between that of Scott [3] and Ghanbari et al. [11] methods with $\upsilon_{soil} = 0.2$, $E_{soil} = 15$ MPa, $E_{str} = 26$ GPa, $\rho_{soil} = 1900$ kg/m^3, $\rho_{str} = 2320$ kg/m^3, H = 7 m and $a_h = 0.3$, $a_v = 0.5a_h$.

4 Conclusion

In this research, considering the failure wedge in the soil backfill behind the retaining structures, the natural frequency of retaining structures was obtained. An analytical model using the Coulomb and Mononobe-Okabe failure wedges, to determine the natural frequency of retaining structures was developed. In addition to the consideration of horizontal and vertical seismic acceleration coefficients, height and thickness of structure, Young modulus and density of the soil and the structure material and also Poisson ratio and friction angle of soil backfill were considered. Considering the earth pressure plays an important role in determining the natural frequency of the retaining structures; each variation of the soil backfill and seismic accelerations changed the system and naturally the natural frequency.

References

1. Matsuo, H., Ohara, S.: Lateral earth pressure and stability of quay walls during earthquakes. In: 2nd World Conference on Earthquake, Tokyo-Kyoto (1960)
2. Wood, J.H.: Earthquake induced soil pressures on structures. Ph.D. dissertation, California Institute of Technology, Pasadena, California (1973)
3. Scott, R.F.: Earthquake-induced earth pressures on retaining walls. In: 5th World Conference on Earthquake Engineering, Rome (1973)
4. Veletsos, A.S., Younan, A.H.: Dynamic soil pressures on rigid vertical walls. Int. J. Earth. Eng. Str. Dyn **23**, 275–301 (1994)
5. Ortiz, L.A., Scott, R.F., Lee, J.: Dynamic centrifuge testing of a cantilever retaining wall. Earth. Eng. Str. Dyn. **11**, 251–268 (1983)

6. Alampalli, S., Elgamal, A.W.: Dynamic response of wall backfill retaining system. Earth. Shock. Vib. **4**, 251–259 (1997)
7. Wu, G.: Dynamic soil-structure interaction: pile foundations and retaining structures. Ph.D. dissertation, University of British Columbia, Vancouver (1994)
8. Wu, G., Finn, W.D.: Seismic lateral pressures for design of rigid walls. Can. Geotech. J. **36**, 509–522 (1999)
9. Hatami, K., Bathurst, R.J.: Frequency response analysis of reinforced soil retaining walls. In: 8th Canadian Conference on Earthquake Engineering, Canada (1999)
10. Hatami, K., Bathurst, R.J.: Effect of structural design on fundamental frequency of reinforced soil retaining walls. Soil Dyn. Earth. Eng. **19**, 137–157 (2000)
11. Ghanbari, A., Hoomaan, E., Mojallal, M.: An analytical method for calculating the natural frequency of retaining walls. Inter. J. Civ. Eng. Geotech. Eng. **11**(1–9), 3 (2013)
12. Coulomb, C.: Essai sur une application des règles de Maximis & Minimis à quelques problèmes de statique relatives à l'architecture. Mémoires de Mathématiques et de physique, Académie Royale des sciences, Paris (1776)
13. Okabe, S.: General theory of earth pressure. J. Japan. Soc. Civ. Eng. **12** (1926)
14. Mononobe, N.: On the determination of earth pressure during earthquakes. In: World Engineering Conference (1929)
15. Sitar, N., Wagner, N.: On seismic response of stiff and flexible retaining structures. In: 6th International Conference on Earthquake Geotechnical Engineering, New Zealand (2015)
16. Winkler, E.: Die lehre von der Elasticitaet and Festigkeit. Dominicus (1867)
17. Kacar, A., Tan, H.T., Kaya, M.O.: Free vibration analysis of beams on variable Winkler elastic foundation by using the differential transform method. Math. Comput. Appl. **16**, 773–783 (2011)
18. Guechi, L., Belkacemi, S.: Effects of soil, structure and seismic parameters on the natural frequency of retaining structures. In: the 2018 Structures Congress, Incheon (2018)

Modelling of Material and Geometrical Nonlinearities of Footing by a New Non-Linear Macro-Element

Mourad Khebizi, Hamza Guenfoud, and Mohamed Guenfoud

Abstract

This paper focused on the development of a new non-linear macro-element for the modelling of soil-foundation interaction. Material and geometrical nonlinearities were taken into account in the present macro-element to examine the response of shallow foundations under monotonic and cyclic loads. An application of soil-foundation systems was studied. The results obtained from this application were in very favorable agreement with those obtained through other numerical models in the literature.

Keywords

Soil-footing • Interaction • Macro-element • Modelling • Non-linear • Material • Geometrical

1 Introduction

The non-linear behavior of shallow foundations is a major objective for civil engineering researchers. On the one hand, the nonlinearities of rigid shallow foundations are related to the soil plasticity under a foundation and, on the other hand, to the uplift of the soil-foundation interface [1, 3]. They are not only related to the nature of the materials (soils + foundations) [6], but also to the type of loading. The latter can be static, cyclic, or dynamic; they cause vertical displacement, horizontal displacement and rotation of the foundations [3]. Foundation systems are also subject to the action of seismic solicitations and are damaged, sometimes with very serious consequences for the structures [6].

M. Khebizi (✉)
Department of Civil Engineering, Mentouri University of Constantine, Constantine, Algeria
e-mail: Mourad_gc@yahoo.fr

M. Khebizi · H. Guenfoud · M. Guenfoud
Civil Engineering and Hydraulic Laboratory, University of Guelma, Guelma, Algeria

A. Kallel et al. (eds.), *Recent Advances in Geo-Environmental Engineering, Geomechanics and Geotechnics, and Geohazards*, Advances in Science, Technology & Innovation, https://doi.org/10.1007/978-3-030-01665-4_74

In this paper, we presented a simplified and reliable numerical modelling method which is able to take into account the soil-foundation interaction by considering all nonlinearities associated with foundation uplift and soil yielding. The method used a "gap" element connected in a series with a "non-linear link that follows Wen plasticity" [7] in order to form a new macro-element [5]. The horizontal behavior of the foundation is controlled by a non-linear horizontal link. The soil-foundation system can be modelled by a single macro-element located in the footing center, as it can be modelled by a rigid beam resting on a set of the vertical macro-elements according to the Winkler approach.

2 Presentation of the Macro-Element

The macro-element developed in this work is a "gap" element connected in a series with a non-linear link (see Fig. 1). This macro-element is able to describe the material and geometrical nonlinearities [5].

The "gap" element is used to simulate the footing uplift. This element carries compression loads only; it has zero stiffness when subjected to tension.

The non-linear link element is used to simulate the soil plasticity under the foundation. The plasticity model is based on the hysteretic behavior proposed by Wen [7].

Fig 1 shows the general structure of the non-linear macro-element. The behavior law associated with this macro-element is not symmetrical (see Fig. 2) and is characterized, on one hand, by a yield load in compression F_y, and by a reduction of the recall force as the foundation is uplifted, on the other hand. The recall force is finally approaches zero.

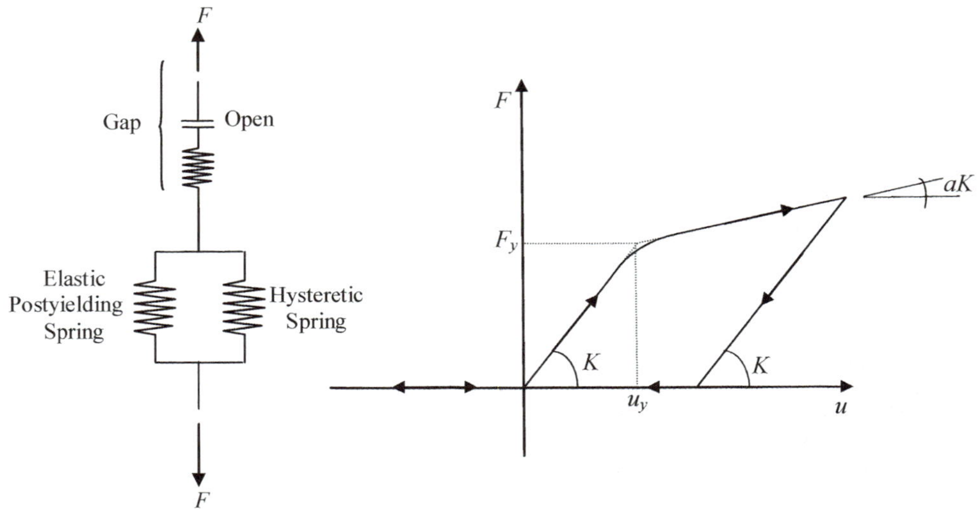

Fig. 1 General structure and behavior of the non-linear macro-element [5]

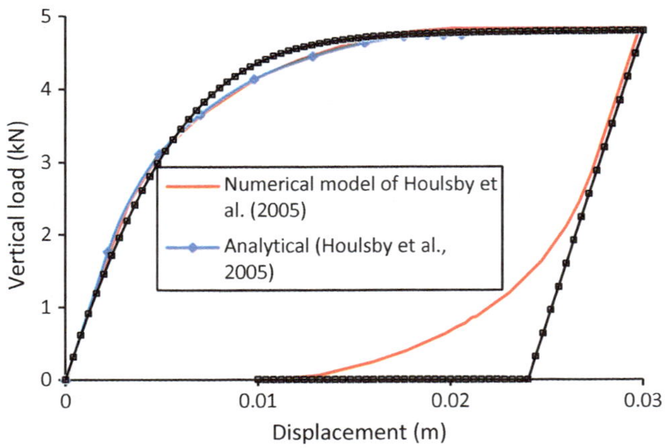

Fig. 2 Vertical load-displacement curve

3 Applications

In order to see if the macro-element is able to reproduce the behavior of a foundation under a monotonic static vertical load, we chose to model a circular footing of diameter D = 1 m, resting on a layer of clay, whose parameters are G = 200 kPa, c = 1 kPa and υ = 0.5.

Houlsby et al. [4] studied analytically and numerically the response of the same foundation. They used the approach of hyper-plasticity theory in order to review the modelling potential of the global behavior of a shallow foundation by adopting Winkler's assumption of decoupled springs.

In our study, the foundation and the subjacent soil were modelled with only one non-linear macro-element located in the center of the foundation on which a cyclic vertical displacement was imposed (only one macro-element is sufficient, since only the translation according to the vertical axis was possible and the foundation was considered infinitely rigid). The elastic vertical stiffness of the macro-element was calculated using the equation developed by Gazetas [2].

The maximum vertical load endured by the foundation F_y for a homogeneous cohesive soil is given by:

$$F_y = 6.06 \times c \times \pi \times D^2/4 = z \quad (1)$$

The numerical parameters of the macro-element are presented in Table 1.

Figure 2 compares the results obtained with our macro-element and those of Houlsby et al. [4]. It can be noted that the non-linear response of the foundation described by the macro-element is practically the same as that described analytically and numerically by Houlsby et al. [4] at the time of the loading phase. Regarding the unloading phase, we noted a discrepancy between the results, as the foundation's response described by the numerical model of Houlsby et al. [4] at the time of the unloading phase after very large displacement produces an unrealistically large uplift of the foundation, as shown in Fig. 2, where there is a very significant decrease of the tangent to the curve in the unloading phase (this is a weakness of Houlsby et al.'s model).

Fig 2 also shows that the macro-element allows a good description of the material non-linearity of the foundation during the loading phase, and the elastic behavior during unloading, as well followed by a good description of the foundation uplift (geometrical non-linearity) during the inversion of the loading sign (cancellation of the load by the activation of the gap element). It can noted that during

Table 1 Macro-element parameters

Vertical elastic stiffness, K	Yield load, F_y	Ratio of post-yield, a	Exponent, Exp
800 kN/m	4.7571 kN	0.001	1

this loading–unloading cycle, the two components of total displacement were clearly observed, namely elastic and plastic displacements.

4 Conclusion

In this paper, we proposed a new macro-element oriented to the application of earthquake engineering, it is able to describe the material nonlinearities associated with soil plasticity, as well as geometric nonlinearities relative to foundation uplifting. A numerical application in the literature was studied in order to validate the response of the proposed macro-element under cyclic loads. According to this application it was noted that:

A good agreement exists between the results obtained from the proposed macro-element and those obtained from other numerical models in the literature.

The nonlinearities of the soil-foundation system are well reproduced by the present macro-element.

References

1. Gazetas, G.: 4th Ishihara lecture: soil-foundation-structure systems beyond conventional seismic failure thresholds. Soil. Dyn. Earthq. Eng. **68**, 23–39 (2015)
2. Gazetas, G.: Formulae and charts for impedance functions of surface and embedded foundations. J. Geotech. Eng. **117**(9), 1363–1381 (1991)
3. Gelagoti, F., Kourkoulis, R., Anastasopoulos, I., Gazetas, G.: Rocking-isolated frame structures: margins of safety against toppling collapse and simplified design approach. Soil. Dyn. Earthq. Eng. **32**, 87–102 (2012)
4. Houlsby, G.T., Cassidy, M.J., Einav, I.: A generalised Winkler model for the behaviour of shallow foundations. Géotechnique **55** (6), 449–460 (2005)
5. Khebizi, M., Guenfoud, H., Guenfoud, M.: Numerical modelling of soil-foundation interaction by a new non-linear macro-element. Geomech. Eng. **14**(4), 377–386 (2018)
6. Khebizi, M., Guenfoud, M.: Numerical modelling of the damaging behaviour of the reinforced concrete structures by multi-layers beams elements. Comput. Concr. **15**(4), 547–562 (2015)
7. Wen, Y.K.: Method for random vibration of hysteretic systems. J. Eng. Mech. Div. **102**(EM2), 249–263 (1976)

Seismic Response of Back-to-Back MSE Walls

Mohamed Djabri and Sadok Benmebarek

Abstract

In the latest Federal Highways Administration guidelines, back-to-back mechanically stabilized earth walls (BBMSEW), considered with complex geometries or significant external loads, are practical and require consideration. From these guidelines, under static condition, the distance between two opposing walls, *D*, is a key parameter used for determining the analysis methods of BBMSEW. The existing design methodologies and numerical investigations do not provide a clear and justified answer to how the tensile strength of reinforcement and the external stability changes under seismic conditions. In this study, the height of the walls was chosen as 6 m in the two-dimensional plane, the strain finite element model was incorporated in the PLAXIS software code and the base acceleration was chosen to be a harmonic motion. This study focused on the effects of the distance between opposite walls, facing type and reinforcement stiffness.

Keywords

Back-to-back walls • Reinforcement • Seismic load • Displacement

1 Introduction

Several research contributions are conceived and executed to improve the understanding of the effects of dynamic loading on mechanically stabilized earth walls. Many of these studies have shown that this type of walls with granular backfill soils exhibit good performance under strong earthquake loading [1].

M. Djabri · S. Benmebarek (✉)
NMISSI Laboratory, Biskra University, BP 145 07000 Biskra, Algeria
e-mail: sadok_benmebarek@yahoo.com

Although the MSEW behavior is widely studied in the literature, back-to-back mechanically stabilized earth walls (BBMSEW) were rarely investigated. Recently, under self-weight, it was shown that two walls perform independently when they are far apart and interact with each other, when they are close [2]. Furthermore, the effect of the reduction of the distance between the opposing walls on internal and external stability of BBMSEW with geosynthetic reinforcement and concrete panel facing under harmonic cyclic was investigated and reported [3].

In this paper, the finite element analysis has been employed using PLAXIS 2D software [4] to investigate the effects of facing systems and reinforcement stiffness on the dynamic behavior of a BBMSEW. Also, the influence of the distance between opposite walls for different walls with various component properties has been studied and discussed.

2 Numerical Modeling

2.1 Finite Element Method and Geometry of Investigated Model

PLAXIS code proved the ability of the finite element method to model the BBMSEW with sufficient accuracy as well as reasonable demand on quality and quantity of input whether under static or dynamic loadings [2, 3]. In this contribution, we considered a reinforced basic BBMSEW of height $H = 6$ m with the width $W = 12$ m, which corresponds to a ratio $W/H = 2.0$. The distance between walls is $D = 3.6$ m. The adopted length of the wall reinforcements was L = 4.2 m (i.e., 0.7 times the height as per FHWA [3]). As shown in Fig. 1, 8 horizontal reinforcement layers, uniformly spaced, were attached to a facing system. The walls and soil regions were supported by a stiff foundation.

A. Kallel et al. (eds.), *Recent Advances in Geo-Environmental Engineering, Geomechanics and Geotechnics, and Geohazards*, Advances in Science, Technology & Innovation, https://doi.org/10.1007/978-3-030-01665-4_75

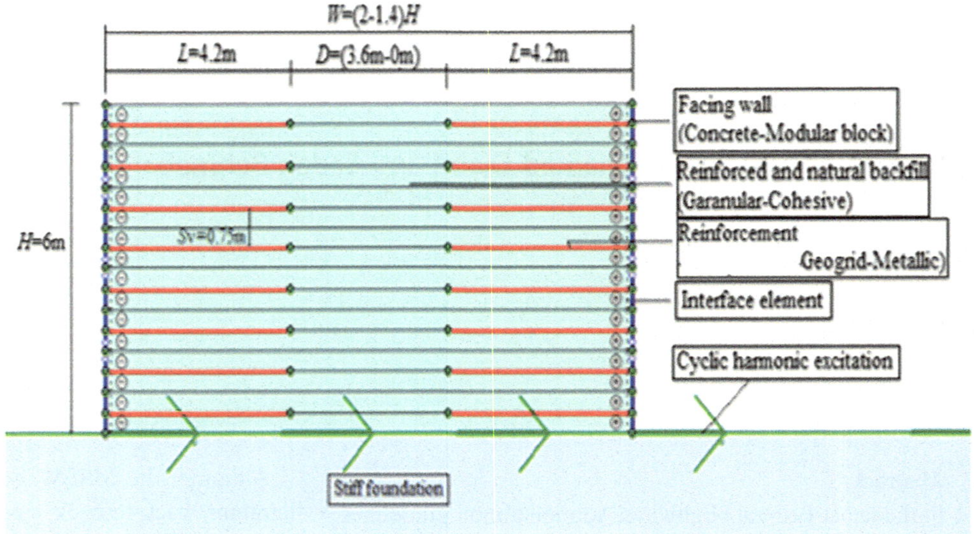

Fig. 1 The geometry of investigated models subject cyclic harmonic load

2.2 Materials Proprieties, Boundary Conditions and Cyclic Load

The elasto-plastic Mohr-Coulomb model was selected to represent the soils in the physical model, which properties were the same as used in several previous studies under static and seismic loads [2, 3]. Tables 1 and 2 summarize the material elements properties assigned to panels and modular block facing, respectively. The elastic axial stiffness (EA) is the main material property of reinforcement, whose values are 20,000 and 69,000 kN/m for geogrids and metal reinforcements, respectively [5].

The interface elements were located between facing systems and backfill soil, and between the backfill soil and reinforcement. For the Mohr-Coulomb model, the PLAXIS code suggests that in the absence of detailed information, a conservative interaction factor $R_{\text{inter}} = 0.67$ should be used [4]. Under self-weight, the boundary conditions were defined by considering total fixity for the bottom, restraining both vertical and horizontal movement. Only horizontal fixities were considered for the sides of the model so that the side boundaries were free to move vertically [2]. Furthermore, during the dynamic analyses, special absorbent boundaries conditions were specified to avoid spurious reflection of the waves [3, 5].

The cyclic load was applied at equal time intervals of 6 s and the accelerogram obtained and given by input ASCII file in PLAXIS software is shown (Fig. 2) [3].

3 Results and Discussion

From the displacement perspective (Fig. 3), it can be seen that the deformations of the wall decreases significantly with the decreasing distance, *D*, from 3.6 to 0 m. This result is in agreement with all the models after applying seismic excitation. For example, when the modular block with stiffer reinforcement is modeled, the maximum displacement diminishes from 41.2 to 14.3 mm by decreasing the distance, *D*, from 3.6 to 0 m (see Fig. 3b). When BBMSEW are closer, the effect of the reinforcement axial stiffness was not detectable on the wall deformation mainly when using a modular block.

The tensile loads in reinforcements are plotted in Fig. 4. It can be seen that the decrease of the distance to zero involves a reduction in the maximum tension load mainly when the concrete panel with metallic reinforcement is selected.

Table 1 Material properties of concrete panels [2, 3]

Material model	Elastic stiffness (*EA*)	Flexural rigidity (*EI*)	Unit weight (γ_c)	Weight of panel	Poisson ratio (ν)
Mohr-Coulomb	4.5×10^6 [kN/m]	8.438 [kN/m^2/m]	24 [kN/m^3]	3.6 [kN/m^2/m]	0.1

Table 2 Material properties of modular block facing elements [5]

Material model	Unit weight	Elasticity modulus (*E*)	Poisson ratio (ν)	Cohesion (*C*)	Internal friction angle (φ)
Mohr-Coulomb	20 [kN/m^3]	30,000 [kN/m^2]	0.1	200 [kN/m^2]	35°

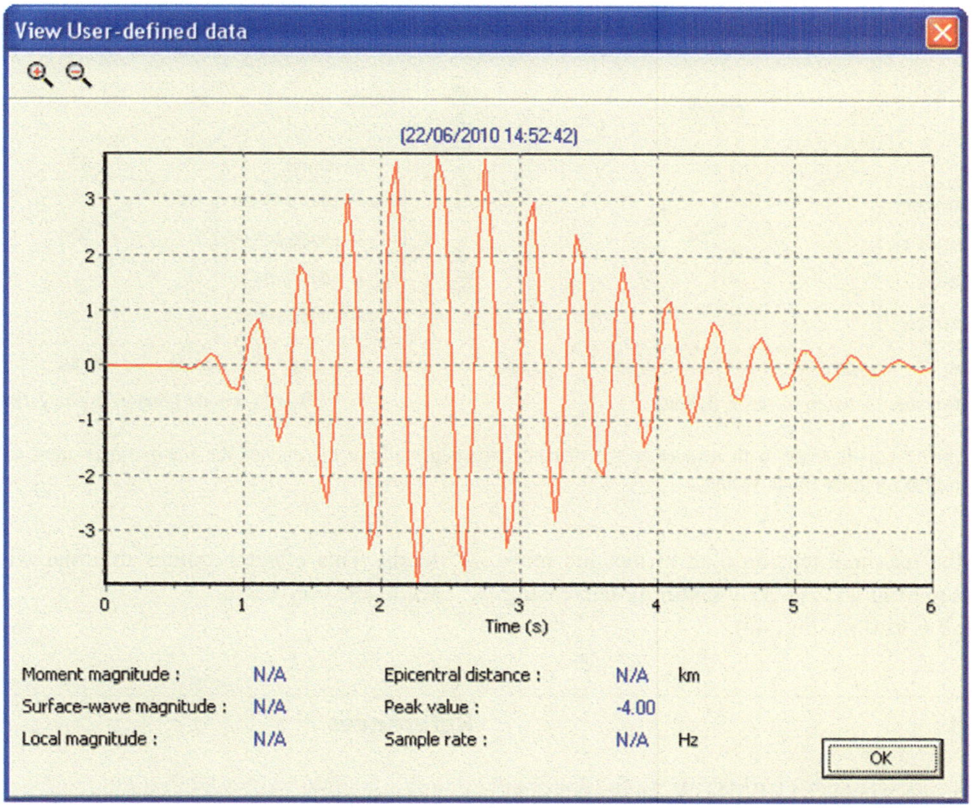

Fig. 2 Accelerogram of cyclic loading for PGA = 0.4 g and f = 1.5 Hz used in analysis [3]

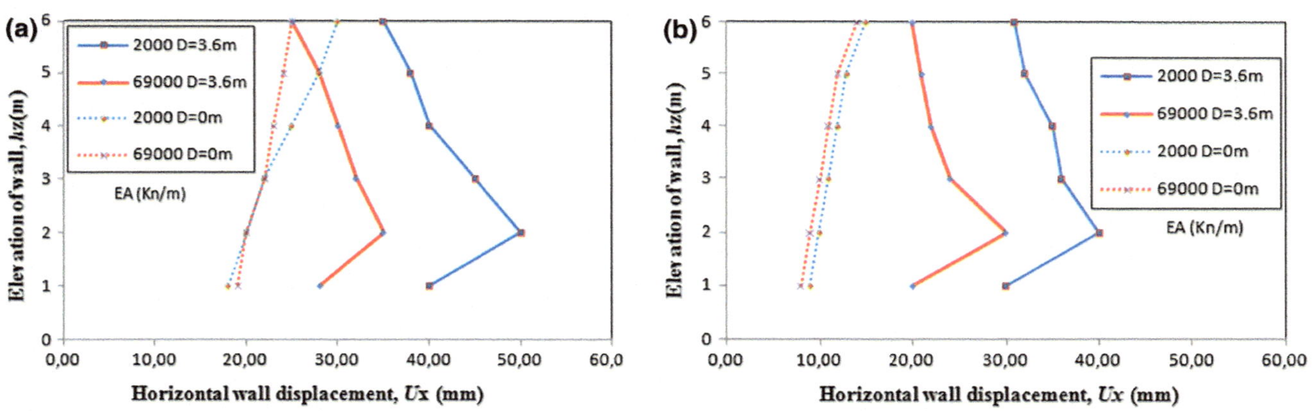

Fig. 3 Horizontal wall displacements with respect of the distance between walls at the end of the seismic excitation for granular backfill: **a** concrete panel facing; **b** modular block facing

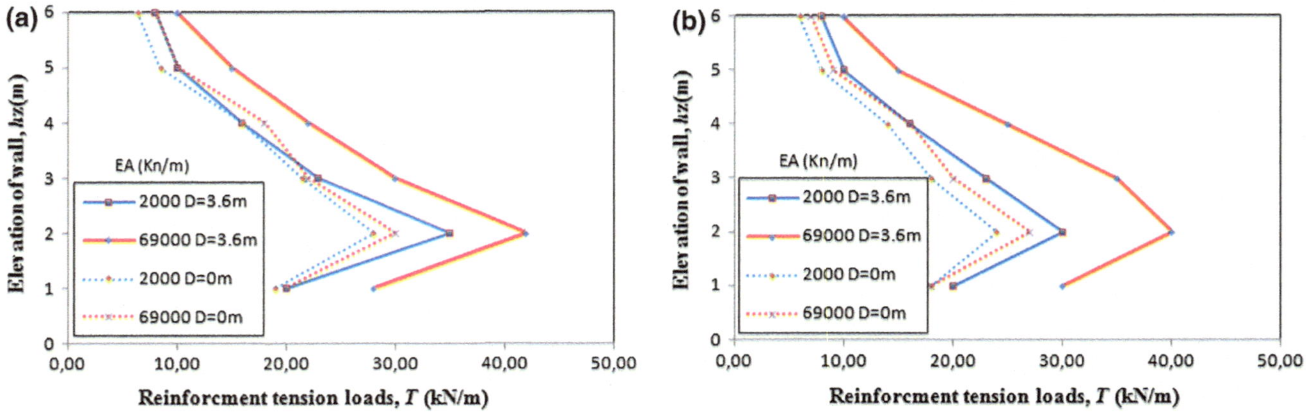

Fig. 4 The reinforcement tensile loads with respect of the distance between walls at the end of the seismic excitation for granular backfill: **a** concrete panel facing; **b** modular block facing

For example, the maximal tension load in metallic reinforcement decreases from 41.2 to 35.1 kN/m by decreasing the distance from 3.6 to 0 m (Fig. 4a).

4 Conclusion

The effects of the distance between BBMSEW, facing type and reinforcement stiffness were investigated under a harmonic motion. From this investigation, we can note the following:

- Decreasing the distance, *D*, to 0 m, involves a significant decrease of the wall displacement;
- For close BBMSEW, the maximum displacement occurred at the top of the wall. However, for wide BBMSEW, it occurred near the base of the wall;
- When BBMSEW are closer, the reinforcement axial stiffness has no effect on the wall deformation particularly when the modular block is used;
- Using stiffer reinforcements (metallic) reduces the wall displacement but increases the reinforcement tensile loads. This effect becomes minimal when the opposite walls are very close.

References

1. Leshchinsky, D., Han, J.: Geosynthetic reinforced multitiered walls. J. Geotech. Geoenviron. Eng. ASCE **130**(12), 1225–1235 (2004)
2. Djabri, M., Benmebarek, S.: FEM analysis of back-to-back geosynthetic-reinforced soil retaining walls. Int. J. Geosynth. Ground Eng. **2**, 26 (2016)
3. Benmebarek, S., Djabri, M.: FE analysis of back-to-back mechanically stabilized earth walls under cyclic harmonic loading. Indian Geotech. J. **48**(3), 498–509 (2018)
4. Brinkgreve, R.B.J., Broere, W.: 2D PLAXIS Finite Element Code (Version 8.0). Delft University of Technology, Delft, Netherlands, PLAXIS bv (2006)
5. Güler, E., Cicek, E., Demirkan, M.M.: Numerical analysis of reinforced soil walls with granular and cohesive backfills under cyclic loads. Bull. Earthq. Eng. **10**, 793–811 (2012)

Numerical Modeling and Parametric Study of Flexible Wall Reinforced with Anchor System

Dounia Amrani

Abstract

This study dealt with the achievement of a diaphragm wall anchored in deep excavations to protect and support the multilayer soil structure. In this research study a reference model was established then a parametric study was carried out. The final obtained model of the case study was of a good approximation to the reality (the inclinometric data of wall displacement). In the parametric study, several calculation parameters (discretisation and precisions) and geotechnical parameters (the interface, water pressure and the anchor system) were varied individually and finally compared from the anchored to the non anchored wall. It is evident from this study that the variation of the parameters in a reasonable interval has considerable effects of horizontal displacement and moments on the wall. In the end it was concluded that the use of numerical analysis procedures can provide more efficient and more economic geotechnical project design.

Keywords

Numerical modelling • Diaphragm wall • Anchors • Multilayer soil • PLAXIS code

1 Introduction

Numerical modeling methods in geotechnical engineering are widely used to solve practical engineering problems, with emphasis on numerical analysis of computation of displacements and stresses in soils and internal forces in structural elements. The finite element method [1, 2] is one of the most powerful numerical methods available to solve stability problems. This study developed a detailed evaluation of the conventional methods of a diaphragm wall with an anchoring system and compared them with Plaxis finite element code [1, 3]. The example taken in this study case was a slope consisting of complex soils. In addition, this work included a contribution on the calculation of safety and stabilization of the cut using diaphragm wall and anchoring systems [1, 2, 4–7].

A strong correlation between wall displacements and bottom uplift stability was found in both field measurements. A finite element parametric study has also been studied in many cases. More recently, a formulation in Boundary Elements has been used by [2, 4–7], who used the solution obtained inside a multi-layer soil in the term of displacements of charges repartition. In this study, the 3% accuracy in the Plaxis wais reduced to 1% to get 3% of the total horizontal displacement and 10% of the moments. The variation of the different parameters (geotechnical parameters, hydrostatic pressure) can give good results making it possible to predict the affection behavior of the flexible wall and its anchoring system.

2 Materials and Methods

The studied case in this research was a project consisting of a flexible wall of 1.20 m of thickness and 40 m of depth footed in the layer of clays in order to stabilize the excavation of the 17 m wide pit. The wall was retained by two anchors not-prestressed slightly inclined, after cuts of 17 m high behind the wall, the layer-by-layer backfilling procedure was considered (Fig. 1).

The aim of this study was to mention the different cases of stability of the wall construction system with the anchor system. Additionally, without the anchor system, and using the numerical modeling with the PLAXIS Code 8.2 each construction phase of the anchored wall was analyzed.

D. Amrani (✉)
Institute of Mines, University of Tebessa, Tébessa, Algeria
e-mail: doniaamrani05@gmail.com

A. Kallel et al. (eds.), *Recent Advances in Geo-Environmental Engineering, Geomechanics and Geotechnics, and Geohazards*, Advances in Science, Technology & Innovation, https://doi.org/10.1007/978-3-030-01665-4_76

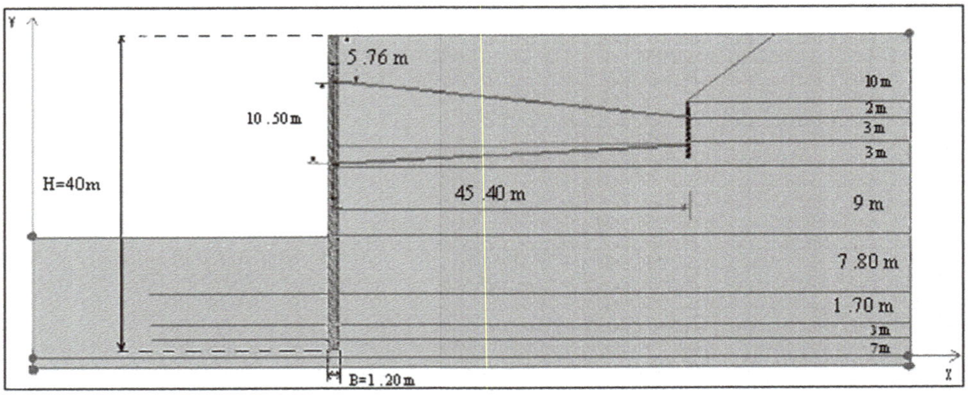

Fig. 1 Geometrical model

3 Results and Discussion

The problem was solved in a several-stage construction sequence in the Plaxis program with an increasing complexity of the anchored flexible retaining wall (Fig. 2).

The obtained results in all model phases prove the precision of calculation in the construction steps (staged construction); however, we can conclude that the numerical procedures gave an acceptable prediction of the wall movements in the whole wall construction phases.

4 Validation Criteria

Forces and displacement predictions present the major objectives of interaction analysis of soil-structure. The above results are presented in terms of total and horizontal displacements; Based on the following aims we obtain the validation step in this work (Fig. 3):

- Graphical presentation of the measured real displacement;
- Graphical presentation of the results obtained from the calculus model;
- Comparing the measured displacements with those obtained from the model.

The calculation obtained from the proposed numerical model of the diaphragm wall movements, indicate that all the displacements are generally close to the real behavior observed in situ (Fig. 4).

5 Conclusion

The obtained results in this study depend on the parameters used; these are; displacement and moments of the diaphragm wall support a multilayer soil, the effect of the different parameters; for the mesh varied from 6 and 15 nodes, the effect of the precision variation from 1 to 3% in the calculation steps was not considered. The geotechnical parameters effect may resume in the interface effects and the rigid interface vary in 1–1.65 can provide the best significance in the calculation. There is a diminution in horizontal displacement and in the bending moments in the case $R_{inter} = 1$,

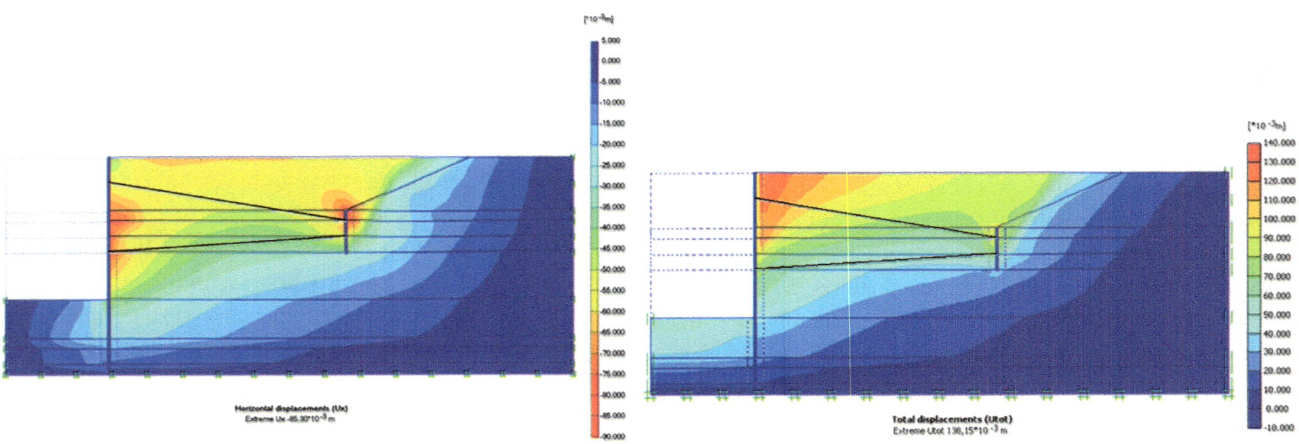

Fig. 2 Total and horizontal displacements in the structure system

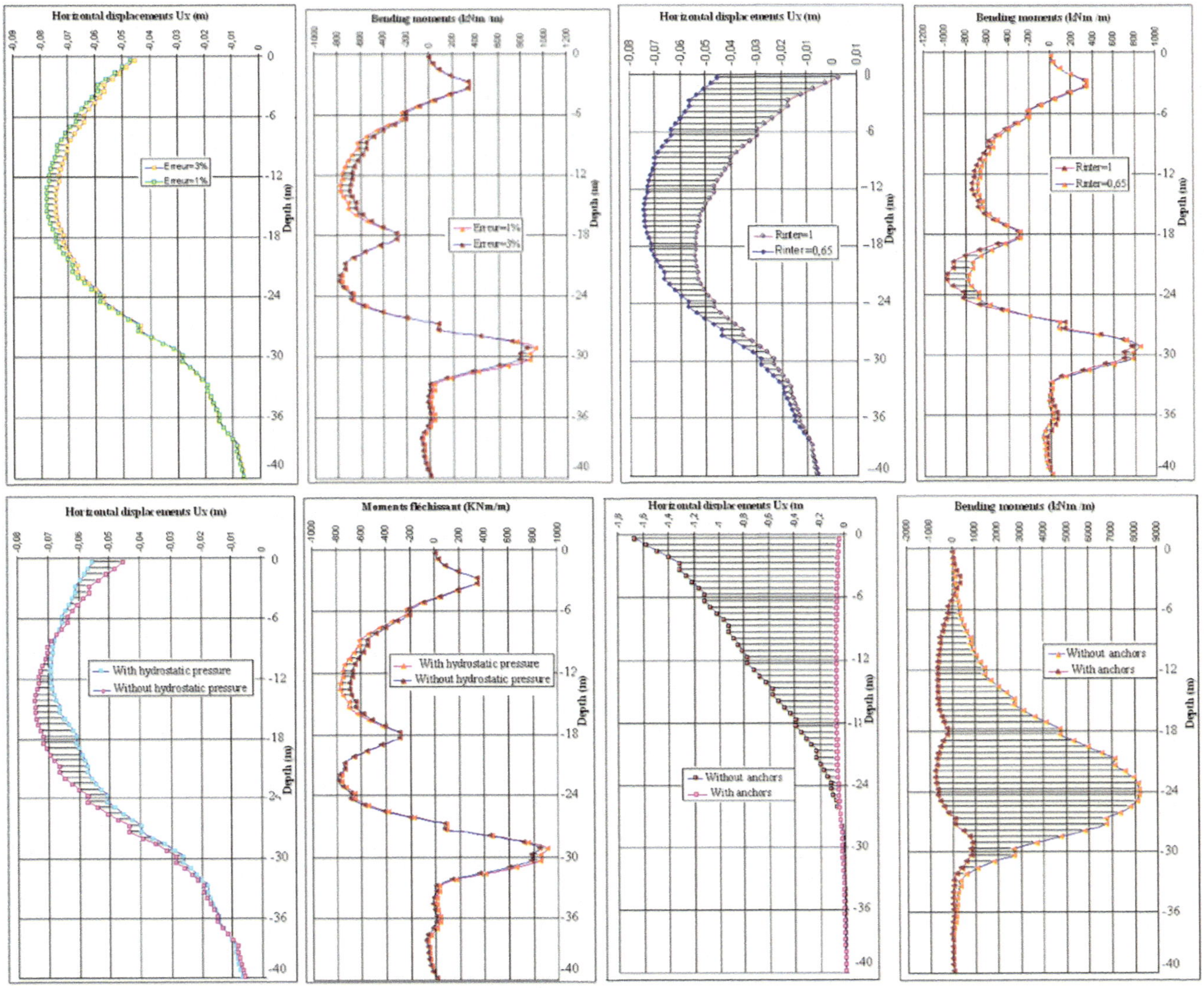

Fig. 3 Graphical presentation of the obtained results

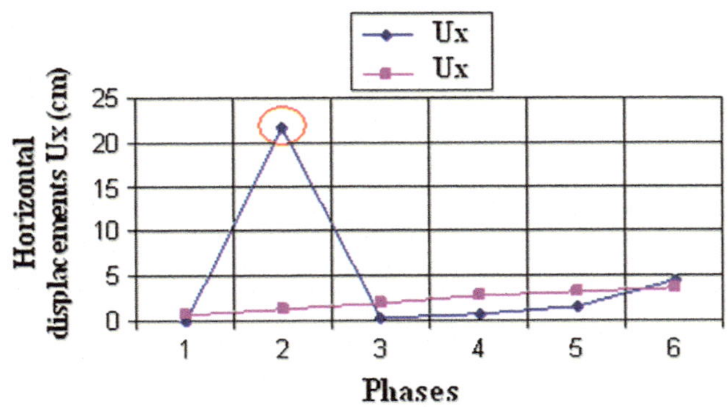

Fig. 4 Measured actual versus numerical results

but when R_{inter}= 1.65 the findings are accurate and give very close results to the real ones. Hydrostatic pressure effect can resume in the diminution in the horizontal displacement to the center of the diaphragm wall but in the head of the wall augmentation of horizontal displacements are observed with variation in the bending moments. Anchor systems, when compared to the obtained results without anchoring, have an important and very high influence on the stability of the diaphragm wall and the total stability of the entire structure.

References

1. Saeterbo, M.G., Nordal, S. Emdal, A.: Slope stability evaluations using the finite element method. NGM. In: XIV Nordic Geotechnical Meeting, Vol. 1, pp. A49–A61 (2004)
2. Kumar, N., Dey, A.: GeoInnovations. Finite Element Analysis of Flexible Anchored Sheet Pile Walls: Effect of Mode of Construction and Dewatering (2014)

3. Brinkgereve, R.B.J., Vermeer, P.A.: PLAXIS Version 8 Material Mode Manual. Delft University of Technology & PLAXIS BV, Pays-Bas (2003)
4. Bilgin, O.: Numerical studies of anchored sheet pile wall behavior constructed in cut and fill conditions. Comput. Geotech. **37**, 399–407 (2010)
5. Bilgin, Ö., Erten, M.B.: Analysis of anchored sheet pile wall deformations. Contemporary topics in ground modification, problem soils, and geo-support (GSP 187). In: Proceedings, International Foundation Congress & Equipment Expo (2009)
6. Boone, S.J.: Design of deep excavations in urban environments. A thesis submitted in conformity with the requirements for the degree of Doctor of Philosophy, Graduate Department of Civil Engineering, University of Toronto (2003)
7. Krajewski, W., Reul, O.: Deep Excavation in Stiff Clay: Comparison Between Numerical Analyses and In-situ-measurements, Ottawa, Canada, 25–27 Aug (2004)

Reliability Analysis of Bearing Capacity of Shallow Foundations

Faiçal Bendriss and Zamila Harichane

Abstract

This paper presents a reliability analysis of the bearing capacity of a shallow foundation by the FORM method and Monte Carlo Simulations. A strip foundation posed on a coherent soil exposed to a concentrated vertical load was assumed. The probabilistic methods were used to estimate the failure probability of the foundation. A numerical example was conducted in order to test the applicability of the two methods in problems dealing with bearing capacity of foundations. The probability of failure of foundations was successfully obtained with both methods.

Keywords

Reliability • Bearing capacity • Strip footing • FORM • Monte Carlo

1 Introduction

Over the past 75 years, several theories have been proposed to obtain the ultimate bearing capacity of shallow foundations [1–4]. The bearing capacity of a shallow foundation can be assessed by deterministic as well as stochastic methods. Deterministically, the existing equations and diagrams are used to evaluate the acceptable load capacity of foundations. The safety factor used in the deterministic approach takes into account the natural variability of statistical uncertainty, measurement errors and analytic model limitations which is a tool to limit deformations. But this factor does not accurately reflect random uncertainty.

Reliability-based analyses are therefore more rational because they consider the natural (or random) uncertainty of each dataset. Numerous studies have been carried out on the reliability analysis of the bearing capacity of superficial foundations over the last twenty years by different researchers [5–7].

2 Probability of Failure

Reliability analyses offer a rational context for capturing uncertainties in a foundation design [8]. Two variables are commonly utilized for assessing the reliability: the reliability index and the failure probability. Generally, a failure is given by a performance function g(·) [9]. The failure probability is given as:

$$p_f = \int_{g(x)<0} I_g(x) f_x(x) dx \tag{1}$$

where

$$\mathrm{I}(x) = \begin{cases} 1 \text{ si } \mathrm{G(x)} \leq 0 \\ 0 \text{ si } \mathrm{G}(x) > 0 \end{cases} \tag{2}$$

3 Monte Carlo Simulations (MCS)

The Monte Carlo simulations have been successfully used in engineering applications. The probability of failure can be written as follows:

$$\bar{p}_f = \frac{1}{m} \sum_{k=1}^{m} I_g\left(x^{(k)}\right) \tag{3}$$

where m is the number of samples.

F. Bendriss (✉) · Z. Harichane
Geomaterial Laboratory, University Hassiba Benbouali of Chlef, Ouled Farès, Algeria
e-mail: fa.bendriss@univ-chlef.dz

A. Kallel et al. (eds.), *Recent Advances in Geo-Environmental Engineering, Geomechanics and Geotechnics, and Geohazards*, Advances in Science, Technology & Innovation, https://doi.org/10.1007/978-3-030-01665-4_77

4 First Order Reliability Method (FORM)

The FORM method makes it possible to write the probability of failure equation in terms of a normal cumulative distribution function $\Phi(\cdot)$:

$$p_f = 1 - \Phi(\beta) = \Phi(-\beta) \tag{4}$$

5 Results and Discussion

A strip footing of width B = 4.51 m and length L = 25 m, embedded in a soil mass at a depth D_f = 1.8 m. The foundation is subjected to a horizontal load Q_h applied at a point 2.5 m above the base and vertical load applied to the center with a value Q_v. The unit weight of the soil equals 21 KN/m^3. The input parameters c, φ, Q_h and Q_v were assumed lognormally distributed random variables with mean values of 15; 25; 300 and 1100, respectively, and with coefficients of variation (Cvs) equal to 0.20; 0.10; 0.15 and 0.10, respectively.

To study the bearing capacity failure, a performance function (PerFunc) is defined as:

$$PerFunc = q_u - q \tag{5}$$

where

$$q = \frac{Q_v}{B'} \tag{6}$$

$$q_u = \frac{1}{2}\gamma B N_\gamma S_\gamma d_\gamma i_\gamma g_\gamma b_\gamma + q N_q S_q d_q i_q g_q b_q + c N_c S_c d_c i_c g_c b_c \tag{7}$$

In these equations, q_u is the ultimate bearing capacity, q the applied load. N_γ, N_q and N_c are the bearing capacity factors. S_γ, S_q and S_c are shape factors. d_γ, d_q and d_c are depth factors. i_γ, i_q and i_c are load inclination factors. g_γ, g_q and g_c are soil inclination factors.

In a simple example, the probability of failure is obtained with FORM method and Monte Carlo simulations by varying the vertical load values Q_v while the other input parameters vary randomly. As it may be observed in Fig. 1, the two methods give similar results. The probability of failure is low for relatively small vertical loads and increases as the vertical load increases. Principally, this example was achieved to make a comparative study between the two probabilistic methods and how to apply them in geotechnical calculations. However several applications may be carried varying the different parameters governing the bearing capacity and according to different models.

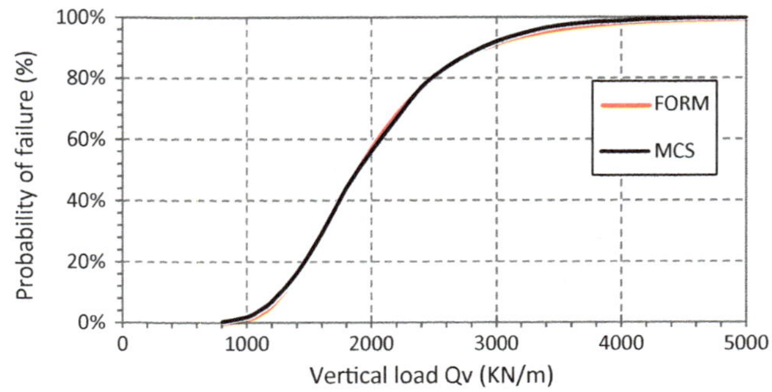

Fig. 1 Probability of failure due to vertical load and random input parameters

6 Conclusion

The reliability analysis of bearing capacity of shallow foundations was carried in this study using two different methods: FORM and MCS. Through the treated example consisting in varying a vertical concentrated load and taking into account the uncertainties in the input parameters (c, φ, Q_h and Q_v), it was found that both methods give similar results. Also, as the vertical load increases the failure probability increases. However a deeper study should be carried in order to make conclusions on the design of foundations in media with random parameters due to uncertainties made during measurements or due to inherent variability of soil properties.

References

1. Terzaghi, K.: Theoretical Soil Mechanics. Wiley, New York (1943)
2. Meyerhof, G.G.: Some recent research on the bearing capacity of foundations. Can. Geotech. J. **1**(1), 16 (1963)
3. Hansen, J.B.: A Revised and Extended Formula for Bearing Capacity. Bulletin No. 28. Danish Geotechnical Institute, Copenhagen (1970)
4. Vesic, A.S.: Analysis of ultimate loads of shallow foundations. J. Soil Mech. Found. Div. **99**(1), 45 (1973)
5. Cherubini, C.: Reliability evaluation of shallow foundation bearing capacity on c', φ' soils. Can. Geotech. J. **37**, 264–269 (2000)
6. Sivakumar Babu, G.L., Srivastava, A., Murthy, D.S.: Reliability analysis of the bearing capacity of a shallow foundation resting on cohesive soil. Can. Geotech. J. **43**(2), 217–223 (2006)
7. Belabed, L., Bencheikh, M.: Analyse semi-probabiliste de la capacité portante des fondations superficielles. Revue Française de Géotechnique n° 124, 61–75 (2008)
8. Nadim, F.: Tools and strategies for dealing with uncertainty in geotechnics. In: Probabilistic Methods in Geotechnical Engineering, pp. 71–95, Italy (2007)
9. Bucher, C.: Computational Analysis of Randomness in Structural Mechanics. Taylor & Francis, London (2009)

Numerical Analysis of Piled Raft Interaction in Soft Clay

Abdelkrim Ferchat, Sadok Benmebarek, and Mohamed Nabil Houhou

Abstract

The piled raft (PR) system was developed to use the load-carrying capabilities of both shallow foundations (rafts) and deep foundations (piles). These components interact with each other through a complex soil-structure interaction scheme. The interaction effects affect the overall load response of PR. In this study, the explicit finite difference code FLAC^{3D} (Fast lagrangian Analyses od Continua) was used to investigate the behavior of PR subjected to vertical loading. Several foundations embedded in soft clay were considered. The results prove that even in the case of soft clay, the load-carrying capacity of the raft and piles, when combined in a PR, becomes different from those of unpiled raft (UR) and group of piles (GP), due to the interactions between raft and piles.

Keywords

Piled raft • Settlement • Interaction • Soft clay • FLAC^{3D}

1 Introduction

The piled rafts are foundations for which the piles heads are connected by a raft which is in contact with the ground surface between the piles and which consequently contributes to the distribution of loads. To this end, the PR system was developed to use the load-carrying capabilities of both components: raft foundations and deep foundations [1]. These components interact with each other through a complex soil-structure interaction scheme, including the raft-soil interaction, pile-soil interaction, pile-pile interaction, and the pile-raft interaction. The interaction effects affect the overall load response of piled raft [2]. The evaluation of load capacity for PR is complicated and requires the consideration of various design parameters, such as dimensions of each foundation components, number of piles (n), piles spacing (Sp) and soil conditions. The load response and load carrying capacity of PR have been investigated experimentally and analytically using different methods. Cooke [3] reported an extensive series of small scale model tests on UR, GP and PR of various sizes, and he observed that the load distribution between piles in PR depends on the number and spacing of piles. de Sanctis and Mandolini [4] carried out, a parametric study, using of 3D FE analysis, to estimate the ultimate bearing capacity of PR loaded vertically from the separate ultimate capacities of its components. They concluded that the load capacity of PR was influenced only by pile-raft interaction. This may be limited for PR embedded in undrained clays. Based on analytical solution, Lee et al. [5] proposed a load-sharing model for considering compaction-dependent variation in clays. To check the validity of the proposed model, they conducted a centrifuge test. They neglected the piled raft interaction effects, where they stated that the piled raft interaction effect is probably not very significant for clay soils. But, the centrifuge test was conducted only for one given pile number and piles spacing. In 2015, Lee et al. [6] performed a series of 3D FE analysis to study the behavior of PR in sandy soil. They state that for sands, the effect of piled raft interaction effect is significant.

In this study, using the code FLAC^{3D} [7], a parametric study has been carried out for various foundations such as PR, GP and UR embedded in soft clay considering the full interaction between the PR components. To check the PR interaction effect for drained soft clays conditions, various piles numbers and piles spacings were considered.

A. Ferchat · S. Benmebarek (✉) · M. N. Houhou
NMISSI Laboratory, Biskra University,
BP 145 07000 Biskra, Algeria
e-mail: sadok_benmebarek@yahoo.com

A. Kallel et al. (eds.), *Recent Advances in Geo-Environmental Engineering, Geomechanics and Geotechnics, and Geohazards*, Advances in Science, Technology & Innovation, https://doi.org/10.1007/978-3-030-01665-4_78

2 Numerical Analysis

The behavior of the PR embedded in soft clays was investigated by carrying out numerical analyses using FLAC^{3D}. Various types of foundations were considered, including PR, UR and GP. In order to minimize the large amount of storage and the time required for numerical computation, small dimensions for the foundations in this study were considered as well as general dimensions and parameters considered by most researchers in the literature. The PR modeled in this study consisted of a square raft with side Br of 10 m and thickness of 1 m, and piles in different configurations. Three cases of pile arrangements, 2 × 2, 3 × 3 and 4 × 4, were investigated corresponding to 4, 9 and 16 piles, respectively. For each pile arrangement, four spacings were considered: 3d, 4d, 5d and 6d where d is the pile diameter. The length (*L*) and diameter (*d*) of all piles were 12 m and 0.5, respectively. The pile head was connected rigidly to the raft. The contact between the soil and pile was described as able to slip. A relatively fine mesh was used near the pile–soil and raft–soil interface while a coarser mesh was used further from the piles and raft.

A roller boundary was used to model zero lateral displacement along the vertical boundaries and to allow downward movement of the soil layers. Based on the preliminary results obtained, the length of the continuum was taken as 8 × B, and a depth of 2 × L were assumed in this present study, beyond which no appreciable stresses variation were observed at the boundaries. Figure 1 plots a typical FLAC^{3D} mesh used in this numerical analysis. Because of the large amount of time required to account for material nonlinearity, the settlement (S) achieved in the analyses is S = 40%d.

The soft clay behavior was assumed to be elastic perfectly plastic following the Mohr–Coulomb failure criterion. The raft and piles were modeled with an isotropic elastic model. The drained characteristics of the soft clay were adopted from reference values as reported by Nguyen [8]. Table 1 summarizes the foundation and soft clay parameters used in the analyses.

3 Computed Results

The load–normalized settlement curves obtained from these analyses of UR, GP and PR are given in Fig. 2a for the case of n = 16 and Sp = 4d. As expected the load-carrying capacity of PR is higher than those of UR and GP, so that installing a limited number of piles to the raft helps to avoid the excessive settlement and enhances the bearing capacity of the foundation. Furthermore, the load carrying capacity obtained from GP was found to be higher than those obtained from UR until a certain settlement level, but after this settlement range, the load-carrying capacity obtained from UR becomes higher than that obtained from GP. It can be noted that the load-carrying capacity of the piles is mobilized earlier than that of the raft due to the smaller foundation size. This result is in good agreement with those in [5]. In contrast, the load–normalized settlement curves of raft component (Rpr), piles component (Ppr) in PR system and PR, obtained from the present analyses are given in Fig. 2b for the case of n = 16 and Sp = 4d. We can see that the load carrying capacity obtained from Ppr shows higher values than those of Rpr for the entire settlement range considered in this study. This is due to the piles and raft interaction effects.

To study the effect of n and Sp on the PR performance, Fig. 3a introduces the load–settlement curves of UR, PR for different n values and for given Sp = 4d. In which we can see that the load-carrying capacity of piled raft increases as the number of piles increases. Similar results and trends were observed for other cases with different Sp values. n has

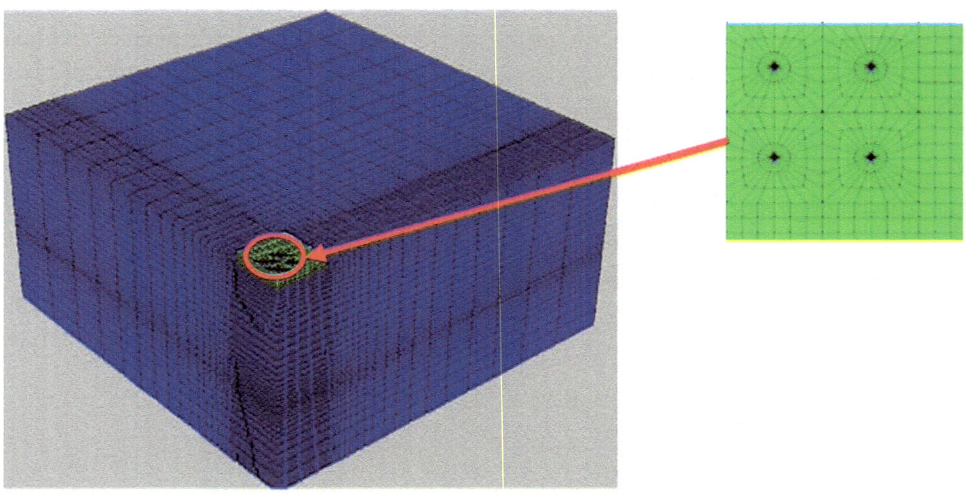

Fig. 1 Mesh used in FLAC^{3D} simulations

Table 1 Material parameters used in this study

	Model	E (MPa)	C′ (kPa)	Φ′ (°)	ν	γ_t (kN/m^3)
Foundation	Elastic	30,000	–	–	0.2	25
Soft clay	Mohr–Coulomb	4.8	15	31	0.3	14

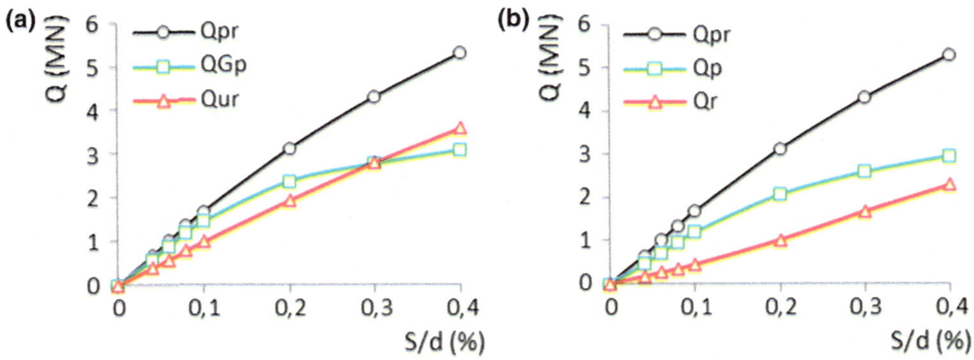

Fig. 2 Load–normalized settlement curves for n = 16 and Sp = 4d, PR, Ur and GP in (**a**) and PR, Ppr and Rpr in (**b**)

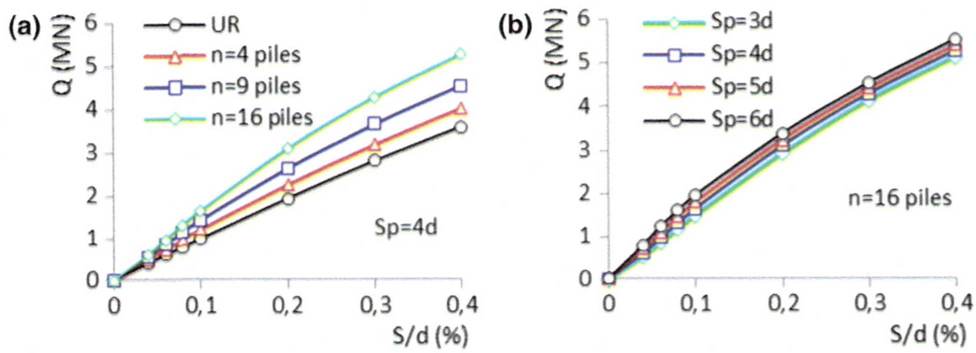

Fig. 3 Load–norrmalized settlement curves of UR and PR, **a** Sp = 4d and n = 04, 09 and 16 piles, **b** n = 16 piles and Sp = 3d, 4d, 5d and 6d

significant effect on load-carrying capacity and load-settlement responses of PR compared with the Sp effect. As shown in Fig. 3b, for given n = 16, increasing of Sp has a negligible effect on load-carrying capacity and load-settlement responses of PR, at least for the cases analyzed in this study.

4 Conclusions

From this investigation, we can note the following:

- The load-carrying capacity of PR is higher than those of UR and GP;
- The load-carrying capacities of Rpr and Ppr, when combined into a PR, become different from those of UR and GP, due to the effects of PR interactions;
- The number of piles has a major influence; however, the pile spacing has little effect on the load-settlement responses of PR.

References

1. Davis, E.H., Poulos, H.G.: The analysis of piled raft systems. Aust. Geotech. J. **2**(1), 21–27 (1972)
2. Horikoshi, K., Randolph, M.F.: Centrifuge modelling of piled raft foundations on clay. Geotechnique **46**(4), 741–752 (1996)
3. Cooke, R.W.: Piled raft foundations on stiff clays—a contribution to design philosophy. Geotechnique **36**(2), 169–203 (1986)
4. de Sanctis, L., Mandolini, A.: Bearing capacity of piled rafts on soft clay soils. J. Geotech. Geoenviron. Eng. **132**(12), 1600–1610 (2006)
5. Lee, J., Donggyu, P., Kyujin, C.: Analysis of load sharing behavior for piled rafts using normalized load response model. Comput. Geotech. **57**, 65–74 (2014)
6. Lee, J., Park, D., Park, D., Park, K.: Estimation of load-sharing ratios for piled rafts in sands that includes interaction effects. Comput. Geotech. **63**, 306–314 (2015)
7. Fast Lagrangian Analysis of Continua in 3 Dimensions. ITASCA Consulting Group, Inc, Minneapolis (2013)
8. Nguyen, N.: Modélisation du comportement des fondations profondes dans les argiles molles. Diss. Laboratoire Sols, Solides, Structures-Risques, IP Grenoble (2008)

Multivariate Assessment of Soil—Building Foundation Interaction Using PLAXIS Software

Ivan A. Shibaev, Ivan E. Sas, Dmitry M. Bagriantcev, and Oleg L. Dudchenko

Abstract

This paper discussed mathematical and physical simulations of the geological and geotechnical conditions of a rock mass disturbed by building construction and soil-building foundation interaction. The structural features of the rock mass are schematized with different levels of details; PLAXIS software was used to simulate soil-building foundation interaction for different schematic diagrams of the rock sequence. The results of the numerical simulations based on the different schematic diagrams differ by no more than 3.6%.

Keywords

Numerical simulation • Stress-strain state • Soils • Foundations • Deformation

1 Introduction

Emergency prevention is very important in the construction industry. Emergency prevention matters should be necessarily addressed at the initial stage of the design process when analyzing the geotechnical conditions of a construction site [1, 2].

The problem of assessing the stress-strain state in a virgin rock mass and choosing appropriate engineering solutions is a very important problem [3–6].

I. A. Shibaev (✉) · I. E. Sas · D. M. Bagriantcev
O. L. Dudchenko
The National University of Science and Technology "MISiS", Leninsky Prospect 4, Moscow 119991, Russia
e-mail: sh1baev@yandex.ru

I. E. Sas
e-mail: ivan_sas@bk.ru

D. M. Bagriantcev
e-mail: dmitry.bagryantsev@mail.ru

O. L. Dudchenko
e-mail: dionis_4444@mail.ru

The purpose of this work was the simulation of the stress-strain state under various schemes of schematization and subsequent comparison of the results.

2 Materials and Methods

The numerical simulation of the geotechnical behavior of rock masses requires that the natural conditions be simplified so as to develop a soil model.

Due to high degree of dissimilarity, schematization of rock masses is individual in each case.

Below are the simplified versions of the geological cross-section shown in Fig. 1.

Quaternary deposits, i.e. the uppermost technogenic deposit (tQ) and lower deposits (aQ4) numbered 1, 2, and 3, are represented by light-gray fine-grained loose sand with admixture of fine gravel and pebbles in places.

The rock mass under study (Fig. 1) consisted of sandy soils; it is considered heterogeneous due to its very variable physical and mechanical properties (Table 1).

The rock mass was schematized by two schematic diagrams (Fig. 2).

1. The first diagram shows the rock mass structure in as much detail as possible: all heterogeneities are taken into account (Fig. 2a);
2. The second diagram describes the rock mass as a layered medium with planar boundaries (Fig. 2b).

Clearly, the schematic diagrams in Fig. 2a, b significantly differ from each other in geometry.

The Mohr-Coulomb model was used to simulate soil behavior. The building was modeled as follows: the foundation was considered a plate with characteristics of B25 concrete. A distributed load of 0.15 MPa was applied to the foundation. PLAXIS 2D software wais used to perform calculations.

A. Kallel et al. (eds.), *Recent Advances in Geo-Environmental Engineering, Geomechanics and Geotechnics, and Geohazards*, Advances in Science, Technology & Innovation, https://doi.org/10.1007/978-3-030-01665-4_79

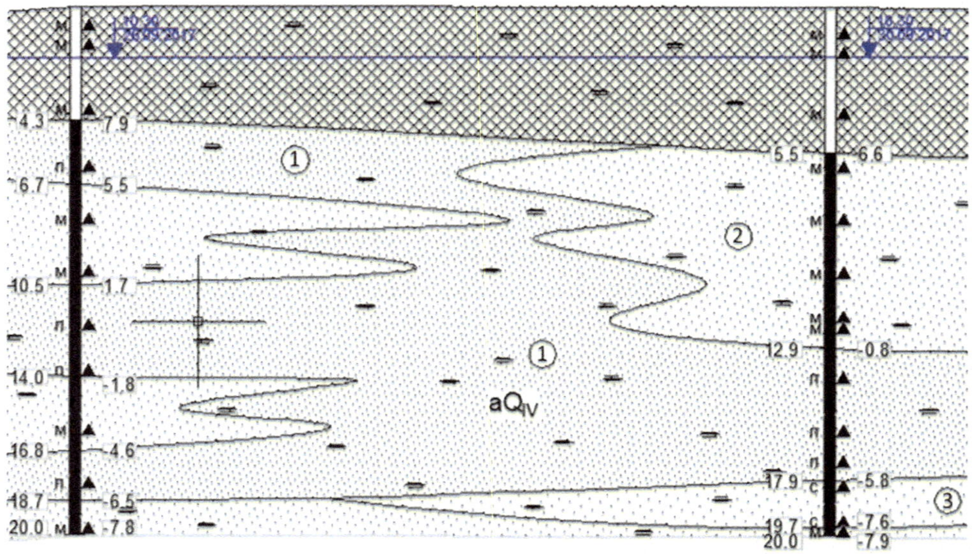

Fig. 1 Engineering geological cross-section

Table 1 Physical and mechanical properties of the designed soil types

Soil type	ρ (g/cm^3)	ρd (g/cm^3)	e	E (MPa)	ν	C (kPa)	ϕ (°)	Description
tQ	1.78	1.65	0.607	32	0.3	15	33	Very fine sand
aQ4 (1)	1.89	1.51	0.801	20	0.3	5	28	Very fine sand, water-saturated
aQ4 (2)	1.93	1.54	0.738	24.9	0.3	4	28	Water-saturated fine sand
aQ4 (3)	1.96	1.61	0.659	31.4	0.3	4	29	Water-saturated medium sand

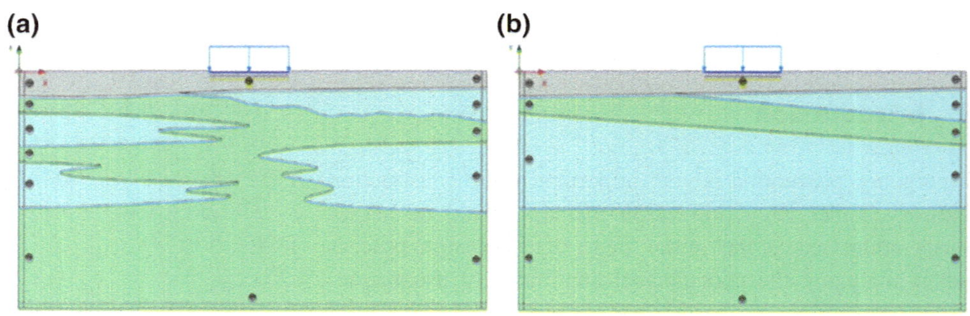

Fig. 2 Schematic diagrams of the rock mass

3 Results

Foundation settlement was estimated for two cases. The first case was a detailed representation of the geological section (Fig. 2a), and the second one was a simplified representation of the rock mass as a layered medium with planar boundaries inclining at certain angles (Fig. 2b).

The maximum settlement of the foundation was found to be 5.4 cm when the rock mass structure is represented in most detail. In the case of the layered medium model, the settlement is 5.6 cm. Thus, the variation in foundation settlement is 0.2 cm.

4 Discussion

It is shown that a more complicated and detailed model of the rock mass does not always mean a more accurate result. This is due to the fact that a great number of factors affect the results of the numerical simulations, including the physical and mechanical properties of soils, the model of their behavior, etc.

Naturally, the level of detailedness of the model should be determined by accuracy requirements for settlement prediction. Also for the simulation of the stress-strain state, the accuracy of the results plays an important role in engineering-geological studies and the use of the soils physico-mechanical properties values.

5 Conclusions

Based on the above results, it may be concluded that the simplification of the rock mass structure affects the numerical simulations accuracy by no more than 3.6%.

Acknowledgements This work was carried out with financial support from the Ministry of Education and Science of the Russian Federation in the framework of the Increase Competitiveness Program of NUST 'MISiS' (no. K2-2017-003).

References

1. Wong, K., Chan, A.P., Yam, M.C., Wong, E.Y., Kenny, T., Yip, K. K.: Findings from a research study of construction safety in Hong Kong: accidents related to fall of person from height. J. Eng. Des. Technol. **7**, 130–142 (2009)
2. Chen, Z., Wu, Y.: Explaining the causes of construction accidents and recommended solutions. In: Proceedings of the International Conference on Management and Service Science, pp. 1–5 (2010)
3. Gibb, A., Hide, S., Haslam, R., Hastings, S., Suraji, A., Duff, A.: Identifying root causes of construction accidents. J. Constr. Eng. Manage. **127**, 348–349 (2001)
4. Sas, I.E., Morozov, D.V., Morozov, N.A.: On calculation of the bearing capacity of self-opening ground anchors using PLAXIS 2D software package. Lect. Notes Mech. Eng. PartF **11**, 104–109 (2017)
5. Melnikov, N.N., Mesyats, S.P., Ostapenko, S.P., Cherepetskaya, E.B., Shibaev, I.A., Morozov, N.A., Kravcov, A.N., Konvalinka, A.: Investigation of disturbed rock zones in open-pit mine walls by seismic tomography. Key Eng. Mater. **755**, 147–152 (2017)
6. Pohl, C., Placzek, D.: Soil-structure-interaction of power plant foundations—monitoring and numerical calculation of an extensive spread foundation. In: The XVI European Conference on Soil Mechanics and Geotechnical Engineering: Geotechnical Engineering for Infrastructure and Development

Probability of Failure Assessment of Building Using Traditional and Enhanced Monte Carlo Simulation Techniques

Badreddine Chemali and Boualem Tiliouine

Abstract

The main objective of this work was to evaluate the influence of the randomness of damping on the peak response of structure under rotating machine vibrations. Two methods: Traditional and Enhanced Monte Carlo Simulation techniques were used for a quantitative analysis of the failure probability for an industrial building in the neighborhood of a resonant frequency. The results show that excellent agreement was obtained using these methods for values of covariance of damping equal or less than 60%. However, for larger values, more important difference in results between both methods was observed for structures with light damping. Moreover, computer time savings were achieved by both methods and time domain solution strategies to evaluate sensitivity functions for dynamic systems with large number of degrees of freedom were discussed.

Keywords

Peak response • Industrial building • Monte Carlo simulation • Failure probability • Sensitivity functions

1 Introduction

Problems involving vibration occur in many areas of mechanical and civil engineering. Quite often, vibration is not desirable and the interest lies in reducing it by dissipation of vibration energy or damping. The difficulty of choosing an adequate damping makes the design of the structure more complex for the engineer, due to the uncertainty generated by several parameters. [1, 2] show that the damping is not only related to the physical phenomenon but also to other parameters such as vibration amplitudes, nature of underlying damping mechanisms … etc.

A wrong estimation of the damping matrix generally induces large errors on the structural response. Such problems were the main interest of several researches showing the importance of damping in dynamic systems for different types of loading using probabilistic methods (e.g. [3, 4]).

In this study, the applicability of a Sensitive Derivative Enhanced Monte Carlo (SDEMC) method was used to predict probability of failure for the response of industrial building with uncertain damping. Failure was considered when the lateral top displacement exceeds the limit level prescribed by current design codes (e.g. H/500 in accordance with IBC (2009) design code). The numerical results obtained by the SDEMC method were independently compared to those generated using traditional MCS technique. The significance of random damping and its implications on the sensitivity of maximum displacement of the industrial building to the resonance were discussed in light of considerable ranges of coefficient of variation.

2 Sensitivity Derivative Enhanced Monte Carlo (SDEMC) Method

For a system with multi degrees of freedom subjected to a dynamic loading $\boldsymbol{F(t)}$ that generates a displacement $\boldsymbol{X(t)}$, the equations of motion can be written in matrix form as follows:

$$[M]\ddot{\boldsymbol{X}} + [C]\dot{\boldsymbol{X}} + [K]\boldsymbol{X} = \boldsymbol{F(t)} \quad (1)$$

The damping coefficients c_i are assumed lognormally distributed with a mean value $\bar{c}_i$ and standard deviation σ_{ci} total number of dampers, in the general case, is $p \geq n$. Arranging the damping coefficients in a p dimensional vector $\boldsymbol{c} = [c_1, \ldots, c_n]^T$, denoted by $\boldsymbol{X^0(t)}$, the value of $\boldsymbol{X}$ when $\boldsymbol{c}$ takes on its mean value.

B. Chemali (✉) · B. Tiliouine
Ecole Nationale Polytechnique, 10 Rue des Frères OUDEK, 16200 El-Harrach, Algeria
e-mail: badar093@yahoo.fr

A. Kallel et al. (eds.), *Recent Advances in Geo-Environmental Engineering, Geomechanics and Geotechnics, and Geohazards*, Advances in Science, Technology & Innovation, https://doi.org/10.1007/978-3-030-01665-4_80

There are several methods for dealing with the effect of uncertainty on the response of structures. The simplest is the traditional MCS technique, which involves solving N deterministic problems. But this process takes a considerable time especially for large systems. So the idea was to improve this method to reduce the time of convergence. In [5], the availability of the SD is exploited to achieve variance reduction via Sensitivity Derivative Enhanced Sampling (SDES) techniques. SDES applied to a traditional MCS is termed as Sensitivity Derivative Enhanced Monte Carlo (SDEMC) method. The SDEMC method employs the calculation of the first-order Taylor series approximation, X(c) expanded about the mean values of the input parameters $\bar{c}$ where the highest order terms were neglected in the present case.

Further analysis (shown in [5]) suggests that by incorporating knowledge of the first order sensitivity derivatives evaluated at the input parameter mean values, one can approximate $\bar{X}(c)$ by applying the MCS. The resulting approximation for mean of the output function X is given as

$$\textbf{FO-SDEMC} \quad \bar{X}(c) \approx X(\bar{c}) + \frac{1}{N}\sum_{i=1}^{n}\sum_{l=1}^{N}\frac{\partial X}{\partial c_i}(c_{i,l} - \bar{c}_i) \tag{2}$$

Introducing sensitivity derivatives

$$\xi_{ij} = (\partial X_i/\partial c_j)_{c_k=\bar{c}_k} \quad \text{for k} = 1\ldots\text{p}$$

The sensitivity functions ξ_{ij} are available by differentiation of (1) with respect to c_j as follows:

$$[M]\ddot{\xi} + [C]\dot{\xi} + [K]\xi = -(\partial[C]/\partial c_j)\dot{X} \tag{3}$$

It should be noted that the Eq. (3) is identical to the equation of motion (1) with a fictitious loading $(-(\partial[C]/\partial c_j)\dot{X})$. Thus, to obtain the vectors ξ_j (j = 1…p), first, we need to obtain the fictitious loading by solving Eq. (1) after which (3) is solved, p times. Thus, the overall number of calculations consists essentially of a computer simulation for the evaluation of the solution of (1) and the solution for ξ_j p times.

3 Numerical Example

To illustrate the method developed previously, and quantitatively demonstrate the influence of the level of uncertainty in damping on the overall dynamic response, we suggested the following simulation. Considering the example of an industrial building with five stories subjected to dynamic excitation resulting from a rotating machine; with frequency corresponding to the second resonant frequency of the structural system; applied at the first story. The parameters of structure are given in Table 1.

In this study, the mean values $\bar{c}$ of inter-story damping constants examined is equal to 197.19, 394.37, 1183.11, and 1971.86 KN/m/s corresponding to 0.5, 1, 3 and 5% of critical damping, respectively. Each value of the damping constant was assigned a coefficient of variation (COV = $\sigma/\bar{c}$) that varied from 20 to 80% with uniform increments of 20%.

In Fig. 1, The SDEMC method was used to obtain the standard deviation of structural response to 2nd story for different values of the mean damping. The results show that the uncertainty in damping increases the structural response. It is also noted that the effects of uncertainty are more pronounced for a light damping and a large COV variability.

In Fig. 2, the probability of failure (Pf) of structural response with light damping based on the exceedance of the limit level set on the performance variable, selected here to be the top displacement were presented for the two methods: SDEMC and MCS. The results show that both methods are in good agreement up to a coefficient of variation of $\boldsymbol{c}$ less than or equal to 60%. It should also be mentioned that a large dispersion was observed between the two methods for structure under dynamic loading with light damping and high COV values of damping, due to the linearization technique.

It should also be noted that the effort to compute the probability of failure of structures subjected to dynamic loading using the traditional MCS method was excessive due to the number of samples required to achieve the convergence of the method. It this study this number was fixed to five thousand samples for traditional MCS technique while for the SDEMC method the required number is two thousand samples.

Table 1 Data for 5 story industrial building example

Parameter	Symbol	Mean (μ)
Lumped mass	m_i	150 t
Inter-story stiffness	k_i	210×10^3 KN/m
Height of story	h_i	3.6 m
Amplitude of dynamic excitation	F	560 KN

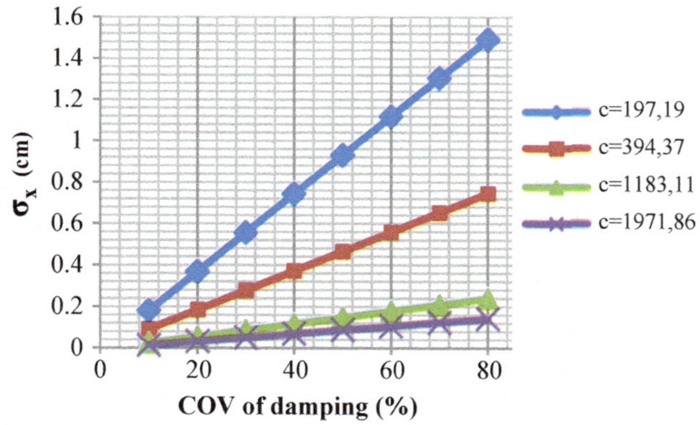

Fig. 1 Standard deviation of response (SDEMC) at the 2nd floor for different COV and damping values

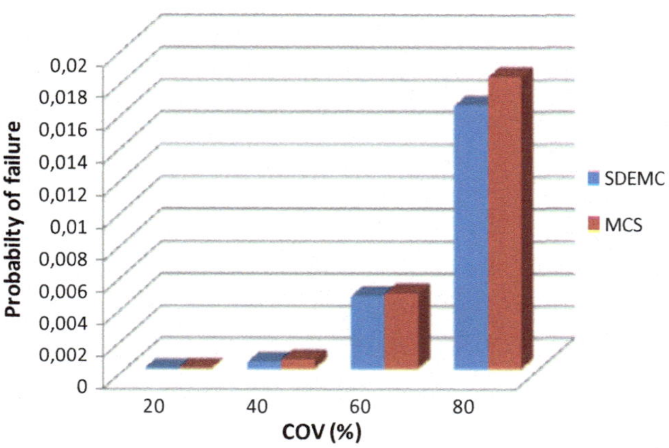

Fig. 2 Comparison Pf of top displacement between SDEMC and MCS of building with light damping ($\xi = 1\%$)

4 Summary and Conclusion

In this study, the probability of failure of an industrial building under resonant conditions is evaluated by considering the randomness of damping. Two methods traditional MCS technique and SDEMC method were used on the maximum top displacement to calculate the probability of failure of the building. The comparison between these methods proved the effectiveness of the SDEMC method for COV of damping values less than or equal to 60%, with efficient computational effort to evaluate the probability of failure.

Furthermore, the numerical results show the importance of damping uncertainty for accurate estimates of dynamic response under resonant conditions. It should also be noted that for height variability of damping the inclusion of higher order terms in the statistical response model is necessary to get more accuracy.

References

1. Li, Q.S., Xiao, Y.Q., Wong, C.K., Jeary, A.P.: Field measurements of wind effects on the tallest building in Hong Kong. Struct. Des. Tall Spec. Build. **12**, 67–82 (2003)
2. Butterworth, J., Lee, J.H., Davidson, B.: Experimental determination of modal damping from full scale testing. In: 13th World Conference on Earthquake Engineering, Vancouver, Canada, 2004, Paper No 310
3. Kareem, A., Gurley, K.: Damping in structures: evaluation and treatment of uncertainty. J. Wind Eng. Ind. Aerodyn. **59**, 131–157 (1996)
4. Tobit, J.: Evaluation of non-stationary damping characteristics of structures under earthquake excitations. J. Wind Eng. Ind. Aerodyn. **59**(2, 3) 283–298 (1996)
5. Cao, Y., Hussaini, M.Y., Zang, T.A.: Exploitation of sensitivity derivates for improving sampling methods. AIAA J. **42**, 815–822 (2004)

Slope Stability Analysis Under External Static Surcharge

Soumia Merat, Lynda Djerbal, and Ramdane Bahar

Abstract

In the assessment of slopes, factor of safety values still remain the primary indices for determining how close or far slopes are from failure. Traditional limit-equilibrium techniques are the most commonly-used analysis methods. Recently, however, the significant computing and memory resources typically available to the geotechnical engineer, combined with low costs, have made the Finite Element Method (FEM) a powerful, viable alternative. This article used a finite element approach to analyze the response of homogeneous and layered slopes. A detailed parametric analysis was presented to study the effect of surcharge (it can be an imposition of building load… etc.) on the stability of the slope.

Keywords

Slope stability • Surcharge • Finite element method • Factor of safety • Layered slope

1 Introduction

Slope stability analysis is one of the most important areas of interest in geotechnical engineering. The embankment slope can be a failure due to its geotechnical properties, the various external conditions such as depth of water, seismic loads, surcharge, etc.

This paper dealt with the assessment of slope stability under static load taking into account a number of influencing parameters such as: ground water level (GWL), slope geometry, soil properties, distance to crest…etc. The effect of each factor was discussed in this paper. Plaxis^{2D} 2015 software was used for finite element analysis which was successfully implemented to calculate Factor of Safety [1, 2, 3] for more than a decade.

2 Materials and Methods

2.1 Slope Stability Analyses by Fem and Boundary Conditions

The geometric model (Fig. 1) was incorporated in the Plaxis^{2D}. After that, the properties of soil were assigned for the specified interface and the material model was set to be the Mohr–Coulomb model. The two lateral boundaries were allowed to move only in the vertical directions (roller type: restrained against horizontal movement) whereas the bottom boundary was completely restrained and fully blocked from movement (Closed: restrained in both horizontal and vertical directions).

2.2 Geometry and Soil Parameters

In this study, a slope with an inclination angle of 26.56° (2/1) to the horizontal plane and a height of 50 m were considered (Fig. 1). The groundwater table was assumed to be at the ground surface. The soil type was specified as «Marl» of Azazga. The soil properties related to the unsaturated soil were taken as given by [4].

The surface soil layer behavior was modeled as linear, elastic-perfectly plastic material using Mohr–Coulomb model and analyzed using soil parameters. The FEM analysis using Mohr-Coulomb material model requires such parameters as modulus of elasticity, poison's ratio, angle of internal friction and cohesion. The properties of soil used in the present study are presented in Table 1. Two types of soils were considered.

S. Merat (✉) · L. Djerbal · R. Bahar
LEEGO, University of Science and Technology Houari Boumediene, BP 32 16111 El Alia, Algiers, Algeria
e-mail: smerat@usthb.dz

A. Kallel et al. (eds.), *Recent Advances in Geo-Environmental Engineering, Geomechanics and Geotechnics, and Geohazards*, Advances in Science, Technology & Innovation, https://doi.org/10.1007/978-3-030-01665-4_81

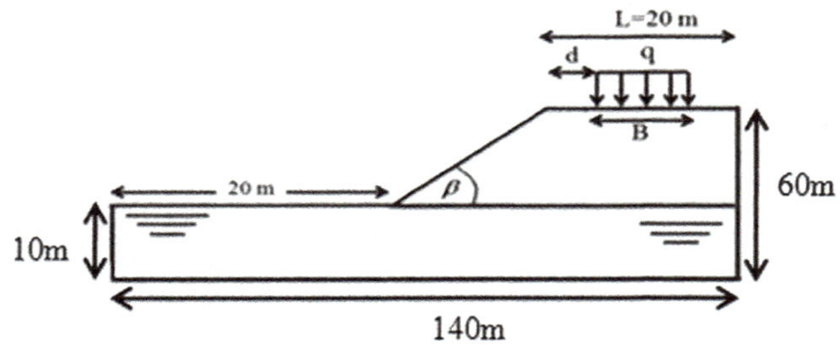

Fig. 1 Geometry of the numerical model of the considered slope in the present study

3 Results and Discussion

3.1 Ground Water Depth Effect

Before applying a surcharge to the model, we tried to understand the effect of ground water level on the slope stability; we decreased the ground water level by 5 m and we tried to find the relation between FS-GW depths. The preliminary model indicated, as obvious, that the stability of the slope is found to increase if the water table is located at deeper locations and it is linearly proportional as shown in Fig. 2.

3.2 Ground Water Level and Surcharge Effect

A combined study was used to show the effect of ground water level and surcharge on the slope stability (Fig. 3). The safety factor appears to be proportional to the increase of ground water depth and load intensity as shown in the figure below. It is also noted that the factor of safety is proportional to the applied surcharge where the safety factor is higher for lower surcharge values as clearly displayed in the figure below.

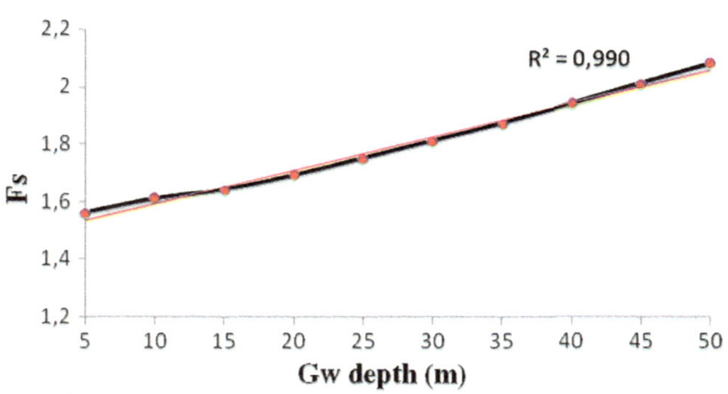

Fig. 2 Variation of factor of safety (FS) for different ground water depth values

3.3 Soil Properties Effect (Cohesion and Friction Angle)

For different values of cohesion and friction angle, a comparison between a model without load and a loaded one reveals that the surcharge decreases the stability of the slope as shown in the figure below (Figs. 4 and 5). From the Table 1, it can be noted that soil-1 represents high properties in comparison with soil-2. This is why the slope failure (critical safety) is reached first if the layer takes the soil2 properties during the application of a surcharge. The initial factor of safety for soil1 (2.08) is greater than that of soil2 (1.29) for initial values of cohesion and friction angle.

3.4 Layered Slope Effect

From the Table 1, it is noted that soil-1 represents high properties in comparison with soil-2. Consequently, the slope failure (critical safety) is reached first if the layer takes soil-2 properties during the application of a surcharge. However, a layered slope as shown in Fig. 6 gave a higher degree of stability in comparison with a slope with only one layer of soil-2.

Because the clayey loam layer (soil-2) is a weak soil layer, the increase in the layer thickness in a layered slope decreases the stability of the slope as shown in Fig. 7.

3.5 Slope Geometry Effect

3.5.1 Slope Height Effect

For the same model as shown in Fig. 1 the slope angle was fixed (26.56°) and the slope height was varied (from 10 to 50 m) (Fig. 8).

3.5.2 Slope Angle Effect

The slope height was fixed, it took 50 m as a value and the slope angle was varied (26,56°; 33,69°; 45°) (Fig. 9) in order to know the behavior of the slope for different slope angle values. The results show that the factor of safety decreased

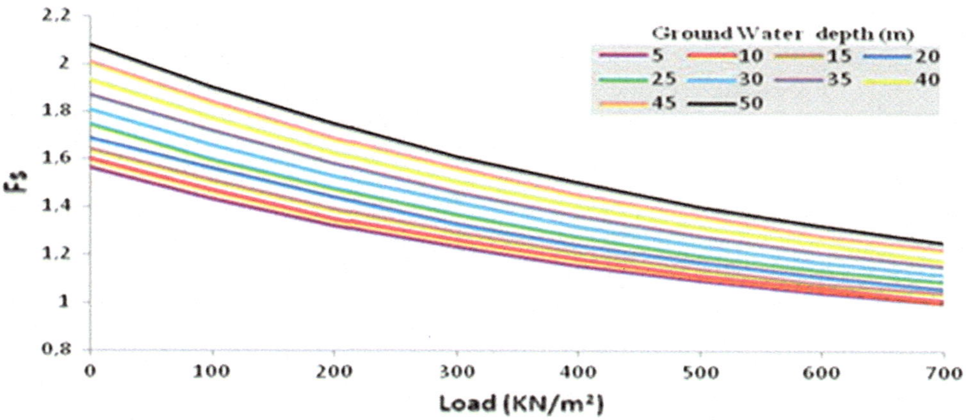

Fig. 3 Variation of factor of safety (FS) with surcharge (q) for different ground water depths

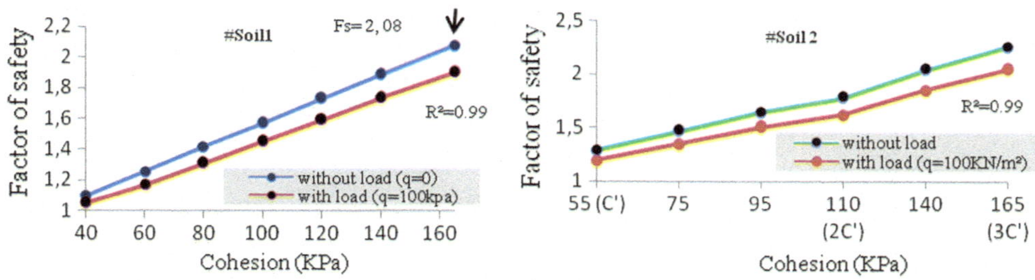

Fig. 4 Variation of factor of safety (FS) with and without surcharge (q) for different values of cohesion

Fig. 5 Variation of factor of safety (FS) with and without surcharge (q) for different values of friction angle

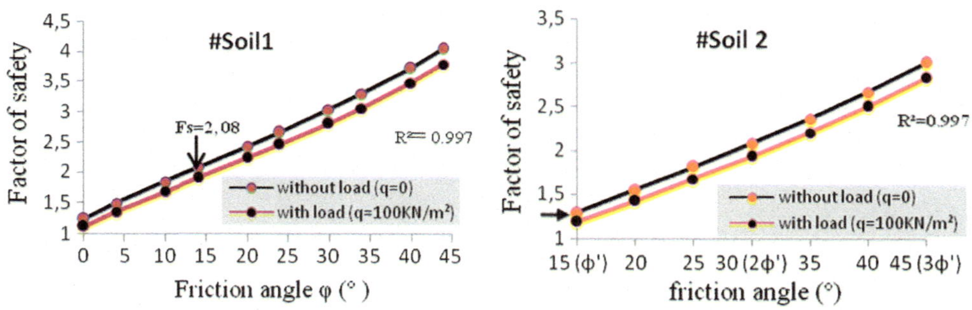

gradually when the slope angle increased during load application (Fig. 9).

3.6 Surcharge Position Effect

A slope profile with L = 80 m was used as an example to show the effect of surcharge position from the crest on the factor of safety. The surcharge width B was fixed (B = 15 m). As shown in Fig. 8, the distance to the crest varied linearly with factor of safety (with $R^2 = 0.988$). Figure 10 shows that the factor of safety increased when the distance of surcharge increased from the crest of the slope.

4 Conclusion

The results analysis allows us to say that the major findings of the study can be stated as follows:

- The factor of safety increases with the increase of soil properties (cohesion, friction angle).

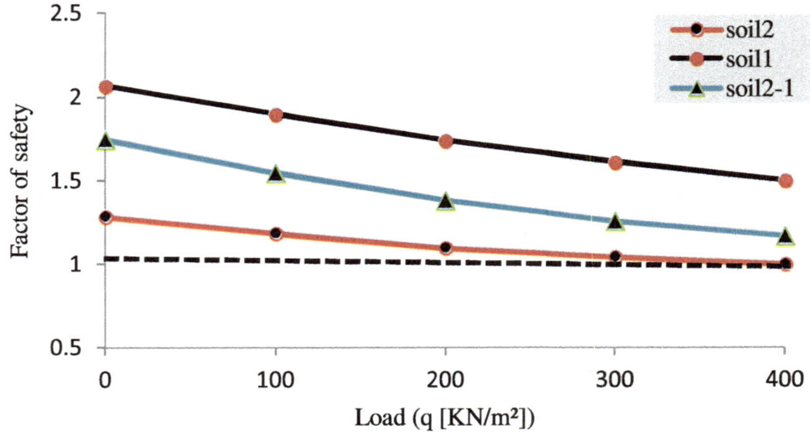

Fig. 6 Variation of factor of safety for the different soil cases

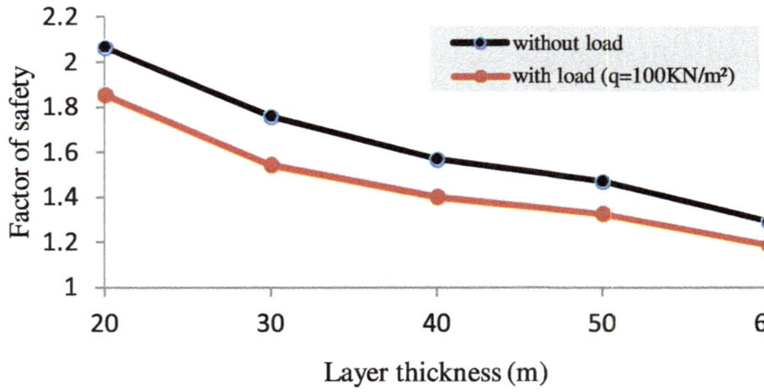

Fig. 7 Variation of factor of safety for the different layer thicknesses

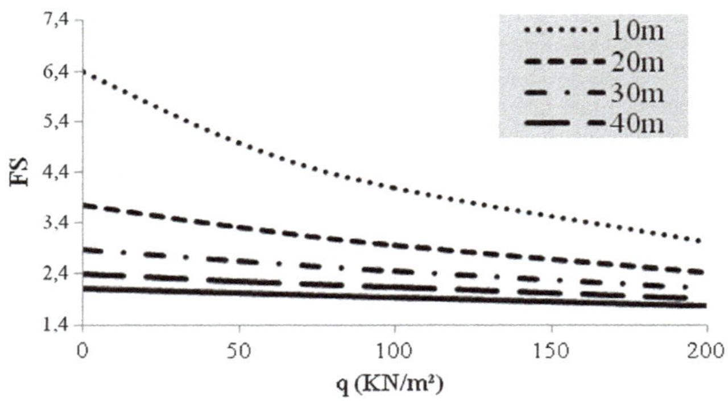

Fig. 8 Variation of factor of safety (FS) with surcharge (q) for different slope heights

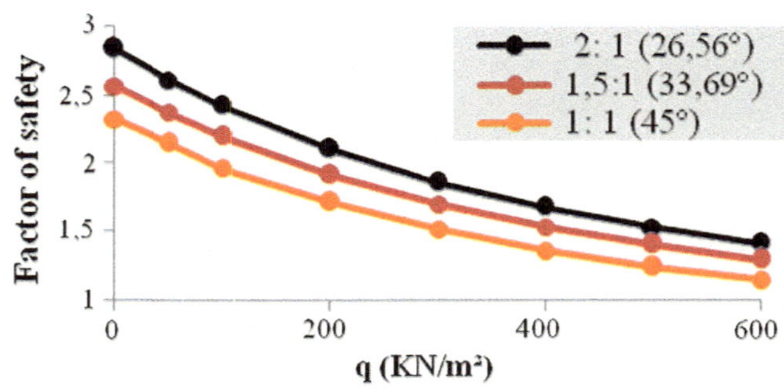

Fig. 9 Variation of factor of safety (FS) with surcharge (q) for different slope angles

- For the range of parameters considered in the present study, the factor of safety of the slope decreases with the increase of surcharge;
- At any surcharge, the factor of safety decreases with the increase of ground water level;
- The slope stability increases with the decrease of both slope angle and slope height;
- The factor of safety of a slope with layered soil decreases with the increase of weak soil layer thickness;
- Factor of safety increases with the increase of the position of surcharge from the crest of slope.

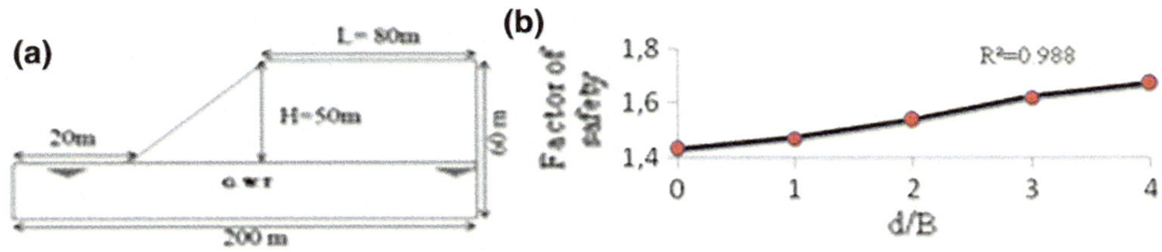

Fig. 10 Study of the effect of load position. **a** Slope profile used for the study, **b** variation of factor of safety in function of distance to crest values

Table 1 Summary of soils parameters of Azazga soil layer used in the FEM

Soil layer	Soil model	γunsat (kN/m³)	γsat (kN/m³)	N	C (kPa)	φ (°)	K (m/min)	E′ (kPa)
Soil 1 (Marl)	Mohr–Coulomb	18	19	0.33	165	14	$Kx = 10^{-5}$ $Ky = 10^{-5}$	4×10^6
Soil 2 Clayey loam	Mohr–Coulomb	15.8	17.8	0.33	55	15	$Kx = 10^{-3}$ $Ky = 10^{-3}$	4.038×10^5

References

1. Dawson, E.M., Roth, W.H., Drescher, A.: Slope stability analysis by strength reduction. Geotechnique **49**(6), 835–840 (1999)
2. Griffiths, D.V., Lane, P.A.: Slope stability analysis by finite elements. Geotechnique **49**(3), 387–403 (1999)
3. Hammah, R.E., Curran, J.H., Yacoub, T.E., Corkum, B.: Stability analysis of rock slopes using the finite element method. In: Proceedings of the ISRM Regional Symposium EUROCK 2004 and the 53rd Geomechanics Colloquy. Salzburg, Austria (2004)
4. van Genuchten, M.T.: A closed-form equation for predicting the hydraulic conductivity of unsaturated soils. Soil Sci. Soc. Am. J. **44**, 892–898 (1980)

Stability Analysis of Slopes Prone to Circular Failures Using Logistic Regression

Mehmet Sari

Abstract

The analysis of stability of slopes is a classical problem for geotechnical engineers. In practice, much user friendly software is available for proper usage. Besides additional techniques capable of providing information useful for decision-making are necessary. In this research, extensive slope failure data collected by Sah et al. [1] were used for the stability analysis of slopes subjected to circular failures. For the purpose, 44 separate slope models were prepared for each slope case given in the original work. The Slide® program produced similar safety factors for the studied cases. Binary logistic regression analysis was applied in the study as an alternative technique to predict stability condition of slopes. The model predicted the stability condition of slopes with 90.9% accuracy.

Keywords

Limit equilibrium analysis • Circular failure • Slope stability • Logistic regression • Slide program

1 Introduction

One of the most challenging problems in geotechnical design is the stability analysis of slopes. The occurrence of slope failures in an area depends on a complex interaction of natural as well as man-made factors, such as rock and soil strength, slope geometry, permeability, precipitation, presence of old failures, proximity to streams and flood-prone areas, land cover patterns, excavation of lower slopes and/or increasing the load on upper slopes, alteration of surface and subsurface drainage, seismic events etc. [2]. Many methods have been developed for analyzing slope stability. The limit equilibrium method is one of the most often used techniques for practical problems. Locating the critical slip surface that has the lowest factor of safety is an important part of analyzing slope stability, and a large number of computer techniques have been developed to automate as much as possible this process.

The main objective of this study was to model some case slopes susceptible to circular failures using slope stability analysis software. Another objective of the study was to search for proper statistical techniques for predicting slope stability conditions. The aim was to find the major contributing elements on the occurrence of circular slope failures.

2 Building Slope Models Using Slide® Program

In this research, slope failure data sets collected from available literature by Sah et al. [1] were used for the stability analysis of slopes subjected to circular failures. 46 cases were reported in the study of Sah et al. [1] for circular or near circular failures in soils or highly fractured rocks. 2 cases were excluded in the current analysis due to duplicate counting in the original study. Basic descriptive statistics for the established database by Sah et al. [1] is given in Table 1.

In the beginning of the study, the collected data sets were carefully evaluated to settle a unique threshold value of safety factor. It was expected that this value could sufficiently classify the slopes as "stable" or "failed" based on the observed slope conditions. For this purpose, the mean of safety factor values given in Table 1 was chosen. It was seen that this value could reasonably classify the condition of slopes with 93.2% accuracy (see Fig. 1). In the original study of Sah et al. [1], the type of limit equilibrium method that was employed to analyze safety factors for the reported cases was not clearly stated. For this reason, the models of slopes studied in this paper were rebuilt in commercially available software, Slide® v5.0 [3], which is a 2D limit

M. Sari (✉)
Mining Engineering Department, Aksaray University, 68100 Aksaray, Turkey
e-mail: mehmetsari@aksaray.edu.tr

A. Kallel et al. (eds.), *Recent Advances in Geo-Environmental Engineering, Geomechanics and Geotechnics, and Geohazards*, Advances in Science, Technology & Innovation, https://doi.org/10.1007/978-3-030-01665-4_82

Table 1 Basic descriptive statistics for the established database [1]

Parameter (unit)	Symbol	Min	Max	Mean	Std Dev
Unit weight (kN/m^3)	UW	12.00	28.44	20.07	3.60
Cohesion (kPa)	C	0.00	150.05	21.41	32.13
Internal friction angle (deg)	IFA	0.00	45.00	27.40	11.22
Slope angle (deg)	SA	16.00	53.00	32.61	10.14
Slope height (m)	SH	3.66	214.0	45.63	49.09
Pore water pressure ratio	PWPR	0.00	0.50	0.18	0.19
Safety factor	SF	0.63	2.05	1.25	0.38

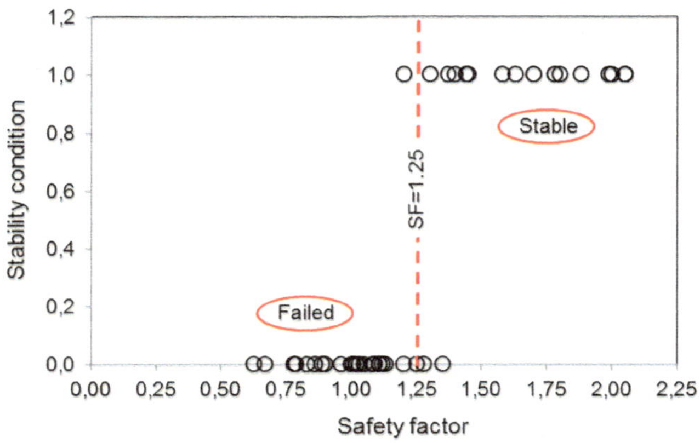

Fig. 1 Classification of the stability condition of slopes based on a threshold safety factor

equilibrium slope stability program for soil and rock slopes. Slide® analyzes the stability of slip surfaces using vertical slice limit equilibrium methods. Individual slip surfaces can be analyzed, or search methods can be applied to locate the critical slip surface for a given slope. In this study, three well-known techniques namely Bishop simplified [4], Janbu simplified [5] and Spencer [6] were used to locate critical circle using grid search technique in the program. In total, 132 separate safety factors were calculated by Slide® program for three modeling methods as given in Fig. 2. It was seen that in general, three modelling methods calculated safety factors for the studied cases in very similar ways.

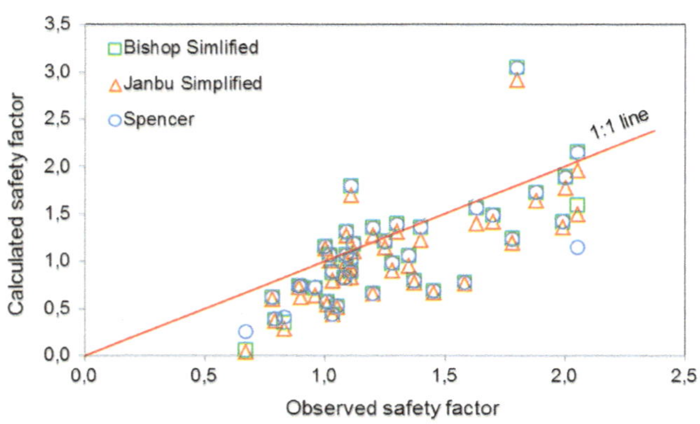

Fig. 2 Comparison of observed and calculated safety factors by three modelling methods

3 Logistic Regression Analysis

In this section, the logistic regression analysis was carried out on 44 data sets using statistical software package SPSS 16.0 [7]. The information about the logistic regression technique can be found in many textbooks in the literature [8–10]. The binary logistic regression is essentially an extension of multiple regressions in situations where the dependent variable (Y) is not a continuous or quantitative parameter. It is a useful tool for analyzing data that included categorical response variables, such as yes-no or live-die or stable-failed, as compared to the regression of quantitative data. Independent variables (Xs) may be in continuous or categorical form. In the logistic regression analysis, the relationship between the dependent variable and independent variables is assumed to be non-linear. The Logistic regression does not model the dependent variable directly, but it is rather based on the probabilities associated with the values of Xs.

In this study, the logistic regression model was developed with six independent variables (Xs) and one dependent variable which is stability (Y) coded as 1 for a stable slope and 0 for a failed one. The logistic regression equation is as follows:

$$p(X) = 1/(1 + \exp-(0.274UW + 0.174C + 0.513IFA - 0.521SA - 0.127SH - 16.386PWPR)) \tag{1}$$

where p(X) is probability of stability of slope, UW is unit weight (kN/m^3), C is cohesion (kPa), IFA is internal friction angle (deg), SA is slope angle (deg), SH is slope height (m), and PWPR is pore water pressure ratio.

Based on the logistic regression model given in Eq. (1), the slopes stability probability can be calculated for each case history. When the outcome of this equation is less than 0.5, the slopes are considered "failed" and when the outcome is equal or more than 0.5, they are considered "stable". The binary logistic regression model classified whether a slope is "failed" or "stable" with a percent accuracy of 90.9. The Wald statistics revealed that the internal friction angle was the most significant factor in the occurrence of circular failures followed by slope angle, pore water pressure ratio, slope height, cohesion, and unit weight, respectively.

4 Conclusion

In this study, the data collected from available literature were used to investigate the stability of slopes susceptible to circular failures. For this purpose, a 2D limit equilibrium slope stability analysis program was performed to model the slopes based on their geotechnical properties and slope geometry. It was seen that the built models produced very similar safety factors for the used methods. Another objective of the study was to obtain the most relevant parameters on the occurrence of circular failures using statistical modeling techniques. The results of logistic regression model explicitly showed that the most and the least significant parameters contributing to stability of slopes were internal friction angle and unit weight, respectively. This model can classify a slope to be stable or not with 90.9% accuracy.

References

1. Sah, N.K., Sheorey, P.R., Upadhyaya, L.N.: Maximum likelihood estimation of slope stability. Int. J. Rock Mech. Min. Sci. Geomech. Abst. **31**(1), 47–53 (1994)
2. Nandi, A., Shakoor, A.: Application of logistic regression model for slope instability prediction in Cuyahoga River Watershed, Ohio, USA. Georisk **2**(1), 16–27 (2008)
3. Slide v5.0: User Manual. Rocscience Inc., Toronto, Canada (2005)
4. Bishop, A.W.: The use of the slip circle in the stability analysis of slopes. Geotechnique **5**(1), 7–17 (1955)
5. Janbu, N.: Slope stability computations. Soil Mechanics and Foundation Engineering Report. The Technical University of Norway, Trondheim, Norway (1968)
6. Spencer, E.: A method of analysis of the stability of embankments assuming parallel inter-slice forces. Geotechnique **17**(1), 11–26 (1967)
7. SPSS v16.0: Statistical Analysis Software (Standard Version). SPSS, New York (2007)
8. Hosmer, D.W., Lemeshow, S.: Applied Logistic Regression. Wiley, New York (2000)
9. Kleinbaum, D.G., Klein, M.: Logistic Regression: A Self-Learning Text. Springer, New York (2002)
10. Leech, N.L., Barrett, K.C., Morgan, G.A.: SPSS for Intermediate Statistics: Use and Interpretation. Lawrence Erlbaum, New Jersey (2005)

Comparative Study on the Influence of the Variation of Initial Stress on Slope Stability

Brahim Lafifi and Mohamed Salah Nouaouria

Abstract

The analysis of slope stability is one of the oldest research subjects in the geotechnical engineering field [1]. The modelling of this stability is complex and remains a significant and unfinished research topic [2, 3]. The goal of this work was to study the influence of the variation of the initial stress field on the slope stability by the finite element method. The stability of a slope built with respect to three different configurations was studied through calculation of the safety factor using the Plaxis software. The first way of creating a slope consists of a progressive loading of the slope weight by a multiplying coefficient, the two other methods, take into account the chronology of the events and consist of performing cut slopes or fill slopes. Finally a comparison was carried out between the results of these studied cases.

Keywords

Slope stability • Initial stress • Finite element method • Safety factor

1 Objective of the Study

The goal of this work was to study, by finite elements modelling, the influence of the stress and strain state on the safety factor of a slope from purely a mechanical point of view [4]. To do so, a slope was achieved by three different ways using PLAXIS software. The first way consisted of a progressive loading of the slope's own weight, using a multiplier coefficient [5]. The two other methods took into account the chronology of the events and consisted of the construction of the slope by cutting or by embankment.

B. Lafifi (✉) · M. S. Nouaouria
Laboratory LGCH, Université 8 Mai 1945 Guelma, 24000 Guelma, Algeria
e-mail: lafifi.brahim@univ-guelma.dz

The calculation of the stability was achieved in two stages:

1. Creation of the slope by one of these three methods;
2. Calculation of the safety factor using the c-φ reduction method.

2 General Model

The analysis was carried out in the context of plane deformations [5], we considered a two-dimensional cut of the slope, established by a homogeneous soil of a height of 10 m, with a width of 60 m. The geometry and boundary conditions of the model are given in Fig. 1.

The massif was discretized by triangular elements with 15 nodes. The behavior of this massif was modelled with the Mohr Coulomb model. The set of parameters of the t soil is presented in Table 1.

2.1 Calculation Phases

The analysis of this model was performed in two steps:

- The application of the soil weight in a progressive way by a multiplier coefficient to simulate the effect of the gravity.
- Calculation of the safety factor using the c-φ reduction method.

3 Analysis of Results

The calculation of the safety factor gives us an average value of $M_{sf} = 1.43$ (Fig. 2).

A. Kallel et al. (eds.), *Recent Advances in Geo-Environmental Engineering, Geomechanics and Geotechnics, and Geohazards*, Advances in Science, Technology & Innovation, https://doi.org/10.1007/978-3-030-01665-4_83

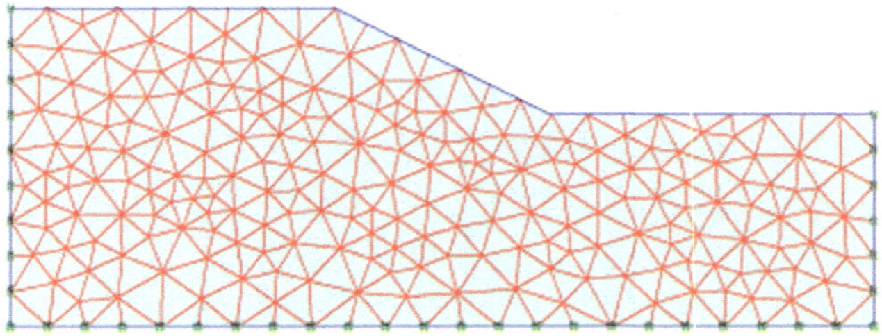

Fig. 1 Mesh and boundary conditions

Table 1 Soil parameters properties

Dry unit weight γ_d (kN/m^3)	Cohesion c′ (kN/m^2)	Friction angle φ' (°)	Dilatancy angle ψ (°)	Young's Module (kN/m^2)	Poison's ratio v
17.0	1.0	34.0	0.0	60,000.0	0.30

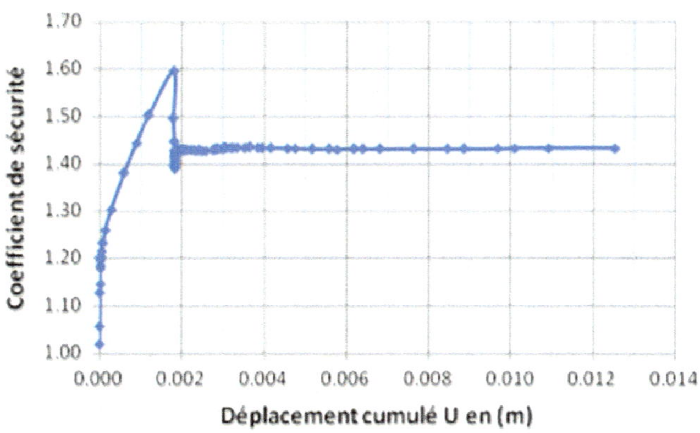

Fig. 2 Variation of the safety factor

3.1 The Slope by Excavation Modelling

This technique consists of an excavation of a massif of soil. We disabled groups of elements to create an embankment. The starting model is represented by a rectangle of dimension 60 m × 10 m and 5 layers of soil are removed in 5 steps (Fig 3).

The representation of the variation of the safety factor, according to the height of the excavation, shows us a sudden decrease of its value at the beginning of the excavation, then its value tends to stabilize at the value of 1.42 at the end of excavation (Fig. 4).

3.2 The Slope by Embankment Modelling

In this method, we considered a layer of soil on which other soil layers were added in order to make an embankment. The initial conditions were fixed by the soil weight. The starting model is represented by a rectangle with dimensions of 60 m × 10 m. This backfill was achieved by adding 5 layers of soil in 5 steps (Fig. 5).

The representation of the variation of the safety factor, according to the height of the embankment, shows us a decrease of its value at the beginning of the construction, then its it tends to stabilize in the value of 1.41 at the end of construction (Fig. 6).

4 Conclusion

Through this work, we studied three methods of creating a slope using the finite element method. Then, safety factors were calculated for each case, and minimal, if not no difference was observed for the different modellings. In conclusion, we notice, that the way of setting up the slope does not influence the safety factor and consequently, we can conclude that the so-called embankment method is already in place, and thanks to its simplicity, it seems the most suitable for studying the stability of slopes.

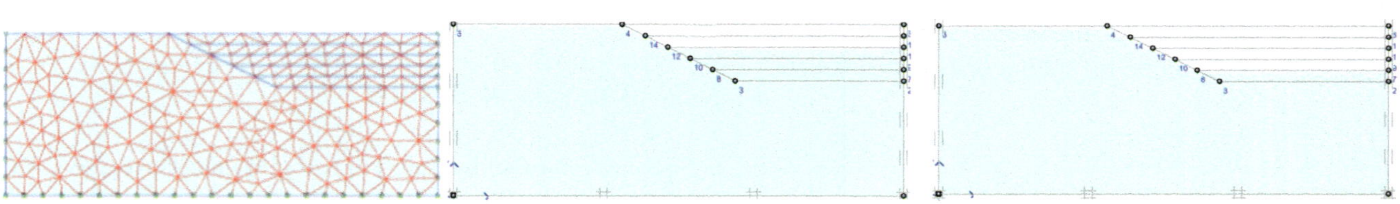

Fig. 3 Construction of the embankment by excavation

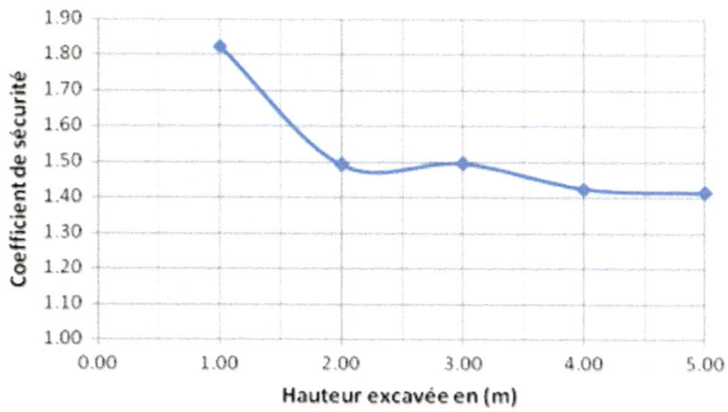

Fig. 4 Variation of the safety factor during the excavation

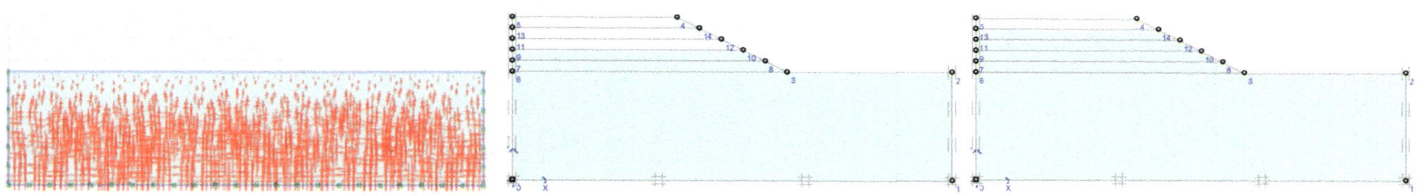

Fig. 5 Construction of the slope by embankment

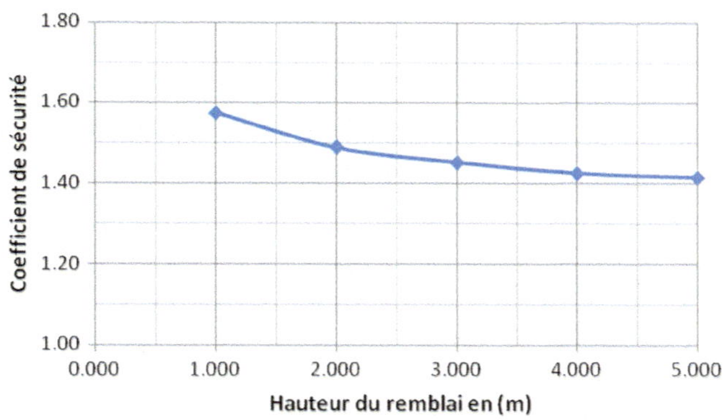

Fig. 6 Variation of the safety factor during the construction of the embankment

References

1. Borja, R.I., Lee, S.R, Seed, R.B.: Numerical simulation of excavation in elastoplastic soils. Int. J. Numer. Anal. Methods Geomech. **13**(3), 231–249 (1989)
2. Brinkgreve, R.B.J., Bakker, H.L.: Non-linear finite element analysis of safety factors. In: Beer, G., Booker, J.R., Carter, J.P. (eds.) Computer Methods and Advances in Geomechanics, pp. 1117–1122, Balkema, Rotterdam (1991)
3. Dawson, E.M., Roth, W.H., Drescher, A.: Slope stability analysis by strength reduction. Geotechnique **49**(6), 835–840 (1999)
4. Duncan, J.M., Wright, S.G.: Soil Strength and Slope Stability. Wiley, Hoboken, New Jersey, (2005). Homepage, http://www.springer.com/ASTI. Last accessed 2016/11/21
5. Griffiths, D.V., Lane, P.A.: Slope stability analysis by finite elements. Géotechnique **49**(3), 387–403 (1999)

The Applicability of the Hybrid Method to Analyze Slope Stabilizing Contiguous Pile Walls

Mehdi Dib, Salim Kouloughli, and Mabrouk Hecini

Abstract

Reinforcing precarious slopes is a very delicate operation, especially when the slope is near an urbanized area. The use of contiguous pile walls can be very effective. This type of reinforcement ensures that soil does not flow between the piles. The present paper represented a numerical investigation regarding the applicability of the hybrid method to analyze and design slope stabilizing contiguous pile walls. Both the coupled and the simplified models were analyzed. It was found that the simplified model gives close results to those given by the fully coupled model with an error of 9.5%. The analyses of pile walls using the hybrid method allow researchers and engineers to perform parametric analyses at a reduced numerical cost compared to the fully coupled numerical analysis.

Keywords

Contiguous pile wall • Hybrid method • Slope

1 Introduction

Stabilizing slopes is a very sensitive operation. When the slope is near an urbanized area, the reinforcement must be efficient, and the soil must not flow between piles. Contiguous pile walls can be very efficient in this case, the very small pile spacing ensures that soil does not flow between piles.

Understanding the behavior of slope stabilizing piles was undertaken by many researchers among whom we can cite: Ito and Matsui [1], Wang et al. [2], Poulos [3], Jeong et al. [4], Ashour and Ardalan [5], Kourkoulis et al. [6], Ho [7], Keawsawasvong and Ukritchon [8].

Kourkoulis et al. [6] proposed a new concept in the analysis of slope stabilizing piles which consists in a hybrid methodology. This concept regroups analytical approaches in slope stability, with the three-dimensional finite element numerical analysis. The philosophy of the hybrid method is firstly to analyze the slope using conventional analytical methods like Bishop, in order to estimate the needed stabilizing force that piles must offer, then, secondly, to perform a numerical simulation on a simplified numerical model to design the adequate reinforcing piles.

The hybrid method was developed for a minimum pile spacing S = 2D (spacing equal twice the pile diameter), therefore, in the present paper, the applicability of the hybrid method in case of contiguous pile wall was investigated.

2 Materials and Methods

In order to validate the applicability of the hybrid method to analyze contiguous pile walls, a fully coupled slope reinforced by a contiguous pile walls was analyzed first, the slope has an inclination of 26.5° with a pre-existing potential sliding interface (PEPSI) (Fig. 1a), piles are made of concrete with a diameter D = 1.2 m. Then secondly, the simplified hybrid model was analyzed (Fig. 1b). The material properties of both slope and piles are summarized in Tables 1 and 2, respectively. The finite element software ABAQUS was adopted, where Mohr-Coulomb failure model was implemented for soils, and piles were simulated as elastic.

3 Results and Discussions

After the analysis of the coupled model, it was found that the pile head deflection after the stabilization is equal to 4.38 mm. On the other hand, the analysis of the slope

M. Dib (✉) · S. Kouloughli
Université des frères Mentouri, 25000 Constantine, Algeria
e-mail: dib.mehdi@umc.edu.dz

M. Hecini
Laboratory of Mechanical Engineering (LGM), Université de Biskra, 07000 Biskra, Algeria

A. Kallel et al. (eds.), *Recent Advances in Geo-Environmental Engineering, Geomechanics and Geotechnics, and Geohazards*, Advances in Science, Technology & Innovation, https://doi.org/10.1007/978-3-030-01665-4_84

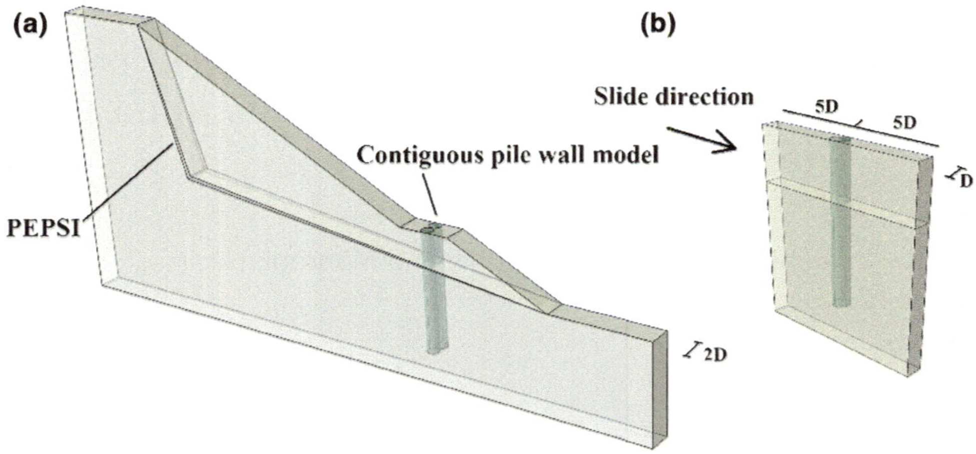

Fig. 1 The analyzed models; **a** the fully coupled model, **b** the hybrid model

Fig. 2 The analyzed models with the soil magnitude and lateral displacements contours; **a** the fully coupled model, **b** the hybrid model (Scale: 1)

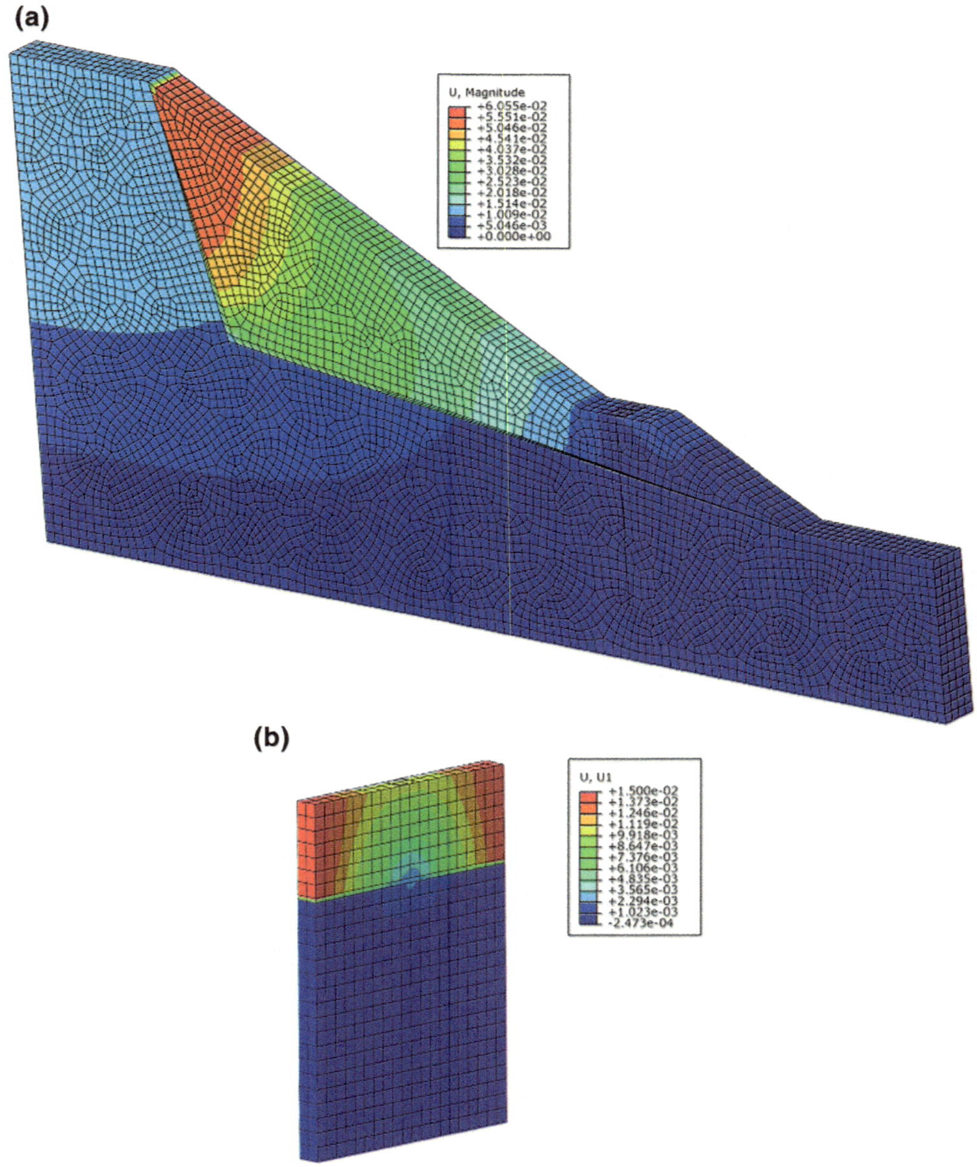

Table 1 Soil properties

	Unit weight (KN/m^3)	Undrained shear strength Cu (KPa)	Young modulus E (KPa)
Unstable part	16	72	48,000
Stable part	19.2	191	482,000

Table 2 Piles properties

	Unit weight (KN/m^3)	Young modulus E (KPa)	Poisson's ratio
Concrete	25	25,000,000	0.2

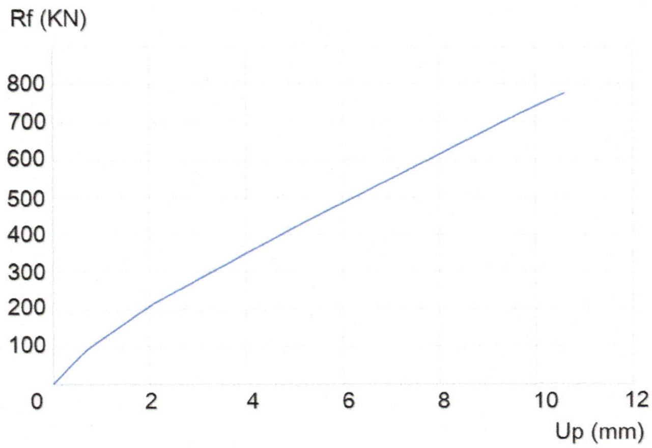

Fig. 3 The resisting force (Rf) given by the simplified model with respect to pile head deflection (Up)

without piles using the method of slices shows that the resisting force needed to stabilize the slope was equal to 410 KN. This was given by the simplified hybrid model at a pile head deflection of 4.8 mm, with an error of 9.5%. The results from the analysis are plotted in Figs. 2 and 3.

4 Conclusions

From the presented research, the following conclusions were reached:

Contiguous pile walls can be very effective in slope reinforcement and give a small pile head deflection.

The hybrid method can be used to analyze contiguous pile walls, an error of 9.5% was observed compared to the coupled model. The use of the simplified hybrid models implies a smaller finite element model size which reduces the simulation cost.

References

1. Ito, T., Matsui, T.: Methods to estimate lateral force acting on stabilizing piles. Soils. Found. **15**(4), 43–59 (1975)
2. Wang, M.C., Wu, A.H., Scheessele, D.J.: Stress and deformation in single piles due to lateral movement of surrounding soils. In: Lunggren, R. (ed.) Behavior of Deep Foundations, pp. 578–591. ASTM Special Technical Publication, 670, ASTM, West Conshohocken (1979)
3. Poulos, H.G.: Design of reinforcing piles to increase slope stability. Can. Geotech. J. **32**(5), 808–818 (1995)
4. Jeong, S., Kim, B., Won, J., Lee, J.: Uncoupled analysis of stabilizing piles in weathered slopes. Comput. Geotech. **30**(8), 671–682 (2003)
5. Ashour, M., Ardalan, H.: Analysis of pile stabilized slopes based on soil–pile interaction. Comput. Geotech. **39**, 85–97 (2012)
6. Kourkoulis, R., Gelagoti, F., Anastasopoulos, I., Gazetas, G.: Hybrid method for analysis and design of slope stabilizing piles. J. Geotech. Geoenviron. Eng. **138**(1), 1–14 (2012)
7. Ho, I.-H.: Numerical study of slope-stabilizing piles in undrained clayey slopes with a weak thin layer. Int. J. Geomech. **06014025**, 1–12 (2014)
8. Keawsawasvong, S., Ukritchon, B.: Undrained limiting pressure behind soil gaps in contiguous pile walls. Comput. Geotech. **83**, 152–158 (2017)

Analysis of the Track Critical Velocity in High-Speed Railway

Wan-Kai Zhang, Li Yuan, and Zhen-Dong Cui

Abstract

A model was developed for analyzing the track critical velocity in this study. It is a double-beam model including rail, rail pad, track slab and cement-asphalt mortar (CA mortar). The model vibration differential equations were transformed by the mobile coordinate system and the Fourier Transform. Then the numerical solution could be easily obtained based on Matlab software. The results from this model were presented for track critical velocities at different rail pad stiffness values. According to these results, there are several track critical velocities in the track structure and the stiffness of rail pad has no obvious effect on the minimum track critical velocity which is the most important velocity in reality.

Keywords

Double-beam • Fourier transform • Track critical velocity

1 Introduction

With the great development of high-speed railway, the track structure dynamics have become more and more important. Analytical and numerical methods have always been applied in the analysis of rail vibration [1, 2]. The former is suitable for some simple problem through the continuous viscoelastic beam whereas the latter is highly adaptable by means of various analysis software.

In this paper, a double-beam model was established for analyzing the dynamic response of the track. The model included rail, rail pad, track slab and cement-asphalt mortar (CA mortar). In this paper, the track critical velocity would be discussed with the help of Matlab software.

2 Simplified Double-Beam Model

In this section, the slab track model for high speed railway was simplified as the double-beam model, which consists of rail, rail pad, track slab and CA mortar as shown in Fig. 1.

The vibration differential equations of the model are given by

$$\begin{cases} E_1I_1\dfrac{\partial^4 w_1}{\partial x^4}+m_1\dfrac{\partial^2 w_1}{\partial t^2}+c_1\left(\dfrac{\partial w_1}{\partial t}-\dfrac{\partial w_2}{\partial t}\right)+k_1(w_1-w_2) \\ =F_0e^{i\omega t}\delta(x-vt) \\ E_2I_2\dfrac{\partial^4 w_2}{\partial x^4}+m_2\dfrac{\partial^2 w_2}{\partial t^2}+c_2\dfrac{\partial w_2}{\partial t}-c_1\left(\dfrac{\partial w_1}{\partial t}-\dfrac{\partial w_2}{\partial t}\right) \\ +k_2w_2-k_1(w_1-w_2)=0 \end{cases} \tag{1}$$

where w_1 and w_2 are the vertical deflection of the rail and track slab, respectively; E_1 and I_1 are the elastic modulus and horizontal inertia moment of rail, respectively; E_2 and I_2 are the elastic modulus and horizontal inertia moment of rail, respectively; m_1 and m_2 are the mass per unit length of the rail and track slab, respectively; k_1 and k_2 are the stiffness of the rail pad and CA motor, respectively; c_1 and c_2 are the damping of the rail pad and CA motor, respectively.

In order to solve Eq. (1), the mobile coordinate system was adopted which means $z=x-vt$. As the load is a harmonic load about time t, we can assume the equations as follows.

$$\begin{cases} w(x,t)=y(z,t) \\ y_1=M(z)e^{i\omega t} \\ y_2=W(z)e^{i\omega t} \end{cases} \tag{2}$$

W.-K. Zhang · L. Yuan (✉) · Z.-D. Cui
State Key Laboratory for Geomechanics and Deep Underground Engineering, School of Mechanics and Civil Engineering, China University of Mining and Technology, Xuzhou, China
e-mail: yuanli-1213@163.com

A. Kallel et al. (eds.), *Recent Advances in Geo-Environmental Engineering, Geomechanics and Geotechnics, and Geohazards*, Advances in Science, Technology & Innovation, https://doi.org/10.1007/978-3-030-01665-4_85

Fig. 1 Double-beam model

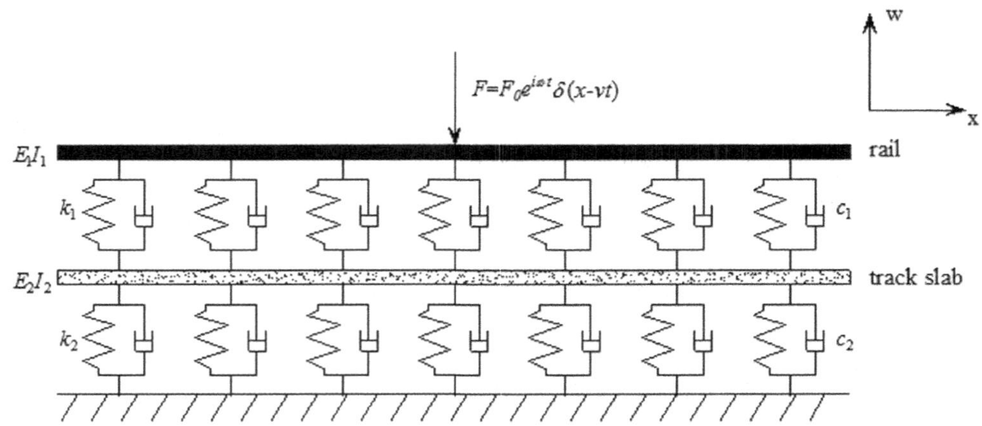

Substitute Eq. (2) by Eq. (1), and then apply the Fourier Transform. So Eq. (1) can be transformed as follow.

$$\begin{cases} \left[E_1I_1\beta^4 + m_1v^2\beta^2 + (2m_1v\omega - ic_1v)\beta \right.\\ \quad \left. + \left(-m_1\omega^2 + ic_1\omega + k_1\right)\right]\tilde{M} \\ \quad + [ic_1v_1\beta - (ic_1\omega + k_1)]\tilde{W} = F_0 \\ \left\{E_2I_2\beta^4 - m_2v^2\beta^2 + [2m_2v\omega - i(c_1+c_2)v]\beta \right.\\ \quad \left. + \left[-m_2\omega^2 + i(c_1+c_2)\omega + (k_1+k_2)\right]\right\}\tilde{W} \\ \quad + [ic_1v_1\beta - (ic_1\omega + k_1)]\tilde{M} = 0 \end{cases} \tag{3}$$

The solution of Eq. (3) is given below.

$$\begin{cases} \tilde{M} = \dfrac{A}{AC - B^2}F_0 \\ \tilde{W} = \dfrac{-B}{AC - B^2}F_0 \end{cases} \tag{4}$$

where

$$\begin{cases} A = E_2I_2\beta^4 - m_2v^2\beta^2 \\ \quad + [2m_2v\omega - i(c_1+c_2)v]\beta \\ \quad + \left[-m_2\omega^2 + i(c_1+c_2)\omega + (k_1+k_2)\right] \\ B = ic_1v_1\beta - (ic_1\omega + k_1) \\ C = E_1I_1\beta^4 + m_1v^2\beta^2 + (2m_1v\omega - ic_1v)\beta \\ \quad + \left(-m_1\omega^2 + ic_1\omega + k_1\right) \end{cases} \tag{5}$$

Then Eq. (4) is transformed with the inverse Fourier transform.

$$\begin{cases} M(z) = \dfrac{1}{2\pi}\displaystyle\int_{-\infty}^{\infty} \tilde{M}(\beta)e^{i\beta z}d\beta \\ W(z) = \dfrac{1}{2\pi}\displaystyle\int_{-\infty}^{\infty} \tilde{W}(\beta)e^{i\beta z}d\beta \end{cases} \tag{6}$$

So the vertical deflection can be obtained.

$$\begin{cases} y_1(z,t) = M(z)e^{i\omega t} \\ y_2(z,t) = W(z)e^{i\omega t} \end{cases} \tag{7}$$

Equation (6) can be solved by the inverse fast Fourier Transform with the help of Matlab software to get the numerical solution.

3 Results

The effect of the stiffness of the rail pad on the track critical velocity was investigated in this section based on the model shown in Fig. 1. The calculation parameters are shown in Table 1 and results are shown in Fig. 2.

Figure 2 shows that there are several track critical velocities in the track structure and the minimum track critical velocities at different rail pad stiffness are 280, 290, 310 and 295 m/s, respectively.

4 Discussion

Although there are several track critical velocities in the track structure, the most important velocity in reality is the minimum one. From the results presented here, the conclusion that can be drawn is that the rail pad stiffness has no obvious effect on the minimum track critical velocity. The minimum track critical velocity is still nearly equal to 300 m/s.

5 Conclusion

In this paper, an investigation on the track structure dynamics using a double-beam model was presented. The vibration differential equations were transformed by the mobile coordinate system and the Fourier Transform. And

Table 1 Model parameters

E_1I_1 (MPa/m^2)	m_1 (kg/m)	k_1 (MN/m^2)	c_1 (kN s/m^2)	E_2I_2 (MPa/m^2)	m_2 (kg/m)	k_2 (MN/m^2)	c_2 (kN s/m^2)	ω (rad/s)
66	120	25	0	40	2500	100	0	0
		50						
		100						
		200						

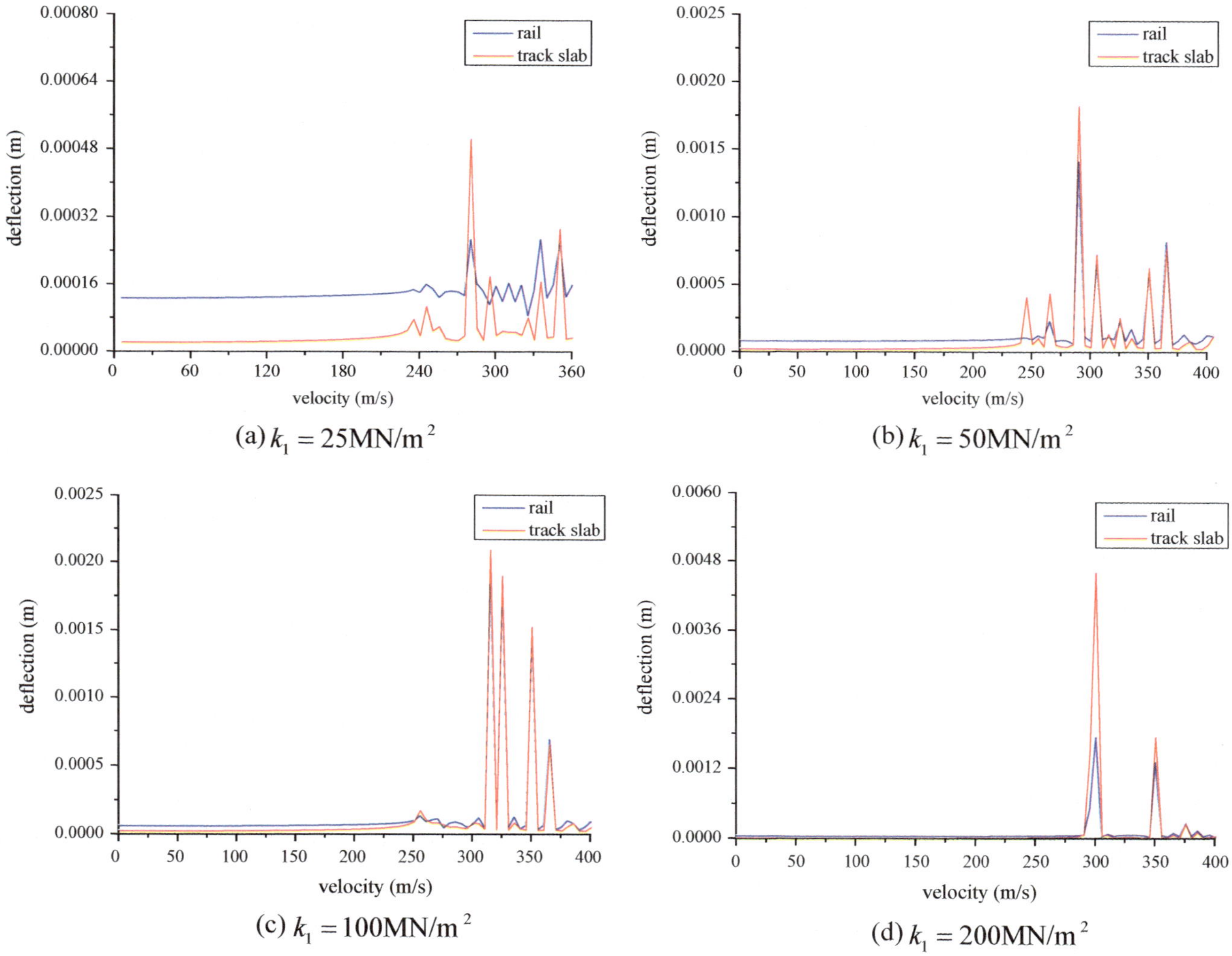

Fig. 2 Track critical velocity at different rail pad stiffness values

then with the help of Matlab software, the numerical solution could be easily obtained.

Based on the example results, the conclusion that can be drawn is that the stiffness of rail pad has no obvious effect on the minimum track critical velocity.

Acknowledgements This work was funded by National Key R&D Program of China (2017YFC1500702).

References

1. Sheng, X., Jones, C.J.C., Thompson, D.J.: A theoretical model for ground vibration from trains generated by vertical track irregularities. J. Sound Vib. **272**(3), 937–965 (2004)
2. Sheng, X., Jones, C.J.C., Thompson, D.J.: A theoretical study on the influence of the track on train-induced ground vibration. J. Sound. Vib. **272**(3), 909–936 (2004)

Flow Filling of a Closed, Circular and Almost Horizontal Pipe

Wahiba Mokrane and Ahmed Kettab

Abstract

When a water supply hydraulic project is achieved, particular attention must be given to the filling process of a discharge pipe. Transient phenomena will take place and the air phase may become an obstacle. In this work, we studied the filling of a closed circular pipe with a weak slope. So, we counted the filling time for three various discharges. We determined the flow type while taking in accounts the air and water phases. From our observation of the flow behaviour, we deduced that the flow is configured as a smooth stratified for the first and second tests. But for a higher value of the discharge speed, we visualized a wavy flow, with simple and elongated bubbles. This was observed until the complete pressurizing of the pipe. In this experimental case, air was completely removed, argued by our weak values of the flow discharge and the pipe size.

Keywords

Filling • Discharges • Bubbles • Pressurizing • Air

1 Introduction

The filling process includes a free surface flow, a transition, a full section no pressurized and a pressurized two-phase flow. Filling a pipe implies removing the air. But frequently, during this operation air seems to be an obstacle. Flow may take different configurations depending on the discharge values. This field of hydraulic situation has been treated through several approaches and methods. Cunge and Wegner [1] used the Preissman slot approach and neglected the air phase. Dong [2] considered both the pipe elasticity and the water compressibility but did not take into account the air phase. Gomez and Achiaga [3] estimated air volume as one percent of the flow in their study. Kerger [4] attracted the attention to use a negative slot approach to simulate the depressurized flow.

On the other hand, in their work, Wright and Vsconcelos [5] showed that a rapidly filling would not be described by a stratified flow.

Thus, we can deduce that this phenomenon remains a difficult problem. We believe that experimental testing is the most appropriate way to understand the phenomenon and help practitioners. Our present work aims to study experimentally the filling of a transparent circular pipe for three flow discharges values.

2 Materials and Methods

To visualize and follow the flow steps of the pipe filling, we used experimental tests. Our set up consists of a plexiglass pipe supplied with water by a pump. The pipe is of 0.05 m diameter, three meters in length and 0.005 m thickness. A valve was placed on each side of this pipe. An electromagnetic flow meter was used to measure the discharge.

We carried out three test series for different values of the flow discharge measured at the upstream of the experimental pipe. These values are respectively 0.43, 0.56, and 1.44 in liters per seconds. For each test, the upstream valve was completely opened and the downstream valve was partially closed. Piezometers tubes, installed at four positions allowed us to measure the pressure values for the first and the second tests. But manometers were used for the third test.

W. Mokrane (✉)
Research Laboratory of Water Sciences-LRS-EAU, National Polytechnic School, 10 Av. Hacene-badi, BP182 16200 El-Harrach, Algiers, Algeria
e-mail: mokranewah@yahoo.fr

W. Mokrane · A. Kettab
Department of Urban Hydraulic, Research Laboratory of MVRE, National High School of Hydraulic, BP31 Blida, Algeria

A. Kallel et al. (eds.), *Recent Advances in Geo-Environmental Engineering, Geomechanics and Geotechnics, and Geohazards*, Advances in Science, Technology & Innovation, https://doi.org/10.1007/978-3-030-01665-4_86

3 Results

3.1 Results of the First Test

For a discharge of 0.43 L per second and for the four positions of the pipe, the filling pressures are shown on the following Fig. 1.

3.2 Results for the Second Test

For a discharge of 0.56 L per second and for the four positions of the pipe, the filling pressures are given on the following Fig. 2.

3.3 Results for the Third Test

For a discharge of 1.44 L per second and for the four positions of the pipe, the pressures filling are shown on the following Fig. 3.

4 Discussion

The experimental results presented in the first figure allow deducing that the filling time was 19 min but we noticed that at the fourth position, free surface flow occurred for few seconds then it became rapidly pressurized with about five times the pipe diameter value pressure. The flow at the other

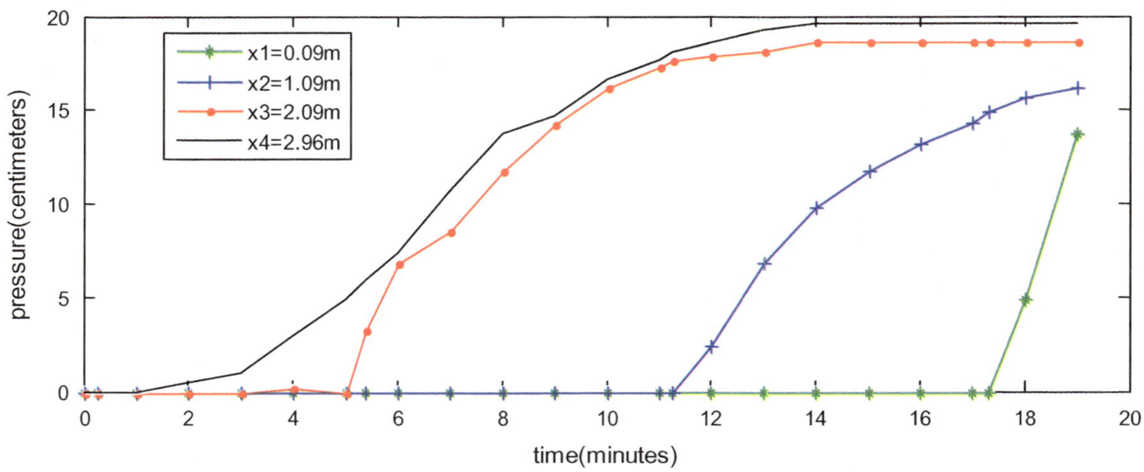

Fig. 1 Pressure evolution at the four positions for the first test

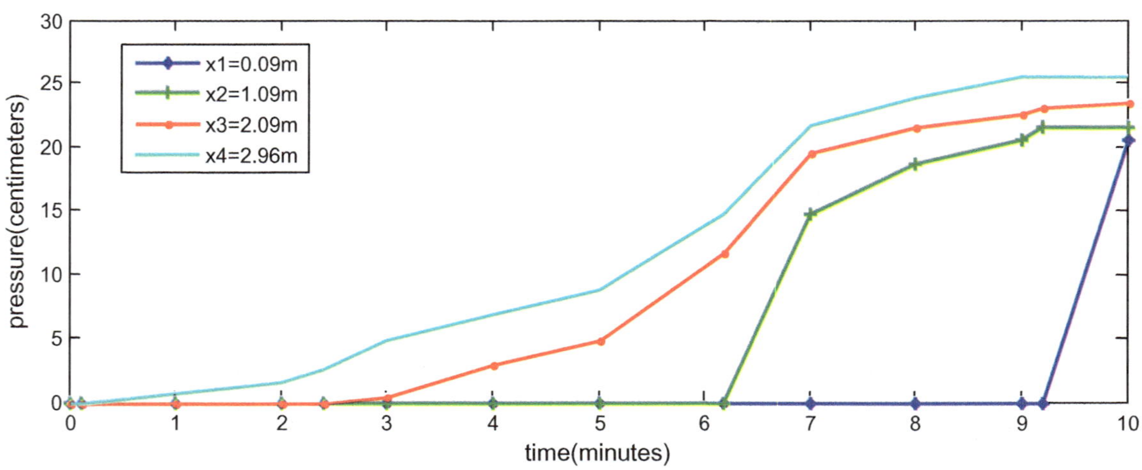

Fig. 2 Pressure evolution, at the four positions and for the second test

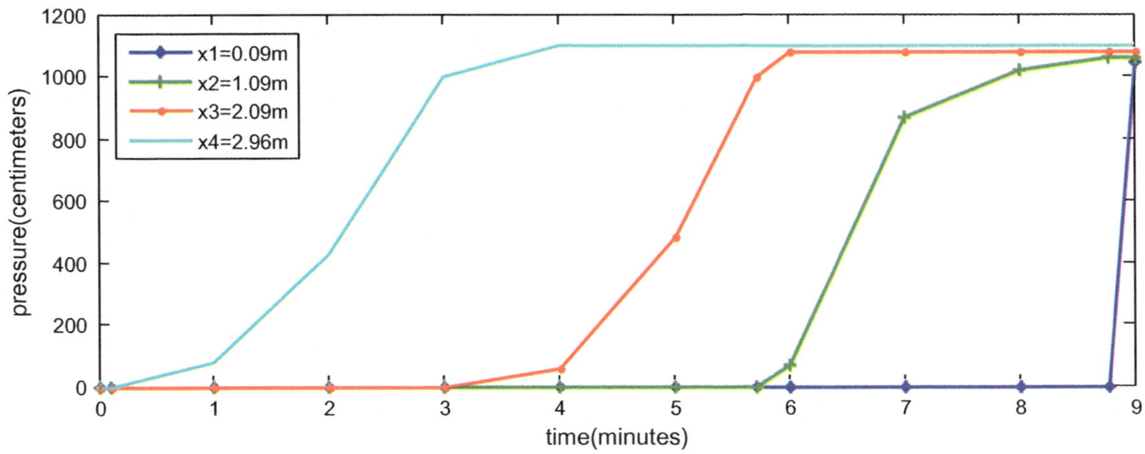

Fig. 3 Pressure evolution, at the four positions for the third test

positions passed from free surface to a full section not pressurized and finally became pressurized when pressure exceeds the pipe diameter. The filling time, for the second test was 10 min. We observed the same flow filling steps as those noted for the first test. The water–air flow configuration detected was the stratified one.

For the third test, the filling time was 9 min. The pipe was rapidly pressurized; we did not detect the full no pressurized section. The flow water-air is configured as wavy, with elongated bubbles and with bubbles before the total pressurizing. In this case pressure rose to two hundred and twenty times of the pipe diameter value. When Gomez and Achiaga [3] reached a value of twenty-five of their pipe diameter which is nine times bigger than ours. In the third test the water-air flow could not be stratified. This is in agreement with the results of Wright and Vasconcelos [5].

5 Conclusion

Filling a discharge pipe must be performed while avoiding important pressures which may induce considerable pipe damage. Also, all of the air must be removed. In this background, we carried out experimental tests for three different discharges on a transparent pipe. For values less than one liter per minute the water–air flow is stratified and pressure reached five times the internal pipe diameter. But for a value of more than one liter per minute the water-air flow is wavy, bubbly and not stratified. In this case, pressure is hundreds times the pipe diameter.

Thus, we conclude that the discharge pipe filling must be carried out with weak flows values. This will protect pipes against air bubbles erosion, air obstacle and overpressure.

References

1. Cunge, J.A., Wegner, M.: Integration numérique des équation de Barré Saint venant par un schéma implicite de différences finies. Houille Blanche **1**, 33–39 (1964)
2. Dong, N.T.: Sur une methode numérique de calcul des écoulements nonpermanents soit à surface libre, soit en charge, soit partiellement à surface libre et partiellement en charge. Houille Blanche **2**, 149–158 (1990)
3. Gomez, M., Achiaga, V.: Mixed flow modeling by pressure fronts from upstream and downstream extremes. Urban drainage modeling, World water and environmental resources, congress, pp. 461–470. SASCE, Florida, USA (2001)
4. Kerger, F., et al.: An exact Riemann solver and a Godunov scheme for simulating highly transient mixed flows. J. Comput. Appl. Math. **235**, 2030–2040 (2011)
5. Wright, S., Vasconcelos, J.: Geysering in rapidly filling storm water tunnels. J. Hydraul. Eng. **132**(6), 555–562 (2011)

Part IX

Geohazards: Assessments of Mass Movement Based Geohazards

Assessment of Geological Hazards Along Alagaba Highway-Red Sea State, Sudan

Esamaldeen Ali and El-Khider Rahamt Allah

Abstract

Khartoum-Portsudan national highway represents the economic and commercial highway that link Sudan capital with main cities in eastern Sudan such as Suakin, Bashair and Digna harbors. The highway crosses the Red Sea Hills through high ridge mountains of Alagaba region. The highway consists of two roads: old and new road that are constructed in different time. Some geo-hazards such as rock falls, debris flows and rock failures represent the main natural geological hazards along the new road in compare with old road. However, the properties of rock mass discontinuities play an important role in the behavior of rock slope stability. This study is aimed to assess the geological hazards, probability of failure, and suggest suitable treatments of unstable rocks to mitigate the road from hazardous phenomena. This study is mainly based on geological observations and structural measurements. Field observations as well as kinematic analyses show that rock fall and debris flow can occur. In addition, toppling and wedge failures can probably happen. Based on the data analysis some suitable reinforcements (e.g. benches, catch fences and meshes) are recommended in the location of hazardous sites to prevent the road against from rock falls, debris flows and failures.

Keywords

Alagaba highway • Sudan • Rock fall • Slope failure • Kinematic analysis

1 Introduction

Generally, the highway systems along the mountain areas may suffer from unstable slopes that pose risks on a daily basis to the traveling public, transportation infrastructure and environment. These may not be considered hazardous unless rocks enter the roadway [1, 3, 4]. However, rock falls and debris flows represent one of the main natural geological hazards along highways in mountain areas.

For instance, in Sudan region Khartoum-Port-Sudan express road represents the main highway passing the Red Sea Hills. This road links capital of Sudan "Khartoum City" with Port-Sudan port passing through eastern cities such as Suakin, Bashair and Digna harbors (Fig. 1a). This road was constructed in 1979 by a Germen Company called STRAPAC. At that time this was the only highway path across Alagaba ridge with two lines. Alagaba area, which is a part of the Red Sea hills, is about 780 m height from the sea level (Fig. 1b). During the last years the Sudan government organized the construction of a new road in the present place through the Sudanese Company to reduce accidents and traffic. The new road of Alagaba highway extends about 25 km passing through a major Khor, while old road extends about 18 km. However, some geo-hazards are observed along the new road in compare with old road. In fact, the nature and properties of rock mass discontinuities (e.g. orientation, spacing, roughness, infilling materials, persistency, joint set, block size and degree of weathering) all together play an important role in the behavior of rock mass strength and controlling rock slope stability [5–8].

However, since its construction, there was no available detailed geotechnical investigation along Alagaba new road. Therefore, this study aims to assess the geological hazards, probability of failure, and suggest suitable treatments of unstable rocks to mitigate the road from hazardous phenomenon. However, characterization of rock discontinuities represents main factors controlling rock slope stability.

E. Ali (✉) · E.-K.R. Allah
Faculty of Petroleum and Minerals, Al-Neelain University, Khartoum, Sudan
e-mail: esamali77@gmail.com

A. Kallel et al. (eds.), *Recent Advances in Geo-Environmental Engineering, Geomechanics and Geotechnics, and Geohazards*, Advances in Science, Technology & Innovation, https://doi.org/10.1007/978-3-030-01665-4_87

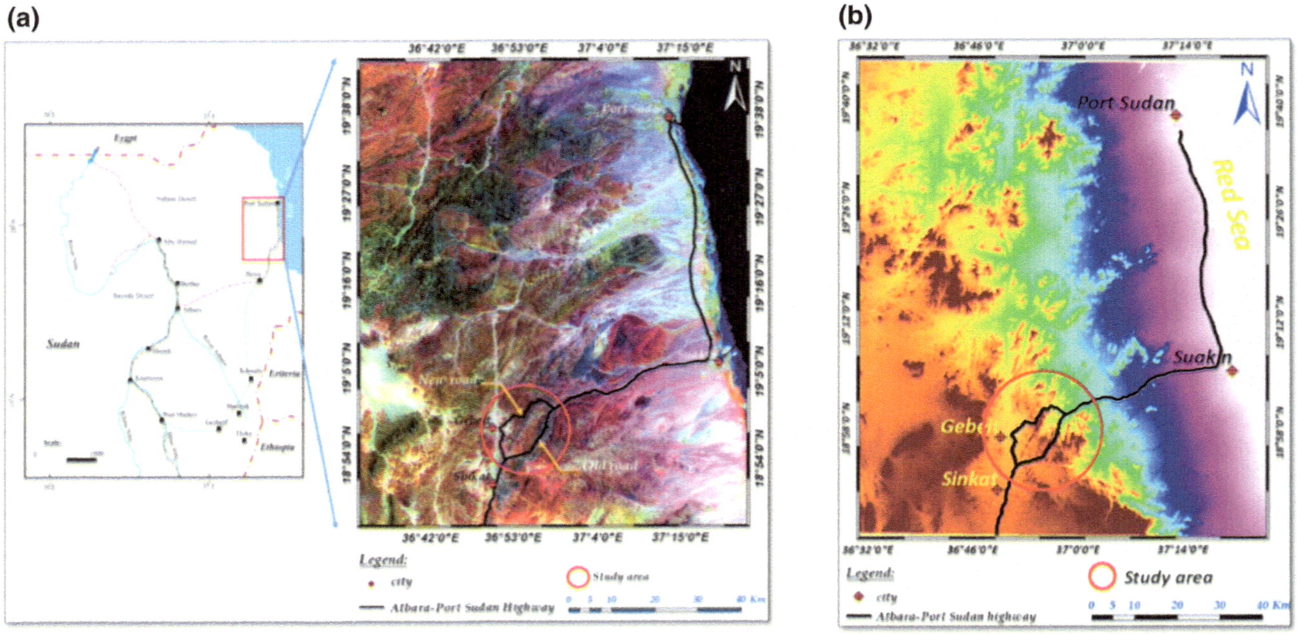

Fig. 1 **a** Location map of Alagaba highway, Red Sea Hills, Sudan and **b** Digital elevation model show the elevation contrast and general topography features of the study area

2 Materials and Methods

In order to accomplish this research, comprehensive geotechnical and structural information are considered along Alagaba new road. In this study Digital Elevation Model (DEM) with 90 m resolution was also used to identify the variations in elevations main streams and gullies. Geological survey, which spends about ten days, were mainly aimed to collect geological and structural measurements of discontinuity planes (e.g.: joints, faults and shears) and their trends in order to identify the Kinematic analysis of rock slope along the new road. A suitable number of representative rock samples were collected for laboratory tests and petrographical investigations. The Schmidt hummer "L-type" was used to determine the preliminary rock strength.

3 Geotechnical Site Investigation

Generally, the geotechnical site investigation started with lithological discrimination along the new road. However, based on the field investigation coupled by the regional geology of the Red Sea Hills, the litho-stratigraphic of Alagaba area is characterized by green schist to low amphibolite facies of metavolcano-sedimentary sequences, which represent the most abundant litho-stratigraphic suite of the study area. They form high rugged topography with deep narrow gorges. The rocks consist of meta-volcanics, meta-sediments and volcano-clastic tuffs. They are moderately sheared, folded and cutting by quartz veins and dykes with varies composition. The above units are intruded by variety of foliated syn to late orogenic and post orogenic igneous intrusions.

For easy site investigation, the two slopes sides of the road are divided into seven localities (Table 1) show general characterizations of rock mass discontinuities along Alagaba new highway that were conducted during site investigation. The different rock falls, debris flow and potential rock failure are well observed along Alagaba new road (Plates 1 and 2). The absence of shoulders and catchment benches play against protection of the road from reaches the rock falls (Plate 2a). There are no obstruction, traffic signs, ditch and fences to protect the cars and auto-trucks from fell down (Plate 3b). Also, many accidents were occurred due to the dangerous road curvatures and higher elevation (Plate 3c). In addition, the unsuitable method of excavation leads to the irregularity face as a result of poor blasting technique.

In this study Kinematic analysis which is a graphical representation was applied, that is required to identify possible modes of slope failure and establish full engineering characterizations of slope design. Rock Slope failures can be classified into four categories based on the geometrical, mechanical nature of the discontinuity and the conditions of the rock masses, plane, wedge, toppling and circular failure.

Table 1 Characterizations of rock mass discontinuities along the new road

Location	1	2	3	4	5	6	7
Persistency (m)	3.7	3.9	4.3	4.9	3.3	3.5	4.2
	The average of persistence = 3–10 m (medium persistence)						
Aperture (mm)	2	21	5.5	15.5	1	8.5	7
	Aperture is ranging from open (0.5–2.5 mm) to very widely open (1–10 cm)						
Joint spacing (cm)	47	41	74	61	53	41	68
Orientation (°)	Many joint sets have different directions (North, NE, NW, NNE and NNW)						
Joint Roughness Coefficient "JRC" [2]	Range from smooth surface to very rough surface						
Weathering degree	According to the classification of weathering scheme the rock masses are classified into slightly weathered, moderately weathered, highly weathered and completely weathered						
Infilling materials	Calcite, epidote and quartz, gypsum, chlorite, talc, graphite and serpentine, in-active clay, swelling clay and cohesion less materials such as sand. Clean discontinuities without filling materials or coatings are also identified						

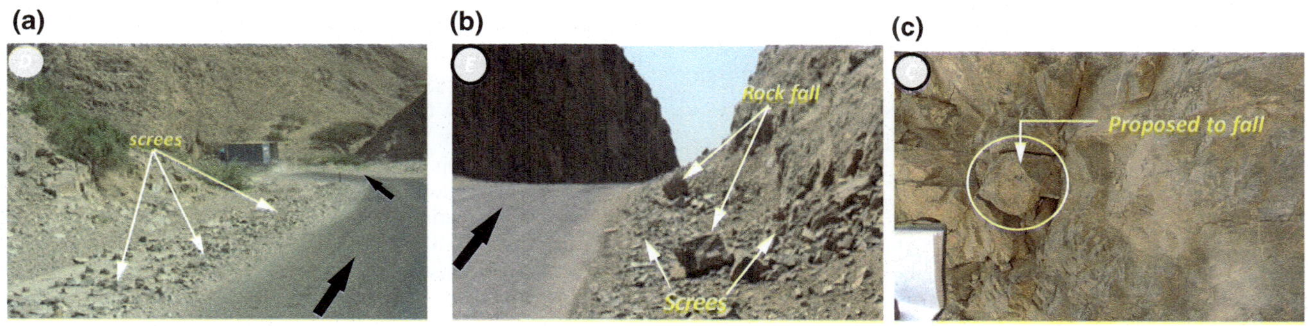

Plate 1 Different types of rock falls along Alagaba new road. **a** Debris flow; **b** Rock falls

Plate 2 Different types of slope failures along Alagaba new road. **a** Plane failure; **b** Wedge failure; **c** Toppling failure

Plate 3 **a** The absences of benches, shoulders, obstructions, fences and traffic signs; **b** Ditch in the new road, **c** Dangerous curvatures of the new road

4 Conclusion and Recommendations

In Sudan region Khartoum-Port-Sudan express road represents the main highway passing through eastern cities of Red Sea Hills such as Suakin, Bashair and Digna harbors. However, potentially unstable slopes present hazards and pose risks to traveling public transportation and infrastructure, which are observed along the new road in compare with old road. The degree and nature of the rock fall hazard depends on the characteristics of the rock mass discontinuities. This study is aimed to assess the geological hazards, probability of failure, and suggest suitable treatments of unstable rocks to mitigate the road from hazardous phenomenon. The discontinuity properties (e.g. orientation, spacing, roughness, infilling materials, persistency, joint set, block size and degree of weathering) all together are effecting rock mass strength and controlling rock stability. Based on the detailed site investigation this study noted that rock falls and debris flows represent one of the main natural geological hazards along road. Also, there are no obstruction, traffic signs, ditch and fences to protect the cars and auto-trucks from fell down. In addition, rock discontinuities characterizations pose high risks along the new road. In this study, kinematic analysis was also applied, in order to identify the possible modes of rock slope failure. However, the study concluded that three types of slope failure potentiality are resulted which represent the main hazards along the road.

To establish full engineering characterizations of slope design, this study recommended that support systems are required depending on the type of rock failure and possible instability. Also, because the road is cutting across many Khors, it must be redesigned and such that catch fences get constructed and benches and meshes get installed especially in deep slopes.

References

1. Ahmed, M.Y., Norbert, H.M.: Slope stability hazard assessment and mitigation methodology along eastern desert Aswan-Cairo highway, Egypt. JKAU Earth Sci. **20**(2), 161–181 (2008) (A.D./1430 A.H.)
2. Barton, N., Choubey, V.: The shear strength of rock joints in theory and practice. Rock. Mech. **10**, 1–54 (1977)
3. Chau, K.T., Wong, R.H.C., Liu, J., Lee, C.F.: Rock fall hazard analysis for Hong Kong based on rock fall inventory. Rock. Mech. Rock. Eng. **36**(5), 383–408 (2003)
4. Chau, K.T., Sze, Y.L., Fung, M.K., Wong, W.Y., Fong, E.L., Chan, L.C.P.: Landslide hazard analysis for Hong Kong using landslide inventory and GIS. Comput. Geosci. **30**, 429–443 (2004)
5. Farmer, I.W.: Rock testing: deficiencies and selection. In: Geomechanics, vol. 91, pp. 3–7. Balkema, Rotterdam (1992)
6. Hoek, E.: Rock mechanics laboratory testing in the context of a consulting engineering organization. Int. J. Rock Mech. Min. Sci. Geomech. **14**, 93–101 (1977)
7. ISRM.: Basic geotechnical description of rock masses (BGD). Int. J. Rock. Mech. Min. Sci. Geomech. **18**, 85–11 (1981)
8. ISRM.: Suggested methods for determining point load strength. Int. J. Rock. Mech. Min. Sci. Geomech. **22**, 53–60 (1985)

Mining Geohazards at the Perimeter of the Amyntaio Open Pit Coal Mine, West Macedonia, Greece

Constantinos Loupasakis

Abstract

Land subsidence phenomena affecting the plain area surrounding the open pit as well as large landslides affecting the open cast slopes are listed among the mining induced catastrophic geo-hazards affecting the perimeter of mines. These large scale geo-hazards are related to both hydro–geological and geotechnical parameters and they cause unrevoked damage. In the present study the investigated site is the area extending at the perimeter of the Amyntaio opencast coal mine at Florina Prefecture, Northern Greece. The overexploitation of the aquifers for the dewatering of the slopes turned the mine to a large diameter well. The surface deformations caused by the land subsidence phenomenon extends 1–3 km around the mine affecting Anargiroi and Valtonera villages. Furthermore, on June 10th, 2017, a massive landslide, of approximately 80 million cubic meters, occurred at the working slopes of the mine, burrying 25 million tons of lignite, resulting in severe damage to large number of mining equipment and causing the evacuation of the nearby Anargyroi village. Apart from presenting the two catastrophic events, the current research aimed to correlate the geological, hydrogeological and geotechnical factors affecting both phenomena and outlining their mechanisms.

Keywords

Mining hazards • Subsidence • Landslide • Opencast mines

C. Loupasakis (✉)
School of Mining and Metallurgical Engineering, National Technical University of Athens, Zographou Campus, 157 80 Athens, Greece
e-mail: cloupasakis@metal.ntua.gr

1 Introduction

Within the current study the surrounding plane of the Amyntaio opencast coal mine was investigated. The deep wells draining network for the protection of the slopes drained the surrounding area forming an extensive depression cone all around the mine. The land subsidence phenomena, extending to 1–3 km around the mine, have been subjected to detailed investigations by numerous past studies [1–6]. Beside the land subsidence phenomena, a massive landslide occurred at the working slopes of the mine on June 10th, 2017. The current research aimed to correlate the geological, hydrogeological and geotechnical factors driving the mechanism of both phenomena.

2 Geological, Geotechnical and Hydrogeological Setting

The Florina-Ptolemaida basin is occupied by crystalline-schist bedrock covered by Neogene and Quaternary deposits. The Neogene deposits are divided in three series. The lower and upper series consist of fine grained deposits, sandy clays to marls, and they surround the lignite deposits constituting the middle series. These formations are interrupted by pre-existing sheared zones and intercalated by low strength clay layers [7]. The Quaternary depositions are divided, from bottom to top, to [8]: (A) The Proastion Formation, consisting of alternating loose sand to clayey sand and conglomerate with red clay. (B) The Perdikas Formation, consisting mainly of intercalations of fine sand with alternating layers of sandy clays, clays and marls. This relatively loose and compressible formation is mainly susceptible to the manifestation of land subsidence phenomena [3]. (C) The Anargiri Formation, including clayey sand or thin sand interrupted by clay layers or lenses with angular fragments. Within the mine slopes, several faults have been identified, some of them active (Vegoritida and Anargiroi

A. Kallel et al. (eds.), *Recent Advances in Geo-Environmental Engineering, Geomechanics and Geotechnics, and Geohazards*, Advances in Science, Technology & Innovation, https://doi.org/10.1007/978-3-030-01665-4_88

fault), presenting NE-SW and secondarily NW-SE to NNW-SSE striking directions [1].

In the wider study area, the Quaternary deposits host semi-confined aquifers. The systematic exploitation of the aquifers started taking place after the redistribution of the land, in 1974. The excavation of the mine, in 1989, changed radically the setting of the aquifers, as large quantities of water pumped out from the slope protection wells, combined with the ground water inflows into the opencast, turned the mine to a large diameter well. The above described mechanism is also enchased by the operation of the farmers' irrigation wells.

The influence of the mining activity on the aquifers piezometry can be easily evaluated by comparing the diachronic changes of the spatial distribution of the isopiezometric contour lines [3, 4, 9]. By means of this comparison, it is clear that the depression cone at the west of the mine extends until the Chimaditida Lake and at the northwest until the Valtonera village [3, 4, 10]. Also, the maximum drawdown of the piezometric surface, next to the mine, is more than 60 m.

Besides the semi-confined aquifers, into the underlay bedrock formations a karstic aquifer extends, developing high piezometric loads. By evaluating the provided data [9], it is practically clear that those loads get transferred by means of the coarse grain layers of the lower Neogene series close to the excavation floor. Obviously, the combination of the high piezometric loads with the occurrence of the pre-existing sheared zone generates negative slope stability conditions.

3 The Mining Induced Geo Catastrophic Events

The land subsidence deformations occurred initially at the Anargiri village, at 2001, and until 2006 the phenomenon affected the entire study area. The strike direction of the surface raptures is parallel to that of the main tectonic lines of NE-SW to ENE-WSW directions (Fig. 1). Also, the

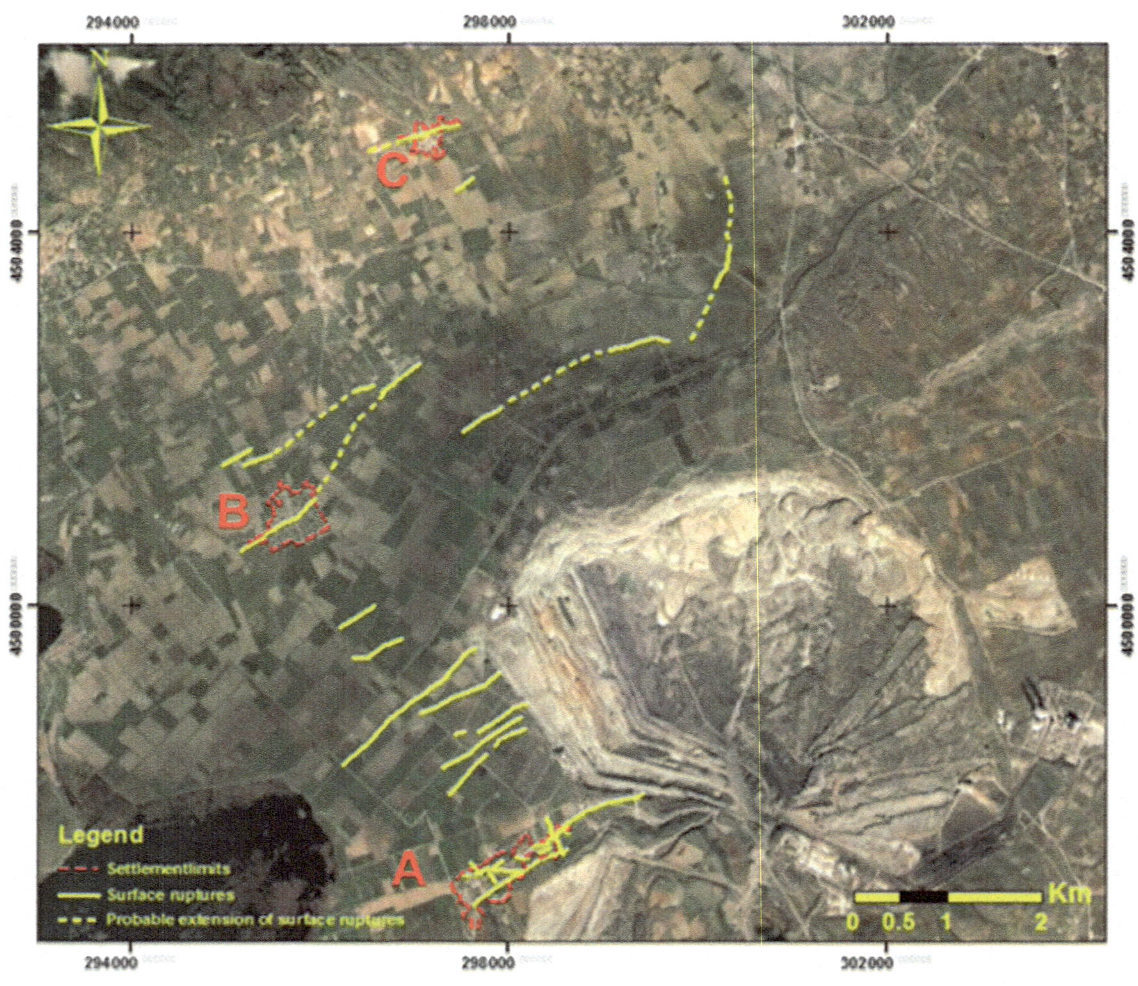

Fig. 1 Spatial distribution of the surface ruptures at the perimeter of the Amyntaio open cast mine. *A* Anargiroi village, *B* Valtonera village and *C* Fanos village [3, 5]

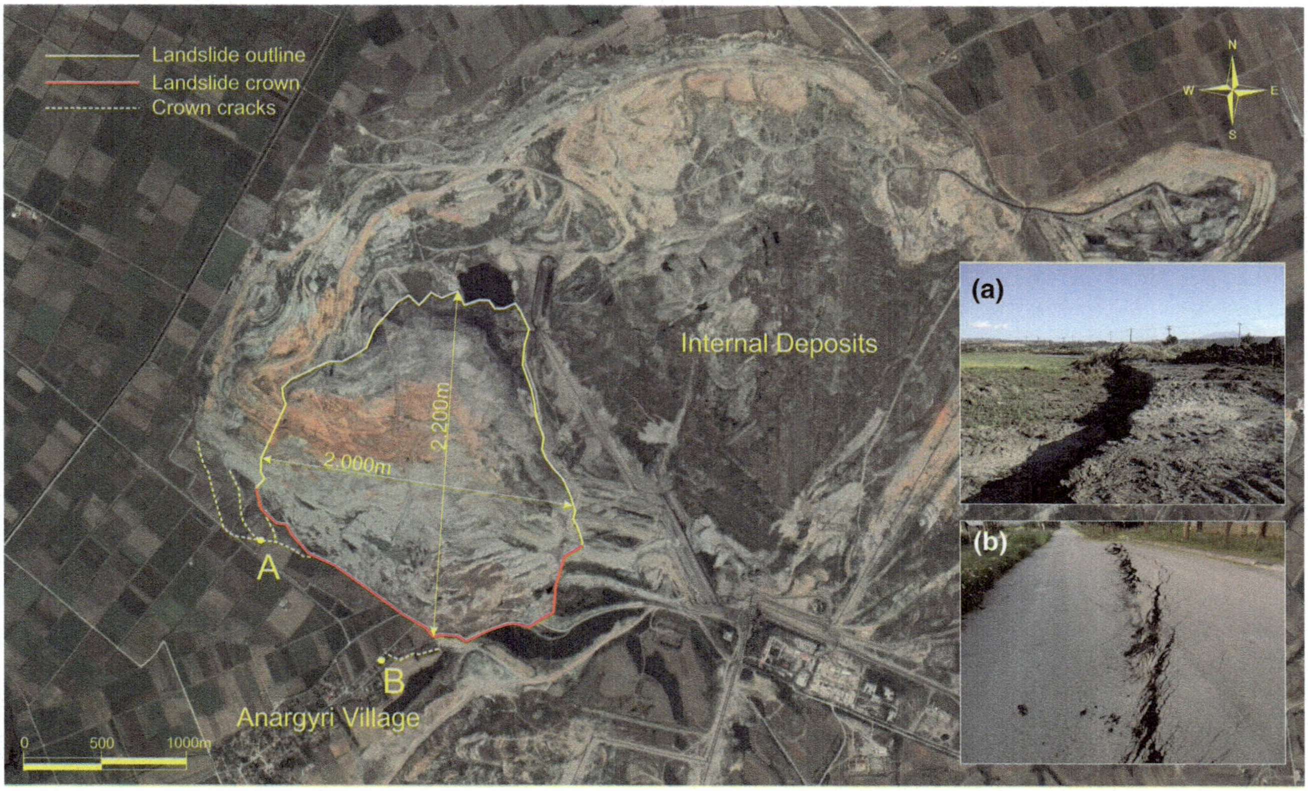

Fig. 2 The June 10th, 2017 massive slope failure at the Amyntaio open pit coal mine

differential vertical displacements identified along the surface raptures are inversely proportional to the distance from the mine.

The strict confinement of all the tension cracks within the limits of the depression cone connects them directly with the overexploitation of the aquifers. Also their gradual debilitation in relation to the distance from the mine and their continuous aseismic slip do not allow any correlation with seismic activity.

The massive landslide affected the entire, 200 m high, working slope of the mine (Fig. 2). The occurrence of the pre-existing sheared zones within the Neogene formations, the extensive faults intersecting the site and the high piezometric loads underneath the floor of the open pit are the main natural preparatory casual factors of the landslide. On the other hand, the partial failure of the dewatering system eliminating the pore pressures established by the semi-confined aquifer and the geometry of the slopes, are the main man made triggering casual factors. All the above described factors were adequately studied but their coaction was not properly evaluated.

4 Discussion—Conclusion

Considering the above described mechanisms it is clear that the occurring catastrophic events were driven by contradictive casual factors. The dewatering of the slopes triggers the land subsidence phenomena while the attenuation of the draining reduces the safety factor of the slopes.

References

1. Loupasakis, C.: Study of the geotechnical conditions of the Amintaio coalmine slopes close to the Anargiri village, Aetos Municipality, Florina Prefecture, Greece, Unpublished technical report, I.G.M.E, Athens (2006)
2. Loupasakis, C.: Geotechnical study in the Anargiroi village, Aetos Municipality, Florina Prefecture, Greece. Unpublished technical report, I.G.M.E., Athens (2010)
3. Loupasakis, C., Angelitsa, V., Rozos, D., Spanou, N.: Mining geohazards—land subsidence caused by the dewatering of

opencast coal mines: the case study of the Amyntaio coal mine, Florina, Greece. Nat Hazards **70**, 675–691 (2014)
4. Tzampoglou, P., Loupasakis, C.: New data regarding the ground water level changes at the Amyntaio basin-Florina Prefecture, Greece. In: Proceedings of the 14th International Congress of the Geological Society of Greece, Thessaloniki, Bulletin of the Geological Society of Greece, No. 2, pp. 1006–1015 (2016)
5. Tzampoglou, P., Loupasakis, C.: Mining geohazards susceptibility and risk mapping: the case of the Amyntaio open-pit coal mine, West Macedonia, Greece. Environ Earth Sci **76**, 542 (2017)
6. Tzampoglou, P., Loupasakis, C.: Evaluating geological and geotechnical data for the study of land subsidence phenomena at the perimeter of the Amyntaio coalmine, Greece. Int J Min Sci Technol (2018). https://doi.org/10.1016/j.ijmst.2017.11.002
7. Leonardos, M., Terezopoulos, N., Rim slope failure mechanism in the Greek deep lignite mines—a case study. Mining Technol (Trans. Inst. Min. Metall.: Sect. A) 112:197–204 (2003). https://doi.org/10.1179/037178403225001656
8. Koukouzas, C., Kotis, Th., Ploumidis, M., Metaxas, A.: Coal exploration of Anargiri–Amynteon area, Mineral deposit research No. 9, pp. 1–69, I.G.M.E., Athens (1979)
9. Dimitrakopoulos, D.: Hydrogeological conditioning of Amyndeon mine. Problems during exploitation and overcoming them. Ph.D. Thesis, NTUA—School of Mining and Metallurgical Engineering, NTUA, Athens, p. 202 (2001)
10. Dimitrakopoulos, D., Koumantakis, I.: Hydrodynamic regime of Amynteon basin. Influence of open lignite mines. In: Proceedings of the 11th International Hydrogeological Congress of Greece, Athens, No. 2, pp. 101–112 (2017)

Geomorphic Mapping Reveals ~NW-SE Extension in NW Himalaya

Afroz Ahmad Shah, Adi Ameza Binti Mohd Addly, and Mohammad Iskandar Bin Abdul Samat

Abstract

The work presented here shows new mapping of an array of ~NE-SW trending normal faults in NW Himalaya. This is achieved by geomorphic mapping of tectonic landforms using a variety of freely available satellite images that include Google maps, global earth, and global multi-resolution topography. The geomorphic mapping is performed by tracing of active fault ruptures, faulted glacial landforms, fault scarps, topographic breaks, deflected drainages, triangular facets, ridge axes offset, and broken Holocene to Recent sedimentary deposits. The results show that the faults are dominantly trending ~NE-SW with a few ~NNW-SSE. The faulting is widespread and extends for >140 km. These faults suggest ~NW-SE extension on a regional scale, and the tectonic transport is—orthogonal to the regional convergence between India and Southern Tibet. The faults are active and the entire region is undergoing regional ~NW-SE extension, and therefore, a new tectonic model is proposed for the formation of these faults, which are interpreted to have formed syntectonic with the regional convergence of the plates in collision, and could potentially host medium to large magnitude earthquakes in the near future. The expected earthquake focal mechanism "beach ball" on each fault is estimated, which suggests what type of earthquake is possible on the newly mapped faults. The new tectonic interpretation of the region supports the observed seismicity, and correlates with the available geodetic and geologic data.

Keywords

Kashmir basin fault • Active fault • Normal fault • Extension

A. A. Shah (✉) · A. A. B. M. Addly · M. I. B. A. Samat
Department of Physical and Geological Sciences, Universiti Brunei Darussalam, Gadong, Brunei
e-mail: afroz.shah@ubd.edu.bn

1 Introduction

The occurrence of normal faulting parallel to the regional convergence vector orientation in convergent tectonic setting is not very common, however, normal faulting oblique to regional convergence is widespread [2]. This paper reported for the first time the large scale normal faulting north of the Kashmir basin in NW Himalaya, and these faults are orientated—oblique to the regional convergence of India and Eurasia plates (Fig. 1). These faults are mapped on freely available Google satellite data, and are supported with geologic, seismological, geodetic, and previously published structural maps (e.g. [4, 5]). The new geomorphic mapping (see below) adds data to the existing literature about the tectonics of NW Himalaya. The previous knowledge about the tectonics of Kashmir Himalaya largely suggests reverse faulting south of the Kashmir basin. The Kashmir basin sits on top of the actively growing reverse fault systems, and this makes it a classic example of a piggyback basin [1]. The basin was formed during the deformation that involved underthrusting of the Indian lithosphere along the active Main Himalayan Thrust (MHT) below the Tibetan Plateau [8]. It was formed ~4 Ma ago during the subsiding phase of the Himalayan orogeny [1]. The previously mapped regional geological structures show a consistent pattern that largely reflects their direct association with the regional convergence that is portrayed through the orientation pattern and slip on geological structures that are mapped in the region.

The actively growing MHT reaches the surface as Main Frontal Thrust (MFT), onto which most of the regional faults root. It is estimated that ~50% of regional convergence is absorbed on the MFT and the rest on the interior faults. The occurrence of active thrust faults in Kashmir basin [4, 5] largely reflects ongoing deformation away from the frontal portions of the Himalaya, and the new mapping that shows normal faulting north of Kashmir basin is important to understand the dynamic nature of brittle deformation in NW Himalaya (Fig. 1).

A. Kallel et al. (eds.), *Recent Advances in Geo-Environmental Engineering, Geomechanics and Geotechnics, and Geohazards*, Advances in Science, Technology & Innovation, https://doi.org/10.1007/978-3-030-01665-4_89

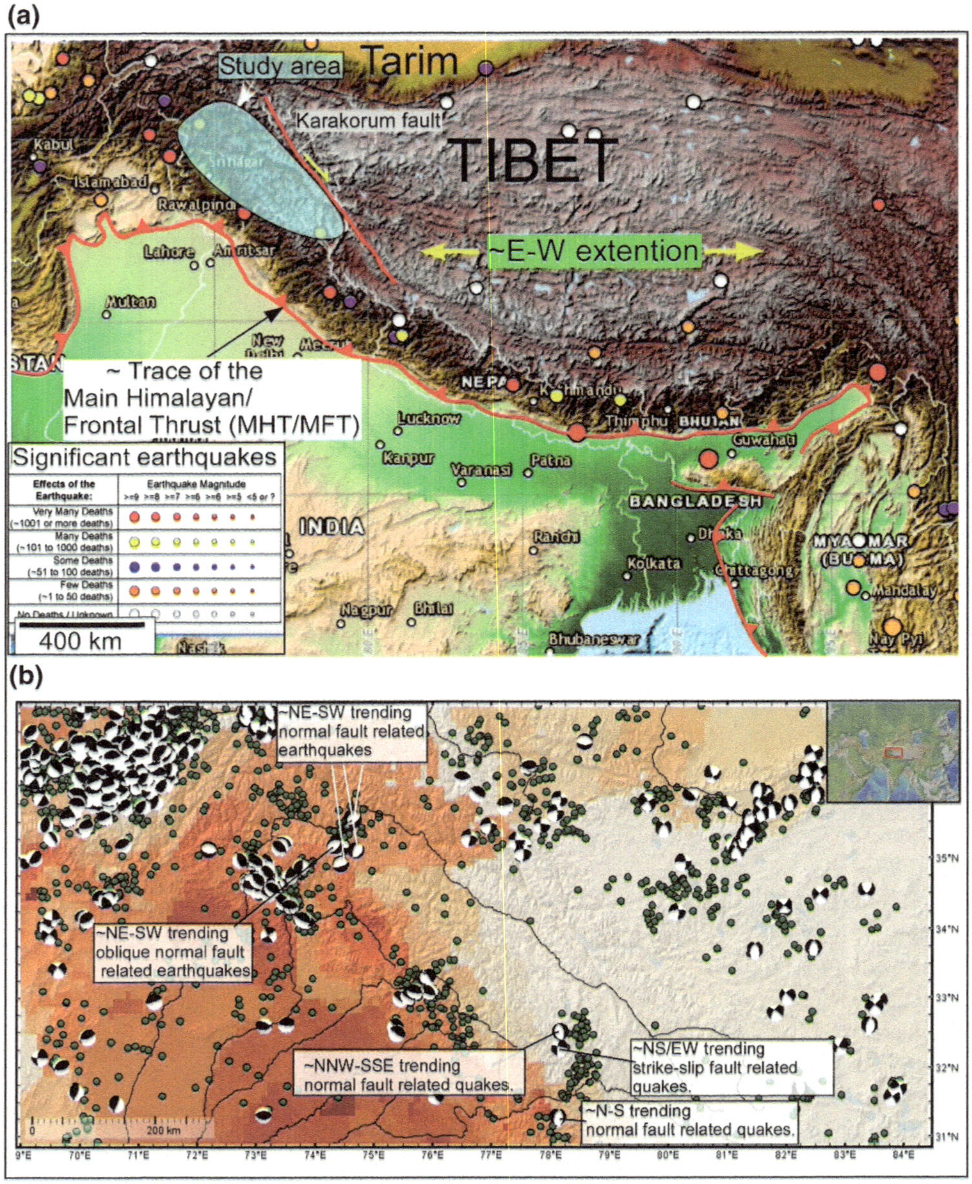

Fig. 1 **a** Shows the significant earthquakes plotted on the satellite image of Himalaya (colored dots). Some of the major active faults are shown as red lines. The polygonal region is the study area. **b** Shows the available data on moment tensor solutions plotted on satellite data. The various types of earthquakes are highlighted. The small green circles are earthquake hypocenters

2 Tectonic Settings

The study area lies to the north of relatively well studied Kashmir basin which is formed ~4.0 Ma ago during the subsiding phase of the Himalayan orogeny [1]. The Kashmir basin was later raised by the actively growing faults that delineate it from the southwest. These faults show a consistent younging towards the Himalayan frontal zone and indicate ~SW tectonic transport [5]. The major faults that were mapped previously show a progressive and sequential tectonic origin, and largely indicate younging towards south-southwest. These are formed as a result of progressive deformation that involves underthrusting of the Indian lithosphere along the active Main Himalayan Thrust (MHT) below the Tibetan Plateau [8]. The surficial expression of MHT is Main Frontal Thrust (MFT), which is exposed at certain portions of the Himalayan arc, and is mostly conceived or blind (Fig. 1). The regional faults (e.g. MCT, MBT) root on this fault, and some are still actively moving, and transporting rocks to upper crustal levels. However, there are a number of regional faults that transport material orthogonal to the regional convergence (~NW and ~SE), and such faults have been mapped in Tibet and south of Karakorum fault (KF), and none in the north of Kashmir basin. Therefore, as shown below, a number of tectonically active normal faults exist in the region.

3 Methodology

Google satellite data allow mapping of large and small scale map large and small scale geological structures, and particularly in regions that are politically controversial or physically inaccessible [4]. This study uses Google terrain mode that offers online contour viewing, and high quality freely accessible data that makes mapping doable on a regional scale. Previously, several works have used a range of satellite data to map active landforms (e.g. [3, 4, 6]).

When satellite image interpretations are coupled with geologic, seismologic, geodetic, and previously mapped structural data it becomes a vigorous methodology for mapping of tectonic landforms. This approach has been adapted below to map the geomorphic expression of faulting where topographic breaks, deflected drainages, deflected/broken ridge axes, incision and erosion patterns, surface ruptures, fault scarps, uplifted/tilted Quaternary to Recent landforms are mapped. The geomorphic mapping is supported by seismological, geological, and geodetic data. The slip on mapped faults is interpreted through the use of available focal mechanisms that is also used to create a set of possible fault plane solutions on each new fault that is mapped. The expected fault plane solutions are created in a freely available structural geology software known as FaultKin 7, by Prof. Rick Allmendinger: http://www.geo.cornell.edu/geology/faculty/RWA/programs/faultkin.html.

4 Results, Interpretation and Discussion

4.1 ~NW-SE Directed Regional Extension

The clear evidence of extensive normal faulting (Fig. 2) above Kashmir basin (Fig. 1) suggests the widespread transitional tectonics in the region. The figure on the left shows an example of regional normal faulting in the study area and the mapped faults trend ~NE-SW, which is oblique to the regional India-Eurasia convergence. This is similar to what Tibet is witnessing today [2]. The mapped normal faults lie west of the major dextral strike-slip fault system, the Karakoram fault (KF), and suggests the need to understand the cause of extension in the region. The relatively well studied KF is a large scale dextral strike-slip fault system that houses a range of extensional basins throughout its length which is ~750 km from the Pamirs in the NW to the Kailas region of SW Tibet in the SE. The Tibetan plateau is located to the east of Karakoram fault, and this region is known for the large scale extension along conjugate strike-slip faults that transport rocks from ~NW to ~SE along well established larger scale strike-slip fault systems [7]. The cause of eastward extension of Tibetan Plateau remains a controversial topic (e.g. [7]). A larger number of studies show extension as a result of gravitational collapse [6] and outward spreading of the plateau, and some argue an oblique convergence along the arcuate plate boundary, while others show the role of mantle upwelling [7]. The region west to KF is generally considered to show a dominant ~S-SW directed tectonic transport via reverse faulting and ~NW-SE trending folds that form most of the Himalayan fold-thrust belt (Fig. 1). The present study however shows that ~NW-SE directed crustal extension is actively taking place in Kashmir Himalaya as well. The mapped ~NE-SW trending normal faults are mostly related to the regional ~NW-SE directed convergence that involves underthrusting of the Indian lithosphere along the active Main Himalayan Thrust (MHT) below the Tibetan Plateau [8]. Their

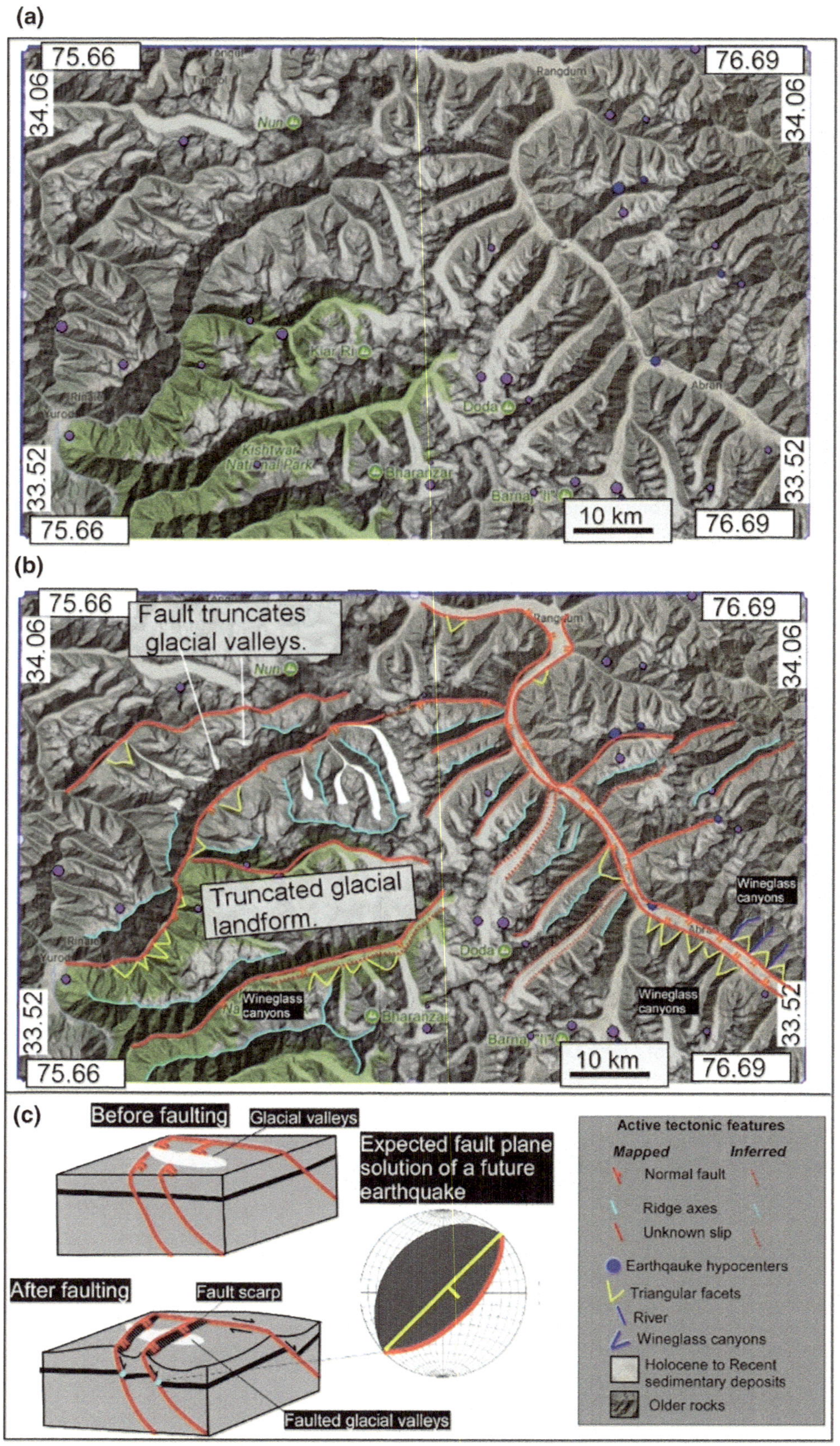

Fig. 2 Shows example of normal faulting where faults cut through the young glacial landforms. The un-interpreted image (**a**) is interpreted (**b**) where a range of active geomorphic features are also shown. It highlights triangular facets, faulted glacial landforms, topographic breaks, and wineglass canyons. The cartoons (**c**) show the landforms before and after the faulting, and the expected fault plane solution of a future earthquake on the mapped faults (dark region shows compressional quadrant and light region shows extensional quadrant). Color filled circles show earthquake hypocenters

occurrence throughout the northern portions of the Kashmir Himalaya reflects the overall ~NW-SE directed extension in the region.

Acknowledgements Authors are very grateful to three anonymous reviewers for improving the quality of the manuscript. The financial help from the National Geographic Society is highly appreciated.

References

1. Burbank, D.W., Johnson, G.D.: The late Cenozoic chronologic and stratigraphic development of the Kashmir intermontane basin, northwestern Himalaya. Palaeogeogr. Palaeoclimatol. Palaeoecol. **43**(3–4), 205–235 (1983)
2. McCaffrey, R., Nabelek, J.: Role of oblique convergence in the active deformation of the Himalayas and southern Tibet plateau. Geology **26**, 691–694 (1998)
3. Sieh, K., Natawidjaja, D.: Neotectonics of the Sumatran fault, Indonesia. J. Geophys. Res. Solid Earth **105**(B12), 28295–28326 (2000)
4. Shah, A.A.: Earthquake geology of Kashmir Basin and its implications for future large earthquakes. Int. J. Earth Sci. **102**(7), 1957–1966 (2013)
5. Shah, A.A., Malik, J.N.: Four major unknown active faults identified, using satellite data, in India and Pakistan portions of NW Himalaya. Nat. Hazards **88**(3), 1845–1865 (2017)
6. Tapponnier, P., Molnar, P.: Active faulting and tectonics in China. J. Geophys. Res. **82**(20), 2905–2930 (1977)
7. Yin, A.: Mode of Cenozoic east–west extension in Tibet suggesting a common origin of rifts in Asia during the Indo-Asian collision. J. Geophys. Res. Solid Earth **105**, 21745–21759 (2000)
8. Zhao, L.S., Xie, J.: Lateral variations in compressional velocities beneath the Tibetan Plateau from Pn traveltime tomography. Geophys. J. Int. **115**(3), 1070–1084 (1993)

Contribution of Morpho-Tectonic Analysis in the Study of Spatio-Temporal Evolution of Land Movements in North-West Tunisia: Example of Balta and Dir El Kef

Radhia Mansour, Mourad El Koundi, Nassira Zouaoui, and Abdessallem El Ghali

Abstract

Gravitational ground movements are a natural event that causes the degradation of the environment and bores the socio-economic development of a population. To study this phenomenon, we adopted a multidisciplinary cartographic approach involving the determining factors of the spatio-temporal appearance and evolution of these gravitational movements. Active faults (seismic genesis), soft lithology, steep slopes and rainfall are the relevant parameters, the combination of which allowed us to highlight the hazard map of the landslide in northwestern Tunisia.

Keywords

Morpho-tectonics • GIS • Spatio-temporal analysis • Land movements

1 Introduction

In recent years, the new perspectives of risk prevention research have focused on gravity movements that are capable of (1) provoking major human and structural disasters, (2) silting hydraulic structures at a rapid rate and (3) degrade the natural environment especially on the slopes of deep incised valleys.

In Tunisia, this phenomenon of ground instability is very developed in the north-west sector known for its very contrasting lithology, rugged morphology and rather complex tectonic structures. The interaction of these three parameters on certain slopes is a catalyst for the initiation and development of landslides, most often affecting the clay layers outcropping on steep slopes and in the vicinity of faults and fractures showing fairly high densities.

In this study we focused on the role of each of the three parameters (lithology, geomorphology and tectonic) and their impact on the initiation and the spatio-temporal evolution of landslides in two geographically distinct areas. The first is located in the Kef region, on the western slope of Dir El Kef and the second in the Balta region, on the northern edge of the plain of Bou Salem also called “plain of the middle Mejrada”.

On the bioclimatic plan, these two study sites belong roughly to the same domain. The statistical study of annual, monthly and daily rainfall over the last decades allowed us to evaluate the degree of impact of this factor on the onset and evolution of landslides in the two localities.

2 Materials and Methods

The present study dealt with the numerical mapping of the ground movement hazard in the north-west of Tunisia, in particular in the region of Kef and that of Bou Salem, both belonging to the most watered bioclimatic field of the country.

The adopted approach to achieve the goal consists in evaluating the ground movement hazard based on the characterization of the lithological and morpho-structural factors determining the instability of the land in the two localities belonging to two distinct geological and structural domains. To do this, the basic data and their analysis results are hierarchical by theme and organized according to a flowchart (Fig. 1). The combination of the resulting analytical maps has led to the development of several thematic maps whose successive crossings have led to a map of the “landslide” hazard synthesis in the two sectors [1–3].

R. Mansour (✉) · N. Zouaoui · A. El Ghali
Faculty of Sciences of Bizerte, Carthage University, Tunis, Tunisia
e-mail: dsmansour.radhia@gmail.com

M. El Koundi
Mining National Office, Tunis, Tunisia

A. Kallel et al. (eds.), *Recent Advances in Geo-Environmental Engineering, Geomechanics and Geotechnics, and Geohazards*, Advances in Science, Technology & Innovation, https://doi.org/10.1007/978-3-030-01665-4_90

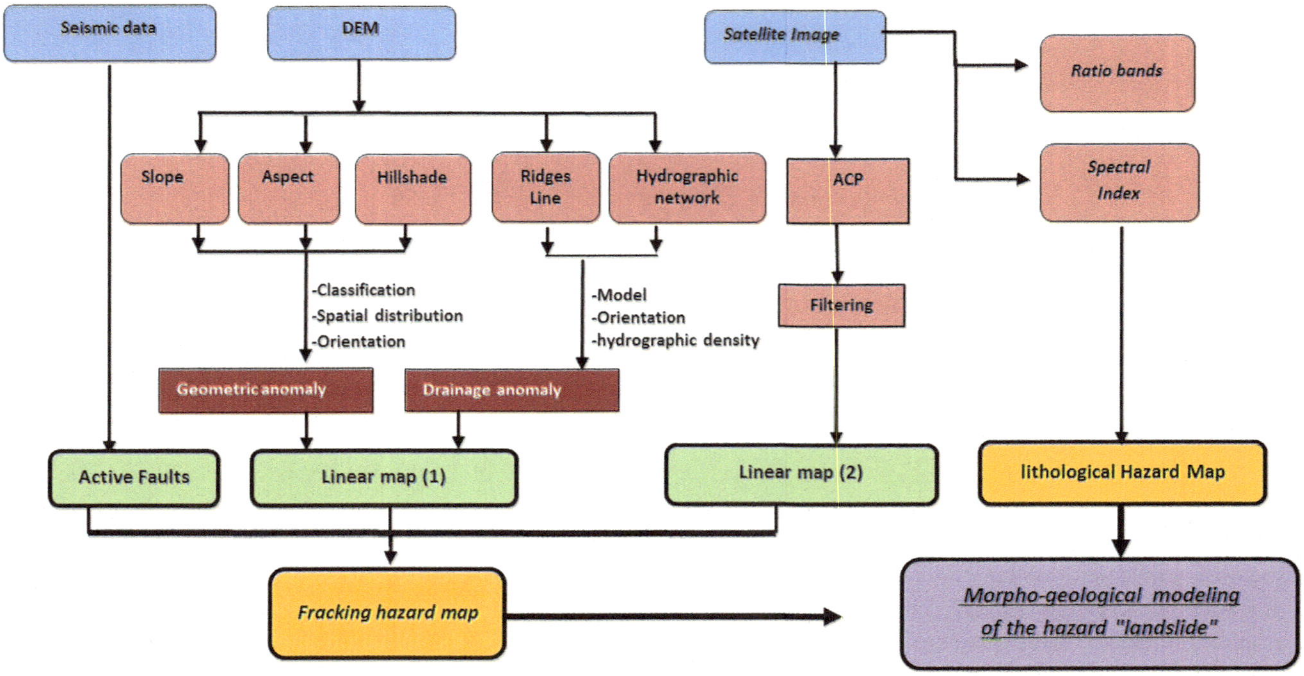

Fig. 1 Methodologic flowchart adopted

3 Results and Discussion

On the lithological plan, the potential lands of landslides are mainly due to the soft and clay-like geological outcrops capped by hard limestone slabs on the backs of the cliffs affected by fairly intense brittle deformations. The irregular rainfall and its torrential nature have only aggravated the situation on the slopes [4].

In general, the areas affected by the movements of land are mountainous regions with steep slopes of which the areas of Kef and Balta are part. However, comparing the slopes relative to the two sites, balta slip is characterized by a much steeper slope and therefore has a much longer movement arrow downstream of the watershed. Thus, the factor "slope" plays a decisive role in the length of the trajectory of the movement of the slipped mass.

As for the factor "fracturing", its importance is especially marked by the limestone slabs capping the clay masses. In fact, the deepest and highest density fractures characterize the most pronounced landslides and the longest ridge ledges and the deepest sliding planes, such as Balta. This can be explained by reaching the roof of the clay layer (sliding seat) by a large amount of vertical and lateral infiltration water through the fractures and karsts affecting the limestone bar capping the clay mass, causing thus its supersaturation and its change towards a more fluid and less resistant physical state.

The interaction of the different triggers, added to the irregularity of the rainfall and its torrential nature destabilized some sectors and reactivated others that were already unstable or have undergone previous shifts.

From the study of the slip cases presented in this work, it is clear that the onset of the ground movements and their evolution in time and space result from the conjunction of several factors dependent on each other but numerically dissociable. The maps derived from them are obtained by photo-interpretation, satellite image processing and verified by direct observations on the ground (ground truth). The selective map combination (weighted or linear) of elementary geological, morphometric, tectonic, climatic, hydrological maps …) highlights the possibility of assessing the susceptibility of terrains to mass movements and allows the elaboration of landslides hazard maps.

4 Conclusion

The various operations of compilation and combination of the basic numerical data with spatial reference: lithological, morphometrical, tectonic and rainfall, organized according to the flowchart presented in this study, allowed the elaboration of the landslides hazard map in several areas of northern Tunisia, particularly Dir el Kef and Balta. On the other hand, this study highlighted an important interaction between several natural factors whose tectonic relationship—topographic slope—lithological nature, seem to be the most determining for the triggering and evolution of landslides during rainy seasons.

References

1. Gyozo, J.: Terrain Modelling with GIS for Tectonic Geomorphology. Geocentrum, Uppsala (2004)
2. Mebarka, T.: Determining the role of lineaments in underground hydrodynamics using Landsat 7 ETM + data, case of the Chott El Gharbi Basin (western Algeria). Arab. J. (2018)
3. Amin, B.: Application of advanced spaceborne thermal emission and reflection radiometer (ASTER) data in geological mapping. Int. J. Phys. Sci. (2011)
4. El Aroui, O.: Contribution to the geomorphological study and mapping of the ground movements on the slopes of the valleys of the exotic rivers of Kroumirie (northern Tunisia). Thesis (2015)

Mapping Geological Risk in Urban Areas (by the Example of Moscow, Russia)

Irina Kozliakova and Olga Eremina

Abstract

The geological risk of economic losses was qualitatively assessed for Moscow, Russia, proceeding from the comprehensive consideration of geohazards and the vulnerability of urban environment. The map of geo-ecological conditions was built based on assessing karst, suffosion, waterlogging and landslide hazards in Moscow; and the schematic map of geological risk (scale 1: 50,000) was compiled for the Moscow territory, representing risk as an integral parameter of probable damage caused by geo-hazards and the anthropogenic load on the urban territory. The mapping procedure included four steps: grading and mapping exogenous geo-hazards; grading and mapping the urban environment vulnerability; distinguishing risk categories; and compiling the resultant risk map. This qualitative approach provides reliable and sufficient data permitting urban planners to optimize investments in Moscow city development.

Keywords

Exogenous geological processes • Geo-hazards • Risk analysis • Urban areas • Urban environment vulnerability

1 Introduction

Transition from evaluation and mapping of exogenous geological hazards to risk assessment is a serious task of urban geology for ensuring sustainable development of cities. Geological risk is usually defined as a measure of geological hazard or a number of hazards determined for a particular object in the form of possible absolute or relative economic losses [1]. This definition suggests that risk is a function of the hazard impact value and the vulnerability of an urban environment. Researchers involved in risk assessment in urban areas agree that the combination of geo-hazard maps with the maps of urban environment vulnerability is the most promising approach to the geo-risk assessment in cities [2]. Qualitative grading and building risk maps for the existing urban infrastructure should be performed at the early stage of urban planning. Mapping risk is aimed at outlining areas, for which special requirements should be set on the engineering survey, and constraints should be imposed on urban development. The aim of this study was to develop the procedure of qualitative geo-risk assessment and mapping in Moscow.

2 Procedure of Qualitative Geo-risk Assessment in Moscow

The investigation procedure included the following steps: (1) ranking and mapping exogenous geological processes (EGPs) according to their hazard degree; (2) ranking and mapping vulnerability of urban environment to EGPs (the urban environment here is taken to mean the city territory with buildings and on-surface engineering facilities); (3) distinguishing risk categories according to hazardous geological processes affecting the city territory and surface urban infrastructure; and (4) compiling the risk map by superposition of the integral map of EGPs hazard and the map of urban environment vulnerability (Fig. 1).

3 Results

Moscow is located within the Eastern European platform; and its geological structure is formed by sedimentary strata of Meso-Cenozoic age. Landslides, karst collapses, suffosion ground subsidence and waterlogging are the most hazardous

I. Kozliakova · O. Eremina (✉)
Sergeev Institute of Environmental Geosciences, Russian Academy of Sciences, Ulansky Lane 13, 101000 Moscow, Russia
e-mail: sci-council@geoenv.ru; o_eremina@mail.ru

A. Kallel et al. (eds.), *Recent Advances in Geo-Environmental Engineering, Geomechanics and Geotechnics, and Geohazards*, Advances in Science, Technology & Innovation, https://doi.org/10.1007/978-3-030-01665-4_91

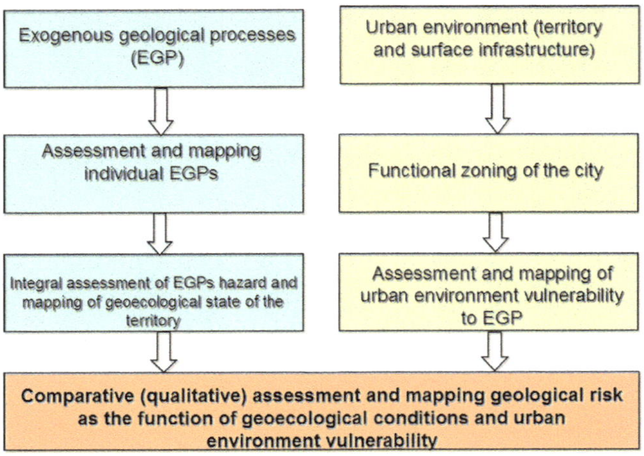

Fig. 1 Algorithm of assessing geological risk in Moscow

EGPs to urban infrastructure. These EGPs were ranked according to their hazard degree and the hazard maps were built at scale 1:50,000 for each EGP. Superposition of these maps resulted in the integral assessment of geo-ecological conditions of Moscow. Five categories of geo-ecological conditions were distinguished ranging from favorable to very unfavorable categories depending on the EGP combination and possible adverse consequences to engineering structures. The resultant map of the geo-ecological conditions of Moscow (based on the EGP assessment) was built using this ranking [3].

The urban environment is taken to mean here the city territory with buildings and on-surface engineering facilities. For a small-scaled study, the vulnerability of urban environment is controlled by density of engineering structures and buildings distribution, which may be obtained from the functional zoning of the city. The functional zoning of Moscow territory includes 5 types of functional zones, and we assumed that the density of territory building-up with on-surface engineering structures increases in the row: recreational—transport—industrial—public—residential zones. By combining the map of geo-ecological conditions and the zones of urban environment vulnerability we come up with the qualitative (schematic) assessment of geological risk.

The geological risk level in the city territory is obtained from superposition of the geo-ecological conditions map (hazardous EGP assessment) and the functional zoning of the territory. Five risk categories were distinguished according to the level of possible loss caused by the manifestation of individual EGPs or their combinations: no risk—low risk—moderate risk—high risk—very high risk (Fig. 2).

The risk increases with intensifying geological hazards and increasing density of building-up in the area (Fig. 3).

Eventually, a schematic map of geological risk in Moscow was compiled at scale 1: 50,000, which shows the risk as a comparative integral characteristics of the probable damage caused by the geological hazards as well as the level and type of the technogenic load of the territory. This map shows that the areas of high and very high risk (that should be of particular attention for the authorities) occupy 14%; the areas of moderate risk—46%; and the areas of low risk—40% of Moscow territory.

Fig. 2 The content of geological risk categories for Moscow area

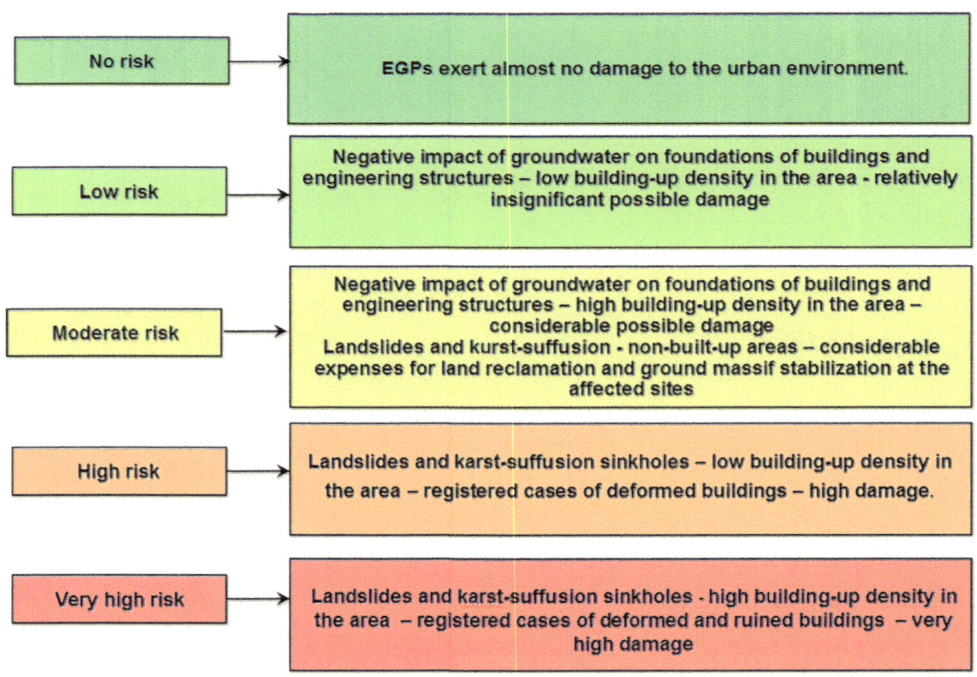

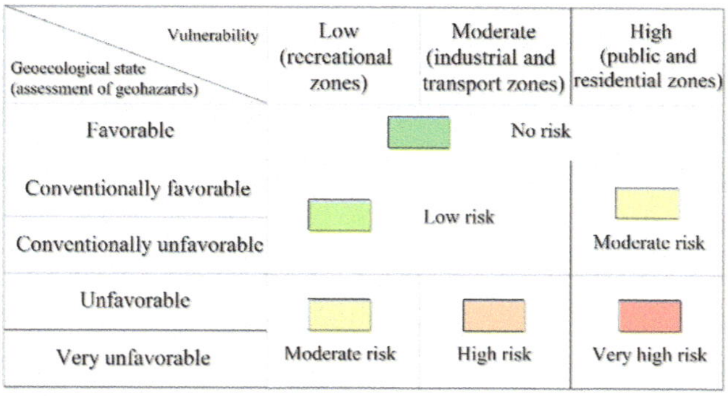

Fig. 3 Legend to the schematic map of geological risk in Moscow

4 Conclusion

The performed studies in the assessment and mapping of geological risk in Moscow prove that risk-analysis according to integral consideration of geo-hazards and vulnerability of urban environment is an adequate and promising tool for the qualitative assessment of possible economic losses from EGPs manifestation in urban area. For the city territory at the early stage of investigation, this risk analysis should consist in zoning territory by the risk level. The qualitative risk assessment permits us to distinguish the city areas, in which the constraints should be imposed on urban development due to a high risk of damage for the existing facilities.

Acknowledgements This study was supported by the Russian Science Foundation, project no. 16-17-00125.

References

1. Ragozin, A.L. (ed.): Natural hazards of Russia. Assess. Manage. Nat. Risks, 320 p. (2003) (in Russian) (Topical vol., Moscow, KRUK)
2. Clayton, C.R.I.: Urban site investigation. In: Culshaw, M.G., Reeves, H.J., Jefferson, I., Spink, T.W. (eds.) Engineering Geology for Tomorrow's Cities, vol. 22, pp. 15–141. Geological Society, Engineering Geology Special Publication, London (2009)
3. Kutepov, V.M., Anisimova, N.G., Eremina, O.N., Kozhevnikova, I.A., Kozliakova, I.A.: Hazardous geological processes and geoecological state in Moscow. In: Proceedings of the 2nd Scientific and Practical All-Russia Conference on Ecological and Geological Problems in Urbanized Territories, pp. 194–196. Yekaterniburg (2009) (in Russian)

Numerical Simulation of Land Subsidence Caused by Both Dewatering and Recharging

Xu-Bing Xu and Zhen-Dong Cui

Abstract

The groundwater recharge was conducted to prevent land subsidence in many major cities such as Beijing, Shanghai and Tianjin in China. Specifically, the use of recharging to prevent ground subsidence in foundation pit projects has become a popular technique. In this paper, the numerical simulation was used to study the influence of the distance between the recharging well and the dewatering well on the land subsidence. The results show that the appropriate distance between the recharging well and the dewatering one is favorable to prevent subsidence.

Keywords

Land subsidence • Numerical simulation • Dewatering • Groundwater recharge

1 Introduction

In the last century, Powrie and Roberts studied the engineering example of using an aquifer recharge in the pit to limit the depth of the water level outside the pit caused by the precipitation in the pit. The results show that the recharge method can effectively prevent the ground settlement in the foundation pit projects [1]. Phien-wej et al. carried out on-site recharge test and soil deformation monitoring, and the results showed that recharging could cause the aquifer and its weak system to rebound [2]. Scanlon et al. conducted comparison and research on the recharge technology. The results of the study showed that appropriate recharging techniques can increase the recharge rate and better improve the recharge effect [3, 4]. The recharge of phreatic aquifer was carried out in the actual projects of Hangzhou, and successfully controlled the settlement of buildings [5]. In Shanghai, the multi-circulation pressure test was carried out with different recharge amounts, different recharge times, natural recharge and pressure recharge [6].

Although recharging is widely used in engineering examples, few people have studied the effect of the distance between the recharging well and the pumping well on settlement. In this paper, the numerical simulation method was used to study the effect of distance between the recharging wells and the pumping ones on land subsidence under 2D conditions.

2 Methods

In the numerical simulation, the set parameters were used to obtain the effect radius R of single-well pumping, and the recharge wells were arranged at $R/3$, $R/2$, $2R/3$ and R in models, respectively. When calculating the influence radius R of a single well with stable groundwater seepage in a phreatic aquifer, the influence radius R can be calculated according to this formula in the absence of experimental conditions:

$$R = 2s_w\sqrt{kH} \quad (1)$$

where R is influence radius of the dewatering well (m); s_w is difference between the static water level and the dynamic water level (m); k is the permeability coefficient of the aquifer (m/d); H is the thickness of phreatic water aquifer (m).

The permeability coefficient is 1.29×10^{-2} m/d and the H is 100 m. In model 1, s_w is 80 m. So the influence radius R, can be calculated as:

$$R = 2 \times 80 \times \sqrt{1.29 \times 10^{-2} \times 100} = 180.93 \text{ m}$$

X.-B. Xu · Z.-D. Cui (✉)
State Key Laboratory for Geomechanics and Deep Underground Engineering, School of Mechanics and Civil Engineering, China University of Mining and Technology, 221116 Xuzhou, Jiangsu, People's Republic of China
e-mail: cuizhendong@cumt.edu.cn

A. Kallel et al. (eds.), *Recent Advances in Geo-Environmental Engineering, Geomechanics and Geotechnics, and Geohazards*, Advances in Science, Technology & Innovation, https://doi.org/10.1007/978-3-030-01665-4_92

According to the above calculation, the distance between the recharge well and the dewatering well in the model 2–5 are $L2 = R/3 = 60$ m, $L3 = R/2 = 90$ m, $L4 = 2R/3 = 120$ m, $L5 = R = 180$ m, the location of the recharging wells are shown in Fig. 1.

3 Results

Figure 1 is the settlement cloud picture, it can be seen that the closer the distance between the recharge well and the pumping well is, the more effective to prevent the development radius of the dewatering well from developing outward. In model-2, it can be seen that when the recharge well is close to the dewatering well, the influence radius of the pump well will continue to expand beyond the recharging well. It is not difficult to imagine that the closer the distance to the pumping well is, the stronger its influence is. If the recharge well is too close to the pumping well, the recharge part will not be enough to make up for the amount of extracted water, and the influence radius will continue to expand outward. In model-3 and model-4, it can be seen that when the recharging well is set in a suitable position, the influence radius can be effectively prevented from expanding outward. In model-5 we find that the recharging well has little effect when it is set remotely. The same conclusions can be drawn about Fig. 2.

Figure 3 shows that with the development of pumping time at 20 m, the settlement curves are largely consistent, and the settlement rate is becoming slower and slower with the development of time, which indicates that the development of settlement near the pumping well is basically not affected by the remote recharge. Compared to Fig. 4, at 80 m the settlement can be clearly found that when the location of the recharge well is 60 m, the settlement is obviously less than that of the recharge well at 90, 120 and 180 m, which indicates that the settlement of the ground can be prevented well in the protection range of the recharge well.

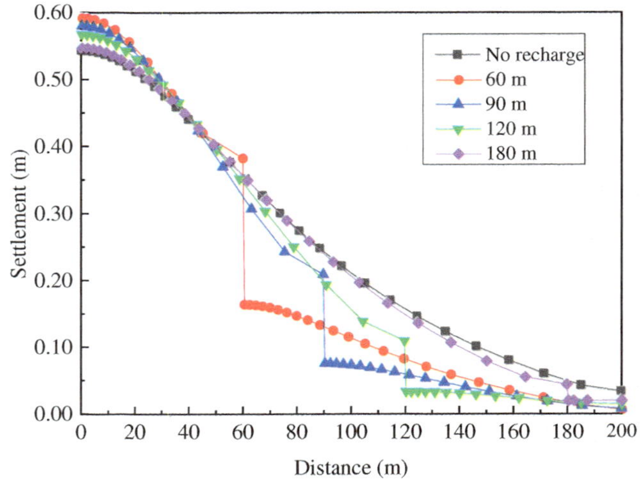

Fig. 2 Different recharge distance settlement distributions

4 Discussion

The reasonable position of the recharge well in the dewatering of the foundation pit projects has a positive effect on preventing the ground subsidence. In the actual project, the combination of water-resisting curtain and recharge well can be used to avoid the harm to the important buildings. This paper discussed the dewatering and recharging simulation of the phreatic layer. In the process of recharging prone to blockage makes it difficult to continue recharging, in the actual project the main and auxiliary wells recharge method is often used to solve the plugging problem.

In many recharge systems, dual-use equipment for the dewatering and recharging is often used, and an intelligent

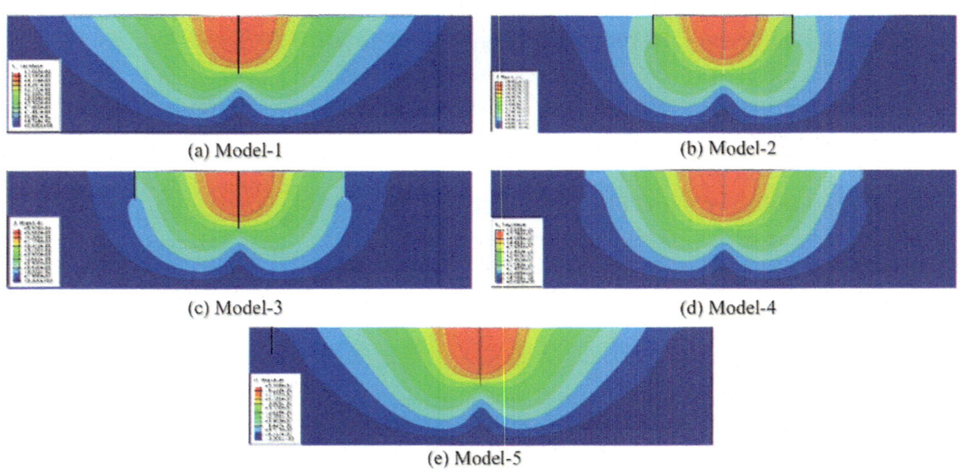

Fig. 1 Settlement cloud picture under different recharging wells

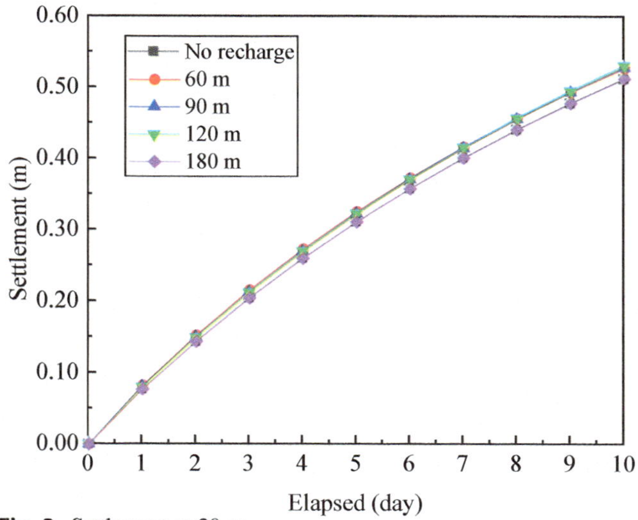

Fig. 3 Settlement at 20 m

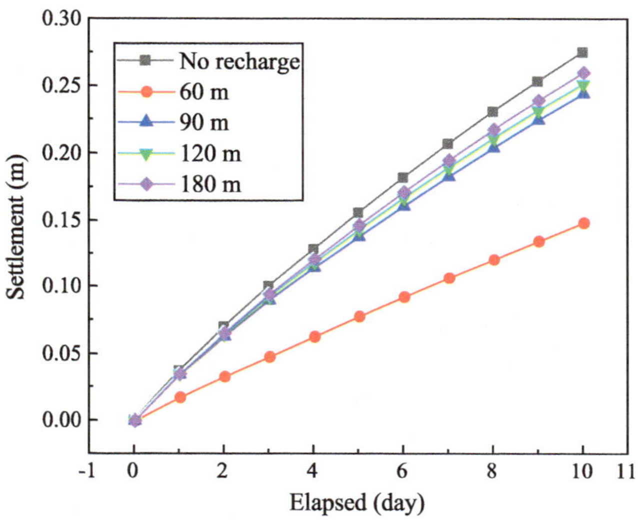

Fig. 4 Settlement at 80 m

equipment can be used to effectively control the recharging process. At present, although recharge has received more and more attention, there is still a lack of theoretical research on recharge. How to efficiently and reasonably perform recharge should be further investigated.

5 Conclusion

The numerical simulation was conducted to study the effect of different recharge distances on land subsidence during recharging. The main obtained conclusions are as follows.

(1) When the dewatering well is close to the recharging well, the influence radius R of the pumping well will continue to expand beyond the recharging well.

(2) The closer the distance between the dewatering well and the recharging one is, the more favorable it is to prevent ground settlement. When the distance of the recharging well exceeds the dewatering influence radius, it will have little effect on preventing the land subsidence. The reasonable position of the recharging well should be at a distance from the pumping well Within $R/2$.

Acknowledgements This work was funded by National Key R&D Program of China (2016YFC0600904).

References

1. Powrie, W., Roberts, T.O.L.: Case history of a dewatering and recharge system in chalk. Géotechnique **45**(4), 599–609 (1995)
2. Phien-wej, N., Giao, P.H., Nutalaya, P.: Field experiment of artificial recharge through a well with reference to land subsidence control. Eng. Geol. **50**(1/2), 187–201 (1998)
3. Scanlon, B.R., Healy, R.W., Cook, P.G.: Choosing appropriate techniques for quantifying groundwater recharge. Hydrogel **10**, 18–39 (2002)
4. Scanlon, B.R., Keese, K.E., Flint, A.L., Flint, L.E., Gaye, C.B., Edmunds, M.W., Simmers, I.: Global synthesis of groundwater recharge in semiarid and arid regions. Hydrol. Process. **20**, 3335–3370 (2006)
5. Yu, J.L., Gong, X.N.: Study on the design and the application of the groundwater recharge system in excavation. J. Build. Struct. **22**(5), 70–74 (2001)
6. Wang, J.X., Wu, Y.B., Zhang, X.S.: Field experiments and numerical simulations of confined aquifer response to multi-cycle recharge–recovery process through a well. J. Hydrol. **464/465**, 328–343 (2012)

Displacement Distribution Caused by Pumping from the Aquifer in Soil

Wen-Hao Guo, Zhen-Dong Cui, and Zhen Li

Abstract

A new phenomenon that the accumulative multi-layered settlement is not equal to the land subsidence was discovered during the engineering dewatering. Displacement distribution induced by pumping from the confined aquifer was investigated using axisymmetric consolidation of multi-layered soils based on the theory of Biot's consolidation. The variations of the displacement showed tensile and shear deformations that appeared in the overlying aquitard when pumping from the confined aquifer. The tensile deformation decreased with the increasing distance away from the pumping well. The shear deformation first increased and then decreased with the growing distance away from the pumping well.

Keywords

Vertical displacement • Horizontal displacement • Anisotropic permeability • Multi-layered soils

1 Introduction

Soil deformation, including the horizontal displacement and the vertical one, often leads to the failure of buildings and infrastructure. Many methods, such as the theory of elastic layered system [1], the Mindlin displacement solution and the method based on Cosserat mechanics [2] have been used to calculate deformations. The land subsidence can be predicted approximately using the above methods. However, a new phenomenon was discovered in Shanghai [3] and Tianjin [4] during the engineering dewatering. The accumulative multi-layered settlement is not equal to the land subsidence, and some strata are not compressed but expansive. Some researchers considered that the phenomenon could be explained by the combination of high-lower seepage characters, dewatering in deep confined aquifer-inversed rebound of overlaying strata, consolidation from deep-coordination of deformation and gradual boundary [3]. Other investigators deemed that the soil-arch effect caused the phenomenon [4]. Nevertheless, there was lack of the computing method that can reflect the phenomenon. In this paper, the soil deformation of different depths induced by the pumping from the confined aquifer was calculated utilizing the method based on the theory of Biot's consolidation with anisotropic permeability.

2 Methods

By using the Laplace-Hankel transforms, Li and Cui [5] get the exact expression of stresses, displacements, excess pore pressure and vertical seepage velocity from the axisymmetric Biot's consolidation governing equations with anisotropic permeability. Equations (1) and (2) are defined to establish a relationship among stresses, displacements, excess pore pressure and seepage velocity, which are vectors made of $A_1 \sim A_6$.

$$\hat{U}(\xi, z, s) = [\hat{u}_{r1}(\xi, z, s), \hat{\sigma}_{z0}(\xi, z, s), \hat{p}_0(\xi, z, s)]^T \quad (1)$$

$$\hat{\Gamma}(\xi, z, s) = \left[\hat{\sigma}_{rz1}(\xi, z, s), \hat{u}_{z0}(\xi, z, s), \hat{V}_0(\xi, z, s)\right]^T \quad (2)$$

where, $\hat{u}_{r1}(\xi, z, s)$ is the horizontal displacement; $\hat{\sigma}_{z0}(\xi, z, s)$ is the vertical stress; $\hat{p}_0(\xi, z, s)$ is the pore pressure; $\hat{\sigma}_{rz1}(\xi, z, s)$ is the horizontal stress; $\hat{u}_{z0}(\xi, z, s)$ is the vertical displacement; $\hat{V}_0(\xi, z, s)$ is the vertical seepage velocity; s is the Laplace transform variable; ξ is the Hankel transform variable. These six factors are listed in Appendix.

The impact of pumping can be defined through Eq. (3)

W.-H. Guo · Z.-D. Cui (✉) · Z. Li
State Key Laboratory for Geomechanics and Deep Underground Engineering, School of Mechan-ics and Civil Engineering, China University of Mining and Technology, Xuzhou, Jiangsu, People's Republic of China
e-mail: cuizhendong@cumt.edu.cn

A. Kallel et al. (eds.), *Recent Advances in Geo-Environmental Engineering, Geomechanics and Geotechnics, and Geohazards*, Advances in Science, Technology & Innovation, https://doi.org/10.1007/978-3-030-01665-4_93

$$T(\xi, z, s) = [Q_2(\xi, \Delta H, s), -Q_3(\xi, \Delta H, s), Q_1(\xi, \Delta H, s)]^{\mathrm{T}} \tag{3}$$

where, Q is the water flow velocity; $Q_1(\xi, \Delta H, s)$, $Q_2(\xi, \Delta H, s)$ and $Q_3(\xi, \Delta H, s)$ are factors influenced by Q; ΔH is the thickness of the confined aquifer.

The relationship can be expressed through Eqs. (4) and (5) which are matrices made of $\mathrm{A}_1 \sim \mathrm{A}_6$.

$$\begin{bmatrix} -\hat{\mathrm{U}}(\xi, 0, s) \\ \hat{\mathrm{U}}(\xi, \mathrm{H}, s) \end{bmatrix} + \begin{bmatrix} -\mathrm{T}(\xi, \Delta\mathrm{H}, s) \\ \mathrm{T}(\xi, \Delta\mathrm{H}, s) \end{bmatrix} = R \cdot [\,\mathrm{A}_1 \quad \mathrm{A}_2 \quad \mathrm{A}_3 \quad \mathrm{A}_4 \quad \mathrm{A}_5 \quad \mathrm{A}_6\,]^{\mathrm{T}} \tag{4}$$

$$\begin{bmatrix} \hat{\Gamma}(\xi, 0, s) \\ \hat{\Gamma}(\xi, \mathrm{H}, s) \end{bmatrix} = S \cdot [\,\mathrm{A}_1 \quad \mathrm{A}_2 \quad \mathrm{A}_3 \quad \mathrm{A}_4 \quad \mathrm{A}_5 \quad \mathrm{A}_6\,]^{\mathrm{T}} \tag{5}$$

Substituting Eq. (5) with Eq. (4)

$$\begin{bmatrix} -\hat{\mathrm{U}}(\xi, 0, s) \\ \hat{\mathrm{U}}(\xi, \mathrm{H}, s) \end{bmatrix} + \begin{bmatrix} -\mathrm{T}(\xi, \Delta\mathrm{H}, s) \\ \mathrm{T}(\xi, \Delta\mathrm{H}, s) \end{bmatrix} = \Phi \cdot \begin{bmatrix} \hat{\Gamma}(\xi, 0, s) \\ \hat{\Gamma}(\xi, \mathrm{H}, s) \end{bmatrix} \tag{6}$$

where $\Phi = R \cdot S^{-1}$, Φ is the stiffness matrix.

Based on continuity condition, the global stiffness matrix of multi-layered soils can be obtained in the form of Eq. (7) shown below.

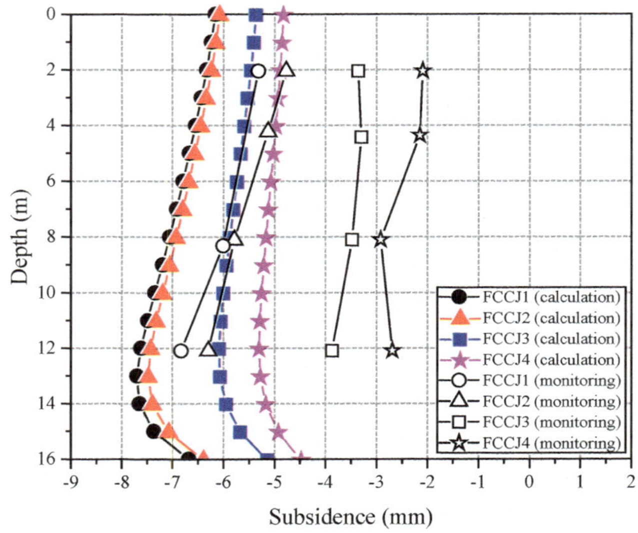

Fig. 1 Comparison of the calculation with the in-site monitoring

$$\begin{bmatrix} -\hat{\mathrm{U}}(\xi, 0, s) \\ 0 \\ \vdots \\ -\mathrm{T}(\xi, \Delta\mathrm{H}_j, s) \\ \mathrm{T}(\xi, \Delta\mathrm{H}_j, s) \\ \vdots \\ 0 \\ \hat{\mathrm{U}}(\xi, \mathrm{H}_n, s) \end{bmatrix} = \begin{bmatrix} \Phi_1 & & & & \\ & \Phi_2 & & & \\ & & \ddots & & \\ & & & \Phi_{(n-1)} & \\ & & & & \Phi_n \end{bmatrix}_{(3n+3)\times(3n+3)} \begin{bmatrix} \hat{\Gamma}(\xi, 0, s) \\ \hat{\Gamma}(\xi, \mathrm{H}_1, s) \\ \vdots \\ \hat{\Gamma}(\xi, \mathrm{H}_{j-1}, s) \\ \hat{\Gamma}(\xi, \mathrm{H}_j, s) \\ \vdots \\ \hat{\Gamma}(\xi, \mathrm{H}_{n-1}, s) \\ \hat{\Gamma}(\xi, \mathrm{H}_n, s) \end{bmatrix} \tag{7}$$

3 Results

3.1 Reliability of the Method

A series of monitoring points of settlement were placed at 2 m (FCCJ1), 5 m (FCCJ2), 21 m (FCCJ3), and 30 m (FCCJ4) away from the pumping well. Comparing the results with the in-site monitoring data in Fig. 1, the vertical displacement calculated by this method is close to the monitoring data near the pumping well. Thus, the method can be used to predict the soil deformation.

3.2 The Displacement of Different Depths

Figure 2a shows the variations of the vertical displacement of different depths with the distance away from the pumping well. Within 10 m away from the pumping well, the vertical displacement of 13 m away from the ground is the biggest, and the surficial vertical displacement is the smallest. When the distance exceeds 10 m, the vertical displacement of 16 m away from the ground becomes the minimum. Figure 2b illustrates the variations of the vertical displacement with the depth the different increasing distances away from the pumping well. When the distance away from the pumping well grows, the vertical displacement decreases at all depths, and the difference of the different depths also decreases. The tensile deformation decreases with the increasing distance away from the pumping well.

The first graph in Fig. 3 illustrates the variations of the horizontal displacement at different depths with the increasing distance away from the pumping well. When the distance away from the pumping well exceeds 40 m, the horizontal displacement decreases with the increasing depth. The second graph in Fig. 3 shows the variations of the horizontal displacement with the increasing depth at different distances away from the pumping well. The variations of the horizontal displacement at 0 and 16 m away from the ground

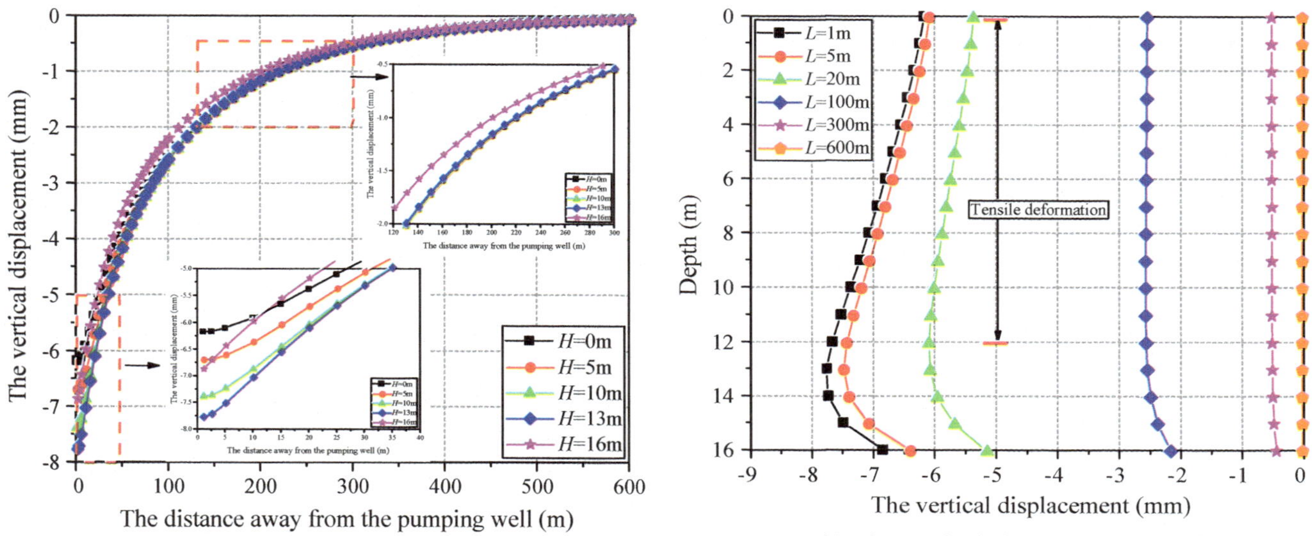

(a) The vertical displacement of different depths

(b) The vertical displacement of different distance away from the pumping well

Fig. 2 Results of the vertical displacement

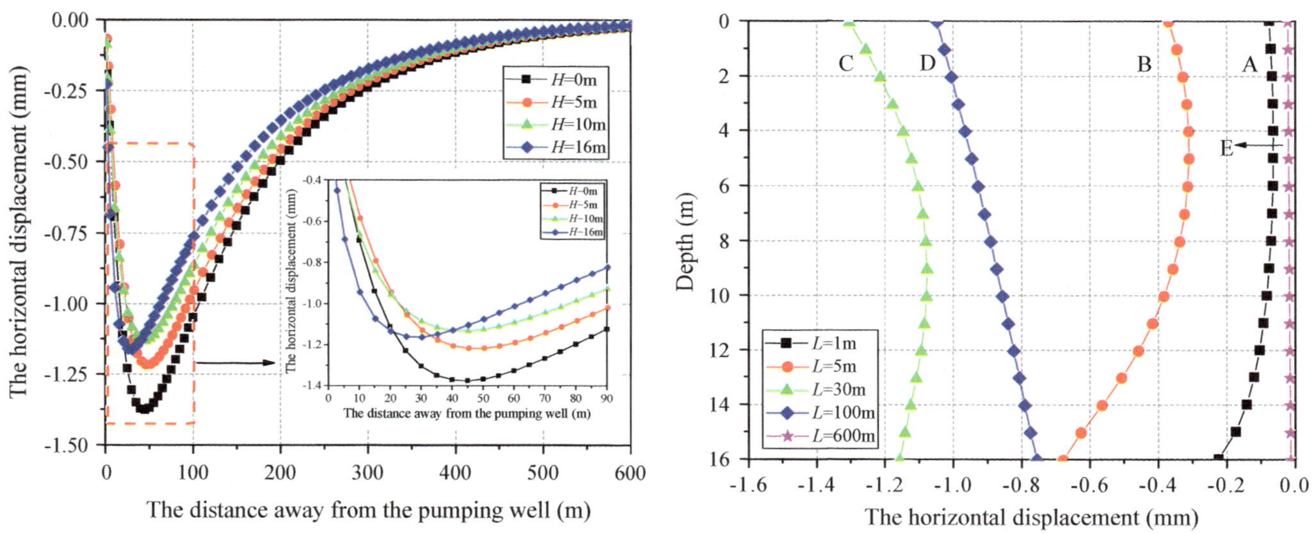

(a)The horizontal displacement of different depths

(b)The horizontal displacement of different distances away from the pumping well

Fig. 3 Horizontal displacement results

are different. At the different distances, the maximum of the horizontal displacement is at different depth.

4 Discussion

The method cannot reflect the influence of no pumping time which has an effect on seepage and consolidation. By considering no pumping time, we can deeply analyze the consolidation behaviors of layered soils.

5 Conclusion

Tensile and shear deformations appear in the overlying aquitard when pumping from the confined aquifer. The tensile deformation decreases with the increasing distance away from the pumping well. The shear deformation first increases and then decreases with the increasing distance away from the pumping well.

Acknowledgements This work was funded by National Key R&D Program of China (2016YFC0600903).

Appendix

$$\hat{p}_0 = A_1 e^{-\xi z} + A_2 e^{\xi z} + A_3 e^{-qz} + A_4 e^{qz} - Q_1(\xi, \Delta H, s)$$

$$\begin{aligned}\hat{u}_{r1} = & \frac{1-(M-G)\varphi}{2G} z e^{-\xi z} A_1 - \frac{1-(M-G)\varphi}{2G} z e^{\xi z} A_2 \\ & + \frac{\xi e^{-qz}}{M(-q^2+\xi^2)} A_3 + \frac{\xi e^{qz}}{M(-q^2+\xi^2)} A_4 \\ & + e^{-\xi z} A_5 + e^{\xi z} A_6 - Q_2(\xi, \Delta H, s)\end{aligned}$$

$$\begin{aligned}\hat{u}_{z0} = & \frac{1 - G\varphi - M\varphi + z\xi(1+G\varphi - M\varphi)}{2G\xi} e^{-\xi z} A_1 \\ & + \frac{G\varphi + M\varphi - 1 + z\xi(1+G\varphi - M\varphi)}{2G\xi} e^{\xi z} A_2 \\ & - \frac{qe^{-qz}}{M(q^2-\xi^2)} A_3 + \frac{qe^{qz}}{M(q^2-\xi^2)} A_4 + e^{-\xi z} A_5 - e^{\xi z} A_6\end{aligned}$$

$$\hat{V}_0 = -\xi k_z e^{-\xi z} + \xi k_z e^{\xi z} - q k_z e^{-qz} + q k_z A_4 e^{qz}$$

$$\begin{aligned}\hat{\sigma}_{rz1} = & (G\varphi - z\xi(1+G\varphi - M\varphi))e^{-\xi z} A_1 \\ & - (G\varphi - z\xi(1+G\varphi - M\varphi))e^{\xi z} A_2 + \frac{2q\xi G e^{-qz}}{M(q^2-\xi^2)} A_3 \\ & - \frac{2q\xi G e^{qz}}{M(q^2-\xi^2)} A_4 - 2e^{-\xi z} G\xi A_5 + 2e^{\xi z} G\xi A_6\end{aligned}$$

$$\begin{aligned}\hat{\sigma}_{z0} = & (-1 + M\varphi - z\xi(1+G\varphi - M\varphi))e^{-\xi z} A_1 \\ & - (1 - M\varphi - z\xi(1+G\varphi - M\varphi))e^{\xi z} A_2 \\ & + \frac{2\xi^2 G e^{-qz}}{M(q^2-\xi^2)} A_3 + \frac{2\xi^2 G e^{qz}}{M(q^2-\xi^2)} A_4 - 2e^{-\xi z} G\xi A_5 \\ & - 2e^{\xi z} G\xi A_6 + Q_3(\xi, \Delta H, s)\end{aligned}$$

where, G is the shear modulus; $M = 2\eta G$; $q^2 = \alpha\xi^2 + \frac{\gamma_w s}{M k_z}$

References

1. Cui, Z.D., Jia, Y.J., Yuan, L.: Distribution law of soil deformation caused by decompression of confined water. Environ. Earth Sci. **75** (18), 1281 (2016)
2. Budhu, M., Adiyaman, I.B.: Mechanics of land subsidence due to groundwater pumping. Int. J. Numer. Anal. Meth. Geomech. **34** (14), 1459–1478 (2010)
3. Wang, J., Wu, L., Zhu, Y., Tang, Y., Yang, P., Lou, R.: Mechanism of dewatering-induced ground subsidence in deep subway station pit and calculation method. Chin. J. Rock Mech. Eng. **28**(5), 1010–1019 (2009)
4. Zheng, G., Zeng, C., Diao, Y., Xue, X.: Test and numerical research on wall deflections induced by pre-excavation dewatering. Comput. Geotech. **62**, 244–256 (2014)
5. Li, Z., Cui, Z.D.: Axisymmetric consolidation of saturated multi-layered soils with anisotropic permeability due to well pumping. Comput. Geotech. **92**, 229–239 (2017)

Observations on Failure Behavior of Cut Slopes in Chalky Limestone, Case Studies in RUS Formation

Abdullah Kallash and Hernan Vigil

Abstract

The Lusail Plaza project CP7B is one of the mega infrastructure projects in the state of Qatar, located in the center of the new Lusail City north of Doha, straight on the coast line. Achieving the project required the construction of an open pit excavation with variable depths between 10 and 30 m. The construction of the cut slopes revealed a limestone member RUS formation underlain by a Simsima Limestone layer at a depth of about 12 m below ground level. The RUS formation consisted of chalky limestone mixed with clay. During the site construction, several local failures were observed in the RUS formation. In this paper, the failure behavior of cut slopes in the RUS formation was evaluated.

Keywords

Slope stability • RUS formation • Clay-stone • Failure behavior • Stand-up time • Qatar

1 Introduction

The enabling works for the CP7B project involved the excavation of an open cut, of approximately 30 m below ground level (mbgl), with a volume of excavated material up to 6 million m^3. The top level of the excavation lies at −0.5 m Qatar National Datum (m QND). The deepest point of the excavation lies at −29. 5 m QND.

During the construction of the cut slopes the failure behavior of the RUS was observed. This behavior was evaluated, taking into account the stand-up time of clay stone, the hydrogeological conditions and the possible triggering factors [1].

2 The Geological, Geotechnical and Hydrogeological Conditions

The geological units in the cross section in the Lusail area can be divided as follows: Weathered Simsima Limestone (WSL) was found at the top, followed by a layer of Simsima Limestone (SL). The SL is followed by Midra shale (MS). The RUS formation is normally found below the MS at depths of between −12 and −15 m QND (Fig. 1a). In most cases, the Midra shale and the Simsima limestone were missing, so the weathered Simsima limestone is followed directly by the RUS formation (Fig. 1b). The RUS formation is described as weak, to very weak chalky limestone mixed with layers of clay stone and inclusions based on BS 5930:1999. The strength of the rock changed between a 1 and 6 MPa.

Due to the location—adjacent to the sea shore—a high groundwater level of +1 m QND must always be taken into consideration. This will produce a hydraulic gradient of up to 30 vertical meters with a significant hydrostatic pressure on the cut slopes. To deal with the related risks a complex drainage system was designed to maintain dry cut slopes until the end of the construction work.

3 Examples of Failure Events and Failure Mechanisms

There were several forms of failure observed. The inclination of the constructed slopes was designed to be in the range of 1:7 to 1:10. Under this design criterion, the stand-up time of these slopes varied between a few hours and several months. The observed stand-up time depends on the water content of the slope. So, slopes under dry conditions have longer

A. Kallash (✉)
DB Projekt Stuttgart-Ulm, Stuttgart, Germany
e-mail: abdullah.kallash@deutschebahn.com

H. Vigil
QDSBG Lusail CP07B Project, Doha, Qatar

A. Kallel et al. (eds.), *Recent Advances in Geo-Environmental Engineering, Geomechanics and Geotechnics, and Geohazards*, Advances in Science, Technology & Innovation, https://doi.org/10.1007/978-3-030-01665-4_94

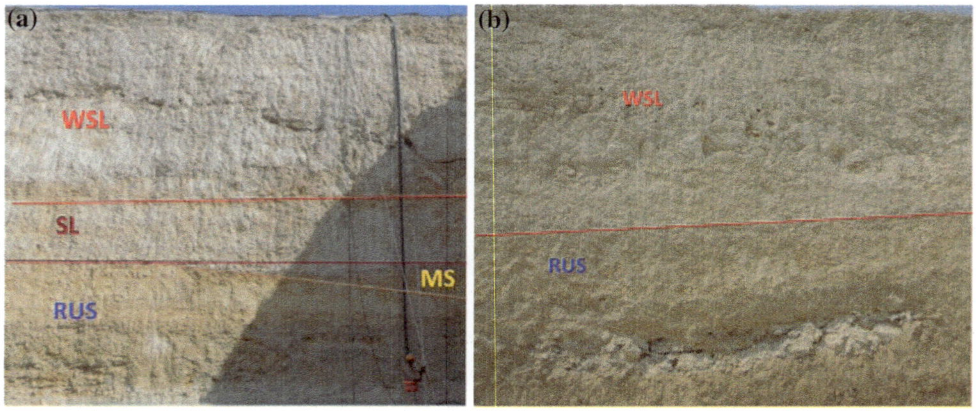

Fig. 1 The geological profile at Lusail area, **a** with all members, **b** RUS formation directly followed by the weathered Simsima limestone

stand-up time in comparison with slopes under wet conditions. Accordingly, the support and protection measures varied from maintaining dry conditions for free standing slopes, up to the use of reinforced shotcrete.

3.1 Case 1, Failures in the Unsupported Chalky Limestone

The slope failures in the first case occurred in cut slopes with a stand-up time ranging from a few weeks to several months. The main factor affecting the stand-up time was the amount of claystone inclusions in the cut slope. The presence of claystone as an impermeable layer, forms areas of hydrostatic pressure on the contact surfaces between the claystone and the limestone.

Figure 2 shows examples of failures where hydrostatic pressure along the contact area rose rapidly in the horizontal joints between the limestone and the claystone, subsequently triggering the failure below the claystone. The thin layer of claystone had, as per its nature a short stand-up time, and was unable to bear the high water pressure and collapsed immediately. Successive failures in the chalky limestone were stopped due to the formation of the arching effect.

3.2 Case 2, Failures Occurring During the Installation of Protection Measures

The failures in the second case took place during the installation of the slope support measures. Figure 3a shows an example of this type of failure. The failure occurred in the weak chalky limestone after a sudden rise of the hydrostatic pressure due to a breakdown of the dewatering pump adjacent to the location, causing the slope to topple down. The failure surface shows striking horizontal fine joints (Fig. 3a, b).

At the time of the failure, the support measures were almost completed with the exception of the designed draining system. The raised water behind the shotcrete wall created high hydrostatic pressure and pushed the shotcrete shale with rock mass behind it to fall down in a successive way. The rock mass fell down until an arch effect was formed. A post simulation of the collapse event proved that the leak of weep holes was the main preparation factor for the failure. Figure 3b shows similar failure mechanism due to non-functionality of the weep holes.

Another example of this type of failure occurred in the very weak chalky limestone, with high amounts of claystone inclusions as shown in Fig. 4. It was observed that the

Fig. 2 Examples of failures in the unsupported RUS formation below the claystone layer

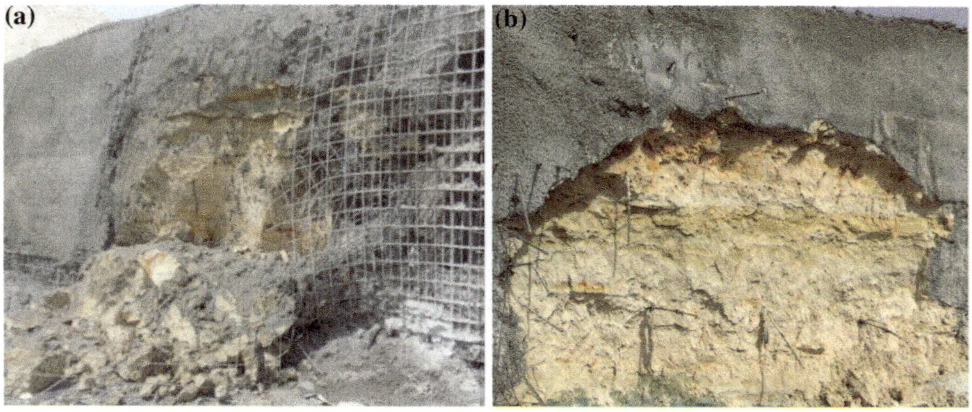

Fig. 3 Examples of failures occurring during the installation of support measures

Fig. 4 Collapses occurred below the installed support measures

stand-up time of such type of rock was at the limit of 4 weeks under the design boundaries. Accordingly, it was necessary to expedite the installation of the designed support measures. The installation of the support measures was progressing in stages, parallel to the installation of the slope draining system. However, during the installation of the support elements in the lower stage, a big block of rock became loose and fell down. A main preparation factor for this type of failure was the dynamic loads, resulting from the drilling activities for the installation of the nailing system.

4 Conclusion

The behavior of the failures in the RUS formation depends on certain factors such as the layering of claystone, or the distribution of the joints in the rock matrix. During the construction of the slopes, there were two main causes that triggered the failures: the raising of hydrostatic pressure and dynamic loading. In comparison to cut slopes in chalky limestone (RUS) with reduced inclination, the observed stand-up time of the above mentioned cases is much shorter.

By controlling these two factors it is possible to improve the stand-up time of the cut slopes under the designated conditions.

Reference

1. Popescu, M.E.: landslide causal factors and landslide remediatial options. In: Slope Stability and Safety of Infra-Structures. Citeseer (2002)

The Landslide of Agrigento Hill (Sicily, Italy)

Liguori Vincenzo

Abstract

This paper illustrated the geological, morphological and hydrogeological studies performed for the analysis and monitoring of the landslide involving the northern side of the hill of Agrigento (335 m a.s.l.) on which the ancient Cathedral was built during the 11th Century. The hill is made of a typical Plio-Pleistocene transgressive succession made of clays (M. Narbone formation), calcarenites, sands and clayey soils (Agrigento formation). The area has been unstable since 1315, involving both the little-welded, very porous and fractured calcarenitic sections (E-W) from Pleistocene and the clay layers interstratified within these sections. Since 1924, from time to time, various typologies of disruption have occurred: falls, flows and more complex phenomena. Fractures of the calcarenitic sections of the hill, pointed out by Grappelli's studies in 1966, have reactivated from March 7, 2005 and until now, thus triggering movements of the side of the hill and damages to the infrastructures. The landslide movement is very slow and it is located in the slope mainly from the bottom of the hill to XXV Aprile Street. The re-activation of the existing discontinuity represents the rotational sliding surface that passes along the sub-vertical crack and through the cracked calcarenites, the silty sands and the altered light-brown clays. The rotational sliding appears to be the main reason for anomalous behavior of the ground Cathedral system and probably the heavy rainfalls, from September 2004 to February 2005, could have caused the movement.

Keywords

Landslide • Cathedral • Agrigento

L. Vincenzo (✉)
Dipartimento di Ingegneria Civile, Aerospaziale dei Materiali (DICAM), Scuola Politecnica, Università degli Studi di Palermo, Viale Delle Scienze, Edificio 8, 90128 Palermo, Italy
e-mail: vincenzo.liguori@unipa.it

1 Introduction

Sicily is a region rich in archaeological and historical sites and any of them are human heritage. In some cases, these areas are in high landform relief with young weak rocks. As results of the erosive processes and low resistance of the rocks, these areas are susceptible to landslides.

In this paper, an attempt has been made to understand the typology of landslide processes affecting a slope in the Agrigento town (37° 18′N–13° 34′E in central southern Sicily) and in particular, the northern slopes leading down from the St. Gerlando Cathedral which since 1951 has elevated to the status of minor basilica. Figure 1 shows the Agrigento cathedral location and its position in relation to Sicily.

Since 2005 the landslide of Agrigento hill has been investigated and monitored by means of geomorphological investigation, geological and geophysical surveys, hydrogeological tests, geotechnical analysis and structural studies. Though the available data are numerous, no unique geological and nor a kinematic model of the landslide has been achieved so far.

The objective of this paper was to establish a relationship between historical and recent data, trying to provide a geological model and describing the landslide in accordance to new update of the Varnes classification [1].

The Agrigento area is located in central-southern Sicily at the edge of the Maghreb-Apennine thrust belt and it is characterized by a geological setting strictly influenced by geodynamic evolution. Along the Maghreb-Apennine thrust front, the Gela Nappe forms a large arcuate salient and, in the area of Agrigento, has a NW-SE structural trend. The tectonic setting of the basin is characterized by large open folds (1–3 km amplitude and 5–10 km half wavelength across the general strike [2]. These structures characterize the geologic-structural layout on the surface, and are kinematically associated with the development of south-verging blind-thrust [3].

A. Kallel et al. (eds.), *Recent Advances in Geo-Environmental Engineering, Geomechanics and Geotechnics, and Geohazards*, Advances in Science, Technology & Innovation, https://doi.org/10.1007/978-3-030-01665-4_95

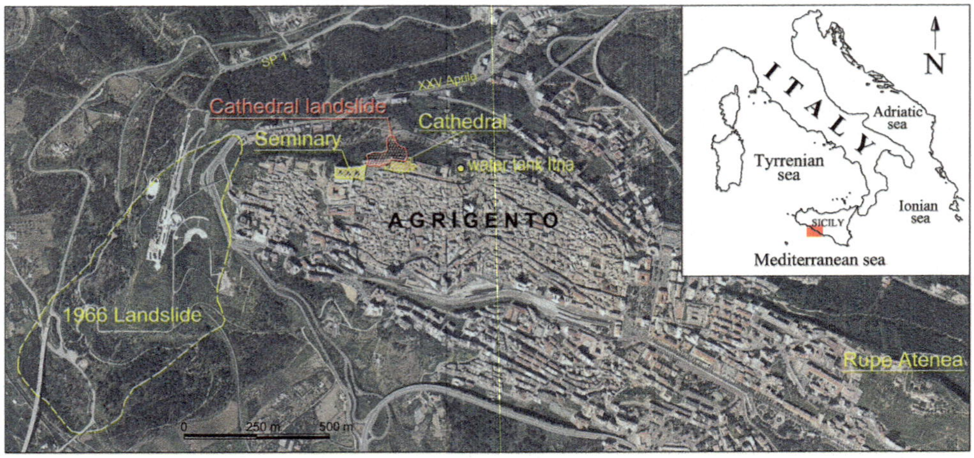

Fig. 1 Location of the Agrigento Cathedral and localization of the landslide in the northern side of the hill

The Agrigento hill is part of a 20°-south sloping monocline. This peculiar morphological condition depends on the geological conditions of the area as well as on the stratigraphic contact between the Mt. Narbone and Agrigento formations (Fig. 2).

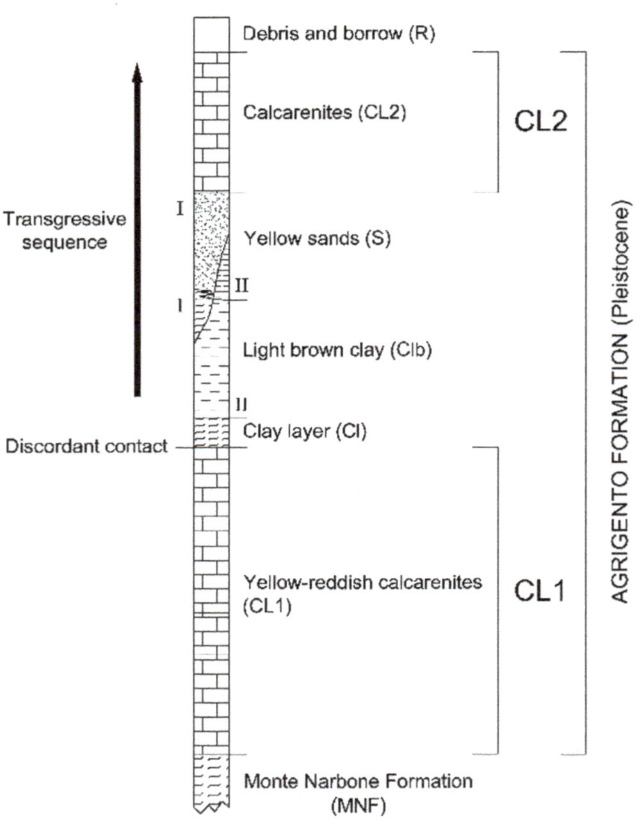

Fig. 2 Stratigraphic column of the northern side of the Agrigento hill

2 Materials and Methods

The main slope of the landslide can be easily identified in the areas of the Archiepiscopal Palace, the Cathedral and the Diocesan Museum, in correspondence to the old superficial crack.

The instability in this area is mainly generated by the geo-structural setting of a fragile lithotype (calcarenite) overlapping a plastic lithotype (sandy clay, silts and marly sands). Such instability is accelerated by the kind of layout of these lithologies, arranged along a regular, large and asymmetric syncline fold limb with variable inclination between 20° and 40°.

2.1 Detail Geology and Stratigraphy

After a proper homogenization and analysis of the data collected from the several surveys carried out from 1966 until now, the detailed stratigraphy of the northern side of the hill was reconstructed. In this way, a unique reference stratigraphic sequence was created including: (a) from the building's floor to −5 m depth, fill material; (b) from −5 to −10 m, porous organogenic yellow-to-reddish calcarenite, sometimes characterized by a bit cementation; (c) from −10 to −30 m, light brown clayey sands, sandy clays, sometimes turning into light brown clays with chalk crystals (altered clays), with fossils; (d) from −30 to −60 m pale-yellow porous organogenic calcarenites, cemented but crumbly, turning in the lowest area into porous reddish calcarenites with fossils (bio-calcarenites); (e) from −60 to −70 m grey clayey clays of Mount Narbone formation (Fig. 3).

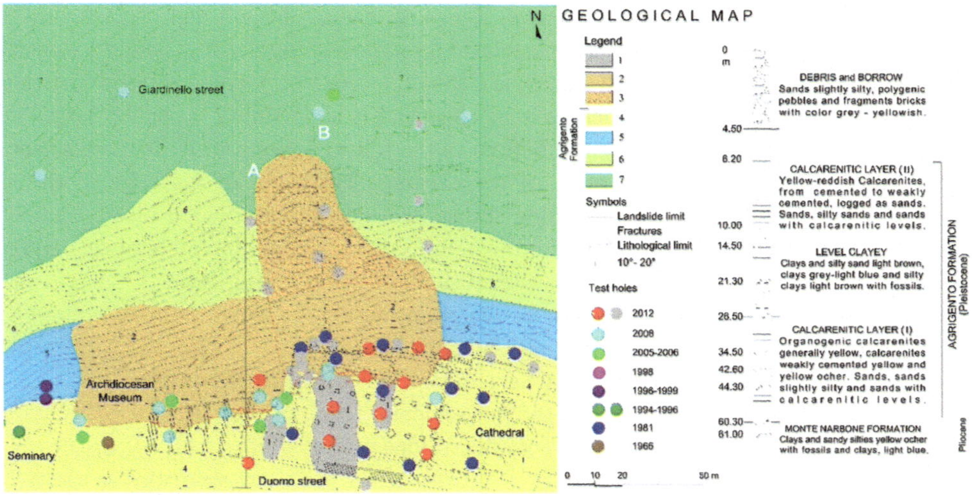

Fig. 3 Geological map and stratigraphy of the Cathedral area with the lacalization of main test hole. **1** backfill; **2** rotational slide zone; **3** earth flow zone; **4** calcarenite layer II—CL2; **5** grey-light blue clays; **6** calcarenite layer I—CL1; **7** Monte Narbone formation

Geotechnical parameters of the foundation soil

The geotechnical properties of the lithologies of the hill has been analyzed.

The calcarenites rocks (CL2 and CL1) have:

porosity (n): 0.36–0.46%; specific weight (γs): 26.9–27.4 kN/m^3; saturated unit weight (γsat): 18.8–20.6 kN/m^3; uniaxial strength (σf): 1.1–4.6 MPa; coefficient of permeability (k): 10–2—10–4 cm/s and Young modulus (E'): 300–700 MPa.

The relevant inclinometer readings, observed between June 2011 and December 2012, indicate different shear surface localized at depths between 10 and 30 m.

3 Results

The geo-morphological aspects characterizing the northern side of the hill include: (a) "creep" type superficial phenomena affecting the clays of Mt. Narbone formation; (b) morphological relicts from complex gravitational phenomena produced by suspected rock collapses involving the edge of the side of the hill; (c) slow shifts of the calcarenitic slope's edge, enhanced by different-size crack systems and residual recesses detecting the preferential directions alongside which such movements can originate. The combination of all this evidence leads to the hypothesis of the existence of a 10–20 m-deep phenomenon with a rotational-slide of movement which was activated during a massive weather event such as that of the March 2005. The landslide area can be divided into two zones: (1) Museum and the Cathedral's staircase; (2) St. Gerlando Cathedral. The instability of the first zone is caused by a rotational-slide movement and is limited to the edge of CL2 calcarenite layer, involving a thick debris layer standing on the sandy clays. The instability of the zone 2, on the other hand, is not limited to CL2 calcarenite layer but also to CL1 calcarenite layer and the Mt. Narbone formation. In this instance, the landslide starts with a rotational slide but then evolves into an earth-flow landslide in the downstream area [1].

As far as the progressive kinematic evolution of the slope is concerned, the clays located beneath the upper calcarenite layer should play a paramount role. These clays have been, in fact, subjected to such flows that involved the calcarenitic blocks upstream. Such instabilities, together with causing cracks in the buildings of Duomo Street, have also caused the collapse of large calcarenitic blocks that reached the area of XXV Aprile Street. The sub-vertical and inclined cracks in the historical building showed a rotational movement with failure in the Northern edge of the hill. The structures crack monitoring highlighted that the rotational movement (North direction) is dominant compared with the failure.

4 Discussion and Conclusion

The paper analyzed the historical and recent data on the Agrigento landslide. Indeed the hill landslide, active since 1315, in the time has damaged a few structures as the St. Gerlando Cathedral. The hill slope, weakly inclined, is susceptible to different landslide phenomena, but one of them determined the damage of the cultural heritage. The numerous landslides, nowadays stabilized or quiescent, that have affected the side of the hill, have left both cracks in the calcarenites and cracks in the historical monuments (Cathedral, Archiepiscopal palace, etc.). All the research carried out from 1966 to 2012 allowed the detection of the rotational slide affecting the slope and the ecclesiastic complex, with a sliding surface that goes from the ground to about −15 m deep (variable from area to area according to the lithology and the alteration status). The kinematic movement has not been identified nor postulated up to now, and consequently the attempts to stabilize the Cathedral were unfounded. The landslide movement is very slow and it is localized in the slope mainly from the bottom of the hill to XXV Aprile Street. The re-activation of the existing discontinuity represents the rotational sliding surface that passes along the sub-vertical crack and through the cracked calcarenites, the silty sands and the altered light-brown clays. The rotational sliding appears to be the main reason for the anomalous behavior of the ground Cathedral system and probably the heavy rainfalls, from September 2004 to February 2005, could have caused the movement. The rotational sliding movement involves the northern slope until the CL1 calcarenite edge.

The two detected collapsing areas subjected to different and not-necessarily-simultaneous shifts. Today the northern slope, under the CL1 calcarenite layer, until XXV Aprile street, even if its morphology is ascribable to ancient landslides (creep, earthflow, collapses etc.) as quoted by several authors, is to be considered quiescent. In conclusion, the main success of this work consisted in detecting a well-defined deformation pattern both of the northern-side of the hill and of Cathedral, confirming a real aggravation of the local instability worsening the risk scenario.

Geological and geotechnical investigations permitted to identify a rather large sliding mechanism involving the North aisle of the Cathedral as well as long adjacent stretches of the edge of the slope. It is now evident that the preservation of the Cathedral to reduce hazard and risk for this cultural heritage, requires the stabilization of the slope as a priority.

References

1. Hungr, O., Leroueil, S.: Picarelli L The Varnes classification of landslide types, an update. Landslides **11**(2), 167–194 (2014)
2. Grasso, M., Miuccio, G., Maniscalco, R., Garofalo, P., La Manna, F., Stareilia, R.: Plio-Pleistocene structural evolution of the western margin of the Hyblean plateau and the Maghrebian foredeep, SE Sicily: Implications for the deformational history of the Gela nappe. Ann. Tecton. **9**, 7–21 (1995)
3. Grasso, M., Lickorish, W.H., Diliberto, E., Geremia, F., Maniscalco, R., Maugeri, S., Rapisarda, F., Pappalardo, G.: Carta geologica della struttura a pieghe di Licata (Sicilia centro-meridionale), scale, vol. 1: p. 50.000, S.EL.CA., Florence (1997)

ROAD Slope Instabilities in Schist Massifs: Case of the National Road Linking Marrakech to Ouarzazate (Morocco)

Abdeltif Bouchehma, Azzouz Kchikach, Imad Kadiri, Hamid Essaidi, Mohammed Himmi, Khalil Bizani, and Hammou Mansouri

Abstract

The national road RN9 connecting Marrakech and Ouarzazate is located in a mountainous area with steep slopes. Several instabilities are recorded along this road including the Tishka massif which rises to 2300 m. These instabilities often hinder the road operation especially during periods of poor weather conditions. A geological and geotechnical study made it possible to better understand the origin of the instabilities in the Tishka massif. The monitoring of these instabilities by a series of inclinometers showed significant displacements (18 mm/month) to 5 and 15 m of depth. This study showed that the instabilities are the result of benches slips according to fractures plans and schistosity or sometimes linked to a regional fault mapped for the first time in this study. A geotechnical model was developed and integrated in the Plaxis software to discuss and validate the different slopes strengthening solutions.

Keywords

Road slopes • Instability • Schist massif • Strengthening • Tishka • Morocco

1 Introduction

The national road RN9 connecting Marrakech and Ouarzazate is the only link with the southeastern part of Morocco. It crosses the High Atlas chain and is characterized by sharp turns at steep slopes, especially in the Tishka massif which rises to 2300 m altitude (Fig. 1). Many serious traffic accidents have been recorded in recent decades, forcing the authorities to implement redevelopment and widening works. This paper dealt with slope instabilities study in the Tishka schist massif carried out within the abovementioned works. After the geological and geotechnical investigations, core drillings, pressuremeter tests and inclinometers measurements, we analyzed the unstable slope strengthening solutions proposed from both stability and cost points of view. The solution chosen for each studied slope was assessed by a digital simulation! [1].

2 Unstable Slopes Geological and Geotechnical Investigations and Strengthening Solutions

The Tishka pass is part of the Hercynian High Atlas Mountains. In this region, the RN9 is located on altered schist soils [2]. A mapping and structural study of the Tishkaa area was performed during the months of May and June 2017 extended by looking into several cored and destructive drillings results showing instability signs (Fig. 2). The rocks mechanical quality was assessed by pressuremeter tests at different depth levels. Figure 2 shows an investigation results example. Inclinometers measured land masses movements between November 2016 and November 2017. Some records are very important especially in sub-surface of altered schists. These movements affect the road pavement as shown in Fig. 1c. Two solutions can be adopted to address excessive displacements: If the healthy schist is less than 10 m in depth with low displacements, we suggest a nailing slope

A. Bouchehma (✉) · A. Kchikach · I. Kadiri
Cadi Ayyad University, B.P. 511, 40000 Marrakesh, Morocco
e-mail: bouchehma@gmail.com

H. Essaidi
Ministry of Equipment and Transport, RN9 Construction Site, 40000 Marrakesh, Morocco

M. Himmi · K. Bizani
Ministry of Equipment and Transport, Regional Office of Marrakech-Safi, 40000 Marrakesh, Morocco

H. Mansouri
Public Laboratory of Tests and Studies, CES, Km7, El Jadida Road, Casablanca, Morocco

A. Kallel et al. (eds.), *Recent Advances in Geo-Environmental Engineering, Geomechanics and Geotechnics, and Geohazards*, Advances in Science, Technology & Innovation, https://doi.org/10.1007/978-3-030-01665-4_96

Fig. 1 Study area location. **a** Identified fault producing landslides examples **b** slope instability signs examples. **c** Location of the slope presented in Fig. 2

stabilization, otherwise, we adopt an underground piles road stabilization combined with nailing.

3 Results and Discussion

3.1 Geotechnical Model Template of Unstable Slopes

Geological mapping revealed two lithological parts: a schisto-sandstone formation topped by green pelites with highly weathered outcrops. The weathered layer thickness may exceed 5 m. Almost all drillings show the same lithological succession of highly weathered pelites with tuffaceous surface resting either on a healthy schist directly or on a green altered layer before reaching the black healthy one (Fig. 2b) [1].

The massif structural survey helped identify faults and fractures partly affecting instabilities. Figure 1 highlights the most important ones. The structural measurements projected in wulf canvas using SteoriWin software show the principal directions, which are N25 for faults and fractures and N160 for schistosity [3]. Pressuremeter tests revealed low pressure limits (PL < 3 MPa) and low pressurometric modulus (EM < 20 MPa) at less than 10 m depths. These values increase beyond this depth. This confirms roughly the lithological descriptions made above. These low values are assigned to the altered rocks.

Most inclinometers reveal significant displacements from 5 to 15 m depths. Sliding surfaces are likely to be affected by lithological contacts between weathered pelites and schist and between altered and unaltered schists. All the inclinometers were sheared certifying the sliding activity at speeds around 17–18 mm/month.

3.2 Modeling and Assessing the Strengthening Models

Two slope instability cases were identified. The first one is about shallow healthy schist. Its stability analysis using the Plaxis software shows that a nailing system with tie-rods is a suitable solution (Fig. 2c). Their orientations are perpendicular to directions and dips of fractures and schistosities. The second case is about deeper healthy schist level exceeding 10 m where stability can be ensured by a road carried on piles in its transversal and longitudinal directions (Fig. 2d).

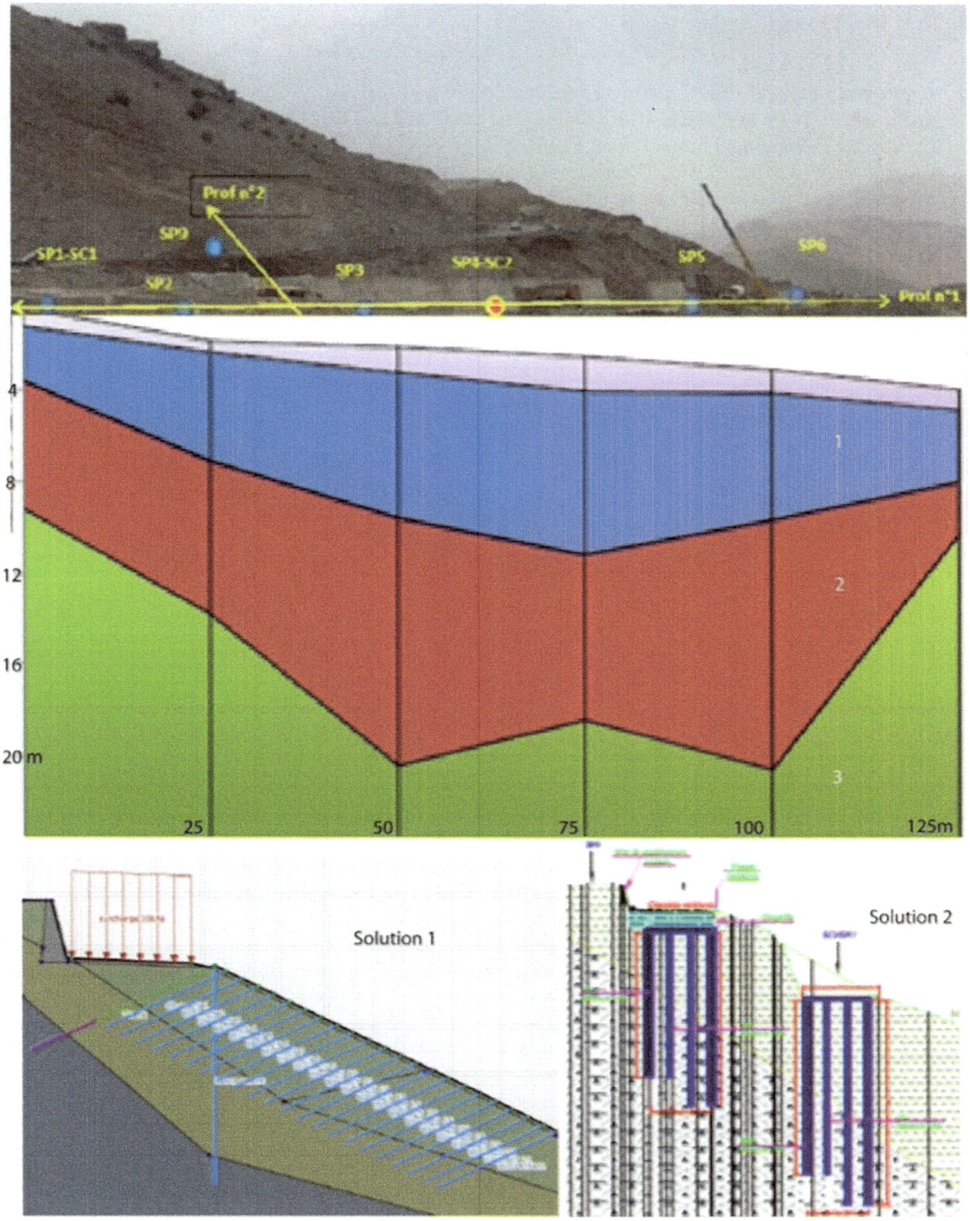

Fig. 2 **a** Core drillings and pressiometer tests locations. **b** Geotechnical model, profile correlation 2. **c** Sol. 1: Nailing stabilization. **d** Sol. 2: Piles stabilization

4 Conclusion

The investigations carried out in the Tishka massif helped define the used geotechnical model. The instabilities recorded are essentially linked to sliding benches according to fracture and stability plans. Major landslides are most likely due to weathered schist contact with healthy schist. Some landslides are linked to a regional fault mapped during this study (see Fig. 1). The developed geotechnical model made it possible to choose the right strengthening solutions. Nails and tie-rods stabilization would minimize displacements within slopes. Underground piles stabilization prevents disorders within road pavements. However, to prevent piles downstream slope displacements, the combination of the two solutions is required for an overall stability. This variant is being assessed and will be detailed in the draft article.

References

1. Engineering firms «LPEE & CID» : Execution design study related to the RN9 service improvement from kilometric point 278 to 349, Ouarzazate, 52 p. Unpublished Report (2014)
2. Gasquet, D., Levresse, G., Cheilletz, A., Azizi-Samir, M.R., Mouttaqi, A.: Contribution to a geodynamic reconstruction of the Anti-Atlas (Morocco) during Pan-African times with the emphasis on inversion tectonics and metallogenic activity at the Precambrian-Cambrian transition. Precambr. Res. **140**, 157–182 (2005)
3. Rafiee, A., Radjaï, F.: Contribution to fractured rock stability study: characterization of in-situ fracturing, geostatistics and mechanics of discrete media. School of Mines of Alès (2008)

Rainfall Effect on Slope Stability Using Numerical Analysis

Soumia Merat, Lynda Djerbal, and Ramdane Bahar

Abstract

Rainfall-induced slope failure is a major geotechnical disaster and has been frequently reported in many regions worldwide. This work aimed to study the rainfall (climate) influence on slope stability. The climate is represented by precipitations rate, precipitation duration and the stability of this unsaturated slope in natural terrain was evaluated based on a key indicator which is "safety factor". The finite elements software (PLAXIS^{2D}) was used to analyze the stability of the slope. The factors of safety were evaluated by means of fully coupled flow-deformation analyses and then compared with the various controlling factors: rainfall intensity, rainfall duration, angles and heights of slopes, soil characteristics and hydraulic conductivity of the soil. The results were discussed in this paper and led to confirm the relationship existing between landslide and climate effects considered in this analysis. In fact, this relation is observed for most part of the north Algerian instabilities.

Keywords

Slope failure • Rainfall intensity • Numerical simulation • Factor of safety

1 Introduction

Slope failure is attributed to many factors including rainfall, earthquake shaking, snow melting, temperature changes, volcanic activity, various human actions and weak soil layers. Rainfall-induced slope failure is a major geotechnical disaster and has been frequently reported in many regions worldwide [1]). The main objective of this study was then to estimate the factor of safety of the slope after predefined time periods of a rainfall event of Azazga city. Factor of safety (FS) has been commonly used to quantitatively assess the stability of a slope. It is used as an index to determine how close or far away a slope is from failure. The analyses were performed employing two-dimensional finite element models using the code (*PLAXIS*^{2D}).

2 Materials and Methods

2.1 Numerical Simulation and Models for Slope Stability Analysis

Researchers as Gasmo, Tsaparas and Rahardjo have performed numerical analyses of rainfall-induced slope failure to study the effects of antecedent rainfall and parameters controlling slope instability [2]. Recent developments in analyses of slope instabilities under rainfall conditions using the limit equilibrium method and coupled with hydro-mechanical modeling are given in some software (such as: Plaxis, Geo-Slope, ABAQUS…etc.).

2.2 Coupled Hydro-mechanical Analysis

The coupled analysis robustly integrates the above two computation procedures together; the soil hydraulic and mechanical responses were calculated simultaneously in the coupled analysis. Oh and Lu commented that the coupled analysis could lead to a more accurate assessment of slope stability under rainfall conditions [1].

2.3 Geometry and Soil Parameters

In this study, a slope with an inclination angle of 26,56° (2:1) to the horizontal plane and a height of 50 m was considered (Fig. 1). The groundwater table is assumed to be at 10 m elevation from the lower ground surface. The soil

S. Merat (✉) · L. Djerbal · R. Bahar
University of Science and Technology Houari Boumediene, BP. 32 El Alia, 16111 Algiers (Leego), Algeria
e-mail: smerat@usthb.dz

A. Kallel et al. (eds.), *Recent Advances in Geo-Environmental Engineering, Geomechanics and Geotechnics, and Geohazards*, Advances in Science, Technology & Innovation, https://doi.org/10.1007/978-3-030-01665-4_97

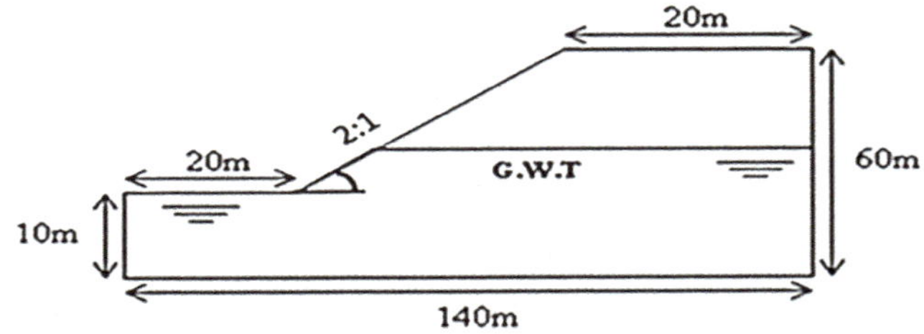

Fig. 1 Geometry of the slope model and initial ground water level

type was specified as «Clayey Loam» of Azazga. The soil properties related to the unsaturated soil were taken as given by van Genuchten [3, 4]. The different values of soil parameters used for the surface layer are recapitulated in Table 1.

The surface soil layer behavior was modeled as linear, elastic-perfectly plastic material by using the Mohr–Coulomb model and analyzed using soil parameters given by undrained tests. The effect of soil dilatancy was not considered in the simulation because the soil dilation angle ψ, was negligible based on the empirical relationship $\psi = -30°$ [1]. So for Undrained (A), it is recommended to use $\psi = 0$ (material models manual, Plaxis^{2D} 2016) [6].

2.4 Initial and Boundary Conditions

The initial stress within the soil body of the slope was generated using the PLAXIS^{2D} gravity loading function for non-horizontal surfaces. For gravity loading it was established by applying the soil self-weight during the first calculation phase and solving for equilibrium (Yang et al. 2016).

The standard fixities were applied as the mechanical boundary. The two lateral boundaries were allowed to move only in the vertical directions whereas the bottom boundary was restrained and fully blocked from movement (Fig. 2b).

Controlling factors for rainfall-induced landslides

The analysis of seepage and infiltration in an unsaturated soil and slope stability can be readily performed using available computer programs. However, the results are sometimes difficult to interpret, because the analysis involves many parameters related to soil properties, slope geometry, groundwater condition, initial conditions and rainfall characteristics. Slopes with a high slope angle, a high slope height and a shallow initial depth of groundwater table form the worst combination of factors for failure, and are more likely to fail as a result of rainfall [7]. The main objective of many research studies was to separate the influence of the various factors, and determine the controlling parameters [8].

According to Rahardjo et al. [7], the slope geometry and the initial water table determine the initial safety factor, and the actual failure conditions are much affected by rainfall characteristics and properties of the soils in the slope. Therefore attention should be paid to soil properties and rainfall characteristics when dealing with rainfall-induced slope failures. In this study, the influence of each controlling factor on the slope during a rainfall was dealt with to

Table 1 Summary of soil parameters of Azazga surface soil layer used in the FEM [5]

Soil layer	Soil model	Drainage type	γunsat (kN/m^3)	γsat (kN/m^3)	v'	C′ (kpa)	φ′ (°)	K (m/min)	E′ (Kpa)
Clayey Loam	Mohr-Coulomb	Undrained A	15,8	17,8	0,33	55	15	Kx = 10^{-3} Ky = 10^{-3}	4,038 * 10^5

v′: Poisson ratio, C′: Cohesion, φ′: Friction angle, K_x and K_y: Horizontal and vertical hydraulic conductivity respectively, E′: Young modulus

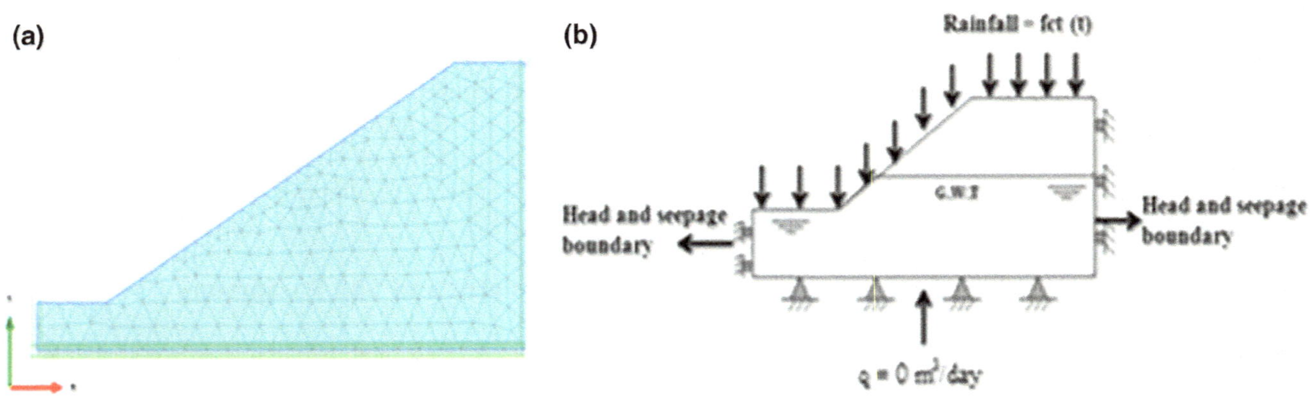

Fig. 2 Slope model used in the analysis **a**—Finite element method based computational model. **b** Hydraulic and mechanical boundary conditions of the slope model

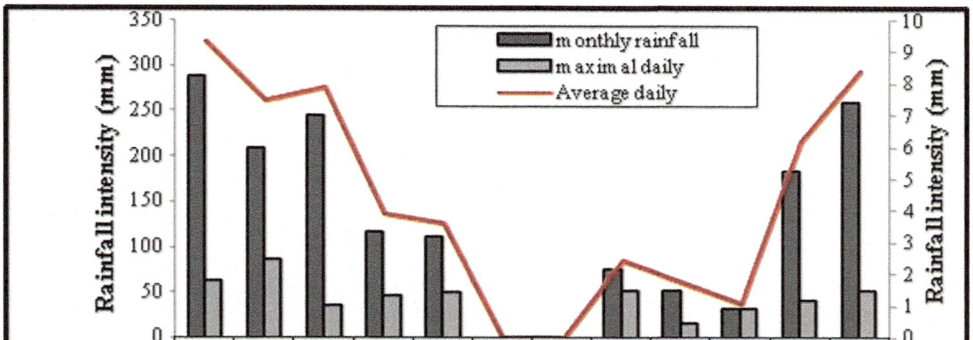

	Jan	Feb	Mar	Apr	May	Jun	Jul	Aug	Sep	Oct	Nov	Dec
monthly rainfall	289	209	245	117	112	0	0	75,5	53	33	184,7	259,5
maximal daily	63	86,5	36	47	50	0	0	53	17	33	42	52
Average daily	9,32	7,46	7,9	3,9	3,61	0	0	2,44	1,77	1,06	6,16	8,37

Fig. 3 Variation of monthly and maximal daily rainfall for the rainy year [9]

understand the mechanism of failure and the response of the slope [8].

2.5 Analysis of Rainfall Data

Figure 3 shows the monthly variation of rainfall of the rainy year. The average annual rainfall is characterized by a value of **973 mm** with a standard deviation equal to **232,8 mm**. The average daily rainfall data were used as input to the model. Rainfall data were obtained from ANRH for the Azazga School station (the station coordinates **X:649,25; Y: 383,9; Z: 430**).

The rainfall intensity of the rainy year from January to December with an accumulated rainfall amount equal to **1577,7 mm** was applied on the slope model. The FoS variation with time due to rainfall conditions can be obtained.

3 Results and Discussion

A number of influencing parameters on slope stability were taken account using series of coupled hydro-mechanical finite element analyses. Each of the studied parameters gave a considerable effect. A slope with a high slope angle and height represents an unfavorable case (Fig. 8). The factor of safety (FoS) and total displacement (U_{tot}) are linearly correlated with the cumulative rainfall intensity (see Figs. 5 and 6).

3.1 Controlling Factors Effect on the Slope

3.1.1 Soil Properties Effect

Different values of cohesion and friction angle parameters were used as soil properties to analyze the slope stability (Fig. 4). The results show that the factor of safety increased gradually with the increase of both cohesion and friction angle. For a friction angle equals to 10°, the factor of safety decreased gradually from 1,11 to 1 (with total cumulative rainfall = 972 mm) until the rainfall stops at the beginning of June and the slope failure was reached, then it remained constant during the next months of the year.

A slope with poor soil mechanical properties (C′ and φ') is very susceptible to reach the failure phenomenon during rainfall as shown in Fig. 4 and it gave a considerable displacement (1,349 m).

For the previous slope model mentioned above in (Fig. 1) with Azazga surface soil properties as a witness case, the

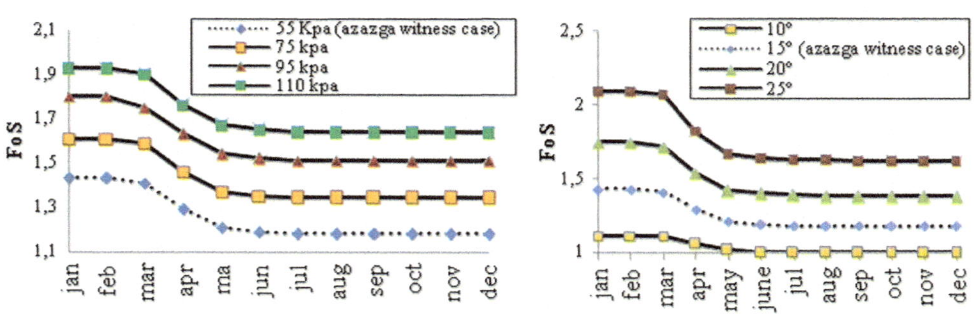

Fig. 4 Effect of soil properties on variation of factor of safety with time: **a**—soil cohesion **b**—friction angle

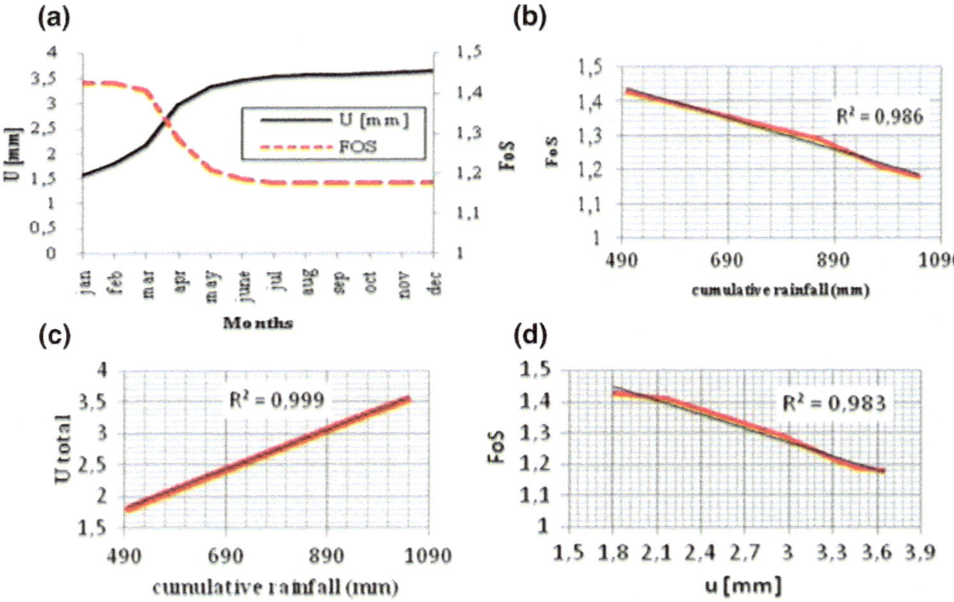

Fig. 5 Variation of total displacement and factor of safety for the adopted Model: **a**—U_{tot} and FoS versus time, **b**—FoS versus cumulative rainfall, **c**—U_{tot} versus cumulative rainfall, **d**—FoS versus U_{tot}

factor of safety decreased gradually with time as cumulative rainfall increased; on the other hand, the total displacement of the corresponding slope increased (the order of a few millimeters) (Fig. 5a). The results confirm that the factor of safety (FoS) and the total displacement (U_{tot}) are linearly correlated with the cumulative rainfall intensity (Figs. 5b and 5c) and they are also correlated among them as shown in (Fig. 5d). So the cumulative rainfall effect is a very important factor for a landslide occurrence.

The variations of total displacement and factor of safety were analyzed for the critical case shown in Fig. 5, for a soil with $\varphi' = 10°$. The linear correlation was also confirmed (Fig. 6).

The total displacement in this case at the end of calculation (derived after the coupled analysis) was considerable (1,349 m) in comparison with the previous case (the order of a few millimeters) (Fig. 7).

3.1.2 Factor of Safety for Various Slope Angle (Tan β) and Slope Heights

The properties of the slope itself such as the geometry (slope angle and slope height) are the internal factors for the rainfall intensity values of the rainy year of Azazga city, the factor of safety remains constant with the rainfall until the third month (March) and thereafter, the factor of safety rapidly decreased then seemed to stabilize with time. The factor of safety tended to decrease gradually with the increase in the slope angle as shown in Fig. 8a. The critical value of factor of safety equals to 1,18 for slope angle ($\beta = 26{,}56°$). For a slope angle equals to 20°, different heights (35, 40, 45 m) were used to analyze the slope stability. The factor of safety seemed to decrease with the increasing slope height. For a 35 m height, the slope was still stable during the rainfall event (the minimum Fos obtained is equal to 1,5). For the other slope heights (40 and 45 m), Fos suddenly decreases from 1,86 to 1,45 and 1,77 to 1,39 respectively) (see Fig. 8b)

3.1.3 Factor of Safety for Various Hydraulic Conductivity Values

Various permeability values were used as input for hydraulic parameters of the soil with the same rainfall conditions. Figure 9 illustrates the variation of the calculated FoS with respect to time for different coefficients of permeability values considered in this study.

The FoS decreased faster with a highest hydraulic conductivity (e.g. $K = 10^{-3}$ m/min and $1{,}4 * 10^{-3}$) due to the

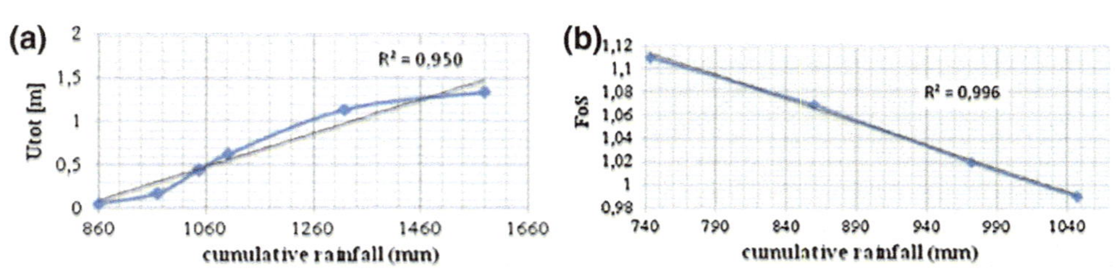

Fig. 6 Variation of total displacement and factor of safety: **a**—U_{tot} versus cumulative rainfall, **b**—FoS versus cumulative rainfall

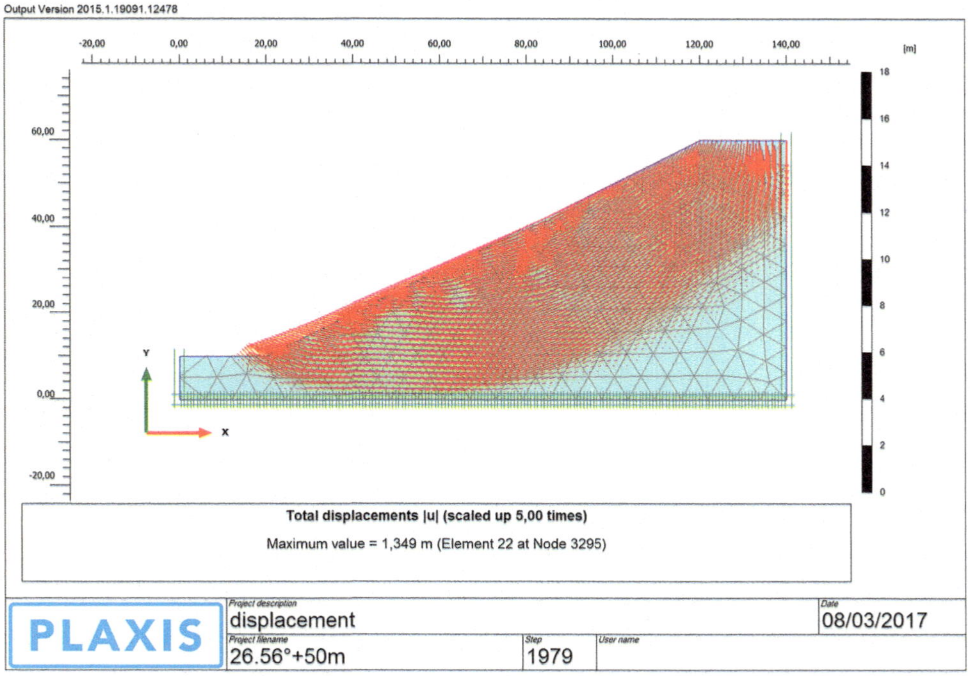

Fig. 7 Total displacement at the end of a rainfall event (Utot = 1,349 m)

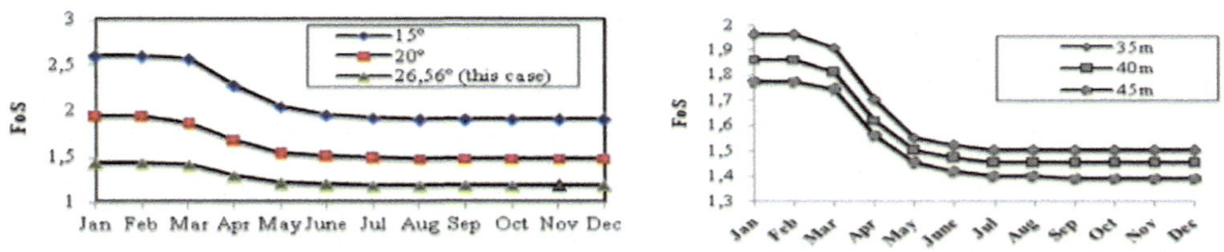

Fig. 8 Variation of safety factor in function of: **a**—slope angle, **b**—slope height

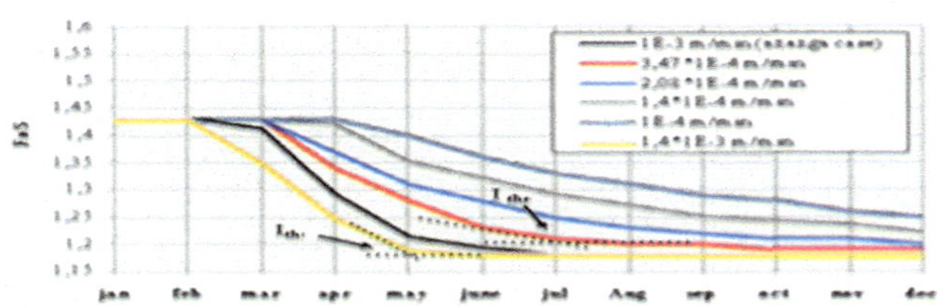

Fig. 9 Factor of safety for various hydraulic conductivity values

higher infiltration ability. For K = 1,4 * 10^{-3} m/min, the FoS decreased rapidly from 1,43 and reached a critical value (1,18) after 4 months with an accumulated rainfall intensity of **860 mm**. In contrast, when the soil has a poor hydraulic conductivity (e.g. K = 10^{-4} m/min), the FoS seemed to decrease slowly with time and took a long period of time to reach the lowest value. Rahardjo et al. [7] suggested that the threshold rainfall intensity can be determined by the maximum reduction of the factor of safety (Fig. 9). In fact, the results in Fig. 9 illustrate that an increase in coefficient of permeability will shorten the time period for slope failure to occur during rainfall event.

4 Conclusion

In this study, numerical analyses were carried out to observe the variations of factor of safety under a rainfall event based on real monitoring data from the study area. The conclusions can be summarized as follows:

- A slope with the highest soil properties (cohesion and friction angle) represents a stable state during a rainfall event;
- During a rainfall event, the factor of safety seemed to decrease gradually while the total displacement tended to increase;

- The factor of safety decreased continually with the increase of the height and slope angles. Slopes with a high slope angle and height constitute the worst combination of factors for failure;
- In the less permeable soil, the effects of the rainfall on slope stability threshold are delayed. In contrast, a high hydraulic conductivity value in soil plays an important role in accelerating the infiltration process and allows the slope failure to occur by reaching rapidly its lowest value.

References

1. Yang, K.H., Uzuoka, R., Thuo, J.N., Lin, G.L., Nakai, Y.: Coupled hydro-mechanical analysis of two unstable unsaturated slopes subject to rainfall infiltration. Eng. Geol. 18 pages (2017)
2. Ering, P., Babu, G.L.S.: Probabilistic back analysis of rainfall induced landslide-a case study of Malin landslide, India, Elsevier. Eng. Geol. **208**, 154–164 (2016)
3. Sun, D.M., Zang, Y.G., Semprich, S.: Effects of airflow induced by rainfall infiltration on unsaturated soil slope stability. Transp Porous Med **107**, 821–841 (2015)
4. Huat, B.B.K., Ali, F.H., Rajoo, R.S.K.: Stability analysis and stability chart for unsaturated residual soil slope. Am. J. Environ. Sci **2**(4) (2006 ISSN 1553-345X), 154–160 (2006)
5. Djerbal, L., Alimrina, N., Melbouci, B., Bahar, R.: Mapping and management of landslide risk in the city of Azazga (Algeria). Landslide Science for a Safer Geo-environment. Springer International Publishing, Heidelberg, pp. 463–468 (2014)
6. Plaxis: Internal Report, Groundwater Flow, Fully Coupled Flow Deformation and Undrained Analysis in Plaxis 2D and 3D. Vahid Galavi (2010)
7. Rahardjo, H., Ong, T.H., Rezaur, R.B., Leong, E. C.: Factors controlling instability of homogeneous soil slopes under rainfall. J. Geotech. Geoenviron. Eng. © Asce, 1532–1543 (2007)
8. Zhang, L.L., Zhang, J., Zhang, L. M., Tang, W.H.: Stability analysis of rainfall induced slope failure: a review. In: Geotechnical Engineering Volume 164 Issue GE5, Proceedings of the Institution of Civil Engineers, pp. 299–316 (2011)
9. ANRH: Données pluviométriques de la station Azazga école, Tizi-ouzou. Agence Nationale des Ressources Hydriques, Algérie (2011)

Temporal Characteristics of the Rainfall Induced Landslides in the Chinese Loess Plateau (China)

Aidi Huo, Jianbing Peng, Yuxiang Cheng, Xiaolu Zheng, and Yiran Wen

Abstract

Prior to assessing the temporal characteristics of the rainfall-induced landslides in Chinese Loess Plateau (CLP), it is fundamental to establish a detailed landslide inventory map. This study documented rainfall data from 2000 to 2015 period and mapped rainfall distribution in the CLP on the basis of yearly, monthly, and the daily precipitation data from 57 rain gauging stations. Geo-statistical analysis and Geographic Information System (GIS) techniques combined with information on geological disaster within the region were used to study the temporal characteristics and trend variations of rainfall and geological disasters distribution in the region. The results indicated that a positive relationship existed between geological disaster and the average annual rainfall, with zonality in the loess gullies, hills (Mao), and ridges (Liang); Based on the field investigations, the main factors affecting loess landslide distribution and development, such as geomorphic conditions, vegetation cover, soil erosion modulus, rainfall, and climate were considered. The lag effect of landside disasters after raining is particularly obvious over the time. These results provide a useful reference for the study of the mechanism of water-induced loess landslides.

Keywords

Geological hazards • Precipitation • Temporal variation • Trend change • Chinese Loess Plateau

A. Huo · X. Zheng · Y. Wen
Key Laboratory of Subsurface Hydrology and Ecological Effects in Arid Region (Chang'an University), Ministry of Education, Xi'an, 710054, Shaanxi, China

A. Huo · X. Zheng · Y. Wen
School of Environmental Science & Engineering, Chang'an University, Xi'an, 710054, Shaanxi, China

J. Peng (✉) · Y. Cheng
College of Geology Engineering and Geometrics, Chang'an University, Xi'an, 710054, Shaanxi, China
e-mail: dicexy_2@126.com

1 Introduction

The Chinese Loess Plateau (CLP) is the major strategic area of Chinese western development. The CLP covers one-twentieth of the Chinese land area and represents one-fifth of the national agricultural land; more than 700 important cities and towns are located in the CLP with one-sixth of the Chinese population. The CLP is also a major base for the development of natural gas, petroleum, and coal resources in China [1]. The main linear projects such as high-speed railways, oil and gas transmission lines, expressways, and power transmission lines cross the CLP [1, 2]. Energy and mineral resource development, major lifeline engineering, and urban construction have to deal with loess and loess-related geo-hazards, which further aggravates major geo-hazard (i.e., landsides) development on the CLP. Therefore, the frangibility of the geo-logical environment of the CLP, the intensity of human activities, and the development of landsides hazards, are related to each other and affect the CLP development [3].

On the basis of a large number of field investigations, comprehensive loess land-slide data caused by rainfall, and an examination of the geological conditions of regional constructions in the CLP, it was recognized that the scientific analysis of the relationship between rainfall and the occurrence of landside disasters is particularly important [2, 4]. Therefore, this paper studied the relationship between the temporal characteristics of rainfall and the mechanism of landslide development in the CLP. Based on the comprehensive analysis, we found that the hydrogeological structure and the rainfall were the internal and external factors to induce the landslide disaster, respectively.

A. Kallel et al. (eds.), *Recent Advances in Geo-Environmental Engineering, Geomechanics and Geotechnics, and Geohazards*, Advances in Science, Technology & Innovation, https://doi.org/10.1007/978-3-030-01665-4_98

2 Materials and Methods

2.1 Study Area

The CPL is located in the middle and upper reaches of the Yellow River, covering an area of 630,000 km^2 (Fig. 1), latitude 33°43′–41°16′, longitude 100°54′–114°33′, which is an important water inflow and main sandy area in the Yellow River. The CLP is the largest loess sedimentary area in the world, and its terrain is complex, with horizontal and vertical gully extending about 1300 km from east to west and about 700 km from north to south, and broken ground. The area covers most of Gansu and Shaanxi Provinces, all of Shanxi Province, and parts of three others (Henan, Qinghai, and Ningxia), among the southern Yinshan Mountain, northern Qinling Mountain, eastern Riyue-Helan Mountain, and western Taihang Mountain. The elevation of the plateau is between 1000 and 1600 m, and the surface is covered by continuous loess on hills, basins, and alluvial plains, with an average thickness in the range from 50 to 80 m, exceeding in some area 100 m, and reaching nearly 200 m in the Dongzhi and the Luocuan plateaus. The loess has been geologically transported from the northwestern Gobi Desert by the wind and has accumulated on the CLP since the beginning of the Quaternary Period. Huangling County is a typical representative area in the CLP. The geological hazard data of Huangling County were obtained from the detailed investigation report of geological disasters in Yan'an City, Shaanxi Province in the CLP.

2.2 Data Used and Methodology

The Global Land Data Assimilations Systems (GLDAS) was jointly developed by the National Oceanic and Atmospheric Administration (NOAA) and NASA Goddard Space Flight Center (GFC) to provide estimates and archives on land surface conditions and fluxes [5]. The processed GLDAS data were accessed online (http://grace.jpl.nasa.gov), using the services of the Goddard Earth Sciences Data and Information Services Center. The GLDAS information used in this study was based on satellite, land surface model, and ground observation data [6]. The GLDAS dataset obtained with the Noah model includes precipitation, air temperature, and soil moisture variables. The spatial resolution of the data is 1.0° × 1.0°, and the time resolution is monthly from 2000 to 2015 [7–9]. Meanwhile, the reference monthly rainfall values were obtained from 57 meteorological stands (stations) in the CLP. Geo-statistical analysis and GIS techniques were used to achieve the temporal characteristics and trend variation research.

3 Results

There is an obvious regularity of the rainfall temporal distribution characteristic of monthly changes in the CLP. The distribution of monthly rainfall time obviously corresponds with Gaussian distribution. In general, the rainfall gradually increases after January and reached the peak in July, August, and September. After September, the rainfall gradually decreases (Fig. 2). There is a trend of overall southward movement in the semi-arid and semi-humid areas.

Figure 3 shows the monthly geological hazard and monthly average rainfall in the central CLP of Huangling County. It can be seen that the monthly rainfall is similar to the annual one; both are closely related to the number of geological disasters. With the increasing frequency of monthly rainfall, the monthly frequency of geological disasters also increases, the basic trend is generally the same.

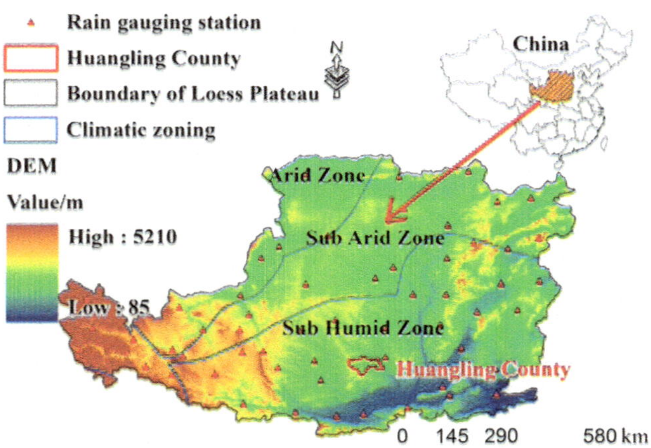

Fig. 1 Geographical position, meteorological stations and climate regions of the CLP

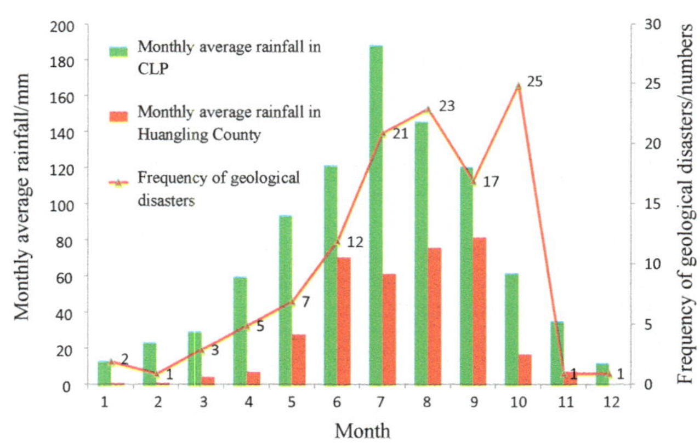

Fig. 2 The relationship between frequency of landslide disasters and the monthly average rainfall in Huangling County and CLP (2000–2015, *Source* GLDAS)

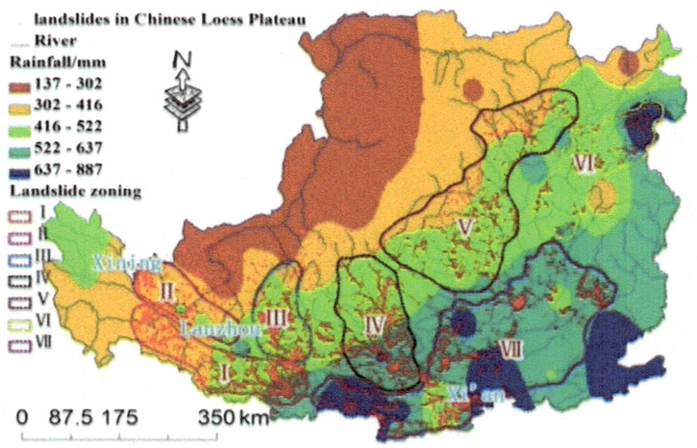

Fig. 3 The combined rainfall distribution-landslide map. (Rainfall: 2000–2015, *Source* GLDAS; Landslides data: the detailed investigation report of geological disasters in CLP)

Disasters were divided into 7 zones (I–VII) according to densities. Compared to the annual rainfall, monthly rainfall has a more obvious lag, that is, the increasing in rainfall in the month, not only with the increasing frequency geological disasters in the month, but also significant increase in the frequency of geological disasters next month. This lag effect is particularly evident, especially in the 6–9 months with more rainfall.

4 Conclusion

This study investigated the temporal characteristics of the rainfall-induced landslides in the CLP. Our results revealed that:

(1) The annual rainfall in the CLP shows a faster decreasing trend. However, the occurrence frequency of geological disasters in the CLP has not been reduced due to the increasing frequency of extreme weather and heavy rains because of global climate change. (2) The geological disaster and the average annual rainfall are positively correlated. In addition to rainfall, the geomorphic structure, stratigraphic lithology, slope structure, and human activities are also important influence factors for the temporal distribution of geological disasters. (3) According to the hydrogeological characteristics of the CLP. The rainfall-induced landslide mechanism can be summarized as follows: rainfall promotes the increase of shear stress on of glide plane of a slope and the decrease of shear strength in the landslide body.

Acknowledgements This study was supported by the following grants: National Natural Science Foundation of China: (Grant No: 41672255, 41302250, 41402255, 41790444, and 41877232)

References

1. Peng, J.-B., et al.: The critical issues and creative concept in mitigation research of loess geological hazards. J. Eng. Geol. **22**(4), 684–691 (2014). (in Chinese with English abstract)
2. Zhu, Y., Zhou, B.: Characteristics of Quaternary phreatic water system in Longdong Loess Plateau. People's Yellow River **33**(10), 51–53 (2011)
3. Xu, L., et al.: Landslides in a Loess platform, North-West China. Landslides **11**(6), 993–1005 (2014)
4. Xu, X.-Z., et al.: Landslides on the Loess Plateau of China: a latest statistics together with a close look. Nat. Hazards **86**(3), 1393–1403 (2017)
5. Anderson, J., et al.: The data assimilation research testbed: a community facility. Bull. Am. Meteorol. Soc. **90**(9), 1283–1296 (2009)
6. Rodell, M., et al.: The global land data assimilation system. Bull. Am. Meteorol. Soc. **85**(3), 381–394 (2004)
7. Huo, A.-D., et al.: Groundwater storage and depletion trends in the Loess areas of China. Environ. Earth Sci. **75**(16), 1167 (2016)
8. Huo, A.-D., et al.: Simulation modeling for water governance in basins based on surface water and groundwater. Agric. Water Manage. **174**, 22–29 (2016)
9. Cui, Y.-F., Nouri, A., Chan, D., Rahmati, E.: A new approach to the DEM simulation of sand production. J. Petrol. Sci. Eng. **147**, 56–67 (2016)

Tempo-Spatial Distribution and Triggering Mechanism of Loess Landslide: A Case Study from South Jingyang Platform, Shaanxi

Zhao Duan, Wen-chieh Cheng, Jianbing Peng, and Wei Chen

Abstract

Landslide hazard is often seen in the boundary of loess platforms in northwest China. There are more than 60 loess landslides recorded in Jingyang platform in the past five decades. Two types of the loess landslide, which are flowslide and slide, have been categorized. In this study, a statistical analysis to characterize the loess landslides was conducted. Additionally, a series of constant shear drained (CSD) triaxial tests were undertaken to reveal the triggering mechanism of the loess landslides. The irrigation-water infiltration and sequence and rule of loess landslide play a crucial role in triggering the loess landslides. These results provide the insights into the countermeasures recommended to resist the loess landslide hazard, thereby reduce the potential of loess landslide in the future.

Keywords

Loess landslide • Jingyang platform • Irrigation-water infiltration • Tempo-spatial distribution

1 Introduction

Due to some internal and external factors, the Loess Plateau is prone to the landslide hazard [1]. Degradation of loess and plateau topography were categorized as the internal factors, while variation of phreatic surface was categorized as the external factors [2–4]. South Jingyang platform, Shaanxi Province, is deemed as one of the typical distress areas in northwest China. Although there have been 62 extensive loess landslides recorded in the south Jingyang platform since its agricultural irrigation began in 1976, the tempo-spatial distribution and triggering mechanism of the loess landslides are still not clear.

The objectives of this study were: (1) to characterize the loess landslides by means of the field geological and topographical investigations and statistical analysis, (2) to reveal the triggering mechanism of the loess landslide by interpreting the results of CSD triaxial tests, and (3) to summarize the countermeasures against loess landslide hazard.

2 Geology and Loess Landslides of South Jingyang Platform

Since the contacting bodies of the Wei River fault started to slide relative to one another and consistent erosion activities associated with Jing River promoted plateau topography, the above factors could be considered to be major contributors to the loess landslides. Figure 1 shows the locations of the loess landslides that are categorized into two types, loess flowslide and loess slide. All the landslides happened in the peripheral areas of the south Jingyang platform clarify the statement above. Two geological boreholes located around Jiangliu indicate that the water contents vary from 10 to 25% and increase with the increasing depth. The higher water contents of loess deeper are due to the fact that the fractures and fissures of loess body provide additional seepage paths to the loess deeper to promote the irrigation-water infiltration.

Z. Duan · W. Chen
College of Geology and Environment, Xi'an University of Science and Technology, Xi'an, 710054, China

W. Cheng (✉)
School of Civil Engineering, Institute of Tunnel and Underground Structure Engineering, Xi'an University of Architecture and Technology, Xi'an, 710055, China
e-mail: w-c.cheng@xauat.edu.cn

J. Peng
School of Geology Engineering and Geomatics, Chang'an University, Xi'an, 710054, China

A. Kallel et al. (eds.), *Recent Advances in Geo-Environmental Engineering, Geomechanics and Geotechnics, and Geohazards*, Advances in Science, Technology & Innovation, https://doi.org/10.1007/978-3-030-01665-4_99

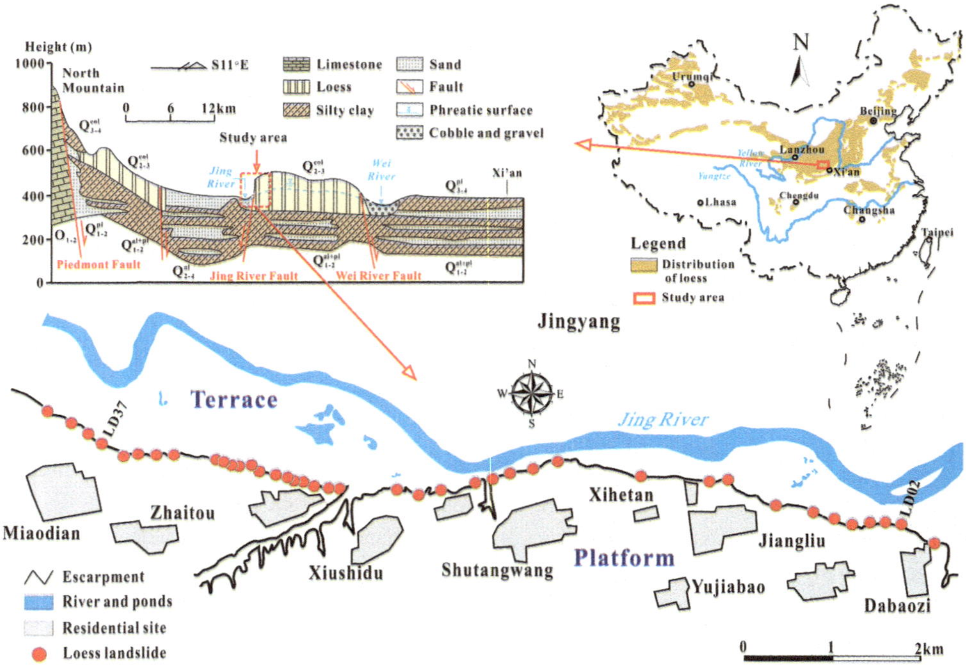

Fig. 1 Geological profile and plan view of the study area

3 Analysis and Discussion

3.1 Statistical Analysis

The statistical results indicate that the relationship between the tanϕ_α and the number of loess landslide followed a bell-shaped curve where mostly the loess flowslides and loess slides were ended up with the tanϕ_α value being equal to 0.25 and 0.45, respectively, as shown in Fig. 2. In other words, the horizontal runout distance for the loess flowslides could reach a value four times the slope height, whereas for the loess slides it could attain a value about two times the slope height. Figure 3 shows the relationship between the tanϕ_α and the volume of loess landslide for the south Jingyang platform. As can be seen, the accumulative sliding volume to differentiate the two types of loess landslides was evaluated to be approximately 100 ~ 150 × 10^3 m^3 according to the landslide influence extents. No any clear relationship between the number of loess landslides and the tanϕ_α value has been revealed.

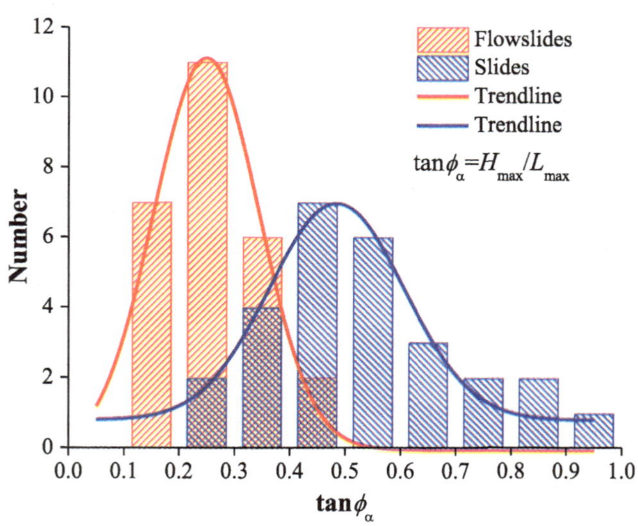

Fig. 2 Relationship between the number of loess landslides and the tanϕ_α value

It can be seen from Fig. 1 that the loess landslides presented a cluster feature. In the case the loess body of both sides has already slid down, the loess body left in between is more easily to slide down due to an inability of mobilizing the side resistance resulting from the normal contacts given by the loess body of both sides, during sliding phase.

3.2 Analysis for Triggering Mechanism of Loess Landslide

The constant shear drained (CSD) triaxial test was employed to investigate the behavior of loess subjected to the rising phreatic surface induced by the irrigation-water infiltration. The cylindrical specimens 39.1 mm in diameter and 80 mm in height were prepared. The CSD tests were performed at three stress levels of the specimens retrieved at 6, 12, and 24 m depths below the platform surface, respectively. CO_2 was introduced in saturation phase to displace air from the specimens allowing for back pressure saturation to be performed. The specimens were then sheared in a drained condition along a constant deviatoric stress path by

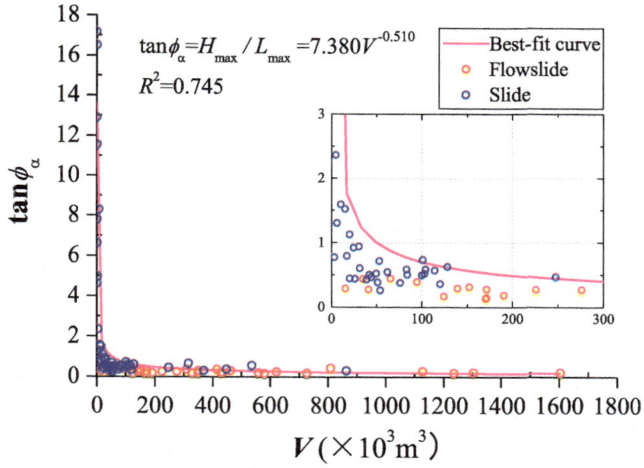

Fig. 3 Relationship between the $\tan\phi_\alpha$ value and the volume of landslide deposit V

increasing the back pressure u at a rate of 1 kPa/hr. Figure 4 shows the results of CSD tests. The axial strain and volumetric strain showed a small change during a steady reduction in the effective mean stress p' while the deviatoric stress remained constant. This may be ascribed to the internal particle structural effects of loess. When the back pressure u was increased to a value where the axial strain and volumetric strain showed a substantial change and the soil stress approached the critical state, the specimen collapsed suddenly. At that time, the ratio of the back pressure u to the cell pressure σ_3' reached a value of 0.7

The triggering mechanism of the loess landslides may be interpreted from the results of the CSD tests that the sudden collapse of the saturated loess due to the rising phreatic surface leads a downward movement of the slope producing a larger undrained loading on the saturated loess at the toe of the slope. Thus, a loess landslide is likely to be triggered as subjected to the generation of excess porewater pressure in a further sliding phase [5–10].

4 Conclusion

This paper presented investigation results of 62 loess landslides in the south Jingyang platform. Based on the analysis results, the following conclusions can be drawn:

(1) The irrigation-water infiltration and sequence and rule of loess landslide seemed to be the major contributor to the loess landslides in the south Jingyang platform.
(2) The sudden collapse of loess induced by the rising phreatic surface led to a downward movement of the slope and the subsequent landslide of loess as subjected to the generation of excess porewater pressure in further sliding phase.
(3) Fieldworks to stabilize the toe of loess slope due to the consistent erosion activities and regional irrigation rather than irrigation over the whole area may reduce the potential of loess landslide hazards. Additionally, more efforts to enhance the resistance of loess against

Fig. 4 Results of constant shear drained (CSD) triaxial tests: **a** ε_a versus p', **b** ε_v versus p', **c** q' versus p', and **d** ε_a versus u/σ_3

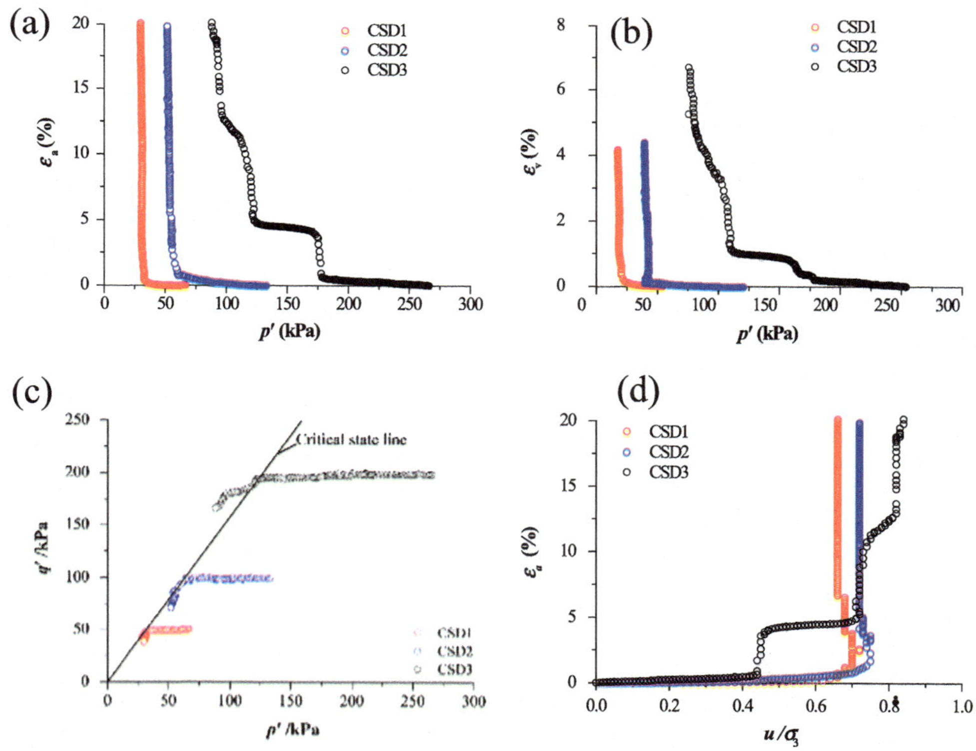

the seepage force should be made to avoid triggering a loess landslide.

References

1. Derbyshire, E.: Geological hazards in loess terrain, with particular reference to the loess regions of China. Earth Sci. Rev. **54**, 231–260 (2001)
2. Zhang, M.S., Liu, J.: Controlling factors of loess landslides in western China. Environ. Earth Sci. **59**, 1671–1680 (2010)
3. Wang, J.J., Liang, Y., Zhang, H., Wu, Y., Lin, X.: A loess landslide induced by excavation and rainfall. Landslides **11**, 141–152 (2014)
4. Peng, J.B., Fan, Z.J., Wu, D., Zhuang, J.Q., Dai, F.C., Chen, W. W., Zhao, C.: Heavy rainfall triggered loess-mudstone landslide and subsequent debris flow in Tianshui, China. Eng. Geol. **186**, 79–90 (2015)
5. Cheng, W.C., Ni, J.C., Cheng, Y.H.: Alternative shoring for mitigation of pier-foundation excavation disturbance to an existing freeway. J. Perform. Constr. Facil. **31** (5), 04017072 (2017)
6. Cheng, W.C., Song, Z.P., Tian, W., Wang, Z.F.: Shield tunnel uplift and deformation characterisation: A case study from Zhengzhou metro. Tunn. Undergr. Space Technol. **79** (83–95), (2018)
7. Cheng, W.C., Ni, J.C., Shen, J.S., Wang, Z.F.: Modeling of permeation and fracturing grouting in sand: Laboratory investigations. J. Test. Eval. **46** (5), 20170170 (2018)
8. Liu, X., Zheng, X., Cheng, W.C., Kong, Q., Wang, J.: The shear strength of the nature loess joint: A case study in shaanxi province. J. Test. Eval. *47* (3), 20170759 (2018)
9. Wang, Z.F., Cheng, W.C., Wang, Y.Q., Du, J.Q.: Simple method to predict settlement of composite foundation under embankment. Int. J. Geomech. **18** (12), 04018158 (2018)
10. Cheng, W.C., Ni, J.C., Arulrajah, A., Huang, H.W.: A simple approach for characterising tunnel bore conditions based upon pipe-jacking data. Tunn. Undergr. Space Technol. **71** (494-504), (2018)

Landslides in the Mila Basin-AGIS Approach

Nadira Bounemeur, Riad Benzaid, and Souad Atoub

Abstract

The study area is part of the Rhumel-Kebir watershed which belongs to the northern zone of the Neogene Constantine-Mila post-nappe basin (Northeast Algeria). In this zone, the phenomenon of landslides manifests in detrital formations of the Mio-Pliocene and seriously threatens homes and basic infrastructure. The main objective of our work was to contribute to the knowledge of the probable causes of these landslides. All these ground movements are becoming more and more important with the abundance of rainfall. However, tectonics and lithology are two predisposing factors that play an important role in their evolution. These factors are precursors of instability that decision-makers have to take into account when implementing any development project. Through this work we hope to contribute to the mapping of areas at risk of landslide in the Mila basin. The maps produced will be of major importance for decision support in future development programs.

Keywords

Watershed • Landslides • Lithology • Tectonics • GIS

1 Introduction

The basin of Mila, located northwest of the city of Constantine, extends west to the border of the region of Setif. The areas subject to land movements are mainly the Tell Atlas [1, 2], which includes the Mila basin, due to the existence of highly contrasted reliefs, friable geological formations (clays, flyschs, marls, etc.) and severe climatic conditions. In such geological and climatic conditions landslides, collapse of roads and natural cavities can be observed in some areas (Fig. 1).

This work is also based on various cartographic documents such as, topographic maps of the study area at the scale of 1:50,000 and geological maps at 1:50,000 scale. These data will be supplemented by GPS measurements and observations on the field. The software Arc Gis10.2.2, Map Info and Global Mapper 17.0 were used as part of image processing and geo-referencing of data [3].

2 Results and Discussions

2.1 Slope Map and Hydrographic Network

The geometric parameters of the slopes were derived from a digital terrain model. Two geometric parameters were used in the model: the slope and the hydrographic network (Figs. 2 and 3). Slope is one of the most important factors in the genesis of landslides [3]. Thus the hydrological study passes mainly, by the treatment of hydro-climatic measures making it possible to establish the water balance and consequently to understand the mechanisms of the circulation and the infiltration of waters inducing the instability of the grounds.

According to the precipitation histogram results (Fig. 4), the rainiest season is winter while the driest season is summer. Thus, almost half of the rain received in the study area occurs in winter, where the potential for risk of instability increases considerably during this period, since the feeding of rivers and groundwater by precipitation is important in winter.

2.2 The Lineaments

The lineaments (Fig. 5) correspond to morphological structures as well as elements of structural origin, visible lines and fractures or zones of breaches of faults [4].

N. Bounemeur (✉) · R. Benzaid · S. Atoub
Laboratoire de Génie Géologique, Université Mohammed Seddik Benyahia-Jijel, B.P. 98, Cité Ouled Issa, 18000 Jijel, Algeria
e-mail: bounemeur_nadira@yahoo.fr

A. Kallel et al. (eds.), *Recent Advances in Geo-Environmental Engineering, Geomechanics and Geotechnics, and Geohazards*, Advances in Science, Technology & Innovation, https://doi.org/10.1007/978-3-030-01665-4_100

Fig. 1 Examples of landslides in the Mila Basin (Northeast Algeria)

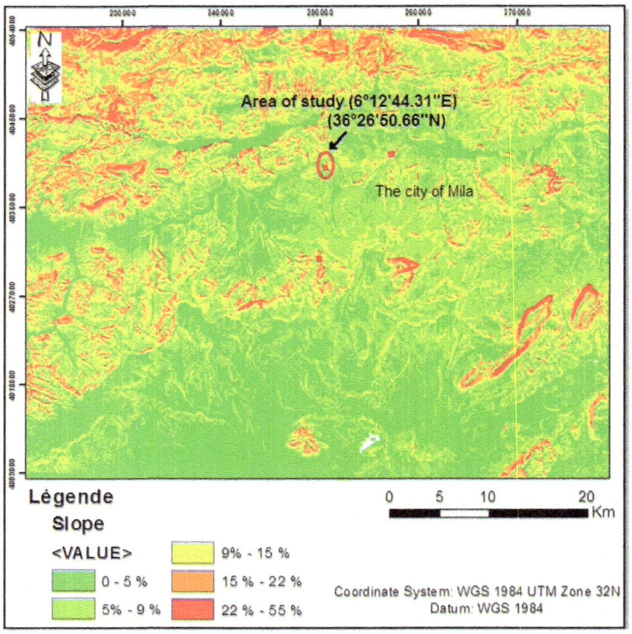

Fig. 2 Map of slopes in the Mila Basin

The results obtained at the study site (6° 12′ 44.31″ E) at (36° 26′ 50.66″ N) demonstrate a high sensitivity to the water of the clay-marly layer, affected by the landslides. For the sake of prevention and forecasting, it is imperative to undertake a thorough investigation of the phenomenon and to ensure a rigorous monitoring of all vulnerable sites. For some landslides, a system of drainage and capture of water sources or simply reforestation can prevent landslides and improve the stabilization of unstable slopes.

3 Conclusion

The study site is mainly characterized by heterogeneous detrital formations, mainly clay-marl, combined with other factors favoring mudslides. The factors generating the ground movements are diverse. We believe that in our case study, the geological nature of the formations characterizing the Mila Basin and the climatic conditions are the main causes of most of the landslides encountered. Nevertheless,

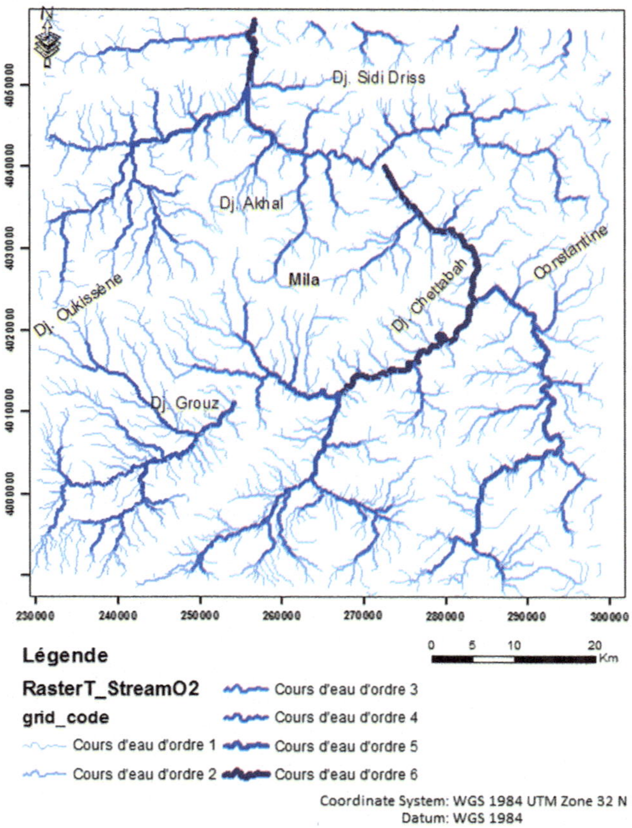

Fig. 3 Hydrographic network map

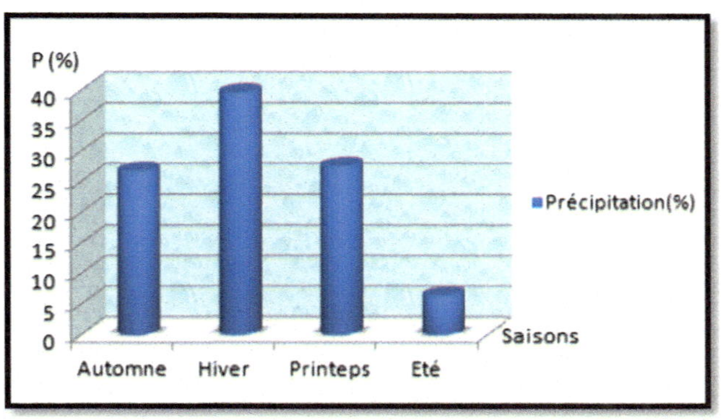

Fig. 4 Histogram of average seasonal precipitation (1990–2003, ANRH-Constantine)

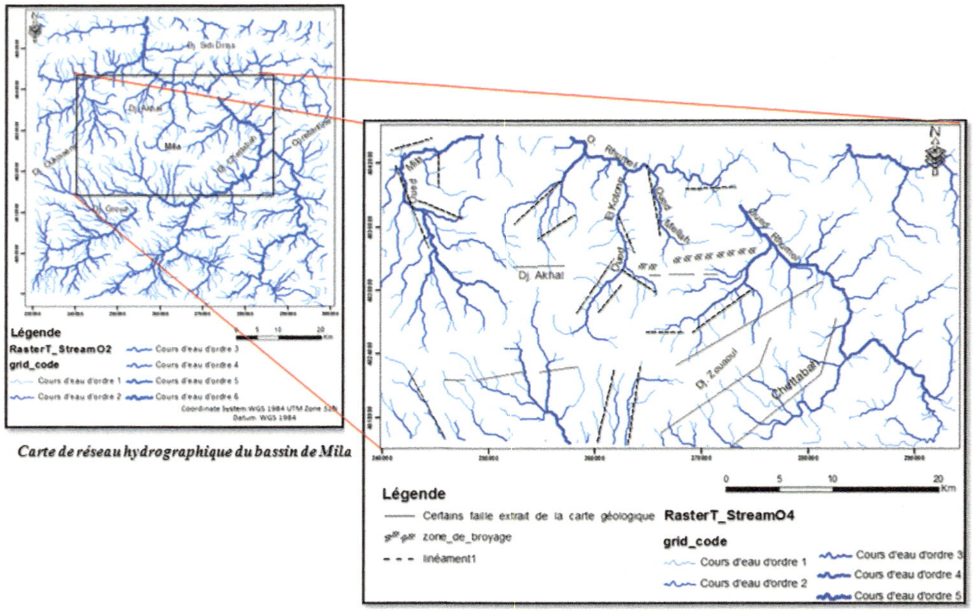

Fig. 5 The lineaments map in the Mila basin (Northeast Algeria)

anthropogenic actions are often the catalyst or the accelerator of these landslides.

References

1. Marmi, R., Kacimi, M., Boularak, M.: Land movements in the Mila (Nord-EastAlgeria)- impact on infrastructure. Revista de geomorfologie **10**, 51–56 (2008)
2. Atmania, D.: Clay mineralogy and shrinkage-swelling phenomenonin the Mila basin (North Constantinois). Thesis in Science, Constantine 1 University (Algeria), 172p (2010)
3. Bennia, A., Saad, A.Z., Mesbah, C.: 1/50.000 geological hazard mapping from high resolution satellite imagery(ALSAT-2A) and geographic information systems (SIG): the case of the pilot area of Bejaia. Geogr. Sci. Bull. N° **30**, 7P (53–59)
4. Belabbas, S.: Inventory and characteristics of active accidents (impact on seismicity of the Constantine Region), Magister thesis, Ferhat Abbas-Sétif University (Algeria), 120p (2012)

Landslide Distribution Analysis and Susceptibility Mapping: A Case Study from Haveli District, Pakistan

Muhammad Basharat, Rizwan Yousaf, and Muhammad Tayyib Riaz

Abstract

Landslide is a frequent and recurrent phenomenon in the northern mountainous terrain of Pakistan and Kashmir region. Where fatalities and severe economic losses occur along with the long time blockage of transport lines. The primary purpose of this research was to establish primary data and information about landslide occurrence in this area. For this purpose, interpretation of SPOT-5 satellite images and field investigations were carried out to develop a landslide inventory of the region. A total 534 landslides were identified and subsequently verified through field visits in a 540 km^2 area. A Digital Elevation Model (DEM) of 12.5 m resolution was used to derive topographic factors such as slope, aspect, curvature elevation and drainage in ArcGIS 10.1. Land cover map was generated using Landsat 8 imageries and processed in ENVI 5.3 software. After that, overlaying the landslide inventory map on these causative factor maps, landslide distributions were assessed in each class of causative factors. Landslide concentration was significantly found to be very high in slope gradient of more than 60° (3.84), followed by water bodies class of land cover (3.35), and the road network (3.07). The landslide susceptibility map was developed using weight of evidence method to rank the region in low, moderate, high and very high susceptibility zones. This map could be helpful for planners to develop a safe hazard plan of the region.

Keywords

Landslide distribution • Himalayas • Landslide susceptibility • Weight of evidence

M. Basharat (✉) · R. Yousaf · M. T. Riaz
Institute of Geology, University of Azad Jammu and Kashmir, Muzaffarabad, 13100, Pakistan
e-mail: basharatgeo@yahoo.com

1 Introduction

Landslides are not only dangerous to human life and property, but also cause the blockage of routine movement of passengers and freight traffic. The Himalayan mountain belt was revealed to be susceptible to landslides due to its topography, fragile geology, climate and large tectonic activities. It has been assessed that 30% of the world mass movements occur in these Himalayan series. Hence, landslides are very common and frequent phenomenon throughout the mountainous area of northern Pakistan, causing damage to roads, buildings, pipelines, and disrupting the movements of the local people.

The landslides hazard depends on various factors like lithology, slope gradient, morphology of the area, land use, rainfall that causes landslide vulnerability and the drainage of the area as well. The hilly zones are prone to slope failures. One of the most important triggering factors of landslides in the mountainous areas is earthquakes. For instance, 2008 Wenchuan earthquake and 2005 Kashmir earthquake have induced thousands of landslides in huge zones [1, 2, 4]. The death toll of 26,000 people was reported directly or indirectly associated with the 2005 Kashmir earthquake induced landslides [3]. Due to adverse effects of mass movements on these socio-economic values, a scientific investigative approach is the need of the time to minimize these natural hazards by effectively investigating the geological and topographical conditions.

In order to minimize the damage due to the mass movement in the future, this study has focused on producing GIS based landslide distribution and susceptibility maps and analysis with respect to many geological and topographical factors for Haveli district. This study is the first attempt in the evaluation of landslide susceptibility of the study area and will contribute as a primary database to the future landslide research in this area. As a result, it will be helpful for planners and other concerned authorities to mitigate the landslide hazards in the region.

A. Kallel et al. (eds.), *Recent Advances in Geo-Environmental Engineering, Geomechanics and Geotechnics, and Geohazards*, Advances in Science, Technology & Innovation, https://doi.org/10.1007/978-3-030-01665-4_101

2 Materials and Methods

The methodological steps adopted in this research are shown in Fig. 1. To generate landslide susceptibility map, all causative factors and these factor maps were rasterized and weights of each class of causative factors were calculated using ArcSDM (Spatial data modular) an extension of ArcGIS (Fig. 1).

3 Results

3.1 Spatial Distribution Analysis

After establishing a brief landslide inventory by field visits and satellite imageries, the analysis of these landslide occurrences was carried out by different causative factors including geological and topographic factors with the aid of distribution maps. The study area comprises mainly debris slides, followed by rock slide, rock fall and debris fall.

The analysis of landslide distribution is shown in Fig. 2. The results indicate that slope greater than 60° has the highest landslide concentration as compared to all other slope classes. Elevation class 1500–2000 m, east facing slope, near water bodies, Paleocene/Eocene rock units, within 500 m distance to fault classes have maximum concentration of landslides as compared to all the other classes of rasters.

3.2 Landslide Susceptibility Mapping

The main objective of this study was to produce a landslide susceptibility map. This study made use of data-driven approach weight of evidence (WofE). Data driven models were easily applied where many landslide locations or occurrences were known and sufficient data existed from which related evidence could be drawn. Using cumulative area posterior probability curve (CAPP) the whole study area was classified into four susceptibility zones as shown in Fig. 3. Accuracy of landslide susceptibility map was also analyzed using prediction rate curve (PRC) and 76% accuracy was achieved.

4 Conclusion

This research work applied landslide spatial distribution analysis to assess landslide characteristics for Haveli district, Azad Jammu and Kashmir, Pakistan. A landslide concentration value of 3.84 was found in slope class of above 60°. Elevation range of 1500–2000 m contained the majority of total landslides i.e. 33.7%. And elevation class of 1000–1500 m had high LC value of 1.66. Land cover classes of water bodies and barren land had very high LC values, 3.35 and 3.10 respectively. And urban land had an LC value 1.79. First 100 m buffer zone around the road network had about 41.02 percent of the total landslide and LC value was also

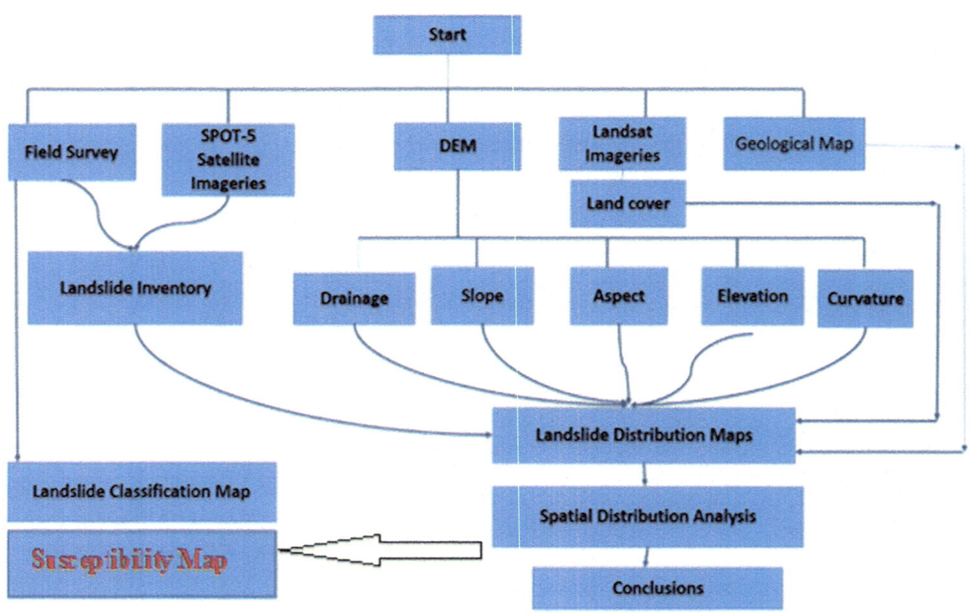

Fig. 1 Flow diagram of the research work

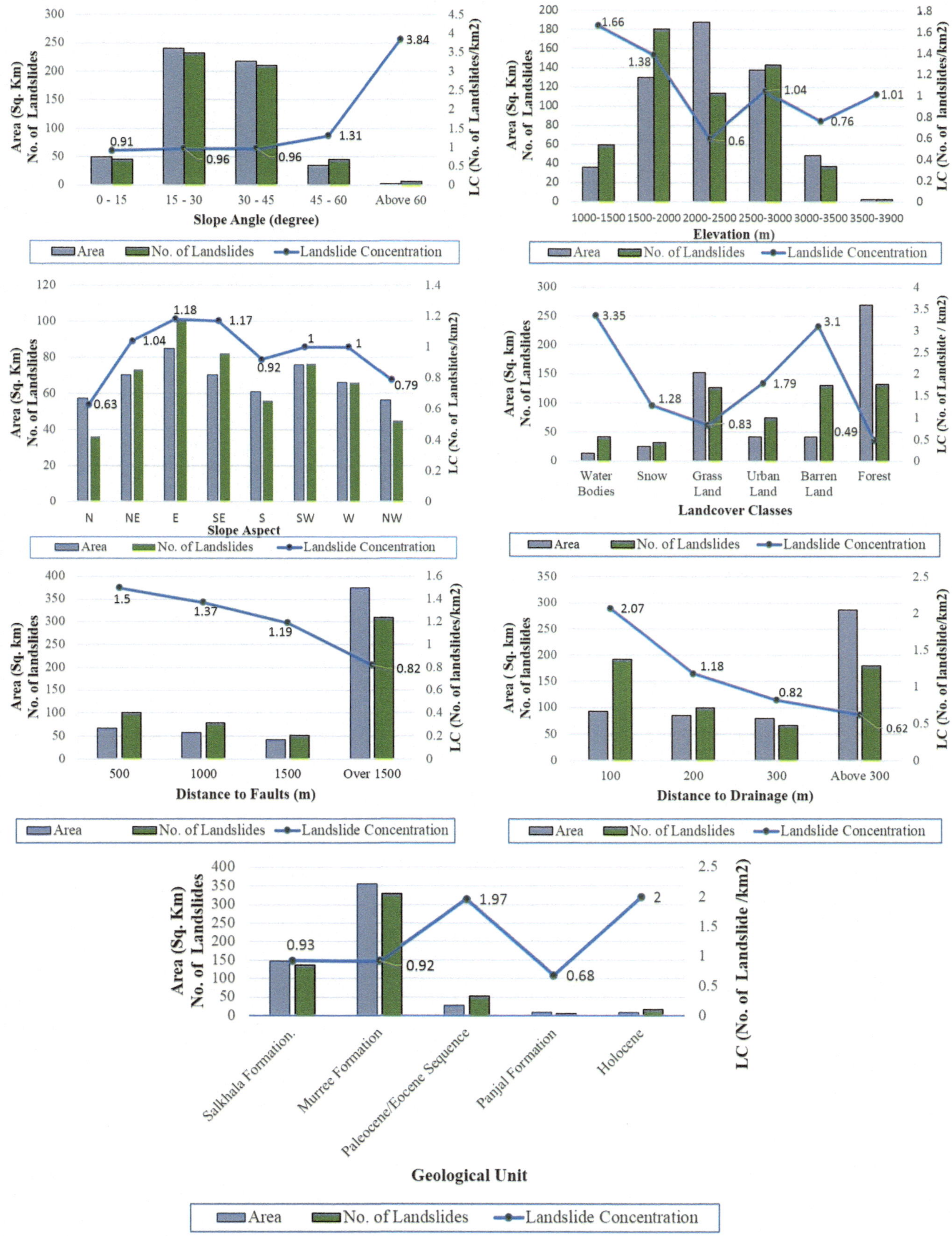

Fig. 2 Spatial distribution analysis of landslides

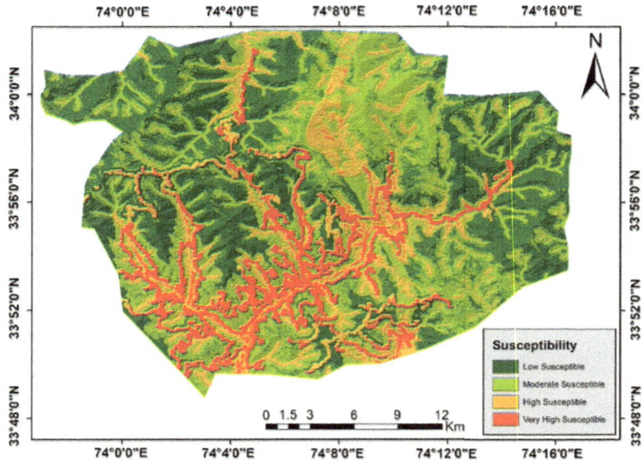

Fig. 3 Landslide susceptibility map of the study area

significantly high (3.07) in this zone. First 100 m buffer zone around drainage held 35.76% of the total landslides and landside concentration was also high (2.07) in this zone. The landslide concentration varied among different lithological units. However, landslide concentration is high in Quaternary alluvium. Landslide frequency and LC values are a bit higher in east direction as compared to other orientations.

References

1. Basharat, M., Rohan, J., Baig, M.S., Khan, M.R.: Spatial distribution analysis of mass movement triggered by Kashmir earthquake 2005 in the NE Himalayas of Pakistan. Geomorphology **206**, 203–214 (2014)
2. Owen, L.A., Kamp, U., Khattak, G.A., Harp, E.L., Keefer, D.K., Bauer, M.A.: Landslides triggered by the 8 October 2005 Kashmir earthquake. Geomorphology **94**, 1–9 (2008)
3. Petley, D., S. Dunning, Rosser, N., Kausar, A.B.: Incipient landslides in the Jhelum Valley, Pakistan following the 8th October 2005 earthquake. Disaster mitigation of rock flows, slope failures and landslides by Universal Academy Press, pp. 1–9 (2006)
4. Sato, P.H., Hasegawa, H., Fujiwara, S., Tobita, M., Koarai, M., Une, H., Iwahashi, J.: Interpretation of landslide distribution triggered by the 2005 Northern Pakistan earthquake using SPOT 5 imagery. Landslide **4**, 113–122 (2007)

Debris Flow Susceptibility Assessment at a Regional Scale Based on Flow-R Model (China)

Yinping Nie and Xiuzhen Li

Abstract

After determining the thresholds value of sources identification and the parameters of propagation, the Flow-R was used to simulate the potential hazard range of Bayi Gully debris flow. Then the simulation results were evaluated using the confusion matrix. Finally, the susceptibility of Bayi gully debris flow was assessed. The results show: (1) The small catchment channels at the top of the gullies provide rich sources for debris flow, (2) The sensitive areas are distributed roughly within the range of 40 m on both sides of channels and account for 20.06% of the study area, (3) The very high areas are distributed in the center of the channels and account for 50.38% of the sensitive areas. The susceptibility gradually decreases from the center of the channel to both sides.

Keywords

Debris flow • Susceptibility assessment • Confusion matrix • Flow-R • Bayi gully

1 Introduction

Located in the meizoseismal areas hit by the China Wenchuan earthquake on May 12, 2008, frequent mountain hazards occurred in Bayi Gully and caused great harm to the safety of the people's lives and property. Over only four years (2008–2011), five large-scale debris flows struck this area. At present, most evaluation methods are usually a debris flow gully corresponding to a susceptibility class. In fact, the susceptibility class is unequal in different parts of a gully. Flow-R model is based on empirical and physical models. This is a method that can assess the susceptibility of different parts of a debris flow gully. Therefore, it is of great practical significance to use the Flow-R to evaluate the regional susceptibility of Bayi Gully debris flow. The location of the study area and the field investigation pictures are shown in Fig. 1.

2 Flow-R Model

Flow-R based on GIS platform mainly consists two parts: the identification of the source areas and the assessment of the propagation. The source areas identification is the process of identifying these grids that simultaneously satisfy all source areas threshold conditions. The assessment of the propagation includes: (1) flow direction algorithms and persistence functions for the spreading assessment, and (2) implemented algorithms for the run-out distance assessment. This model is described in detail by Horton et al. [1]. There are three outputs of Flow-R (Fig. 2).

3 Susceptibility Assessment of Debris Flow

3.1 Parameters Selection

The required parameters of the Flow-R can be divided into two classes: source identification thresholds and propagation parameters. 10 $\times$ 10 m DEM, flow accumulation, slope, lithology and plan curvature were selected as sensitive factors to identify the sources areas. By comparing the river network extracted from different drainage area thresholds, 0.01 km^2 was finally selected as the upslope contributing area. For slope threshold, 15° is the general value recommended by Horton et al., while 25° is the empirical value of Bayi Gully 15° with a higher efficiency of confusion matrix was selected as the slope threshold. The Controlling

Y. Nie · X. Li (✉)
Institute of Mountain Hazards and Environment, Chinese Academy of Sciences, Chengdu, 610041, China
e-mail: lxzljt@sina.com

Y. Nie
University of Chinese Academy of Sciences, Beijing, 100049, China

A. Kallel et al. (eds.), *Recent Advances in Geo-Environmental Engineering, Geomechanics and Geotechnics, and Geohazards*, Advances in Science, Technology & Innovation, https://doi.org/10.1007/978-3-030-01665-4_102

Fig. 1 **a** The location of Bayi Gully; **b** full view of the debris flow in Bayi Gully; **c** loose deposits in the gully

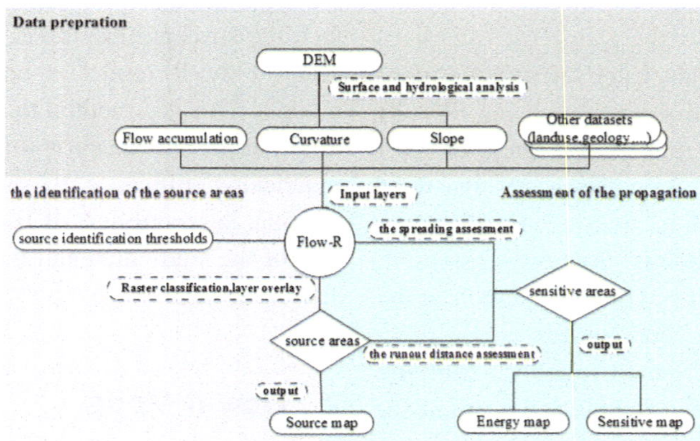

Fig. 2 Flow chart of main processes within Flow-R

persistence function was the only variable to simulate the local area, and the cosines method with highest accuracy was selected to simulate Bayi Gully. Other parameters are shown in Table 1.

3.2 Debris Flow Simulation, Assessment and Susceptibility Assessment Results

The simulation results of Flow-R are shown in Fig. 3. This figure reveals that: (1) the source areas are distributed in the gullies, (2) the energy from the sub-channels to the main channels integrally changes from high to low, and (3) the susceptibility gradually decreases from the center of the channel to both sides.

The simulation results were evaluated by the confusion matrix (Fig. 4; Table 2) [2]. Affected by factors such as vegetation and weathering, the historical debris flow paths don not really reflect the hazard ranges of actual debris flow, so the statistics of the confusion matrix cannot be completely accurate and it might be generally higher.

The results of susceptibility assessment are shown in Table 3 and Fig. 5. The debris flow susceptibility in different parts of the gully is different. And the susceptive areas

Table 1 Parameters table

Source identification thresholds			Propagation parameters (SFLM)	
Flow accumulation	Slope	Plan curvature	Travel angle	Maximum velocity
0.01 km^2	15°	$-2/100\ m^{-1}$	7°	16 m/s

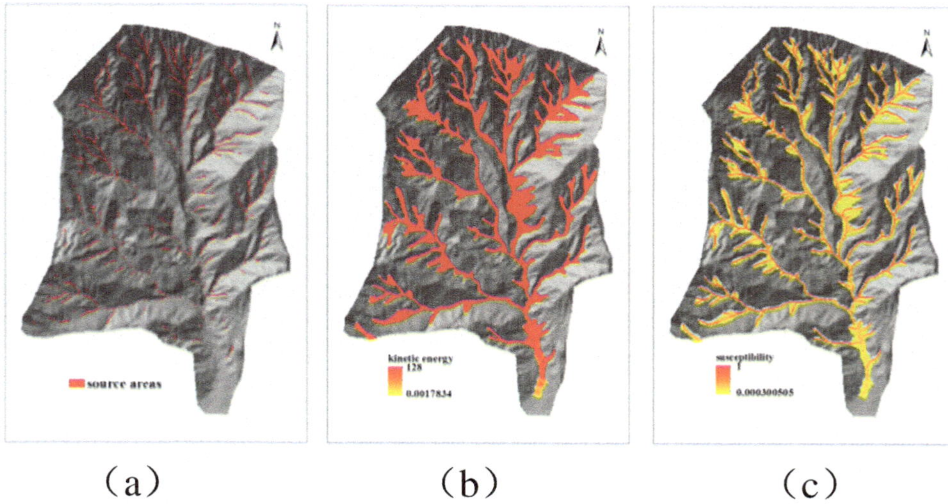

Fig. 3 **a** Sources areas map; **b** Energy map; **c** The sensitive areas map

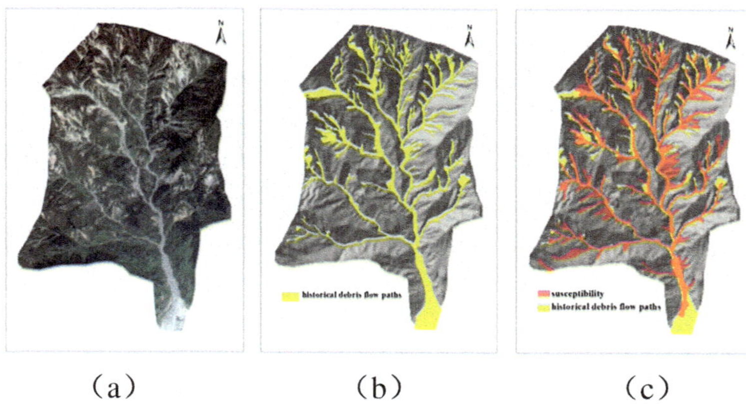

Fig. 4 **a** Bayi Gully Google earth satellite image (recording date 2011.4.26); **b** Historical debris flow paths extracted from Google earth satellite image; **c** Overlay analysis of sensitive areas and historical debris flow paths

Table 2 The evaluation results of the confusion matrix

Efficiency	Positive prediction power	Sensitivity
84.18%	47.99%	62.83%

Table 3 Susceptibility assessment results

Susceptibility class	Low	Moderate	High	Very high
Flow-R values	0.0003–0.004	0.004–0.032	0.032–0.11	0.11–1
The ratio in sensitive area (%)	16.00	18.27	15.35	50.38
The ratio in study area (%)	3.21	3.66	3.08	10.10

controlled by topography and landform are obviously distributed along the channels. Analyzing the spatial distribution characteristics of debris flow has given significance for debris flow prevention and control decision.

The model does not consider the influence of inducing factors such as rainfall and earthquake. In addition, the loose materials on the slope outside the channel can also become the sources of debris flow, which should be considered, too. These problems need to be solved in future studies.

4 Conclusion

Our main conclusions can be stated as follows:

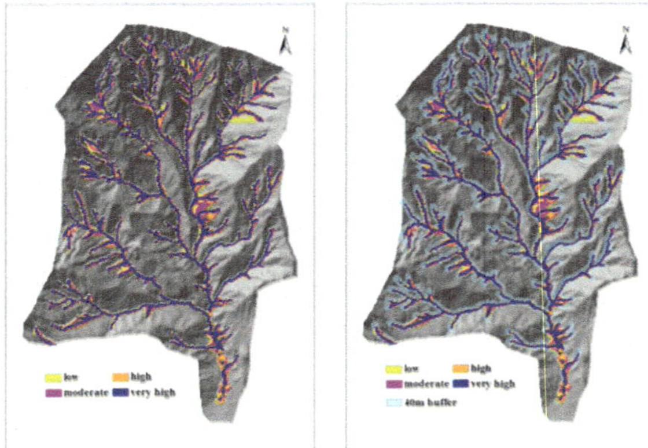

(a) (b)

Fig. 5 **a** The susceptibility zoning map of Bayi Gully debris flow, **b** the 40m distribution map of Bayi Gully debris flow

(1) The debris flow source areas are distributed in the gullies. Most of them are distributed in the slope range from 20° to 50° and the elevation range from 1400 to 1600 m, (2) the entire sensitive areas are distributed roughly within the range of 40 m on both sides of channels and divided into four classes: very high, high, moderate and low. In addition, the entire sensitive areas account for 20.06% of the study area, (3) the evaluation results of the confusion matrix show that efficiency is 84.18%, the positive prediction power is 47.99%, and the sensitivity is 62.83%. The simulation can basically reflect the sensitive characteristics of Bayi gully debris flow.

References

1. Horton, P., Jaboyedoff, M., Rudaz, B., et al.: Flow-R, a model for susceptibility mapping of debris flows and other gravitational hazards at a regional scale. Nat. Hazards Earth Syst. Sci. **13**(4), 869–885 (2013)
2. Park, D.W., Lee, S.R., Vasu, N.N., et al.: Coupled model for simulation of landslides and debris flows at local scale. Nat. Hazards **81**(3), 1653–1682 (2016)

Vegetation Dynamics on Clay Landslides After Bioengineering Works: Three Case Studies in North Apennines, Italy

Donatella Pavanelli, Antea Gennari, Lorenzo Sulpizi, and Claudio Cavazza

Abstract

Landslides in Italy and especially in the Emilia Romagna region represent a characteristic component of the Apennines landscape, characterized by predominantly argillaceous lithology of sedimentary origin. The total landslide area in Emilia Romagna is about 2717 km^2, there are about 758 km^2 of active landslides, mainly found on agricultural land. The paper examined three representative landslides of the Emilia Apennines (province of Bologna), characterized by anthropogenic abandonment during the last 70 years. The three case studies, Vimignano, Marano and Ca′ dei Ricci landslides, are located in the Reno River Mountain Basin and are united by lithological and climatic features that have caused the recurring and simultaneous land movements that occurred in the ′90 s. The study analyzed the effectiveness and status of stabilization and bioengineering intervention about 15–20 years after their execution and the ability of the natural vegetation to colonize and grow on the landslides in the calanchive basins (badlands) after stabilization, despite the unfavorable climatic and edaphic conditions. We also determined the standing vegetation which developed on the landslides on field sites.

Keywords

Bioengineering • Vegetation dynamics • Landslides • Badlands • Italy

D. Pavanelli (✉)
Department of Agricultural and Food Sciences, University of Bologna, Bologna, Italy
e-mail: donatella.pavanelli@unibo.it

A. Gennari · L. Sulpizi
School of Agriculture and Veterinary Medicine, University of Bologna, Bologna, Italy

C. Cavazza
Reno Catchment Authority, Emilia-Romagna Region, Bologna, Italy

1 Introduction

In the Emilia-Romagna Region (northern Italy), over 32,000 landslides bodies were identified. The high number of large and periodically reactivated landslides depends primarily on geological causes and the rock mass lithology. Landslide reactivation generally occurs following periods of intense and/or prolonged precipitation events; it has been demonstrated that another triggering cause may consist in snow melting. The paper examines three representative clayey landslides of the Emilia Apennines located in the Reno River Mountain Basin (RMB). Their similar lithological and climatic features have probably caused the recurring and simultaneous land movements that occurred in the ′90 s. Since then several works have been made in order to control the mass movement dynamics and secure the rivers, the road network and nearby villages and towns. Most of the control works dealt with bioengineering practices: palisades piles, geotextiles, seedings, surface flow control works and drainage ditches. The lithological and climatic characteristics of the median Apennine are quite challenging for the survival of bioengineering works, especially for the living materials used. The main causes of the said issue are the clayey substrate and the rainy winter and the dry summer, which cause runoff. According with Giupponi et al. [1] measuring the success of bioengineering works shows problems similar to those that affect the evaluation of the success of environmental restoration intervention: there are still great difficulties in identifying valid evaluation criteria. The aims of the present study were: (1) to analyze the effectiveness and status of stabilization and bioengineering works 10–20 years after their achievement, and (2) to evaluate the ability of the spontaneous species to colonize and grow in the degraded soil upon completion of the bioengineering works.

A. Kallel et al. (eds.), *Recent Advances in Geo-Environmental Engineering, Geomechanics and Geotechnics, and Geohazards*, Advances in Science, Technology & Innovation, https://doi.org/10.1007/978-3-030-01665-4_103

2 Materials and Methods

The three landslides are located in the northern Apennines, about 30 km south of Bologna, on the eastern slope of the Reno River Valley (Fig. 1). The RMB area is 1061 km^2 with outlet at Casalecchio di Reno and its mean riverflow is 26 m^3/s . The highest and mean altitudes of the catchment are 1945 m a.s.l. and 639 m a.s.l., respectively. The relatively easily erodible clayey rocks of the study area of RMB represents the intact Ligurian nappe and its wedge-top catchment deposits collectively referred to as Epiligurian rocks. The impermeable and erodible lithology and the heavy rainfall causing runoff, soil erosion and landslides easily start in badlands (Calanchi). The Reno catchment falls within the Apennine climatic zone, characterized by two periods of maximum rainfall, in autumn and spring, and one period of minimum in summer.

The yearly average rainfall is 1307 mm yr^{-1} [2]. The mean temperature is 12.4 °C . The three landslides examined, classified as complex landslides, and located in badlands areas (Calanchi basin) were : Vimignano (L1), Marano (L2) and Ca' dei Ricci (L3). The activity of the landslides is known since 1948 (L1) and 1900 (L2, L3), when they moved damming part of the Reno (L2, L3) and Limentra (L1) Rivers. The landslides slope is about 20 and 22% for L2 and L3 respectively, and 31% for L1, with the surface area of 65,000 m^2 (L1), 70,000 m^2 (L2) and 130,000 m^2 for the L3 landslide. The methodology applied to the three case studies in the Bologna Apennines is meant to evaluate the effectiveness of bioengineering works with focus on the plant species involved and their dynamic. In addition to that, it focused on the identification of the responses of the slope and the ground to the techniques and of the works themselves over time, in terms of maintenance requirement and vitality of the plant material used. The methodology used in order to evaluate the reestablishment of vegetation in the landslide areas, consisted in the integration of the ArcGIS software with satellite images and aerial photographs and identify plant species in transects on the landslides. The monitoring required the use of the following digital ortophotos: 1954 IGM-GAI b/n aerial photo, AIMA 1996 and AGEA 2003, 2008, 2011, TeA 2014 and Quickbird 2017. The vegetation was classified considering these categories: forest/scrublands, herbaceous, ground-rocks. The transects (Fig. 2) were identified in a similar position in each landslide body.

3 Results

The works made at the foot of the landslides responded positively to the stresses of the slope, the structural state is discrete, while the response of the live material was more problematic. The use of willow cuttings in the bioengineering works has led to some problems of environmental adaptation that penalized their development. Willows often have good results during the first two, three years of planting but on longer periods suffer lack of water.

The spontaneous vegetation dynamic shows a reduction of the bare soil /rock outcropping (clays) of about 25% for L1 and L2 landslides. The wood-shrubs vegetation remains almost unchanged after the landslide settlement. With regard to the herbaceous vegetation, there is an increase from 5% to 22% of the surface of the L1 and from 37% to 48% of landslide L2 area. A separate case is the L3 landslide which had a high grass cover (40–43%) and around 50% of the wood-shrubs vegetation, from the beginning of the observations, so it remained almost unchanged after the landslide settlement (Fig. 3).

Pioneer species native to environments with xerothermic affinity, such as those detected thanks to the transects: blackthorn (*Prunus spinosa*), rose hip (*Rosa canina*), juniper (*Juniperus communis*), Spanish broom (*Spartium junceum*), flowering ash (*Fraxinus ornus*), seaberry (*Hippophae rhamnoides*).

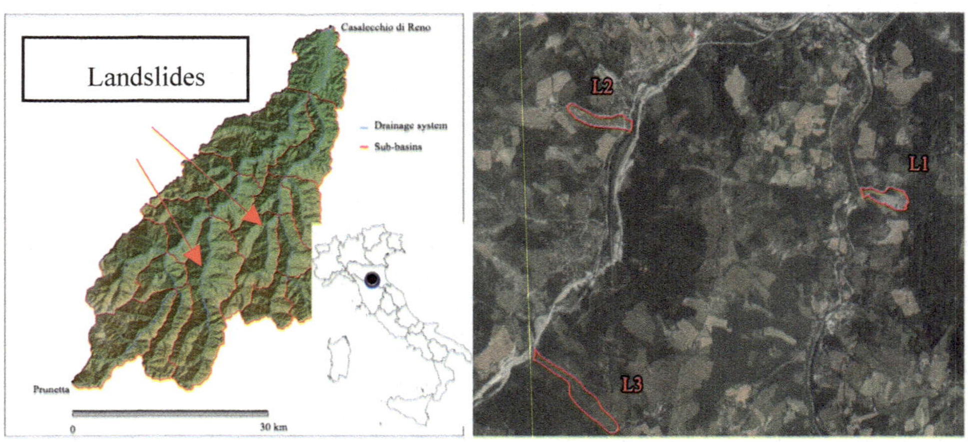

Fig. 1 Reno mountain basin (south Bologna) and the image with landslides (Google hearth)

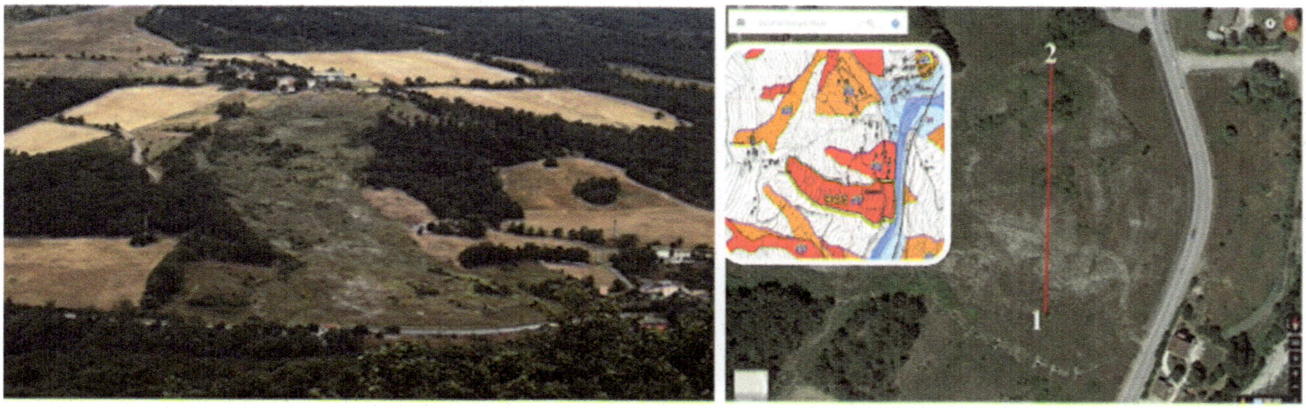

Fig. 2 Marano landslide (L2) and Regional Geomorphologic map. Right: transect on the landslide foot

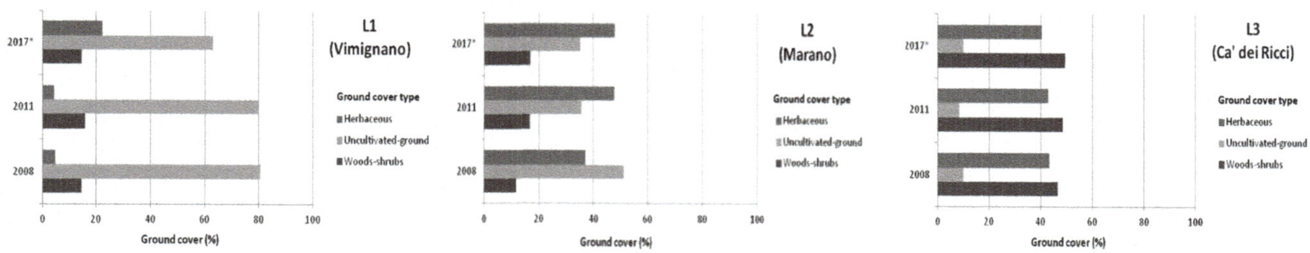

Fig. 3 The spontaneous vegetation dynamic (% area) of the landslides, after bioengineering works (* Based on satellite images)

4 Discussion

This study aimed to be a contribution to the possibilities of improving the performance and durability of naturalistic engineering intervention in extreme conditions, in particular: the clay lithology, the high slopes and the critical climatic events that are increasing due to climate change (cold winters and warm, arid summers). Concerning the structural part of the interventions the survey confirmed the good results and durability of the materials. The survey showed that the actual vegetation, after 10–20 years from works, was not related to the planted species in the bioengineering works. Willow cuttings in the bioengineering works has led to some problems of environmental adaptation that penalized their development In this case the action of planted species has been limited to the years after the works. The spontaneous plants colonized the landslide, out-competing the planted ones. At present, the majority of the plant species has spread from the nearby areas, and is mainly composed of pioneer species typical of the secondary successions. The typical shrubs can be used in arid slopes characterized by cold winters and hot summers and can give better results compared to willows, in these mountainous area. The L3 landslide has a high grass cover and around 50% of the wood-shrub vegetation from the beginning of the observations; such behavior can be explained by the type of stabilization intervention carried out, which did not include the reshaping of the soil slope.

5 Conclusions

According to these results, Bochet [3] concluded that understanding the fate of seeds in overland flow is also a critical issue for the successful restoration of severely eroded slopes. In conclusion, an interdisciplinary approach involving scientists from different fields related to plants, soil, geomorphology, hydrology, ecological restoration and modelling should broaden our understanding of seed fate in overland flow as well as its ecogeomorphological consequences in vegetation structure and function to help fill the aforementioned gaps [3].

References

1. Giupponi, L., Bischetti, G.B., Giorgi, A.: A proposal for assessing the success of soil bioengineering work by analysing vegetation: results of two case studies in the Italian Alps. Landscape Ecol. Eng. **13**, 305–318 (2017)
2. Pavanelli, D., Capra, A.: Climate change and human impacts on hydroclimatic variability in the Reno River catchment, Northern Italy. In: CLEAN - Soil, Air, Water, WILEY-VCH GmbH&Co. Germany (2013)
3. Bochet, E.: The fate of seeds in the soil: a review of the influence of overland flow on seed removal and its consequences for the vegetation of arid and semiarid patchy ecosystems. SOIL **1**, 131–146 (2015)

Training of Sensors for Early Warning System of Rainfall-Induced Landslides

Naresh Mali, Pratik Chaturvedi, Varun Dutt, and Venkata Uday Kala

Abstract

Landslides have been a major issue in the Himalayan region where slopes are cut and reformed for construction practices for infrastructure development, deforestation, and many other human activities. In lieu of the mitigation measure for rainfall-induced landslides to improve the factor of safety against failure, several warning techniques have been suggested. However, they are quite expensive, resulting in an only limited application for infinite slopes. In lieu of the existing conditions, early warning systems (EWS) for detecting slope failure using the sensors have been found to be handy to control the fatality of the disaster. But, the various sensors have been used for these warning systems are not unique. Hence, they need to be trained for each type of soil and other favorable conditions. For the proposed study, Micro-Electro-Mechanical Systems (MEMS) based sensors have been used to predict the slope failures under rainfall conditions at controlled laboratory scale prototype and to perform a series of flume tests in order to develop the threshold for moisture levels and movement that can trigger the slope failure.

Keywords

Slope-instability • Flume test • Sensors • Early warning system

N. Mali (✉) · V. U. Kala
School of Engineering, Indian Institute of Technology Mandi, Kamand Campus, Mandi, 175005, Himachal Pradesh, India
e-mail: malinareshmudhiraj@gmail.com

P. Chaturvedi
Scientist 'D' with DTRL, DRDO, New Delhi, India

V. Dutt
School of Computing and Electrical Engineering, Indian Institute of Technology Mandi, Mandi, Himachal Pradesh 175005, India

1 Introduction

In Himalayan region of India; slope failures predominantly occur during or immediately after rainfall [1–4], which leads to increase in piezometer levels such as rainwater infiltration, thereby, triggering slope failures. Several mitigation techniques have been proposed in order to decrease the effect due to landslides. However, where the slopes are steep and extend to greater heights, most of these may be neither applicable nor economical. In such situations, the early warning systems for detecting slope failure using the sensors have been found to be handy to control the fatality of the disaster. But these warning systems employing various sensors are not unique. Hence, efforts have been made to train the sensors for each type of soil and other conditions to retrieve the data from the field. During the course of the study, the multi-disciplinary involvement for bringing out the sensors, assembly, applications, calibration, testing, placing, data retrieving and model-based predictions were developed [1, 3, 4].

In the proposed study, Micro-Electro-Mechanical Systems (MEMS) based sensors were used to predict the slope failures under rainfall conditions at controlled laboratory scale prototype and to perform flume tests in order to develop the threshold for moisture levels and movement that can trigger a slope failure.

2 Materials and Methodology

Most of the slope failures in and around Mandi region of Himachal Pradesh are shallow failures. The index properties of soil (Table 1) were determined (IS-2720). The present methodology developed by performing flume tests (Fig. 1.) considering various conditions such as dry state and also by increasing the amount of moisture levels [1, 3, 4] within the slope. However, while performing the tests, the amount of moisture content, displacement, acceleration and velocity

A. Kallel et al. (eds.), *Recent Advances in Geo-Environmental Engineering, Geomechanics and Geotechnics, and Geohazards*, Advances in Science, Technology & Innovation, https://doi.org/10.1007/978-3-030-01665-4_104

Table 1 Physical properties of sand

Symbol	Description	Value
G_s	Specific gravity	2.62
%	Gravel size	4
%	Sand size	49
%	Silt size	44
%	Clay size	3
USCS	Soil Classification	SP (poorly graded sand)
C (kPa)	Cohesion	5
Φ	Angle of internal friction	28°

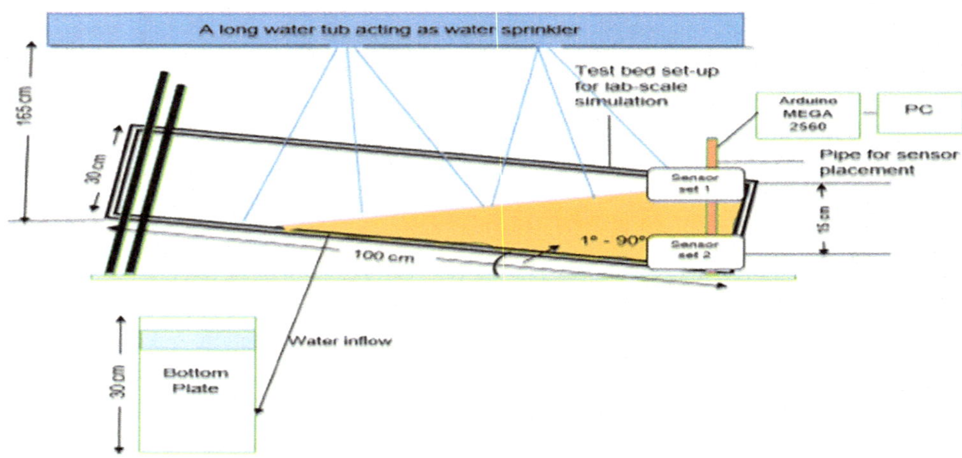

Fig. 1 Line diagram of experimental setup (Not to scale)

were monitored using sensors and identifying the response of soil slope failure. Based on the experimental threshold sensors data, the SMS alert will be generated.

3 Results and Discussion

Tri-axial accelerometer: It is capable of measuring acceleration forces (static and dynamic) by providing simultaneous measurements in three orthogonal directions x, y and z. Thus, the sensor could be used for the analysis of different vibrations experienced by a structure. This sensor expends two capacitors formed by a moveable plate held between two fixed plates. Under zero net force, the two capacitors are equal but a change in force causes the moveable plate to shift closer to one of the fixed plates, increasing the capacitance.

Soil-moisture sensor: It is usually used to detect the soil humidity. When the soil is wet, the output voltage decreases but it increases when the soil is dry.

Force sensor: It is a piezo-resistive conductive polymer, which changes resistance in a predictable manner following application of force to its surface. Like all resistive sensors, this requires a relatively simple interface and can operate satisfactorily in moderately hostile environments.

Tilt sensor: Tilt sensors are devices that produce an electrical signal, which varies with angular movement. These sensors are used to measure slope and tilt within a limited range of motion. They are usually made by a cavity and a conductive free mass inside, such as a blob of mercury or rolling ball. One end of the cavity has two conductive elements (poles). When the sensor is oriented so that its end is downwards, the mass rolls onto the poles and shorts them, acting as a switch.

3.1 Soil Moisture Sensor

For sensing the soil-moisture content in percentage, we used YL-69 module (Fig. 2a). Analog readings for soil-moisture sensors in dry and completely wet states were 395 (0% moisture) and 1022 (100% moisture), respectively. To represent the analog values from the soil-moisture sensor in percentage, and hence the moisture percentage was calculated by the following equation.

$$\text{Moisture Percentage} = \frac{1022 - \textit{Analog value}}{1022 - 395} \times 100 \quad (1)$$

The results depicted that, based on the performing of flume tests, initially the tests were conducted for more than

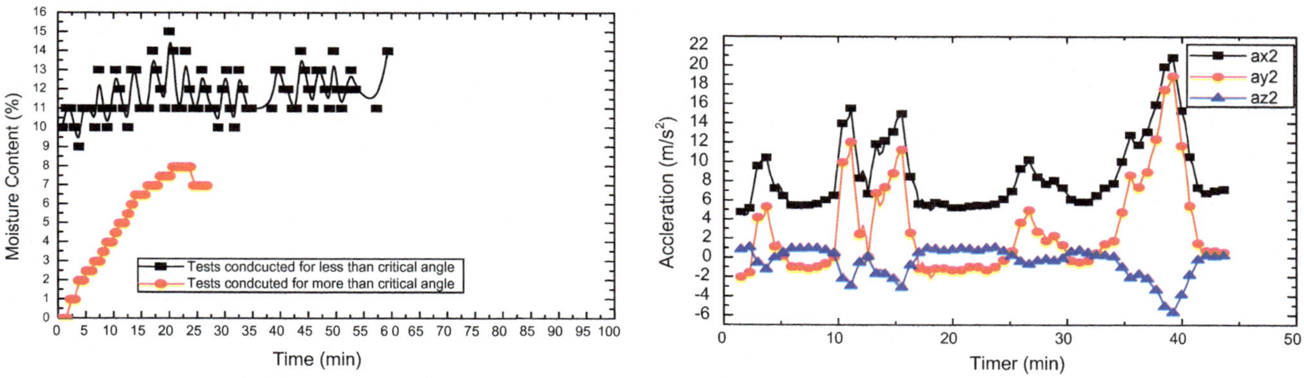

(a) Plot showing the moisture levels with respect to time period

(b) Plot showing acceleration (m/s^2) with respect to time period

Fig. 2 **a** Plot showing the moisture levels with respect to time period. **b** Plot showing acceleration (m/s^2) with respect to time period

the critical angle the failure is observed at around 12% moisture contents whereas, the tests conducted for less than the 17% moisture levels. According to [1, 3, 4], moisture content after reaching the threshold value causes the failure of the slope leading to debris flow.

The failure of slope also depends upon several factors such as soil type, angle of inclination, ramp surface texture and others.

3.2 Accelerometer

For the determination of acceleration, GY-61 accelerometer module was employed. It measures the tri-axial accelerations in three orthogonal axes (Fig. 2b). This accelerometer is capable of measuring acceleration in the range of ± 3 g (where, g = 9.81 m/s^2), in each of the three orthogonal axes. For the accelerometer placed at the top and base of the pipe, a_x, a_y, and a_z refers to the axis perpendicular to the base of the ramp (pointing upwards), along the width of the ramp (from right to left), and sloping at an angle with the soil away from the ramp (θ is the angle of the ramp with horizontal), respectively. Based on the predefined threshold values, the modules outputs LOW, otherwise, it outputs HIGH. The threshold value for the digital signal can be adjusted using the built-in potentiometer. The integration of different sensors to a microcontroller and then the microcontroller's connection to an Internet cloud was carried out. The microcontroller receives data from sensors and it is connected to a GSM modem for transmitting the sensors' readings to the web server. The values from sensors were logged onto an in-house developed web server http://www.landslidemonitoring.esy.es/ with the help of GSM Module.

After the successful reception of the data from soil moisture sensors and accelerometer sensors, the landslide probability was computed with the weights assigned to individual sensor values. When the probability value crosses a prefixed threshold (in this study, 85), an alert would be triggered to the registered users.

3.3 Experiences During the Training Sensors While Performing the Tests

1. Soil moisture sensor needs to be calibrated frequently for calculating the soil-moisture percentage.
2. The probes of the soil moisture sensor are at a distance of 37 mm apart, hence the resistance will be created at the probe, but not in the gap.
3. Calibration should be achieved before placing these sensors in the soil at the time of testing.
4. Calibration values may not be the same for all the sensors
5. Flex sensors should be supported to avoid getting detached from the one end of the node.
6. Any change in voltage would affect the tilt sensor.

4 Conclusion

Early Warning System (EWS) architecture is in place for the people to be alerted about the oncoming disaster. The threshold values are evaluated from the analysis of logged soil-moisture and soil-movement values, by performing the set of flume tests on the ramp. Once the soil-moisture or soil-movement thresholds are reached, the above-mentioned alert generation unit generates landslide alerts. As soon as any activity in the area under surveillance of sensors crosses a pre-determined limit, an alert would be triggered to inform the concerned people to take necessary steps. Hence, as soon as any value of the database crosses the threshold level, an alert is sent via an SMS.

Acknowledgements The authors would like to thank the **State Council for Science, Technology & Environment, Himachal Pradesh, India**, for providing the financial support to pursue this study.

References

1. Huang, C.-C., Lo, C.-L., Jangand, J.-S., Hwu, L.-K.: Internal soil moisture response to rainfall-induced slope failures and debris discharge. Eng. Geol. **101**, 134–145 (2008)
2. Fell, R.: Landslide risk assessment and acceptable risk. Can. Geotech. J. **31**, 261–272 (1994)
3. Lourenco, S.D.N., Sassa, K., Fukuoka, H.: Failure process and hydrologic response of a two layer physical model: implications for rainfall-induced landslides. Geomorphology **73**, 115 – 130 (2006)
4. Wu, L.Z., Huang, R.Q., Xu, Q., Zhang, L.M., Li, H.L.: Analysis of physical testing of rainfall-induced soil slope failures. J. Environ. Earth Sci. **73**(12), 8519–8531 (2014)

A Photogrammetric Surface Comparison for a Dam Reservoir in a Landslide—Prone Area in Eastern Anatolia, Turkey

Sultan Kocaman, Eray Sevgen, and Candan Gokceoglu

Abstract

The purpose of the study was to analyze the slope deformations of a dam reservoir by comparing digital elevation models (DEMs) produced before and after data sets. The DEMs were generated automatically from large format aerial images acquired in 2012 and Gokturk-2 satellite images acquired in 2018. Both DEMs were compared and analyzed to obtain slope deformations. The results showed that the reservoir waters affected the slopes and some displacements were seen clearly in northeast slopes of the reservoir. The slope deformations can be monitored accurately with the proposed approach, with a note on the ground resolution of the images, which affected the accuracy and the density of the DEMs directly. The methodology provided promising results for automatic, low-cost and practical monitoring for slope deformations in large mountainous regions.

Keywords

Landslide • Dam • Photogrammetric surface analyses • Digital surface model • Digital terrain model

1 Introduction

Slope deformations in landslide-prone areas are often monitored using in situ equipment, such as inclinometers, and/or conventional surveys. However, the use of these methods are limited in case of very steep topography and can only be used in relatively small areas due to operational reasons and high costs. Digital elevation models (DEMs) are often used for the determination of morphological features and landslides. Monitoring slope deformations is also possible by analyzing the obtained DEMs at different times.

Conventionally, DEMs have been mostly generated by terrestrial surveys, manual contour measurements in aerial stereo images and digitizing contour lines in existing maps. With these methods, DEMs can be obtained with low density and mainly comprise the terrain elevation. With the development of new image acquisition and multi-image matching techniques, digital surface models (DSMs) and digital terrain models (DTMs) can be produced in large areas with higher density, accuracy and reliability. Especially with the help of dense information, the terrain characteristics can be modeled better. Data collection devices vary from airborne LiDAR (Light Detection And Ranging) to optical imaging cameras operating on satellite and airborne platforms. In this study, a multi sensor approach was used to obtain DSMs and DTMs at different times to monitor slope deformations by comparing DEMs.

Landslides are one of the major problems in dams and their reservoirs. Dam sites are often selected in deep valleys having steep slopes. Such sites are generally prone to landsliding and dam reservoirs even cause their occurrence due to increasing pore water pressure and decreasing shear strength parameters. In this study, we proposed a surface comparison approach for monitoring the slope deformations in a dam reservoir constructed in a mountainous area with very steep slopes. Additionally, the selected area has very complex geological and tectonic characteristics with high seismicity. The area is located in Bingol Province, Eastern Anatolia, Turkey. An overview of the study area is provided in Fig. 1. The methodology described here is suitable for the selected area because the conventional methods are not applicable due to access limitations and study area size.

S. Kocaman · E. Sevgen
Geomatics Engineering Department, Engineering Faculty, Hacettepe University, 06800 Beytepe, Ankara, Turkey

C. Gokceoglu (✉)
Geological Engineering Department, Engineering Faculty, Hacettepe University, 06800 Beytepe, Ankara, Turkey
e-mail: cgokce@hacettepe.edu.tr

A. Kallel et al. (eds.), *Recent Advances in Geo-Environmental Engineering, Geomechanics and Geotechnics, and Geohazards*, Advances in Science, Technology & Innovation, https://doi.org/10.1007/978-3-030-01665-4_105

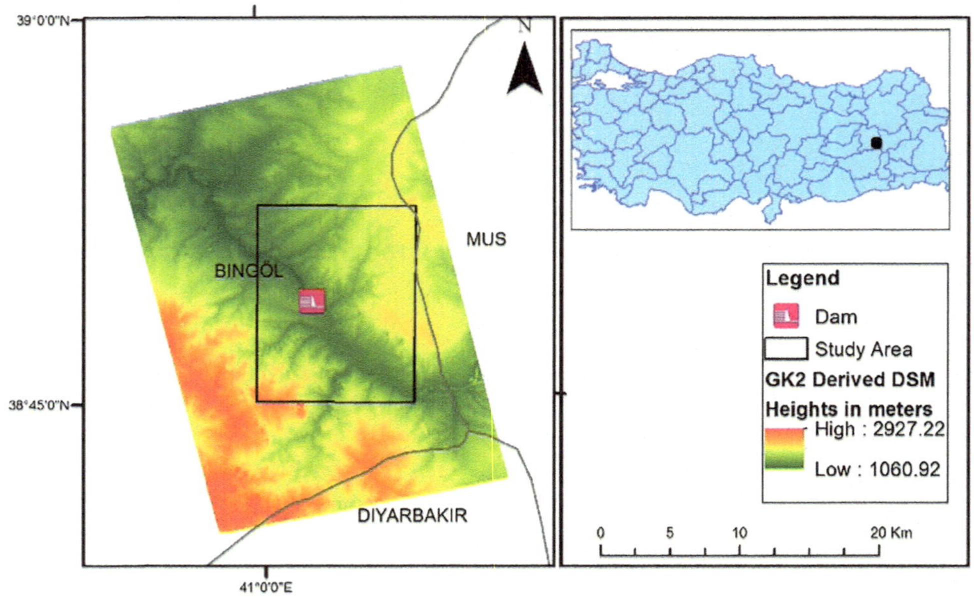

Fig. 1 The location of the study area and colored altitude map generated from the stereo Gokturk-2 satellite images acquired in April 2018

2 General Characteristics of the Study Area

Turkey is an orogenic collage created by some collided continental fragments and the remnants of oceanic environments separating them [1]. Consequently, the complex geology of Turkey results in faulting, folding, shear zones, high horizontal stresses and tectonized rock masses.

According to Turkey's Landslide Inventory Map, the study area has some large – very large active and inactive landslides (Fig. 2). The main conditioning factors of these landslides are the highly jointed and weathered rock masses, young and steep topography. The large earthquakes and toe incision are the main triggers of the landslides in the study area. A real extensions of the landslides are large, and depths of failure surfaces are high. The general failure mechanisms of the landslides are circular, and typical hummocky topography can be observed easily in the field (Fig. 3). When considering these adverse conditions, the landslides should be monitored carefully for the economic life time of the dam and the safety of properties in close vicinity of the reservoir. However, it is impossible to monitor such a large area by conventional methods.

Fig. 2 Landslide inventory map with inactive (yellow) and active (red) landslides [2]

3 Data, Methods and Results

Aerial photographs with 45 cm ground resolution were taken by using Ultracam Eagle camera by the General Command of Mapping in 2012. A total of 34 images were provided together with camera calibration and exterior orientation parameters for this study. A dense point cloud was produced by multi-image matching (matching at every 2 pixels) and a ground filtering was applied to generate a DTM by removing trees and buildings. In addition to the DSM and DTM, an orthomosaic was also generated from this dataset. On the other hand, stereo Gokturk-2 images with 2.5 m spatial resolution (panchromatic band) of the area were ordered and acquired in April 2018 to model the terrain surface after the reservoir was filled by water in February 2018. The difference in the water level can be observed in Fig. 4. Gokturk-2 images were oriented by using control points extracted from the aerial orthophotos and the DSM. After the orientation, a DSM was also generated from the

Fig. 3 A general view of a large landslide in the study area (typical hummocks are seen clearly)

Fig. 4 Aerial orthophoto from 2012 (left) and one Gokturk-2 panchromatic image from 2018 (right). The triangles in the images denote the location of the dam

Gokturk-2 images. Subsequent comparisons between the aerial and Gokturk-2 DSMs were performed using the open source CloudCompare [3, 4]. The residuals were analyzed in terms of elevation differences between the reference and search surfaces. Color coded residual maps are provided in Fig. 5.

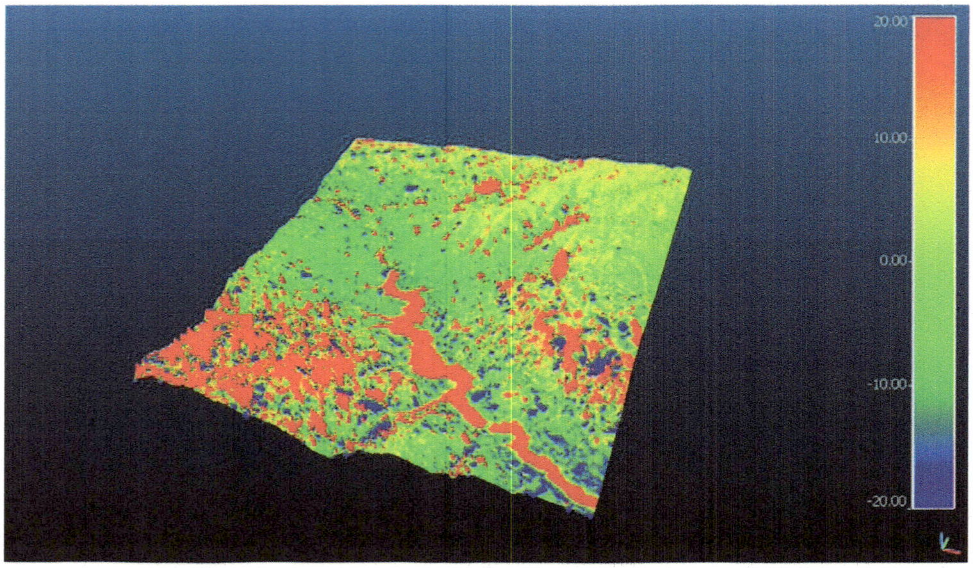

Fig. 5 The residuals after the comparison of aerial DTM from 2012 and Gokturk-2 DSM from 2018 (before and after filling the dam reservoir)

4 Conclusion

In the present study, a practical and comprehensive methodology was introduced for detecting slope deformations in a dam reservoir constructed in a landslide—prone area. The comparison showed that the reservoir waters affected the slopes and some displacements were seen clearly in northeast slopes of the reservoir. Depending on the ground resolution of images, the slope deformations can be monitored accurately.

The further steps of the study include validation of the approach introduced here by comparing previous data sets and improving of the Gokturk-2 DSM by manual editing. Additionally, field checks will be conducted to assess the methodology.

References

1. Okay, A.I., Tüysüz, O.: Tethyan sutures of northern Turkey. In: Durand, B., Jolivet, L., Horváth, F., Séranne, M. (eds.) The Mediterranean Basins: Tertiary extension within the Alpine orogen, pp. 475–515. Geological Society, London, Special Publication 156 (1999)
2. Turkey General Directorate of Mineral Research and Exploration: http://yerbilimleri.mta.gov.tr/anasayfa.aspx. Retrieved 30 Apr 2018
3. CloudCompare: http://www.danielgm.net/cc/. Retrieved 30 Apr 2018
4. Kocaman, S., Ural, S., Karakas, G., Bakici, S.: 3D Processing of Gokturk-2 imagery. In: 38th Asian Conference on Remote Sensing, New Delhi, India, 23–27 Oct 2017

Part X

Geohazards: Characterization and Impacts of Different Geohazards

Evaluation of Natural Radioactivity Levels of Pre-Cambrian Basement Rocks from the South-Western Margin of Arabian-Nubian Shield, Sudan

Mohammed Abdallsamed, Mushaal Salih, and Asim El Mansour

ABSTRACT

This study aimed to investigate the activity concentration levels of ^{238}U and ^{40}K for the basement rocks using in situ gamma-ray spectrometry to set a base data to be used as a reference in case of any radiological accident. In average, the activity concentration levels of ^{40}K are 1226 Bg\kg and 860 Bq\kg in migmatite gneisses and micaschist, respectively. The higher levels in the migmatite gneisses are due to high concentration of feldspar minerals. These average values are either analogous or higher than the standard concentration levels of continental crust. The activity concentration levels of ^{238}U from migmatite gneisses show an average value of 42.4. The micaschist displays an average value of 45.6 Bq\kg. These values are greatly elevated than the average values recorded from basement rocks worldwide.

Keywords

In situ gamma-ray spectrometry • Natural radioactivity • Basement rocks • North kordofan state • Sudan

1 Introduction

Natural radioactive geological materials are the most important sources of the activity concentration levels in the environment. In the past years, a great deal of research has been achieved in many countries to evaluate the hazards linked to the natural radioactive materials (NRM) [1, 2, 9]. In Sudan, a lot of research has been carried out for occurrences of natural radioactive materials (NRM) on the diverse environmental radioactivity since late 1980, just after Chernobyl accident, aiming to produce radiation map for the country to be used as a reference in case of any radiological accident of global dimension (e.g., [3, 4]). However, the current study assessed for the first time the natural background radioactivity levels of ^{238}U and ^{40}K from Basement rocks. It is hoped that the information obtained will allow for the construction of a radiation map of the State which provides a useful reference for the natural background radioactivity levels.

2 Geological Features

The selected study area is situated in Sheakan locality, North Kordofan State in central Sudan. It's characterized by a semi-desert type of climate. Geologically, the area is formed of a part of the Pre-Cambrian Basement rocks complex of North Kordofan. This complex is characterized by the existence of high-grade metamorphic rocks of amphibolite facies as the most interesting features and associated superficial deposits. The metamorphic rocks are composed of migmatite gneisses (constitutes most of the basement complex in Sudan), micaschist, quartzite and amphibolite [5–7] (Fig 1).

3 Sampling and Measurements

Sampling areas were chosen on the basis of field work. Two localities (Jebel Kordofan and Jebel Abu Uroag) were chosen, and a total of 9 rock samples were collected from migmatite gneisses and micaschist. The activity concentration of ^{238}U and ^{40}K of metamorphic rocks samples were measured using gamma-ray spectrometry system equipped with high purity germanium detectors (HPG), assuming that the activities of ^{214}Bi and ^{214}Pb in equilibrium with their parents represent the ^{238}U activity. ^{238}U was determined using photo peak of ^{214}Bi (19.9 min half-life) (609 keV) and ^{214}Pb (28.8 min half-life) (352 keV). ^{40}K was determined

M. Abdallsamed (✉) · M. Salih · A. El Mansour
Department of Geology, Faculty of Science,
University of Kordofan, B160, 51111 Elobeid, Sudan
e-mail: 2778414071@qq.com

A. Kallel et al. (eds.), *Recent Advances in Geo-Environmental Engineering, Geomechanics and Geotechnics, and Geohazards*, Advances in Science, Technology & Innovation, https://doi.org/10.1007/978-3-030-01665-4_106

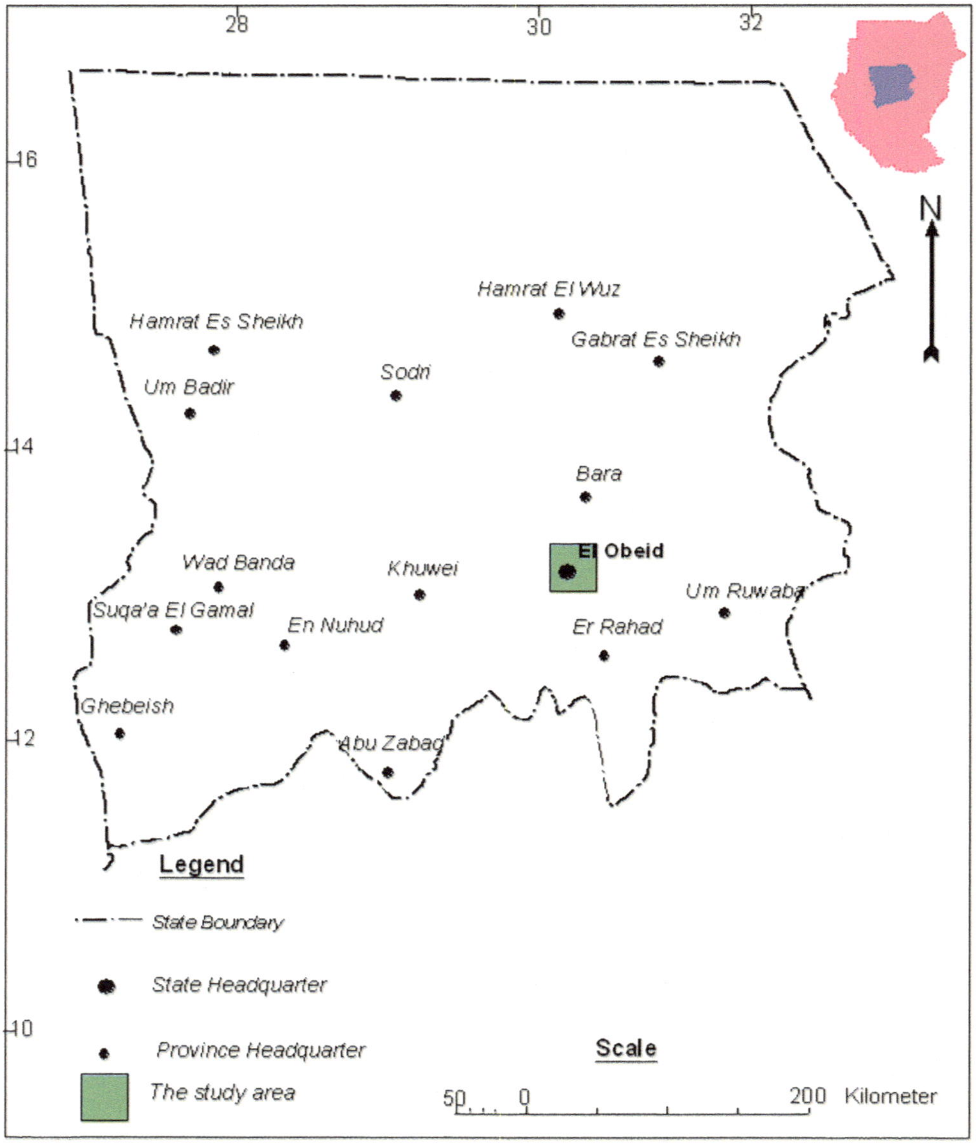

Fig. 1 Location map of the study area

directly by (1461 keV). The certified reference material supplied by International Atomic Energy Agency (IAEA) was used as a quality control sample and the results show good agreement between measured and certified values with a relative determined error of < 1%.

4 Results and Discussion

4.1 ^{40}K Activity Concentration Levels

The measured activity concentration levels of ^{40}K ranged from 278 to 2389 Bg\kg with average values of 1226 Bg\kg in migmatite gneisses and about 631 Bq\kg to 1315 Bq\g with an average value of 860 Bq\kg in micaschist. The average value of migmatite gneisses is clearly higher than the activity concentration levels of continental crust average (i.e. 850 Bg\kg), while the average value of micaschist is within the range [8], which implies that study area falls within the range of normal background radiation. The higher activities for ^{40}K noted in samples migmatite gneisses can be explained in relation to mineralogical compositions.,

4.2 ^{238}U Activity Concentration Levels

The activity concentration levels of ^{238}U from migmatite gneisses ranged from 6.24 to 84.5 Bg\kg with an average value of 42.4. The corresponding values were 30.3 Bq\kg to 65.9 Bq\kg with an average value of 45.6 Bq\kg in micaschist. These values are much higher than the worldwide average reported, which is 25 Bq/kg [9]. However, no effect of radioactivity levels on the environment and human health in the study area was remarked.

5 Conclusion

In situ measurements carried out in typical rocks showed that the activity concentration of 40 K measured from migmatite gneisses is clearly higher than the activity concentration levels of the average continental crust, while the micaschist displays an average value comparable to the continental crust. ^{238}U displays much higher values than the average values reported in different rocks worldwide.

Acknowledgements The authors wish to express their deep appreciation and gratitude to the team of SAEC for analyzing the collected samples.

References

1. UNSCEAR: Source and affects of ionizing radiation. Report to General Assembly, with Scientific Annexes, United Nations, New York (1993)
2. El-Mageed, A.I., El-Kamel, A.H., Abbady, A., Harb, S., Youssef, A.M.M., Saleh, I.I.: Assessment of natural and anthropogenic radioactivity levels in rocks and soils in the environments of Juban town in Yemen Radiation. Phys. Chem. 80(2011), 710–715 (2011)
3. Sam, A.K., Holm, E.: The natural radioactivity in phosphate deposits from Sudan. Sci. Total Environ., 162, 173Ð178 (1995)
4. Sam, A.K., et al.: Assessment of terrestrial gamma radiation in Sudan. Radiat. Prot. Dosimetry. **71**, 141–145 (1997)
5. Vail, J.R.: Outline of The Geological of The Nuba Mountains and Vicinity. Southern Kordofan Province, Sudan (1973)
6. Vail, J.R.: Outline of the geology and mineral deposit of the democratic Republic of the Sudan and adjacent areas. Overseas Geology and Mineral Resources, 49 (1978)
7. IFAD: Re-Appraisal Report, vol. II. Annexes, Elobeid, Sudan (1999)
8. Eisenbud, M., Gesell, T.: Environmental Radioactivity from Natural, Industrial and Military Sources, pp. 134–200. Acad.press, SanDiego, CApp (1997)
9. UNSCEAR: Source and affects of ionizing radiation. Report to General Assembly with Scientific Annexes, United Nations, New York.Mukhtar, A. (2003): Environmental Study for Radionuclides at Miri Lake. Ph.D. Thesis, Faculty of Science, Department of Physics, University of Khartoum (2000)

Temporal Shallow Water Tidal Analysis at Sharm Obhur, the Red Sea

V. R. Shamji

Abstract

The shallow water tidal variation in the Sharm obhur plays temporally more dynamic and more active role in the physical and biological processes. Comprehensive studies on the shallow water tidal analysis were carried out at the Sharm obhur with one complete data in 2013 using different data analysis techniques. The harmonic analysis technique IOS was used for deriving shallow water tidal constituents and temporal analyses were also carried out. The selective amplification was identified for the shallow water constituents, Zo and the same sample was analyzed seasonally along with other important shallow water tidal constituents like MSf, M2, K1 and N2 etc. The study reveals that tidal range in the Sharm obhur is mainly shaped by the shallow water constituent Z_0, can be used for the long-term sea level variation study in this Red Sea region.

Keywords

Shallow water tidal constituents • Harmonic analysis • IOS method • Sea-level

1 Introduction

A detailed tidal analysis study in the Red sea is lacking [1–3]. Most of the studies in the red sea focus on mean sea level changes and local tide characteristics. This may be due to different factors like spatial unavailability of data over the Red sea coast and lack of historical data resources. The tidal analysis and study in the Jeddah has more practical importance. The temperature and salinity variation during low and high tide time at sharm obhur is the evidence of the influence of tidal activity in the Sharm obhur [4]. The present investigations have more significance in the field of academic research and developments. The historical data blank in this region was also one of the limitations for the present investigations. The study aimed to carry out shallow water tidal constituents analysis and its applications.

2 Methodology

The Red sea is a typical of a micro-tidal tropical environment. The data were collected near the maritime college, Sharm Obhur, Jeddah. The study area is located north of Jeddah and is approximately 10 km long 500 m wide [4]. Sharm obhur is shallow with irregular topography and the average depth is 30 m [5]. The Sharm Obhur inlet is protected with coralline limestone [6]. The tide gauge station is located at 21° 42′ 33″N 39° 5′ 45″E. The data were collected along with the other atmospheric parameters. The raw data were in simple water pressure data. The data were processed and used for analysis. The harmonic analysis were carried out with the processed data to extract major tidal constituents and IOS method (Institute of Ocean Sciences) used for deriving tidal constituents including 45 astronomical main constituents and 24 shallow water constituents.

3 Results

The IOS method was adopted to derive 45 astronomical main constituents and 24 shallow water constituents including major and minor constituents. The derived tidal constituents can be used for tidal predictions and hindcast. The different techniques adopted for the derivation of major and minor constituents were discussed in the following sections.

V. R. Shamji (✉)
Faculty of Maritime, King Abdulaziz University, Jeddah, Saudi Arabia
e-mail: vrshamji@gmail.com

A. Kallel et al. (eds.), *Recent Advances in Geo-Environmental Engineering, Geomechanics and Geotechnics, and Geohazards*, Advances in Science, Technology & Innovation, https://doi.org/10.1007/978-3-030-01665-4_107

3.1 IOS Method

The field data during the year 2013 was used to calculate major tidal constituents via the IOS method. The shallow water constituent Z_0 is the dominant component among the resolved 30 tidal constituents and it shows selective amplification in this region. The Z_0 is associated with mean sea level variation and used for sea level variation study. The study shows that, among the derived tidal constituents Z_0, MSf, M_2, K_1 and N_2 are the most prominent components. The seasonal variations of constituents are given in Fig. 1. The predominant shallow water constituents Z_0 and MSf show selective amplifications compared to other constituents. The Z_0 shows more amplifications than other prominent constituent MSf, but a maximum amplification during February, November and December 2013 and shows less amplification during June to September, i.e., during the summer seasons. Hence the amplifications of Z_0 is related to seasonal change and other environmental forces. The other significant constituent MSf also shows seasonal variations in the amplifications with fortnightly periodicity. Even though the Msf tidal motions are relatively weak ($\sim$1 cm at most coastal locations, [7], it is considered the most important constituent in the shallow water areas. The Msf is very weak during summer and shows much amplification during April. The spatial study of Msf over Sharm obhur helps to identify the variation of amplifications in the upstream and downstream areas. The other constituents M_2, K_1 and N_2 also show a seasonal variation in amplitudes (Fig. 2). The analysis also shows the least importance of the major semidiurnal constituents S_2, K_2 and major diurnal constituents O_1 and P_1 respectively in the Sharm obhur. Most of the constituents show different phases except the Z_0 (Fig. 2). Msf also shows seasonal variations in the phase. The major semi diurnal components, M_2 and N_2 show the same trend in phase variation throughout the year. The meteorological factors affect the seasonal variations of the major semi diurnal and diurnal components in the Sharm obhur.

3.2 Shallow Water Constituent and Sea-Level Change Study

The study shows well-defined prominent shallow water constituent Z_0 at the study area. The shallow water constituents Z_0 is related to mean sea level variation and can be used for sea level variation studies in the Red Sea. The studies conducted by Ahmad and Sultan [8], Sultan et al. [9, 10], and Maghrabi [11] show sea level displaying more variations during the summer in central Red sea. The wind stress component in the long-shore has a more significant role than cross-shore stress for sea level variations during the summer [12]. The study shows that among the constituents, Z_0 and MSf components exhibit more selective amplifications. The tidal amplification in the Sharm obhur

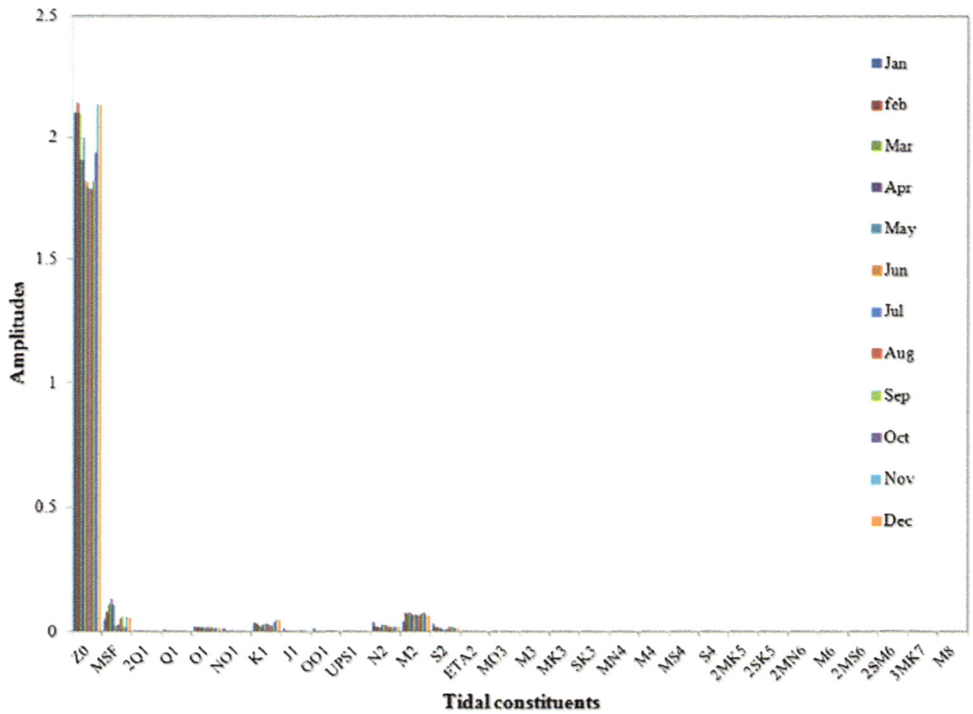

Fig. 1 Monthly distributions of tidal constituents at Sharm obhur during the year 2013

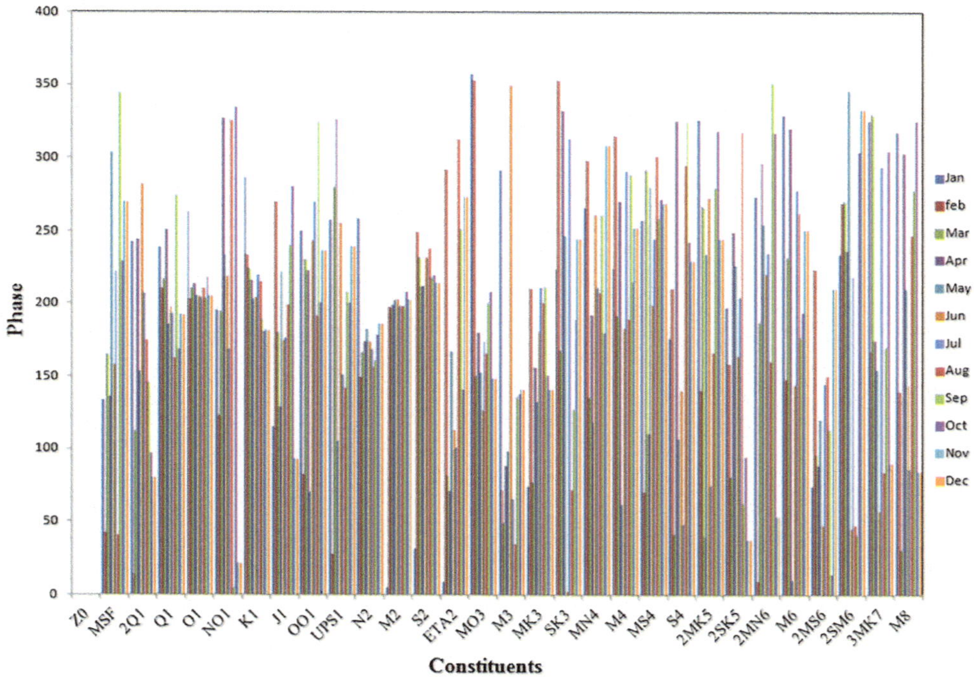

Fig. 2 Distributions of phases of constituents at Sharm obhur during the year 2013

mainly shaped by the shallow water constituent Z_0 and the same can be used for long-term sea level change study.

4 Conclusions

This paper discussed the tidal analysis and predictions at Sharm Obhur using the tidal analysis technique. The tidal constituents were derived using different techniques and the derivations of shallow water tidal constituents have more practical importance. The Z_0 and MSf are mainly shaping the tidal range in the Sharm obhur and the constituent Z_o can be used for the long-term sea level change study in the Red Sea. The derived tidal constituents have more significance for tidal predictions and hindcast in the region.

References

1. Madah, F., Mayerle, R., Bruss, G., Bento, J.: Characteristics of tides in the red sea region, a numerical model study. Open. J. Mar. Sci. **5**, 193–209 (2015). https://doi.org/10.4236/ojms.2015.52016
2. Saad, M.A.: Seasonal flucuation of mean sea level at Jizan, Red Sea. J. Coast. Res. **13**, 1166–1172 (1997)
3. Monismith, S.G., Genin, A.: Tides and sea level in the Gulf of Aqaba (Eilat). J. Geophys. Res. **109**, C04015 (2004)
4. Al-Barakati, Alaa, M.A.:Water exchange of Sharm Obhur, Jeddah, Red Sea. JKAU: Mar. Sci. **20**, 49–58 (2009 A.D. / 1430 A.H.) (2009a)
5. Behairy, A.K.A., Al-Kholy, A.A., Hashem, M.T., El-Sayed, K.H.: Preliminary study on the geology and fisheries of the coastal area between Jeddah and Yanbu. J. Fac. Mar. Sci. **2**, 1–47 (1983)
6. Basham, A.S., El-Shater, A.: Textural and mineralogical characteristics of the surficial sediments of Sharm Obhur, Red Sea coast of Saudi Arabia. JKAU: Mar. Sci. **5**, 51–71 (1994)
7. Woodworth, P.L., Blackman, D.L., Pugh, D.T., Vassie, J.M.: On the role of diurnal tides in contributing to asymmetries in tidal probability distribution functions in areas of predominantly semi-diurnal tide. Estuar. Coast. Shelf. Sci. **64**, 235–240 (2005)
8. Ahmad, F., Sultan, S.A.R.: Tidal and sea level changes at Jeddah, Red Sea. Pak. J. Mar. Sci. **2**(2), 77–84 (1993)
9. Sultan, S.A.R., Ahmad, F., El-Hassan, A.: Seasonal variations of the sea level in the central part of the Red Sea. Estuar. Coast. Shelf Sci **40**, 1–8 (1995b)
10. Sultan, S.A.R., Ahmad, F., Elghribi, N.M.: Sea level variability in the Red Sea. Oceanol. Acta **18**(2), 607–615 (1995a)
11. Maghrabi, S.O.: Variations of Mean Sea Level in the Red Sea. Master Thesis. King Abdulaziz University. 98 p (2003)
12. Zubier, M.: Sea level variations at Jeddah, Eastern Coast of the Red Sea. JKAU: Mar. Sci. **21**(2), 73–86 (2010 A.D. / 1431 A.H.) (2010). https://doi.org/10.4197/mar2.1-2.6

Sinkhole Morphologies from Photogrammetry and Distinct Element Modeling—An Example from the Dead Sea

Djamil Al-Halbouni, Eoghan P. Holohan, Hussam Alrshdan, Ali Sawarieh, and Torsten Dahm

Abstract

Sinkholes are a high impact natural hazard common in karstic rocks worldwide. They show different morphologies depending on formation mechanisms and involved geologic materials. We performed high-resolution photogrammetric analysis at the sinkhole site of Ghor Al-Haditha, Jordan, and Distinct Element Modeling to investigate such morphological differences for two characteristic sinkhole types common at the Dead Sea. By geomechanical modelling we proved that a low-strength cover material promotes wide and flat sinkholes as observed in the former Dead Sea lakebed. A high-strength cover material favours narrow and deep sinkholes as observed in the alluvial fans of the former Dead Sea shorelines. The combination of both field measurements and numerical simulation improved our conceptual understanding of sinkhole formation important for assessment of this natural hazard.

Keywords

Sinkholes • Dead Sea • Distinct element modeling • Natural hazard • Photogrammetry

1 Introduction

Sinkholes are a common natural hazard in karst areas worldwide [1]. In geologic materials rich in evaporite or carbonate minerals, voids are likely to form in the underground by chemical (i.e. dissolution) and physical removal of subsurface materials [1]. Dependent on the material types involved, sinkholes form either by sudden collapse of the overburden, or by continuous and relatively slow subsidence and sinking of the overburden into the voids or by a mixture of both [2, 3].

Over the last three decades, related to the lowering of the Dead Sea level, the numbers of sinkholes at the Dead Sea have been increasing [4]. A pronounced sinkhole area lies at the SE border of the Dead Sea, near the alluvial fan area of Ghor Al-Haditha (Fig. 1), with more than 1000 sinkholes recorded so far. This intense appearance of sinkholes has been related to the dissolution of evaporite minerals (halides, carbonates, and sulphates) in subsurface deposits of the Dead Sea [4]. Physical erosion by ground water of the weak mud-flat and alluvial fan sediments has also been invoked [2, 5].

In this paper, we outline the main sinkhole morphologies revealed by a photogrammetric survey conducted in 2014. We present numerical models of sinkhole formation that relate differences in morphology and collapse style to variation in material strength.

2 Materials and Methods

2.1 Close-Range Photogrammetry

In 2014, a 16 MP camera fitted to a helium filled balloon was used to record images of the Ghor Al-Haditha sinkhole area from an average altitude of ca. 100 m above ground (Fig. 1). Over 60 ground control points were measured with a dual-frequency GPS recorder. Using the structure-from-motion technique in close-range photogrammetry [6] and

D. Al-Halbouni (✉) · T. Dahm
Helmholtz Centre, German Research Centre for Geosciences (GFZ), Telegrafenberg, 14473 Potsdam, Germany
e-mail: halbouni@gfz-potsdam.de

E. P. Holohan
UCD School of Earth Sciences, University College Dublin, Belfield, Dublin 4, Ireland

H. Alrshdan · A. Sawarieh
Ministry of Energy and Mineral Resources, Mahmoud Al Moussa Abaidat Street, Amman, 140027, Jordan

A. Kallel et al. (eds.), *Recent Advances in Geo-Environmental Engineering, Geomechanics and Geotechnics, and Geohazards*, Advances in Science, Technology & Innovation, https://doi.org/10.1007/978-3-030-01665-4_108

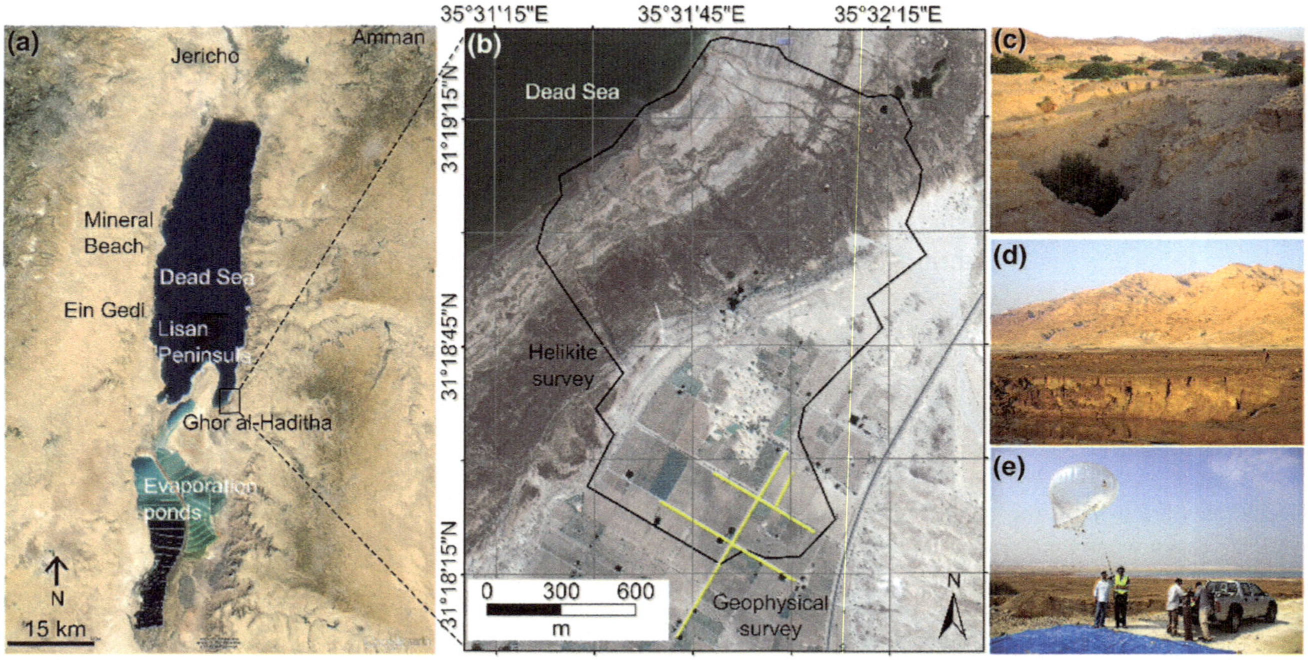

Fig. 1 Location of the sinkhole area of Ghor Al-Haditha in Jordan. **a** Digital globe satellite image from 2014 of the Dead Sea region. **b** Pleiades satellite image from 2015 of the part of the sinkhole area photogrammetrically surveyed in 2014. Yellow lines are a geophysical profiles conducted simultaneously in 2014 (cf. [5]). Photos show: **c** a large depression zone formed by cumulative and nested sinkhole collapses in the alluvial fan area; **d** a typical sinkhole in the mud-flat area, the former Dead Sea lake-bed; and **e** the equipment used for the helium balloon survey

GIS software, a high resolution and high accuracy Digital Surface Model (DSM) and an Orthophoto of the area have been assembled and analysed [2].

2.2 Distinct Element Modeling and Rock Parameter Calibration

The Distinct Element Method (DEM) [7, 8] was used for 2D mechanical simulations of sinkhole formation with the commercially available software PFC5. Here, disk shaped particles interact via a linear elastic force-displacement law and the Newton/Euler laws of motion. Beam-like elastic parallel bonds between particles break when their strength is exceeded, leading to fracturing within a bonded particle assembly. This intrinsic formation of cracks and fractures enables simulating even large strain deformations, the main advantage towards classical FEM methods [9].

The model dimensions are H $\times$ W = 400 $\times$ 400 m with uniformly distributed particles of a mean radius of 0.32 m. For simulation of sinkhole formation, a growing void area within the assembly simulated by constant particle removal was implemented. It produces realistic material failure and overburden collapse under gravitational loading [10]. Two different material assemblies were tested, low-strength limestone-carbonates (mud) and high strength sandy-gravel sediments (alluvium) with a mud interlayer. The micro-properties were translated by simulated compression and tension tests on samples of the materials into classical bulk rock macro-properties (Table 1, [10]).

3 Results

3.1 Photogrammetric Analysis

Figure 2 shows representative topographic cross-sections for sinkholes, sinkhole clusters and streams in the alluvial fan and the mud-flat. The sinkhole depth to diameter (De/Di) ratios are significantly different in each material type, being lower in the mud-flat clay/silt (De/Di $\sim$ 0.13) and higher in the alluvial sand/gravel (De/Di $\sim$ 0.39).

3.2 Distinct Element Models

Figure 3 shows modelled void space growth stages for the two investigated materials: low strength "mud" and high strength "alluvium" with a mud interlayer, cf. Table 1. The weaker mud-flat end-member shows significant early surface subsidence before continuously sagging to form a sinkhole of De/Di $\sim$ 0.21. The stronger alluvium end-member shows

Table 1 DEM Micro- and Macroproperties derived by simulated rock tests based on a Mohr-Coulomb failure criterion

Property/material	Mud	Alluvium
Youngs' Modulus E [MPa]	100	200
Parallel bond tensile strength Pb_t [MPa]	0.1	0.5
Parallel bond cohesion Pb_c [MPa]	0.5	0.02
Parallel bond friction angle ϕ [°]	2.4	34
Density ρ [kg/m^3]	2715	2750
Particle packing porosity n	0.2	0.2
Bulk modulus E [MPa]	84	174
Bulk Poisson ratio ν	0.2	0.31
Bulk unconfined compressive strength UCS [MPa]	0.25	0.5
Bulk tensile strength T [MPa]	0.2	0.24
Bulk friction angle ϕ [°]	6	22
Bulk density ρ [kg/m^3]	2145	2200

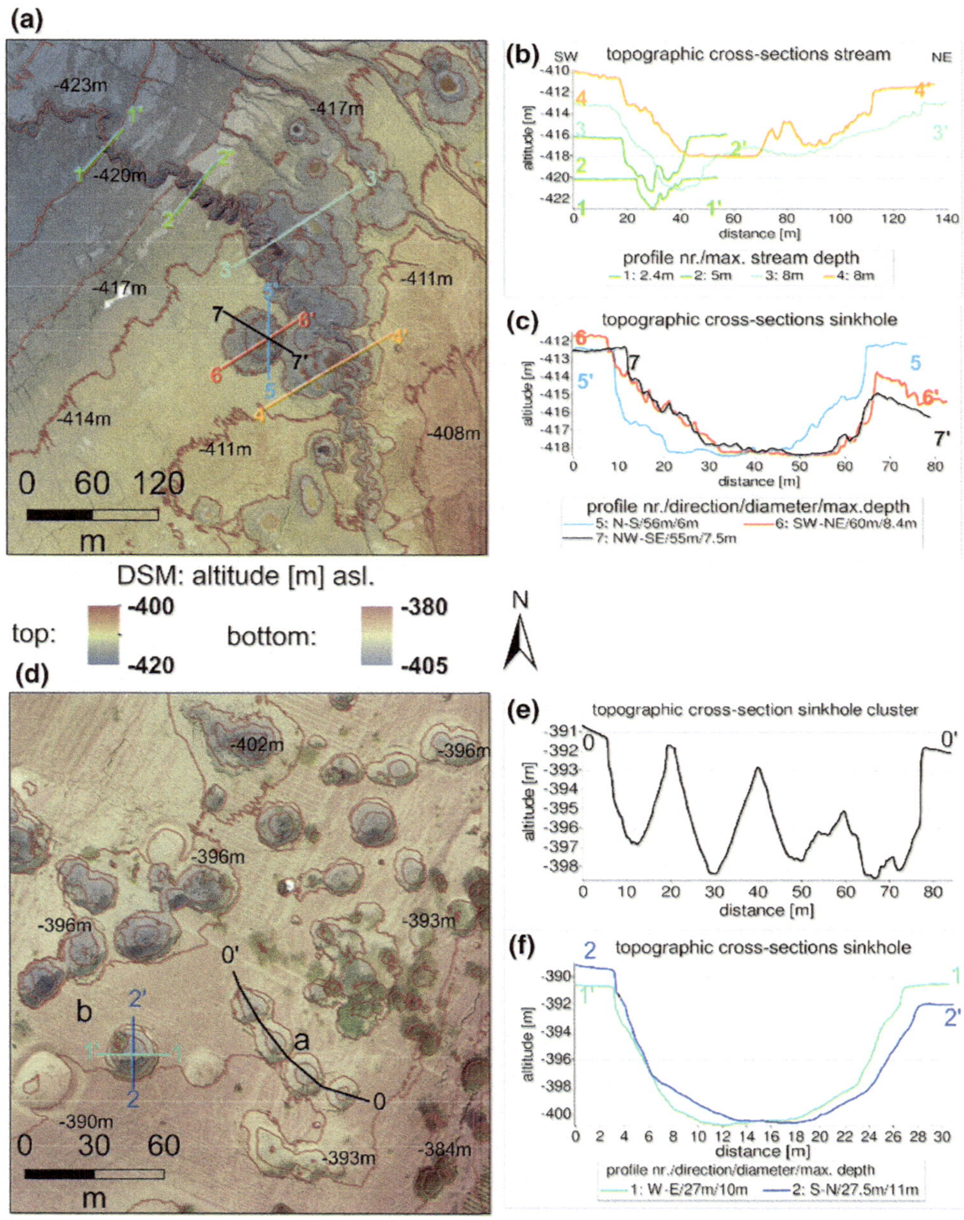

Fig. 2 Topography of sinkholes at Ghor Al-Haditha. **a** and **d** are orthophotos overlain on a hill shaded DSM of 2014 with 3 m topographic contours and marked topographic profiles for: **b** mud-flat stream channel, **c** and individual sinkhole, **e** alluvium aligned sinkhole cluster and **f** individual sinkhole

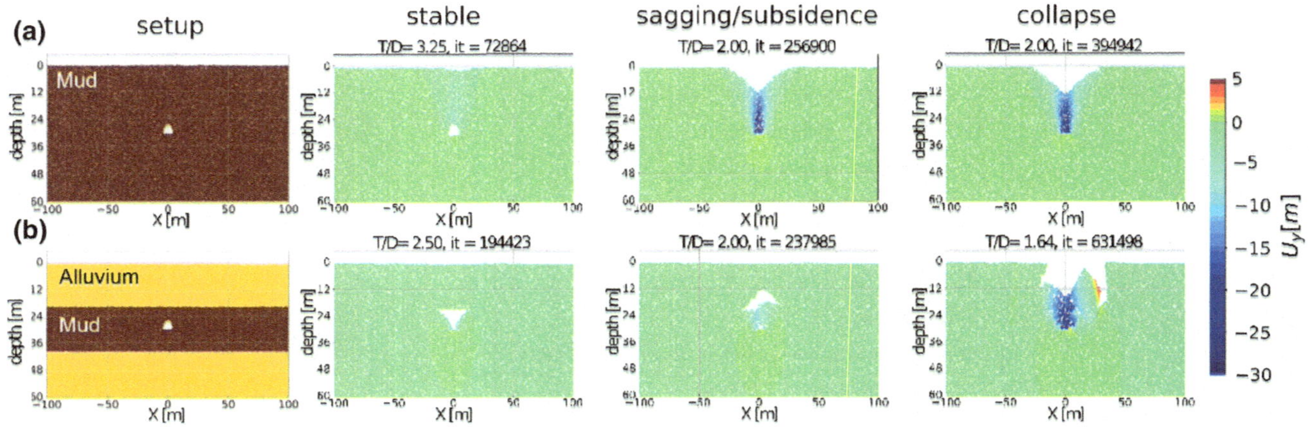

Fig. 3 End-members of DEM sinkhole development for setups **a** ("mud") and **b** ("alluvium"). Colour scale shows vertical displacement (U_y). T/D is the ratio of overburden thickness to void diameter, it the iteration number (~time)

an initially stable void space with little surface subsidence before suddenly collapsing to form a sinkhole of De/Di ~ 0.4.

4 Discussion

From the photogrammetric analysis, we observed a clear difference in the morphology of sinkholes formed in mud or in alluvial overburden (cf. [2]). From DEM geomechanical modelling, similar variation in morphological characteristics can be found, and the collapse process of the sinkholes is also different in such materials. A brittle failure occurs in the alluvial sediments, while a rather brittle-ductile failure occurs in the mud-flat sediments. We can therefore deduce that the mechanical strength of the cavity overburden strongly controls the sinkhole morphology and collapse process. This behaviour enables statements about the sinkhole hazard estimation. A substantial but continuous subsidence can happen in the mud-flat sinkhole end-member, merging into a sagging type style. A rather slight subsidence may be detectable by e.g. remote sensing techniques before a sudden and most hazardous collapse in alluvial fan sinkholes. Also, recognizable cracks may appear at the subsurface and strongly disturbed subsurface layers may be detectable by shallow geophysical methods [5].

5 Conclusions

Geomechanical modelling by the Distinct Element method successfully reproduced sinkhole morphologies measured by a photogrammetric survey at Ghor Al-Haditha in Jordan. The mechanical strength of the materials involved in the collapse process is a key parameter in sinkhole morphology. Low-strength material promotes wide and flat sinkholes as observed in the former Dead Sea lakebed. A high-strength material favours narrow and deep sinkholes as observed in the alluvial fans of the former Dead Sea shorelines. The analysis of numerical models provide important information about the sinkhole phenomena, complementary to field measurements, and are important for the assessment of this hazard.

References

1. Waltham, A.C., Fookes, P.G.: Engineering classification of karst ground conditions. Q. J. Eng. Geol. Hydrogeol. **3**(1), 1–20 (2005). https://doi.org/10.1144/1470-9236/2002-33
2. Al-Halbouni, D., Holohan, L., Saberi, Eoghan P., Alrshdan, H., Sawarieh, A., Closson, D., Walter, T.R., Dahm, T.: Sinkholes, subsidence and subrosion on the eastern shore of the Dead Sea as revealed by a close-range photogrammetric survey. Geomorphology **285**, 305–324 (2017). https://doi.org/10.1016/j.geomorph.2017.02.006
3. Gutierréz, F., Parise, M., De Waele, J., Jourde, H.: A review on natural and human-induced geohazards and impacts in karst. Earth-Sci. Rev. **138**, 61–88 (2014). https://doi.org/10.1016/j.earscirev.2014.08.002
4. Abelson, M., Yechieli, Y., Baer, G., Lapid, G., Behar, N., Calvo, R., Rosensaft, M.: Natural versus human control on subsurface salt dissolution and development of thousands of sinkholes along the Dead Sea coast. J. Geophys. Res. Earth Surf. **122**(6), 1262–1277 (2017). https://doi.org/10.1002/2017JF004219
5. Polom, U., Alrshdan, H., Al-Halbouni, D., Dahm, T., Sawarieh, A., Atallah, M. Y., Krawczyk, C. M.: Shear wave reflection seismics yields subsurface dissolution and subrosion patterns: application to the Ghor Al-Haditha sinkhole site, Dead Sea, Jordan. Solid Earth, **9** (Environmental changes and hazards in the Dead Sea region), 1079–1098 (2018). https://doi.org/10.5194/se-9-1079-2018.
6. Luhmann, T., Robson, S., Kyle, S., Boehm, J.: Close-range Photogrammetry and 3D Imaging, 2nd edn., Walter De Gruyter (2014)
7. Potyondy, D.O., Cundall, P.A.: A bonded-particle model for rock. Int. J. Rock Mech. Min. Sci. **41**(8), 1329–1364 (2004). https://doi.org/10.1016/j.ijrmms.2004.09.011

8. Schöpfer, M.P.J., Childs, C.: Walsh, W: Two-dimensional distinct element modeling of the structure and growth of normal faults in multilayer sequences: 1. Model calibration, boundary conditions, and selected results. J. Geophys. Res. **112**(B10), B10401 (2007). https://doi.org/10.1029/2006jb004902
9. Jing, L., Stephansson, O.: Fundamentals of Discrete Element Methods for Rock Engineering, Elsevier, (2007)
10. Al-Halbouni, D., Holohan, E. P., Taheri, A., Schöpfer, M. P. J., Emam, S., Dahm, T.: Geomechanical modelling of sinkhole development using distinct elements: Model verification for a single void space and application to the Dead Sea area. Solid Earth Discuss (2018), in review. https://doi.org/10.5194/se-2018-62.

An Assessment of Sensitivity to Desertification in Western High Atlas of Morocco: An Application to Ain Asmama Site

Adnane Labbaci, Belkacem Kabbachi, Abdelkarim Ezaidi, and James Thorne

Abstract

In Morocco, desertification affects large areas, and has become more pronounced as conditions become more arid with increasingly a long drought, and on poor soils, highly vulnerable to erosion. In addition, the precarious living conditions of rural populations grows to overuse natural resources to meet their growing needs, which amplifies further environmental degradation [1]. Concerning this big Site of Biological and Ecological Interest (SBEI) (West of Morocco), its bioecological qualities especially its very specific and original flora has attracted naturalists over a long time. This region seems to be still relatively preserved despite the erosion and anthropogenic advanced threats. This article presented a study undertaken to map the sensitivity of Ain Asmama's SBEI to desertification via MEDALUS (Mediterranean Desertification and Land Use) method [2]: This approach is based on a classification system developed in the parametric algorithm. It is worth noting that the sensitivity of soils to desertification depends on the quality of soil [3], climate, vegetation and management system. More than 90% of the territory has a medium climate quality. More than 63% are arid. The vegetal cover has a high quality to protect against desertification. 77% of the site's area represents a low fire risk. More than 14% of soils have a soft parental material. More than 70% have a high slope degree. In conclusion, the largest part of Ain Asmama is very vulnerable to desertification.

Keywords

Morocco • Desertification • MEDALUS • Vulnerability • SBEI of Ain Asmama

1 Introduction

Desertification is the result of land degradation in arid, semi-arid and dry sub-humid areas. It is mainly due to human activities and climatic variations [4]. The methodology adopted in this study is based on the MEDALUS approach. This approach was adapted by the OSS (Sahara and Sahel Observatory) within the framework of the DISMED (Desertification Information System to support National Action Programs in the Mediterranean) project [5] and used in the countries of the North Africa and Egypt for mapping sensitivity to desertification. To contribute to this objective, this study was undertaken to map the sensitivity of Ain Asmama's SBEI to desertification via four layers: climate, soil, vegetation and management quality. Each of them was determined by a specific methodology. The objective was to build a GIS model to define the critical area of desertification impact. To develop each factor will help to understand all factors' impact and then would help in combating the impacts of those drivers

2 Materials and Methods

Between 9.03° and 9.33° longitude East and between 30.74° and 31° North, the SBEI of Ain Asmama occupies an area of approximately 23,564 ha in Northwestern Morocco. It is an important site of biological values, landscape, aesthetics and culture. Its rugged environments, reliefs and powerful varied vegetation, dense, continuous and well preserved, offer a wide variety of habitats and ecosystems important for maintaining wildlife communities against desertification. Morocco has signed the CCD (Convention Combating

A. Labbaci (✉)
High Commission on Water Forests and Combating Desertification, Quartier Administratif, Challah Rabat, Morocco
e-mail: labbaciadnane@gmail.com

B. Kabbachi · A. Ezaidi
Faculty of Sciences of Ibn Zohr, Geology Department, Geomaterials Laboratory, Ibn Zohr University, Agadir, Morocco

J. Thorne
Environmental Science and Policy Department, University of California Davis, Davis, CA, USA

A. Kallel et al. (eds.), *Recent Advances in Geo-Environmental Engineering, Geomechanics and Geotechnics, and Geohazards*, Advances in Science, Technology & Innovation, https://doi.org/10.1007/978-3-030-01665-4_109

Desertification) and developed a national program to combat desertification. The sensitivity of Ain Asmama area to desertification was determined by mapping the quality of soil, climate, vegetation and management system.

2.1 Soil Quality

The soil quality was determined using the map of parental materials, texture, depth and slope. To make those maps, it was necessary to synthesize most part of the geological and pedology studies undertaken all over the SBEI area. Each soil was affected by an index calculated using the formula:

$$SQI = (PMi * Di * Ti * Si)^{1/4}$$

where:

PMi Parent Material index,
Di Depth index,
Ti Texture index,
Si Slope index

2.2 Vegetal Quality

Vegetation is at the same time the expression and the index of the natural balance of the ecosystems. In fact, the vegetal cover is the first component enduring anthropogenic impacts and its state is related to the process of natural degradation: desertification. It shows indications related to: Resistance to drought, Fire risk and ability to protect, Protection to erosion; and Plant cover. The vegetal quality index is a combination of four other indexes: Fire risk FRi, Resistance to drought RDi, Protection against erosion PEi, Plant cover PCi:

$$VQI = (FRi * RDi * PEi * PCi)^{1/4}$$

2.3 Climate Quality

The climate can be considered as the major cause of land degradation. In fact, soil erosion can be due to heavy and concentrated rainfall in space and time, affecting primarily the vegetal cover.

The map of climate quality was produced on the basis of the aridity index, rainfall and aspect. The data (P mm, Aridity, Aspect) of eight stations were used. The climate quality index is a combination of the three cited indexes, aridity, rainfall and aspect:

$$CQI = (Ai * Ri * Asi)^{1/3}$$

2.4 Management Quality

To map the management quality index (MQI), we have used two indexes: The management quality index (MQI) was assessed as the product of land use intensity and poverty index using the following algorithm:

$$MQI = (LUIi * Pi)^{1/2}$$

3 Results

3.1 Soil Quality Index (SQI)

According to the used criteria, about 1% of the territory of Ain Asmama's SBEI is slightly sensitive to desertification. More than 36% has high sensitivity (index > 1.45). This means that about all soils of the Ain Asmama's SBEI soils are vulnerable to desertification, even the other factors, vegetal cover, climate and management, are also favorable.

3.2 Vegetal Quality Index (VQI)

Relying on the obtained results, the zones with low and medium quality represent respectively 22 and 78%.

3.3 Climate Quality Index (CQI)

The climate quality index is a combination of the three cited indexes, aridity, rainfall and aspect. Contrary to two previously calculated indices, we notice that zones with medium climate quality are dominant (91%) followed by the range with good quality zone (8.14%), while zones with low climate quality represent only 1.14%. In other words, the SBEI of Ain Asmama has an arid to semi-arid climate convenient to the desertification.

3.4 Management Quality Index (MQI)

The management quality index (MQI) was assessed as the product of land use intensity and poverty index. The MQI index so established shows that the majority of the SBEI (78%) present a medium quality of management. On the other hand, the low management range is represented with 22% of site area in particular because of the intensification of the agricultural practices and the pressure of the population on natural resources.

3.5 Sensitivity Desertification Index (SDI)

The Sensitivity Desertification Index (SDI) is determined by the intersection of the multiplicative maps according to the following equation:

$$SDI = (SQI * VQI * CQI * MQI)^{1/4}$$

In addition to this phenomenon, especially in mountainous areas, erosion is still an important factor of degradation, particularly in areas where the vegetation has undergone significant degradation due to the combined impact of a harsh climate and samples as well as for grazing for the various needs of the rural population, including fuel wood and firewood.

4 Discussion

Desertification is one of the most serious environmental and socioeconomic problems in Morocco. Monitoring and assessing desertification has been a key and an extremely important task, for combating desertification in threatened areas to evaluate the impacts and provide the best indicators for decision makers to improve projects impacts in terms of sustainability. Using GIS and Remote Sensing has become an important data source and technology to have a big picture of the phenomenon. In this article, we tried to use several keys to understand the desertification issue and its impacts. However, some problems were met including: (1) the reference scale of each parameter usually related to the data and its availability, (2) impact of each factor, we had to use mathematical models to evaluate the weight of each parameter in the equation, (3) underemphasized temporal and spatial scales of desertification and the scale effect, as well as (4) those extracting and assessing methods of real desertification information are in need for improvement.

5 Conclusion

The methodology used to develop the sensitivity desertification map is based on the collection of a considerable amount of textual and graphical information. A coherence scale through a spatial geo-referenced allowed to use the data within a GIS. The interpretation of satellite imagery and ground truth collection helped complete the available spatial and thematic data, especially the cover land. The result of the sensitivity map of desertification has shown two main areas classified into sensitive and very sensitive areas to Desertification.

References

1. Ghanam, M.: La désertification au Maroc-Quelle stratégie de lutte. In: 2nd FIG Regional Conference Marrakech, Morocco, pp. 2–5, December 2003
2. Kosmas, C., Kirkby, M.J., Geeson, N.: The Medalus Project: Mediterranean Desertification and Land Use: Manual on Key Indicators of Desertification and Mapping Environmentally Sensitive Areas to Desertification; Directorate-General Science, Research and Development (1999)
3. Badraoui, M., Stitou, M.: Status of soil survey and soil information system in Morocco. Options Mediterr. Ser. B **34**, 193–204 (2001)
4. Johnson, P.-M., Mayrand, K., Paquin, M.: Governing Global Desertification: Linking Environmental Degradation, Poverty and Participation. Ashgate Publishing, Ltd. (2006)
5. Brandt, J., Geeson, N.: Desertification information system to support National Action Programmes in the Mediterranean (DISMED). DIS4ME, Desertification Indicator System for Mediterranean Europe. Eur. Environ. Agency (2005)

Erosion Sensitivity Mapping Using GIS-Based Multicriteria Analysis – Case Study of the Semiarid Macta Watershed, North-West of Algeria

Mohammed Amine Hamadouche, Fatima Zohra Daikh, Mohammed Chrair, Djamel Anteur, Youcef Fekir, and Miloud Driss

Abstract

The exponential rise in the natural risk related to the soil erosion and important impacts on the human activities threatening the biological richness of different regions in Macta watershed located at the northwest of Algeria. This serious and complex situation has been an environmental concern in Algeria. Soil erosion is mainly due to a combination of factors such as climate (rainfall), topography, vegetation cover, type of soil and land use (agricultural activities). The study aimed to develop a methodological approach, based on the combined use of geographic information system (GIS), remote sensing and multicriteria analysis method (MCA), dedicated to the soil erosion sensitivity mapping. The different factors are involved in a geographical database as layers in order to make it easier to implement the decision aid dataset such as alternatives and criteria. The obtained results show that the most threatened areas by the soil erosion phenomena are situated, in general, in the south and the east. Areas of high vulnerability correspond to some sub-watershed with different degrees, i.e., Mezoua, Taria, Mekerra, and Fergoug.

M. A. Hamadouche (✉) · M. Chrair · Y. Fekir
Laboratory of Research on Geomatic, Ecology and Environment, University Mustapha Stambouli of Mascara, Mascara, Algeria
e-mail: m.hamadouche@univ-mascara.dz

F. Z. Daikh
University Mustapha Stambouli of Mascara, Mascara, Algeria

D. Anteur
Laboratory of Water and Environment, University of Saida, Saida, Algeria

M. Driss
Laboratory of Sciences and Water Techniques, University Mustapha Stambouli of Mascara, Mascara, Algeria
http://www.univ-mascara.dz/fst/index.php/fr/recherchescientifique/laboratoires-de-recherche/2-non-categorise/54-labolste

Keywords

Soil erosion • GIS • Multicriteria analysis • Semiarid • Macta watershed • Algeria

1 Introduction

Algeria is among the most erodible regions in the world [1]. Rain affects 28% of the land in northern Algeria and go in the form of runoff, which is the main factor of water erosion [2, 3]. However, the Tellian steep mountains are the most affected with a rate of 45% representing 12 million hectares. The Macta watershed represents a specific ecosystem on the western coast of Algeria by its geographic situation and the richness of its biodiversity. However, it is highly threatened by the different manmade activities. Since 2001, the Macta watershed has been classified as wetland to be protected in the frame of the Ramsar convention [4].

The integration of Multicriteria Decision Analysis approaches (MCDA) in a Geographical Information System (GIS) provides a powerful spatial decision support system, which offers the opportunity to efficiently map the sensitivity of soil to erosion risks by taking into account its different factors. Because of their spatial aggregation functions, MCDA methods can facilitate decision making in situations where multiple decision-makers preferences are to be considered and multiple conflicting criteria (risk factors) are to be explicitly evaluated [5].

2 Materials and Methods

2.1 Study Area

The study area is located in Mascara state of Algeria (see Fig. 1). Particularly, Macta watershed located in north-western Algeria (Mascara state) between 34° 18′ N to 35° 54′ N Latitude (length approximately 166.76 km) and

A. Kallel et al. (eds.), *Recent Advances in Geo-Environmental Engineering, Geomechanics and Geotechnics, and Geohazards*, Advances in Science, Technology & Innovation, https://doi.org/10.1007/978-3-030-01665-4_110

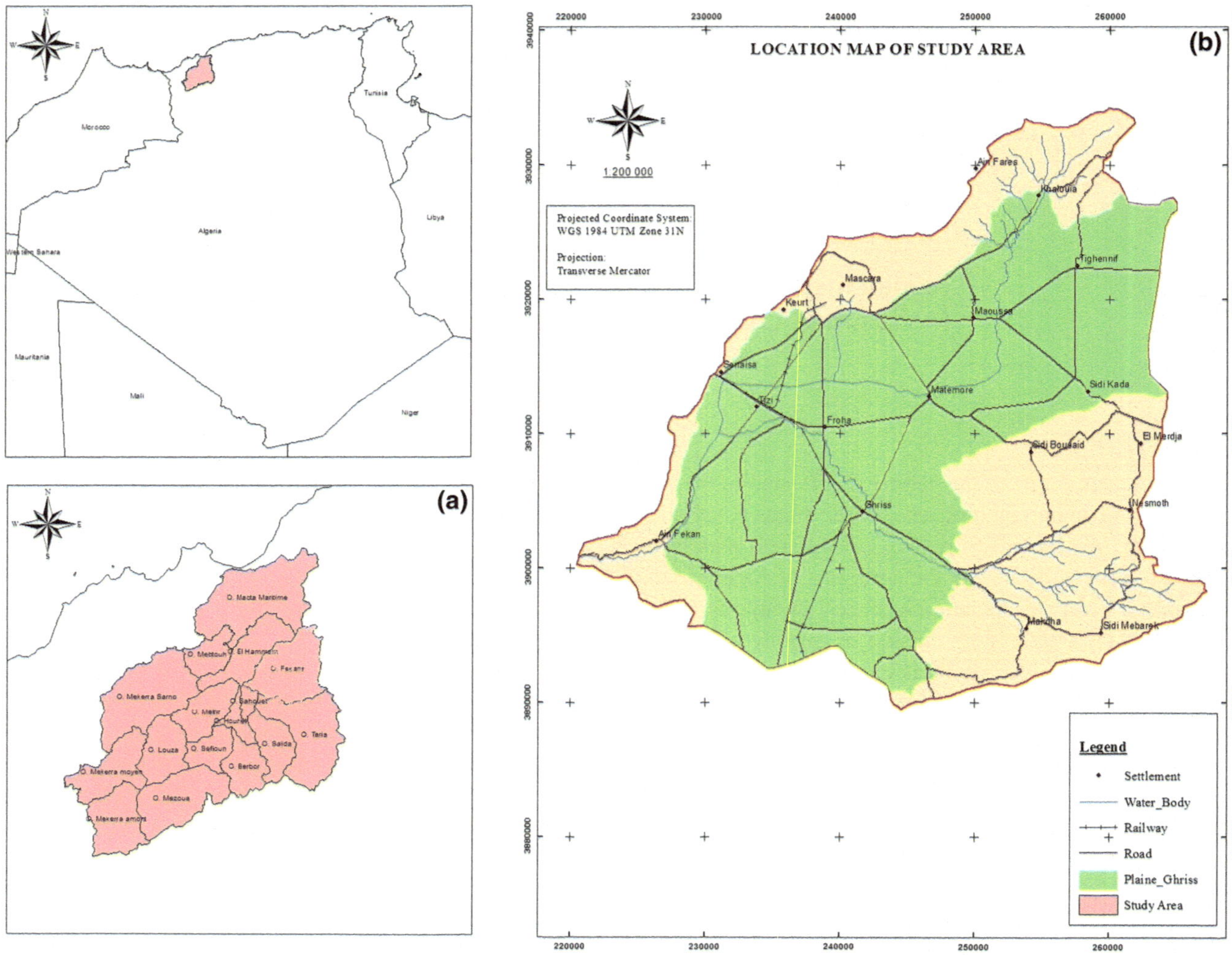

Fig. 1 Presentation of the study area

1° 12′ W to 0° 34′ E Longitude (width approximately 159.59 km) thereby covering a total geographical area of approximately 260,000 hectares. This study area is characterized by its important agricultural production. It also includes the plain of Ghriss town, which covers an agricultural area of more than 72,000 hectares.

2.2 GIS-Based Evaluation and Soil Erosion Risks

The capabilities of GIS in handling spatial aspects to analyze and map the eventual risk of soil erosion have boosted its use in the criteria-based evaluation for the determination of exposed areas to soil erosion phenomena. This is because most of the criteria (risk factors that promote soil erosion) are spatial data. GIS can be used for areas measurement, testing different theoretical assumptions and developing useful models in order to find new resources based on statistical analysis of the relationship with environmental and cultural data. The integration of remote sensing data within a GIS has a significant potential in regional applications for the management of a big set of data. Various researchers worldwide use these tools today. Some of these applications include the location of new features such as archaeological sites, road network, automated extraction of hydrology network, slope and aspect maps, land use and land cover.

2.3 GIS-Based Evaluation and Multicriteria Analysis

Spatial decisions problems are referred to as heterogeneous systems, which interact with many different factors. Controlling the complexity of these problems requires the use of methods, techniques and powerful analysis tools, which must not only manage but also analyze different spatial data [6]. GIS is often recognized as a spatial decision support system, but it lacks the mechanisms of integrating the decision makers' preferences to make a choice in a context of objective evaluation and conflicting criteria [7]. These restrictions, in terms of decision-maker preferences modeling like the different thresholds of preference and selection procedures mean that the scope of GIS as a tool for spatial decision support is currently limited. Multicriteria decision aid (MCDA) is suitable for structuring decision problems, designing, evaluating, and prioritized alternative decisions. It is also appropriate to the decision-making processes related to the choice situation where points of views are often contradictory. Therefore, GIS-based Multicriteria analysis method constitutes a powerful way to develop spatial decision support system.

3 Results

The methodology developed in this study uses qualitative rules, assessments, and a hierarchy of parameters in relationship with water erosion: land cover, land use, topography, climate conditions, and type of soil. The set of data were integrated in a GIS as layers (vector and raster). All preferences of the decision process were implemented on the GIS as layers and attributes tables according to their types.

The combination of these maps following the multicriteria decision process, has allowed producing of thematic map which presents the different categories land fragility

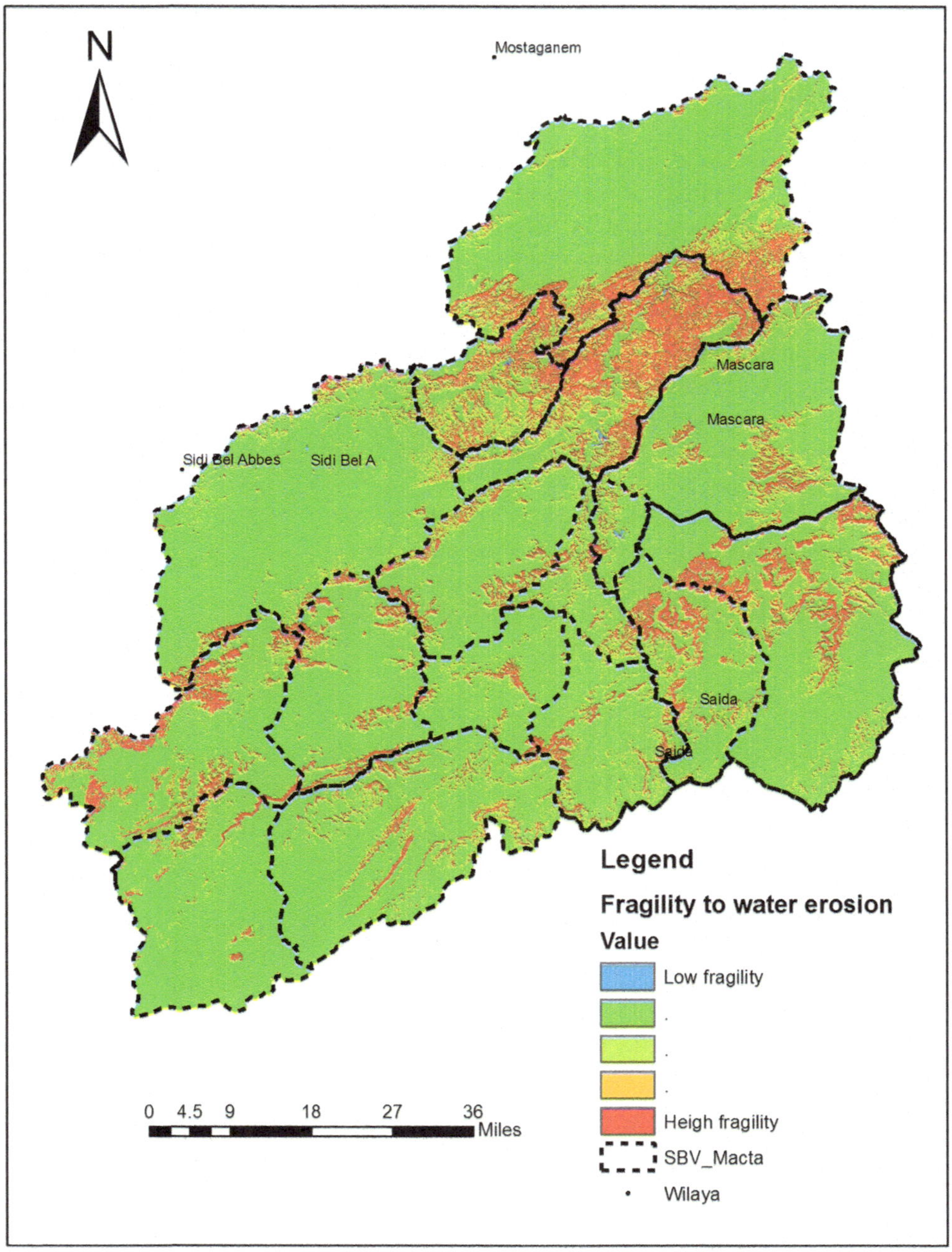

Fig. 2 Fragility to water erosion map

(see Fig. 2). It includes four classes: very fragile, fragile, moderately fragile and little fragile. Then an important area of Macta watershed is exposed to the erosion risk.

4 Discussion

The obtained results show that dominant classes of fragility to erosion correspond to low and middle levels. These areas correspond to areas of low to medium slopes. Consequently, we have to retain that the sensitivity to water erosion of the study area is strongly in relationship with the degree of slope, the vegetation cover density and the type of soil, in addition to the human action caused by agricultural practices which have already impacted the fragility of these lands.

5 Conclusion

Water erosion presents a great challenge to preserve the environment and the surface water resources. Mapping the vulnerable area to erosion, enables reducing and limiting the consequences of this phenomenon. In this paper, we have contributed to map the water erosion fragility of a large semi-arid basin of Macta in the northwest of Algeria. In this study, multi-criteria methods integrated into GIS tools were shown to play an important role in the detection of high vulnerability areas to water erosion. Indeed, this allows the determination of causes in favor of increasing the water erosion risk for specific areas.

The analysis of the maps obtained under GIS environment for different parameters shows that the areas with a high vulnerability to erosion are located in the north east and some are in the south of the watershed. Mapping the risk of water erosion constitutes an efficient and fast way to evaluate and combat soil degradation. The final product of this work is a decision support map for forest managers presenting the priority intervention zones in Macta watershed.

References

1. Probst, J.L., Suchet, P.A.: Fluvial suspended sediment transport and mechanical erosion in the Maghreb (North Africa). J. Sci. Hydro. **37**(6), 621–636 (1992)
2. Banasik, K., Górski, D., Popek, Z., Hejduk, I.: Estimating the annual sediment yield of a small agricultural catchment in central Poland. In: Erosion and Sediment Yields in the Changing Environment. Proceedings of the Chengdu Symposium. IAHS Publ. 356, pp. 267–275 Oct. 2012
3. Benchetrit, M.: L'érosion actuelle et ses conséquences sur l'aménagement de l'Algérie, p. 216. PUF, Paris (1972)
4. Ghodbani, T., Amokrane, K.: La zone humide de la Macta: un espace à protéger sur le littoral ouest de l'Algérie. Physio-Géo [En ligne], vol. 7, mis en ligne le 02 mars 2013, consulté le 01 mai 2018. (2013) URL: http://journals.openedition.org/physio-geo/3228; https://doi.org/10.4000/physio-geo.3228
5. Hamadouche, M.A., Mederbal, K., Larid, L., Regagba, Z., Fekir, Y., Anteur, D.: "GIS-based multicriteria analysis: an approach to select priority areas for preservation in the Ahaggar National Park. Algeria. Arab. J. Geosci. **7**, 419 (2014)
6. Hamadouche, M.A., Mederbal, K., Khaldi, A., Fekir, Y., Anteur, D., Driss, M., Oulha, R.: Gomatic tools and multicriteria analysis for managing and conserving the biodiversity of the national park of ahaggar (algeria). Oral presentation at the 1st First International Geomatics Symposium in Saudi Arabia, Jeddah (2011)
7. Molines, N., Chevalier, JJ,: Système d'information géographique et analyse multicritère – une association novatrice au service du processus d'évaluation des grandes infrastructures linéaires. Centre de Recherche sur l'environne-ment et l'aménagement (CRENAM.) Saint-Etienne le Centre de Recherche en Géomatique Université Laval, pp. 195–211 (2002)

Experimental Investigation of Several Different Types of Soil Erosion Protection Systems

Hossein Moayedi, Ramli Nazir, Loke Kok Foong, Mansour Mosallanezhad, and Biswajeet Pradhan

Abstract

Most unprotected slopes face stability problems due to erosion. Generally, an unstable slope issue arises when erosion begins on its surface. Several erosion protection methods provide a solution to protect the slope surface by dividing a large slope area into many small cells, hence increasing the stiffness of the soil surface. In this research, a total of 964 tests were performed on unconfined slopes and slopes with three types of confinement systems varying in three different sizes. The experiments were conducted based on various rainfall intensities, rainfall durations, and slope angles. Furthermore, the experimental results for the unconfined and confined slopes were compared to indicate the effectiveness of the soil confinement system. The confined slope showed significantly lower soil loss compared to the unconfined slope. Additionally, the results revealed that the triangular type was the most effective confinement system, as the lowest soil loss mass was recorded. In general, the introduced erosion protection system shows a promising way of preventing slope failure due to erosion at an early stage.

Keywords

Soil protection • Erosion • Pollution prevention • Sediments • Sustainability

H. Moayedi (✉)
Centre of Tropical Geoengineering (Geotropik), School of Civil Engineering, Faculty of Engineering, Universiti Teknologi Malaysia, Johor, Malaysia
e-mail: hossein.moayedi@gmail.com

R. Nazir · L. K. Foong
Faculty of Civil Engineering, Centre of Tropical Geoengineering, Universiti Teknologi Malaysia, Johor Bahru, Johor, Malaysia

M. Mosallanezhad
Department of Civil and Environmental Engineering, Shiraz University, Shiraz, Iran

B. Pradhan
Faculty of Engineering and IT, School of Systems, Management and Leadership, University of Technology Sydney, Ultimo, NSW, Australia

1 Introduction

Most slope failures are commonly related to the rainfall and seepage processes. The situation is a triggering factor that even leads to the slope failure. Various types of soil characteristics, slope formations, slope drainage systems, and types of constructions with future developments are considered as main causes of soil erosion [1, 2]. The protection materials are very well known in the soil erosion treatment systems. This has become more popular worldwide, as is evidenced by the increasing amount of research [3–5]. A cellular protection system is the first stage of prevention. The standard design of cellular protection systems could reduce costs in remedying slopes failure.

Slopes with different lengths, inclinations, and soil properties can be protected against erosion by choosing a suitable kind of soil confinement [6]. In terms of slope failure rehabilitation applications, this type of system is still in its infancy stage. Besides providing slope stability, an attractive feature of different types of soil confinement systems means they enable the visible surface to remain natural and green. The main objective of the present research was to study various types of fine soils and their vulnerability to soil erosion hazards. Furthermore, the influence of slope angle and rainfall characteristics, that could significantly increase the rate of soil erosion, were assessed on the soil erosion. Through a small scale laboratory model, different types of confinement systems were introduced to reduce the erosion rate and the results were compared to find the most applicable confinement type. Finally, the performance of the soil confinement systems for a selected site was evaluated in comparison to slopes without any confinement system.

A. Kallel et al. (eds.), *Recent Advances in Geo-Environmental Engineering, Geomechanics and Geotechnics, and Geohazards*, Advances in Science, Technology & Innovation, https://doi.org/10.1007/978-3-030-01665-4_111

2 Materials and Methods

Three clayey soil samples were collected from a particular site located at Johor Bahru, in Peninsular Malaysia. All of the used soils in the small scale modeling were obtained from a mixture of three collected soil samples. The depth of the confinements was constant at 10 mm throughout the experimental test. Small confinement models with different confinement sizes of 50, 100, and 150 mm width were used. A 10 mm thickness also was selected for the confinement according the scaling selected for the laboratory model.

The soils were placed inside the confinement while the cellular confinement installed in different slope angles of 30°, 40°, 50°, and 60°. As mentioned earlier, each type of confinement was made up of three sizes i.e., 50, 100, and 150 mm with a thickness of 10 mm. Series 1, 2, and 3 were then subdivided into 12 sub-series to explore the effect of various values of rainfall intensity (20–75 mm/h), slope angle (30°–60°), and rainfall duration. Table 1 summarizes the 964 laboratory tests performed in this study.

3 Results

The rate of soil erosion obtained in a non-confined system can summarized as follows: the soil loss mass increases with (i) steeper slopes, (ii) higher rainfall intensities, and (iii) prolonged rainfall duration. The steeper slope will induce a higher gravitational force when soil particles break up, while prolonged rainfall saturates the soil. The broken up soil particles at the top of the slope will move downwards and push the adjacent soil particles. This is an early process of surface erosion. The retrogressive process occurs via a transportation element such as water. Besides, a steeper slope reduces the internal frictional force between the soil particles. It is important to note that, the initial amount of water splashing on the surface of the soil will break up the soil particles. An increase in the amount of water flowing on the surface will transport the soil particles to the bottom of the slope. As the slope becomes steeper, the soil particles will be transported faster due to the effect of gravity, as mentioned before. The results illustrated the influence of slope angle on slope confinement systems of 50, 100, and 150 mm embedded. In all of the three soil confinement systems, the soil loss mass increased with an increment in the slope angle. The variation of soil loss mass also appeared to increase with increments of the slope angle with respect to the confinement size. For the confinement size of 50 mm and slope angles of 30°–60°, the soil loss mass increases, gradually. The maximum soil loss mass increased with increments of confinement size. The maximum soil losses for the slope angle of 60°, rainfall duration of 60 min, and rainfall intensity of 75 mm/h with the triangular confinement systems of 50, 100, and 150 mm were 40.9, 68.5, and 97.5 g, respectively. In particular, smaller soil loss masses were observed for smaller confinement sizes, lower slope angles, and lower rainfall durations.

Table 1 Summary of laboratory experiments for every test series

Confinement size (mm)	Slope angle	Type of confinement and number of tests			
	Degree	WCF	CF	TF	SF
Non-confinement	30	24	–	–	–
	40	24	–	–	–
	50	24	–	–	–
	60	24	–	–	–
50	30	–	24	24	24
	40	–	24	24	24
	50	–	24	24	24
	60	–	24	24	24
100	30	–	24	24	24
	40	–	24	24	24
	50	–	24	24	24
	60	–	24	24	24
150	30	–	24	24	24
	40	–	24	24	24
	50	–	24	24	24
	60	–	24	24	24
Sub-total		96	288	292	288

In the cases of a confined slope, the results of soil erosion for various slope angles, confinement systems, rainfall intensity, and durations (10–60 min) are as follows. For the same slope angle condition of 60°, a confinement size of 100 mm, a rainfall duration of 60 min, and rainfall intensity of 75 mm/h, the soil loss masses for the triangular (TF), circular (CF), and rectangular confinement systems (SQF) were 68.5, 81.9, and 87.35 g, respectively. As the triangular-shaped confinement systems (TF) had the lowest soil loss mass followed by the circular- and rectangular-shaped confinement systems.

The results unveiled the variation of soil loss mass versus slope angles of 30–60° for three types and sizes of confinements. The soil loss mass of the unconfined slope is much higher than that of the confined slopes. The soil loss mass under the conditions of a slope angle of 60°, rainfall intensity of 75 mm/h, and rainfall duration of 60 min for the unconfined slopes and the 50 mm-sized triangular, circular, and rectangular confinements were 961.7, 40.9, 46.2, and 54.8 g, respectively. For example in a slope angle equal to 60°, the rainfall intensity is 75 mm/h and 60 min rainfall duration time. Different confinement type specimens such as WCF-LAB, CF-50-LAB, CF-100-LAB, CF-150-LAB, SQF-50-LAB, SQF-100-LAB, SQF-150-LAB, TF-50-LAB, TF-100-LAB, and TF-150-LAB results in 961.7, 46.23, 81.89, 111.7, 54.84, 87.36, 125.8, 40.86, 68.49, and 97.51 g soil loss mass, respectively. Consequently, a huge difference between the unconfined and confined slopes was found. The difference increases largely with the increment of the slope angle. This finding confirmed that the confinement system is effective for the slope angles tested in this research.

4 Conclusion

The effectiveness of soil confinement systems was determined by performing small scale model tests in the laboratory. The tests on the slope confinements with sizes of 50, 100, and 150 mm showed that they had a similar trend to the unconfined slope. The confined and unconfined slopes were compared. It was observed that the slope angle (i.e., for slopes between 30 and 60 angle, the soil loss mass was differed up to 300%) and rainfall intensity (i.e., up to 9 times more when rainfall intensity changed from 20 to 75 mm/h) of a certain amount make a significant contribution toward increased soil loss. The soil confinement system largely reduced the rate of erosion, allowing the slope stability rehabilitation. In this regard, the slope can be permanently protected by the soil confinement system. It was found that a smaller confinement element will apply smaller forces to resist the detachment of the soil particles. The relationship of soil loss mass with slope angle and rainfall intensity proved that the triangular shape presented less soil loss than the circular and square shapes.

References

1. Haghighi, I., Chevalier, C., Duc, M., Guédon, S., Reiffsteck, P.: Improvement of hole erosion test and results on reference soils. J. Geotech. and Geoenviron. Eng. **139**, 330–339 (2013)
2. Pradhan, B., Chaudhari, A., Adinarayana, J., Buchroithner, M.F.: Soil erosion assessment and its correlation with landslide events using remote sensing data and GIS: a case study at Penang Island Malaysia. Environ. Monit. Assess. **184**, 715–727 (2012)
3. Blaikie, P: The Political Economy of Soil Erosion in Developing Countries. Routledge, (2016)
4. Nazir, R., Ghareh, S., Mosallanezhad, M., Moayedi, H.: The influence of rainfall intensity on soil loss mass from cellular confined slopes. Meas. **81**, 13–25 (2016)
5. Posthumus, H., Deeks, L.K., Rickson, R.J., Quinton, J.N.: Costs and benefits of erosion control measures in the UK. Soil Use Manage. **31**, 16–33 (2015)
6. Nazir R., Moayedi, H.: Soil mass loss reduction during rainfalls by reinforcing the slopes with the surficial confinement: World academy of science, engineering and technology. International Journal of Environmental, Ecological, Geomatics, Earth Science and Engineering. **8**(6), 331–334 (2014)

Land Use Sensitivity Map for Impact of Land Management on Extreme Flood Events at the Northeast Coast of Peninsular (Malaysia)

Nader Saadatkhah, Jafar Rahnamarad, Shattri Mansor, Zailani Khuzaimah, Arnis Asmat, Nor Aizam Adnan, and Siti Noradzah Adam

Abstract

In the recent decades, the alteration in the land cover has had severe hydrological effects disturbing the flood occurrences at the east coast of the peninsular Malaysia. The Kelantan River (KR) basin has been extremely influenced because of the quick variations in the usage of land between 1984 and 2013, due to deforestation and the conversion of the agricultural area. The purpose of the current study was to evaluate the influence of deforestation and the potential impact of agricultural area transformation on the hydrological functions such as surface water, lost water, and soil moisture on the KR basin. The Hydrological and Geotechnical Grid-based Regional Flood Model (HGGRF) was exploited to calculate the infiltration rate and consequent modifications in the discharge volume involving hourly rainfall occurrences over the last ten days of December 2014. The implementation of the HGGRF model revealed that the modifications in land cover brought about substantial alterations in hydrological response to surface water.

Keywords

Flood occurrence • Kelantan river basin • HGGRF model

N. Saadatkhah · J. Rahnamarad (✉)
Department of Geology, Zahedan Branch, Islamic Azad University, Zahedan, Iran
e-mail: Rahnamarad@iauzah.ac.ir

N. Saadatkhah
e-mail: N.saadatkhah@iauzah.ac.ir

S. Mansor · Z. Khuzaimah · S. N. Adam
Department of Civil Engineering, Faculty of Engineering, Geospatial Information Science Research Centre (GISRC), University Putra Malaysia, Serdang, Selangor, Malaysia
e-mail: shattri@upm.edu.my

Z. Khuzaimah
e-mail: zailani@upm.edu.my

S. N. Adam
e-mail: snadzah@upm.edu.my

A. Asmat
Faculty of Applied Sciences, School of Chemistry and Environmental Studies, Universiti Teknologi MARA (UiTM), Shah Alam, Selangor, Malaysia
e-mail: Rnis_annis@salam.uitm.edu.my

N. A. Adnan
Centre of Studies Surveying Science and Geomatics, Universiti Teknologi MARA (UiTM), Shah Alam, Selangor, Malaysia
e-mail: nor_aizam@salam.uitm.edu.my

1 Introduction

Alterations in land cover/land use (LULC) by deforestation, population upsurge and agricultural growth in the last few decades was investigated as the prevailing case of augmented flood occurrence by Fritsch [3], De Roo et al. [1], and Saadatkhah et al. [5]. The evaluation of the effect of changes in LULC on flood occurrences showed that several types of land cover and land use have momentous functions and effects on water resources, runoff generation and flood [1, 5, 6].

The KR basin, at the east coast of the peninsular Malaysia, is among those areas that are liable to floods. This area is subjected to the intense and prolonged monsoon rainfall under tropical climates. The land cover alterations led to higher runoff volume and peak flow by uncontrolled deforestation in KR basin throughout the preceding thirty years [2]. The target of current research was to reduce the risk of flooding in the river catchment areas, by assessing the effect of LULC changes (such as deforestation and urbanization) on increased rate of flood at the KR basin using hydrological regional modeling of rainfall-induced runoff occurrence that were used to calculate rate of infiltration, and succeeding variations in the discharge volume.

A. Kallel et al. (eds.), *Recent Advances in Geo-Environmental Engineering, Geomechanics and Geotechnics, and Geohazards*, Advances in Science, Technology & Innovation, https://doi.org/10.1007/978-3-030-01665-4_112

2 Materials and Methods

This paper studied the implementation of the HGGRF model as explained by Shattri et al. [6] accompanying evident techniques and tools for flood management considering land cover and its modifications in the KR basin. The complete research flow diagram is given in Fig. 1.

The Richards equation has been exploited, as also reported by Lawrence and Chase [4], in the enhanced model to explain unsaturated vertical flow in response to the infiltration/loss of water over the ground surface accompanying the rainfall interception loss approach and precipitation distribution data relating to experimental equation. Keeping in mind the hydrological and geotechnical characteristics of the examined area, the improved model was performed by exploiting both GIS framework system and Microsoft Excel®. The existing model was implemented relying on the three key constituents namely rainfall infiltration analysis, runoff simulation model and rainfall interception loss model.

3 Results and Discussion

According to Shattri et al. [6], the 69.96% of total Kelantan state area (956,139.9 ha) consists of forests and jungle reserves. Moreover, most of the agricultural area (395,156.8 ha) is situated in the upstream of Kelantan river basin, which covers an area of 26.32% of Kelantan area, 23.27% of the area (349,365.4 ha) is enclosed by palm oil and rubber, whereas only 1.60% (14,616.5 ha) consists of urban and development area. During 1984 and 2002, there was an increased conversion of forests into less canopy coverage areas that decelerated between 2002 and 2013.

Accordingly, the crucial runoff occurrences were principally dispersed over the unoccupied areas and the areas under progress (e.g., urban areas). Figure 2 showed the sensitive level of land use sensitivity map divided into equal classes based on the overall number of classes i.e., critically sensitive, very high, high, moderate, low and very low level. In this regard, forest transformation into less canopy coverage areas (e.g., rubber, diverse horticulture areas and palm oil) results in high to extremely high level of runoff occurrences in the examined area. Furthermore, the runoff and destructive debris flow were instigated by the blending of the transformation of forests into agricultural plantations in thin granitic soils, extreme rainfall, and destruction of natural forest cover (manmade).

4 Conclusion

The current study targeted the evaluation of the land cover effect on the runoff processing with the help of HGGRF modeling during rainfall occurrences and alleviation plan through the recent 30 years in KR basin. Through the outcomes of this study, the subsequent conclusions could be deduced:

- Land cover alterations instigated substantial dissimilarities in hydrological response to surface water by the implementation of HGGRF model. The rising runoff volume at KR basin mainly depends on the transformation of forests into agricultural land (rubber and mixed-agriculture), as well as on urbanization and deforestation.
- Deforestation and transformation of forests into agricultural land brought about upsurges in flood volume because of evapotranspiration capacity and lower interception loss. Conversely, urban development areas lead to a superior impervious ground surface resulting in surplus runoff volume owing to the lower rate of infiltration.

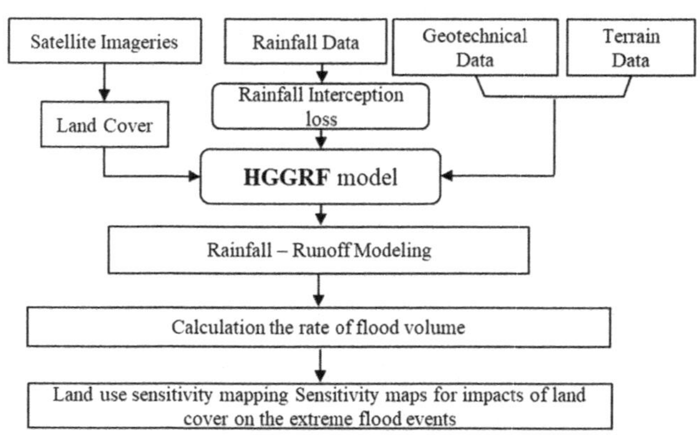

Fig. 1 Overall framework

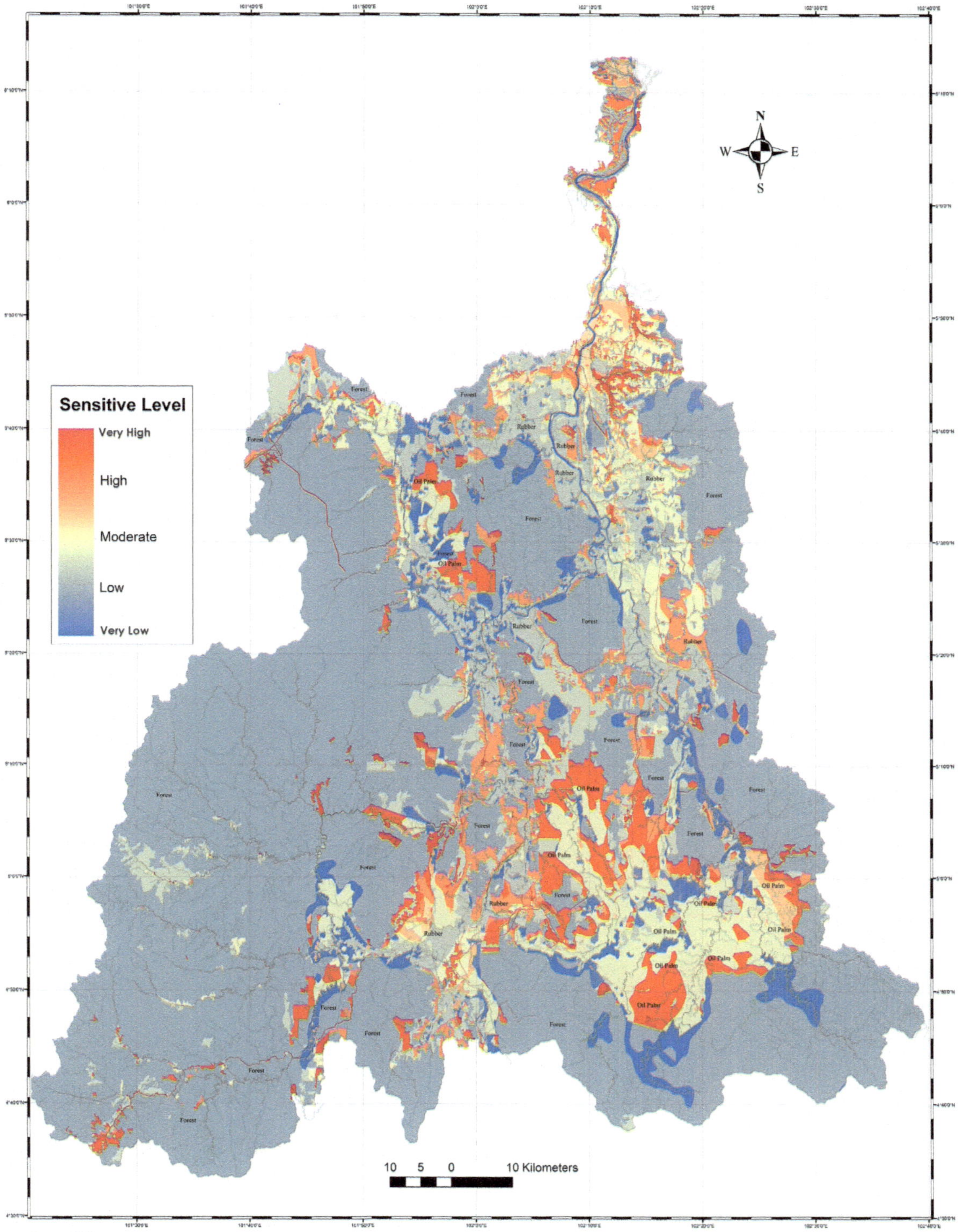

Fig. 2 Land use sensitivity map

- Forests act as an important factor in regulating water flow, consequently reducing the runoff volume in the examined area. On the other hand, an escalated flood magnitude was predicted in case forests are substituted by distinctive land cover varieties, for example, agricultural land and development area, due to the reduced loss/infiltration.

References

1. De Roo, A., Schmuck, G., Perdigao, V., Thielen, J.: The influence of historic land-use changes and future planned land-use scenarios on floods in the Oder catchment. Phys. Chem. Earth, Parts A/B/C, vol. 20, pp. 1291–1300 (2003)
2. Drainage and Irrigation Department (DID): Annual flood report of DID for Peninsular Malaysia. Unpublished report, Kuala Lumpur (2004)
3. Fritsch, U.: Entwicklung von Landnutzungsszenarien fur die okologische Modellierung. Brandenburgische Umweltberichte 12, pp. 1–132. (2002) Downloadable on: http://opus.kobv.de/ubp/volltexte/2005/398/pdf/vol12.pdf
4. Lawrence, P.J., Chase, T.N.: Representing a new MODIS consistent land surface in the Community Land Model (CLM3.0). Journal of Geophys. Res.-Biogeosciences, vol. 112(G1) (2007)
5. Saadatkhah, N., Tehrani, M.H., Shattri, M., Zailani, Kh, Kassim, A., Saadatkhah, R.: Impact assessment of land cover changes on the runoff changes on the extreme flood events in the Kelantan River basin. Arab. J. Geosci. **9**, 687 (2016)
6. Shattri, M., Saadatkhah, N., Zailani, Kh., Arnis, A., Noraizam, A. and Noradzah, A: Regional modelling of rainfall-induced runoff using hydrological model by incorporating plant cover effects: case study in Kelantan, Malaysia. Nat. Hazards, (2018) https://doi.org/10.1007/s11069-018-3322-z

Influence of Anthropogenic Activity on the Development and Spreading of Flood Hazardous Events in Madeira Island (Portugal)

Davide Baioni

Abstract

Over the last few decades, the island of Madeira has become an important offshore tourism and business centre, with a rapid economic and demographic development that have caused changes to the landscape due to human activity. In Madeira's recent history, there has been an increase over time in the frequency of occurrence of damaging flood events. As a result, the costs of restoration work due to damage caused by flood events have become a larger and larger component of Madeira's annual budget. Flood phenomena on Madeira deserve particular attention because they represent the most serious hazard to human life, property, and the natural environment and its important heritage value. The work reported in this paper involved the analysis of historical data regarding damaging flood events on Madeira in particular for the period from 1941 to 1991, together with data on geological characteristics, topographic features, and climate, and from field observations. This analysis showed that the increase in the number of damaging flood events recorded on Madeira Island, especially in recent times, seems to be mostly related to human activity, specifically to economic development and population growth, rather than to natural factors.

Keywords

Madeira Island • Human activity • Flood events • Hazards

1 Introduction

The island of Madeira is part of an archipelago of volcanic origin situated in the Atlantic Ocean (from 32° 23′ N to 33° 07′ N and from 16° 15′ W to 17° 15′ W) approximately 900 km south-west of Portugal and 700 km west of the coast of Morocco.

Madeira is part of Portugal but it is an autonomous region with its own government that can freely deliberate and regulate many aspects of its own territory. It has a resident population of approximately 245.000 people concentrated almost exclusively in the coastal areas, dominantly on the southern side of the island where the majority of the important cities lie and in which the most important part of the economic life of Madeira is located.

Madeira in the last decades has been turned into an important off-shore tourism-business centre making a rapid economic growth. The desire to increase tourism business produced important human development activity which in turn led to an explosion of building in a very short period of time.

Human activity due to urban growth because of the lack of suitable space made profound changes to the existing morphology. In Madeira's recent history there is an increase over time in the occurrence of flood hazardous events [1, 2] and there is a clear increase in the costs of restoration work, protection measures and damage caused by flood phenomena that have progressively become a growing feature in the public economy of Madeira.

Flood phenomena in Madeira deserve particular attention because they represent the main and the most serious hazard for human life, property and the natural system and its important heritage [3, 4].

D. Baioni (✉)
Urbino University, Campus Scientifico Sogesta, 61029 Uribino, Italy
e-mail: davide.baioni@uniurb.it

A. Kallel et al. (eds.), *Recent Advances in Geo-Environmental Engineering, Geomechanics and Geotechnics, and Geohazards*, Advances in Science, Technology & Innovation, https://doi.org/10.1007/978-3-030-01665-4_113

2 Data and Methods

In this work a detailed analysis of flood hazardous phenomena was carried out through the study of historical data about flood damaging events that occurred in Madeira during the period 1941–1991, to determine their distribution in space and time, as well as possible variations in frequency and the recent evolutionary trend [2].

In the data recorded, supplied by the archive of the regional institutions of Madeira, numerous attributes are often described for each individual event surveyed, such as the type of event, the location, the date, the meteorological conditions previous to the event described, as well as the effects on the population and on structures. The available information provides an excellent description of each individual event.

Historical records analysis has been integrated with geological characteristics, topographic features, precipitation data and field investigations. In particular, the analysis of the evolutionary trends of rainfall and flood events was carried out with the aim of checking if the climatic factor is the only or main cause of the increase in the frequency of flood events encountered in Madeira.

3 Results

The flood damaging events recorded all over the island during the period 1941–1991 indicate that (i) they are mainly flash floods; (ii) there is an increase in frequency over time; (iii) the areas with the higher number of events are those drainage basins where the main urban centres are located; (iv) there is a completely different incidence to flooding which is a countertendency as regards the climatic conditions and the rainfall data of the different areas of Madeira.

Rainfall records of the period 1941–1991 show no increase over time in the annual values, and no increase of the number of high-intensity rainfall events (Fig. 1).

A detailed field investigation has observed that urban development caused the reduction and occupation of the mouths of the rivers and led to the occupation of the flood plains areas of the main rivers [3]. Urban growth also made changes in the fluvial morphology by narrowing the original riverbeds, reducing the original section (Fig. 2), shaping the original routes, altering the length and reducing the bed roughness, increasing the flow velocity. Moreover, the expansion of cities and the construction of related infrastructures has increased impermeable surfaces raising the runoff coefficient, leading to a more rapid concentration of runoff with a consequent increase in the magnitude of flash floods.

4 Discussion

The results of this study highlight that there is a geographical correspondence between the incidence of hazardous flood events and human activity within the territory of Madeira.

The increase in flood events mainly occurs in the areas related to human activity and there seems to be no spatial and temporal relationship between hazardous flood events and rainfall.

The increasing in Madeira of damaging flood events seems to be related mainly to the rapid increase in the exposure and vulnerability as a consequence of the modifications on the territory caused by the human activity due to the rapid economic and demographic development.

5 Conclusion

Comparative analyses of all the data highlight positive territorial and timing relationship between anthropogenic opera, land-use change and flood events.

The study carried out in this paper shows that the increase and the growing-up trend in the recent time of hazardous flood events in Madeira island are highly related to human activity and to the environmental modifications, rather than to changes in climatic conditions.

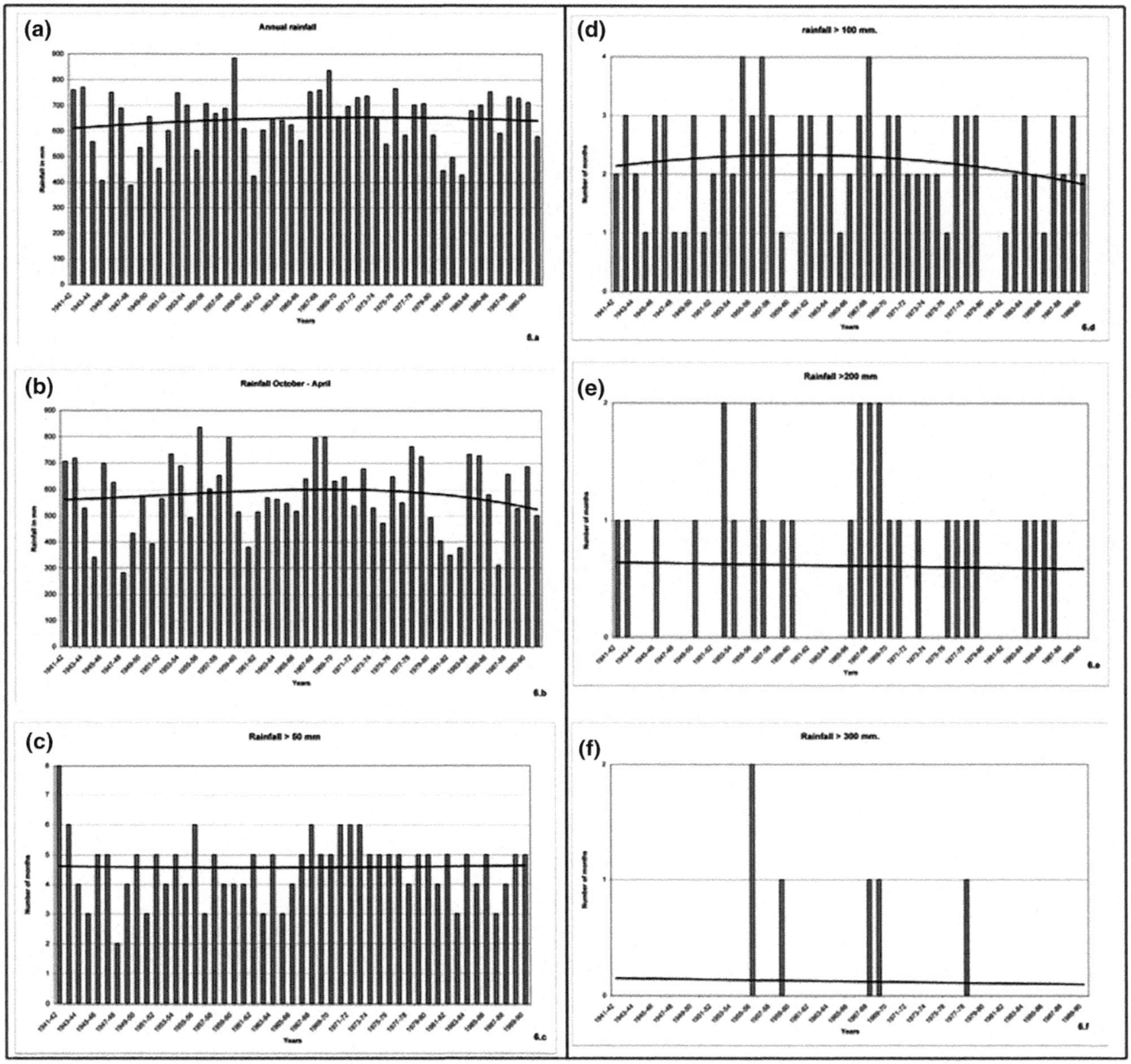

Fig. 1 Rainfall recorded in Santa Catarina for the period 1941–1991; **a** annual values; **b** total values for the period October–April for each year; **c** months with rainfall over 50 mm.; **d** months with rainfall over 100 mm.; **e** months with rainfall over 200 mm.; **f** months with rainfall over 300 mm

Fig. 2 Example of changes in the fluvial morphology by narrowing the original riverbeds, and reducing the original section of the "Ribeira de Santa Luzia" river within Funchal city

References

1. Rodrigues, D., Ayala-Carcedo, F.: Georisk: a natural hazard database of historic events in Madeira Island. 25th General Assembly EGS, Nice, vol. 286 (2000)
2. Fragoso, M., Trigo, R.M., Pinto, J.G., Lopes, S., Lopes, A., Ulbrich, S., Magro, C.: The 20 February 2010 Madeira flash-floods: synoptic analysis and extreme rainfall assessment. Nat. Hazards Earth Syst. Sci. **12**(3), 715–730 (2012)
3. Almeida, A.B., et al.: Risk assessment of flash floods in Madeira Island. Funchal Secretaria Regional do Equipamento Social e Transportes (2010)
4. Fernandes, F.: Built heritage and flash floods: biking trails and tourism on Madeira Island. J. Heritage Tourism **11**(1), 88–95 (2016)

Towards Development of Risk-Based, Soil and Groundwater Screening Values (Threshold Values) for Petroleum Hydrocarbon (PHC) in Libya and Tunisia by Considering Climate and Regional "Geographical" Factors

Salahadein Ahmed Alzien, Roger Brewer, Olfa Ben Said, and Rafig Azzam

Abstract

Petroleum Hydrocarbon Contamination (PHC) is a widespread environmental problem that can be found in many countries. Several contaminated land management approaches have been developed so far, all of which include some form of site investigation, risk assessment and remediation processes. Most of these approaches are often based on risk-based land management concepts. Most countries have a common framework for a risk assessment procedure for contaminated sites which endanger human health, involve ecological risk, and risks to water resources and construction materials. Usually, risk assessment of contaminated land is triggered by suspicious soil or groundwater contamination which is identified by Screening Values (SVs) approach. SVs specify generic quality standards for contaminated land. The application of SVs varies from adjusting long-term quality objectives, through making further investigations, to applying remedial actions. SVs derivation methods have scientific, geographical, socio-cultural, regulatory and political categories. They therefore differ from country to country. However, a "one-size-fits-all" approach is often adopted and a single SV might be applied to all areas within a country. In this paper, the authors demonstrated how differences in climatic and environmental conditions within a country may require the use of region-specific SVs. Here, Exposure SVs for PHC in Libya and Tunisia were used.

S. A. Alzien (✉) · R. Azzam
Department of Engineering Geology and Hydrogeology, RWTH Aachen University, Lochnerstr. 4-20, 52064 Aachen, Germany
e-mail: alzien@lih.rwth-aachen.de

R. Brewer
Hawaii Department of Health, 2385 Waimano Home Road, Uluakupu Building 4, Pearl City, HI 96782, USA

O. Ben Said
Environment Biomonitoring Laboratory LBE, Faculty of Sciences, Bizerte, Tunisia

Keywords

Risk assessment • Screening values • Contaminated land • Climate conditions • Libya • Tunisia • USA

1 Introduction

Screening Values (SVs) or Soil Screening Levels (SSLs) are generic quality standards accepted in many countries for the management of contaminated land. They are commonly presented as concentration thresholds of contaminants in soil above which certain actions strongly recommended and completed (e.g., milligrams of chemical per kilograms of soil in dry weight, mg/kg). When SVs are exceeded, the actions to be taken depend on national regulatory frameworks and vary from the need for further investigations to the need for remedial actions [2]. The Soil Screening Guidance: User's Guide [7] focuses on the application of a simple, site-specific approach by providing a step-by-step methodology for calculating site-specific SSLs and planning the necessary sampling to apply them for future residential use [7, 8]. Figure 1 shows the scope of the risk management concept at a contaminated land, with a spectrum of soil contamination concentrations that range between the below levels (Fig. 1) with no harm to human health, is considered and levels of very high contamination concentrations that need remedial action are set.

Recommended, potential pathways of exposure to pollutants in the soil include direct and indirect exposure via incidental ingestion of soil, dermal absorption and inhalation of volatiles and fugitive dust. This paper focused on the development of screening levels or SVs to address these exposure risks. [1] discuss the influence of climate-based, regional factors on the development of SVs for vapor intrusion. The USEPA document also provides guidance for

A. Kallel et al. (eds.), *Recent Advances in Geo-Environmental Engineering, Geomechanics and Geotechnics, and Geohazards*, Advances in Science, Technology & Innovation, https://doi.org/10.1007/978-3-030-01665-4_114

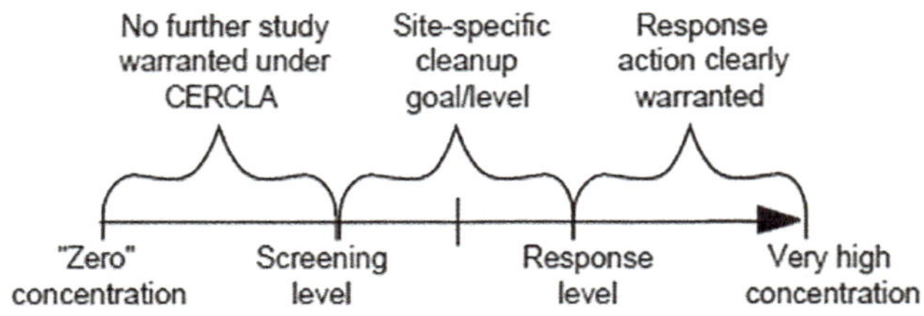

Fig. 1 Conceptual risk management spectrum for contaminated soil [7]

the ingestion of contaminated groundwater caused by migration of chemicals through the soil to an underlying potable aquifer and ingestion of homegrown product that has been contaminated via the uptake of chemicals from the soil. Although potentially important, climate-specific factors associated with these exposure pathways are beyond the scope of this paper.

2 Selection of Screening Values (SVs) for "Libya and Tunisia" from International Approaches Screening Approach

Screening values developed in the United States (US) typically include future residential land use assumptions and related exposure scenarios. The USEPA Soil Screening Guidance (SSG) introduces a framework for screening contaminated soils and developing SVs [7]. Generic soil SVs are calculated by using a conservative hypothesis, and they are, therefore, less robust compared to SVs based on site-specific characteristics. Site-specific or even region-specific SVs are calculated with the same equations used for the generic SVs. However, the site-specific calculations are more representative and applicable, because hydrological, soil, and meteorological data are used. Hydrological characteristics such as rainfall, infiltration rate and groundwater flow rate and soil characteristics such as organic carbon content and permeability are common factors considered for site-specific development of SVs. Consideration of a broader spectrum of meteorological data and the resulting effects on key parameters of the SVs models, in particular the Volatilization Factor (VF) for emissions of vapors from PHC-contaminated soil and the Particulate Emission Factor (PEF) for non-volatile chemicals in dust.

Affect of Climate on SV Volatilization Factor and Particulate Emission Factor:

Climate is one of the important exposure variables. The PEF (m^3/kg), was used to address intake from inhalation of contaminated soil-derived particulates. Climate conditions in Libya and Tunisia and the USA are almost similar. Climate zones for Libya and Tunisia are summarized in Fig. 2. The Köppen climate classification [3, http://koeppen-geiger.vu-wien.ac.at/shifts.htm, 6] was employed to identify countries with similar climatic conditions to Libya and review methods applied in those areas for use is these countries. This suggests that the same climatic condition parameters and model algorithms as for those areas of the USA can be applied to similar areas of Libya and Tunisia.

A principal factor used for deriving both of PEF and VF is the air dispersion factor (Q/C). This is defined as the inverse of the mean dust or vapor concentration at the center of a 0.5 acre square source (g/m^2 s per kg/m^3).

3 Results

Table 1 describes the Q/C value for Las Vegas-Nevada, Phoenix-Arizona, Albuquerque-New Mexico (Climatic Zone III) and Los Angeles (Climatic Zone II) presented in

Fig. 2 Climate conditions in the USA, Libya and Tunisia [5]

Table 1 Recommended Q/C values for calculation PEF and VF for Libya and Tunisia

Climate Zone	Correlative US Region	Selected Q/C Value (g/m^2-s per kg/m^3)		Calculated PEF (m^3/kg)		Calculated VF (m^3/kg)	
		LY	TU	LY	TU	LY	TU
Mediterranean "Csa"	Los Angeles, CA	68.82	68.82	$5.5575*10^9$		X	Y
Steppe-Hot "Bsh"	Phoenix, AZ	64.06	64.06	$5.1715*10^9$		X	Y
Desert "Bwh"	Las Vegas, NV	95.51	95.51	$7.7128*10^9$		X	Y
Steppe-Cold "Bsk"	Albuquerque, NM	–	84.18	–	$6.7979*10^9$	–	Y

X,Y: These calculated values depend on the chemical and site specific properties

Appendix E of the USEPA SSL guidance [8]. They were selected for application to Libya and Tunisia. These were then used to calculate PEF and VF values for the different climate regions of Libya and Tunisia. These values can subsequently be used to develop generic or site-specific SVs for these individual regions.

4 Discussion

Soil characteristics are also important parameters which include soil texture, dry soil bulk density, moisture content, organic carbon, PH, moisture retention exponent, saturated hydraulic conductivity, and average moisture content. Another significant parameter, which reflects the geographical characteristics, is the meteorological data where the air dispersion factor (Q/C) is a site-specific value reflecting the corresponding area (temperature, wind).

5 Conclusion

Climatic conditions and contaminant properties (physio-chemical constants), such as vapor pressure, Henry's law constant, solubility in soil water, solubility in soil organic matter are very important parameters particularly regarding their significant effects on the volatilization factor and therefore have a significant influence on outdoor environment.

References

1. Brewer, R., Nagashima, J., Rigby, M., Schmidt, M., O'Neill, H.: Estimation of Generic Subslab Attenuation Factors for Vapor Intrusion Investigations. Groundwater Monit. Rem. **34**(4), 79–92 (2014). https://doi.org/10.1111/gwmr.12086
2. Carlon, C.: Derivation methods of soil screening values in europe. A review and evaluation of national procedures towards harmonisation. Eur. Comm., Jt. Res. Centre, Ispra, EUR 22805-EN, 306 pp (2007)
3. Chen, D., Chen, H.W.: Using the Köppen classification to quantify climate variation and change: An example for 1901–2010. Environ. Dev. **6**, 69–79 (2013). https://doi.org/10.1016/j.envdev.2013.03.007
4. http://koeppen-geiger.vu-wien.ac.at/shifts.htm. World Maps of Köppen-Geiger climate classification
5. Köppen-Geiger. Main Köppen-Geiger Climate Classes for US countie (2006)
6. Kottek, M., Grieser, J., Beck, C., Rudolf, B., Rubel, F.: World Map of the Köppen-Geiger climate classification updated. Meteorol. Z. **15**(3), 259–263 (2006). https://doi.org/10.1127/0941-2948/2006/0130
7. USEPA. Soil screening guidance: technical background document (1996)
8. USEPA. Supplemental guidance for developing soil screening levels for Superfund sites (2002)

Understanding Preparedness Insufficiency in the Context of DRRM: A Case Study in Malinao, Albay, Philippines

Ana Marie R. Abante

Abstract

This paper presents a preparedness insufficiency conceptual model relative to the actual land utilization and location information relating the DRE raster surface after interpolation of points using an inverse distance weighted technique inferring preparedness insufficiency. With ArcGIS Geostatistical Analytical Tool, this study brings out about 259.87 hectares of land utilized for economic, social, institutional, housing and critical infrastructure needs sensitive spatial planning and appropriate risk reduction activities to achieve the preparedness sufficiency level or balanced physical developments in Malinao. This study infer risk reduction changing needs add up to disasters likely to happen happen again and again.

Keywords

Preparedness insufficiency • Risk reduction • GIS • DRE

1 Introduction

Prevention [1–7] is the 1st level of receptiveness where risk is always expected headed for corrective measures and continual improvement to reduce undesired effects of climate changes such as stronger typhoons carrying more rainwater transporting lahar, sand, soils from Mayon Volcano, Mt. Masaraga and Mt. Malinao triggers Quinali 'B' river bank erosions and overflows submerging the alluvial flats mostly utilized for rice production in Malinao [1, 7, 8]. Next is mitigation [5, 9–12] taking in the improvement of existing irrigation canals into flood control structures to which it shall also serve as farm-to-market roads and when prompt with flood hazards it shall be used as escape routes to the three evacuation centers as the case of Malinao. It is critical to Malinaons as the altogether utilizing the downstream or flood plain areas for agricultural use mixed with dwelling and other socioeconomic land uses. The third stage of receptiveness is the preparedness [13–18] to which it emphasizes the need for a safe, comfortable and accessible evacuation centers to ensure the zero-casualty strategy of the Malinaons or Albayanos [3]. In this stage, local resiliency action plan where the responsibilities and accountabilities on a share reduction environment among the disaster reduction managers in Albay are in place [3]. The responsiveness [19–21] composes of the disaster and post disaster phase wherein the disaster phase is caring about the emergency response plan as well as the public safety of the people. The work was designed to analyze preparedness insufficiencies occurring barangays of Malinao. Multiple hazards, landscape condition, physical features and land use of the municipality were processed and converted into attributes of point feature distributed 2-km apart within the entire municipality. The attribution was used to analyze polygon objects featuring the estimated disaster risk per land use along with location information using Inverse distance weighted (IDW) technique to understand preparedness insufficiency concept model that is exactly right for Malinao remained vulnerable to changes. The term 'Preparedness Sufficiency' is defined as the conceptual border between receptiveness and responsiveness in the context of DRRM. In contrast, 'Preparedness Insufficiency' is the conceptual space setting the people or critical infrastructure unprotected, unsafe, uncomfortable because of not meeting the appropriate spatial plans or risk reduction actions and programs of the government.

1.1 Preparedness Insufficiency Benchmarking

The Preparedness Insufficiency Benchmark in this study attempt to provide a basis for the five circumspectial stages,

A. M. R. Abante (✉)
Bicol University, Legazpi City, 4500, Philippines
e-mail: anamarie.abante@bicol-u.edu.ph

A. Kallel et al. (eds.), *Recent Advances in Geo-Environmental Engineering, Geomechanics and Geotechnics, and Geohazards*, Advances in Science, Technology & Innovation, https://doi.org/10.1007/978-3-030-01665-4_115

these are: (i) state of alertness, (ii) state of emergency, (iii) state of calamity, (iv) disaster phase and (iv) post disaster phase. The researcher tried to analyze the borderline between receptiveness and responsiveness in relation with extrinsic hazards that happen again and again in the study area. By pairing the stages of DRRM with the Circumspectial Stages, the study disclose the 'state of calamity'-'preparedness' isometric pair define the line for 'preparedness sufficiency' and space for 'preparedness insufficiency'. Figure 1 illustrates the adjustment from the Disaster Risk Estimate (DRE) [1] to DRE* [2] to put in matrix form the five DRRM cycle: prevention, mitigation, preparedness, response and recovery stages paired with the 5 stages of DRRM with the circumspectial stages.

This matrix was arranged to make simpler analyzing the interconnectivity of elements and sub-elements of risk. The black line denotes the 'Line for preparedness sufficiency. [22–29] By pairing the stages of DRRM with the Circumspectial Stages, the study discloses the 'state of calamity'-'preparedness' isometric pair define the 'Borderline for Preparedness Sufficiency' and 'space for preparedness insufficiency'. The 'Borderline for Preparedness Sufficiency' refers to the black line bordering the receptiveness and responsiveness occurring right after the state of emergency and before the state of calamity having and RLQ equal to 1.0 as shown in Fig. 2. The 'Space for Preparedness Insufficiency' refers to the space generated as black borderline moving downward to the left direction before it passes over the disaster stage.

This study comprises 5 descriptions of location quotients pertaining to the DRE [1], these are: (i) 5/5 or 1.0 Location Quotients (LQ) in which the immediate information needs should be initiated in re-development spatial planning to provide significant information on space allocation such as permanent safe relocation sites, potential safe settlements sites, and so on obligatory in revising development and rehabilitation plans to implement adequate action for desired re-development; (ii) 4/4 LQ wherein the urgent information needs should be initiated in Emergency Response Plans to provides a clear communication relative to safe emergency routes, safe evacuation sites, safe pre-positioned food, medicines and emergency response resources, safe location of incident command center, responders and rescuers, and so on. The National Government thru Office of the Civil Defense in cases where 2 or more regions are affected; (iii) 3/3 LQ wherein the direct information needs should be initiated in a DRRM Plan wherein spatial planning on safe and accessible evacuation routes, safe and sufficient evacuation sites, evacuation protocols for zero casualty strategy of Albay, moderate to very high vulnerability and exposure register, GIS database, and so on are necessary in local action planning; (iv) 2/2 LQ wherein the high priority information needs should be initiated in a comprehensive development and investment plan resource needs on mitigation phase; and (v) 1/1 LQ wherein the pressing information needs should be initiated in a land use plan and zoning ordinance as mandatory spatial planning tool to provide information such as space allocation for permanent

DISASTER RISK ESTIMATES (DRE)			CIRCUMSPECTIAL STAGES: Responsiveness			Receptiveness	
			Post Disaster Phase	Disaster Phase	State of Calamity	State of Emergency	State of Alertness
DRRM STAGES			Multi-Hazards	Vulnerable to Multi-hazards	Exposed Location	Far-flung Evacuation Shelter	Unaffordable land for Resettlement Projects
Receptiveness	Prevention	Unsuitable LU	125 & above	64	27	8	1.0 (Bal.)
Receptiveness	Mitigation	Unsafe	15.63	8	3.36	1.0 (Bal.)	0.13
Receptiveness	Preparedness	Uncomfortable	4.63	2.37	1.0 (Bal.)	0.3	0.04
Responsiveness	Emergency Response & Public Safety	Insensitive Actions/ Spatial Plans	1.95	1.0 (Bal.)	0.42	0.13	0.02
Responsiveness	Recovery	Repetitive Evacuation Exercises	1.0 (Bal.)	0.51	0.22	0.08	0.01

Fig. 1 Disaster risk estimates (DRE*) matrix

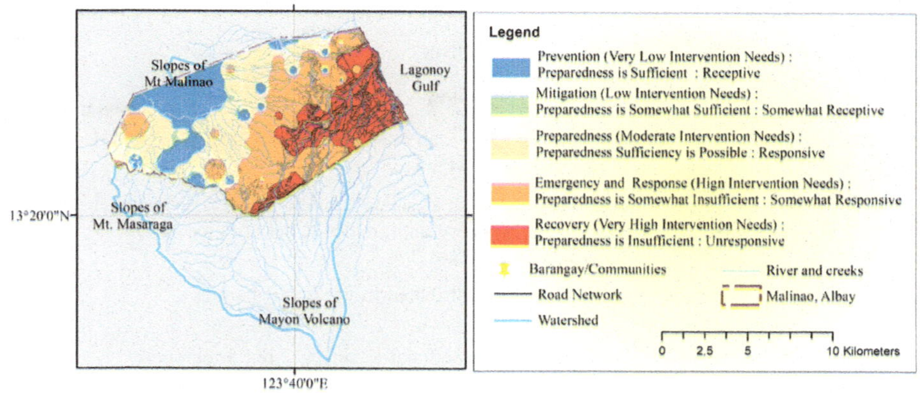

Fig. 2 Preparedness insufficiency model

relocation sites & safe housing sites, internal-external access road, policy on protection, production, infrastructure and settlements, and so on. [1, 2].

2 Discussion

With ArcGIS geostatistical analyst tool, it was possible to generate the desired output as shown in Fig. 2 to demonstrate the input point data that were distributed 2 km apart (within the 11,338.53 Hectares total land area of Malinao) to interpolate the raster Multi-hazard levels surface generated from these points using an inverse distance weighted (IDW) technique tool.

Then the values of the multi-hazard raster surface were used to calculate the DRE level along with the following assumption: (i) vulnerability (human) [1, 2] used value equal to 3 denoting the class of the municipality; (ii) exposure [1, 2] take values multi-hazard values intersecting the surface with settlements or urban use areas; (iii) preparedness take the values 3 denoting a fair DRRM Plan was prepared in harmony with the approved Comprehensive Land Use Plan and Zoning Ordinance was approved though it is hard to monitor the implementation since the publication of the said ordinance; and the competency [20, 21, 24] and coping capacity [24, 30, 31] values equal to 3 to signify the 3rd Class Municipality. With the Abante DRE formula written as $\frac{Multi-hazard}{Preparedness*} x \frac{Vulnerability}{Competency} x \frac{Exposure}{CopingCapacity}$, wherein the attributions of the risk elements were calculated and the result presented display the 5 DRE levels. The Preparedness [23] for this work match up the 5 stages of DRRM, in which it resembles where the needs to achieve preparedness sufficiency. The desired output on preparedness insufficiency per land use as presented below comprise the lack of prevention and mitigation measure to protect the following: (i) economic sector utilized for commercial, industrial and other mixed uses with an aggregated land area of 7.2 hectares; (ii) social facilities constructed within 7.89 hectares of land; (iii) housing and residential area with an aggregated land area of 154.90 Hectares; and (iv) Institutional and critical infrastructures with an aggregated land area of 89.867,194 hectares. To further understand preparedness insufficiency in the Malinao, the DRE Map overlaid with the proposed land use plan to visualize where receptiveness along with responsiveness needs lies in the study area. The DRE for Malinao in 2018 is 119.04 location quotient denoting the following: (i) RLQ is between 119.04 fall within 64 ≥ but <125; (ii) Level of Significance is between ≥95% but <99% level of Significance; (iii) Significance Rendering is Risk is High to very High; (iv) Implication may be catastrophic, environmental damaging, and economic activity disruption; and (v) Mitigation is obligatory, production is allowed but must protect environmental critical areas, and permanent relocation is obligatory for the people are lives in a no-dwelling-zone. With the DRE tool and by setting aside the swaying intervention needs and temporal data the resulting 119.04 RLQ in the Municipality of Malinao, Albay for the year 2018 infer an enormous gap in DRRM. Keeping the people safe and comfortable, a balanced physical space for development in harmony with the environment requires to eliminate if not reduce the 119.04 RLQ or 11,904% to 1.0 or 100% (Balance state of line between receptiveness and responsiveness infer sense of balance reckoning preparedness sufficiency) by applying this risk location quotient as a multiplicative factor while allowing land development but maintaining the mandatory zero-casualty strategy of the Province of Albay. As the neutrality at different pairs of DRRM-CIRCUMSPECTIAL Stages hinged at 1.0 RLQ, with the 119.04 RLQ for locality remained at the circlet of disaster phase and/or emergency response and public safety. The gap (see Fig. 2) is nearing the 125 RLQ pointing towards the post-disaster and/or recovery phase to which it implies the following: disaster is always imminent; catastrophic and environmental damaging; and economic activity disruption. Table 1 is the attribute table of the polygon objects featuring the DRE levels by land use along with

Table 1 Preparedness insufficiency Model

Preparedness assessment result	Land use per barangays	Exposed areas (Hectares)	DRRM enhancement needs
5— Insufficient	Economic activities such as commerce and industries located in 7 Barangays of Malinao	1.96	Full Recovery
4–5—Somewhat insufficient to Insufficient	Social Facilities such as schools, religious sites, recreational areas, parks, cemetery and protective services located in 7 Barangays of Malinao	7.83	Emergency, Response & Recovery
4–5—Somewhat insufficient to Insufficient	Institutional Use distributed in the municipalities	1.00	Emergency, Response & Recovery
1–3—Somewhat sufficient to sufficient	Residential in Burabod and Estancia. 2 out 3 temporary evacuation shelters are in these 2 barangays	1.44	Protection and Land reservation
4—Somewhat insufficient	Residential Areas across 13 Barangays of Malinao	37.43	Allocation of Suitable, safe, comfortable, accessible Housing/Residential areas
5—insufficient	Residential areas in 19 Barangays of Malinao	103.02	Relocation to Suitable, safe, comfortable, accessible sites
2–5—somewhat sufficient to insufficient	Utilities connecting all barangays by land and communication facilities and other infrastructures	87.11	Mainstreaming DRRM into Spatial and development plans

location information. The DRE levels present raster surface after interpolation of points using an inverse distance weighted (IDW) technique inferring preparedness insufficiency.

3 Conclusions

The imaginary space for 'preparedness insufficiency' remained pressing consequences in Malinaons, to which it is undeniably logical to changing intervention needs and corrective measures thru spatial planning and risk reduction actions towards balance state of development in harmony with the physical environment. While this imaginary space denotes unsafe and uncomfortable place to live, socio economic infrastructure at risk, and environment critical areas remained likely to changes such as river morphological changes due to lahar and sand siltation, bank erosions and overflows submerging the ecological buffers and adjacent settlements or communities or urban areas attributable to reoccurrences of severe rainfall and stronger typhoons the mitigation intervention needs varies and temporal. The government directive to set aside the 5% for the Local DRRM Fund from the local budget cannot counter the 11,904% need multiplicative factor for intervention to become stable. The standby Calamity and Quick Response Fund (QRF) allocated for public works and highways, national defense, education, social welfare, agriculture, and civil defense departments. However, these funds are accessible only after declaration of state of calamity to assist areas stricken by catastrophes and crises. Therefore, preparedness insufficiency likely to increase due to reoccurrences of multiple hazards prompting the non-conforming utilization of every hectare of land. Accordingly, the temporality of nature and the changing needs in DRRM convey an imaginary space (denoting the gap called preparedness insufficiency) between receptiveness and responsiveness. In contrast, risk reduction is pictured as trimming the topological space (polygon features) until a borderline (denoting preparedness sufficiency) is pulled off.

References

1. Abante, A.M.R., Abante, C.G.R.: Sensitive land use planning in Malinao, Albay, Philippines. IOP Conf. Ser.: Earth Environ. Sci. **123** (2018)
2. Abante, A.M.R.: Bicol University Completed Research Study entitled "Risk Modeling: A Case Study of Malinao, Albay, Philippines (2017)
3. Daep, C.D.: Bicol University Graduate School Unpublished Dissertation entitled "The Implementation of the Disaster Risk Reduction Management Program in the Province of Albay" (2011)
4. Tabayag, S.G.: Disaster risk reduction through land use planning in lahar-devastated footslopes of Mayon Volcano in Albay [Philippines]. *J. ISSAAS [Int. Soc. Southeast Asian Agric. Sci.] (Philippines)* (2010)
5. Ondiz, R.L.: Bicol University Graduate School Unpublished Dissertation entitled "Flood Disaster Preparedness and Mitigation Program Implementation in Quinali 'A' River System" (2006)
6. Candelaria, A.P., Revale, I.F.H.: Scaling up coastal resource management in an agrarian reform community [in Dimasalang, Masbate, Philippines]. J. of ISSAAS [Int. Soc. Southeast Asian Agric. Sci.] (Philippines) (2010)
7. de Jesus, S.C., Dioneda, R.R., Revale, I.H., Doloiras, A.D., Nolial, A.L., Ocampo, A., Alcazar, D.S.: Assessment of the Ecological Habitats of Bacon District, Sorsogon City in the Philippines. 黒潮圏科学 **4**(1), 43–52 (2010)

8. Punay, J.P., Perez, G.J.P.: Evaluation Of modis cloud products-derived rainfall estimates
9. Mascarinas, A.M., Baas, S., Koksalan, N., Amano, L.O., Nieves, P.M., Binoya, C.S., … & Torrente, E. C.: Mainstreaming Disaster Risk Reduction into Agriculture: a Case Study from Bicol Region, Philippines. Environment and Natural Resources Management Series. FAO (2013)
10. Demelletes, R.D. Jr.: Bicol University Graduate School Unpublished Dissertation entitled "Responsiveness of Department of Environment and Natural Resource—National Capital Region in Attaining the Environmental Security". Bicol University Library, Legazpi City, Philippines (2017)
11. Diaz-Pavilando, L.: Showcasing integrated farming (crop-livestock) technologies as coping mechanisms for lowland and upland typhoon devastated coco-based farms: a rehabilitation strategy [in Albay, Bicol, Philippines]. J. ISSAAS [Int. Soc. Southeast Asian Agric. Sci.] (Philippines) (2010)
12. Barrameda, J.T.: Bicol University Graduate School Unpublished Dissertation entitled "Development Practices on Disaster Management for State Universities and Colleges". Bicol University Library, Legazpi City, Philippines (2015)
13. Bradecina, R.G., Shinbo, T., Nieves, P.M.: Exploring local coastal recreational tourism as potential strategy to support changing agriculture in a Typhoon-Prone Small Island of San Miguel. Tabaco City, Albay, Bicol Region (2011)
14. See, E.S., Ables, L.R., De Guzman, M.D., Bartolata, J.L., See, M. M.A.M.: Electric energy utilization in the households of Albay Province, Philippines: contexts, conservation practices, and future efficiency strategies. ASIAN RURAL SOCIOL. IV, **367** (2010)
15. Areola, E. L.: Bicol University Graduate School Unpublished Dissertation entitled "Comparative Analysis of Maintenance of National Road Network in Region V". Bicol University Library, Legazpi City, Philippines (2013)
16. Evasco, K.E., Rafaelito IV, A.B.: Disaster preparedness measures in the high-risks barangays [villages] of the province of Albay [Bicol, Philippines]. J. of ISSAAS [Int. Soc. Southeast Asian Agr. Sci.](Philippines) (2010)
17. Gonzales, A.B.: Sustainable decommissioning and integrated closure planning of selected mine sites in the Bicol Region, Philippines. Eur. J. Environ. Sci. **8**(1), 76–82 (2018)
18. Lauraya, F.L.P.M., Pavilando, L.D., Bartolata, J.L., Mamansag., M.B.: The Bicol University community outreach services strategy: educating to build resilient communities. ASIAN RURAL SOCIOL. IV (2010)
19. Demelletes, R.D. Jr.: Bicol University Graduate School Unpublished Dissertation entitled "Responsiveness of Department of Environment and Natural Resource – National Capital Region in Attaining the Environmental Security". Bicol University Library, Legazpi City, Philippines (2017)
20. Aguilar, A., Imran, M.A.: Land for the landless. WIT Trans. Built Environ. **150**, 291–297 (2015)
21. Nieves, P.M., Pelea, N.R., Bradecina, R.G., Pereyra, M.A., Morooka, Y., Shinbo, T., & Rivero, M.C.P.: Socio-economic conditions, the status of fisheries and agriculture and the adaptive capacities of households and communities in San Miguel Island, Albay, Philippines in the Kuroshio sphere of influence (2009)
22. Carreño, M.L., Cardona, O.D., Barbat, A.H.: A disaster risk management performance index. Nat. Hazards **41**(1), 1–20 (2007)
23. Schlobinski, S., Zuccaro, G., Scholl, M., Meiers, D., Denzer, R., Guarino, S.,… & Frysinger, S.: Decision making and strategic planning for disaster preparedness with a multi-criteria-analysis decision support system. In: International Symposium on Environmental Software Systems, pp. 178–186. Springer, Cham (2015, March)
24. Taylor, S.S., Ehrenfeld, J.M.: Electronic health records and preparedness: lessons from Hurricanes katrina and Harvey. J. Med. Syst. **41**(11), 173 (2017)
25. Boholano, C.A.: Bicol University Graduate School Unpublished Dissertation entitled "The Disaster Risk Reduction and Management Capability of the Local Government Unit of Malilipot in the 1st District of Albay". Bicol University Library. Legazpi City Philippines (2013)
26. Borch, F.L.: Comparing pearl harbor and" 9/11": intelligence failure? American Unpreparedness? Military Responsibility? J. Military Hist. **67**(3), 845–860 (2003)
27. Mulyasari, F., Inoue, S., Prashar, S., Isayama, K., Basu, M., Srivastava, N., Shaw, R.: Disaster preparedness: looking through the lens of hospitals in Japan. Int. J. Disaster Risk Sci. **4**(2), 89–100 (2013)
28. Briggs, D.J., Forer, P., Järup, L., Stern, R. (eds.): GIS for Emergency Preparedness and Health Risk Reduction (Vol. 11). Springer Science & Business Media (2012)
29. Levac, J., Toal-Sullivan, D., OSullivan, T.L.: Household emergency preparedness: a literature review. J. Community Health **37**(3), 725–733 (2012)
30. Kallos, R.C.: Bicol University Graduate School Unpublished Dissertation entitled "The Organizational Competency and Performance of Tabaco City". Bicol University Library,. Legazpi City, Philippines (2010)
31. Banua, A.S.: Bicol University Graduate School Unpublished Dissertation entitled "Phychosocial Care Activities for the Evacuees During Disasters and their Coping Mechanisms". Bicol University Library, Legazpi City, Philippines (2011)

Transposition of the Genius Paraseismic Knowledge at Elements of Architectural Design; Case of Algerian Architects

Mohamed Benabdelfattah and Youcef Kehila

Abstract

Since 1954, the code for earthquake-resistant design was one of the priorities of the public authorities in order to take into account the seismic risks. The earthquake-resistant rules have steadily evolved (AS, PS 69, RPA 1983, RPA 1988, RPA 1999, RPA 2003). In fact those rules do not dictate any architectural prerequisites. All they offer is just literary guidelines answering the question: «what is it to be done?». But they do not answer the esthetic question: «how to do it?». The aim of this study was to establish a methodological framework that enables the transposition of the genius earthquake-resistant knowledge in architecture at elements of design for the project. In this work, usage needs assessment method is recommended to assist the architect in the earthquake-resistant design project. This last consists in applying a model of seismic vulnerability assessment throughout the different design stages of the architectural project. The Validation of utilized tool was achieved to check its practical, economic and creative benefits in terms of exchanging information and seismic vulnerability assessment during the early stages of architectural design.

Keywords

Design process • Paraseismic design • Needs analysis • Vulnerability index • Algerian context

1 Introduction

Several analyses approaches were adopted in order to propose a mechanism which is able to find ways to overcome the constraints when providing the necessary knowledge and skills. The needs analysis method is one of these approaches, which seems to be similar to the act of architectural design [1]. Indeed, it can be used as an assistant tool for architectural specific purposes. It can be used to assist the architect during the architectural earthquake-resistant design since he usually lacks the knowledge and information of such problems.

In the Algerian context, the earthquake-resistant rules are evolving constantly, yet despite this the buildings conceived with regard to old rules can demonstrate a high degree of vulnerability if compared to those which comply with the recent rules [2]. In fact, those rules do not dictate any architectural prerequisites answering the semantic questions [3]. The lesson brought by the last seismic events demonstrates that a building can be earthquake-resistant only if it is the product of a close cooperation between an architect and a civil engineer as of the early phases of the design. However, in the Algerian case neither cooperation is provided nor device tools are used by architects during the earthquake designing process [4]. For this reason, a reasoned and adopted earthquake-resistant design strategy is highly required.

In this paper, mixing fuzzy case based reasoning to the seismic vulnerability index was proposed. The ultimate aim of this contribution was to set a cornerstone of Algerian earthquake-resistant BIM software to be used during the different schematic design phases of an architectural project.

2 Methods

This contribution targeted the Algerian architects, which was made up of 7808 physical persons registered in the national table of the Architects [5]. Their geographic distribution

M. Benabdelfattah (✉) · Y. Kehila
Laboratoire LAE, Ecole polytéchnique d'architecture et d'urbanisme EPAU d'Alger, Oued Smar, Algeria
e-mail: benabdelfattah_med@univ-adrar.dz

M. Benabdelfattah
Université Africaine Ahmed Draia, Adrar, Algeria

A. Kallel et al. (eds.), *Recent Advances in Geo-Environmental Engineering, Geomechanics and Geotechnics, and Geohazards*, Advances in Science, Technology & Innovation, https://doi.org/10.1007/978-3-030-01665-4_116

regarding province seismicity is concentrated at 70.02% in the areas of high seismicity.

In our case, the term needs analysis refers to the process of identifying the different aims and objectives of particular learners from tracking a specific support or training. Three different types of needs were identified: present situation analysis (it means what the people concerned know in terms of the subject field), target situation analysis (what they ought to achieve by support using or taking the course), and learning needs analysis (the preferable and effective methods of achieving the objective, in other terms bridging the gap between the two stages) (Fig. 1).

The validation was achieved through the comparison between the achieved results and those obtained by experts of National Centre for Applied Research on Earthquake Engineering CGS-Algeria (referenced results of a seismic vulnerability index method obtained by experts in the field). This step was carried out by comparing the assessment results obtained by the seismic vulnerability index method and those obtained through the proposed fuzzy inference system. The seismic vulnerability analysis of these buildings was carried out by a classification method called "Vulnerability index method" (Iv). It indicates the state of the structure before and after the occurrence of the seismic event. The Iv represents the balanced sum of the numerical values expressing the seismic quality of the structural and nonstructural elements [6].

3 Results and Discussion

Among the collected remarks, it was noticed that there are two attitudes the two main actors of the architectural earthquake-resistant design can adopt:

The practice in series: In which engineers intervene as amenders, being satisfied to check the feasibility of forms and structures with regard to the current standards. They intervene only on stages where the architectural party is already established. This habit frequently leads to a contradictory situation between the two project parts; the architectural and the structural one, as confirmed by Mr. Slimani Y., a civil engineer in APC Adrar: «*…We intervene after the achievement of the architectural designing phase*».

The practice in common way allows a kind of reconciliation between the architect and the technical designer. The two actors can proceed with a common and simultaneous design of the form which allows a great creativity and a larger territory of exploration by mixing both know-hows. The current architectural practices which separate between the architectural design and the technical phase make us as insurance agents of the project stability. As a result, they have only one limited and focused intervention but very essential for the project as it is mentioned by Mr. Foudou A., civil engineer in an architectural office: «*…the tight cooperation with the architect allows a great architectural freedom and the project is more interesting…*»; as also confirmed by Bekraui K. (a civil engineer, Assistant Professor at African university of Adrar, Algeria) «*…When the work is done independently, the architect sometimes gets stuck because the structural provisions must be validated from the technical point of view…*».

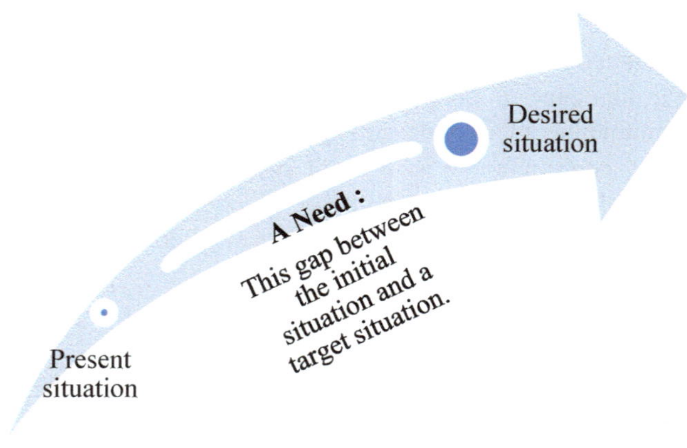

Fig. 1 Definition of a need

4 Conclusions

To assist the architect in his/her architectural earthquake-resistant design, a fuzzy case based reasoning combined with the seismic vulnerability index was proposed. The ultimate aim of this contribution was to set a cornerstone of Algerian earthquake-resistant BIM software to be used during the different schematic design phases of an architectural project. It allows a kind of reconciliation between the objective and the subjective design parts in order to facilitate the exchange of communication between the different actors, and hence ensure a tight collaboration, which is a fundamental principle in the earthquake-resistant design.

Encouraging precisions are noted from the obtained results, which indicate an excellent correlation between the designed procedure and referenced results, considering the predefined attributes. The obtained results were in conformity with the those obtained by experts of National Centre for Applied Research on Earthquake Engineering CGS-Algeria, and the registered comments in the report of the samples projects. Proposal tool can be considered as a tool of dialogue and decision-making aid during the development process of the architectural project, and even a tool of communication with the common public.

References

1. Benabdelfattah, M., Kehila, Y., Makhloufi, A.: Meeting paraseismic knowledge needs of Algerian architects. In: Kallel, A., Ksibi, M., Ben Dhia, H., Khélifi, N. (eds) Recent Advances in Environmental Science from the Euro-Mediterranean and Surrounding Regions. EMCEI 2017. Advances in Science, Technology & Innovation (IEREK Interdisciplinary Series for Sustainable Development). Springer, Cham (2018)
2. Guide de la conception parasismique des bâtiments, Association Française de Génie Parasismique, AFPS, publié avec le soutien du Ministère de l'Ecologie et du Développement Durable (MEDD/ DPPR/ SDPRM). (2002). ISBN n° 2-911709-13-6
3. Abdelkrim, Y.: La sismicité de la région Nord Algérienne. Presented at Rencontre Nationale Sur la Révision des Règles Parasismiques Algériennes RPA, CRAAG, Alger, 10 Oct 2010
4. Benabdelfattah, M., Kehila, Y., Makhloufi, A., Bouhania, B.: Towards an Algerian earthquake-resistant BIM software used during the very early schematic design phase of an architectural project. In: IEEE—The 4th International Conference on Optimization and Applications (ICOA 2018), Morocco (2018)
5. CNOA, Conseil National de l'Ordre des Architectes: Tableau national des architectes. CNOA., Algiers, Algeria (2017)
6. Abed, H., Rezoug, N.: Intégration de la logique floue dans le raisonnement à base de cas: application dans le domaine du bâtiment, Online. (2009) http://www.univ-msila.dz/ar/wp-content/uploads/2009/12/STIC09/articles/paper_65-1.pdf

Protecting Heritage Structures Against Liquefaction: Recent Developments

Amir Tophel and G. V. Ramana

Abstract

The occurrence of liquefaction of loose, saturated and cohesionless soils during major earthquakes is a reasonably common problem and has a more significant potential for destruction. Landmark research regarding the prevention of liquefaction using traditional techniques such as air injection, biogas generation, soil densification and other procedures are well documented. These techniques are energy consumptive and ecologically destructive. However, for protecting the heritage structures, a non-destructive method should be used to preserve the already existing structures. One of the recent research interests is using a novel technique called Microbially Induced Calcite Precipitation (MICP) to prevent liquefaction. This test is not only non-destructive but also green and has long-term effects overcoming all the limitations of the conventional practices. An attempt was made to compile the ongoing research for liquefaction resistance using MICP.

Keywords

Liquefaction • Earthquake • Densification • Non-destructive • MICP

1 Introduction

Heritage structures according to UNESCO [1], are the memories of the golden historical era of over a thousand years. They should be protected and preserved, and the legacy should be passed on to future generations. Liquefaction is known to cause catastrophic damage to the heritage structures [2]. In simple terms, liquefaction is a phenomenon in which the cyclic stress on the soil exceeds the cyclic strength of the soil due to the build-up of excess pore pressure by rapid vibration caused by the earthquake. Some case histories from different parts of the world are covered in [3–6]. The conventional techniques to increase the liquefaction resistance of soil include desaturation of soil using air injection and biogas generation, strengthening the soil using soil densification, electro-osmosis, mixing stabilising agents like lime, cement and materials like tire-chips [7–11]. These techniques, however, are expensive, can only be implemented before the construction of any structure, and are destructive if they are used below an existing structure. Methods which use air and gas injection to mitigate the liquefaction have a short-term effect and thus have a significant drawback with their implementation [12, 13]. A recent interdisciplinary biogeochemical technique called MICP has the potential to overcome all the drawbacks and limitations these conventional techniques have.

2 Methodology

MICP is a well-established biochemical method where calcium carbonate (calcite) is formed as a consequence of microbial metabolic activity. This calcite is produced using two methods; (a) by ureolysis using ureolytic bacteria and (b) using denitrification. The calcium carbonate increases the strength and stiffness of the soil [14] and thus increases the liquefaction resistance of the soil.

2.1 MICP by Ureolysis

In this method, the calcite precipitation takes place by converting urea into ammonium and carbonate. Bacteria such as Bacillus Sphaericus and Bacillus Pasteurii are some examples of ureolytic bacteria. This method consists of two chemical reactions:

A. Tophel (✉) · G. V. Ramana
Indian Institute of Technology Delhi, Delhi, India
e-mail: amirtophel@gmail.com

A. Kallel et al. (eds.), *Recent Advances in Geo-Environmental Engineering, Geomechanics and Geotechnics, and Geohazards*, Advances in Science, Technology & Innovation, https://doi.org/10.1007/978-3-030-01665-4_117

The Hydrolysis of Urea to carbonate ion:

$$CO\,(NH_2)_2 + 4H_2O \rightarrow CO_3^{-2} + 2NH_4^+ + 2H_2O \quad (1)$$

Precipitation of calcite in the presence of calcium ions.

$$Ca^{+2} + CO_3^{-2} \leftrightarrow CaCO_3 \quad (2)$$

2.2 MICP by Denitrification

Microbial denitrification of nitrate using denitrifying bacteria produces carbon dioxide and nitrogen gas, which in turn uses organic compounds, calcium salts of fatty acids as an electron donor and carbon source, to produce calcium carbonate.

Both of the above methods of producing MICP have an equal potential for liquefaction remediation.

2.3 Liquefaction Prevention Using MICP

Researchers found a new novel, environment-friendly technique to remediate the soil against liquefaction. Cyclic tests were preferred to study the behaviour of liquefaction under earthquake load. Montoya et al. [15] carried out centrifuge tests to compare the liquefaction resistance of cemented and untreated sand and found that the cyclic stress ratio was smaller in case of untreated sand showing more excellent resistance to liquefaction as shown in Fig. 1a. O'Donnell et al. [16] tested on semi-stagnant and continuous-flow soil columns and showed that even 0.4% of $CaCO_3$ precipitation could increase the cyclic shear resistance of the sand.

The monuments are not built on the same type of soil everywhere, to see the effect of different ground conditions such as confining pressure, saturation, researchers have tried to account for these factors in their studies. Simatupang et al. [17] conducted an undrained cyclic shear test with samples at different saturation degrees during curing and reported that significant amelioration against liquefaction took place mainly at a low saturation degree. Xiao et al. [18] experimented on a series of undrained cyclic triaxial tests to investigate liquefaction potential of treated and untreated calcareous sand. The test results were; (a) a decrease in the liquefaction resistance was reported with an increase in confining pressure for both clean and bio-cemented specimens as shown in Fig. 1b and (b) the number of cycles to liquefaction increased with increasing cementation content and showed that MICP-treatment could significantly reduce liquefaction potential.

3 Conclusion

Liquefaction due to earthquakes has occurred long before human development and will continue to occur in future. Protection of heritage structures against this phenomenon has become one of the major concerns for today's generation. The conventional techniques for liquefaction remediation are exorbitantly expensive, have a short-term effect and are destructive. To mitigate a land with the already existing structures, we would want a process which has a long-term mitigation effect and is non-destructive. The soil stabilised using MICP is now a well-established technique which takes care of all the drawbacks that the conventional techniques had. MICP is gaining popularity as it is an eco-friendly and cost-effective solution for mitigating the liquefaction potential over a large area.

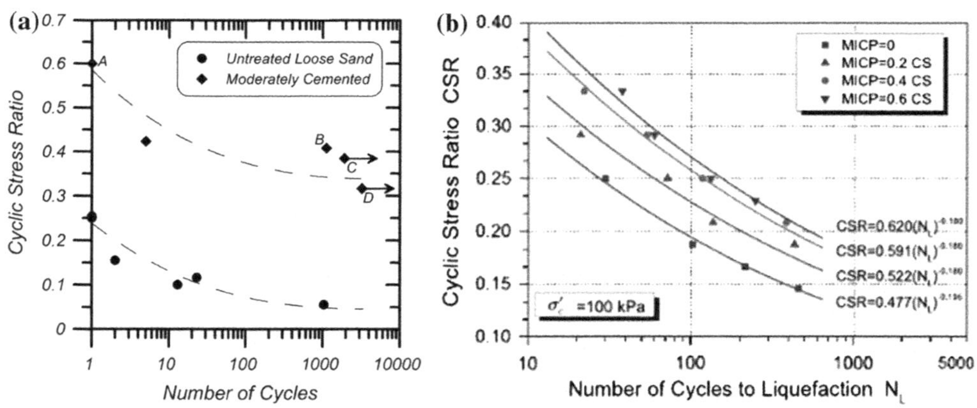

Fig. 1 a, b Cyclic stress ratio versus no. of cycles to achieve liquefaction for cemented and untreated sand [modified after 15, 18]

References

1. UNESCO Homepage. https://whc.unesco.org/en/about/. Last accessed 28 Apr 2018
2. Lo, R.C., Wang, Y.: Lessons learned from recent earthquakes—geoscience and geotechnical perspectives. Adv. Geotech. Earthq. Eng. Liq. Seism. Saf. Dams Monum. InTech, 1–43 (2011)
3. Berardi, R., Margottini, C., Molin, D., Parisi, A.: Soil liquefaction: case histories in Italy. Tectonophysics **193**(1–3), 141–164 (1991)
4. K. Wakamatsu.: Liquefaction history, 416–1997, in Japan. In: 12th World Conference on Earthquake Engineering, pp. 1–8 (1997)
5. Huang, Y., Yu, M.: Review of soil liquefaction characteristics during major earthquakes of the twenty-first century. Nat. Hazards **65**(3), 2375–2384 (2013)
6. Idriss, I.M., Boulanger, R.W.: Soil liquefaction during earthquakes. Earthq. Eng. Res. Inst. **136**(6), 755 (2008)
7. Hazarika, H., Hyodo, M., Yasuhara, K.: Investigation of tire chips sand mixtures as preventive measure against liquefaction. Gr. Improv. Geosynth. **GSP** 207, 338–345 (2010)
8. He, J., Chu, J., Ivanov, V.: Mitigation of liquefaction of saturated sand using biogas. Géotechnique **63**(4), 267–275 (2013)
9. Seed, R.B., Cetin, K.O., Moss, R.E.S., Kammerer, A.M., Wu, J., Pestana, J.M., Riemer, M.F., Sancio, R.B., Bray, J.D., Kayen, R.E., Faris, A.: Recent advances in soil liquefaction engineering : a unified and consistent framework. In: Report no.EERC 2003-06, Earthquake Engineering Research Center, pp. 1–72 (2003)
10. Youd, T.L., Idriss, I.M.: Liquefaction resistance of soils: summary report from the 1996 NCEER and 1998 NCEER/NSF workshops on evaluation of liquefaction resistance of soils. J. Geotech. Geoenvironmental Eng. **127**(10), 817–833 (2001)
11. Okamura, M., Ishihara, M., Tamura, K.: Degree of saturation and liquefaction resistances of sand improved with sand compaction pile. J. Geol. Geoenviron. Eng. **132**(2), 258–264 (2006)
12. Kavazanjian, E., O'Donnell, S.T.: Mitigation of earthquake-induced liquefaction via microbial denitrification: a two-phase process. Ifcee **GSP** 256, 2286–2295 (2015)
13. O'Donnell, S.T., Rittmann, B.E., Kavazanjian, E.: MIDP: liquefaction mitigation via microbial denitrification as a two-stage process. I: desaturation. J. Geotech. Geoenviron. Eng. **143**(12), 4017094 (2017)
14. DeJong, J.T., Mortensen, B.M., Martinez, B.C., Nelson, D.C.: Bio-mediated soil improvement. Ecol. Eng. **36**(2), 197–210 (2010)
15. Montoya, B., DeJong, J., Boulanger, R.: Dynamic response of liquefiable sand improved by microbial-induced calcite precipitation. Géotech. **470**(4), 302–312 (2013)
16. O'Donnell, S.T., Kavazanjian, E., Rittmann, B.E.: MIDP: liquefaction mitigation via microbial denitrification as a two-stage process. II: MICP. J. Geotech. Geoenviron. Eng. **143**(12), 4017095 (2017)
17. Simatupang, M., Okamura, M.: Liquefaction resistance of sand remediated with carbonate precipitation at different degrees of saturation during curing. Soils Found. **57**(4), 619–631 (2017)
18. Xiao, P., Liu, H., Xiao, Y., Stuedlein, A.W., Evans, T.M.: Liquefaction resistance of bio-cemented calcareous sand. Soil Dyn. Earthq. Eng. **107**, 9–19 (2018)

Printed by Printforce, the Netherlands